Biodegradable Waste Management
in the Circular Economy

Biodegradable Waste Management in the Circular Economy

Challenges and Opportunities

Edited by

Malgorzata Kacprzak
Warsaw University of Technology
Faculty of Civil Engineering, Mechanics and Petrochemistry
Institute of Civil Engineering, Płock, Poland
Czestochowa University of Technology, Department of Environmental Engineering and Biotechnology, Częstochowa, Poland

Eleonore Attard
University of Pau, Pau, France

Kari-Anne Lyng
Kråkerøy, Norway

Helena Raclavska
VSB Technical University of Ostrava, Ostrava, Czech Republic

BalRam Singh
Norwegian University of Life Sciences, As, Norway

Eyob Tesfamariam
University of Pretoria, Pretoria, South Africa

Franck Vandenbulcke
Université de Lille, Lille, France

WILEY

This edition first published 2022

Registered Offices
John Wiley & Sons, Inc., 111 River Street, Hoboken, NJ 07030, USA
John Wiley & Sons Ltd, The Atrium, Southern Gate, Chichester, West Sussex, PO19 8SQ, UK

Editorial Office
The Atrium, Southern Gate, Chichester, West Sussex, PO19 8SQ, UK

For details of our global editorial offices, customer services, and more information about Wiley products visit us at www.wiley.com.

Wiley also publishes its books in a variety of electronic formats and by print-on-demand. Some content that appears in standard print versions of this book may not be available in other formats.

Library of Congress Cataloging-in-Publication Data

Names: Kacprzak, Malgorzata, editor.
Title: Biodegradable waste management in the circular economy : challenges and opportunities / edited by Malgorzata Kacprzak Czestochowa University of Technology, Częstochowa, Poland, Eleonore Attard, University of Pau, Pau, France, Kari-Anne Lyng, Kråkerøy, Norway, Helena Raclavska, VSB Technical University of Ostrava, Ostrava, Czech Republic, BalRam Singh, Norwegian University of Life Sciences, As, Norway, Eyob Tesfamariam, University of Pretoria, Pretoria, South Africa, Franck Vandenbulcke, Université de Lille, Lille, France.
Description: First edition. | Hoboken, NJ : John Wiley & Sons, Inc., 2022. | Includes bibliographical references and index.
Identifiers: LCCN 2021050904 (print) | LCCN 2021050905 (ebook) | ISBN 9781119679844 (hardback) | ISBN 9781119679813 (adobe pdf) | ISBN 9781119679851 (epub)
Subjects: LCSH: Organic wastes–Recycling. | Waste management. | Compost.
Classification: LCC TD804 .B54 2022 (print) | LCC TD804 (ebook) | DDC 363.72/88–dc23/eng/20211202
LC record available at https://lccn.loc.gov/2021050904
LC ebook record available at https://lccn.loc.gov/2021050905

Cover Design: Wiley
Cover Image: © New Africa/Shutterstock

Set in 9.5/12.5pt STIXTwoText by Straive, Chennai, India
Printed and bound by CPI Group (UK) Ltd, Croydon, CR0 4YY

C9781119679844 _080622

Contents

Preface

According to the definition of the European Union, the bioeconomy encompasses the production of renewable biological resources and the conversion of these resources and waste streams into value-added products, such as food, feed, bio-based products, and bioenergy. Biodegradable waste contains a wide group of waste: not only a part of municipal solid waste (sometimes called BMW, biodegradable municipal waste) or green waste, food waste, paper, cardboard, and biodegradable plastics, but also human waste, manure, sewage sludge, and slaughterhouse waste. Strictly, bio-waste means biodegradable garden and park waste, food and kitchen waste from households, restaurants, caterers, and retail premises, and comparable waste from food processing plants and agricultural waste. Trends in biodegradable waste generation show that in most countries volumes of such waste produced increase and it is clear that the trend will continue. This should be followed by correct management of this fraction in terms of quality and quantity of the collected flow and selection of sutainable management models. It requires a new approach to management and disposal, including not only the identification of streams or new technologies but also environmental safety or LCA analysis.

Many research groups from different countries are focused on activities aimed at assessing different technologies/strategies for more effective recovery of matter and recourses and/or energy from biodegaradable waste. Ecotoxicologists and environmentalists pay attention to the risks associated with the presence of pollution in such waste. The question arises about the possibility of renewable recycling organic matter and many others compounds. Answering this broad topic requires a holistic approach and cooperation between academia and R&D. Only this approach will allow a holistic view of secure biodegradable waste management, and provide data to fill gaps. Scientific and technological modifications will lead to: development of new technologies processing, reducing greenhouse gas emissions, limiting the use of the most sensitive resources, and reducing the consumption of fossil fuels. All these activities fit very well into the idea of sustainable development. Activities aimed at creating the foundations of new biodegradable waste economy policies will cease to perceive such substances as waste, because they will become a valuable product with a definitely positive added value. This approach implies the development of new strategies for processing biodegradable waste and makes the issues related to their management become one of the biggest challenges in the next decade. However, all biodegradable waste processing technologies generate new waste and this problem should be taken into consideration in the context of a clear circular economy, because every, even new, technology causes waste production. Collaboration between highly interdisciplinary and international experts provides excellent scientific and complementary skills, a unique idea that will be presented in this book.

The book covers the very important and actual point of a circular economy – the management of biodegradable waste. The reader will find a complete book that reflects the latest trends in the rapidly changing management of biodegradable waste. The monography shows the different approaches used depending on the country and region, as the authors are experts from different countries and even continents.

The book was designed within EnviSafeBioc project, financed by the Polish National Agency for Academic Exchange.

FACULTY of INFRASTRUCTURE AND ENVIRONMENT CZESTOCHOWA UNIVERSITY OF TECHNOLOGY

ENVISAFEBIOC ENVIRONMENTAL SAFETY OF BIOSOLIDS IN THE CIRCULAR ECONOMY

NAWA POLISH NATIONAL AGENCY FOR ACADEMIC EXCHANGE

contract no PPI/APM/2018/1/00029/U/001

1

Circular Bioeconomy and Sustainability

Işıl Aksan Kurnaz[1,2], *Elif Damla Arisan*[1,2], *and M. Levent Kurnaz*[2,3]

[1] *Gebze Technical University, Institute of Biotechnology, Gebze, Kocaeli, Turkey*
[2] *Original BioEconomy Resources Center of Excellence (OBEK), GTEAV, Gebze, Kocaeli, Turkey*
[3] *Center for Climate Change and Policies & Department of Physics, Bogazici University, Bebek, Istanbul, Turkey*

1.1 Introduction

The UN projections indicate that by the end of the century world population will total 11.2 billion people, and considering some calculations indicate that the Earth's carrying capacity is around 7 billion people, it becomes clear that the resources are becoming increasingly scarce while demand constantly increases. In order to feed, fuel, and heal the world, a different approach is required to maximize the efficiency of resource use and to minimize the waste thereof. The circular economy was formulated as such an approach, whereby production and consumption of materials are optimized. The European Parliament defines **circular economy** as "a model of production and consumption, which involves sharing, leasing, reusing, repairing, refurbishing and recycling existing materials and products" [1].

The bioeconomy, on the other hand, is defined by the European Commission (EC) as "the production of renewable biological resources and conversion of these resources and waste streams into value added products, such as food, feed, bio-based products and bioenergy" [2]. The societal challenges that drive this need are food security and sustainability of natural resources, and hence the need for reducing our reliance on non-renewable resources but instead increasing use of renewable resources, and of finally the climate change issue that we are all facing.

In order to achieve that, the EC has set a Bioeconomy Strategy and outlined three major action plans:

1. To develop new technologies and processes for the bioeconomy,
2. To develop markets and competitiveness in the bioeconomy, and
3. To push policymakers and stakeholders to cooperate.

The bioeconomy relies on biological resources (such as animals, plants, microorganisms, or organic waste) and involves all production and economic sectors that use biological resources (such as forestry, agriculture, aquaculture, etc., including those in food, feed, bio-based products, and bioenergy). Interestingly, although medical and aromatic plants are heavily used in the pharmaceutical sector, it is not explicitly stated within some definitions of bioeconomy, which is sometimes interpreted as the biopharmaceutical sector not being included in the bioeconomy. However, all sectors involving biological resources in one way or another, and those creating

Biodegradable Waste Management in the Circular Economy: Challenges and Opportunities, First Edition.
Edited by Malgorzata Kacprzak, Eleonore Attard, Kari-Anne Lyng, Helena Raclavska, BalRam Singh, Eyob Tesfamariam, and Franck Vandenbulcke.

organic waste, has to be considered as an inherent component, and all bioeconomies need to have sustainability and circularity in their focus, not only to make efficient use of resources but also to make sure we protect and enhance biodiversity.

1.2 Overlaps and Differences – Bioeconomy, Circular Economy, Circular Bioeconomy

Linear economies rely on renewable or non-renewable raw materials that enter the production line, where a product is generated and marketed for user consumption, at the end of which there is waste that is thrown away. More recently, a recycle-based system has been established whereby the linear economy has an additional feedback loop, where some aspects of the product that the consumer uses, such as plastic or paper packaging, is recycled back to the production line, thereby to a certain extent reducing the waste that is deposited to the environment or used as landfill (Figure 1.1a and b). However, as neither system regenerates the input material, the challenge of finite and diminishing natural resources remains unresolved.

The circular economy essentially describes an economic system [3] where the business models divert from an "end-of-life" model for production toward a "re-use, refurbish, repurpose, re-cycle and recover" model to achieve sustainable production, reducing negative environmental effects while maximizing economic prosperity and social equalities (Figure 1.1c).

Although there is no worldwide agreed-upon definition, bioeconomy focuses on the sectors that are involved with renewable biological resources, including agricultural and forestry products, fish, animals, and microorganisms, for the food, feed, materials, and energy sectors, and essentially can be linear or circular, and includes only bio-based products and services, usually with the use of biotechnology [3].

The circular bioeconomy is a relatively recent concept and is at the intersection of the circular economy and the bioeconomy, aiming to improve resource efficiency, reducing the demand for fossil fuels, valorization of waste material and such; however, the bioeconomy has unique aspects that are beyond the circular economy, including reutilization of by-products and bio-waste, new production processes that minimize toxicity to humans and environment, protection of biodiversity, healthy bio-based products, etc., which are incorporated into circular bioeconomy [4].

Waste is a central concept in the circular bioeconomy since it provides "sustainable biomass" from which new bio-based products can be generated, in addition to being available for compost production (Figure 1.1d). Another important tenant of the circular bioeconomy is the biorefinery concept, which can be used to generate a single bio-based product or in more recent versions integrated biorefineries can handle multiple co-products from the same biomass through efficient design and innovative conversion technologies. Waste biorefining is different from classical waste management and the recycling concept, in the sense that "waste" of one process may be used as a "resource material" for another process, generating a completely different bio-based product in circular bioeconomies, whereas waste recycling in the classical schemes simply reuse or recycle the waste in some aspect of the same production line [5]. The most common waste materials used in bio-based production include lignocellulosic wastes, mostly from rice or wheat straw or corn stover, followed by municipal solid waste and food wastes. Among these, whey – cheese waste – is a by-product of cheese manufacturing and the worldwide production is estimated at around 200 million tons year^{-1}, which could be valorized through biotechnology and redirected to the generation of a number of different value-added bio-based products including lactic acid, polylactic acid used in bioplastics, and bio-based fertilizers. This would not only prevent pollution and

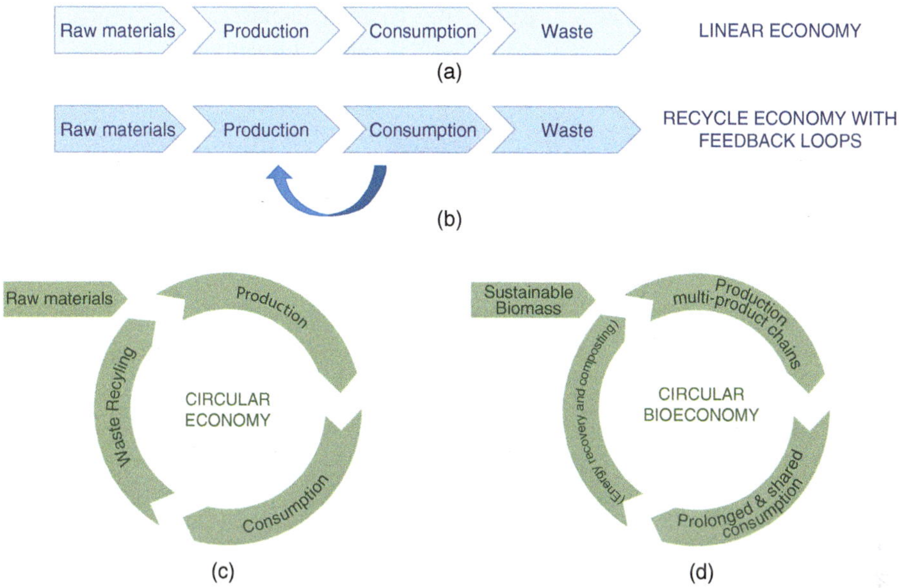

Figure 1.1 The linear economy with and without recycling feedback, circular economy vs circular bioeconomy. (a) The linear economy starts with raw materials that enter the production chain, which is then marketed and consumed, eventually ending in waste materials. (b) The recycle economy with feedback loops is essentially a linear economy where raw material enters the production chain and is marketed; but, after consumption some products will be recycled and entered to the production chain, such as plastic or paper packaging, but eventually ending in waste materials. (c) The circular economy relies on raw materials entering the production circle, but the principle is the sustainable recycling of waste material back to the production chain, which eliminates (as much as possible) waste materials and minimizes raw material requirements. (d) The circular bioeconomy relies on sustainable biomass (including residues and waste) as raw material, which enters the integrated production circle that maximizes multiple bio-based production (e.g. biofuel, biodiesel, biomaterials, food, feed, biopharmaceuticals, etc.) through biorefineries, encourages prolonged and shared use of products so as to minimize over-consumption, and instead of generating recyclable waste material, relies on energy recovery from waste as well as use of waste in composting, thereby renewing and supporting long-term sustainability of biological resources.

relieve negative environmental pressure but also create new circular value chains and innovative manufacturing ecosystems [5].

Similar to the Circular Economy Action Plan adopted in 2015, which aims to turn Europe's economy into a more sustainable economy, a number of policies and strategy papers indicate that it is possible to boost global competitiveness of countries or regions via promoting sustainable economic growth. Circularity of waste management provides minimal generation of waste and maximizes the recycling and reuse.

Biomass cascading is a relatively recent concept in waste management of the circular economy or bioeconomy, where biomass is exploited for high value-added product generation in succession, where residual biomass left after each production step is utilized for generation of another product (and/or co-product), until all value-added products possible have been exploited and the remaining residue is ultimately used as an energy source (Figure 1.2 shows the cascading concept within the context of the circular bioeconomy).

Bottlenecks in such processes include the high cost of biomass resources, enzymes, and chemicals for the conversion process, as well as lack of optimization for bioprocesses, as experienced in early bioethanol and biodiesel research and development (R&D). In fact, contrary to common belief

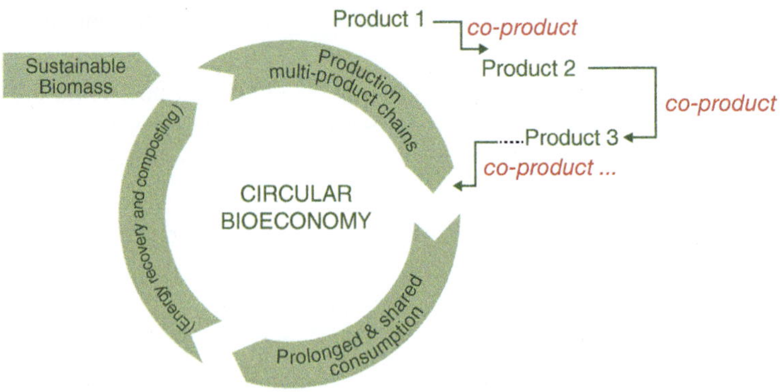

Figure 1.2 Product cascading. Use of biomass can be cascaded to include parallel or serial bioprocess conversion steps so as to generate multiple products and co-products that can be utilized in different industry sectors.

bio-based production is not necessarily more environment-friendly than fossil-based production schemes: when inherent inefficiencies of bioprocesses in terms of titer, yield, and productivity are considered, these inefficiencies may in fact result in higher consumption of other natural resources so that the overall environmental effects or resource efficiency may not be greatly improved over classical manufacturing technologies. Hence, high-technology R&D is required for the design of more efficient and more productive bioprocesses.

Plastic waste is also an issue in bioeconomies, and hence many EU and other funds have been diverted to bio-based plastics that are biodegradable and recyclable. In fact, bioplastics are considered a new chain value for the development of a sustainable plastics economy and promotion of biodegradable bioplastics for packaging industries are greatly encouraged.

Agriculture and forestry are one of the first sectors to incorporate circular bioeconomy in their workflows, since they are unique in their reliance on natural sources and cycles as their primary inputs [6]. The circular bioeconomy emphasizes intrinsic natural recycling and feedback loops and cycles and requires high technical know-how. Agriculture and forestry are particularly important in that not only do these sectors ensure food and nutrition security but they also present an opportunity for carbon sequestration and storage, thereby compensating for emission-related climate effects. It should be noted that agriculture is the largest consumer of the world's freshwater supplies and almost a quarter of the global energy usage goes to food production.

Our growing population needs a robust and sustainable food system. Presently this system has many flaws, as has been described by the UN Sustainable Development Goals (SDG) 2, no hunger. It is clear that our food system is not satisfactory to meet humanity's long-term needs. Food production has helped the population growth in the past 100 years, partly with the help of synthetic fertilizers and pesticides, high-yielding crop varieties, and advanced farm equipment. However, we are all aware of the linear nature of this system, which also has many negative consequences. Presently we have a huge amount of food waste and, in parallel, the global population is suffering from malnutrition, micronutrient deficiency, and obesity [7].

The health costs of malnutrition, micronutrient deficiency, and obesity reach nearly 5.7 trillion USD annually. About the same amount has also been caused by the negative social costs of producing food in a linear nature using modern food production systems. By this, we mean that the present food production system:

- *Is wasteful*. While about 821 million people go to bed hungry every night, about one-third of all the food produced goes to waste. In addition to this, only 2% of all the valuable nutrients in the uneaten food and human waste generated in cities is returned to the food cycle to generate a circular production pattern.

- *Uses finite natural resources.* Valuable mineral resources like phosphorus and potassium are used in synthetic fertilizers. These elements are finite and, especially for phosphorus, humanity might have passed the production peak. Most of the activities in our food production system are powered by fossil fuels.
- *Degrades natural capital.* Soil is our most important natural capital. In many parts of the globe, we have been performing agriculture for many millennia. Because of this, the soil is now tired. We are trying to compensate for this by using excess synthetic fertilizers, pesticides, and water. These practices further degrade about 39 million hectares of arable soil each year.
- *Pollutes the environment.* Pesticides and synthetic fertilizers used in agriculture and the mismanagement of manure can contaminate soils, cause air pollution, and let chemicals into the surface and underground water supplies.

Degraded soil is our most important problem in terms of a sustainable bioeconomy. Our present way of producing food damages the soil, causes health problems both for the workers and consumers, and creates large amounts of harmful greenhouse gases. A regenerative approach to farming promises humanity a healthy and sustainable food supply.

Regenerative farming systems depend on healthy and biologically active farm ecosystems. We should not take soil as a given property for agriculture but should start by producing the soil as well. Industrial-scale composting is important for soil fertility and long-term sustainability of the finite resources of the Earth. Our tired soils benefit enormously from the addition of compost and biodigester as organic additives. There are quite a few techniques considered as regenerative, like rotational grazing, natural system restoration, agroecology, agroforestry, and conservation agriculture. When we keep in mind that in Nature there is no waste, we can farm the land sustainably, which decreases the use of synthetic inputs, especially pesticides, fertilizers, and growth hormones.

In a circular farming system, the main focus is on healthy soils that have improved soil organic material, water-holding capacity, and a microbial population. Permaculture farming techniques significantly help by improving the diversity of crops, animal species on the farm, and the general biodiversity of the local environment.

Unfortunately, synthetic additives have been used for farming for nearly a century and because of economies of scale, they offer many financial advantages. However, to achieve the sustainability of our food supply in the near future, a circular approach together with the addition of end products like compost and biodigestates should be made affordable.

Besides producing calories and micronutrients, the by-products of the food industry can easily be used to create new and exotic food products, supplementary inputs for agriculture, and new materials for the industry in general. When we allow organic materials to rot without intervention, methane is produced. Methane is a greenhouse gas, which is about 25 times more potent than carbon dioxide. Therefore, in our production circle, the most important point is not letting this methane seep through into the atmosphere. Fortunately, as there is already a production system in place, biogas is readily produced from organic waste. However, this step should be considered as a last resort in the circular farming practices as biogas is also organic material and we should not take away this organic additive in the circular production to support another industry.

One additional factor in agriculture is the tilling of the fields. This is the traditional method in farming but tilling is one of the major factors of carbon dioxide emissions in agriculture. In a regenerative agricultural system, the soil is not tilled and so the organic carbon content of the soil is constantly increased. The circular nature of regenerative farming suggests that we should not take more than we give to the soil. Synthetic fertilizers are not materials within the cycle of production and hence they should be used only sparingly. To be able to attain a sustainable food supply, we should take care of all of our environmental systems. Therefore, none of our production practices should be harmful to the environment in general.

1.3 Sustainability Strategies, Climate Action, and the Bioeconomy

In the 1970s, the increasing growth of the world population where natural resources were finite in a finite Earth has brought to our attention the fact that "business as usual" is not a sustainable practice, as quite bluntly stated in "The Limits to Growth" [3, 8]. Not surprisingly, 12 of the UN Sustainable Development Goals (SDGs) can be considered to be involved in the sustainability of natural resources and manufacturing/production capacities, of which almost half include biotechnology and bioeconomy-related issues (SDGs 2 Zero Hunger, 3 Good Health and Well-Being, 6 Clean Water and Sanitation, 7 Affordable and Clean Energy, 9 Industry, Innovation and Infrastructure, 12 Responsible Consumption and Production, 13 Climate Action, 14 Life Below Water, and 15 Life on Land [9]).

Sustainability refers to the balanced existence of human civilization within the boundaries of a finite planet with diminishing resources, without compromising the needs of future generations. From a biological perspective, it can be defined as a homeostatic balance between world population and the Earth, protecting nature, climate, and humans. In the current status, on the one hand, our planet's resources are not sufficient to feed and sustain an ever-growing human population, even at the expense of other ecosystem elements, creating food and water security problems, hunger and social inequality problems, and by 2050 the natural resources required to feed and sustain a human population estimated to be the equivalent of three planets; on the other hand, it is estimated that each year approximately one-third of all food produced goes to waste [10].

In terms of sustainability strategies, waste management (of, for example, lignocellulosic wastes for bio-based production), integrated biorefineries for the production of multiple bio-based products from the same biological resource, and composting appear as pillars of circular bioeconomy [5]. However, global capacity for cellulosic biorefining is still limited, with one plant operational in Canada, one in Finland, one in Italy, one in China, two in Brazil, and four in the US as of 2018 [5].

In terms of biomass and bioenergy policies, resource efficiency strategies are on the front page. Indicators of best practices (using parameters such as input material, design, production, consumption, and end-of-life for products) show that some countries have efficiently established bioeconomy strategies to that effect: Denmark has been a pioneer in circular economy since 1972 and established a National Bioeconomy Panel in 2013 [5, 11]; Finland's National Bioeconomy Strategy was published in 2014; Italy established its Bioeconomy Strategy in 2016 [12] and updated it recently [13]; Spain set up a strategy emphasizing utilization of biological wastes and residues in 2016; [14]; while Scotland initiated a Biorefinery Roadmap in 2015 and developed an integrated approach for biorefineries such as utilization of whisky co-products and waste [15]. Whisky waste is a good example for bioeconomies: the whisky industry in Scotland generates biological waste that has a high biological oxygen demand containing yeast and organic and inorganic compounds, and can be utilized for novel bio-based production chains. Each country, therefore, could have its unique approaches to an effective integration of circular bioeconomy platforms in line with the national manufacturing capacities and priorities.

The US Department of Energy published a Strategic Plan for a Thriving and Sustainable Bioeconomy in 2016, which particularly focuses on renewable biofuels (such as cellulosic ethanol, for which 88 million gallons of commercial cellulosic ethanol capacity was built in 2015) [16]. Together with the Federal Activities Report on the Bioeconomy, published in 2016 by the Biomass Board of the US Department of Agriculture, the Bioeconomy Vision of the US was presented as expanding the sustainable use of biomass resources while maximizing economic, social, and environmental outcomes [17].

The strategic goals and take-home messages of these strategies or roadmaps of countries are mostly similar:

(i) ensuring a competitive environment for bioeconomies to grow,
(ii) ensuring accessibility and sustainability of biomasses and biological resources,
(iii) preventing biodiversity loss,
(iv) creating new economic growth and new jobs in line with sustainable development.

Therefore, a number of reports published by different countries show regional awareness for bio-based products and related bioeconomies. These are the initial steps of the implementation of knowledge arising from bio-based industries into other disciplines to promote a sustainable future.

1.4 The Bio-based Industry Clusters Increased the Impact of Innovative Bioeconomies for a Healthy Future

Public health demographics indicate unstable and sustainable health issues all over the world due to increased industrial progress and irregular urbanization patterns. Since the increasing world population is more evident, we face serious public health problems. As one of the remarkable events, the COVID19 pandemic has increased public attention on bio-based industries. According to a recent report published in Forbes, the annual bio-economy market share is estimated at about 4 trillion USD in the US economical records, with market estimates of the synthetic biology sector alone at 5 – 15 billion USD [18, 19]. Although the definition of the bio-economy points out the validation and standardization of bio-based products for a healthy future with a sustainable environment, the additional definition is given by the US White House Office of Science and Technology unit, which has adopted a broad definition: "The infrastructure, innovation, products, technology, and data derived from biologically related processes and science that drive economic growth, promote health, and increase public benefit."

This definition explains a future strategic position of bio-based economies on different economies with the power of data science. On the other hand, OECD defines the bioeconomy as the set of economic activities relating to the invention, development, production, and use of biological products and processes from a broader perspective. Major socioeconomic contributions of the increased number of bioeconomies in OECD and non-OECD countries potentially improve the health outcomes via boosting health, clean food, and energy.

Although existing abilities in different disciplines with the power of data science is fascinating, the transformation of existing knowledge with the benefits of bioeconomies is not easy. Country-based need assessments for the powerful transformation of knowledge with an innovative perspective needs different skills and know-how. A well-trained human workforce is important for a rapid transformation of knowledge. The education of basic sciences and related disciplines should be supported by the authorities under the science, technology, engineering, art, and mathematics (STEAM) umbrella. Quality in a STEAM education will increase the powerful transformation strategies for a sustainable future. Additionally, educationally prepared communities should be supported by cultural knowledge according to local need, which may also increase novel products for global markets.

Today, the major threats are a sustainable food and clean water supply with an increased perspective of green economies for urbanization [19]. The rural challenges of a bioeconomy also include increasing demand for natural resources as a result of high utilization of existing biomass and environmental pressures created by waste, consumption, and behavioral patterns toward new and sustainable products such as synthetic meat or biofuels, and the need for new economic and growth models that are sure to attract serious opposition from existing production chains.

Therefore, the authorities indicate that more investments are required for infrastructure, know-how on a biological perspective to rapidly overcome COVID19 or similar disease-related alterations in the normal life routines of everyone. COVID19 has dramatically altered economies globally, including a number of bioeconomies. According to Baltic blue economy reports, the lockdown of the cities and distance rules negatively affect small businesses and related supply chains. Hence, there are some roadblocks in establishing effective circular bioeconomies [20]:

1. Sustainable production or access to biological resources while maintaining biodiversity is one challenge to be dealt with.
2. To achieve circularity in bioeconomies, innovative and cost-competitive conversion technologies must be developed and rapidly adopted.
3. Distribution and supply chains as well as the whole production systems have to be changed and ready for such a large-scale adoption.
4. Workforces must be re-trained for a transition to a circular bioeconomy system, from project design to production and transport, market analysis to financial management areas.
5. To achieve that, public and private collaboration must be established across sectors, including business, industry, academia, public sector, as well as policy-makers.
6. Lastly, the consumer must be educated in positive aspects of biomass and bio-based products.

These roadblocks have to be addressed and mitigated through investment in scientific knowledge, innovation and skills management, biorefinery infrastructure, cohesive policies that integrate production systems across various sectors, informed dialogue with the society, and public awareness of the importance of bioeconomy and sustainability issues [21]. To that end, research and innovation funding must be aligned with National Bioeconomy Policies and Sustainability Roadmaps and education programs must be established to train multidisciplinary high- and low-skilled workforces to achieve highly integrated circular bioeconomy systems.

Such national and regional circular bioeconomy policies should, therefore, take into consideration national and regional manufacturing capacities and priorities, national and regional development programs, pressure on land and sea, the environment, and optimize waste valorization strategies and create new circular value chains and innovative manufacturing ecosystems. In a specific perspective globally, aging societies underline the unmet need of innovative products for sustainable bioeconomies. Today's concept is to boost the existing product portfolio, where a number of leading clusters on bioeconomies are remarkable. New products could be possible with the help of innovative technologies rising from STEAM knowledge and a multidisciplinary working environment.

One of the leading examples is RoHealth, a Romanian non-governmental organization that supports the "clustering process" of different types of health-related entities in order to increase competitiveness in the health sector. RoHealth aims to increase the economic competitiveness of the members of the cluster, to enhance the bilateral cooperation between different organizations with the support of research, and to provide sustainable development of the health sector. In a similar perspective, Bloom and BioM are the promising clusters in the European Union, which aims to increase bio-based products and their value in global markets. A report published by Wageningen Research for Bloom cluster was on mapping bioeconomies [22]. They suggested that innovative bioeconomies boost new types of jobs. Biom is the other leading biocluster, which provides opportunities for health-based start-ups at Munich and other locations in Germany. The BioM working model is very promising and hass produced innovative rapid products. Transformation of knowledge from basic sciences to products with the support of industrial partners in a cluster triggers an innovative cycle of the production pipeline. The BioM cluster brings university, research, and SMEs together with industrial leaders on a focused biotechnology production facility. This is critical because of the complexity of the bioeconomy, where gaps between current practices and biobased solutions could be closed by strategic parties for the focused targets. Additionally, the well-structured network will be critical to executing new strategies and policies for the globalization of bio-based solutions for a healthy future for everyone.

As indicated in the bioeconomy mapping report by Wageningen research, social integration of the bioeconomy is crucial for sustainable growth in bio-based sectors. For this issue, they figured out six fundamental concepts, which interact with each other and are relevant for encouraging biobased innovations and transitions. Clusters with different stakeholders, connected platforms, integrated value chains, the readiness level of technology, awareness for a circular economy, and an innovation pipeline are able to address bioeconomy social issues. Sustainable growth of bio-based economies depends on a strong supply chain between different industrial stakeholders with the power of transformative knowledge of the basic sciences. Increased supply incentives for companies may lead to a generation of innovative work outcomes or new supply resources to increase capital and competence, with different actors to deploy resources. Additionally, recognition of technological possibilities and vital solutions in economic perspectives are important to facilitate the exchange of information and knowledge. This is relevant to functional innovation systems.

The active networking strategies with bio-based clusters are critical to building a functional infrastructure with various actors from government, civil society, and business. One of the outputs of this network mechanism is stimulation or creation of new supply chains and markets for different bio-based industries. Additionally, the presence of active information hubs reduces social uncertainty of a bio-based industrial transformation. In a similar approach, a platform action entitled Research and Innovation Strategies for Smart Specialization, or RIS3, initiated by the European think tank (ERA), optimally used regional strengths, abilities, and opportunities to enlighten regional comparative advantages [23]. The smart specialization provides a deeper analysis of socioeconomic innovation engines for regional growth with a competitive advantage. Furthermore, regional clusters on bio-based industries encouraged new actors to proceed with sustainable strategies for rapid transformation of different industries. RIS3 has objectives for four fundamental steps, business, research, public administration, and civil society, for different actors. The most important part of the innovation is entrepreneurship skills of regional bio-based industrial hubs. Instead of STEAM abilities, again a workforce with twenty-first-century skills is the leading factor for sustainable growth of bioeconomies. Resilience, negotiation, problem-solving, critical thinking, or a growth mindset are the crucial abilities needed for the required workforce for a

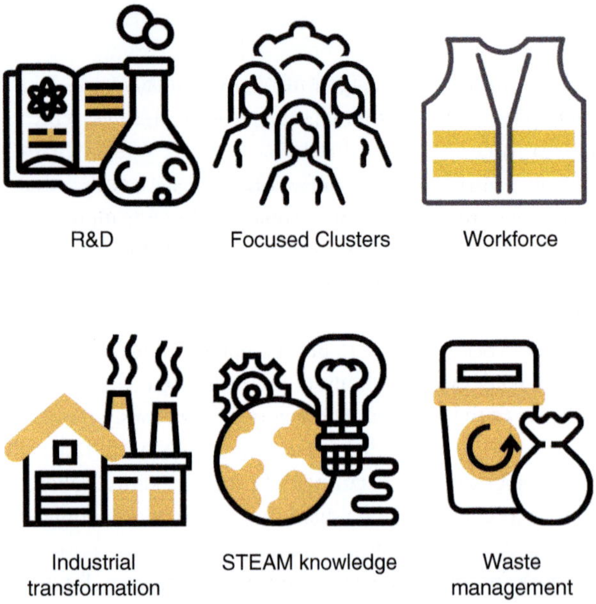

Figure 1.3 A sustainable healthy future with bio-based economies depends on active R&D capacities of different industries with the power of basic sciences to promote innovation, focused industrial clusters for circular knowledge, need and materials between partners, a well-educated workforce for the implementation of bio-based industrial capabilities at different perspectives, elasticity of industrial skills for knowledge transformation and STEAM knowledge for innovative production, biomass preservation at every step of production via organized, structured waste management strategies, and policies to prevent climate-based problems and not to overcome planetary boundaries.

strong innovation cycle of bio-based industries [24]. The most frequently cited industries for RIS3 are health, energy, agro-food, advanced materials, eco-innovation, creative industries, advanced materials, services/tourism, and ICT. One of the examples of agro-food is higher-added-value food, which is closely linked to health and security.

In conclusion, the sustainable bioeconomies have the potential to promote existing technologies in different industries. The innovative new products arising from circular economy working strategies are promising. The existing problems, such as the climate crisis, underline the unmet need for bio-based products [24]. Knowledge is the main required element needed to transform the current strategies for a sustainable future. Strong policies that provide a deeper understanding and awareness for bio-based economies are important in order to implement the bio-based knowledge in the transformation of existing technologies (Figure 1.3).

The alliance of bio-based industries empowers the impact of the bioeconomies. Bioeconomy clusters are critical in the rapid development and implementation of innovative methods of bioeconomies. STEAM knowledge and twenty-first-century abilities will be critical for a workforce that directs the bio-based economies and transformation of existing industries for a sustainable and healthy future.

Acknowledgement

This work was supported by an East Marmara Development Agency program grant for the project toward establishment of the "Original BioEconomy Resources (OBEK) Center of Excellence."

References

1 Circular economy: definition, importance and benefits. Available from: https://www.europarl.europa.eu/news/en/headlines/economy/20151201STO05603/circular-economy-definition-importance-and-benefits (accessed 3 July 2020).

2 EC Report. (2012). Innovating for sustainable growth – A bioeconomy for Europe; ISBN 978-92-79-25376-8. doi: 10.27777/6462. Available from: https://op.europa.eu/en/publication-detail/-/publication/1f0d8515-8dc0-445-ba53-9570e47dbd51 (accessed 12 March 2020).

3 Giampietro, M. (2019). On the circular bioeconomy and decoupling: implications for sustainable growth. *Ecological Economics* 162: 143–156.

4 European Commission. (2020). The "Circular Bioeconomy" – Concepts, opportunities and limitations. Knowledge of Policy 2018 report. Available from: https://ec.europa.eu/knowledge4policy/publication/circular-bioeconomy-concepts-opportunities-limitations_en (accessed 3 July 2020).

5 OECD Report. (2018). Realising the Circular Bioeconomy, 2018 Policy Paper No. 60. Available from: http://www.wcbef.com/assets/Uploads/Publications/Realising-the-Circular-Bioeconomy.pdf (accessed 10 November 2019).

6 The Circular-Bioeconomy in agriculture and forestry: Sustainable bioresource pathways. Dr. Ben Allen's presentation, Institute for European Environmental Policy (IEEP) OECD CRP Governing Board, Paris, 30 November 2018.

7 Cities and Circular Economy for Food, Ellen MacArthur Foundation. Available from: https://media.sitra.fi/2019/01/24094741/cities-and-circular-economy-for-food.pdf. (accessed 18 May 2020).

8 Meadows, D.H., Meadows, D., Randers, J., and Beherens, W. III, (1972). The Limits to Growth. A report for the Club of Rome's projects on the predicament of mankind. New York: Universe Books. ISBN-13: 978-0876631652.

9 UN Sustainable Development Goals (2020). Available from: https://sustainabledevelopment.un.org (accessed 3 July 2020).

10 UN Sustainable Development Goals. (2020) Goal 12: Ensure sustainable consumption and production patterns. Available from: https://www.un.org/sustainabledevelopment/sustainable-consumption-production (accessed 3 July 2020).

11 National Bioeconomy Panel Report of Denmark. Available from: http://naturerhverv.dk/fileadmin/user_upload/NaturErhverv/Filer/Indsatsomraader/Biooekonomi/Fact_sheet_the_National_Bioeconomy_Panel.pdf (accessed 22 April 2020).

12 Finnish Bioeconomy Strategy. Available from: https://www.bioeconomy.fi/facts-and-contacts/finnish-bioeconomy-strategy (accessed 25 April 2020).

13 Bioeconomy in Italy Strategy Report. Available from: http://cnbbsv.palazzochigi.it/media/1774/bit_en_2019_02.pdf (accessed 2 May 2020).

14 Spanish Bioeconomy Plan. Available from: https://ec.europa.eu/knowledge4policy/publication/2016-spanish-bioeconomy-action-plan_en (accessed 2 May 2020).

15 Scotland Bioeconomy Plan. Available from: www.sdi.co.uk/media/2092/biorefinery-roadmap-for-scotland-building-a-sustainable-future.pdf (accessed 2 May 2020).

16 US Department of Energy. (2020) Strategic Plan for a Thriving and Sustainable Bioeconomy. Available from: https://www.energy.gov/sites/prod/files/2016/12/f34/beto_strategic_plan_december_2016_0.pdf (accessed 3 July 2020).

17 Biomass Research & Development Board Report. (2020) Federal Activities Report on the Bioeconomy. Available from: https://www.biomassboard.gov/pdfs/farb_2_18_16.pdf (accessed 3 July 2020).

18 Cumbers, J. (2020). New McKinsey report sees a $4 trillion gold rush in this one hot sector. Who's selling picks and shovels? Forbes, 30 May 2020. Available from: https://www.forbes.com/sites/johncumbers/2020/05/30/mckinsey-report-4-trillion-gold-rush-bioeconomy-synthetic-biology/#255ff53b4cfd (accessed 20 June 2020).

19 Gropp, R.E. (2020). COVID-19 and the bioeconomy; Editorial. *BioScience* 70 (6): 443. https://doi.org/10.1093/biosci/biaa056.

20 Blue Economy Report 2020. Available from: https://ec.europa.eu/commission/presscorner/detail/en/ip_20_986 (accessed 1 June 2020).

21 Review of the European Bioeconomy Strategy 2012. Available from: https://op.europa.eu/en/publication-detail/-/publication/c2f36c72-2e59-11e8-b5fe-01aa75ed71a1/language-en (accessed 3 July 2020).

22 Wageningen Research Report on Bioeconomy Challenges and Implementation. Available from: https://www.wur.nl/en/Research-Results/Research-Institutes/Economic-Research/Research-topics-WEcR/Bio-based-economy.htm (accessed 18 June 2020).

23 (2020) Research and Innovation Strategies for Smart Specialization or RIS3. Available from: https://s3platform.jrc.ec.europa.eu/s3-guide (accessed June 2020).

24 EEA Report 2017. Circular by design. Products in the circular economy. European Environmental Agency Report No. 6/2017. ISBN 978-92-9213-857-8; doi: https://doi.org/10.2800/860754.

2

Cradle to Cradle Marketplace

Malgorzata Kacprzak[1,2]

[1] *Warsaw University of Technology, Faculty of Civil Engineering, Mechanics and Petrochemistry, Institute of Civil Engineering, Płock, Poland*
[2] *Czestochowa University of Technology, Department of Environmental Engineering and Biotechnology, Częstochowa, Poland*

2.1 Introduction

The **2030 Agenda** cannot be achieved without a systemic transformation from various point of views. One of these approaches is systemic changes in production and consumption cycles. Remaining within the paradigm of the linear system and implementing mere incremental changes will not be sufficient. The profound challenges that confront us underline the urgent need for a comprehensive transformation of existing system of production and consumption, way of living, and relationship to products and services.

The **circular economy (CE)** is a significant lever to enable such a transformation and help achieve the 2030 Agenda. It provides a holistic framework for system-level redesign, a new paradigm to set us on a path toward a more sustainable future that achieves prosperity and human well-being for all within planetary boundaries.

Within the linear system of production and consumption, recycling and eco-efficiency are two of the many strategies that could be incorporated to achieve greater sustainability. Eco-efficiency seeks to minimize the ecological impact of production without reducing the output and consumption. The principle of the 3Rs – reduce, reuse, and recycle – is certainly an important strategy that has been made to reduce the impact of the externalities through more efficient processes. Recycling is applied as an end-of-pipe solution, often resulting in downcycling, a reduction in material quality. In contrast to recycling, CE offers a systemic perspective. Systems Thinking at the Core CE aims to close all material loops and create a material round-flow instead of through-flows. In a CE, there is no such thing as waste. The waste of one product or industrial process is used as an input into another one [1].

In this chapter, the history, principles, and current approach to the "cradle to cradle" concept will be presented. Moreover, the same cases of the certified products will be described.

2.2 Cradle to Cradle Concept

The Cradle to Cradle design (also referred to as Cradle to Cradle, C2C, cradle2cradle, or regenerative design) is a biomimetic approach to the design of products and systems. The Cradle to Cradle approach is the opposite of developing a material flow based on the in-line production principle

Biodegradable Waste Management in the Circular Economy: Challenges and Opportunities, First Edition.
Edited by Malgorzata Kacprzak, Eleonore Attard, Kari-Anne Lyng, Helena Raclavska, BalRam Singh, Eyob Tesfamariam, and Franck Vandenbulcke.

known as "*cradle to grave*." According to this concept, the flow of all materials takes place without any conservation of resources and there is an irretrievable loss of such resources, both for people and for the economy. Each life cycle of a new product must start with new resources. The products created according to the model, the final stage of which is the disposal of the product, intentionally transform valuable raw materials into waste, thus destroying valuable resources in accordance with the principle of "*take – make – waste*."

This suggests that industry must protect and enrich ecosystems and nature's biological metabolism, while also maintaining a safe, productive technical metabolism for the high-quality use and circulation of organic and technical nutrients. Put simply, it is a holistic economic, industrial, and social framework that seeks to create systems that are not only efficient but also essentially waste free. The model in its broadest sense is not limited to industrial design and manufacturing; it can be applied to many aspects of human civilization, such as urban environments, buildings, economics, and social systems.

Living things thrive on the energy of the current solar income. Similarly, human constructs can utilize clean and renewable energy in many forms – such as solar, wind, geothermal, gravitational energy, and other energy systems being developed today – thereby capitalizing on these abundant resources while supporting human and environmental health.

Around the world, geology, hydrology, photosynthesis, and nutrient cycling, adapted to locale, yield an astonishing diversity of natural and cultural life. Designs that respond to the challenges and opportunities offered by each place fit elegantly and effectively into their own niches.

The cornerstone of the Cradle to Cradle product design concept is the transition from efficiency to eco-efficiency. The product is designed from the outset to follow resource cycles. The materials used to create the product are suitable for a safe and complete return to the environment (*biological cycle*) or are suitable for high-quality technical recycling that allows their quality to be maintained (*technical cycle*) (Figure 2.1).

In the future, products should be designed in such a way that as many materials as possible are recycled and that reusing the raw material produces products of at least the same

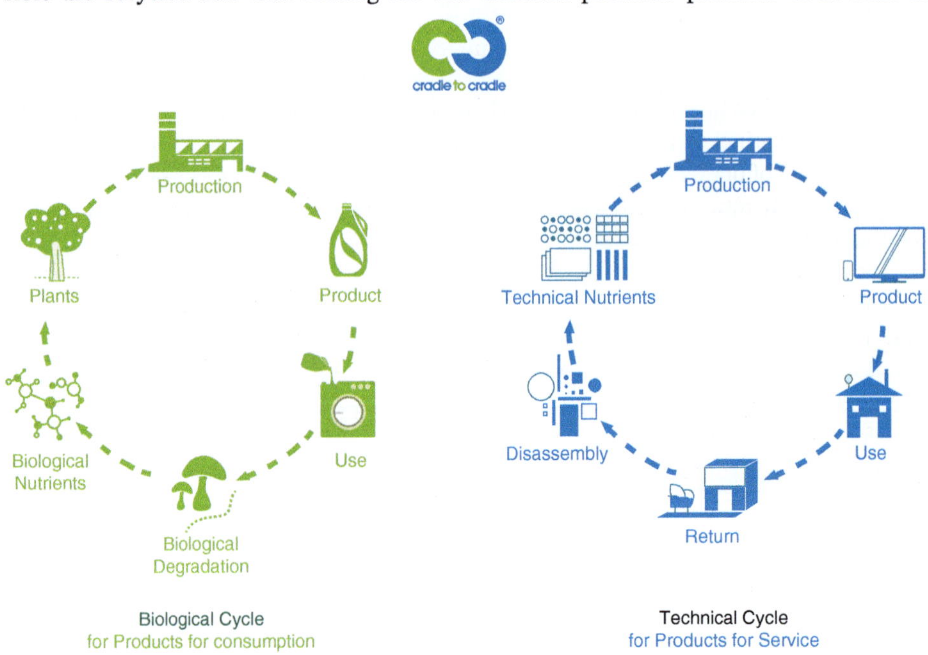

Figure 2.1 Graphical design of the Cradle to Cradle concept working within human activity. Source: Cradle to Cradle Products Innovation Institute (http://wp.auburn.edu/sustainability/tag/cradle-to-cradle, 26 July 2020).

quality as when they were first used. Technical nutrients are limited to non-toxic, non-harmful materials that will not be able to damage the environment and can be re-used without losing quality. This means that these materials can be used over and over again for their original purpose without being "downcycled" and ultimately becoming waste. The biological nutrients are organic materials that will decompose naturally without adversely affecting the natural environment.

Ingredients, including pigments and other additives, should be selected so as to exclude toxicity during use or at other stages of the product life cycle, such as manufacturing, recovery, and reuse. For example, according to the Cradle to Cradle concept, crude oil, once used to make plastics, remains in the cycle and is not burned irretrievably. Energy needed for recycling, as in other production processes, should be obtained from renewable energy sources in a way that protects the environment and climate. A good example of this is the production of a new PET bottle from an old PET bottle.

2.3 Cradle to Cradle History

The term Cradle to Cradle Marketplace was originally introduced by Walter R. Stahel in the 1970s and made popular by William McDonough and Michael Braungart in their book *Cradle-to-Cradle, Remaking the Way We Make Things*, published in 2002 [2]. McDonough is an American architect and Braungart a German chemist. They presented an integration of design and science that provides enduring benefits for society from safe materials, water, and energy in circular economies and eliminates the concept of waste. Earlier, in 1992 they drafted the Hannover Principles, Design for Sustainability. This document was created for planning Expo 2000, a World's Fair in Hannover, Germany, in the year 2000. There are nine Hannover Principles. Number six states: "Eliminate the concept of waste. Evaluate and optimize the full life-cycle of products and processes, to approach the state of natural systems, in which there is no waste." This new approach for designing intelligent products, processes, and systems takes into account the entire life cycle of the product, optimizing material health, recyclability, renewable energy use, water efficiency and quality, and social responsibility.

The book *Cradle-to-Cradle, Remaking the Way We Make Things* put forward a design framework characterized by three principles derived from nature: everything is a resource for something else. In nature, the "waste" of one system becomes food for another. Everything can be designed to be disassembled and safely returned to the soil as biological nutrients, or re-utilized as high-quality materials for new products as technical nutrients without contamination.

In 2010, McDonough and Braungart established the Cradle to Cradle Products Innovation Institute, "... a global non-profit dedicated to transforming the safety, health, and sustainability of products through the Cradle to Cradle Certified Product Standard ...."

In their follow-up book, *The Upcycle: Beyond Sustainability – Designing for Abundance* (2013), McDonough and Braungart go well beyond a "do no harm"/stewardship of the planet aspiration. Their vision for humans is positive, bold, and inspiring: "We should not just protect the planet from ourselves but should redesign our activity to improve the planet. And that goal is well within our reach."

2.4 Cradle to Cradle Design Criteria

Based on the Cradle to Cradle approach and the dual cycle principle, design criteria can be derived for the creation of new products, processes, and objects. The following 10 design criteria are regarded as crucial [3]:

1. Design all products, processes, and objects so that after their initial use all materials can be fully re-used in a biological or technological cycle. Avoid hybrid material streams that are very difficult to separate. Define the materials and their associated life cycle usage paths.
2. Make sure that no harmful or toxic substances are released or used in the production process and during the usage phase of the product or object.
3. Make sure that the production process and the use of the product or object create added value for the stakeholders.
4. Design on the basis of a Triple E (Economy, Ecology, and Equity) approach (Figure 2.2).
5. Use renewable energy sources such as the sun, wind, and (ground) water.
6. Respect the diversity of the location, species, innovation, and culture.
7. Protect and maintain the quality of water reserves.
8. Carry out the production process with social responsibility. That means no child or forced labor, no unhealthy workplaces, no discrimination, and with freedom of association.
9. Follow a local approach in the production process.
10. Make intentions transparent and translate them into measurable targets. Concrete targets can be reached by drawing up a roadmap with milestones.

2.5 Cradle to Cradle Certification Procedure

In 2005, McDonough Braungart Design Chemistry (MBDC, LLC) company created the Cradle to Cradle Certified™ Products Program to recognize high levels of sustainability achieved by its clients and to inspire others to optimize their products and "rethink the way they make things." In 2010, MBDC donated to the Cradle to Cradle Products Innovation Institute an exclusive license for the certification program and methodology.

According to the Cradle to Cradle Certified product standard, product materials and processing steps are evaluated in five categories:

- Material health of used ingredients. The material health category helps to ensure products are made using chemicals that are as safe as possible for humans and the environment by leading designers and product developers through a process of inventorying, assessing, and optimizing material chemistries. As a step toward full certification, manufacturers may also earn a separate Material Health Certificate for products that meet Cradle to Cradle Certified material health requirements.

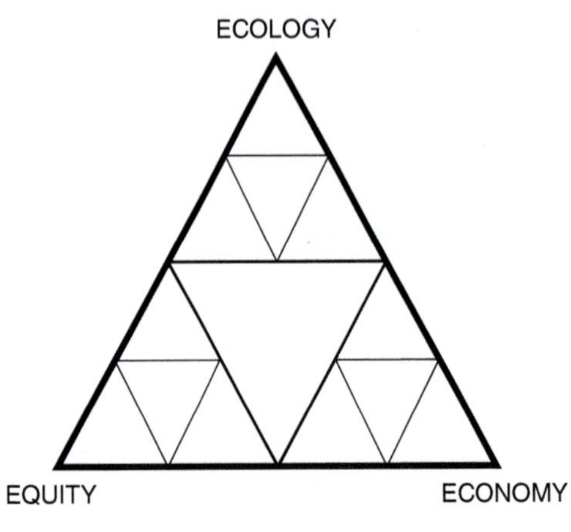

Figure 2.2 Three dynamics of sustainable communities: economy, ecology, and equity [1].

- The material reutilization category aims to eliminate the concept of waste by helping to ensure products remain in perpetual cycles of use and reuse from one product use cycle to the technical or biological cycle.
- Use of the renewable energy and carbon management category helps to ensure products are manufactured using renewable energy so that the impact of climate changing greenhouse gases due to the manufacturing of the product is reduced or eliminated.
- The water stewardship (responsible water management) category helps ensure water is recognized as a valuable resource, watersheds are protected, and clean water is available to people and all other organisms.
- Compliance with social standards. The aim of this category is to design business operations that honor all people and natural systems affected by the manufacture of a product.

A product is assigned an achievement level (Basic, Bronze, Silver, Gold, Platinum) for each category. A product's lowest category achievement also represents its overall certification level. The standard encourages continuous improvement over time by awarding certification on the basis of ascending levels of achievement and requiring certification renewal every two years.

The Cradle to Cradle Products Innovation Institute evaluates products for certification through a network of assessment companies: ToxServices LLC, Eco Intelligent Growth, EPEA GmbH – Part of Drees & Sommer, EPEA Switzerland GmbH, Bilfinger Tebodin, SGS Search Consultancy, a division of SGS Search Ingenieursbureau B.V., ARCHE Consulting, C2C Platform, OmniCert Umweltgutachter GmbH, and Vugge til Vugge ApS.

2.6 Cradle to Cradle Certified Products in the "Biological Nutrient" Cycle

Waste = food is a basic concept of organic waste materials in the Cradle to Cradle concept, becoming food for bugs, insects, and other small forms of life who can feed on it, decompose it, and return it to the natural environment, which we then indirectly use for food ourselves.

Figure 2.3 The Jiffy Pot DK2C pots (https://www.c2ccertified.org/products/scorecard/jiffypot-jiffystrip-jiffyinternational; 2 August 2020).

One of the samples is the *Jiffy Pot DK2C*, pots manufactured from peat, wood pulp, and a little lime in order to adjust the pH of the peat (Figure 2.3). The pots are designed to be planted directly into the earth and to disappear after some time.

2.7 Conclusion

The application of the Cradle to Cradle continuous-cycle principles into practical management is taking off rather slowly. One of the important reasons is the lack of government policy to promote the use of materials in closed cycles, without harmful effects on the environment. Although some argue that C2C philosophy is even a utopian concept and it actually offers some concepts that already exist in sustainable design practices, it offers some insights to think about.

References

1 Henk, M. (2014). *Ecology, Economy, Equity*. American Library Association.
2 Braungart, M. and McDonough, W. (2002). *Cradle to Cradle: Remaking the Way We Make Things*. North Point Press.
3 Van de Westerlo, B. (2011). Sustainable development and the Cradle to Cradle® approach. In: *A literature study of the opportunities to apply the Cradle to Cradle® approach in the built environment*. Enschede: University of Twente https://doi.org/10.3990/1.9789036531818.

3

New Products from Old Wastes Concept – Analysis of the Current State of CE in Biodegradable Waste Management

Biodegradable Waste Management in the Circular Economy: Challenges and Opportunities, First Edition.
Edited by Malgorzata Kacprzak, Eleonore Attard, Kari-Anne Lyng, Helena Raclavska, BalRam Singh, Eyob Tesfamariam, and Franck Vandenbulcke.

3.1

Valorization to Energy

Ewa Neczaj and Anna Grosser

Czestochowa University of Technology (CUT), Czestochowa, Poland

3.1.1 Introduction

A circular economy (CE) is an economic concept in which the value of products and materials is maintained for as long as possible and resource use and waste generation are minimized. The CE takes into account the life cycle of the product, starting from its design, through production, consumption, waste collection, to its management and disposal [1–3]. On 11 March 2020, the European Commission adopted a new circular economy action plan based on an implementation report of a previous plan adopted in 2015. The document emphasizes that to achieve climate neutrality by 2050 “the EU needs to accelerate the transition towards a regenerative growth model that gives back to the planet more than it takes, advance towards keeping its resource consumption within planetary boundaries, and therefore strive to reduce its consumption footprint and double its circular material use rate in the coming decade.” It is therefore obvious that the recovery of raw materials from waste, including their variation into energy, will be a key element of the CE economy in the future [4, 5]. According to the World Bank (2018), annual waste generation will increase by 70% by 2050. Whether this scenario will come true depends on the implementation of the Zero Waste Policy [6–8], in which the bioenergy sector with organic waste as an energy resource will play a very important role. This requires a change of approach to the waste hierarchy (prevention, preparing for reuse, recycling, recovery, and disposal) proposed in the Waste Framework Directive [6]. Zero Waste Europe together with Zero Waste International Alliance developed Zero Waste Hierarchy [9]. Out of seven levels, five relate to waste, while bioenergy production is possible in three of them (Figure 3.1.1).

Different types of feedstock including among others municipal solid waste (MSW), sewage sludge, industrial waste, and biomass residues can be used as biofuels resources. Organic waste conversion products such as biogas, biodiesel, bioethanol, biohydrogen, and biocoal are the principal renewable energy sources that may limit the use of coal, natural gas, oil, and other fossil fuels [11–15].

3.1.2 Feedstock for the Bioenergy Sector in the EU

Biomass residues and waste are huge groups of feedstock that can be used for energy production [16]. We distinguish the first, second, and third categories of biomass residues. Primary residues such as steams, straw, and leaves are generated during the plantation of target food and forest

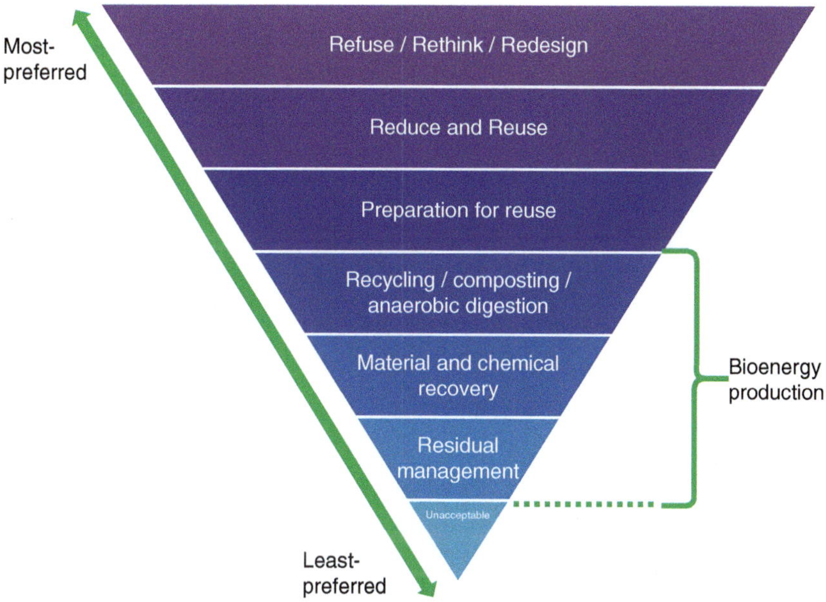

Figure 3.1.1 Place of the bioenergy system in the Zero Waste Hierarchy (Source: adopted from [10]).

crops. During the transformation of food crops into final products, the second categories of biomass residues are generated, e.g. woodchips, rice hulls, and coffee husk. Tertiary residues are produced after biomass-derived consumption by human or animals and are presented as an organic fraction of MSW (OFMSW) [17]. This chapter covers the most important groups of waste that can be used to produce bioenergy.

3.1.2.1 Municipal Solid Waste

Figure 3.1.2 shows waste generation by economic activities and households in EU-27 by 2018. MSW represents about 10% of the total waste generated in Europe on average [18] and as much as 8.2% of all waste generated is household waste.

Factors such as lifestyle, economic level, and the existing legal framework have the greatest impact on both the composition of the MSW and the ways of its management [19]. The predominant fraction (46%) of MSW is OFMSW. Other fractions are paper (17%), plastics (10%), glass (5%), and others (18%), respectively [20]. Due to legal requirements [21] Member States were obligated to reduce the amount of OFMSW going to landfill, which resulted in implementation of different strategies in biodegradable waste management, such as composting (including fermentation), incineration, and pre-treatment (e.g. mechanical-biological treatment). This resulted in an increase in the amount of waste recycled (material recycling and composting) to the level of 104 million Mg (37 million Mg in 1995) and an increase in the share of MSW recycled to 47% (19% in 1995). According to the European Environment Agency, municipal waste consists mainly of food waste (60%), garden waste (35%), and waste classified as "other" (5%). About 43% of biodegradable waste was collected separately in Europe (EU-28) in 2017, which means that 57% was lost for recycling [22]. With regard to the resources recovery from biodegradable waste, its segregation at source is crucial. The quality of the feedstock has an impact on the treatment efficiency and the possibility of the final product disposal. Although composting is more often used for the treatment of biodegradable

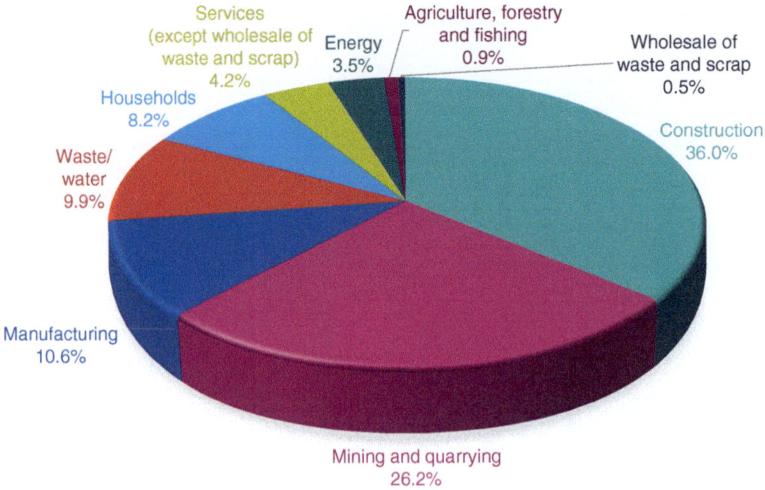

Figure 3.1.2 Waste generation by economic activities and households in EU-27 by 2018 [18].

waste, anaerobic digestion (AD) is the more preferred treatment method because it allows both materials and energy to be obtained [23–25].

3.1.2.2 Wood Residues and Agriculture Biomass

Agricultural biomass is a very broad category that includes [26–28]:

- crops, e.g. sugarcane, corn, beets
- oilseeds, e.g. rapeseed, soybeans, sunflower
- agricultural residues, such as:
  - herbaceous crop residues, e.g. cereal straw, rice straw, corn stover
  - permanent crop residues, e.g. orchard prunings
  - agro-industrial by-products, e.g. olive cake, sunflower husks
  - grassy and woody energy crops
  - animal waste, e.g. manure.

Agricultural residues and dedicated energy crops were identified by Bioenergy Europe as the most relevant for the bioenergy sector due to their high market potential as well as economic and environmental benefits [28].

The waste generated in the wood processing through lumber processing and sawmills can also be used as feedstock for bioenergy production. Wood residues from paper and mills can be used as boiler fuels as well as feedstock for bioethanol production. Currently, a high level of use of mill residues is observed (e.g. for particleboard production), which probably means that it will not be an important source of bioenergy in the future. Biomass residues have great potential for production of different kinds of bioenergy, such as biogas, bio-oil, bioethanol, biodiesel, biochar, and biohydrogen [29–33].

3.1.3 Bioenergy from Biodegradable Waste

3.1.3.1 Waste to Energy Technologies

A holistic approach in assessing the possibility of using biodegradable waste for energy production is essential. The waste feedstock in modern technologies should be use to integrate production of power and fuel chemicals [34]. This approach is necessary for more effective use of organic waste as a renewable resource and reduction in the eco-footprint. Further technology development should be based on a holistic model focused on sustainable resource management. Figure 3.1.3 shows the holistic biorefinery approach proposed by Mohan et al. [35]. Implementation of the future biorefinery design will result in limiting the use of non-renewable sources [36].

Waste-to-energy (WtE) technologies that can be potentially used in velorization of biodegradable waste can be divided into three groups: Thermo-chemical (e.g. Combustion, Gasification, Pyrolysis, and Liqefecation), Biochemical (e.g. Anaerobic Digestion and Photobiological Hydrogen Production), and Chemical (e.g. Transestrification) (Figure 3.1.4) [39–41].

A large group of biodegradable waste requires preliminary treatment prior to its conversion into energy and materials. As an example, Figure 3.1.5 shows possible technologies for the conversion of sewage sludge into fuels and other valorized co-products, including pre-treatment methods for intensification of process efficiency, e.g. ultrasonic disintegration and thermal hydrolysis [43–46].

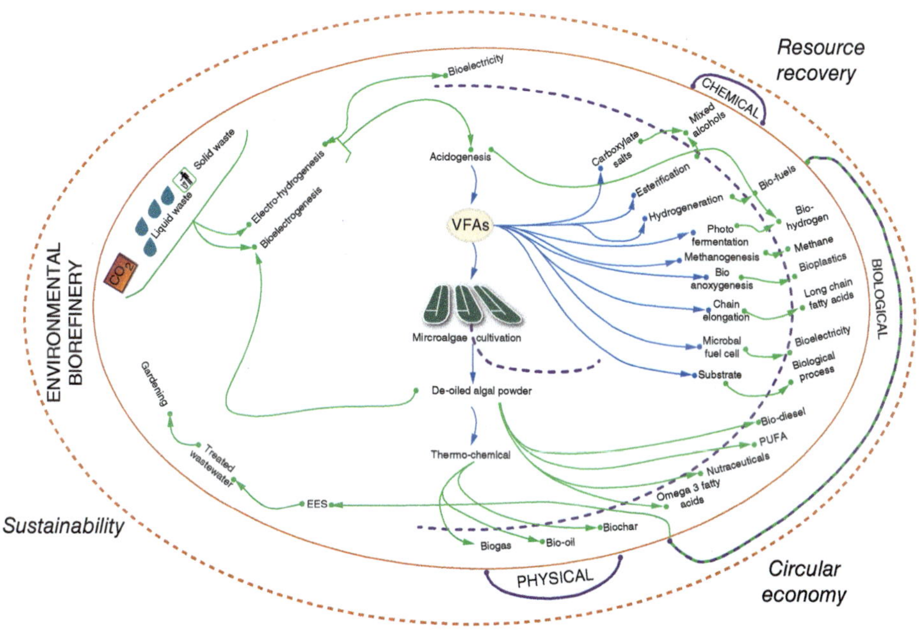

Figure 3.1.3 Holistic biorefinery approach (Source: adopted from [35]).

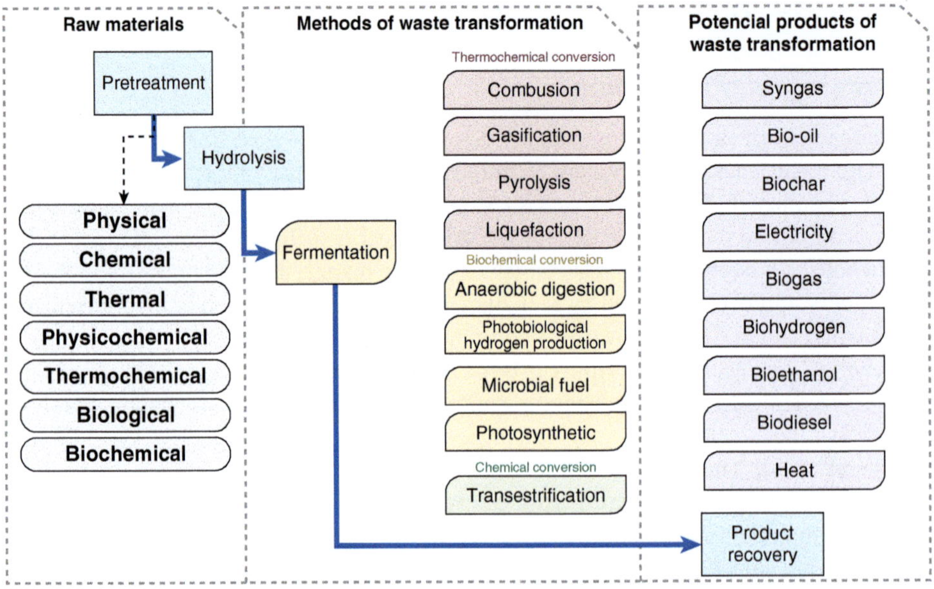

Figure 3.1.4 Potential WtE technologies for valorization of biodegradable waste [37, 38].

3.1.3.2 Types of Bioenergy

According to the US Department of Energy's Oak Ridge National Laboratory's (ORNL) Bioenergy Feedstock Network definition, bioenergy is "useful, renewable energy produced from organic matter – the conversion of the complex carbohydrates in organic matter to energy" [12]. Bioenergy needs a continuous carbon cycle between the atmosphere and the earth in which CO_2 is taken from the atmosphere and converted into biomass via photosynthesis. Additional carbon dioxide generated during the biological process is converted into energy [47]. Forms of bioenergy include heat, power, and solid, gas, and liquid fuels. Biopower is traditional bioenergy generated during combustion of solid biomass sources such as wood and other biomass pellets in biopower plants. Biomass can be burned alone or with natural gas, coal, and other fuels. The electricity produced in power plants can be used for residential and industrial purposes. Moreover, steam generated during biomass combustion can be used in situ to directly power a mechanical process. For advanced bioenergy production, conversion of feedstock into liquid or gas form is required (Figure 3.1.6) [49, 50].

Solid fuels such as wood and pellets are typically used for space heating via combustion. Gas and liquid fuels are used for industrial process and transportation. Conversion of feedstock into liquid and gas biofuels takes place via such processes as fermentation, transesterification, gasification, torrefaction, and pyrolysis [50].

Biofuels are categorized into three groups as follows [51–53]:

- First-generation biofuels produced through simple technologies where oils, sugars, and starches originating in food crops are used as a feedstock.
- Second-generation biofuels are produced from non-food crops, e.g. perennial grasses, woody materials, non-food parts of food crops. Many of the technologies that will be used for their production are in development.
- Third-generation biofuels produced from algae have multifold times fuel yield. Cost-effective technologies for commercial scale in this area are still under development.

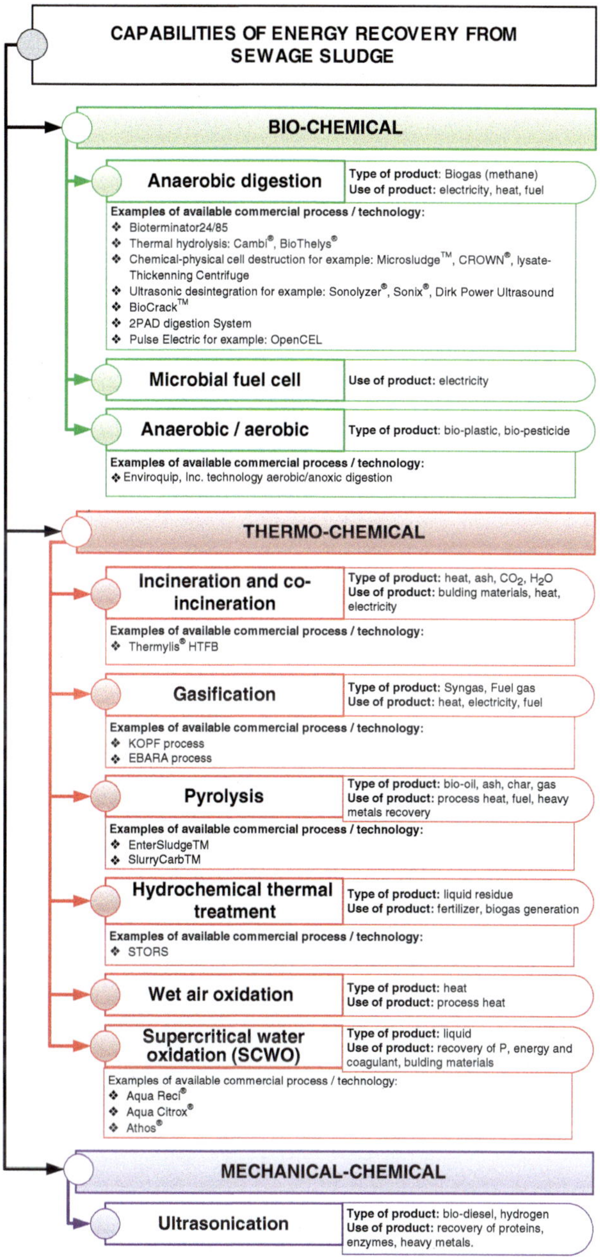

Figure 3.1.5 WtE technologies for sewage sludge valorization [42].

3.1.3.3 The Generation and Use of Bioenergy in the EU

Primary production of energy within the EU in 2018 was 1.1% lower than in 2017 and achieve value of 635 million tonnes of oil equivalent (Mtoe). While for renewable energy an increase of 2% was registered (Figure 3.1.7), which had a 34% share in primary energy production. The share of renewable energy in the gross inland energy consumption in the EU has also been growing in recent years (Figure 3.1.8). A rise of over 200% in renewable energy consumption was registered

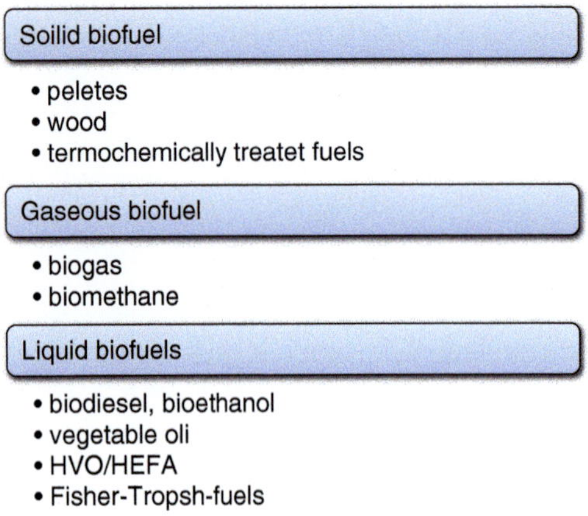

Figure 3.1.6 Types of biofuels [48].

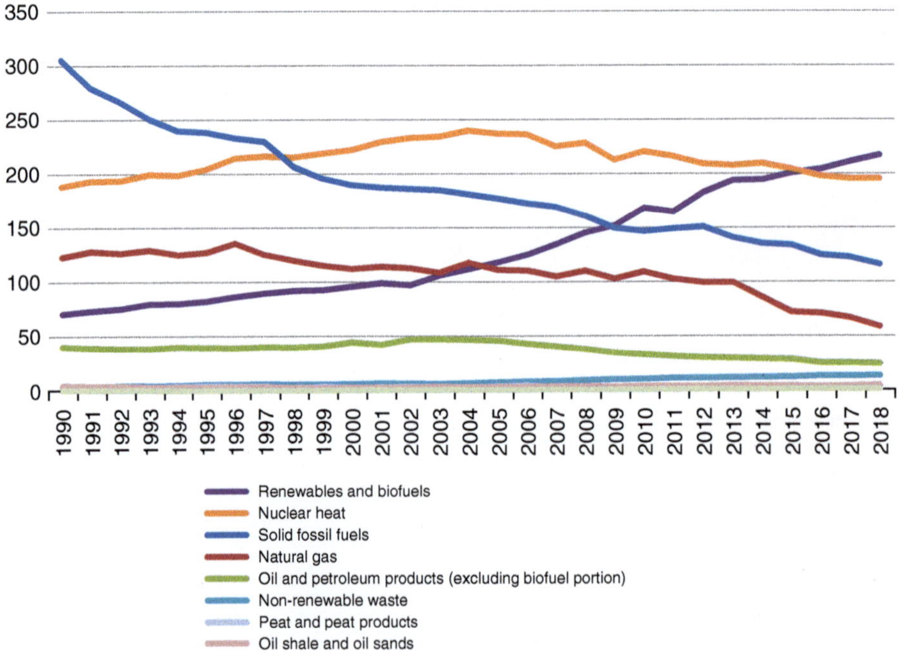

Figure 3.1.7 Primary energy production by fuel, EU-27, in selected years, 1990–2018 (million tonnes of oil equivalent) [54].

compared with 1990. Biomass for energy (bioenergy) is the main source of renewable energy in Europe, with a share of about 60%. Most of the bioenergy is used by the heating and cooling sector (87%) and represents 18% of European gross final consumption of energy in this sector [56].

Almost half of total bioenergy for heating (49.6%) is used for residential consumption where wood, woodchips, or pellets are used as a feedstock for stoves or boilers. Industry and "district heating" represent 26 and 17% of biomass consumption, respectively. These sectors have still great potential for development. Bioenergy represents 5.3% of the overall EU-27 power generation and 17% of the EU-27 renewable electricity production (Figure 3.1.9) [18].

More than 71% of biomass electricity generated comes from combined heat and power plants (CHP), while conventional thermal source CHP plants represent only 28%.

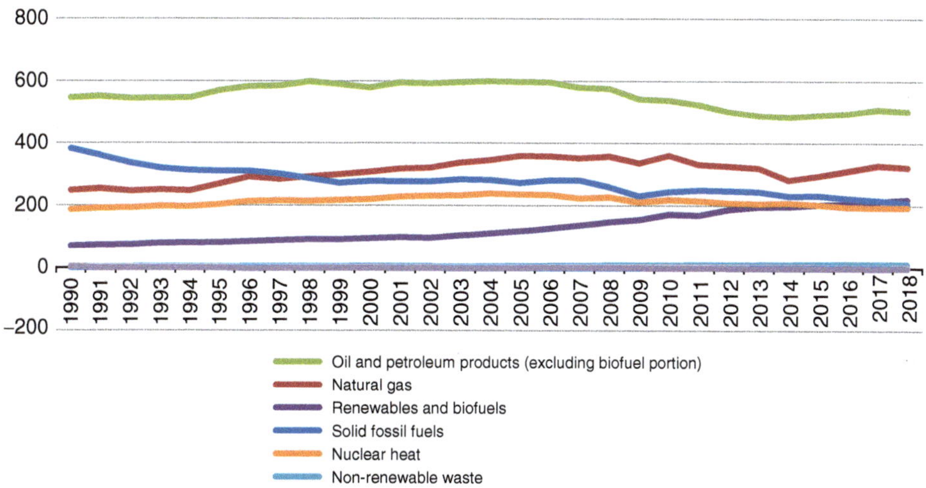

Figure 3.1.8 Gross inland energy consumption by fuel, EU-27, 1990–2018 (million tonnes of oil equivalent) [55].

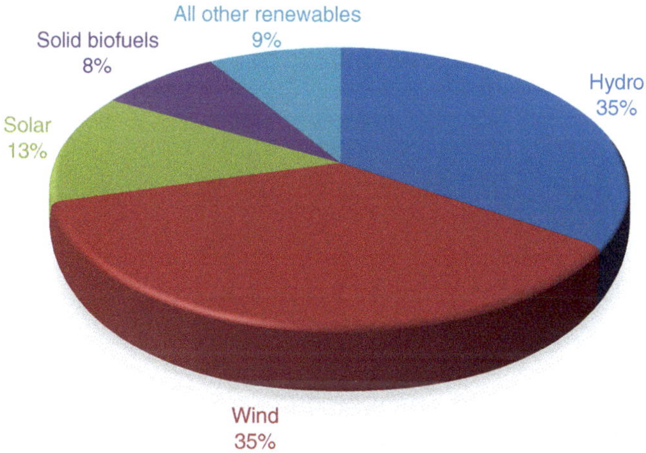

Figure 3.1.9 Electricity generation from all renewable sources in 2019 (EU-27) [18].

In 2018, the total indigenous biogas production in EU-27 was more than 580 thousand TJ (161 TWh) including the share of "other gases from anaerobic fermentation from methane fermentation at the level of 82% (131 TWh)". The biogas from landfills accounted about 9% (15 TWh) and biogas from sewage sludge was 8% (13 TWh). The remaining 1% was generated during thermal processes (2 TWh). In 2018, around 55% of the total EU biogas production was in Germany (around 90 TWh), 13% in Italy (21 TWh), 7% in France (11 TWh). In 2019, the total biogas production reached 590 thousand TJ (164 TWh).

In 2018, renewables represented 8.0% of EU-27 total energy consumption (22.473 Ktoe) in transport, 68% of which was provided by biofuels (biodiesel, bioethanol). EU-27 was the world's largest biodiesel producer and consumer in 2018 (13.303 Ktoe). Due to the Covid-19 pandemic, global transport biofuel production decreased in 2020 as compared to 2019. In particular, the biggest decline was in US and Brazilian ethanol and European biofuel [57]. Table 3.1.1 presents examples of plants for bioenergy production from different types of feedstock.

Table 3.1.1 Examples of plant energy recovery.

Company or owners	Operation year and country	Feedstock	Product	Quantity per year	References
COFCO/Sinopec	2011, China	Corn stover	Ethanol	50 000 tonnes	[58]
Diamond green diesel	2013, USA	Animal fats, used cooking oil, and inedible corn oil	Diesel	275–675 MG	[59]
Mascoma	2014, USA	Wood waste	Ethanol	20–80 MG	[37]
BlueFire renewables	2013, USA	Sorted MSW, wood waste, and agricultural residues	Ethanol	19 MG	[37]
Enerkem and Alberta Innovates Edmond	2015, Canada	Sorted MSW	Methanol, Ethanol (2017)	38 million liters of biofuels	[60]
Abengoa	Under construction, USA	Solid urban waste	Jetfuel	10 000 000 gallons	[61]
	2010, Netherlands	Wheat and corn	Ethanol	480 000 m^3	
	2010, Brazil	Bagasse straw	Electricity and steam	70 MW	
	2009, Spain	Vegetable oils	biodiesel	225 000 tonnes	
	2003, Spain	Cereal straw	Electricity	29 MW	
	2000, United Kingdom	Poultry manure	Electricity	13 MW	
CalBio	2019, USA	Dairy waste	Electricity	320 MW	[37]
VärmlandsMetanol AB	2006, Sweden	Forest residue	Methanol	130 000 m^3	[62]
BioMCN	2009, Netherlands	Biogas from, e.g., solid waste landfills and anaerobic digestion plants	Methanol	65 000 Mg	[63]
Södra	2020, Sweden	Wood raw material	Methanol	5250 m^3	[64]
MefCO project	2019, Germany	Captured CO_2	Methanol	1 Mg day^{-1}	[65]
Carbon Recycling International (CRI)	2012, Iceland	CO_2 and hydrogen	Methanol	Five million liters	[66]

Advanced Plasma Power Ltd., BioSNG pilot plant	2016, United Kingdom	Refuse derived fuel (100 kg h^{-1})	SNG Power (electricity)	SNG (4 kg h^{-1}) power (electricity) (0.06 MWel)	[67]
AustroCel Hallein	Under construction, Austria	Sulfite spent liquor (SSL, 33% dry content) from spruce wood pulping (600 000 tonnes y^{-1})	Ethanol	30 000 tonnes	[68]
ArcelorMittal, Ghent Steelanol	Under construction, Belgium	Steel flue gas	Ethanol	16 000 tonnes	
Agnion Technologies GmbH	2001, Germany	Waste Wood (80 000 Mg y^{-1})	Heat, power	Heat (28 MWth) power (electricity) (6.1 MWel)	
Borregaard Industries AS	1938, Norway	SSL (33% dry content) from spruce wood pulping (400 000 tonnes y^{-1})	Ethanol	15 800 tonnes	
Sindal District Heating Company	2018, Denmark	Forestry by-products, wood processing industry by-products, garden and park waste (20–60% moisture content) (15 000 tonnes y^{-1})	Heat, power	Heat (5 MWth) power (electricity) (0.8 MWel)	

MG, million gallons

3.1.4 Conclusion

Biomass residues and waste can be converted into biofuels and bioelectricity using chemical and biochemical pathways. The choice of process technology for bioenergy generation depends on the feedstocks end product desired. The waste feedstock in modern technologies should be used to integrate the production of fuel, power, and chemicals. Further technology development should be based on a holistic model focused on sustainable resource management. Bioenergy can play a very important role in achieving the EU's renewable energy targets for 2030. However, biomass transformation for energy must be carried out in a sustainable and efficient way in order to reduce greenhouse emissions and maintain ecosystem services without causing deforestation, land degradation, and loss of biodiversity.

References

1. Ng, H.S., Kee, P.E., Yim, H.S. et al. (2020). Recent advances on the sustainable approaches for conversion and reutilization of food wastes to valuable bioproducts. *Bioresource Technology* [Internet]. 302: 122889. https://doi.org/10.1016/j.biortech.2020.122889.
2. Venkata Mohan, S., Dahiya, S., Amulya, K. et al. (2019). Can circular bioeconomy be fueled by waste biorefineries – a closer look. *Bioresource Technology Reports* [Internet]. 7 (July): 100277. https://doi.org/10.1016/j.biteb.2019.100277.
3. Geissdoerfer, M., Savaget, P., Bocken, N.M.P., and Jan, E. (2017). The circular economy – a new sustainability paradigm? *Journal of Cleaner Production* 143: 757–768. https://doi.org/10.1016/j.jclepro.2016.12.048.
4. Lieder, M. and Rashid, A. (2016). Towards circular economy implementation: a comprehensive review in context of manufacturing industry. *Journal of Cleaner Production* [Internet]. 115: 36–51. https://doi.org/10.1016/j.jclepro.2015.12.042.
5. Kumar, G., Kim, S.H., Lay, C.H., and Ponnusamy, V.K. (2020). Recent developments on alternative fuels, energy and environment for sustainability. *Bioresource Technology* 317 (August): 2019–2021.
6. European Parliament and of the Council. (2008). *Directive 2008/98/EC of the European Parliament and of the Council of 19 November 2008 on waste and repealing certain Directives* [Internet]. Available from: https://eur-lex.europa.eu/legal-content/EN/TXT/?uri=celex%3A32008L0098.
7. Iqbal, M.W., Kang, Y., and Jeon, H.W. (2020). Zero waste strategy for green supply chain management with minimization of energy consumption. *Journal of Cleaner Production* 245: 118827.
8. Kurniawan, T.A., Avtar, R., Singh, D. et al. (2020). Reforming MSWM in Sukunan (Yogjakarta, Indonesia): a case-study of applying a zero-waste approach based on circular economy paradigm. *Journal of Cleaner Production* 284: 124775.
9. Singh, S. and Hussain, C.M. (2021). Zero waste hierarchy for sustainable development. In: *Concepts of Advanced Zero Waste Tools* (ed. C.M. Hussain), 123–142. Elsevier.
10. Simon JM. A. (n.d.) *Zero Waste hierarchy for Europe. New tools for new times. From waste management to resource management* [Internet]. Available from: https://zerowasteeurope.eu/2019/05/a-zero-waste-hierarchy-for-europe.
11. Hassan, S.S., Williams, G.A., and Jaiswal, A.K. (2019). Moving towards the second generation of lignocellulosic biorefineries in the EU: drivers, challenges, and opportunities. *Renewable and Sustainable Energy Reviews* 101: 590–599.

12 Dahiya, A. (2020). *Part 1 Bioenergy – biomass to biofuels: an overview* [Internet]., 2e, vol. 1, Bioenergy., 1–4. Elsevier https://doi.org/10.1016/B978-0-12-815497-7.02001-7.

13 Bhatia, L., Bachheti, R.K., Garlapati, V.K., and Chandel, A.K. (2020). Third-generation biorefineries: a sustainable platform for food, clean energy, and nutraceuticals production. *Biomass Convers Biorefinery*: 1–16.

14 Karmee, S.K. (2016). Liquid biofuels from food waste: current trends, prospect and limitation. *Renewable and Sustainable Energy Reviews* [Internet]. 53: 945–953. http://dx.doi.org/10.1016/j.rser.2015.09.041.

15 Ahorsu, R., Medina, F., and Constantí, M. (2018). Significance and challenges of biomass as a suitable feedstock for bioenergy and biochemical production: a review. *Energies* 11 (12): 3366.

16 Speight, J.G. and Singh, K. (2014). *Environmental Management of Energy from Biofuels and Biofeedstocks*. Wiley.

17 Ho, D.P., Ngo, H.H., and Guo, W. (2014). A mini review on renewable sources for biofuel. *Bioresource Technology* 169: 742–749.

18 Eurostat. 2018. *Waste statistics*. [Internet]. Available from: https://ec.europa.eu/eurostat/statistics-explained/index.php/Waste_statistics#Waste_generation_excluding_major_mineral_waste.

19 Edjabou, M.E., Jensen, M.B., Götze, R. et al. (2015). Municipal solid waste composition: sampling methodology, statistical analyses, and case study evaluation. *Waste Management* [Internet]. 36: 12–23. https://doi.org/10.1016/j.wasman.2014.11.009.

20 Hoornweg, D. and Bhada-Tata, P. 2012. *What a waste: a global review of solid waste management*.

21 Union C of the E. Council Directive 1999/31/EC of 26 April 1999 on the landfill of waste [Internet]. 1999. Available from: https://eur-lex.europa.eu/legal-content/EN/ALL/?uri=CELEX%3A31999L0031.

22 EEA. 2020. Bio-waste in Europe – turning challenges into opportunities. pp. 1–56.

23 Bharathiraja, B., Sudharsanaa, T., Bharghavi, A. et al. (2016). Biohydrogen and biogas – an overview on feedstocks and enhancement process. *Fuel* [Internet]. 185: 810–828. https://doi.org/10.1016/j.fuel.2016.08.030.

24 Chand Malav, L., Yadav, K.K., Gupta, N. et al. (2020). A review on municipal solid waste as a renewable source for waste-to-energy project in India: current practices, challenges, and future opportunities. *Journal of Cleaner Production* 277: 123227. https://doi.org/10.1016/j.jclepro.2020.123227.

25 Demichelis, F., Piovano, F., and Fiore, S. (2019). Biowaste management in Italy: challenges and perspectives. *Sustainability* 11 (15): 4213.

26 Maity, S.K. (2015). Opportunities, recent trends and challenges of integrated biorefinery: part i. *Renewable and Sustainable Energy Reviews* 43: 1427–1445.

27 Singh, L. and Kalia, V.C. (2017). *Waste Biomass Management – A Holistic Approach*. Springer International Publishing. 392 pp.

28 Bioenergy Europe Raport (2019). Bioenergy Europe factsheet: biomass for energy – agricultural residues & energy crops. *Biomass for Energy*. Available from: https://bioenergyeurope.org/article/204-bioenergy-explained-biomass-for-energy-agricultural-residues-energy-crops.html.

29 Khoshgoftar Manesh, M.H., Rezazadeh, A., and Kabiri, S. (2020). A feasibility study on the potential, economic, and environmental advantages of biogas production from poultry manure in Iran. *Renewable Energy* [Internet]. 159: 87–106. https://doi.org/10.1016/j.renene.2020.05.173.

30 Ervasti, S., Vainio, M., and Tampio, E. (2019). Use of local resources as co-substrates in a farm-scale biogas plant. *Open Agriculture* 4 (1): 650–660.

31 Vu, H.P., Nguyen, L.N., Vu, M.T. et al. (2020). A comprehensive review on the framework to valorise lignocellulosic biomass as biorefinery feedstocks. *Science of the Total Environment* 743: 140630.

32 Callegari, A., Bolognesi, S., Cecconet, D., and Capodaglio, A.G. (2020). Production technologies, current role, and future prospects of biofuels feedstocks: a state-of-the-art review. *Critical Reviews in Environmental Science and Technology* 50 (4): 384–436.

33 Nikkhah, A., Assad, M.E.H., Rosentrater, K.A. et al. (2020). Comparative review of three approaches to biofuel production from energy crops as feedstock in a developing country. *Bioresource Technology Reports* 10: 100412.

34 Soetaert, W. (2009). *Defining biorefineries and different concepts.* In: Presentation in BioreFuture 2009 Conference, Brussels, Belgium.

35 Venkata Mohan, S., Nikhil, G.N., Chiranjeevi, P. et al. (2016). Waste biorefinery models towards sustainable circular bioeconomy: critical review and future perspectives. *Bioresource Technology* [Internet]. 215: 2–12. https://doi.org/10.1016/j.biortech.2016.03.130.

36 Venkata Mohan, S., Modestra, J.A., Amulya, K. et al. (2016). A circular bioeconomy with biobased products from CO_2 sequestration. *Trends in Biotechnology* [Internet]. 34 (6): 506–519. https://doi.org/10.1016/j.tibtech.2016.02.012.

37 Dessie, W., Luo, X., Wang, M. et al. (2020). Current advances on waste biomass transformation into value-added products. *Applied Microbiology and Biotechnology* 104 (11): 4757–4770.

38 Lee, S.Y., Sankaran, R., Chew, K.W. et al. (2019). Waste to bioenergy: a review on the recent conversion technologies. *BMC Energy* 1 (1): 1–22.

39 Pan, S., Alex, M., Huang, I. et al. (2015). Strategies on implementation of waste-to-energy (WTE) supply chain for circular economy system: a review. *Journal of Cleaner Production* [Internet]. 108: 409–421. https://doi.org/10.1016/j.jclepro.2015.06.124.

40 Funk, K., Milford, J., and Simpkins, T. (2020). *Waste not, want not: analyzing the economic and environmental viability of waste-to-energy technology for site-specific optimization of renewable energy options* [Internet]., 2e. Bioenergy, 385–423. Elsevier https://doi.org/10.1016/B978-0-12-815497-7.00019-1.

41 Kadier, A., Jain, P., Lai, B. et al. (2020). Biorefinery perspectives of microbial electrolysis cells (MECs) for hydrogen and valuable chemicals production through wastewater treatment. *Biofuel Research Journal* 7 (1): 1128–1142.

42 Grosser, A. and Celary, P. (2019). Biogas (methane production) and energy recovery from different sludges. In: *Industrial and Municipal Sludge*, 705–740. Butterworth-Heinemann and Elsevier.

43 Grosser, A. and Neczaj, E. (2017) *Pretreatment methods as a means of boosting methane production from sewage sludge and its mixtures with grease trap sludge.* In B. Kaźmierczak, M. Kutyłowska, K. Piekarska, H. Jouhara, and J. Danielewicz (eds.), E3S Web Conference [Internet]. 7 November 2017; 22:00058. Available from: .http://www.e3s-conferences.org/10.1051/e3sconf/20172200058.

44 Neumann, P., Pesante, S., Venegas, M., and Vidal, G. (2016). Developments in pre-treatment methods to improve anaerobic digestion of sewage sludge. *Reviews in Environmental Science and Biotechnology* 15 (2): 173–211.

45 Wang, B., Strömberg, S., Nges, I.A. et al. (2016). Impacts of inoculum pre-treatments on enzyme activity and biochemical methane potential. *Journal of Bioscience and Bioengineering* 121 (5): 557–560.

46 Carrere, H., Antonopoulou, G., Affes, R. et al. (2016). Review of feedstock pretreatment strategies for improved anaerobic digestion: from lab-scale research to full-scale application.

Bioresource Technology [Internet]. 199: 386–397. Available from: http://dx.doi.org/10.1016/j.biortech.2015.09.007.

47 Naik, S.N., Goud, V.V., Rout, P.K., and Dalai, A.K. (2010). Production of first and second generation biofuels : a comprehensive review. *Renewable and Sustainable Energy Reviews* 14: 578–597.

48 Thrän D, Ponitka J, Arendt O et al. 2015. Konversionspfade zur energetischen Biomassenutzung im 21. Jahrhundert. Focus Bioenergie-Technologien – Fokush Energ Biomassenutzung.

49 Alalwan, H.A., Alminshid, A.H., and Aljaafari, H.A.S. (2019). Promising evolution of biofuel generations. Subject review. *Renewable Energy Focus* [Internet]. 28 (00): 127–139. https://doi.org/10.1016/j.ref.2018.12.006.

50 Ong, H.C., Chen, W.H., Singh, Y. et al. (2020). A state-of-the-art review on thermochemical conversion of biomass for biofuel production: a TG-FTIR approach. *Energy Conversion and Management* [Internet]. 209 (February): 112634. Available from: https://doi.org/10.1016/j.enconman.2020.112634.

51 Bauer, F., Coenen, L., Hansen, T. et al. (2017). Technological innovation systems for biorefineries: a review of the literature. *Biofuels, Bioproducts and Biorefining* 11 (3): 534–548.

52 Castilla-Archilla, J., O'Flaherty, V., and PNL, L. (2019). Biorefineries: industrial innovation and tendencies. *Biorefinery*: 3–35.

53 Ferreira, J.A., Agnihotri, S., and Taherzadeh, M.J. (2019). Waste biorefinery, Chapter 3. *Sustainable Resource Recover and Zero Waste Approaches*: 35–52.

54 Eurostat. (n.d.) *Primary energy production by fuel, EU-27, in selected years, 1990–2018* [Internet]. Available from: https://ec.europa.eu/eurostat/statistics-explained/images/0/07/Primary_energy_production_by_fuel%2C_EU-27%2C_in_selected_years%2C_1990-2018_%28million_tonnes_of_oil_equivalent%29.png.

55 Eurostat. (n.d.) Gross inland energy consumption by fuel, EU-27 [Internet]. Available from: https://ec.europa.eu/eurostat/statistics-explained/index.php?title=File:Gross_inland_energy_consumption_by_fuel,_EU-27,_1990-2018_(million_tonnes_of_oil_equivalent)_F5.png&oldid=492000.

56 European bioenergy day. (2020). What is the EU Bioenergy Consumption? [Internet]. Available from: http://www.europeanbioenergyday.eu/bioenergy-facts/bioenergy-in-europe/what-is-the-eu-bioenergy-consumption.

57 IEA (International Energy Agency). (2020). Renewables 2020 Analysis and Forecast to 2025 [Internet]. Available from: https://www.iea.org/reports/renewables-2020.

58 Bioenergy International. Ethanol Development in China [Internet]. Available from: https://bioenergyinternational.com/biofuels-oils/ethanol-development-in-china.

59 Diamond Green Diesel. North America's Largest Renewable Diesel Production Facility [Internet]. Available from: https://www.diamondgreendiesel.com.

60 LP EAB. Edmonton's Waste to Biofuels and Chemicals Facility [Internet]. Available from: https://www.edmonton.ca/programs_services/garbage_waste/biofuels-facility.aspx.

61 Abengoa. Energy Recovery from Waste and Biomass [Internet]. Available from: http://www.abengoa.com/web/en/negocio/energia/residuos.

62 VÄRMLANDSMETANOL. About VÄRMLANDSMETANOL [Internet]. Available from: http://www.varmlandsmetanol.se/index.htm.

63 BioMCN. Bio-Methanol [Internet]. Available from: https://www.oci.nl/operations/biomcn.

64 Södra. Inauguration of the Biomethanol Plant in Mönsterås [Internet]. Available from: https://www.sodra.com/en/global/Bioproducts/biomethanol/inauguration-of-sodras-plant-for-biomethanol-in-monsteras.

65 Project M. MefCO_2 Methanol Fuel from CO_2 Synthesis of Methanol from Captured Carbon Dioxide Using Surplus Electricity [Internet]. Available from: http://www.mefco2.eu.

66 Chemicals-technology. George Olah CO_2 to Renewable Methanol Plant, Reykjanes [Internet]. Available from: https://www.chemicals-technology.com/projects/george-olah-%0Arenewable-methanol-plant-iceland/%0A.

67 Smith, S. Advanced Plasma Power: Leading Waste to Energy Fuels Technology [Internet]. Available from: http://www.advancedplasmapower.com.

68 ETIP Bioenergy European Technology and Innovation Platform. Production Facilities [Internet]. Available from: https://www.etipbioenergy.eu/databases/production-facilities.

3.2

Valorization to Matter

Kari-Anne Lyng[1] *and Aneta Kowalska*[2]

[1] *NORSUS Norwegian Institute for Sustainability Research, Stadio n 4, 1671, Kråkerøy, Norway*
[2] *Faculty of Infrastructure and Environment, Czestochowa University of Technology, Institute of Environmental Engineering and Biotechnology, Czestochowa, Poland*

3.2.1 Material Recycling of Bio-waste

Valorization of bio-waste to matter can be defined as using bio-waste as an input to produce physical products. According to the Ellen McArthur Foundation [1], the circular economy "entails to gradually decoupling economic activity from the consumption of finite resources, and designing waste out of the system." One can distinguish between biological and technical cycles, where the biological cycles involve returning the organic materials to the bioeconomy, with the purpose of regenerating natural systems [1].

According to The Circularity Gap Report the total resources entering the global economy is 92.8 Gt yr^{-1} globally, of which 28.7 t are from biomass. The remaining 64.1 Gt are minerals, metal ores, and fossil fuels. In total only 8.4 t are from recycled resources, implying a circularity of only 9% [2].

In contrast to inert materials such as plastics and metals, bio-waste cannot be recycled into the same product directly. While the material from a plastic bottle may be remelted and recycled into a new bottle, and an aluminum can may be remelted and transformed into a new can, it can be harder for an average consumer to envisage that food waste can be recycled into new food. Food waste will degrade over time, become smelly and unpleasant, change its properties, and potentially result in greenhouse gas emissions and leakages to water and soil. Bio-waste does, however, contain building blocks, such as carbon and hydrogen, necessary to produce materials and important nutrients for food production, such as nitrogen and phosphorus. These components can be valorized to matter and thus potentially increase resource efficiency and decrease the need for fossil-based materials globally.

This subchapter describes the characteristics of bio-waste, possible conversion pathways for valorization of bio-waste to matter, and the elements that can be recycled. In addition, the barriers and way forward for material recycling of bio-waste are discussed.

According to the waste hierarchy, avoiding waste is the most preferable option, followed by reuse and material recycling [3]. Edible food that can be distributed and eaten by humans or animals is not defined as waste, and food waste prevention and distribution of surplus food will thus not be discussed further in this chapter.

3.2.2 Characteristics of Bio-waste and the Natural Cycles of Carbon, Nitrogen, and Phosphorus

Bio-waste is a broad term, ranging from food waste from household and industry to sewage sludge, garden waste, and organic side streams from manufacturing other products. The properties and composition of bio-waste will be dependent on the source. The properties are often described in terms such as *dry matter content* (DM), *volatile solids* (VS), and *biochemical methane potential* (BMP). The properties of the waste can affect which treatment processes are relevant.

The composition of the waste describes which components are in the waste, such as carbon, nitrogen, phosphorus, and potassium content. The composition varies for different types of bio-waste, which in turn will affect the properties and the potential application areas of the recycled physical products. Chicken manure incineration ash contains over six times more potassium in comparison to sewage sludge ash, and a similar amount of phosphorus oscillating within limits of about 80 000 mg kg^{-1} [4]. Compost from fish waste contains over 11 000 mg_N kg^{-1} and, moreover, consists a valuable source of phosphorus, and compost from beef cattle fed wheat and corn dried distillers' grain with solubles contain over 23 000 mg_N kg^{-1} [5, 6].

Carbon is considered to be the most important substance for life due to the fact that carbon builds all organic matter [7]. In the atmosphere, carbon occurs mainly as carbon dioxide (CO_2) and methane (CH_4), two major greenhouse gases. Carbon is present in all organic waste, but the content is dependent on waste type. For example, the carbon content in cattle manure can be almost twice as high as that in sewage sludge and compost [8]. Soil organic carbon is used as an indicator of soil quality, and it has been estimated that the carbon present in the soil on earth is two times higher than its concentration in the air [9]. Soil is therefore an important carbon sink, and can play an important role in decreasing carbon concentration in the atmosphere, representing an important greenhouse gas mitigation strategy. The natural terrestrial carbon cycle includes several main processes: photosynthesis, plant respiration, and microbial respiration. Plants throughout photosynthesis catch atmospheric carbon to produce energy and build the biomass. During mineralization carried out by decomposers and respiration of organisms, carbon returns into the atmosphere [10]. By developing circular pathways for bio-waste treatment, these natural cycles give support instead of disturbing them.

The nitrogen content in bio-waste is also essential due to the important role of nitrogen in protein building. The nitrogen content in the end product from bio-waste treatment will depend on the treatment method. For example, it has been noticed that composting of bio-wastes causes much higher losses of nitrogen in comparison to anaerobic digestion, because small amounts of nitrogen evaporate in the form of NH_3. Incineration causes the highest losses of nitrogen from bio-waste compared with other treatment methods due to the fact that almost all of the nitrogen evaporates in gaseous form as N_2 and NO. Nitrogen residue in ash from incinerated bio-wastes reaches only a few hundred g kg^{-1} of biomass [11].

The importance of nitrogen as a resource has been underlined when looking at the current position of the seven planetary boundaries within which it is expected that humanity can operate safely, as proposed by Röckstrom et al. [12]. The results showed that the threshold for the rate of interference with the nitrogen cycle is largely exceeded, which indicates that the nitrogen resources on the planet are wasted and are not being used efficiently. Due to extensive fertilization of agricultural fields, the excessive concentration of reactive nitrogen in the environment is a challenge, which can have a negative influence on human health, ecosystem stability, and the economy. High concentrations of nitrogen in the soil are leaching into the groundwater as well as terrestrial water reservoirs [13]. Excessive concentration of nitrogen in terrestrial waters may result in eutrophication [14]. Thus, management of nitrogen use should take into account both maximizing of crop yields caused by fertilization and, on the other hand, minimalizing the effect of nitrogen release.

Nitrogen is essentially important in all ecosystems and participates in all important biochemical processes. The demand for nitrogen fertilizer is expected to increase significantly. In addition, earth's atmosphere consists of 78% of nitrogen. Due to its building role in nucleic acids (DNA, RNA) and peptides, nitrogen constitutes a significant macronutrient that is the fourth most abundant compartment of biomass [15]. Nitrogen occurs in nature in various forms – as gaseous N_2 and as "reactive nitrogen," which is included in compounds involved in cellular metabolism. In its free form the biggest reservoir of N occurs in the atmosphere and in the lithosphere. The nitrogen cycle consists of three major processes: an N_2 fixation, a nitrification, and a denitrification – in all of them, microbes play a key role [15]. The nitrogen fixation leads to ammonium production, which may be assimilated into biomass as well as being further respired by aerobic and anaerobic ammonia-oxidizing microbes. Nitrification consists of two processes: reduction of nitrogen compounds (primarily ammonia) and oxidation to nitrite and nitrate. In turn, denitrification refers to the respiration of anaerobic microorganisms with nitrite, nitric oxide, and nitrous oxide as a substrate to the free form of nitrogen [15–17].

Bio-wastes also consist of other key macroelements, such as phosphorus and potassium. Sewage sludge and sewage sludge ashes may represent a source of phosphorus, not only in the context of soil fertilizer but also through phosphorus recovery and its further treatment. A more sustainable phosphorus management is essential due to predicted depletion of P sources at the end of the twenty-first century [18]. It has been noticed that food wastes such as bones contain a high amount of concentrated phosphorus [19]. Such composition of bio-waste makes them valuable for the fertilizer industry, and thus for the food industry, being a part of the circular economy [20].

According to its essential biochemical role, phosphorus is essentially important for living organisms. Due to the increasing global population, which is estimated to be at the rate of a 33% increment by 2050 in comparison to 2011, the global food production and the need for phosphorus will increase significantly [21]. Due to the essential role of phosphorus fertilizers in cereal cultivation, it is considered that phosphorus demand to 2050 will increase by 2.4 times in comparison with the beginning of the twenty-first century [22]. In turn, in modeling future phosphorus demand, it has been suggested that 23 MMT yr^{-1} will be needed in 2050, which means a 1.64-fold increase [23].

The natural phosphorus cycle indicates that phosphorus is in a continuous circulation in nature. The whole phosphorus cycle is a biogeochemical cycle that include numerous environmental reservoirs and mutual interactions between physical and biological processes. The phosphorus cycle encompasses four major components. First is tectonic uplift and exposure of phosphorus-bearing rock to the forces of weathering. The second one constitutes physical processes of erosion. The third part of the phosphorus cycle is phosphorus transport thought rivers into lakes and oceans. The last component consists of processes of phosphorus sedimentation bounded with organic and mineral matter [24].

Phosphorus enters the soil during organic and inorganic fertilization, and also as an effect of mineralization of organic material thought microbial activity catalyzed by the phosphate enzyme. Phosphorus removal from the soil is mainly carried out thought plant uptake, erosion, and leaching. Control of phosphorus bioavailability takes place through sorption processes. Phosphorus leaching from the soil into rivers and then into a lakes and oceans constitutes a major flow of phosphorus from the continent into water reservoirs. In comparison to other substances, phosphorus may be present in the atmosphere in the form of dust, but does not occur in gaseous forms due to the fact that it is mainly liquid at normal temperatures and pressures [24–26].

Despite increasing troubles with phosphorus sources, the problem with environmental pollution with phosphorus from different sources seems to be more and more important, especially in surface waters. Excessive concentration of phosphorus in water reservoirs may cause an inordinate growth of phytoplankton, macroalgae, and higher plants, leading, in a further perspective, to the eutrophication and imbalance of ecosystem homeostasis [27].

Irresponsible human management of phosphate rock, which is non-renewable, has made losses of almost 0.5 billion tonnes of phosphorus [28]. According to the newest predictions, the actual sources of phosphorus may be depleted by the end of the twenty-first century [18]. One of the major agents driving phosphorus demand is the growing need for livestock food products caused by the continuing increase in the world population and problems with hunger worldwide [29]. The wide usage of phosphorus in different branches of industry makes it a necessity to find a new way to acquire it. To meet the rules of the circular economy, bio-wastes rich in phosphorus seems to be a suitable source for its recovery.

3.2.3 Conversion Pathways to Recycling of Bio-waste

As described in the previous section, carbon, phosphorus, and nitrogen run in natural cycles in the environment. Linear value chains that use virgin material without recycling the materials after use can contribute to disturbing these natural cycles. Developing more circular systems and utilizing bio-waste is a resource to produce physical products that can create more sustainable value chains. There are many possible conversion pathways that can be used to contribute to recycling the valuable components present in the bio-waste resource, as shown in Figure 3.2.1. These pathways range from traditional treatment methods for waste to more advanced methods for utilizing the elements and substances as building blocks for producing advanced materials. The figure is only intended to give examples of pathways and thus is not exhaustive.

Each of the conversion pathways and the recycled physical products that can be generated are described below: valorization to matter by composting, anaerobic digestion, pyrolysis, and biorefinery. Furthermore, the barriers, conclusion, and way forward are discussed.

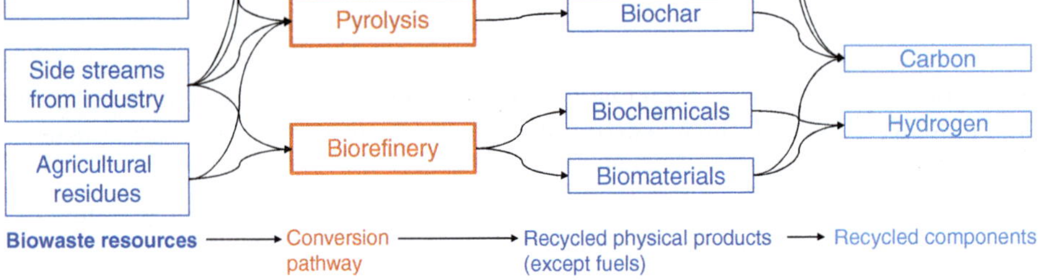

Figure 3.2.1 Conversion pathways for valorization of bio-waste to matter.

3.2.4 Valorization to Matter by Composting

Composting is perhaps the most traditional conversion pathway for food waste. Composting is a natural process where the organic material during a period of time decomposes into humus. Composting can be done on a small scale (e.g. garden composting) and an industrial scale. Compost can be used as a soil amendment, biofertilizer, and for growing media for plants. The quality of the product is dependent on the composition of the input material (the components in the waste resource) and the composting process. For example, structural materials such as wood chips and sand are often added to obtain an amorphous structure. Emissions of greenhouse gases and ammonia may occur in the process but can be minimized by improving access to air by optimizing the composting process [30].

The use of compost can increase the carbon content in soil, and thus serves as a soil amendment product. Organic carbon in soil significantly improves water retention and the physical properties of the soil, which allows plants to easily penetrate the soil. This can result in more fertile land and thus higher crop yields, which in turn can reduce the need for a mineral fertilizer and potentially contribute to improving the global food supply [31]. Soil organic carbon can also contribute by obtaining optimal soil hydration, biodiversity, and a proper pH value of the soil. It has been noticed that fertile soil is less toxic due to the immobilization of pollutants (such as heavy metals) in the organic matter structure [32].

For example, a Danish study investigating changes in soil organic matter composition after repeated applications of organic waste found that compost-amended soils and manure-amended soils had a greater lignin content with a lower degree of oxidation and a greater contribution of bacterial amino sugars relative to fungal amino sugars compared to soils treated with NPK fertilizer [8].

Through soil carbon sequestration of compost, carbon present in the waste that would have been emitted to air if treated by incineration is captured in the soil structure. This can in turn contribute to reducing the concentration of carbon in the atmosphere, decreasing the greenhouse effect. Storage of carbon in the deeper parts of soil for many years thus represents a promising way to mitigate climate change [33, 34].

In addition, compost can be used as a growing media as a substitution for peat [35–37], which may result in reduction of greenhouse gas emissions [38, 39]. The reduction is related to the carbon sequestration in soil for compost and emissions of carbon during the use of the growing media. One Danish study assumed that for the compost 14% of the initial carbon would be left in the soil after 100 years, while for the peat alternative it was assumed that all carbon in the peat would be mineralized and emitted [38]. In addition, the reduced environmental impacts from using compost instead of peat are associated with preparation, excavation, and transportation.

3.2.5 Valorization to Matter by Anaerobic Digestion

Anaerobic digestion is another process that occurs naturally. In the absence of oxygen, microorganisms present in the waste break down biodegradable material and produce biogas. The process is normally conducted by feeding substrates (organic resources) into a heated tank for a defined period of time, called the hydraulic retention time (HRT). The biogas consists mainly of methane (approximately 60%). The remaining components of the gas are mostly carbon dioxide, but small amounts of hydroxide can also be present. Biogas is normally used for generating heat and/or electricity, or as a fuel. An upgrading process where the methane content is increased to over 97% is required to be able to use biogas as a fuel for transport or in a gas grid. The production and use of biogas from bio-waste can be a sustainable measure to substitute fossil energy carriers, but, as it is consumed during its use phase, it does not contribute to recycle nutrients in the waste and is thus not an example of valorization to matter. The anaerobic digestion process does, however, result in another product, which is the digestate. Untreated digestate can be used as a fertilizer, depending on the input material. The use of digestate produced from sewage sludge as a biofertilizer is restricted by regulation in many countries due to the heavy metal content. The digestate may also be separated into a wet and dry fraction, where the wet fraction can be used as a nitrogen fertilizer. The dry fraction can be used for soil improvement or sent to composting.

Substitution of a mineral fertilizer by compost from digestate can result in a reduction in greenhouse gas emissions [40]. In addition to high energy consumption, the production of conventional fertilizers is based on fossil fuels or fossil ore deposits [20]. The use of bio-waste to produce fertilizers therefore also represents a means of circulating nutrients such as nitrogen and phosphorus instead of using non-renewable and scarce resources.

The nutrient balance of the digestate as a biofertilizer can be adjusted by applying different treatment techniques. By separating the digestate in a wet and a dry fraction, one can produce nitrogen fertilizer (as most of the nitrogen follows the wet fraction) and phosphorus fertilizer (dry fraction). This can be applied to even out balances between regions with high and low phosphorus concentration in the soil [41].

3.2.6 Valorization to Matter by Pyrolysis

Pyrolysis is a chemical process where the material under treatment is rapidly heated above its decomposition temperature. The products from pyrolysis include gases, biochar, and biooil, and the primary motivation is often the production of liquids for fuel or chemical production. The biochar consists primarily of carbon (approximately 85%), but it can also contain some oxygen and hydrogen and some inorganic ash [42]. Storage of biologically stable pyrolysis char in soils has been identified as a possible low-tech below-zero emission technology option which can have positive side-effects [43]. Increased carbon in soil is not only a means of improving soil quality and avoiding erosion but is also a means of reducing climate change (removing carbon from the atmosphere). Increasing carbon content in soil can also lead to improvement of soil quality, as described in the previous subchapter.

After the pyrolysis process, it has been estimated that about 54% of the primary carbon content in the input material can be expected to end up in the char. The remaining amount will be released to the atmosphere after conversion of the volatile pyrolysis fractions [43]. In addition to the input materials, the properties of the biochar will depend on the pyrolysis temperature, some technical measures such as the quenching process, and steam reformation [44, 45].

A review that systematically compared the environmental and economic benefits of applying crop residue with applying biochar to soil found that the crop residues can enhance the mineralization rate of soil carbon, while biochar can increase or decrease soil depending on the types of biochar and the soil properties, in addition to duration. The conclusion was that converting crop residues to biochar may be more efficient for sequestering soil carbon, but it is not necessarily more cost-effective [46]. Some studies on the effects of biochar on soil quality and yield indicated that the observed improvements could be short-lived [46].

The impact on long-term yield, the availability of nitrogen for microorganisms, and for crop uptake as a consequence of adding biochar to the soil is complicated and requires in-depth studies, as the results are dependent on individual experimental parameters. In most cases, however, biochars seem to sequester more carbon than relevant alternatives, and in some cases, biochar has been proven to increase crop productivity [46].

3.2.7 Valorization to Matter by Biorefining

Biorefineries separate biomass resources into their building blocks (such as carbohydrates, proteins, triglycerides), which can be transformed into valuable products, biofuels, and chemicals. In a biorefinery, almost all types of biomass feedstocks can be converted to different classes of biofuels and biochemicals through jointly applied conversion technologies [47].

The biorefinery concept represents a more advanced technology than composting and anaerobic digestion and can potentially produce higher-value products than fertilizers and bioenergy. A biorefinery can potentially produce a large range of products, from biofuels such as bioethanol, biodiesel, or biohydrogen, to high value-added bioproducts such as biooil, wax, bioplastics, lactic acid, tri- and di-carboxylic acids, biosurfactants or biolubricants, isolated or extracted products with important applications in the pharmaceutical and food industries, and enzymatic bioremediation, nanoparticles, or foam [48].

Many bio-wastes are non-homogenous and have variable properties, which may be challenging when applied as a feedstock for biorefineries. This may be the reason why the biorefinery concept based on bio-waste is less developed than applying homogeneous feedstock from industry and agriculture [49]. While there are examples of biorefineries with timber and wood chips as feedstocks [50], biorefineries for bio-waste is less common. A review of biorefineries based on organic fraction of municipal solid waste did not identify one single industrial scale biorefinery. Several studies focusing on the concept of producing bioplastics by synthesizing organic acids from waste were, however, identified [49]. The results in these studies are promising, but the process needs to be further optimized.

Biorefineries based on waste streams are characterized by homogenous and predictive waste types as feedstocks, such as residues from potato waste [51, 52] and orange waste [52] and agricultural residues such as straw [53]. Before starting a new biorefinery, it is important to gather information about the composition of the feedstock in order to be able to choose the most adequate processing and desired products [48]. Knowledge about characterization of incoming waste feedstocks is the key to applying the correct technology for product manufacture [52].

3.2.8 Conclusion and the Way Forward

Due to large amounts of bio-waste that is not separated and goes to landfill or to incineration, valuable nutrients are lost. The conversion pathways described in this chapter show that there are many possibilities of valorizing bio-waste to matter and to recycling the valuable components in the waste and to produce physical products. The conversion pathway and type of end product will depend not only on the property of the waste but also on the market for the bio-waste and the various end products. The environmental impacts from valorization of bio-waste to matter are dependent on two main factors: which products the recycled material will replace and the alternative treatment method for the bio-waste and the potential end product(s) for that system.

Despite the many possibilities, there are some barriers that need to be overcome to be able to unleash the full potential. As the bio-waste starts to decompose, dependent on the ambient temperature, the properties of the bio-waste may be difficult to predict. Unlike inert materials, bio-waste cannot be stored and there may be odor problems.

Dependent on the waste source, there may also be large variations in composition. This is a challenge for a potential market of bioresources, as potential users of bio-waste will be interested in predictability related to the properties of the material. When possible, one should try to obtain as pure streams as possible and avoid mixing different types of bio-waste in order to obtain control of the properties and composition of the waste and to exploit the waste at the highest possible value.

It is also important to be aware that some types of bio-waste contain pollution and hazardous materials and should be taken out of circulation, or they may be unwanted in food production. Additionally, impurities may hamper the production process. For example, metals and other impurities in food waste may affect biological processes (mostly enzymatic activity) or the oxidative reductive potential. Moreover, the presence of heavy metals, i.e. Cd, Pb, or Zn, may inhibit amylase, cellulase, or urease synthesis [48]. This underlines the importance of knowing, not just about the waste properties, but also about the waste treatment processes and which types of waste are suitable for the different conversion pathways.

As a conclusion, by regarding bio-waste as a valuable material and not a problem, increased resource efficiency and decreased environmental impacts can be obtained. There are several technologies available for converting bio-waste into useful physical products, but some of the technologies are still immature. More research and innovation is required to obtain a more radical change toward a circular bioeconomy.

References

1 Ellen McArthur Foundation. 2019. Circularity Indicators. An Approach to Measuring Circularity. Methodology.

2 Circle Economy. 2019. The Circularity Gap Report 2019. Closing the Circularity Gap in a 9% World.

3 Papargyropoulou, E., Lozano, R., Steinberger, K.J. et al. (2014). The food waste hierarchy as a framework for the management of food surplus and food waste. *Journal of Cleaner Production* 76: 106–115.

4 Kaikake, K., Sekito, T., and Dote, Y. (2009). Phosphate recovery from phosphorus-rich solution obtained from chicken manure incineration ash. *Waste Management* 29 (3): 1084–1088.

5 Radziemska, M., Vaverková, M.D., Adamcová, D. et al. (2019). Valorization of fish waste compost as a fertilizer for agricultural use. *Waste and Biomass Valorization* 10 (9): 2537–2545.

6 Thomas, B.W., Li, X., Nelson, V., and Hao, X. (2017). Anaerobically digested cattle manure supplied more nitrogen with less phosphorus accumulation than undigested manure. *Agronomy Journal* 109 (3): 836–844.

7 Johnston, A.E., Poulton, P.R., and Coleman, K. (2009). Soil organic matter: its importance in sustainable agriculture and carbon dioxide fluxes, Chapter 1. In: *Advances in Agronomy*, vol. 101 (ed. D.L. Sparks), 1–57. Academic Press.

8 Peltre, C., Gregorich, E.G., Bruun, S. et al. (2017). Repeated application of organic waste affects soil organic matter composition: evidence from thermal analysis, FTIR-PAS, amino sugars and lignin biomarkers. *Soil Biology and Biochemistry* 104: 117–127.

9 Mondini, C., Cayuela, M.L., Sinicco, T. et al. (2018). Storage potential of exogenous organic matter at regional level (Italy) under climate change simulated by RothC model modified for amended soils. *Frontiers in Environmental Science* 6 (144).

10 Schimel, D., Stephens, B.B., and Fisher, J.B. (2015). Effect of increasing CO_2 on the terrestrial carbon cycle. *Proceedings of the National Academy of Sciences* 112 (2): 436.

11 Godbout, S., Palacios, J., Brassard, P. et al. (2012). Combustion of agricultural biomass: mass balance of principal elements. *Valencia: CIGR-EurAgEng* 44: C-1646.

12 Rockström, J., Steffen, W., Noone, K. et al. (2009). A safe operating space for humanity. *Nature* 461 (7263): 472–475.

13 Ahmed, M., Rauf, M., Mukhtar, Z., and Saeed, N.A. (2017). Excessive use of nitrogenous fertilizers: an unawareness causing serious threats to environment and human health. *Environmental Science and Pollution Research International* 24 (35): 26983–26987.

14 Le Moal, M., Gascuel-Odoux, C., Ménesguen, A. et al. (2019). Eutrophication: a new wine in an old bottle? *Science of the Total Environment* 651: 1–11.

15 Stein, L.Y. and Klotz, M.G. (2016). The nitrogen cycle. *Current Biology* 26 (3): R94–R98.

16 Niu, S., Classen, A.T., Dukes, J.S. et al. (2016). Global patterns and substrate-based mechanisms of the terrestrial nitrogen cycle. *Ecology Letters* 19 (6): 697–709.

17 Stevens, C.J. (2019). Nitrogen in the environment. *Science* 363 (6427): 578.

18 Christel, W., Bruun, S., Magid, J., and Jensen, L.S. (2014). Phosphorus availability from the solid fraction of pig slurry is altered by composting or thermal treatment. *Bioresource Technology* 169: 543–551.

19 Wyciszkiewicz, M., Saeid, A., Malinowski, P., and Chojnacka, K. (2017). Valorization of phosphorus secondary raw materials by *Acidithiobacillus ferrooxidans*. *Molecules* 22 (3): 473–485.

20 Chojnacka, K., Moustakas, K., and Witek-Krowiak, A. (2020). Bio-based fertilizers: a practical approach towards circular economy. *Bioresource Technology* 295: 122223.

21 Chowdhury, R.B., Moore, G.A., Weatherley, A.J., and Arora, M. (2017). Key sustainability challenges for the global phosphorus resource, their implications for global food security, and options for mitigation. *Journal of Cleaner Production* 140: 945–963.

22 Tilman, D., Fargione, J., Wolff, B. et al. (2001). Forecasting agriculturally driven global environmental change. *Science* 292 (5515): 281.

23 Bouwman, A.F., Beusen, A.H.W., and Billen, G. (2009). Human alteration of the global nitrogen and phosphorus soil balances for the period 1970–2050. *Global Biogeochemical Cycles* 23 (4): 1–16.

24 Ruttenberg, K.C. (2003). The global phosphorus cycle. *Treatise on Geochemistry* 8: 682.

25 Reinhard, C.T., Planavsky, N.J., Gill, B.C. et al. (2017). Evolution of the global phosphorus cycle. *Nature* 541 (7637): 386–389.

26 Kipp, M.A. and Stüeken, E.E. (2017). Biomass recycling and Earth's early phosphorus cycle. *Science Advances* 3 (11): eaao4795.

27 Farmer, A.M. (2020). Phosphate pollution: A global overview of the problem. 2018 [cited 9 July 2020]. In: *Phosphorus: Polluter and Resource of the Future – Removal and Recovery from Wastewater* [Internet]. IWA Publishing [cited 9 July 2020]; [0]. Available from: https://doi.org/10.2166/9781780408361_035.

28 Cordell, D., Rosemarin, A., Schröder, J.J., and Smit, A.L. (2011). Towards global phosphorus security: a systems framework for phosphorus recovery and reuse options. *Chemosphere* 84 (6): 747–758.

29 Helin, J. and Weikard, H.-P. (2019). A model for estimating phosphorus requirements of world food production. *Agricultural Systems* 176: 102666.

30 Cao, Y., Wang, X., Bai, Z. et al. (2019). S, et al. Mitigation of ammonia, nitrous oxide and methane emissions during solid waste composting with different additives: a meta-analysis. *Journal of Cleaner Production* 235: 626–635.

31 Rabot, E., Wiesmeier, M., Schlüter, S., and Vogel, H.J. (2018). Soil structure as an indicator of soil functions: a review. *Geoderma* 314: 122–137.

32 Grobelak, A. and Kowalska, A. (2020). Heavy metal mobility in soil under futuristic climatic conditions, Chapter 16. In: *Climate Change and Soil Interactions* (eds. M.N.V. Prasad and M. Pietrzykowski), 437–451. Elsevier.

33 Stockmann, U., Adams, M.A., Crawford, J.W. et al. (2013). The knowns, known unknowns and unknowns of sequestration of soil organic carbon. *Agriculture, Ecosystems and Environment* 164: 80–99.

34 Lal, R. (2015). Sequestering carbon and increasing productivity by conservation agriculture. *Journal of Soil and Water Conservation* 70 (3): 55A.

35 Ribeiro, H.M., Romero, A.M., Pereira, H. et al. (2007). Evaluation of a compost obtained from forestry wastes and solid phase of pig slurry as a substrate for seedlings production. *Bioresource Technology* 98 (17): 3294–3297.

36 Ghanbari Jahromi, M. and Aboutalebi, A. (eds.) (2011). *Garden Compost as a Substrate for Vegetable Transplant Production*. Leuven, Belgium: International Society for Horticultural Science (ISHS).

37 Clark, S. and Cavigelli, M. (2005). Suitability of composts as potting media for production of organic vegetable transplants. *Compost Science & Utilization* 13 (2): 150–155.

38 Boldrin, A., Hartling, K.R., Laugen, M., and Christensen, T.H. (2010). Environmental inventory modelling of the use of compost and peat in growth media preparation. *Resources, Conservation and Recycling* 54 (12): 1250–1260.

39 Telnes, L.G., Lyng, K.-A., Valente, C. et al. (2017). Erstatningsmaterialer for torv. Kartlegging av klima- og miljøeffekter. *Utført på oppdrag fra Miljødirektoratet. Østfoldforskning AS OR* 17 (2017): 27.

40 Lyng, K.-A., Modahl, I.S., Møller, H. et al. (2015). The BioValueChain model: a Norwegian model for calculating environmental impacts of biogas value chains. *The International Journal of Life Cycle Assessment*: 1–13.

41 Hanserud, O.S., Lyng, K.-A., Vries, J.W.D. et al. (2017). Redistributing phosphorus in animal manure from a livestock-intensive region to an arable region: exploration of environmental consequences. *Sustainability* 9 (4): 595.

42 Basu, P. (2010). Pyrolysis and torrefaction, Chapter 3. In: *Biomass Gasification and Pyrolysis* (ed. P. Basu), 65–96. Boston: Academic Press.

43 Pröll, T., Afif, R.A., Schaffer, S., and Pfeifer, C. (2017). Reduced local emissions and long-term carbon storage through pyrolysis of agricultural waste and application of pyrolysis char for soil improvement. *Energy Procedia* 114: 6057–6066.

44 Huang, H. and Tang, L. (2007). Treatment of organic waste using thermal plasma pyrolysis technology. *Energy Conversion and Management* 48 (4): 1331–1337.

45 Pariyar, P., Kumari, K., Jain, M.K., and Jadhao, P.S. (2020). Evaluation of change in biochar properties derived from different feedstock and pyrolysis temperature for environmental and agricultural application. *Science of the Total Environment* 713: 136433.

46 Majumder, S., Neogi, S., Dutta, T. et al. (2019). The impact of biochar on soil carbon sequestration: meta-analytical approach to evaluating environmental and economic advantages. *Journal of Environmental Management* 250: 109466.

47 Cherubini, F. (2010). The biorefinery concept: using biomass instead of oil for producing energy and chemicals. *Energy Conversion and Management* 51 (7): 1412–1421.

48 Carmona-Cabello, M., Garcia, I.L., Leiva-Candia, D., and Dorado, M.P. (2018). Valorization of food waste based on its composition through the concept of biorefinery. *Current Opinion in Green and Sustainable Chemistry* 14: 67–79.

49 Vea, E.B., Romeo, D., and Thomsen, M. (2018). Biowaste valorisation in a future circular bioeconomy. *Procedia CIRP* 69: 591–596.

50 Modahl, I.S., Brekke, A., and Valente, C. (2015). Environmental assessment of chemical products from a Norwegian biorefinery. *Journal of Cleaner Production* 94: 247–259.

51 Ekman, A. and Börjesson, P. (2011). Environmental assessment of propionic acid produced in an agricultural biomass-based biorefinery system. *Journal of Cleaner Production* 19 (11): 1257–1265.

52 Matharu, A.S., de Melo, E.M., and Houghton, J.A. (2016). Opportunity for high value-added chemicals from food supply chain wastes. *Bioresource Technology* 215: 123–130.

53 Ekman, A., Wallberg, O., Joelsson, E., and Börjesson, P. (2013). Possibilities for sustainable biorefineries based on agricultural residues – a case study of potential straw-based ethanol production in Sweden. *Applied Energy* 102: 299–308.

3.3

Valorization of Food Waste to Food

Kari-Anne Lyng

NORSUS Norwegian Institute for Sustainability Research, Stadion 4, Kråkerøy, Norway

3.3.1 Food Waste as a Resource and as a Challenge

Valorization of *food waste to food* can lead to increased resource efficiency and reduction of environmental impacts by creating closed-loop systems for food value chains. Closed-loop recycling can be defined as using a discarded product as input to generate a new similar product. Open-loop recycling, on the other hand, signifies using a discarded product as input to the production of a different type of product, often with a lower value or quality. This chapter focuses on the possibilities of valorizing food waste to food, which implies creating more *circular food systems*, using food waste as a resource for producing new food.

Food waste is generated in all parts of the value chain in food production, from cultivation and production to retail and consumer. According to the Food and Agriculture Organisation (FAO) of the United Nations, one-third of the food produced for human consumption is lost or wasted globally, amounting to about 1.3 billion tonnes of edible food per year [1]. A review of studies documenting food waste generated along the supply chain showed that the amount of food waste generated per person each year is estimated to be between 158 and 298 kg in Europe and between 194 and 389 kg globally [2]. A 50% reduction of food waste by 2030 is defined as a target under sustainable development goal 12, *Ensure sustainable consumption and production patterns* [3].

Developing methodology for monitoring the amount of food waste generated in each step of the value chain is key for obtaining a progress toward the reduction targets [4]. Furthermore, the distinction between edible and inedible food waste is crucial in the work of preventing food waste [2]. While edible food waste is waste that could have been eaten, inedible food waste is waste that can occur during preparation of food such as peel, coffee grinds, and eggshells. A Norwegian household food waste composition study showed that on average 2.17 kg of a total of 3.76 kg of food waste generated per household per week was edible food and 0.6 kg of the edible food waste was in original packaging [5]. The same study documented that the majority of the food waste was bread, fruit, and vegetables, as well as leftovers from dinner meals. Another study found that documented food waste in the hospitality sector in Sweden, Norway, Finland, and Germany showed that 50–192 g per portion or around 20% of served food were wasted [6].

This shows that the potential for food waste prevention is large. Additionally, it shows that generation of some inedible food waste is unavoidable, which means that finding ways of utilizing food waste as a resource to create high value products is of importance. The waste hierarchy ranks waste management alternatives in the following order: prevention, reuse, recycling, recovery, and disposal. For food waste, this implies the following ranking: food waste prevention, re-use of surplus food for human consumption, recycling of food waste to animal feed or compost/biofertilizer, energy recovery, and landfill [7]. In the European Union, The Landfill Directive obliges Member States to reduce the amount of biodegradable municipal waste sent to landfill [8].

Food production requires the input of many different resources, such as energy, water, nutrients, and feed. Food waste is rich in nutrients, can have a high water content, and may also be a potential source for renewable energy. It can thus in principle provide all the inputs necessary to produce new food. This shows that matching the waste resources available with the required inputs to produce food and developing *food waste to food* systems can be an important step toward a circular economy. In the following section, an overview of possible pathways for food waste as a resource for producing new food is given.

3.3.2 Possible Paths for Circular Food Systems

Figure 3.3.1 illustrates pathways for obtaining a circular food system where food waste is valorized into food products. The figure is intended to give examples of potential pathways and is thus not exhaustive. For example, valorization of food waste to fish is not included. The figure does not include redistribution, as redistributed food is outside the definition of food waste.

Through different conversion pathways, intermediate products can be developed from food waste, which can be used in different farming practices to produce food products such as meat, milk, cereals, fruit, and vegetables.

Mixed waste from households and industry may be sent to composting or anaerobic digestion, generating compost or biofertilizer that can be used as input to cultivation of cereals, fruits, and vegetables, or to produce forage for animal husbandry. In the retail sector the waste that is treated is much more homogenous. A large part of the waste generated is bread, fruit, and vegetables [9, 10] which in some cases are used as feed for animals and is thus a potential resource for meat and milk production. In the following section the three farming practices of livestock husbandry, crop cultivation, and greenhouse farming are described in more detail.

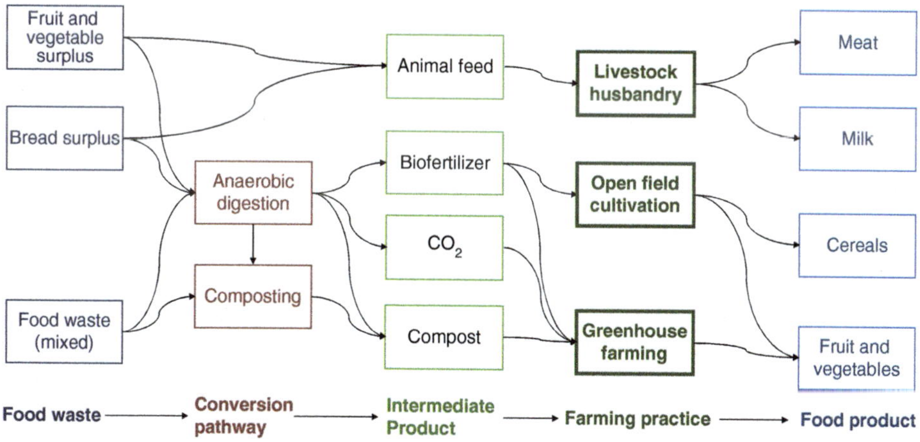

Figure 3.3.1 Pathways for valorization of food waste to food.

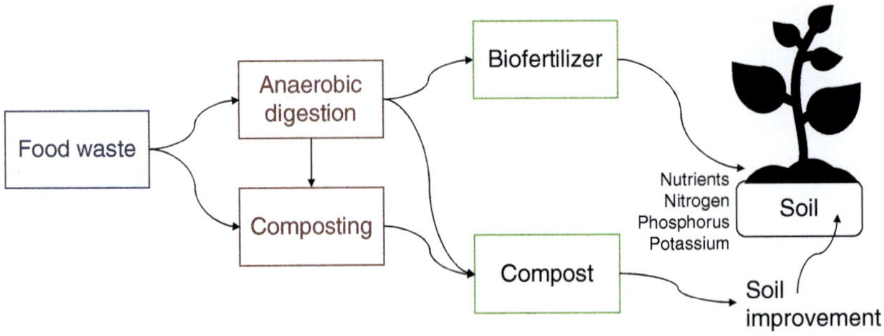

Figure 3.3.2 Conversion pathways for food waste as a resource to produce cereals, fruit, and vegetables.

3.3.3 Food Waste as Resource in Open Field Cultivation

Food such as cereals, fruit, and vegetables produced in open field cultivation will often require inputs such as nutrients and soil improvement to avoid erosion. Food waste can be used as a resource to produce the necessary inputs, as illustrated in Figure 3.3.2.

Modern agriculture can result in a decrease in carbon stocks in the soil and an increase in soil erosion, leading to runoff of nutrients. Applying compost to soil can contribute to improving the carbon content in the soil and thus increase soil quality. The compost can be produced from food waste directly or it can be produced from digestate after anaerobic digestion of the waste. A study from the United States showed that high rates of application of compost had showed significant increases in carbon stock in soil compared with low rates of compost application and control soils [11]. In addition to improvements in total carbon, the study also found reduced bulk density and increased microbial activity, total nitrogen (in comparison to control soils), water holding capacity, and water infiltration rate.

The digestate that is generated in addition to biogas when food waste is treated by anaerobic digestion of food waste contains the necessary nutrients for food production, such as nitrogen, phosphorus, and potassium, and can be used without any further treatment [12]. It can thus replace mineral fertilizer, which can result in a reduction of greenhouse gas emissions [13]. The use of digestate from food waste as a biofertilizer can also improve resource efficiency and reduce the need for extracting scarce resources. The EU has listed phosphorus as a critical raw material [14] and previous studies have shown that the disturbances of natural cycles of phosphorous and nitrogen due to human activities are unsustainable [15].

The digestate can also undergo different treatment techniques to obtain wanted properties and nutrient composition. For example, separation technologies can be used to separate the phosphorus resources, which is likely to end up in the dry fraction, from nitrogen resources, which will follow the liquid phase. Separation of digestate and manure can be an effective measure to a better distribution of phosphorus resources by exporting phosphorus from high livestock density regions to regions with arable farming [16].

Despite its potential, the use of digestate as a fertilizer is limited in many countries due to unfamiliarity with the product and insufficient confidence in its quality and safety [12]. The digestate may contain physical impurities, such as plastics, stones, and glass, chemical impurities such as heavy metals, organic pollutants, and pathogens, and other unwanted biological matter [12]. The impurities will depend on the input material into the digestate. Digestate produced from food chains are in most cases low in chemical impurities [17], but impurities of plastics due to human errors during separation or plastic waste collection bags may occur. Development of a quality standard and improving the communication of the benefits is of great importance to ensure food safety and acceptance of products.

3.3.4 Food Waste as Resource in Livestock Husbandry

The ruminant nutritional cycle is naturally circular. Manure from the animal fertilizes fields of grass, which are food for the animal. In many cases, however, modern meat and milk production are based on linear economy concepts. Fodder is produced from other protein sources and often transported long distances, and grass is cultivated using mineral fertilizer, produced from scarce resources.

In meat production, environmental impacts relating to the raw materials required, such as production of fodder and other nutritional components, have a critical role. A Spanish study found that raw materials contributed more than 46.9% of the environmental impacts, regardless of the impact category considered [18]. Meat production is known to cause considerable environmental impacts and to be resource intensive [19, 20], both in terms of energy and water use, and the need for phosphorus resources. These impacts may potentially be reduced by creating more circular food systems. Figure 3.3.3 illustrates how food waste can be used as feed directly or used as input in a biorefinery to produce meat and milk.

A Spanish study compared the potential environmental impact of a circular economy-based system for pork production with a traditional linear pork production, applying life cycle assessment methodology. The conclusion was that moving toward a fully circular value chain would lead to a reduction in all the environmental impact categories studied [9]. The largest reductions were from substituting fodder ingredients imported from other counties. Additional reductions were obtained if pig manure was used as the fertilizer in the cultivation of fodder.

Food waste can be used as feed for animals as a substitute for conventional feed, which has been explored in several studies [21, 22]. To maintain health and safety standards, the food waste must be without impurities and should not contain meat from the same species. The safety standards can be maintained by using homogenous waste with a known origin, such as surplus bread and other baked goods from bakeries and surplus fruit and vegetables from retailers. Using food waste as feed for animals may result in lower environmental impacts than anaerobic digestion and composting [10].

Treatment of food waste and manure through anaerobic digestion can generate both renewable energy for the farm and digestate that can be used as fertilizer for grass as a substitute for mineral

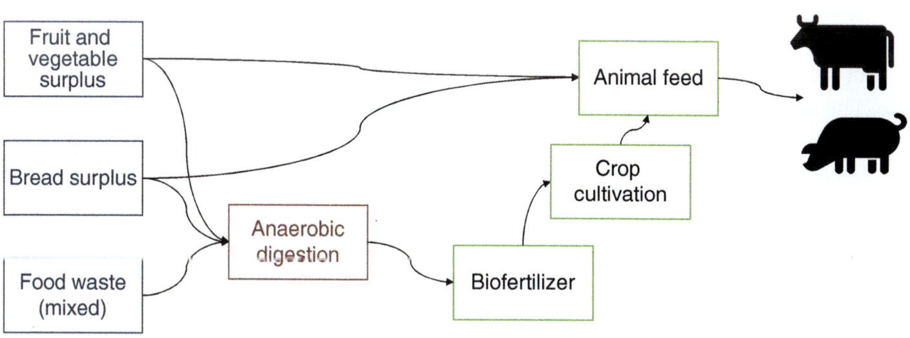

Figure 3.3.3 Conversion pathways for food waste as a resource to produce meat and milk.

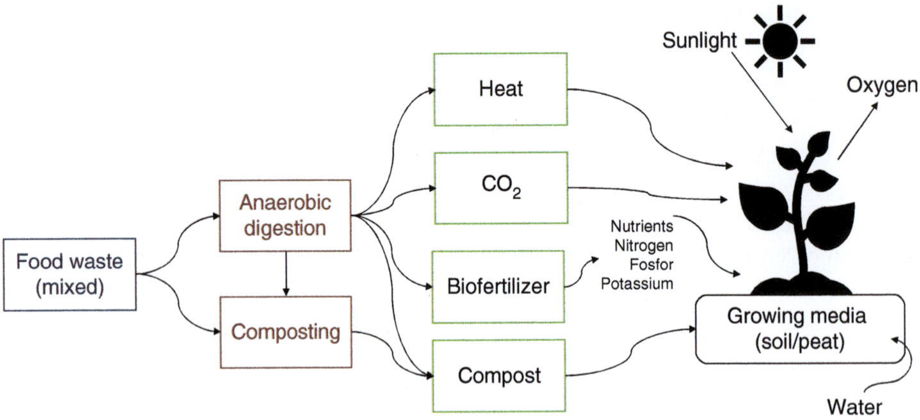

Figure 3.3.4 Recycling of food waste into resources as input to food production in greenhouse cultivation.

fertilizer [13]. The energy required on the farm for meat and milk production varies greatly [23]. The economic viability of investing in a farm-scale biogas plant will depend on the energy required on the farm and the energy prices saved [24].

3.3.5 Food Waste as a Resource in Greenhouse Farming

In open-field cultivation farmers provide water and nutrients in the form of fertilizer. Nature does, however, provide the important basis necessary for obtaining yields: soil containing carbon and nutrients, rain, sunlight, and carbon dioxide present in the atmosphere. In contrast, greenhouse cultivation is a closed system that requires more inputs.

Green plants capture light and carbon dioxide during photosynthesis, which enables them to grow and generate oxygen. In greenhouses, the growing conditions are optimized by adding heat and increasing the concentration of CO_2 in the air. Traditional greenhouse gas production is often based on non-renewable resources such as propane for heating and an increase of CO_2 concentration in the air, growing media produced from peat, mineral fertilizers, and energy for heat and light.

Food waste treated in anaerobic digestion plants generates products that can provide most of the necessary inputs for greenhouse cultivation, and has proved to be technically possible [25]. This is illustrated in Figure 3.3.4.

Biogas can be used to produce heat and electricity. If the biogas is upgraded to biomethane, CO_2 is removed from the gas and can be extracted and compressed and distributed or fed by pipes into a greenhouse. The digestate from anaerobic digestion can serve as a growing media, nutrient fertilizer, and water source, dependent on the properties and treatment method.

The liquid fraction after separation of digestate has been proven to be a satisfactory substitute for mineral fertilizer for short cycle crops (e.g. lettuce, herbs) [25]. The solid fraction of digestate can be used as a growing medium, but must be aged, composted, or preferably vermicomposted to be successfully used in intensive cultivation with traditional substrates [25]. Digestate can also successfully use substituted chicken manure in straw composts in mushroom production [26].

3.3.6 Conclusion and Way Forward

As described in Chapter 3.2, food waste decomposes and changes properties over time, which can make it harder to envisage closed-loop recycling systems for food waste than of other types of wastes such as plastics or metals. Agricultural systems are, however, based on the natural ecosystem and make use of nature's cycles of elements such as water, carbon, nitrogen, and phosphorus. The loop of nutrients related to the food system can thus, in principle, be closed [27].

There have, however, been few attempts of assessing systematically the resources needed to produce food and match these with the food waste resources available. In addition, the products generated from food waste treatment are often seen as a consequence of waste treatment, and not as valuable products that can be further developed and targeted toward a market. For example, most anaerobic digestion plant operators have to date focused on the input side of their business, concentrating their resources on energy production and sometimes regarding digestate as waste [28]. Consequently, the market potential for digestate has not been properly analyzed. In some cases, the digestate represent a cost and a barrier toward increased source separation and increased biogas production, while in other cases prices for biofertilizer are strongly positive [28]. Digestate does, nevertheless, represent a renewable system that utilizes recycled nutrients that would otherwise be lost. Mineral fertilizers, on the other hand, have traditionally relied on non-renewable resources. For this reason, it is regarded as unlikely that the costs of mineral extraction will decline in the future [28]. Increase in prices for competitive products will be an advantage for the products produced from food waste.

This subchapter has illustrated that there is a range of possibilities of using food waste as a resource for food consumption, depending on the composition of the waste. Through different conversion pathways, intermediate products can be produced, such as animal feed, compost, biofertilizer, and CO_2. These products may be used in different farming practices to produce food products such as meat, milk, cereals, fruit, and vegetables. By using the organic waste resources, it seems to be possible to supply remote locations with locally produced vegetables [25]. Obtaining the necessary safety standards and increase knowledge in agriculture about the alternatives and their advantages are, however, of large importance to obtain more circular food systems.

References

1 FAO. 2013. *Food wastage footprint. Impacts on natural resources.* Summary Report.

2 Corrado, S. and Sala, S. (2018). Food waste accounting along global and European food supply chains: state of the art and outlook. *Waste Management* 79: 120–131.

3 United Nations. (n.d.) Webpage: About the sustainable development goals. Available from: https://www.un.org/sustainabledevelopment/sustainable-development-goals.

4 Corrado, S., Caldeira, C., Eriksson, M. et al. (2019). Food waste accounting methodologies: challenges, opportunities, and further advancements. *Global Food Security* 20: 93–100.

5 Hanssen, O.J., Syversen, F., and Stø, E. (2016). Edible food waste from Norwegian households—detailed food waste composition analysis among households in two different regions in Norway. *Resources, Conservation and Recycling* 109: 146–154.

6 Malefors, C., Callewaert, P., Hansson, P.-A. et al. (2019). Towards a baseline for food-waste quantification in the hospitality sector – quantities and data processing criteria. *Sustainability* 11 (13): 3541.

7 Papargyropoulou, E., Lozano, R., Steinberger, K.J. et al. (2014). The food waste hierarchy as a framework for the management of food surplus and food waste. *Journal of Cleaner Production* 76: 106–115.

8 Union E. Council Directive. 1999. 1999/31/EC of 26 April 1999 on the landfill of waste.

9 Filimonau, V. and Gherbin, A. (2017). An exploratory study of food waste management practices in the UK grocery retail sector. *Journal of Cleaner Production* 167: 1184–1194.

10 Cicatiello, C., Franco, S., Pancino, B. et al. (2017). The dark side of retail food waste: evidences from in-store data. *Resources, Conservation and Recycling* 125: 273–281.

11 Brown, S. and Cotton, M. (2011). Changes in soil properties and carbon content following compost application: results of on-farm sampling. *Compost Science & Utilization* 19 (2): 87–96.

12 Al Seadi, T. and Lukehurst, C. 2012. Quality management of digestate from biogas plants used as fertiliser.

13 Lyng, K.-A., Modahl, I.S., Møller, H. et al. (2015). The BioValueChain model: a Norwegian model for calculating environmental impacts of biogas value chains. *The International Journal of Life Cycle Assessment* 20 (4): 490–502.

14 European Commission. COM. (2017). 490 final Communication from the Commission to the European Parliament, the Council, the European Economic and Social Committee and the Committee of the Regions on the 2017 list of Critical Raw Materials for the EU.

15 Rockström, J., Steffen, W., Noone, K. et al. (2009). A safe operating space for humanity. *Nature* 461 (7263): 472–475.

16 Hanserud, O.S., Lyng, K.-A., Vries, J.W.D. et al. (2017). Redistributing phosphorus in animal manure from a livestock-intensive region to an arable region: exploration of environmental consequences. *Sustainability* 9 (4): 595.

17 Govasmark, E., Stäb, J., Holen, B. et al. (2011). Chemical and microbiological hazards associated with recycling of anaerobic digested residue intended for agricultural use. *Waste Management* 31 (12): 2577–2583.

18 Noya, I., Aldea, X., González-García, S. et al. (2017). Environmental assessment of the entire pork value chain in Catalonia – a strategy to work towards circular economy. *Science of the Total Environment* 589: 122–129.

19 Baroni, L., Cenci, L., Tettamanti, M., and Berati, M. (2007). Evaluating the environmental impact of various dietary patterns combined with different food production systems. *European Journal of Clinical Nutrition* 61 (2): 279–286.

20 Swain, M., Blomqvist, L., McNamara, J., and Ripple, W.J. (2018). Reducing the environmental impact of global diets. *Science of the Total Environment* 610–611: 1207–1209.

21 Salemdeeb, R., zu Ermgassen, E.K.H.J., Kim, M.H. et al. (2017). Environmental and health impacts of using food waste as animal feed: a comparative analysis of food waste management options. *Journal of Cleaner Production* 140: 871–880.

22 San Martin, D., Ramos, S., and Zufía, J. (2016). Valorisation of food waste to produce new raw materials for animal feed. *Food Chemistry* 198: 68–74.

23 Kythreotou, N., Florides, G., and Tassou, S.A. (2012). A proposed methodology for the calculation of direct consumption of fossil fuels and electricity for livestock breeding, and its application to Cyprus. *Energy* 40 (1): 226–235.

24 Lyng, K.-A., Bjerkestrand, M., Stensgård, A.E. et al. (2018). Optimising anaerobic digestion of manure resources at a regional level. *Sustainability* 10 (1): 286.

25 Stoknes, K., Scholwin, F., Krzesiński, W. et al. (2016). Efficiency of a novel "food to waste to food" system including anaerobic digestion of food waste and cultivation of vegetables on digestate in a bubble-insulated greenhouse. *Waste Management* 56: 466–476.

26 Stoknes, K., Beyer, D.M., and Norgaard, E. (2013). Anaerobically digested food waste in compost for *Agaricus bisporus* and *Agaricus subrufescens* and its effect on mushroom productivity. *Journal of the Science of Food and Agriculture* 93 (9): 2188–2200.

27 Jurgilevich, A., Birge, T., Kentala-Lehtonen, J. et al. (2016). Transition towards circular economy in the food system. *Sustainability* 8 (1): 69.

28 Dahlin, J., Herbes, C., and Nelles, M. (2015). Biogas digestate marketing: qualitative insights into the supply side. *Resources, Conservation and Recycling* 104: 152–161.

4

Definitions and Procedures for Characterization of Biodegradable Waste

Helena Raclavská, Jana Růžičková, Marek Kucbel, Barbora Švédová, and Konstantin Raclavský

VŠB – Technical University of Ostrava, Centre ENET, Ostrava, Czech Republic

4.1 Definitions

The framework Directive 2008/98/EC [1] on Waste is the main legislative document on waste at the EU level. As it is a directive, it is incorporated into the national law of the Member States as separate legal acts. This Directive defines waste as follows: "Waste is any substance or object which the holder discards or intends or is required to discard." On 2 July 2014, the European Commission adopted a legislative proposal to review waste-related targets in the "Landfill Directive" [2] as well as recycling and other waste-related targets in "Directive 2008/98/EC [1] on Waste" and "Directive 94/62/EC on Packaging and Packaging Waste" [3]. According to this proposal, disposal of recyclable waste (paper, plastics, metals, glass, and bio-waste) in non-hazardous waste landfills should decrease by 2025 to a maximum landfilling rate of 25%.

Under aerobic conditions, in the presence of oxygen, practically all biological wastes are biodegradable. Most biological materials are also biodegradable under anaerobic conditions – without oxygen [4].

"Biodegradable waste" is defined as a broader concept in the "Directive 1999/31/EC on the Landfill of Waste" [2] as any waste that can undergo aerobic or anaerobic decomposition, e.g. paper, cardboard, food, and garden waste (Figure 4.1).

4.1.1 Types of Biodegradable Waste

Biodegradable fractions of municipal solid waste (MSW):

- *Biodegradable Municipal Waste* (BMW) represents the biodegradable component of municipal waste, not including bio-stabilized residual waste. BMW is typically composed of food and garden waste, wood, paper, cardboard, and textiles.
- Other biodegradable wastes. Other biodegradable wastes that may be composted on their own or together with the biodegradable fraction of MSW include mainly the following items:
  - commercial food waste, not collected as part of MSW, including waste from markets, catering waste;
  - forestry residues including bark and wood residues;
  - waste from agriculture including animal husbandry excrements (solid and liquid manure), straw residues, sugar beet and potato haulm, residues from growing beans, peas, flax, vegetables, and spent mushroom compost;

Biodegradable Waste Management in the Circular Economy: Challenges and Opportunities, First Edition.
Edited by Malgorzata Kacprzak, Eleonore Attard, Kari-Anne Lyng, Helena Raclavska, BalRam Singh, Eyob Tesfamariam, and Franck Vandenbulcke.

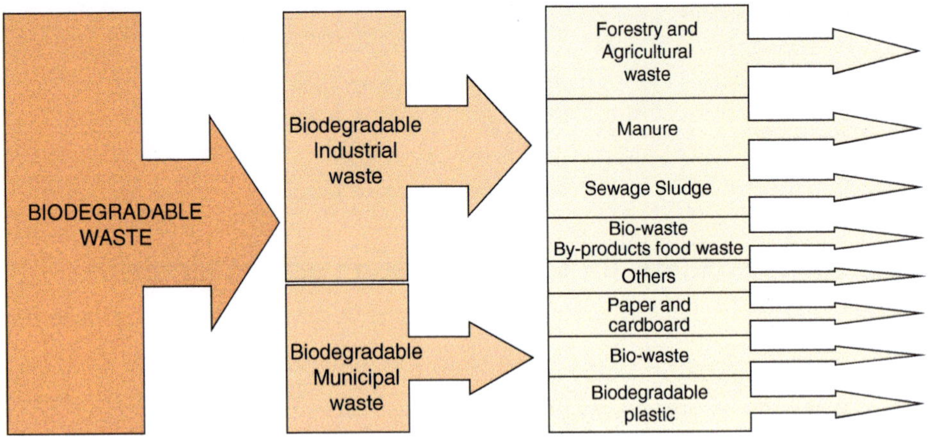

Figure 4.1 Composition of biodegradable waste modified [5].

- wastes from the food and beverage industry, including breweries and malt houses, wineries, the fruit and vegetable production industry, the potato industry, including starch, residues of beet sugar production, meat production, slaughterhouse residues, and whey;
- sewage sludge (derived from biological treatment of municipal wastewater).

"Bio-waste" is defined as "biodegradable garden and park waste, food and kitchen waste from households, restaurants, caterers and retail premises, and comparable waste from food processing plants according to Directive 2008/98/EC [1] on Waste and Repealing Certain Directives." It does not include forestry or agricultural residues, manure, sewage sludge, or other biodegradable waste such as natural textiles, paper, or processed wood. It also excludes those by-products of food production that never become waste. The term "bio-waste" is not used in the same way in different EU countries and sometimes refers only to kitchen waste and excludes green waste (Figures 4.2 and 4.3).

"Municipal solid waste" (MSW) includes waste from private households and other similar waste that is collected together with household waste by municipalities. The composition of MSW varies

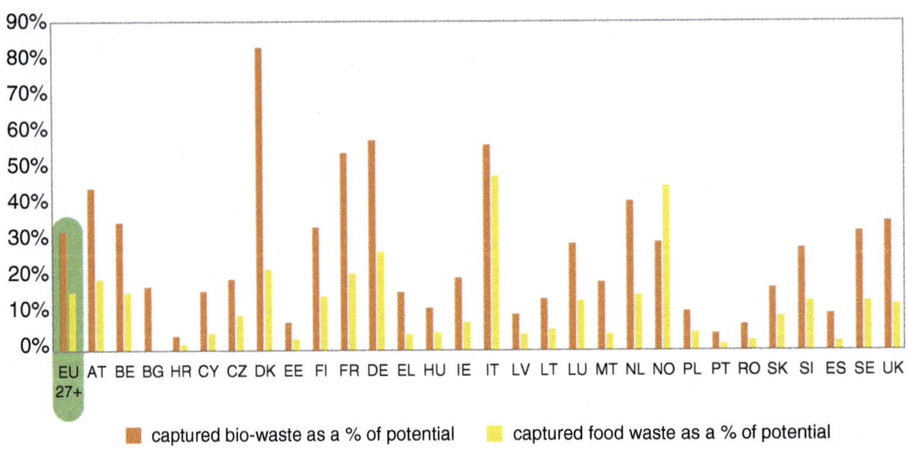

Figure 4.2 The comparison of the theoretical potential for the production of bio-waste and food waste with real collected waste [6].

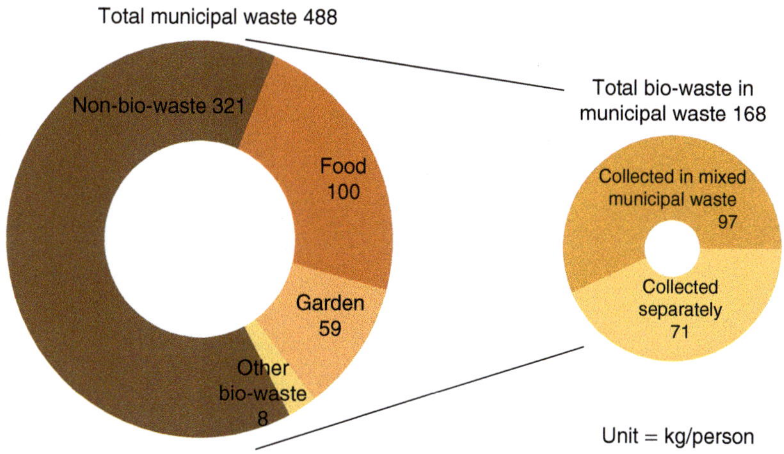

Figure 4.3 Bio-waste in municipal waste and how it is collected, EU-28, 2017 [7].

significantly from city to city and state to state, but it always has an important part of biological material. Food waste and "green" waste from gardens and parks make up from 30 to 50% of the total weight of MSW.

4.2 Production of Biodegradable Waste

Approximately 88 million metric tons (Mt) of food are wasted in the EU every year, which means 173 kg per citizen. That represents significant wasting of resources, but it also influences climate change. The European Parliament is preparing new measures to decrease food waste in the EU by 50%. In the EU, most of the food is wasted in households (53%) and processing (19%) [8]. Minimization of food waste is significant for a sustainable food system, which should provide food for the growing population of the world and at the same time preserve natural systems [9].

The production of municipal food waste (households and food service) in an EU27 average is 113 ± 12 kg capita^{-1} year^{-1}. The lowest amount of food waste is produced in Norway (78.8 kg of food waste per capita), and the highest amount is generated in Spain (114 kg capita^{-1} year^{-1}). Waste production was calculated according to the method developed by Bräutigam et al. [10]. Green waste produced in the regions with the northern and continental climate in cities reaches 40 kg person^{-1} year^{-1}, town, and suburban areas 160 kg person^{-1} year^{-1} and in rural areas 200 kg person^{-1} year^{-1}. In Mediterranean climatic conditions, garden production is lower: in cities 10, in towns + suburban areas 50, and in rural 100 kg person^{-1} year^{-1} [6]. The number of inhabitants being 513, 481, and 690 in EU27, the production estimate of food waste generation is 59 Mt and bio-waste 114 Mt, which corresponds to approximately 52% of food waste in bio-waste. The direct assessment of the food waste content in bio-waste provided a considerably lower value, approximately only 30% compared with the value from the estimate [6].

Municipal solid waste (MSW) in the EU, usually contains between 30 and 40% of bio-waste but it can range from 18% up to 60%. Bio-waste (green waste and food waste) accounts for 41.2% of all MSW sources segregated in Italy in 2016. Intensive collection schemes for bio-waste in Italy divert about 6.5 Mt of bio-waste from disposal to recycling.

It can be considered that the MSW management represents a strategic supply chain problem because it involves waste production, collection, transport, separation, processing, distribution, and disposal [11]. The solution of recycling and recovery of FW from MSW is key in terms of meeting the objectives of the European Directive (EU) 2018/851, known as the "New Waste Framework Directive (WFD)" [12]. Achieving the target of 10% landfill is only possible by sustainable management of bio-waste [13]. Designing an optimal solution for MSW management includes mitigation between costs, disposal of waste, and efficient use of produced waste [11].

4.2.1 Methods for Determination of the FW and BMW Production

The identification of food waste produced in the individual EU Member States was a problem within the non-uniform methodology used for quantification [14]. By implementation of the EC guidelines, published in May 2018 in the form of the "Commission Delegated Decision (EC) supplementing Directive 2008/98/EC [1] of the European Parliament and of the Council as regards a standard methodology and minimum quality requirements for the uniform measurement of the food waste levels," the data is objectified. "Commission Delegated Decision (EU) 2019/1597 of 3 May 2019 supplement Directive 2008/98/EC of the European Parliament and of the Council as regards a common methodology and minimum quality requirements for the uniform measurement of the food waste levels" [15]. The adoption of a uniform and consolidated methodology for measuring food waste at the EU level is an important step in addressing food losses and wastage and generating food waste. Within this decision, direct and indirect methods are recommended for measuring food waste production [16]. Direct methods include direct physical access to food waste and their physical measurements. These methods represent direct measuring, waste composition analysis, scanning, counting, and maintaining records. If the direct analysis is not possible, the methodology recommends the use of indirect methods, which consist of calculation of mass balance with application of coefficients. The methodology for calculating food waste for individual stages of FC, including households, is stated by [17].

Design of technology for the BMW management is only possible if its share in MSW is known. The standardized ASTM D5231-92 [18] method is available for determining the biodegradable waste content in MSW. In addition to the direct determination method, BMW uses EPA Approved Factors, which take into account the 1, 2, or 3 Bin System – segregated collection [19].

4.2.2 Reasons for Decreasing the Biodegradable Waste Production

In accordance with the "Landfill Directive (1999/31/EC)" [2], Member States are required to decrease the amount of BMW landfilled. Landfilling of food waste under anaerobic conditions contributes to the increasing generation of landfill gases – methane and carbon dioxide [20]. Up to 73% of greenhouse gas emissions related to food waste come from food production, 6% from food processing, 7% from retail and distribution, and 8% from food preparation and consumption, with only 6% from food waste disposal. Food waste is a source of unpleasant odors, and seepage from landfills can contaminate groundwater [21]. The presence of food waste in landfills affects the occurrence and multiplication of rodents, birds, and other undesirable animals that can be carriers of various diseases.

Excess food production increases water and energy consumption, which contributes negatively to climate change. According to the FAO (the UN's Food and Agricultural Organization), food waste is a source of approximately 8% of all anthropogenic greenhouse gas emissions. Every kilogram of food produced causes the release of 4.5 kg of CO_2 into the atmosphere [22].

4.2.3 Goals of the EU in the Field of Biodegradable Waste Management

The European Commission (2018) proposed "new targets for municipal waste – 55% is intended to be recycled and prepared for reuse (by 2025). This percentage is supposed to rise to 65% (by 2035)" [23]. The policy rules for the management of bio-waste in the EU may be summarized as follows:

- European Directive (EU) 2018/851, known as the "New Waste Framework Directive (WFD)," included in the "Circular Economy Package," requires the implementation of separate collections of bio-waste as of 1 January 2024 [12].
- Quality recycling is a key issue. The calculation of recycling rates to assess compliance with the EU targets (65% "preparation for reuse and recycling, i.e., net recycling including organic recycling, by 2035") will have to subtract rejects, which are strictly related to impurities included in separated fractions; this puts the emphasis on collection schemes that can ensure a high quality of collected materials.
- Other environmental drivers, aside from the Circular Economy vision and strategy, are propelling interest in separation, processing, and recovery of bio-waste: the production of compost that can enrich soils with bioavailable organic carbon. Reuse of compost will improve soil structure, microbial diversity, and water retention capacity [6].
- Use of bio-waste streams as a resource for the products with value-added applications. Introduction of cascading utilization of biodegradable waste refers to sequential recycling of a material into another type of product at its end of life [24].
- New quality standards and quality assurance of compost, digestate, fertilizers, and soil additives ensuring the minimization of environmental risks in the application of these products ("Regulation EU 2019/1009 of the European Parliament and of the Council of 5 June 2019 laying down rules related to making fertilizing products available on the EU market, and amending Regulations EC No. 1069/2009 and EC No. 1107/2009 and repealing Regulation EC No. 2003/2003") [25].

The introduction of separate collections of bio-waste usually requires significant initial operational investment from the public sector and the cooperation of citizens to ensure the overall efficiency of the system. In many EU countries, there are separate collection systems, particularly of green waste. Significant differences exist in the facilities for separated collection and treatment of bio-waste across Europe. Countries such as Austria, Switzerland, Germany, the Netherlands, Flanders (Belgium), Sweden, and Norway had already decided for separate bio-waste collection and processing systems 15 years ago. Other countries, including the UK, Italy, Finland, Ireland, Slovenia, Estonia, and France achieved significant improvements during this period. Actual costs depend on many factors, including the collection system (door-to-door collection or containers on the roadside), population density, collection frequency, and weather conditions.

4.3 The Selection of Technology

In order to use bio-waste as a source for the production of a high-quality fertilizer or soil adjuvant to improve soil properties, this waste must be collected separately at source, which ensures low contamination. For separately collected bio-waste, composting is used more often than anaerobic digestion. Biological treatment (anaerobic digestion and composting) of bio-waste belongs to recycling if compost or digestate is used on soil or for the production of growing substrates. More than 90% of garden and food waste is processed in the form of compost. In case of separate collections, food waste can be used for the production of high-value products such as bio-based chemicals – bioplastics, colorants, acids, enzymes, and biopesticides [26–29].

The EU Member States follow different approaches when determining the status of compost, i.e. whether it is considered a waste or not. It represents end-of-waste defined by national regulations under waste law or other national environmental regulations. In fact, it is an end-of-waste when registered as a fertilizer. In some countries, compost is not considered waste if the following criteria are met [4]:

- The list of waste from which the compost may be produced. Biodegradable wastes that are currently regarded as suitable for composting or digestion.
- Specifications of the product quality (heavy metal threshold values).
- Temperature–time profile during composting to achieve hygienic safety.
- Quality control provisions on the input materials and the product.
- Mandatory record-keeping (for five years) of batch-wise information on input materials and products, including details of who receives the compost.
- Obligations for registering and notifying the authorities.
- Analytical methods [4].

The List of Waste (LoW) provides EU-wide common terminology for waste classification for easy waste management (Table 4.1). The LoW or Waste Classification Code or EWC code is the code used to identify waste as listed in the European Waste Catalog (EWC). All waste produced will have a corresponding EWC code. LoW in the version set out in "Decision 2000/532/EC as amended by Decision 2014/955/EU" has applied since 1 June 2015 [30].

Waste management in the EU varies greatly from one Member State to another. Three methods are most commonly used: incineration, biological treatment, and landfilling. Bio-waste biological treatment is most widespread in Germany and Austria. Countries that reduce bio-waste by combined heat and power generating incineration include Denmark, Sweden, The Netherlands, Luxembourg, and France. According to the European Compost Network, approximately 30 Mt of bio-waste is composted or digested at the present time. There are more than 3500 treatment plants with 2000 composting plants in Europe. Nevertheless, some EU countries choose other bio-waste disposal techniques such as incineration [31] or landfill. Less frequently used thermal techniques include pyrolysis and gasification. The right choice of bio-waste management in line with environmental and economic benefits is very often complicated by a number of factors: population density, infrastructure, the bio-waste collection system, bio-waste composition and quality, climatic conditions, and the economic potential of using different products obtained from bio-waste such as electricity, gas, heat, and compost. Anaerobic digestion is suitable for moist biological waste, including kitchen waste containing fats. The limiting process during anaerobic digestion is hydrolysis [32].

4.3.1 Criteria for Selection of Technology

To establish an efficient and integrated waste management system, it is usually not enough to morphologically characterize the waste fraction. It is also necessary to determine its characteristics through analysis of the physical and chemical properties as well as by determining its biodegradability. These properties are of great importance for further managing and utilization of biodegradable components, primarily as raw materials for energy recovery. During the

Table 4.1 Positive List of Waste suitable for compost and digestate production [4].

Group No.	Type of waste material	01	02	03	04	07	10	15	16	17	19	20
		EWC Code										
Waste for biological treatment from an exclusively vegetable origin (NO animal by-products or meat)												
1 1.1	Organic vegetable waste from garden and parks and other greens			03 01 01, 03 01 05								20 02 01, 20 01 38
1.2	Vegetable waste, from the preparation and consumption of food, luxury food, and beverages		02 01 03, 02 03 04								19 05 02	20 02 01, 20 03 02
1.3	Organic residues from commercial, agricultural, and industrial production, processing and marketing of agricultural and forestry products – purely of vegetable origin		02 01 03, 02 03 04, 02 03 05, 02 04 03, 02 05 02, 02 06 03, 02 07 01, 02 07 02, 02 07 04, 02 07 05	03 01 05, 03 03 10, 03 03 11								
1.4	Other organic residues – purely of vegetable origin		02 01 03, 02 01 07, 02 01 99, 02 03 04	03 03 01, 03 03 10,	04 02 10	07 05 14		15 01 01, 15 01 03, 15 01 09	16 03 06	17 02 01	19 05 01, 19 05 02, 19 05 99	20 01 01, 20 01 25
1.5	Digestion residues from anaerobic digestion of waste materials – pure vegetable origin		02 03 04			07 01 99					19 06 03, 19 06 04, 19 06 06	

(Continued)

Table 4.1 (Continued)

Group No.		Type of waste material	EWC Code 01	02	03	04	07	10	15	16	17	19	20
Waste for biological treatment with parts of animal origin													
2	2.1	Animal waste, especially waste from the preparation of foodstuffs		02 02 01, 02 02 02, 02 03 04, 02 07 05									20 01 08, 20 01 99
	2.2	Organic residues from commercial, agricultural, and industrial production, processing and marketing of agricultural and forestry products – with parts of animal origin		02 01 02, 02 01 06, 02 02 03, 02 02 01, 02 02 04, 02 02 99, 02 05 02								19 05 03, 19 05 99	
	2.3	Digestion residues from anaerobic treatment of waste materials, which may contain parts of animal origin		02 02 02, 02 02 03, 02 02 99, 02 05 01, 02 06 01								19 06 03, 19 06 04, 19 06 05, 19 06 06	
Further waste for biological treatment													
3	3.01	Wastes from the leather and fur industry'				04 01 01							
	3.02	Municipal sewage sludge										19 08 05	
	3.04	Biodegradable packaging and biodegradable plastics and biobased plastics					07 02 13		15 01 02, 15 01 05				

Additives for composting added in minor quantities (up to 15% of the composting feedstock, expressed in fresh weight) in order to improve the composting process, humification, and maturation or environmental performance of the processes

4	4.01	Rock dust	01 03 08, 01 04 09			
	4.02	Limestone dust		02 04 02		
	4.03	Bentonite				
	4.04	Ash from combustion of plant tissue (e.g., wood, straw)			10 01 01	
	4.05	Excavated soil		02 04 01		17 05 04, 17 05 06
	4.06	Washing soil from sugar beet and potato processing				
	4.07	Biochar				

Explanations: Numbers – EWC code used in the List of Waste.

determination of biodegradability of municipal solid waste and other types of waste, valuable information is gathered that facilitates the decision-making process within the entire waste management system, especially during the design and construction of waste management centers.

The most common methods used for determination of the biodegradability of waste materials are:

- Determination of the biogas potential. There is a lack of standardized procedures for bio-waste. A standardized procedure only exists for sludges from the wastewater treatment plants (ISO 11734).
- Various methods of determination were published for bio-waste [33–35].
- Determination of organic (TC – total carbon, TOC – total organic carbon) and inorganic carbon (IC) content in waste.
- Determination of loss on ignition (LOI) and the biomass content of waste is a measure of the quantity of organic matter in the sample that can be combusted at the temperature of 550 °C. The loss in weight during combustion is equal to the mass of organic matter in the sample. LOI is determined according to the European Standards EN 15169:2007 [36] and EN 14346:2006 [37].
- Determination of aerobic biological stability of waste materials (e.g., RDRI, AT4). The standard ASTM D5929 [38] defines the method to use to analyze and determine the effect of materials on the composting process and perform the assessment of the feasibility of the composting process as a method for managing organic solid waste material. Key parameters of process performance, including theoretical oxygen uptake (ThOU) and theoretical carbon dioxide production (ThCO2P), are determined according to the "ASTM D5929-18 Respirometry Methods" [38], "ASTM D5338-11", and "The Standard Test Method for Determining Aerobic Biodegradation of Plastic Materials under Controlled Composting Conditions Incorporating Thermophile Temperatures" [39].

The application of the principles of circular economics in the management of single-type biodegradable waste from direct production (food production, wood waste) can bring economic and environmental benefits by ensuring the production of products with a higher added value and by saving primary resources. Biodegradable municipal waste (BMW), as part of municipal solid waste (MSW), remains a problem. The information on the quality of "clean kitchen waste collected separately" will be available only after the introduction of separate collections after 2025. In terms of valorization of such waste, a limiting factor for its further use may be newly monitored pollutants (e.g. microplastics).

References

1 The European Parliament. 2008. The Council of the European Union. Directive 2008/98/EC of the European Parliament and of the Council of 19 November 2008 on waste and repealing certain Directives (Text with EEA relevance).

2 The Council of the European Union. 1999. Council Directive 1999/31/EC of 26 April 1999 on the landfill of waste.

3 The European Parliament. 1994. *The Council of the European Union*. European Parliament and Council Directive 94/62/EC of 20 December 1994 on packaging and packaging waste.

4 Saveyn, H. and Eder, P. 2014. Institute for Prospective Technological Studies. End-of-waste criteria for biodegradable waste subjected to biological treatment (compost and digestate): technical

proposals. [Internet]. Luxembourg: Publications Office. Available from: http://dx.publications.europa.eu/10.2788/6295.

5 Ponsa, S., Bolzonella, D., Colon, J. et al. (2017). Recovering energy from sludge. In: *Innovative wastewater treatment and resource recovery technologies: impacts on energy, economy and environment [Internet].* (eds. J.M. Lema and S. Suarez), 325–354. International Water Association. Available from: https://iwaponline.com/ebooks/book/302/chapter/660172/Recovering-energy-from-sludge.

6 Favoino, E., Giavini, M., del Parco di Monza, SA. 2020 Bio-waste generation in the EU: current capture levels and future potential [Internet]. Bio-based Industries Consortium (BIC). Available from: https://zerowasteeurope.eu/wp-content/uploads/2020/07/2020_07_06_bic_zwe_report_bio_waste.pdf.

7 European Environment Agency (EEA). 2020. *Bio-waste in Europe-turning challenges into opportunities [Internet].* Luxembourg: European Environment Agency. 56 pp. Available from: https://www.eea.europa.eu/publications/bio-waste-in-europe.

8 European Parliament. 2017. *Food waste: the problem in the EU in numbers [Internet].* European Parliament. Available from: https://www.europarl.europa.eu/news/en/headlines/society/20170505STO73528/food-waste-the-problem-in-the-eu-in-numbers-infographic.

9 Willett, W., Rockström, J., Loken, B. et al. (2019). Food in the anthropocene: the EAT–Lancet commission on healthy diets from sustainable food systems. *The Lancet* [Internet]. 393 (10170): 447–492. Available from: https://linkinghub.elsevier.com/retrieve/pii/S0140673618317884.

10 Bräutigam, K.R., Jörissen, J., and Priefer, C. (2014). The extent of food waste generation across EU-27: different calculation methods and the reliability of their results. *Waste Management & Research* [Internet]. 32 (8): 683–694. Available from: http://journals.sagepub.com/doi/10.1177/0734242X14545374.

11 Mohammadi, M., Jämsä-Jounela, S.L., and Harjunkoski, I. (2019). Optimal planning of municipal solid waste management systems in an integrated supply chain network. *Computers & Chemical Engineering* [Internet]. 123: 155–169. Available from: https://linkinghub.elsevier.com/retrieve/pii/S0098135418302813.

12 European Parliament. 2018. Council. Directive (EU) 2018/851 of the European Parliament and of the Council of 30 May 2018 amending Directive 2008/98/EC on waste (Text with EEA relevance)

13 Razza, F., D'Avino, L., L'Abate, G., and Lazzeri, L. (2018). The role of compost in bio-waste management and circular economy. In: *Designing Sustainable Technologies, Products and Policies* [Internet]. (eds. E. Benetto, K. Gericke and M. Guiton), 133–143. Cham: Springer International Publishing. Available from: http://link.springer.com/10.1007/978‐3‐319‐66981‐6_16.

14 Yu, Y. and Jaenicke, E.C. (2020). Estimating food waste as household production inefficiency. *American Journal of Agricultural Economics* [Internet]. 102 (2): 525–547. Available from: https://onlinelibrary.wiley.com/doi/abs/10.1002/ajae.12036.

15 European Commission. 2019. *Commission Delegated Decision (EU) 2019/1597 of 3 May 2019 supplementing Directive 2008/98/EC of the European Parliament and of the Council as regards a common methodology and minimum quality requirements for the uniform measurement of levels of food waste (text with EEA relevance).*

16 Łaba, S., Niedek, M., Szczepański, K., and Łaba, R. (2019). Regulation of the food waste measuring in the EU in the light of the need of counteracting the food wastage. A. Kamińska-Dwórznicka, editor. *Environmental Protection and Natural Resources* 30 (4): 1–7.

17 Tostivint, C., Östergren, K., Quested, T. et al. (2016). *Food Waste Quantification Manual to Monitor Food Waste Amounts and Progression.* Paris: FUSIONS EU.

18 ASTM International. 2016. ASTM D5231-92 Standard test method for determination of the composition of unprocessed municipal solid waste. West Conshohocken: ASTM International. Report No.: ASTM D5231-92.

19 Campbell, R., Moriarty, J., Derham, J. et al. 2011 Protocol for the evaluation of biodegradable municipal waste sent to landfill. pp. 1–33. Available from: https://www.epa.ie/pubs/advice/waste/municipalwaste/EPA_Protocol_For%20BMW%20Final.pdf.

20 Moult, J.A., Allan, S.R., Hewitt, C.N., and Berners-Lee, M. (2018). Greenhouse gas emissions of food waste disposal options for UK retailers. *Food Policy* [Internet]. 77: 50–58. Available from: https://linkinghub.elsevier.com/retrieve/pii/S0306919217309168.

21 Yasin, N.H.M., Mumtaz, T., Hassan, M.A., and Abd Rahman, N. (2013). Food waste and food processing waste for biohydrogen production: A review. *Journal of Environmental Management* [Internet]. 130: 375–385. Available from: https://linkinghub.elsevier.com/retrieve/pii/S0301479713006002.

22 Food and Agriculture Organization (FAO). 2013. *Food wastage footprint: impacts on natural resources: summary report [Internet].* Rome: FAO; 61 pp. Available from: http://www.fao.org/3/i3347e/i3347e.pdf.

23 The European Commission. 2018 *Environment. Commission reviews implementation of EU waste rules, proposes actions to help 14 Member States meet recycling targets [Internet].* Available from: https://ec.europa.eu/info/news/commission-reviews-implementation-eu-waste-rules-proposes-actions-help-14-member-states-meet-recycling-targets-2018-sep-24_en.

24 Carus, M. (2017). Biobased economy and climate change – Important links, pitfalls, and opportunities. *Industrial Biotechnology* [Internet]. 13 (2): 41–51. Available from: http://www.liebertpub.com/doi/10.1089/ind.2017.29073.mca.

25 The European Parliament. 2019. *The Council of the European Union.* Regulation (EU) 2019/1009 of the European Parliament and of the Council of 5 June 2019 laying down rules on the making available on the market of EU fertilising products and amending Regulations (EC) No 1069/2009 and (EC) No 1107/2009 and repealing Regulation (EC) No 2003/2003 (Text with EEA relevance).

26 Vea, E.B., Romeo, D., and Thomsen, M. (2018). Biowaste valorisation in a future circular bioeconomy. *Procedia CIRP* [Internet]. 69: 591–596. Available from: https://linkinghub.elsevier.com/retrieve/pii/S2212827117308387.

27 Mirabella, N., Castellani, V., and Sala, S. (2014). Current options for the valorization of food manufacturing waste: a review. *Journal of Cleaner Production* [Internet]. 65: 28–41. Available from: https://linkinghub.elsevier.com/retrieve/pii/S0959652613007440.

28 Machmudah, S., Zakaria, W.S., Sasaki, M. et al. (2012). Lycopene extraction from tomato peel by-product containing tomato seed using supercritical carbon dioxide. *Journal of Food Engineering* [Internet]. 108 (2): 290–296. Available from: https://linkinghub.elsevier.com/retrieve/pii/S0260877411004572.

29 Galanakis, C.M. (2012). Recovery of high added-value components from food wastes: Conventional, emerging technologies and commercialized applications. *Trends in Food Science & Technology* [Internet]. 26 (2): 68–87. Available from: https://linkinghub.elsevier.com/retrieve/pii/S0924224412000532.

30 The European Commission. 2014. 2014/955/EU: Commission Decision of 18 December 2014 amending Decision 2000/532/EC on the list of waste pursuant to Directive 2008/98/EC of the European Parliament and of the Council Text with EEA relevance.

31 Scarlat, N., Fahl, F., and Dallemand, J.F. (2019). Status and opportunities for energy recovery from municipal solid waste in Europe. *Waste and Biomass Valorization* [Internet]. 10 (9): 2425–2444. Available from: http://link.springer.com/10.1007/s12649‐018‐0297‐7.

32 Paritosh, K., Yadav, M., Mathur, S. et al. (2018). Organic fraction of municipal solid waste: overview of treatment methodologies to enhance anaerobic biodegradability. *Front Energy Research* [Internet]. 6: 75. Available from: https://www.frontiersin.org/article/10.3389/fenrg.2018.00075/full.

33 Jingura, R.M. and Kamusoko, R. (2017). Methods for determination of biomethane potential of feedstocks: a review. *Biofuel Research Journal* [Internet]. 4 (2): 573–586. Available from: https://www.biofueljournal.com/article_46479.html.

34 Filer, J., Ding, H.H., and Chang, S. (2019). Biochemical methane potential (BMP) assay method for anaerobic digestion research. *Water* [Internet]. 11 (5): 921. Available from: https://www.mdpi.com/2073-4441/11/5/921.

35 Zheng, W., Phoungthong, K., Lü, F. et al. (2013). Evaluation of a classification method for biodegradable solid wastes using anaerobic degradation parameters. *Waste Management* [Internet]. 33 (12): 2632–2640. Available from: https://linkinghub.elsevier.com/retrieve/pii/S0956053X13003887.

36 European Standards. 2007. *EN 15169:2007 Characterization of waste – Determination of loss on ignition in waste, sludge and sediments.* p. 16.

37 European Standard. 2006. *EN 14346:2006 Characterization of waste – Calculation of dry matter by determination of dry residue or water content.* p. 24.

38 ASTM International. 2018. *ASTM D5929-18 Standard test method for determining biodegradability of materials exposed to source-separated organic municipal solid waste mesophilic composting conditions by respirometry.* West Conshohocken: ASTM International: p. 6.

39 ASTM International. 2011. *ASTM D5338-11 Standard test method for determining aerobic biodegradation of plastic materials under controlled composting conditions. Incorporating thermophilic temperatures.* West Conshohocken: ASTM International: p. 7.

5

Biodegradable Waste Streams

Helena Raclavská[1], Jana Růžičková[1], Barbora Švédová[1], Marek Kucbel[1], Michal Šafář[1], Konstantin Raclavský[1], and Elton Luis Da Silva Abel[2]

[1] *VŠB-Technical University of Ostrava, Centre ENET, Ostrava, Czech Republic*
[2] *CEDAE, Rio de Janeiro, Brazil*

5.1 Agriculture

The most important step for the circular economy is the introduction and use of biological materials while ensuring their higher form of use in terms of the value chain [1]. Table 5.1 provides examples of directions for ensuring better utilization of the input raw material in the value chain for selected industrial activities that use mainly material of plant origin.

Agricultural waste is defined as residues of plants or animals that are not used (or are not further processed) to obtain food or feed. Agricultural activities produce a significant amount of waste in the form of animal manure and sludge; residual biomass from cultivated crops and non-tradable products due to poor quality [2] accounts for up to 30% of agricultural production [3].

It is estimated that up to 121 million metric tonnes (Mt) of agricultural produce residues, 46 Mt of wood biomass residues, and 31 Mt of grass could be produced annually in Europe [4]. Up to 1400 Mt of cow and swine manure are generated annually in the EU, the most important producer being France [5].

5.1.1 Manure

The primary source of nutrients for growing crops was organic material obtained by livestock production or from plant residues. As a result of the intensification of crop production and the separation of arable and livestock farms, interest in the use of solid manure declined in the 1970s. The direct use of organic matter has certain limitations related to the variability of the composition, the high water content, and the nutrient ratio N : P : K, which is different for different crops. However, organic manures are a source of humic and semi-humic C, which helps to sequester C [6]. The required carbon in bioavailable form enters the soil. The highest amount of dissolved organic carbon is leached from poultry manure ($37\,mg\,l^{-1}$), while from cow manure it is only $1.2\,mg\,l^{-1}$ [7]. The direct use of manure for crop production can also bring secondary health and environmental risks. Veterinary medicinal products (tetracyclines, doxycyclines, sulfonamides, fluoroquinolones) and their transformation compounds may also persist in environmental matrices, may affect the microbial community, and can be ecotoxic [8]. After switching to mineral fertilizers, manure often remained unused and causes local environmental burdens. The global production of

Biodegradable Waste Management in the Circular Economy: Challenges and Opportunities, First Edition.
Edited by Malgorzata Kacprzak, Eleonore Attard, Kari-Anne Lyng, Helena Raclavska, BalRam Singh, Eyob Tesfamariam, and Franck Vandenbulcke.

Table 5.1 Possibilities of application of bio-based materials – directions of development for the increased added value of products [1].

Sector	Value chain
Chemicals	Cellulose to biosolvents
Disposable food packaging	Starch to bioplastics food packaging
Agriculture	Starch to bio-base mulch film
Fabrication	Starch to bioplastics for fabrication
Automotive	Vegetable fats to bio lubricants
Agriculture/waste management	Solid biomass to fine chemicals
Textiles	Cellulose to fabric
Food packaging	Cellulose to plastic paper cups
Construction	Waste biomass to insulation materials, waste biomass to wood–plastic composites
Agriculture	Polysaccharides to crops health inducer
Animal husbandry	Plant-based chemicals to fine chemicals

nitrogen from animal manure is 100 Mt annually. Only 20–40% of this amount was used as fertilizer, while the remaining 60–80% was released into the environment [9]. As part of the "pressure" to obtain P (phosphates as a critical raw material) from sewage sludge, a number of nutrient extraction technologies have been developed that can be partially applied to manure. The disadvantage is the price and the small spatial concentration of manure-producing entities, which entails increasing costs for storage and transport. Life cycle assessment (LCA) analysis of manure from the point of view of nutrient distribution and environmental impacts was described by Leip et al. [10].

Manure is a mixture of feces, urine, bedding, wasted feed, and additional water. The composition of manure is influenced by livestock type and age, feeding practice, the amount of bedding and water, storage and storage time, and weather conditions [11]. The content of trace elements (Cu, Mn, and Zn.) is also significant. Animal waste from agriculture (manure, slurry) is usually characterized by a high content of nutrients (phosphorus, nitrogen, potassium), biodegradable organic carbon, pesticide residues, and fecal coliform bacteria. Animals consume nutrients to ensure metabolic functions and growth. Unused nutrients become part of manure. Depending on the animal kind, growth phase, dose and feed composition, livestock excrete 50–90% of nutrients. Excessive nitrogen excretion can be prevented by feeds with sources of high-quality protein and a balance of amino acids. Animals excrete nitrogen in the soluble (urea and uric acid) and insoluble form, organically bound in feces. Phosphorus is excreted in the urine in soluble form; organically bound phosphorus is excreted in feces. Water is used to water the animals. The amount of water used affects the total amount of water in the manure. It is dependent on the animal kind, the growth phase of the animal, and the feed intake (Table 5.2). Livestock bedding affects the manure form (liquid, solid, or semi-solid) and thus the possibility of its further use. Wheat straw, flax straw, sawdust, wood chips, rice hulls, peanut hulls, and recycled paper products are used for bedding. During storage, the concentration of nutrients in the manure changes [15]. There are significant differences between sheep and cattle manure in the content of lignin components and cellulose. Sheep manure [16] has a lower lignin content (7.37%) than cattle manure (12.91%). The cellulose concentration in sheep manure is about 11.60%, while cattle manure contains

Table 5.2 Carbon and nutrient content in manure [12].

	pH	C ($g\,kg^{-1}$)	N ($g\,kg^{-1}$)	C/N	NH_4-N ($g\,kg^{-1}$)	NO_3-N ($g\,kg^{-1}$)	P ($g\,kg^{-1}$)	K ($g\,kg^{-1}$)
Swine	6.63	374.6	46.8	8.0	0.09	4.9	32[a)]	
Cattle	7.86	308.2	28.7	10.7	0.02	0.7	7.89[b)]	38.45[b)]
Chicken	7.73	270.2	53.0	5.1	0.10	2.7	27[c)]	25[c)]

Explanation:
a) [13]
b) [12].
c) [14]

15.30% [17]. Rabbit manure is an excellent source of ammonium sulfate: 108.50 kg ton^{-1} of manure, superphosphate: 109 kg ton^{-1} of manure, and potassium sulfate: 17.85 kg ton^{-1} of manure [18]. Daily production of poultry feces per 1000 birds is highest in geese and ducks, from 190 to 200 kg/1000 birds/day, turkey 160 kg/1000 birds/day, adult hens 150–155 kg/1000 birds/day, and chicken broilers 65 kg/1000 birds/day [19]. Pigs produce 84 kg of total manure per 1000 kg of live mass per day [20]; for cows, the amount is 7.3 kg/day [13].

The dynamic development of meat consumption in poultry farming increases the need for a quick solution for waste. Poultry litter is interesting in terms of macro- and micronutrient content. Before direct use, they must be pre-processed to reduce odor and eliminate pathogens, including bacteria, fungi, helminths, parasitic protozoa, and viruses [21]. Thermal conversion, hydrogen production, and power generation technologies can be used for poultry litter processing. The economic profitability of technology depends on the selling price of the final products: biochar, diesel product, and H_2 [22].

5.1.1.1 Technology for the Production of Fertilizers

The main technological processes used for obtaining nutrients are shown in Figure 5.1. There are various technologies for the production of manure fertilizers, which can be used to obtain NH_4SO_4, NH_4NO_3, as well as N and K concentrate, K fertilizer, and fertilizers containing P: struvite ($NH_4MgPO_4{\cdot}6H_2O$), Ca/Mg-phosphate, and P-rich ashes. An overview of technologies for obtaining P in a full-scale operation unit and the pilot plant is given in Table 5.3. Struvite crystals can be obtained by precipitation from swine effluent, applying sonification using the equipment

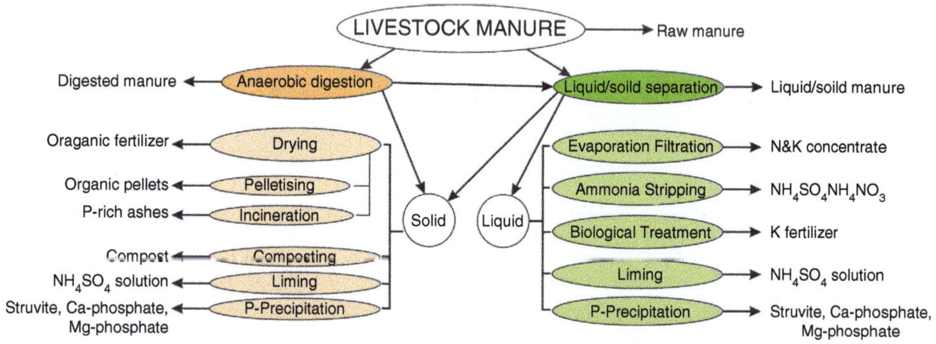

Figure 5.1 Technological possibilities of livestock manure processing for fertilizer production [23].

Table 5.3 Technological processes of phosphorus recovery from manure (full-scale operators and pilot plant).

Process	Input	Output products	Process description	Operating status
Hitachi Zosen	Manure, solid/liquid separated and dried to 60% DM	Stabilized and sanitized biochar, rich in phosphorus and can be used as organic fertilizer and/or soil improver	Pyrolysis at 400–500 °C under oxygen-limiting conditions in the reactor	Pilot scale plants operating in Japan on manure (capacity 720 kg d^{-1}, operating since 2012). Planned full scale 18 t d^{-1}
Agro America VP Hobe	Liquid or dry pig manure	Biochar	Belt-press for solid/liquid separation, drying of solid and the pyrolyzed	250 000 t yr^{-1}, full-scale operation in The Netherlands from 2015
GENIAAL (Nijhuis)	Liquid manure digestate	N–K fertilizer solution, P-rich organic fertilizer	Solid/liquid separation by decanter, flotation, two-stage membrane filtration to produce N–K fertilizer solution and purified water, solid fraction is phosphorus-containing organic fertilizer (1.3–2.6% P DW)	Full-scale plant operating since 2019 at Groot Zevert Digestion, Beltrum, The Netherlands (100 000 t yr^{-1} manure wet weight)
BioEcoSim (Suez)	Liquid manure or liquid digestate	Recovery of precipitated phosphate salts, ammonium sulfate, organic soil amendment (biochar)	Dissolving of nutrients into liquid fraction, solid-liquid separation, steam drying, pyrolysis, phosphate precipitation from liquor, gas permeable membrane ammonia stripping, palletization of biochar	Pilot plant Kupferzell, Germany and from 2019 pilot plant Zorbau, Germany (10.000 t yr^{-1})
N2-Applied	Manure slurry, biogas digestate	Combines N in manure with N fixed from air to produce a stabilized, ammonium nitrate based, liquid fertilizer product	Using renewable electricity and air, a plasma reactor fixes nitrogen by generating nitrogen oxides, which react with ammonia in manure or digestate to form ammonium nitrate, so lowering pH and stabilizing the nitrogen, reducing ammonia and greenhouse gas emissions during storage and field application. After solid/liquid separation the liquid fraction of manure or digestate can be managed as a liquid nitrogen fertilizer and most of the phosphorus will remain in the solid fraction	Several mobile trailer pilot units up to 500 l d^{-1} tested on farms in UK, Norway, Sweden, Finland, Denmark, and South Africa, on fresh cow and pig manure and on a variety of biogas digestates (varying in feedstock). Planned scale-up to 5–15 m^3 (tonnes) liquid per day in 2020
PARFORCE	Sewage sludge, bone ash from slaughterhouses	Phosphoric acid	Acid digestion using HCl or HNO_3 to generate raw phosphoric acid, solid–liquid separation, membrane electrodialysis to separate HCl, Ca, and Mg	Batch pilot (with continuous electrodialysis, capacity 1 t d^{-1} input material), tested for several different materials in 2018 and 2019 at TU Bergakademie Freiberg, Germany

from Hielscher Ultrasonics [24]. The use of VP Systems air flotation (with polymer flocculant) has proved successful in the USA. With a belt filter press, a yield of 66% of the solid phase is obtained, in which 83% of the total content is concentrated [25].

Fertilizer production technologies are composed of a combination of the following processes: liquid and solid separation, followed by evaporation and filtration, liming, ammonia stripping, phosphorus precipitation, biological treatment or anaerobic digestion followed by drying, precipitation of phosphorus, pelletizing, liming, composting, and incineration [9].

Solid–liquid separation of manure is a physical separation process used for liquid swine manures to separate solid and liquid fractions. As a result of the separation, the nutrients are separated. Na^+, Cl^-, $N{-}NH_4{}^+$, and K^+ remain in the water, while organically bound N and P and calcium phosphate accumulate in the solid component. Centrifugation or rotary press is used for separation, using a flocculant to improve phosphorus separation [26]. The cheapest variant is passive separation by gravitation in a settling tank [27].

At the present time, several processes using membrane technologies (nanofiltration, membrane distillation – ceramic membranes, and reverse osmosis), air stripping, and chemical precipitation are utilized for ammonium fertilizer recovery from manure [28].

Obtaining nutrients from dairy farm manure is possible by a system of related technologies; processes of the second and third degree have been introduced only since 2015 [29]. In the "Partial Advanced Nutrient Separation" process, a physical, chemical, or biological process is added to achieve a higher nutrient yield and to achieve the removal of very fine-grained suspended particles with an efficiency of >90% (Figure 5.2). A flocculant, most often polyacrylamide, is added to form large particles with sedimentation ability and easier dewatering. The volume of the liquid phase is not reduced in this process. The amount of manure to be stored will not be significantly reduced before reaching the third stage. By the introduction of filtration, water, which can be used as animal drinking water and for irrigation, will be obtained. The concentration of TSS (Total Suspended Solids) is less than $30\,mg\,l^{-1}$ and $N{-}NH_4 < 15\,mg\,l^{-1}$. The third stage includes ultrafiltration, which yields 96% phosphorus and 88% organically bound phosphorus, while ammonia/ammonium, nitrogen, and potassium remain in the solution [30].

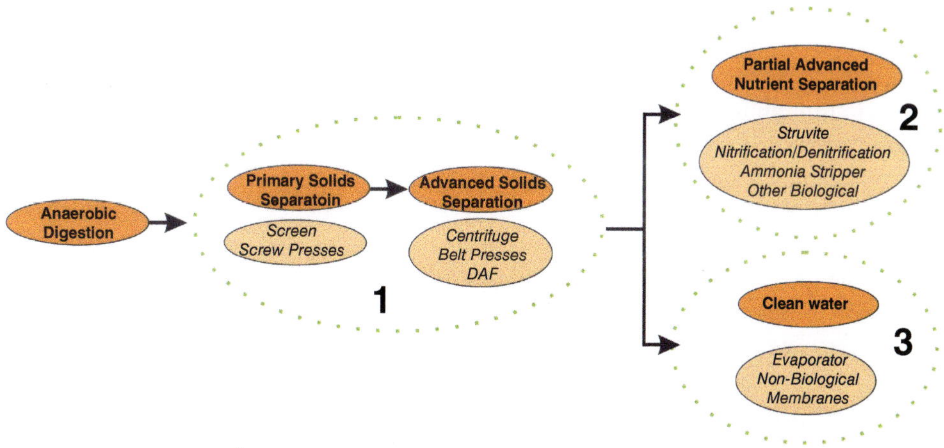

Figure 5.2 Three-step system of nutrient recovery from dairy manure, following anaerobic digestion. DAF, dissolved air flotation [30].

5.1.2 Agricultural Waste – Crops

At present, natural resources are being used unsustainably in many cases, and the European Union has implemented a circular economy and bioeconomy strategies in response. Combining both concepts with the application of specific principles within a system approach would improve resource efficiency and reduce environmental impacts. The bioeconomy is based on the approach of the primary use of the input raw material for bioproducts with higher added value and only subsequent energy use [31]. Extractable phytochemicals (PIC) present in lignocellulosic biomass of perennial industrial crops obtained by supercritical carbon dioxide extraction technology can be used as raw materials for the chemical, pharmaceutical, cosmetics, food, and feed industries [31].

Waste biomass is one of the most energy-rich and unused renewable raw materials suitable not only for the production of alternative energy sources (biomethane, bioethanol, pyrolysis oil, synthesis gas, and biohydrogen) but also for the production of valuable chemicals (cellulose fibers, oligosaccharides, organic acids, polyhydric alcohols, furans, natural antioxidants, and oils), which can be used, for example, in the production of eco-innovative materials (bioplastics or composites with a biocomponent). Agricultural biomass wastes are most often made up of lignocellulosic materials. These materials are composed of insoluble components: cellulose, hemicellulose, and lignin, as well as a soluble component consisting of sugar, organic acids, and amino acids. The decaying part of plants is an optimal source of carbon in soils, which improves soil fertility [32]. Wastes from agricultural activities can be divided into two groups: fibrous (low and high digestibility) and bran. Crops with low degradability include cereal, corn, cotton and soy straw, peanut and cotton husk, and harvest remnants of forage grass seeds. Crops with high degradability are corn, gluten bran, and soy husk. An integral part of agricultural crops are pesticides, which are used for crop protection and can, together with the released nutrients, cause contamination of the aquatic environment [33].

The direct use of the resulting agro-waste products (composts) can be affected by the presence of toxic compounds. Mycotoxins represent a large group of toxic compounds contained in agro-residues. Mycotoxins are secondary fungal metabolites. The most important mycotoxins that contaminate agricultural products, agro-residues, and animal feed during storage and processing include alpha toxins, trichothecenes, zearalenones, fumonisins, and ochratoxins. Important techniques for the elimination of mycotoxins in agro-residues and animal feed include biotransformation, i.e. transforming mycotoxin into different non-toxic substances [34]. The introduction of technologies respecting the conditions of the circular economy and bioeconomy eliminates problematic toxic compounds and organisms present in agro-waste.

The development of technologies for the use of agricultural residues is undergoing intensive research into the conversion of residues into chemical substances and value-added materials. Knowledge of the molecular structure and characteristics of agro-waste needs to be improved. Another direction of technology development is the introduction of advanced processes enabling selective extraction or conversion, such as enzymatic supercritical fluid extraction or depolymerization [35]. These processes would make it possible to obtain bioactive molecules (proteins, hydroxy-acids, and polyphenols) that could be utilized as building components for polymers or other chemicals. These products generally create the added value more than twice higher compared to the production of electricity, fuel, or animal feed [36]. The main obstacles to the recovery and conversion of agricultural residues into organic products and biomaterials are caused by energy requirements of these processes, the large variability of the chemical composition of waste materials, and the presence of contaminants. However, the production of organic acids,

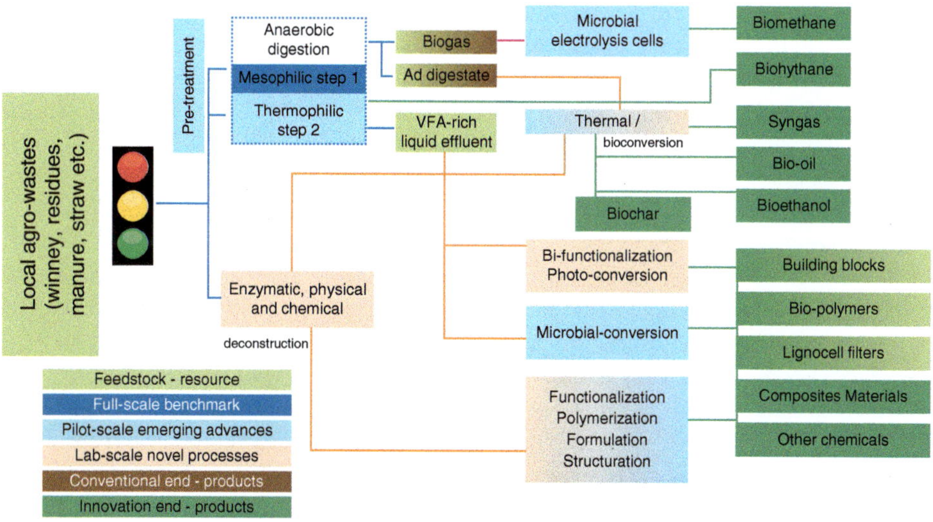

Figure 5.3 Possibilities of processing using the system of cascading technologies utilized at laboratory, pilot plant, or industrial scale levels [35]. VFA, volatile fatty acid.

bioplastics, and enzymes from waste biomass sources creates up to twice the added economic value compared to the production of electricity, feed and fuels [36].

The most common uses are fermentation and thermochemical conversion [37] and anaerobic digestion (AD). AD is considered one of the most advanced and efficient technologies for converting agricultural waste into energy and other bioproducts [38]. AD, combined with the pyrolysis of digestate, will provide up to a 42% increase in energy production [39]. The production of biohythane, a mixture of H_2 and CH_4 produced in two-stage AD, which can replace CNG, is also important [40]. Certain hygienic and environmental risks connected with the use of digestate as a fertilizer [41] can be eliminated by processing it with biochar. Biochar recovered from this process can be used to improve soil properties and carbon sequestration [42].

Microbial electrolytic cells (MECs) use electrochemical hydrogenation to directly convert biodegradable material to hydrogen (Figure 5.3). A microbial electrolytic cell is a modified microbial fuel cell (MFC), the principle of which consists in the microbial decomposition of organic matter under the anaerobic conditions of the anode. This process produces protons, electrons, and carbon dioxide. Protons pass through the electrolyte toward the cathode through the membrane while electrons flow through the circuit with the resistor to the cathode to produce current. Protons then combine with electrons at the cathode to form pure H_2 [43, 44], CH_3COOH, or CH_4 [45, 46]. Although the bacteria are able to produce a certain current on their own, the reaction does not start spontaneously without supplying the necessary voltage difference (0.2 V). This small voltage is applied to the circuit and bacteria of *Geobacter*, *Shewanella*, or *Rhodoferax* species function as ferrireducens [35].

5.1.3 Biorefineries

Many by-products in organic farming can be obtained by recycling agro-energy waste in green biorefineries [2, 47, 48]. Only technologies for conversion of waste biomass to biofuels, biomaterials, and biochemicals together with the generation of electricity or heat can be referred to as biorefineries. Green biorefinery (GBR) is defined as "a dedicated integrative, multifunctional concept which uses biomass from a range of raw material sources for sustainable generation of a range of different products and intermediate products (chemicals, materials, bioenergy including biofuels) while exploiting the biomass to the fullest extent possible" [49, 50]. In 2017, only 2% of bio-based chemicals were produced from the global annual production of 2300 Mt [51]. LCA analysis can provide answers to the usefulness of introducing "green biorefineries" and their future prospects for the production of bio-based chemicals, which, however, encounters a number of problems due to the lack of information [51].

Green biomass has become a surplus raw material in many countries of the EU due to substantial changes in dairy farming and meat production. Interest in the possible use of biomass from grasslands is, therefore, in the present time a significant issue for agriculture. In the EU-27, it accounts for 33% of utilized arable areas of permanent grassland [52]. Technology of GBR provides a novel approach for new ways of using green biomass.

The input raw materials for GBR I are cereal grains, sugar beet, sugar cane, and maize. The output product is only bioethanol, i.e. first-generation biofuel. The input raw materials for GBR II are identical to the first, but the output products are, in addition to ethanol, starch, fructose syrup, carbon dioxide, and distillery slops. However, these technologies use raw materials that can be used in the production of food or feed. Second-generation biofuels should be produced mainly from waste forest biomass, including logging residues, agricultural waste, energy crops, and biodegradable household waste. Thus, second-generation biofuels do not compete with food crops and therefore do not put upward pressure on food prices. Algae are used to produce third-generation biofuels [53] and the production of fourth-generation biofuels is based on the synthesis of carbon dioxide emissions with hydrogen into higher hydrocarbons. The transition to the second and higher generation of biofuels, the so-called 2+ generation of biofuels, and the related second and higher generation of biorefineries is, therefore, a key step. Otherwise, there is a risk of a continuing sharp rise in the prices of cereals, maize, and, consequently, the prices of meat and dairy products. Physical and chemical properties of different types of biomass are the main factor for optimizing the yield of bioproducts in the decision-making process of a sustainable way of biomass conversion [54]. In 2017, only 2% of bio-based chemicals were produced from the global annual production of 2,300 Mt [51].

From the technological point of view, GBRs use four basic technologies (Table 5.4) in a combination of different processes and their interconnection according to the requirements for the type of output product.

Table 5.4 Processes utilized in the technologies of green biorefineries [49].

Physical (+ mechanical)	Thermochemical	Chemical	Biotechnological
Basic operations to change material properties (*milling, drying, heating, cooling, compacting*)	Combustion Gasification Pyrolysis Thermolysis Hydrothermal processes	Basic operations for material transformation (*oxidation, hydrogenation, esterification, etherification, isomerization, hydrolysis, polymerization*)	Enzymatically catalyzed conversions
Cleaning and separation processes (*filtration, distillation, extraction, crystallization, adsorption, sieving*)		Chemically catalyzed conversions	Fermentation and decomposition processes
Extraction processes			

Biorefineries can be divided into conventional, lignocellulosic, oleochemical, green, gasification, and algal according to the nature of the input waste raw material. Currently, conventional biorefineries, which process grain processing waste, are the most technologically advanced and industrially implemented ones. Other types of biorefineries are still under development. For other types of input raw materials, processing technologies have not yet been developed that would guarantee the possibility of energy-efficient, economically profitable, and environmentally friendly complex large-scale processing of biomass into liquid and energy-valuable biofuels and bioproducts. Many studies were performed worldwide on the production of biofuels and bioproducts from various wastes, both on a laboratory and a pilot scale (Table 5.4). At present, there is still a large gap between projected and actual industrial production of biofuels, mainly for economic reasons. Therefore, biofuels and bioproducts cannot compete economically with traditional fuels and petrochemical products. GBR technologies still have opportunities to achieve further reductions in environmental effects, mainly by reducing energy intensity and increasing conversion efficiency [54]. Biorefineries have been developing in Europe from 2015, mainly in Germany, France, The Netherlands, and Scandinavia.

5.1.3.1 Green Biorefinery (GBR)

The input raw material for GBRs is fresh green biomass (from cultivated permanent grassland), green crops: potato leaf, sugar beet leaf, clover, and alfalfa. Primary fractionation is necessary because high water and fiber content complicate further technological stages in a biorefinery. Water and fibers are removed by pressing. The possibility of thermomechanical dewatering or hydrothermal conditioning is verified. Pressing results in two main streams. The first liquid fraction of nutrient-rich organic juice contains organic acids, proteins, amino acids, hormones, enzymes, and inorganic elements. The second fraction consists of a lignocellulosic press cake rich in fiber and containing cellulose, starch, pigments, and dyes. Valorization pathways – separation and fermentation – are used. Separation from press juice (decantation, centrifugation, membrane technology, electrodialysis, purification, and drying,) allows carbohydrates, proteins, enzymes, amino acids, flavors, and colorings to be obtained. After separation of the proteins, the supernatant is used for fermentation. Fermentation produces ethanol, lactic acid and its derivatives, and other organic acids. Pellets of pressed materials can be used as a feed or as fuel and for the manufacture of insulating fibrous materials, substrates for horticulture, and bio composites. They are also the input raw material for lignocellulosic biorefineries, where biofuel, organic acids, and biopolymers can be produced by hydrolysis and fermentation, as well as the input raw material for thermochemical conversion (gasification, pyrolysis) – biofuels – or for anaerobic digestion. The system of the GBR function is shown in Figure 5.4. Currently, five facilities operate as a commercial, demonstration, and pilot plant in Europe: commercial-scale factory (BIOWERT, Brensbach, Germany), demonstration plant Green Biorefinery of the Havellenad type (Haveland, Germany), Green Biorefinery (Utzeinach, Austria) as a pilot plant, Grassa (mobile unit, the Netherlands), Gramitherm® process (Austria) for the production of thermal insulating boards and the juice for AD and animal food, and three facilities in the USA [48]. GBR "Organofiner" are targeted to address the current shortage of protein-rich feeds, especially at the regional level [55].

An LCF (Lignocellulose Feedstock Biorefinery) type biorefinery uses dry biomass and cellulosic and lignocellulosic waste. These biorefineries use physico-chemical processes, such as thermal pressure hydrolysis in an acidic environment or enzymatic hydrolysis or fermentation processes decomposing cellulose into glucose and hemicelluloses into pentoses (xylose). The end products are, for example, bioethanol, furfural, and pure lignin.

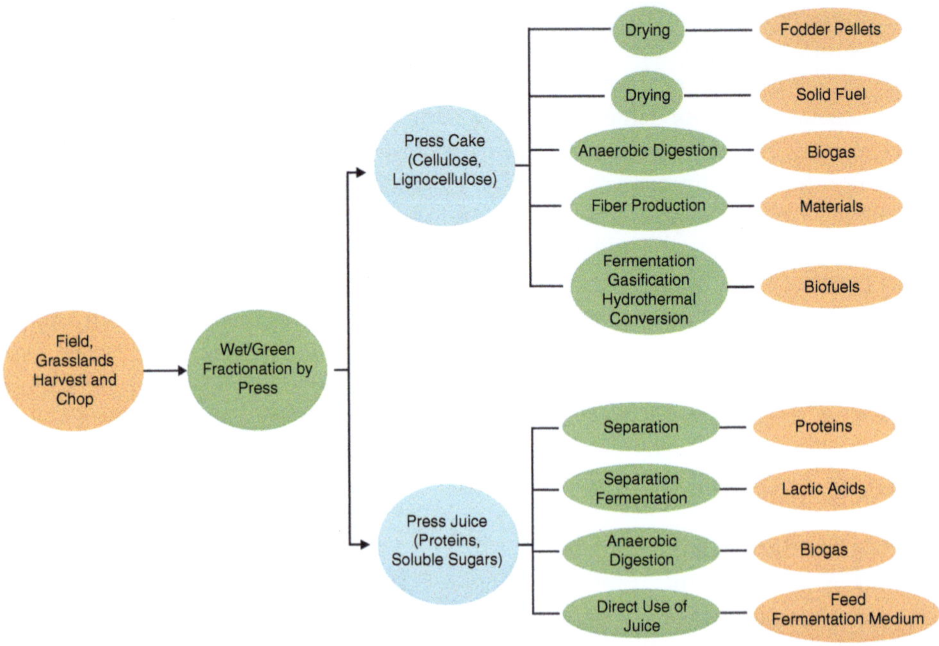

Figure 5.4 Technological processes and products in the framework of the GBR function [48].

Cereal biorefineries CCB (Cereale Com Biorafinery) uses the starchy parts of plants (Figure 5.5). The end products are, for example, bioethanol and lactic acid. Depending on the market situation, fructose syrup from the processing of wet maize grain may be used by confectionery or bioethanol producers. The above concepts of biorefineries are also combined in various ways.

Electricity and heat used in the technology are usually obtained by anaerobic digestion of various waste intermediates. An important step in the success of biorefineries is the maximum increase in the yield of target products. It has been verified in the laboratory that steam distillation of barley straw can achieve better selective isolation, as well as higher yields of polysaccharide-derived products, such as levoglucosane, which improves the economy of the process [57].

5.1.3.2 Triglyceride Biorefinery (TGB)

A biorefinery uses vegetable oils, animal fats, oil obtained from algae, and waste cooking oil as raw materials. The output product is biofuels. Triglycerides are converted by transesterification reactions with methanol in the presence of an acid, base, or enzyme acting as a catalyst. Glycerol is formed as a by-product, which is used commercially and also increases the added value of the whole process [58].

The EU Bioconsortium has received information on the operation of 224 biorefineries producing oil. Fat-based biorefineries produce mostly biodiesel (64) or oleochemicals (54). Biorefineries based on sugar and starch (63) mainly produce bioethanol but also products for food or feed and biochemicals. Wood-based biorefineries (25) produce cellulose, pulp (excluding pulp for paper only), and bioethanol. Five biorefineries produce cellulosic fiber or bioethanol from other lignocellulosic sources such as wheat straw, *Miscanthus*, or switchgrass. Bioethanol or biodiesel is produced from bio-waste in 13 biorefineries [56].

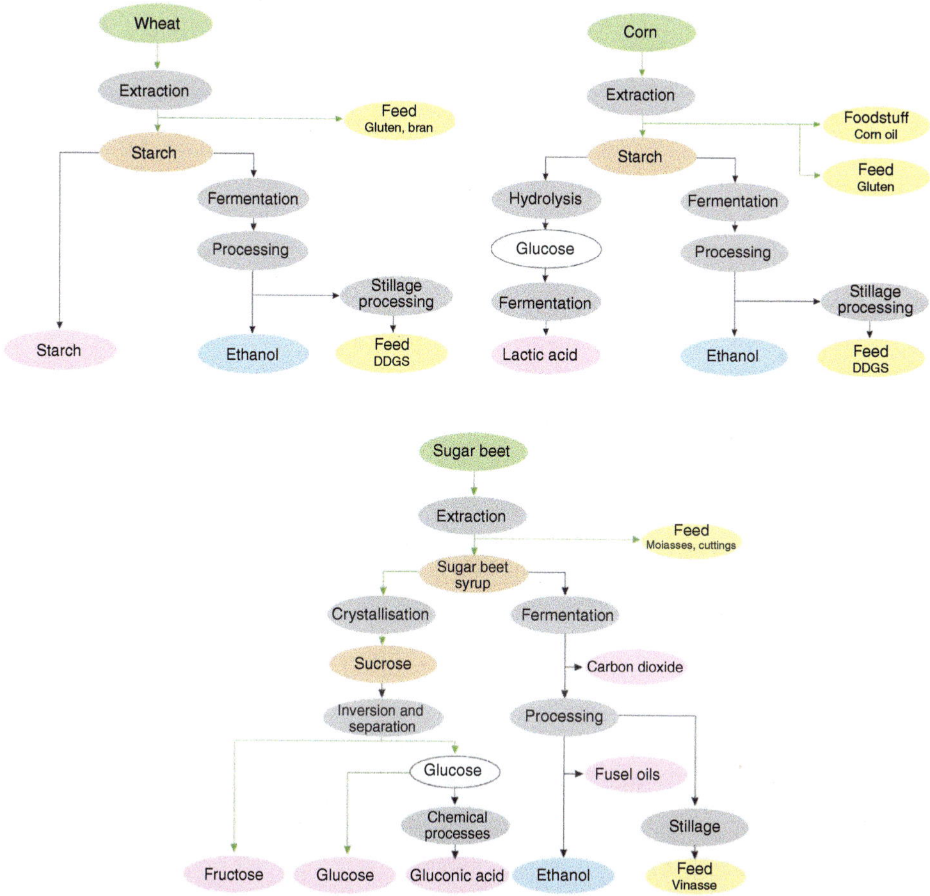

Figure 5.5 CCB Biorefinery and modified GBR refinery for sugar beet [56].

5.1.4 Traditional Ways to Use Agro-waste – Without an Added Value

5.1.4.1 Combustion

Agro-waste incineration accounts for about 9% of global energy production, up to 35% in developing countries [59]. An example of the energy use of agro-waste is rice husk (RH), which represents the largest amount of agro-waste. RH contains up to 20% of amorphous SiO_2. Combustion at <700 °C produces amorphous silica, which is characterized by a high reactivity and a large surface area. At a temperature of >800 °C, a crystalline modification of SiO_2 is formed – cristobalite.

Ash obtained from RH combustion has a wide range of uses: glass production, SiC, silica gel, a component in the production of Li-ion batteries, energy capacitors, filling for thermoplastics and rubber, but mainly for the production of silicon nitride (Si_3N_4), magnesium silicide (Mg_2Si), and other advanced materials. It is also used in the construction industry for lightweight construction materials and production of high strength concrete [60].

5.1.4.2 Production of Biofertilizers

Fertilizers containing living microorganisms are referred to as biofertilizers. They influence the soil ecosystem and produce substances utilized by the plants [32]. Microbial fertilizers work by two mechanisms [61]. Microbes symbiotically attack the plant and increase its ability to absorb nutrients and convert them to a form that is more bioavailable to the plant. The second group acts as biostimulants. Biostimulants contain microorganisms that in plants or in the rhizosphere stimulate natural processes in order to increase nutrient absorption and resistance to abiotic stress, and to improve crop quality. The group of biostimulants includes protein hydrolysates, a mixture of amino acids and peptides, which can be obtained by chemical and enzymatic protein hydrolysis from plant sources and animal wastes. In particular, the plant-based peptides are the most interesting of the biostimulants due to their multifunctional effects.

5.1.4.3 Solid-State Fermentation

Solid-state fermentation (SSF) is a simple technology that only needs input material, simple substrate pre-treatment, and the addition of microorganisms. SSF is a biotechnological process in which organisms grow on an insoluble substrate without the presence of free water [62]. SSF is used to enrich lignocellulosic residues with proteins, increasing the total content of lipids and fatty acids [63]. SSF agro-industrial wastes with fungi and bacteria, such as *Xanthomonas campestris* (potato peel), *Bacillus licheniformis* (wheat straw, maize straw, sugar cane bagasse), *Aspergillus niger* (wheat and rice bran, soybean), *Aspergillus oryzae* (soybean meal), *Streptomyces* sp. (household kitchen waste), accelerate fermentation processes [64].

For SSF, cereal grains (wheat, rice, corn, and barley), legume seeds, wheat bran, lignocellulose materials (straw, wood chips, and sawdust), and many plant and animal materials are most often used as substrates [64]. SSF processes were already known in ancient Egypt, where this principle was used to prepare bread. SSF works in the humidity range of 12–70%; the typical content is about 60%. Over time, this technology has been used to produce various products (enzymes, pharmaceuticals, bioethanol). It has been used as a pre-treatment process for biorefinery processes since 2000 and is expected to be used for the production of industrial biochemicals in the future [65]. The advantages of the technology are mostly higher product output, lower energy requirements, less waste water produced, elimination of the need for solubilization of nutrients from solid substrates, elimination of the need for checking many parameters strictly during the fermentation, no foam generation, and relatively easy recovery of end products.

SSF technology is suitable for the production of various value-added bioproducts, including organic acids, enzymes, biopesticides, biofertilizers, bioethanol, biosurfactants, animal feed, pigments, fragrance compounds, vitamins, and antibiotics. An overview of the individual valuable products that can be obtained from agro-waste by various processes within the SSF is given in Figure 5.6.

SSF technology with various types of microorganisms, such as *Candida utilis* and *Aspergillus niger*, can be used to prepare feed from apple processing waste, where the protein content in the final product is increased by 17–20%. The nutritional value of feeds prepared from agricultural waste and their digestibility can be increased by depolymerization enzymes [66], which are added or obtained in SSF. The enzymes that can be used are xylanase (catalyzed compound of hemicellulose), amylase (starch), cellulase, pectinase (pectic substances), chitinase (glycosidic bonds in chitin), ligninases (lignin degradation), and phytases (phytate – inositol hexaphosphate). Enzymes such as amylase, xylanase, b-glucanase, and phytase are added to cereal-based food to increase the utilization of dietary phosphorus, β-glucans, arabinoxylans, and starch.

5.1.5 Waste from Forestry

Forests and other wooded land occupy around 43% of the European Union's land and are a source of timber and non-timber products [67]. More than a third of primary biomass from

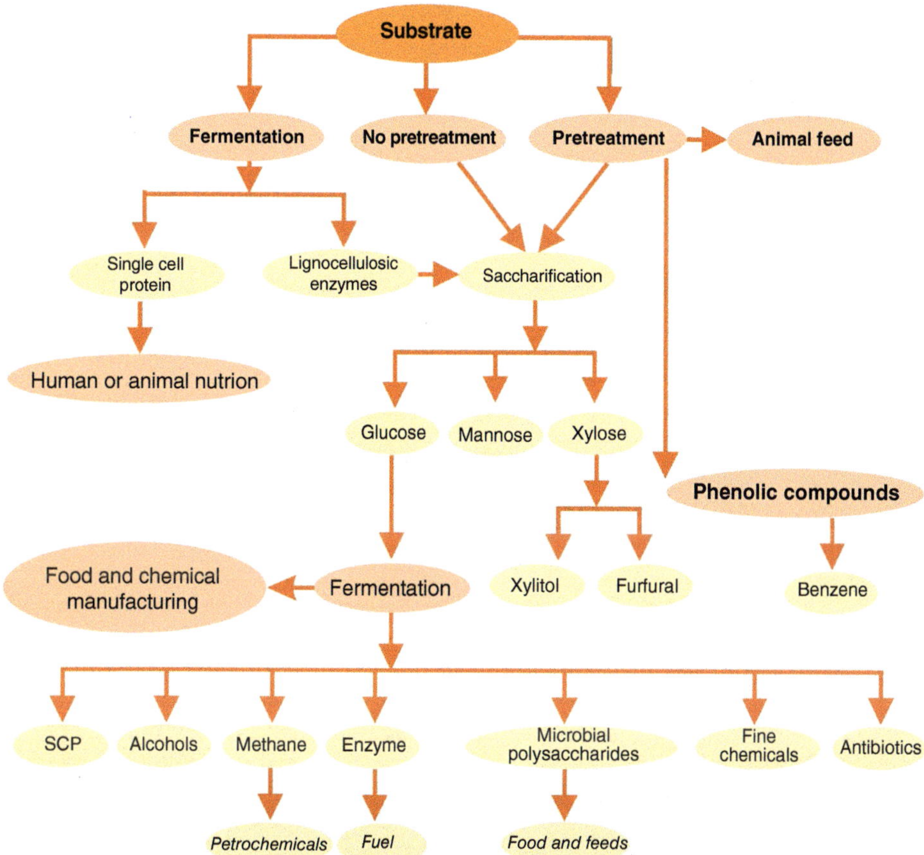

Figure 5.6 SSF technologies – production of value-added products [64].

forests is used directly for energy production. Secondary sources from industrial (pre-consumer) and post-consumer recycling (26% or 21%) cover 47% of the EU's need for wood products, pulp, and paper. Sustainable forest management in Europe is elaborated in the document "EU Strategy for Forests: for Forests and the Forestry Sector" and is in accordance with the objectives of the circular economy [68]. In forestry, the circular economy focuses on technologies aimed at reducing the amount of waste and promoting the use of secondary raw materials [69]. The recycling rate of wood products is relatively low, which is influenced both by the long life of furniture or timber and by their use for energy purposes.

Waste generated during forest harvesting consists of wood, bark, branches, and treetops. The main waste is bark [70] because wood processing (wood-based panels, pulp, and paper) requires debarking [71]. Global annual bark production is estimated at 300 and 400 million m^3 [72]. The bark makes up 2–10% of the stems. Its amount varies depending on the age of the tree, climatic conditions, and the metabolism of the plant [73]. Wood fiber recycling has developed into a cascading industrial system in which wood fibers are reused until the last phase – energy recovery [74]. The Swedish study, The Working Forest, 2019, revealed that the production of "new" paper has a lower carbon footprint than the processing of recycled paper if it comes from places where forests are managed in a sustainable way and the environmental impact of producing electricity used to produce it is low [75].

The use of wood residues in a cascade system is limited by the technical problems associated with the decontamination of waste wood and the economy of the process affected by the relatively low price of raw wood due to the costs associated with the collection, sorting, cleaning, and recovery. Bioenergy subsidies limit the introduction of more efficient wood reuse technologies [76]. Economic conditions are changing rapidly in connection with the surplus of wood after the bark beetle calamity or, conversely, with the demand associated with the transition to renewable electricity sources. There is no obligation to separate recyclable wood at source throughout Europe.

Wood has a high value as the primary raw material. Wood residues created by cutting and trimming can be used for the production of chipboard or pellets. Sawdust and shavings can be used as filling material for packaging, animal bedding, composting (e.g. dry toilets that save water) or in energy production (Figure 5.7). In smaller logging companies, the heat obtained from the incineration of wood waste enables the drying of lumber, which makes it possible to increase the accuracy in sizing and refurbishing wood products. Another possibility of using wood residues is the production of charcoal, pellets, and briquettes, which can be produced with relatively low investments [75]. The current possibilities of using forest waste as a raw material for technological utilization are shown in Figure 5.8.

The basic building blocks of forest waste are cellulose, hemicellulose, and lignin. Minor components include extractives, terpenes, proteins, and pectins. The content of major and minor components varies depending on the type of the tree (deciduous × coniferous), the age of the tree, and the part (branches × wood × bark). Lignin is a macromolecular substance with a three-dimensional structure. It contains a benzene ring with hydroxyl and methoxyl groups and with a side propane chain. Each type of plant has its own lignin (protolignin). Differences in lignin structure are due to the combination of phenylpropane units such as p-hydroxyphenylpropane, guajacylpropane, and syringylpropane units [77]. The lignin content of temperate trees in Europe is higher in conifers, where lignin is about 30%, while in deciduous trees it is about 20% [78]. The lignin content in forest waste is significant in terms of pellet and briquette production. Lignin is involved in plasticization, which helps to form bonds between particles and thus maintains the mechanical stability of pellets/briquettes [79].

The most important polysaccharide in the plant is cellulose. Together with pectin, hemicellulose, and proteins, it participates in the cell wall construction. From a chemical point of view, cellulose is

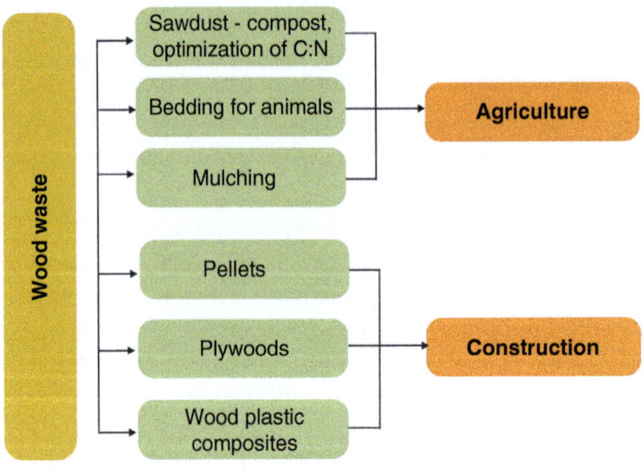

Figure 5.7 Direct utilization of forestry waste in construction and agriculture (only with and without mechanical pre-treatment).

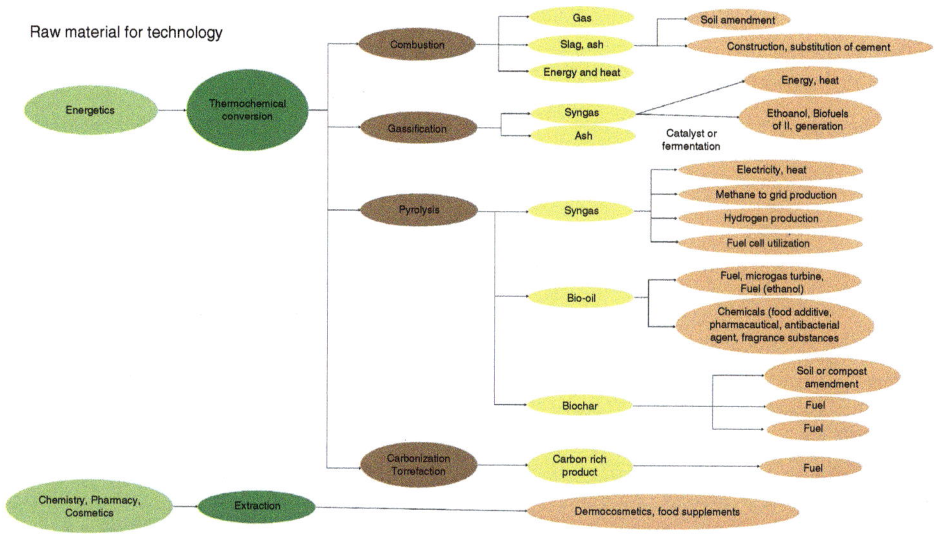

Figure 5.8 Forest waste as a raw material for technology and utilization of technological products.

formed by β-D-glucose units that are linked by a β-glycosidic bond. Hardwood has a higher cellulose content than softwood (Table 5.5). Hemicellulose belongs to a group of polysaccharides that form an amorphous component of the cell wall.

Hemicellulose consists of pentosans (xylan and arabane), methylpentosans, hexosans (mannan, galactan, levulan), and polyuronic acids. Hemicelluloses perform starch-like functions – they are reserve substances and are enzymatically broken down into simple sugars. The chemical composition of biomass, in particular, the content of cellulose, hemicellulose, lignin, and extractives, affects the combustion process. Individual parts of woody plants contain different amounts of major and minor components.

The bark contains significantly more total extractives (15–20%) than wood (4–6%) [83]. The higher proportion of extractives was also confirmed by Sette et al. [73]. Three to five times higher amounts of polar compounds (tannins, polyphenols, and glycosides) than non-polar constituents (fats, waxes, terpenes, steroids, etc.) are contained in bark extractives [81]. The bark has a higher fixed carbon content and has a higher HHV [73] than wood.

5.1.5.1 Technological Possibilities of Forest Waste Processing

The current possibilities of using forest waste as a raw material for various technologies are shown in Figure 5.8.

5.1.5.1.1 Wood-Plastic Composites (WPC)

Lignocelullosic materials are used to make wood-plastic composites (WPCs) [84]. In 2019, worldwide WPC production reached 4 Mt. It is a hybrid material created by blending wood products (bamboo, pulp, bark) with thermoplastic (PE, PVC, PP) or polymers. Some authors state that the use of barks in the production of WPC reduces mechanical properties [85]. As these products are easy to process (drilling and planing) and are created using biodegradable, recycled, and renewable plastic materials, further production growth can be expected in the future. No chemical that is dangerous or toxic to the environment is used in the production process. WPCs have the properties of both wood and plastic. Their biggest advantage is the possibility of use even in temperature

Table 5.5 Content of cellulose, hemicellulose, and lignin in softwood and hardwood [80].

Wood	Cellulose	Hemicellulose	Lignin	Extractives (%)	Ash	C	H	O	N	S	HHV (MJ kg^{-1})
Softwood[a)]	42.5±10.61	27.5±3.54	28.5±2.12		0.5	50.0	6.0	43.5	0.22	ND	
SW-inner bark[a)]					4.0	51.5	5.7	38.8	ND	ND	
SW-outer bark[a)]	3–17		> 20		3.8	44.4	6.4	45.4	ND	ND	
Softwood – wood[b)]	66–72		25–30	2–9	0.2–0.6				ND	ND	
Softwood – bark[b)]	30–48		40–55	2–25	>20				ND	ND	
Hardwoods – wood[b)]	74–80		18–25	2–5	0.2–0.6				ND	ND	
Hardwood – bark[b)]	32–45		40–50	5–10	>20				ND	ND	
Hardwood	47.5±3.53	22.5±3.52		22.5±3.54					ND	ND	
Poplar wood[c)]	42.3	31.0	16.2		1.20	49.40	6.00	43.10	0.23	0.05	19.74
Pine wood[c)]					0.35	45.92	5.27	48.24	0.22	0.01	18.98

Explanations: SW, soft wood

a) [72].

b) [81],

c) [82]

conditions in which wood cannot be used. Their use in the furniture industry is growing (kitchen production, furniture). Otherwise, these materials are most often used in construction and the automotive industry. WPCs formed from wood bark particles and rapeseed oil-based binding material are an example of building materials that exhibit sufficient thermal insulation, have good structural and sound-absorbing properties, and a low environmental impact [86].

5.1.5.1.2 *Tannins as a Plywood Adhesives*

Tannins (flavonoid compounds) derived from bark and wood are used commercially as adhesives [73]. By aqueous extraction of tannins from bark or wood, adhesives suitable for plywood production can be obtained [87]. Plywood adhesives can also be made from fibrillated bark, phenol-formaldehyde (PF) resin, and water in different ratios [88]. The advantage of this procedure is the complete utilization of the bark without prior extraction of tannins. The presence of other major components (hemicellulose, cellulose, and lignin in the form of nanoparticles) in the fibrillated bark creates high glue bond quality.

5.1.5.1.3 *Agriculture*

The most common way of using the bark in agriculture is mulching, which reduces evaporation from the soil surface and helps the reproduction of microorganisms. The presence of bark in the soil prevents compaction, improves water absorption, and reduces soil erosion. The disadvantage is the high cost associated with shipping costs [72].

The direct use of the bark in agriculture is problematic, as bark has an extremely high C : N ratio (200–400), is lacking in nitrogen, as well as other nutrients (potassium and phosphorus), and also contains phenolic acids, tannins, and growth-inhibiting substances. The use of the bark for composting is only possible after several years of storage. Dried chicken droppings and urea can be added to improve the nutrient balance in the raw material for composting [89].

5.1.5.1.4 Thermochemical Conversion

The goal of thermochemical conversion is the decomposition of biomass into gases, which are then synthesized into chemical compounds or used directly for energy production. Thermochemical conversion processes include combustion, carbonization/torrefaction, pyrolysis, gasification, and liquefaction [90].

5.1.5.1.5 Combustion

The bark is chemically distinguished from wood by the amount of ash and extractives. The impurities present in the ashes significantly affect the energy behavior of the bark during the combustion process [73]. The problematic element of biomass in terms of combustion processes is potassium, which occurs in the bark and other parts of woody plants in a concentration of 0.16–0.5%. Problems with fouling and slagging due to higher alkali contents in the fuel are reported by Melissari [91] and Mudgal et al. [92].

High-quality graded wood pellets complying with ISO 17225-2 [93] should be used for combustion in local fireplaces and in small commercial and public buildings (for non-industrial use in combustion plants <1000 kW). Combustion conditions are not ideal in small power plants, so better-quality fuel must be used. A distinction is made between class A1, A2, and B pellets. Differences in ash content (A1 < 0.7%, A2 < 1.2%, B < 2.0%) and total N, S, and chlorine regulate the possible addition of bark to the pellets for classes A2 and B. Class B pellets may contain chemically treated by-products of the wood industry, and these pellets are intended for industrial use. For pellets, the content of trace elements is also defined, which is the same for all classes. The physical and chemical properties of ash from biomass combustion, including the possibility of its use as a soil conditioner, are reported by Lanzerstorfer [94]. Ash from wood combustion as well as crop biomass contains silicates or aluminosilicates of Ca, Mg, and CaO, which have pozzolanic properties and can be used as filler or as a binder for road construction in asphalt, concrete, and pavements [95].

According to the European Pellet Council [96], 26.1 Mt of wood pellets were consumed in the EU28 in 2018, of which 65% was used for residential heating. Italy had the highest consumption of pellets (3.3 Mt). Also in other standards, "Graded Wood Briquettes ISO 17225-3" [97], "Graded Wood Chips ISO 17225-4" [98], and "Graded Firewood ISO 17225-5" [99], the key parameters of fuel quality are the ash content, calorific value, N, S, and chlorine.

5.1.5.1.6 Pyrolysis and Torrefaction

It is a process in which biomass decomposes into syngas, pyrolysis oil, and biochar in the temperature range of 300–650 °C in the absence of air. Biomass syngas consists of a mixture of non-condensable compounds: CH_4 (26%), C_2H_2 (3%), H_2 (15%), CO (35%), CO_2 (17%), and N_2 (4%) with a heating value of 17 MJ Nm^{-3}. The presence of gases in the syngas is influenced by the composition of the input raw material, for example, the amount of lignin [100]. The possibilities of producing hydrogen-rich synthesis gas from biomass with catalytic reforming using microwave and activated carbon made it possible to obtain a gas containing 55 vol% H_2 compared to 30 vol% H_2 without catalytic reforming [101].

Bio-oil can contain up to hundreds of organic compounds, which belong to the group of alkanes, aromatic hydrocarbons, phenol derivatives, and small amounts of sugars, ketones, amines, alcohols, esters, and ethers. Bio-oil can be used as a fuel for a micro gas turbine [102] or a source of chemical compounds [82]. The most important chemical compounds that can be obtained from bio-oil include levoglucosan (food additive, pharmaceuticals), ethanol (biofuel), furfural (pharmaceuticals), acetic acid (speciality chemicals), formic acid (antibacterial agent and preservative), and hydroxyacetaldehyde (fragrance substance).

Biochar is used for agricultural purposes to improve soil properties: increases water holding capacity and cation exchange capacity, improves nutrient management, changes soil microbial populations and their functions, and increases the content of humic acids [103]. If the pyrolysis technology is not well implemented, or low-quality raw materials enter the technology, biochar can be a source of environmental risks related to phytotoxicity and the release of VOCs into the environment [104]. The quality of the obtained biochar can be verified in terms of its use for agricultural purposes according to the "European Biochar Certification Guidelines of the European Biochar Certificate – Version 9.0E, updated on 1st June 2020" [105]. It is also possible to produce biochar from pure wood biomass, which is suitable as a supplement for animal feed ("Guidelines for EBC-Feed Certificate, Biochar in Animal Feed") [106]. The EBC Guidelines now include the requirement that biochar that has not been EBC-certified should be used in industrial processes only as part of building materials, as a component in the production of plastics, or as a substitute for charcoal for grilling. "Draft Proposals for EU Criteria for Recovered Struvite and Phosphate Salts, Recycling of Ashes and for Biochars, as CE Fertilizers under the Revision of the EU Fertilizers Regulation (STRUBIAS)" [107] are currently being prepared in the EU. According to this proposal, it will be possible to classify biochar into categories for PFC 1 (fertilizer), PFC 3 (soil improver), PFC 4 (growing medium), and PFC 6 (non-microbial plant bio-stimulant) in the revised fertilizer regulation.

During torrefaction, lignocellulose materials decompose according to their thermal properties. At 275 °C, the hemicellulose almost completely decomposes while the cellulose remains unchanged (except bark) and the amount of lignin increases [108]. The main positive of torrefied biomass is the increase in fixed carbon, calorific value, reduction of moisture content, and grindability, as well as the fact that it easier to store and transport [109].

Quality requirements for virgin woods and chemically untreated wood residues (TW) and non-woody, chemically untreated herbaceous, fruit, and aquatic biomass (TA – "classes are used for specification of pellets and briquettes from non-woody biomass; TA1, TA2, and TA3 represent non-woody, chemically untreated herbaceous, fruit, and aquatic biomass") are also defined for the thermal use of torrefied biomass (prepared at 200–300 °C) in local fireplaces and for non-industrial use in the standard "ISO/TS 17225-8" [110]. The basic quality requirement is the net calorific value on a dry basis>21 MJ kg^{-1}. The individual classes differ mainly in the ash content (1.2–5%) for TW and for TA <10%, but no limit value is defined in class TA3. Other differences are in the limits for N, S, and chlorine. The carbon value is not specified.

5.1.5.1.7 Conversion of Lignocellulose to Ethanol

Lignocellulose resources are used both for energy production and for chemical compounds and materials. Ethanol produced from lignocellulose (second-generation biofuel) is expected to have the potential to replace fossil fuels [111], and its production will increase further in connection with the use of waste lignocellulose. It is estimated that 110 billion liters were traded on the world market in 2018 [60]. Most bioethanol is produced in the USA (56%), Brazil (28%), the European Union (5%), China (4%), and Canada (2%). Biochemical conversion of lignocellulose wastes into bioethanol (e.g. fermentation) is more affordable than thermochemical conversion (pyrolysis). In order to increase the yield, optimization of the main technological processes must be achieved, including pre-treatment (alkali pre-treatment, as well as a combination with H_2O_2 for lignin removal, ionic liquid pre-treatment, organo-solvents, other delignification treatment, and steam explosion), detoxification of lignocellulose hydrolysate (membrane technology and adsorption on charcoal), enzymatic hydrolysis (additives addition), and fermentation [111].

Bark or wood residues with bark represent potential raw material for the production of ethanol from a lignocellulose source. Different physical properties and chemical composition of the bark in relation to the standard wood chips negatively affect the bioconversion process. Compared to wood chips, it is much more difficult to hydrolyze conifer bark to monomer sugars because the bark has a low enzymatic hydrolyzability. The bark hydrolysate contains a high amount of glucose, 60–70%, as well as 5–15% xylose, 5–10% arabinose, and 3–4% galactose, as well as mannose. Spruce bark contains up to 8% mannose and pine bark 9% arabinose [81]. Nevertheless, in the process of simultaneous saccharification and fermentation (SSF) with the addition of bark, the yield of ethanol decreased from 85 to 59%, similar to the combination of hydrolysis and fermentation [112].

5.1.5.1.8 Products with Higher Added Value

Tree bark is a versatile source of raw materials for the pharmaceutical, cosmetic, and food industries. The bark contains compounds with antimicrobial and antioxidant effects that could be used as preservatives. The strong antioxidant activity of flavonoid compounds in the bark extract of French maritime pine (*Pinus pinaster*, synonym *P. pitit*) and radiata pine was used for the commercial production of nutritional supplements: pycnogenol or enzogenol [87]. Some compounds contained in the bark enhance the taste of food and can be used in the food industry.

Bark contains alkaloids, essential oils, carbohydrates, vitamins, and antibiotics [72]. Some compounds are used to fix odors. In pharmacology, plant components obtained from the bark serve as components of patches, ointments, massages, hair fluids, and other mixtures. Methanol extracts of *Quercus robur*, *Larix decidua*, *Picea abies*, and *Alnus glutinosa* barks suggest their promising use as active components in the dermo-cosmetic industry [113].

5.2 Food Industry

Inhabitants in Europe and North America produce 95–115 kg $year^{-1}$ of food waste per capita [114], but food waste across EU-27 is 179 kg per capita $year^{-1}$ according to Jörissen et al. [115]. In contrast, in Australia and the United States, food waste per capita reached 361 kg and 278 kg $year^{-1}$, respectively [116]. These significant differences in the disposal patterns of households are the result of different ways of food consumption as well as various methodological approaches to data collection [115]. Approximately 129 Mt of food waste is produced in the food supply chain (FSC) from a total input of approximately 638 Mt of primary food commodities [117].

Due to the amount produced worldwide, FWs represent a priority area of the action plan of the European circular economy [118]. The European Commission introduced the EU Platform on Food Losses and Food Waste (FLW) in the framework of the Circular Economy Action Plan [119]. As part of this activity, four technical reports were prepared by the Joint Research Center – JRC [118, 120–122]. In the EU, the FW reduction is supported by maximizing material recycling in several documents as the Circular Economy Action Plan [119], the Bioeconomy Strategy [123], and the Waste Framework Directive [124]. Similarly, the United Nations 2030, in the framework of the Agenda for Sustainable Development, defined the goal of "reducing global food waste at retail and consumer level per capita [125] to one half, reducing food losses throughout the production and supply process, including post-harvest losses."

Individual waste streams of food processing waste may have different recovery/utilization methods. The most significant amount by weight of waste in the processing and manufacturing of food is produced during fruit processing (6.1 Mt), meat (2.9 Mt), vegetables (2.6 Mt), cereals (2.5 Mt), and potatoes 2.1 Mt [117]. The largest source of waste in the entire FSC is represented by fish, which produces up to 51% of FW (food waste), vegetables (46%), fruit (41%), oil crops (36%), potatoes (22%), and cereals (20%). The lowest FW production is shown by dairy (4.5%) and sugar beet (4.3%). Within the whole FSC, including consumption, households are the largest producer (Figure 5.9).

In the consumption phase, 46.3% of FW is generated. In the food supply chain [125], food waste is produced at households (85%) and food services (15%). The variable composition of streams from FW arising in the consumption phase does not allow its use for the recovery of chemical compounds but is suitable as a source of nutrients in biotechnological processes or for the production of biofuels [126].

Valuable compounds in FW include macronutrients – carbohydrates, sugars, proteins, lipids, starch, polyphenols, cellulose, fibers (Table 5.6), and micronutrients or bioactive compounds, which can be further transformed into new added-value products (Table 5.7). Food waste can be more frequently utilized in the production of biofuels, biodegradable plastics, nanoparticles, alcohols, and enzymes [129].

The waste management hierarchy, according to the US EPA and for the principles of application of circular bioeconomy in the EU, is shown in Figure 5.10. The amended Waste Framework Directive (2018/851/EC) [124] requires the Member States to monitor the production of food waste and take measures for its reduction, with the preference for food waste prevention. The material recycling in the circular bioeconomy reflects the future food waste biorefineries. The recovery of energy and nutrients represents two different categories [130]. Waste streams from fruit and vegetable production are used as a source of antioxidants (Table 5.7, Figure 5.11), fibers, phenols, polyphenols, and carotenoids; waste from the meat and dairy industry, where vinasse, grains from distillate production, and fish products contain large amounts of regenerable proteins [126].

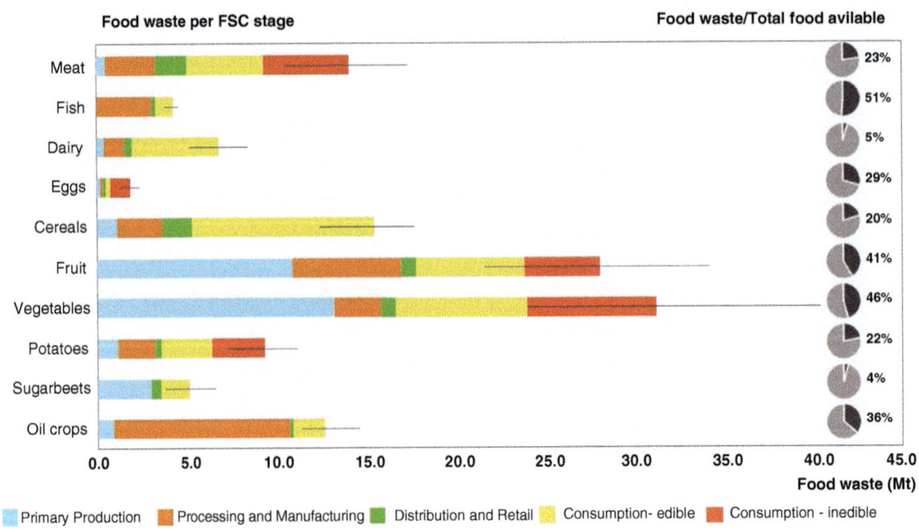

Figure 5.9 FW production in individual FSC sequences [117].

Table 5.6 Components of industrial food-waste streams [127].

Industry	Resource	TS	VS	Carbohydrate	Cellulose	Hemicellulose	Pectin	Proteins	Lipids	Lignin	Antioxidant
Potato processing	Potato peels	9	90	63	–		–	17	1	10	–
Coffee roasting and serving	Coffee grounds	20	98	–	9	37	–	14	17	–	–
Pomegranate juice	Pomegranate peels	92	95	–	5–20.0	–	5.0–11	–	–	9	32
	Pomegranate seeds	13–19	98	–	19	–	–	–	–	21	10
Citrus processing	Citrus peels	19	95–97	1	22–37	5–17	23	16–23	1–4	7–9	–
Apple processing[a)]	Apple pomace	–	–	–	7.2–43.6	4.2–24.4	3.5–14.3	2.9–5.7	1.2–3.9	15.3–23.5	–
Fish processing	Fish tails, skin, heads, and bones	20–50	75–81	–	–	–	–	15–30	0–25	–	–
Cheese processing	Whey	6	92	78	–	–	–	11	7	–	–

Explanations: TS, percentage of total solids (dry solids), VS, volatile solids, Anti, antioxidans
a) [128]

Table 5.7 Industrial food waste and potential products [127].

Industry	Resource	Global mass of resource (t/a)	Potential product from resource	Yield of potential product (kg(product)/kg (resource))	Mass of potential product (t/a)	Technology
Potato processing	Potato peels	25 954 386	Lactic acid	0.140	3 633 614	Fermentation and separation
Cucumber pickling	Pickling brine	2 074 888 925	Lactic acid	0.010	24 898 667	Electrodialysis
Coffee roasting and serving	Coffee grounds	25 570 924	FAMEs	0.170	4 295 915	Dimethyl ether extraction of oil and esterification to FAME
Pomegranate juice	Pomegranate peels	1 302 500	Phenolic microcapsules	0.070	91 175	Ultrasound extraction, emulsification, spray drying
	Pomegranate seeds	370 000	Antioxidant oil microcapsules	0.100	37 000	Ultrasound extraction, emulsification, spray drying
Citrus processing	Citrus peels	24 204 988	Ethanol	0.250	6 051 247	Cellulose hydrolysis, fermentation, distillation
			Pectin	0.23	5 567 147	Hot dilute acid treatment, alcohol precipitation
			D-limonene	0.005	139 179	Steam distillation and alkali treatment
Fish processing	Fish tails, skin, heads, and bones	540	Bioactive peptides	0.730	394	Hydrolysis and electrodialysis with membrane filtration
Cheese processing	Whey	150 000 000	Butanol	0.010	1 950 000	Fermentation and separation
			Acetone	0.001	150 000	Fermentation and separation
			Ethanol	0.001	150 000	Fermentation and separation

Explanations: yield = kg-product/kg-resource, FAMEs, fatty acid methyl ester.

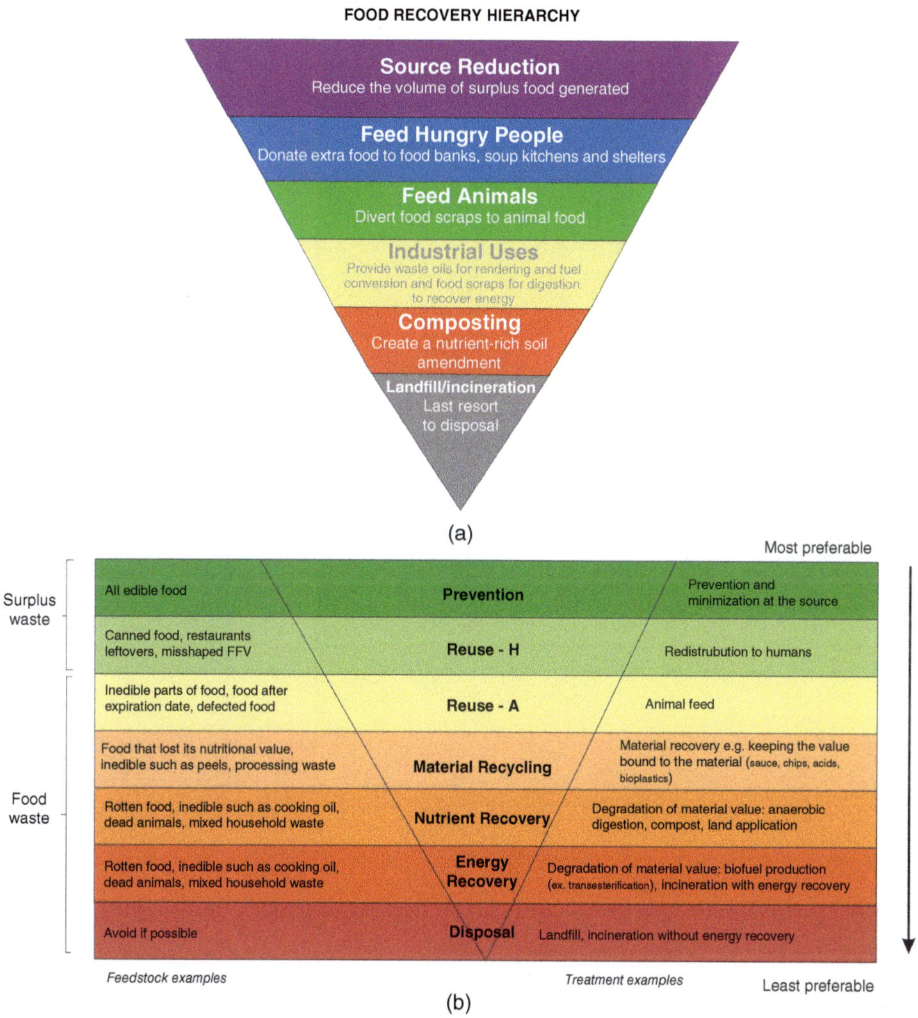

Figure 5.10 Left: Food recovery hierarchy (US EPA). Right: Food recovery hierarchy – circular bioeconomy [130].

A review of valorization/utilization of the main food waste streams is shown in Table 5.8.

5.2.1 Fruit and Vegetable Waste

The world produces a huge amount of fruit and vegetables globally. The highest quantity represents 145 Mt of citrus fruits, 111 Mt of bananas, 83 Mt of apples, 73 Mt of grapes, 45 Mt of mangoes, mangosteens, and guavas, 27 Mt of pineapples, and 292 Mt of other fruits [133]. Production of vegetables [133] includes potatoes (389 Mt), tomatoes (181 Mt), onions (98 Mt), beans (31 Mt), peas (16 Mt), and other vegetables (952 Mt). The composition of fruit and vegetable waste depends on the type of fruit or vegetable, the time of harvest, and the processing technique. The generation of waste from the processing of fruit and vegetables is different. Peel, pitch residues, and seeds represent inedible parts [134]. The smallest amount of waste is generated during the processing of apples (10%) (pomace, peels, and seed), while the most waste is produced from citrus fruits (up to

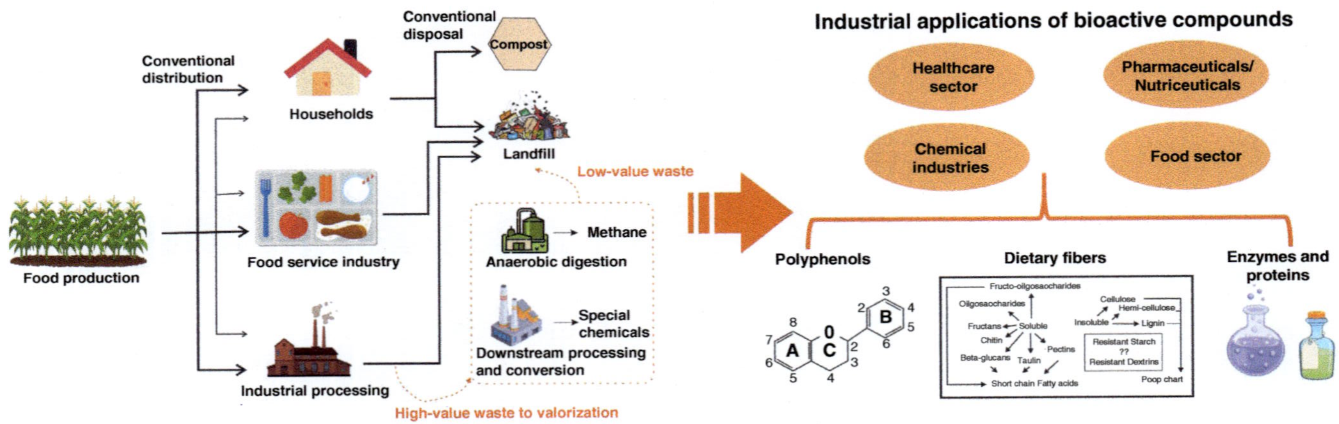

Figure 5.11 Simplified flow diagram – the genesis of food waste and its possible utilization with access to high-value waste valorization [127] production of bioactive components [131].

Table 5.8 Main food waste streams – appropriate for valorization [132].

Food product	Waste stream	Current management	Reason for selection
Ales, lagers & spirits	Spent grains	Animal feed, composting, anaerobic digestion.	HV solid wastes widespread across EU-28.
Lamb, beef, pork, & poultry	Blood	Production of foodstuffs (raw, plasma, albumin). Production of pharmaceuticals, production of animal feed (blood meal, raw, albumin), anaerobic digestion.	Production & processing widespread across EU-28.
Lamb, beef, pork, & poultry	Bones	Production of animal feed (fat, bone meal). Production of chemicals (glue, detergent), production of foodstuffs/pharmaceuticals (gelatin), composting, anaerobic digestion, production of low gel.	Production & processing widespread across EU-28.
Lamb, beef, pork, & poultry	Hair, feathers, hooves, & feet. White and red offal including guts & giblets Proteinaceous matter incl. Category 3 material from slaughter plus carcass fat	Production of chemicals (glue, gelatin, collagen, glycerine, soap). Production of pharmaceuticals (Ca, P, gelatin, collagen, fat, insulin, heparin, pepsin, steroids, cholesterol). Production of animal feed (meat meal, fat). Production of foodstuffs (sausage casing, catalase, additives). Anaerobic digestion. Composting only.	Production & processing widespread across EU-28.
Light wines	Pomace (skin and seeds)	Production of ethanol. Extraction of antioxidants & pigments. Production of grapeseed oil (cooking oil & beauty ingredient) & grapeseed flour (food ingredient). Production of resveratrol. Production of bio-based packaging.	HV produced within the EU (approx. 70% of global wine production; Martí et al., 2014), regionally important for Italy, Spain, France, & Germany.
Spirits	Organic wastes, mash from grain, fruit, or potato	Animal feed. Composting.	HV, regionally important for the UK, France, Germany, Poland & Italy.

(Continued)

Table 5.8 (Continued)

Food product	Waste stream	Current management	Reason for selection
Sugar	Sugar beet pulp	Marketed in fresh/ensiled form as pressed pulp or blended with molasses to give molassed sugar beet feed (MSBF).	HV, regionally important for France, Germany, the UK & Poland.
Tomatoes	Pomace (skin, pulp & seeds)	Animal feed.	HV, regionally important for Spain, Italy, Greece & Portugal, rich source of lycopene.
Vegetable oil & margarine	Crude & extracted press cake or spent meal	Production of fuels. Industrial uses (kernel oil, wood, activated carbon).	HV of meal produced across Europe (~30 Mt, FEDIOL 2016), particularly in Germany, France, Spain, The Netherlands, & the UK.
Vegetable oil	Olive stones	Production of fuels. Industrial uses (kernel oil, wood, activated carbon).	HV regionally important for Spain, Italy, Greece & Portugal.
Wheat milling products	Wheat feed/wheat middlings	Feed for use by cattle, sheep, and pigs.	Production & processing widespread across EU-28.

Explanation: HV, High volume.

50%) (rag, peel, seeds), bananas (35%) (peel), and pineapples (33%) (skin and core) [131], while for vegetables, onions produce 10% (outer leaves), peas 40% (shell), and potatoes 15% (peel) [135].

5.2.1.1 Industrial Processing of Fruit and Vegetables

During the processing of fruit and vegetables, fruit and vegetable waste is generated, which consists of edible and inedible parts. The type of waste produced from the processing of fruit/vegetables/herbs is determined by the technology (liquid, solid phase). Solid parts of the waste include husks, seeds, pitch residue, and pulp.

Recycling processes for fruit and vegetable waste (FVW) can be divided into technologies in which all waste is recycled (composting) and technologies from which chemical compounds are recovered by extraction [136]. The composition of the waste is important for the selection of an optimal processing technology. Wastes from vegetable/fruit processing can be processed by fermentation, anaerobic digestion, and composting. Prior to this processing, economically interesting organic compounds (essential oils, starch, pectin, and acids) can be obtained. An example of zero-waste technology is citrus peel processing (Figures 5.12 and 5.13). Essential oils are produced as an added-value component from the waste of citrus peel waste (CPW) [138].

5.2.1.2 Basic Properties of Fruit and Vegetable Waste

Citrus waste (oranges, grapefruits, lemons, limes, tangerines, and mandarins) consists of husks, seeds, segment membranes plus juice sacs and seeds. The amount of CPW from processing varies between 49 and 69% depending on the type of citrus fruit, the amount of processing water, and the technology used. The most important components of waste are peels, which make up 40–55%, 30–35% internal tissues and less than 10% seeds [139]. In orange peel suitable for fermentation for bioethanol production, the total sugar content is 29–44%. The high content of pectins, phenolic compounds, and dietary fiber is also significant. The minor components of CPW include volatile

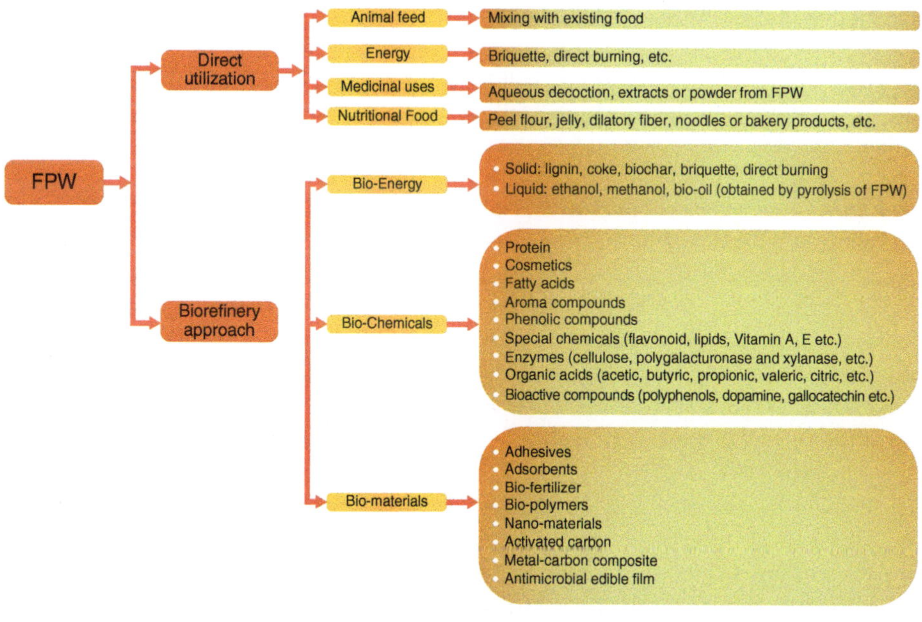

Figure 5.12 Possibilities of fruit peel waste (FPW) processing in relation to the principles of circular economics [137].

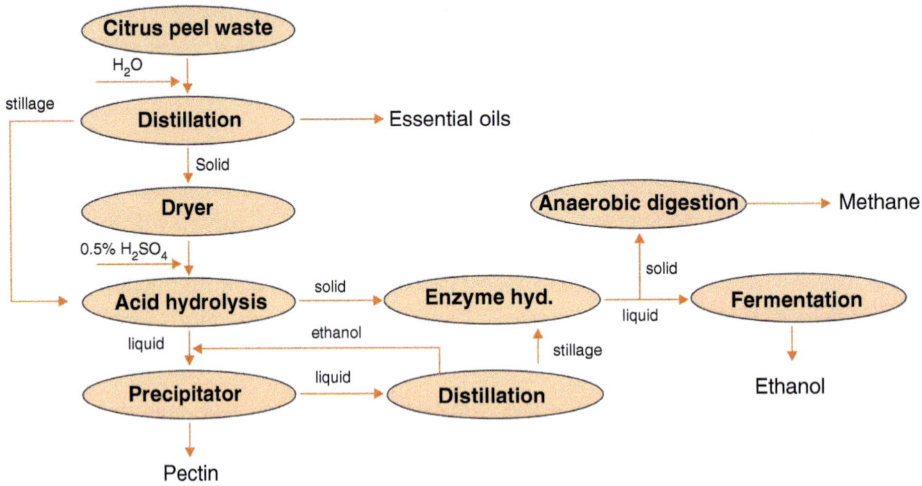

Figure 5.13 Diagram of the zero-waste biorefinery for citrus peel waste (CPW) valorization [138].

compounds (aldehydes, ketones, alcohols, and hydrocarbons), inorganic substances, flavonoids (flavones, flavanones, and anthycanins), essential oils-EO (D-limone), enzymes, pigments, carotenoids, polyphenolics, vitamins, and nitrogen compounds [139]. Removal of compounds like D-limonene is necessary for efficient production of bioethanol [140].

Vegetable waste (peels, pods, seeds) makes up 10–60% of the total weight of a fresh production. Carrot pomace and peels are the main by-products of juice processing. Pomace is rich in fiber, carotenes (9.40–10.80 mg/100 g), β-carotene and ascorbic acid (9.90–11.60 mg/100 g) [141]. The chemical composition of various types of vegetable waste in terms of major components is reported by Bakshi et al. [142].

Wastes from potato processing are composed of skins and marrow residues. In the industrial sector, 0.07–0.14 Mt of peels are generated annually [143]. Residues are rich in starch and reducing sugars. Potato peels are rich in phenolic compounds (1.02–2.92 g/100 g), especially phenolic acids (p-coumaric acid, chlorogenic acid, vanillic acid, gallic acid, caffeic acid, p-hydroxybenzoic acid) and glycoalkaloids, which are important for the pharmaceutical industry [144].

5.2.1.3 Processing Fruit and Vegetable Waste Essentials Oils (EO)

EOs (essential oils) that are used in the food, cosmetic, pharmaceutical, and chemical industries can be obtained from some fruit wastes. EOs are used as a fragrance in soaps, perfumes, creams, lotions, and detergents [145], or as fungicides and antibacterial and antifungal agents [146].

Most EOs are contained in citrus fruits (average content in raw biomass is 5.4 kg of oil per ton). Citrus EOs consist of terpenic hydrocarbons together with their derivatives, such as aldehydes, alcohols, and esters [147]. D-limonene (monoterpene) makes up 90–98% of the volatile substances contained in citrus essential oils [148]. Limonene is obtained from EOs from citrus peels at a rate of 50 million kg year^{-1} [149].

EOs are obtained by cold pressing, centrifugation, steam distillation, and liquid extraction with solvent [148]. Prior to extraction, pre-treatment is performed using thermochemical and thermophysical processes [150] to increase EO recovery. Cold pressing and steam distillation are performed with hexane or carbon dioxide. Cold pressing is applied on the skins followed by separation by centrifugation of a mixture of oil, hexane, and water [134]. During steam distillation, citrus wastes are exposed to hot steam/water and oils are released by evaporation. Newer EO extraction techniques include microwave-assisted extraction (MAE), ultrasound waves (SAE), and hydro-distillation (MAHD). Both MAE and MAHD reduce extraction time [151] and solvent consumption [152]. By creating UV waves, SAE helps to destroy waste cells and thus increases the release of EOs.

If EOs are not recovered from the waste and the waste is treated in another way, it must be disposed of. It is a condition for optimizing other technologies (ethanol fermentation) but also composting [148]. EOs usually have potent antimicrobial, antifungal activities that significantly inhibit microbial processes. Citrus wastes are not suitable for composting because of the relatively low pH value, the content of essential oils that slows the composting process and the biodegradation.

5.2.1.4 Xanthan Gum and Pectin Production

Xanthan gum is an important polysaccharide that is commercially produced by a microbial procedure [153]. It is formed by the action of *Xanthomonas campestris* bacteria from heteropolysaccharide and is used to improve viscosity and acts as a stabilizer in the food production, pharmaceutical, and oil industries [154]. Xanthan gum contains acetyl and pyruvate, and their proportion varies depending on the culture conditions and microorganisms used [155]. The conditions (temperature, pH, airflow) must be carefully controlled to ensure the quality of the product, yield, and acceptable production cost [153, 156].

Pectin represents another valuable compound that can be obtained from waste in the processing of fruit and vegetables. Pectin is currently produced mostly (85%) from citrus peels (56% from lemons, 30% from limes, and 13% from oranges) and from apple pomace (14%), with a minor percentage obtained from sugar beet and other vegetables (Figure 5.14).

Pectin is a complex compound composed of 17 different monosaccharides contained in the cell walls of non-woody plants [158]. It is utilized as a hydrocolloid in food production, with growing applications in other sectors of industry [159]. Pectin used in the food industry as a polymer must contain at least 65% galacturonic acid units [157].

Pectin polysaccharides include homogalacturonan, rhamnogalacturonan-I, and rhamnogalacturonan-II. Pectin is used as a jellifying agent in jam and jellies and as a thickener, emulsifier, and a stabilizer in many foods. Gelation by pectin in fruit concentrates provides stabilization

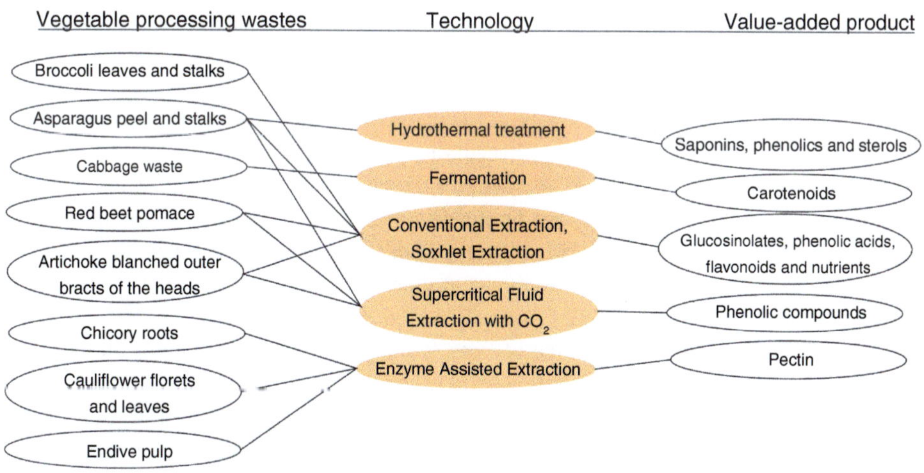

Figure 5.14 Valorization pathways for vegetable processing wastes [157].

emulsions, suspensions, and foam [160]. In pharmacy, it is used as a detoxifying agent and also has an anti-diarrheal effect [161]. Physical methods of pectin recovery include extrusion-cooking or MAE [162], supercritical water extraction [163], and enzymatic extraction [164]. A hydraulic press or centrifugation is used for separation.

5.2.1.5 Enzymatic Production of Oligosaccharides and Flavonoids

Oligosaccharides are carbohydrates, consisting of up to 20 monosaccharides connected by glycosidic bonds [165]. Oligosaccharides are particularly abundant in fruits (pear, nectarine, watermelon, and blueberry), and vegetables (white onion, leek, scallion, and garlic). These foods could be used as sources for commercial production of pre-biotics, which can form ingredients in functional food products [166].

Oligomers designated as prebiotics have the ability to control the gut microbiota by supporting beneficial indigenous flora and suppressing or not affecting non-desirable bacteria such as proteolytic Bacteroides and Clostridia [167]. Important pre-biotics include inulin, fructooligosaccharides (FOS), galactooligosaccharides, and lactulose [168]. Prebiotics are produced by autoclave or microwave extraction [169]. New technology for production of oligosaccharides is represented by membrane reactors where enzymes are concentrated at the reaction side and reused continuously. The main disadvantage of membrane reactors in the industrial scale is fouling and polarization, which decrease permeate flow by accumulation of the produced oligosaccharides [165].

Flavonoids are natural substances with phenolic structures of various kinds. They are contained in fruit, vegetables, grains, stems, flowers, roots, bark, tea, and wine [170]. Flavonoids are important plant pigments. These naturally occurring polyphenolic compounds cannot be artificially synthesized. They have a function of antioxidants in biological systems and have other various biological properties [171]. They have significant antioxidation, anticancerogenic, anti-inflammatory, antiaging, and cardio-protective properties. They are an essential part of many pharmaceutical, nutritional, and cosmetic products. Flavonoids, their classification, and rich dietary sources are reported by Panche et al. [170]. The most important flavonoids include rutin, quercetin, hesperidin, hesperetin, luteolin, naringin, naringenin, tangeretin, eriodictyol, neoeriocitrein, diosmin, and apigening [172]. Flavonoids are usually obtained by extraction with organic solvents or enzyme-assisted extraction: glucanase, a pectinase, and cellulase [173].

5.2.1.6 Production of Dietary Fibers

Extraction of juice is performed from many fruits, e.g., apple, peach, orange, and other citruses. This procedure produces a by-product from which various compounds can be recovered, among which, the fiber fraction is remarkable as it is utilized for the production of functional foods. The waste is generated during the processing of vegetables, such as the carrot, green pea, onion, potato, pepper, asparagus, and artichoke [174]. Dietary fibers can be divided into water-soluble fiber (SDF) or insoluble fiber (IDF). Most foods produced from plants contain a mixture of soluble and insoluble fibers in their composition [175]. Dietary fibers consist of non-starch polysaccharides such as cellulose, pectin, hemicellulose, lignin, β-glucans, and gums [176]. The properties of dietary fiber are affected by the SDF/IDF ratio. For food processing use, the SDF/IDF ratio approaching 1 : 2 is suitable. The main source of fiber is potato peel (9.7–68%), as well as peels from green peas, grapes (50%), berry, and citrus (65%) [177]. Dietary fiber is most commonly used in the food industry. Moreover, fiber improves the sensory characteristics, texture, and shelf-life of foods because of its water-binding ability, gel-forming properties, and antisticking, anticlumping, texturizing, and thickening effects [176]. It is used in the production of pastries, beverages, soup, and pasta.

5.2.1.7 Bioethanol Production

Ethanol is produced from various low-cost agricultural raw materials like banana waste, potato waste, molasses, citrus peels, mango peels, and apple pomace [178]. Bioethanol production involves the process of converting more complex organic compounds of carbohydrates, hemicellulose, and pectins with the participation of microorganisms and their enzymes into simple carbohydrates, which are then fermented to ethanol. Ethanol can be used as an efficient fuel. Pectinase and cellulase are used as enzymes. The structure of more complex compounds in lignocellulosic biomass can also be disrupted by ammonia fiber expansion, hot water, alkaline hydrolysis, extraction by organic solvents, decomposition of lignin by oxidation, and biological methods [179]. The main steps in the production of ethanol from lignocellulosic wastes include pre-treatment of biomass (sulfuric acid), enzymatic hydrolysis at a temperature of 50–60 °C, fermentation (saccharomyces), separation, and effluent treatment.

5.2.1.8 Organic Acids

Organic acids such as citric acid (CA), lactic acid [180], acetic acid [181], and succinic acid can also be obtained by fermentation of FW [138]. The conditions of fermentation by *Aspergillus niger* and fermentation technology for various food wastes are reported [182].Citric acid is used in the food industry (preservative, beverage production), pharmacy, and textile industry [183]. Lactic acid is widely used in food, pharmaceutical, cosmetic, and industrial products [184]. In order to meet requirements for large-scale acid production, it is important that the production process is environmentally friendly by utilizing inexpensive and available products obtained from agro-industrial waste (molasses, corn-starch, orange waste, pineapple waste, apple pomace, carrot waste, coffee-husk) while maintaining high production yields.

5.2.2 Waste from Cereals Manufacturing

The high content of lignocellulose and starch in cereals allows cereal processing wastes to be used for the production of sugar monomers and xylo-oligosaccharides (Figure 5.15). Other compounds that can be extracted from the manufacturing waste of cereals include peroxidases, phenolic compounds, glucuronoarabinoxylans, and β-glucan. The number of studies confirmed that cereals (Food Waste at Manufacturing, FWm) have also been used as a carbon source in the production of various enzymes, unicellular proteins, and organic acids. The processes included either simultaneous saccharification and fermentation or a separate hydrolysis step with a following submerge or SSF [157]. Wheat bran is an important input for biorefineries because it contains carbohydrates (starch, cellulose, and hemicellulose), a relatively high protein content, phenolic compounds (ferulic acid), and phytic acid [185]. Mussatto [186] performed a technological and economic assessment of the possibility to use spent grain from breweries for the production of products such as phenolic compounds, xylitol, lactic acid, and activated carbon [157].

5.2.3 Sugar Crops and Tubers

Potato is the fourth important crop globally [187]. Potato peels are used for the production of biogas, extraction of phenolic acids, lactic acids, and ultrasonic extraction of steroidal alkaloids (glycoalcaloids). They may have great potential use in pharmaceutical applications [184]. Phenolic acids are widely used for food preservation, feeds, and packing materials. Potato peels can be utilized for the production of ethanol, extraction of antioxidants, and production of biocomposites containing cellulose nanocrystals.

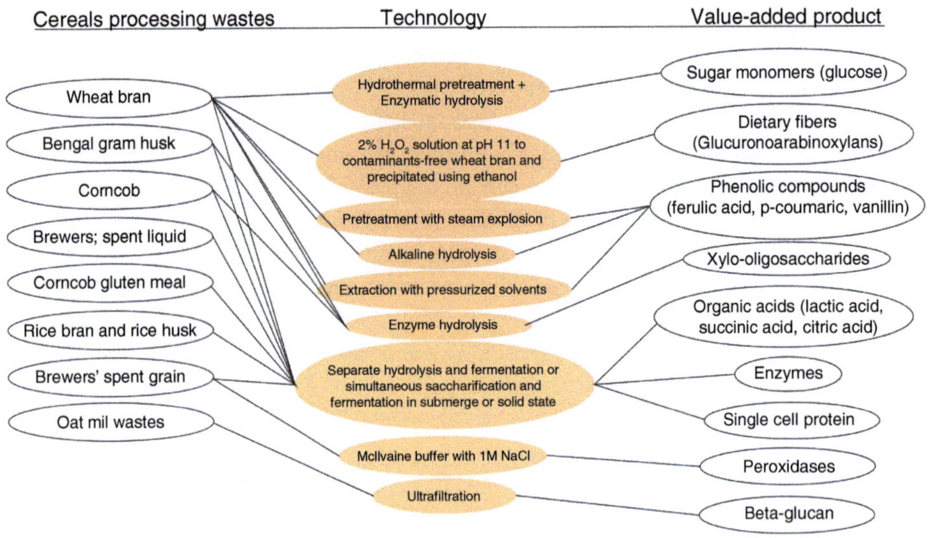

Figure 5.15 Possibilities of waste treatment using waste from cereal processing for products with higher added value [157].

Potato waste can be used as a source of xanthan obtained by fermentation. Xanthan is used in food products as a thickening agent and currently is produced from sugar at a high cost. Polylactic acid (PLA), polymer produced from petroleum, can also be alternatively produced from potato waste [188].

Several fermentation products, e.g., enzymes like amylase, can be derived from sugar crops and tubers [189]. The main components of the sugar bagasse are glucan (39%), lignin (25%), and xylan (22%) [188]. Sugarcane bagasse is a residue produced in large volumes and is a favorable substrate for biotechnological processes because of its high content of carbohydrates (cellulose and hemicellulose) and its low ash concentration. It can be used to produce high value-added products such as pectin, phenolic compounds, and succinic acid. Sugar beet processing plants produce sugar beet pulp, which is currently mainly used as animal feed [190]. Sugarcane consists of cellulose and lignin and thus represents a source for the production of ethanol fuel. Its high content of carbohydrates (sucrose, cellulose, and hemicellulose) and pectin makes it an attractive source of high value-added compounds. Biosourced plastics (such as Bio-PE, PLA, and others) that are currently marketed use food resources such as corn or cane sugar [191].

5.2.4 Utilization of Fish Waste

Fish processing waste should be disposed of as soon as possible. However, this waste provides an important source of relevant biomaterials (Figure 5.16). In aquaculture farming, it is necessary to provide proteins for aquafeed in the form of fishmeal and fish oil. An alternative is to use fish proteins (Food Waste at manufacturing, or FWm), which are a cheap source of protein. Fish protein hydrolysates (FPH) produced by heat processing of fish FWm can effectively replace up to 20% of a conventional fish meal [192]. Hydrolyzed phosphorus can also be obtained from FWm fish, which is also useful in the fish feed industry to support the growth of bones and teeth [193]. Enzymatic, chemical, and physical processes are used for extraction of bioactive proteins from fish FWm. The course of these processes depends on pH, temperature, and extraction time. Fish oil is

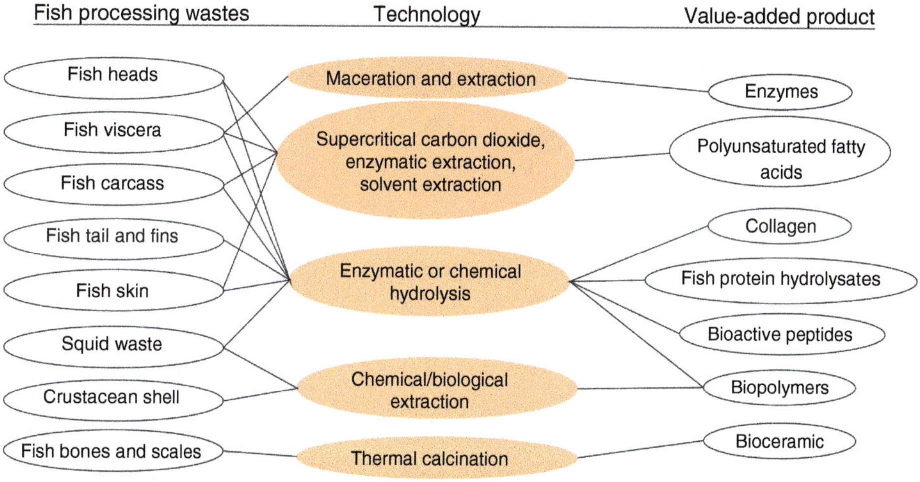

Figure 5.16 Valorization flows for wastes from fish processing with marking of paths for recovery of fish FWm [157].

mainly used for the aquafeed industry; only a small portion is used for human consumption in the form of nutraceuticals [194]. The most important bioactive substances in fish oil are polyunsaturated fatty acids (PUFAs) – Omega–3, –6, –7, and – 9 [157].

The production of bioplastics as new packaging materials from FWm fish waste represents one of its most interesting uses. Proteins extracted from FWm fish have the favorable properties for the production of bioplastics, characterized by excellent elasticity and plasticity. The use of fish FWm in the form of gelatin [195] or myofibrillar proteins [196] provides an inexpensive raw material that is able to incorporate additional substances (anthocyanin and chitosan) extracted from fish. This can prevent the oxidation of lipids in packaged foods, thereby significantly extending their shelf life [195]. Peptides extracted from fish protein hydrolysates and fish food waste at gelatin production exhibit antioxidant effects that can be used to produce anti-inflammatory and anti-cancer agents [197]. The similar antioxidant activity is shown by chitin (protein) and chitosan (polysaccharide produced in chitin processing). These substances are obtained mainly from waste from the processing of crustaceans (i.e., shrimp and crabs) and are used for important applications in medical and other industries [157].

5.2.5 Meat Waste

Animal by-products, as animal waste is rightly called, fall into several categories: hazardous waste (for example, infected with dangerous viruses such as BSE (bovine spongiform encephalopathy – mad cow disease), carcasses of dead animals, and parts of animal carcasses from slaughterhouses (which cannot be utilized in any way). This animal waste is generated mainly during the slaughter of animals in slaughterhouses, production of products of animal origin, disposal of dead animals, and implementation of disease control measures.

Common processing methods include, for example, heat-pressure sterilization. The sterilized material is dried and meat-and-bone meal (MBM) is formed using other technological procedures. MBM has a relatively low content of volatile matter (54.1% d.m.), fixed carbon of 15.0%, and a high amount of ash (25.1% d.m.). Energy use in cement plants or for pyrolysis technology is reported by Kantorek et al. [198]. The disadvantage is the high content of nitrogen, up to 8.7%, and chlorine at 0.8% [199], but the carbon content of about 58% is interesting [198]. MBM is also used as a phosphorus-containing fertilizer (da Silva et al. 2019) or as an input for biogas plants [200].

The global meat industry produces 130 Mt of animal bone residues annually. Countries of the EU produce more than 10% of this waste. After slaughter, one animal produces approximately 18 wt% of bone residues [201]. The value of bones is also given by their nutritional content. They contain nutritionally valuable natural collagen protein and fat that could be used as ingredients for nutritious foods, pet foods, pharmaceutical and cosmetics, and raw materials in biodiesel and fertilizer industries. Given the expected increase in global meat consumption, it is essential to minimize the waste produced during slaughter and industrial meat processing by finding opportunities for its further economic use.

The published data indicate that by-products (organs, lard or fat, skin, feet, abdominal and intestinal contents, bones, and blood) of cattle, lambs, and pigs represent 66%, 68%, and 52% of the live weight, respectively [202]. Non-edible parts of slaughtered animals can be collected and processed to recover fats and proteins for oleochemical and animal feed applications, respectively [203]. Depending on the kind of animal, meat makes up 60–70% of slaughtered bodies, of which 40% are edible and 20% inedible [204]. Proteins can be obtained from bovine lung, porcine lung, and boneless chicken meat using conventional extraction [205], with a protein yield of about 50%.

In a green extraction pilot operation [206], keratin was obtained from pig hair with a protein purity of up to 70% and a yield of approximately 50% [207]. Keratin is one of the most abundant proteins in epithelial tissue and forms so-called intermediate filaments in cells, which participate in the formation of the cell skeleton (cytoskeleton). Bioactive peptides can be extracted from meat production waste by enzymatic hydrolysis [208]. With subcritical hydrolysis at a minimum temperature of 180 °C and a minimum pressure of 1200 kPa, a cycle time of 40 minutes is verified for use on specified risk material [203]. Bovine lungs, intestines, and blood [209], and porcine blood and liver are suitable for protein production [210].

The possibilities of recovery of waste from meat production to create value-added products are listed in Figures 5.17 and 5.18. Waste keratin materials (e.g., feathers after hydrolysis) represent a mixture of amino acids. They can be digested with sulfuric acid and as a source of nitrogen directly used as fertilizers for plants [212]. Similarly, wool introduced into the soil without treatment releases valuable nutrients and absorbs moisture. Wool can also be transformed into a fertilizer using hydrolysis by superheated water without any pre-treatment [213].

5.2.6 FWm from Dairy Production

The main type of FWm produced in the dairy sector is whey, which is the result of the coagulation of proteins in milk during cheese production. Whey is a source of protein, lactose, and salt. Historically, whey was considered as a waste by-product of dairy processing and especially of cheese making, while nowadays, the intrinsic value of its components is widely recognized [214]. It is used for the production of a high-protein medical beverage and whey protein powder (a powder-based soluble protein designed for dietary management). Other possibilities of whey processing include the separation of specific protein components (isolation and purification of α-lactalbumin (α-la) and specific peptides) and production of lactic acid, lactose, and bioethanol [215].

5.2.7 Wastes from the Production of Egg Products

In the industrial production of egg products, considerable amounts of shells are collected. Eggshells can be used as fertilizer, animal feed (low-investment processes) up to heavy metal

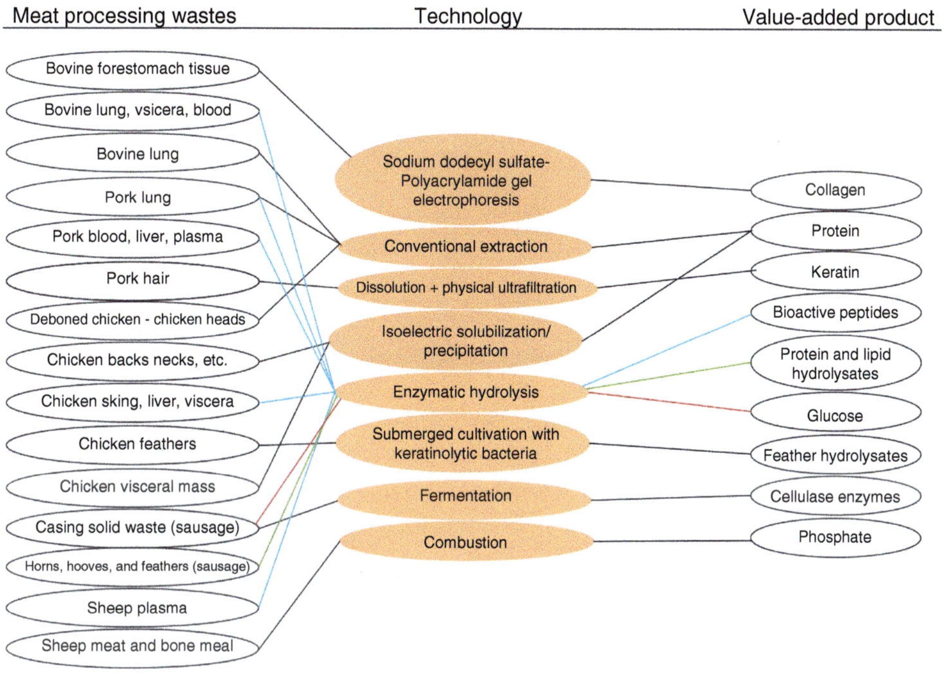

Figure 5.17 Valorization pathways for meat processing wastes. The colors of lines mark the relationship between the waste components and the value-added products [157].

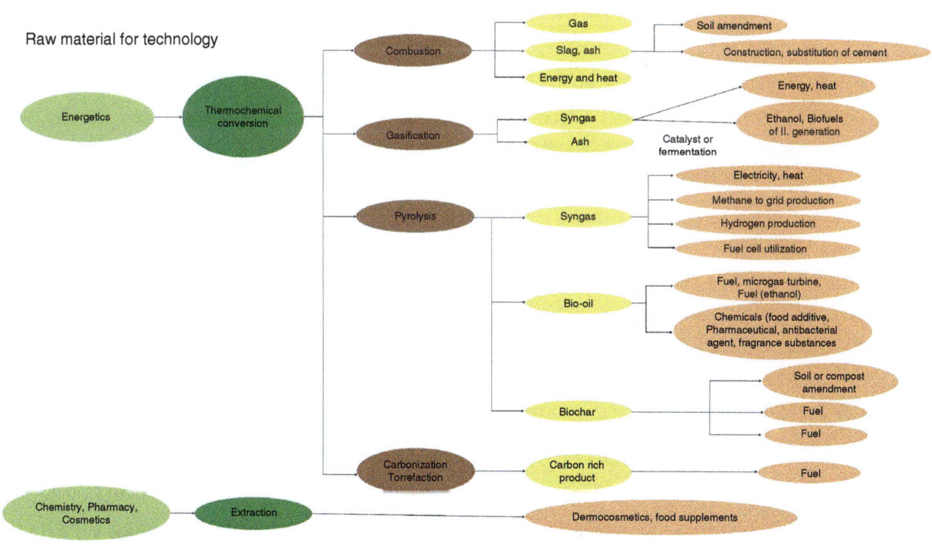

Figure 5.18 Traditional approaches to meat processing wastes [211].

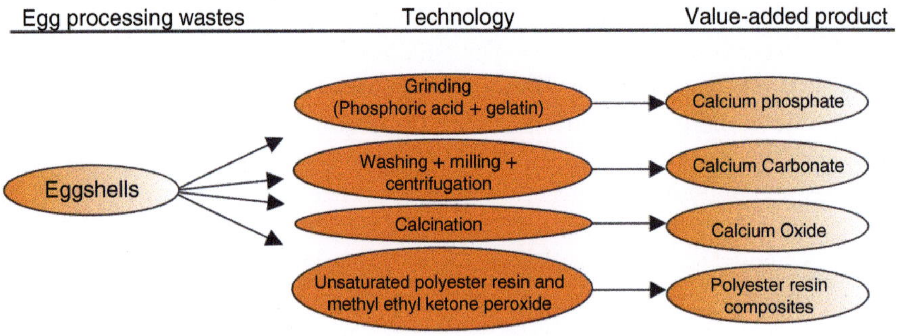

Figure 5.19 Possible valorization for egg waste [157].

absorption products, paper treatment, or catalysts for biodiesel production [216]. Their use was also conditioned by the high cost of landfilling them [132]. Calcium carbonate and calcium oxide, as well as calcium phosphates, can also be obtained from eggshells [157]. Another utilization of eggshells is the production of polyester resin composites for use as artificial carbonaceous rocks [217]. Possible ways of valorization for eggs FWm are illustrated in Figure 5.19.

5.2.8 Food Waste as Animal Food

The growing demand for animal products is expected to result in a further need for feed, especially from coarse grains, such as maize, and protein meals, which also defines the differences between human and animal nutrition requirements [218]. Ensuring sustainable sources of feed will be a challenge in the conditions of food–fuel–feed competition, water shortage, land degradation, biodiversity loss, and climate change [142]. The processing technologies that are available today can lead to the transformation of food waste to safe animal feed products, which contain nutrients of high quality [219]. The pressure on the optimal composition of feed products and the current possibilities of using food waste in obtaining a product with higher added value influences the market behavior in food waste management. Despite the ban on the use of direct feeding of food waste in the EU, changes are slowly taking place. The required quality of feed can be ensured with the utilization of solar energy and use fractionation schemes for food sources. This process produces low carbon emissions and is energy efficient because solar energy is utilized for drying and sanitation of the food waste. Treatment of food waste includes dehydration, pelletizing, extrusion, cooking, probiotic treatment, and ensiling [218]. Safety analysis of transformation food waste to animal feed from three typical treatment processes (i.e., fermentation, heat treatment, and coupled hydrothermal treatment with fermentation) showed that preserved nutritional content and balanced feed value could be obtained for all three conversion processes [220]. Vegetable and fruit waste can be used as animal feed in wet, ensiled, or dried form. Carrot pomace is often used as animal feed due to the content of bioactive components with antioxidant properties [221]. Potato peel is one of the important food wastes that can be used as an animal feed because it represents sources of energy and fiber with low levels of protein [222].

5.3 Municipal Solid Waste

Municipal solid waste (MSW) is defined as the materials managed traditionally by municipalities that are categorized as unwanted and useless solids generated by various activities [223].

MSWs include waste originating from households, markets, shops, cafes or restaurants, office buildings, educational institutions, industrial sites or buildings, agricultural land and farming activities, and waste produced by waste water treatment plants (WWTPs) [224]. MSWs consist of biodegradable (BDMSW) and non-biodegradable components. BDMSWs include food leftovers, kitchen waste, garden waste, paper, and paperboard. MSWs may contain ingredients that increase the level of environmental pollution and affect human health [225]. These are mainly sanitary wastes, which can account for up to 24% of sorted MSWs in urban areas [226].

Population growth combined with rapid urbanization leads to an increase in MSW production. Significant factors influencing the generation of MSWs include socioeconomic conditions of a country, education level, population size, age structure, urban area, climatic conditions, seasonality, nature of heating, electricity sources, number of vehicles operated [227], and the national environmental legislation [228]. A higher standard of living contributes to an increase in the amount of food waste and plastics produced in MSWs [229]. The world's annual waste production in 2017 was about 17 000 Mt. In 2050 the production is expected to be 27 000 Mt [230]. The highest annual amount of MSWs [231] originates in China (300 Mt), United States (225 Mt), India (220 Mt), and in the EU (248 Mt) [232]. The average annual production of MSWs/capita in the EU was 489 kg in 2018. In 2018, 47% of MSWs produced in Europe was used for material recycling together with composting [233].

MSWs most often contain paper (packaging and non-packaging paper), paperboard (packaging), cardboard, glass (packaging and non-packaging glass), plastics (packaging and non-packaging bags and wraps), wood (wood packaging, treated wood, untreated non-packaging wood), textiles, shoes and bags, batteries, electronic stuff, and sanitary textiles. In Europe, the main components are 25–35% of paper, 25–35% of food waste, 10–15% of green waste, 7–10% of plastics, 5–10% of glass, 3–5% of ferrous metals, and 1–2% of ceramic and hazardous waste [234]. The composition of MSWs is influenced by the nature of urban development and the season. Biodegradable waste with a share of up to 34% is the most significant single component of municipal waste in the EU. Some authors report up to 50% share in MSWs and about 40% of recyclables [235]. BDMSW contains 30–70% of polysaccharides (starch, cellulose, hemicellulose), 5–10% of proteins, and 10–40% of lipids [236]. The main characteristics of BDMSV are high moisture, high content of volatile solids, and biodegradability.

5.3.1 Utilization and Disposal of MSWs

In 2018, 247 Mt of MSWs were produced in the EU. There are significant differences in the ways in which MSWs are handled in individual countries (Figure 5.20). On average, in Europe, 23% is landfilled, 30% is used for recycling, 17% has been composted, and 47% has been incinerated, only partially with energy recovery [232].

The main goal in the EU is to minimize the share of waste that is landfilled and to promote recycling as a matter of priority, which is also supported by legislative measures in the EU, such as "The European Union Action Plan for the Circular Economy – Directives 2018/850851," and making the separate collection of bio-waste obligatory by 2023 [119].

5.3.1.1 Incineration

In 2016, 1618 WtE (Waste to Energy) plants were used for waste incineration worldwide, 512 in Europe, 822 in Japan, 88 in the USA, and 166 in China [237]. In Europe, WtE is mostly used by Sweden, Finland, Norway, Switzerland, Germany, France, and the UK. Some countries do not have WtE capacity built at all. In Europe, 93 Mt of waste were incinerated at these facilities, which

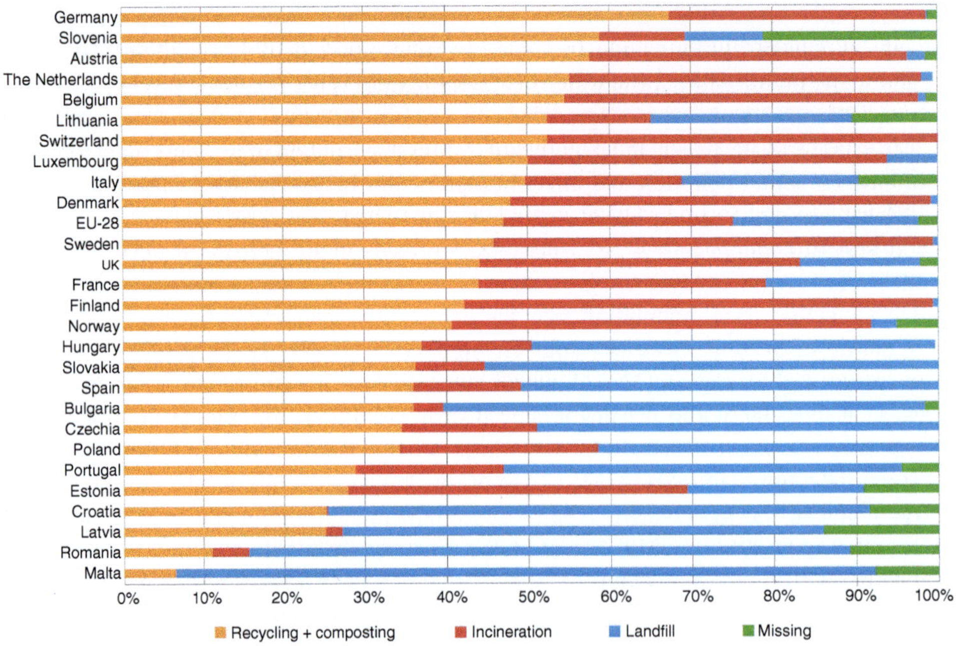

Figure 5.20 Processing, utilization, and disposal of waste in EU28+ [232].

were used for combined heat and power (251), electricity only (161), and 94 heat only [238]. In a combined WtE plant, 2/3 MWh of heat and 1/3 MWh of electricity is produced from 1 ton of MSW with a lower calorific value of 10 MJ kg^{-1}. If the output is only electricity, 0.7–0.75 MWh is produced from 1 ton of waste (10 MJ kg^{-1}) [239]. The results of the analysis performed by Scarlat et al. [238] show that Europe has the potential to build additional facilities with a capacity of 37–50 Mt. The basic requirements for building new WtE capacities include the amount of MSWs (at least 100 000 t yr^{-1}), and the lower calorific value must be at least 7 MJ kg^{-1} and never decrease below 6 MJ kg^{-1} [239]. The use of MSWs in WtE facilities increases the renewable energy percentage in the total production of energy and contributes to the reduction of greenhouse gas production. A very important aspect is the possibility of using fly ash from WtE, which can have the character of hazardous wastes. At present, there are already possibilities to modify this fly ash for further use [240]. Ashes from MSW incineration have pozzolanic properties, so they can be partially used in construction as a substitute for cement [241], and thus contribute to saving primary resources (limestone) and reducing CO_2 emissions. The replacement of conventional fly ash from fossil fuel combustion for these purposes is becoming increasingly relevant due to the rapid retreat from fossil fuel combustion.

The combustible part of MSWs can be used to produce refuse-derived fuel (RDF) and solid-recovered fuel (SRF). Fuel quality is standardized according to EN 15359 [242], which specifies the technical and environmental parameters of the fuel (calorific value, chlorine, and Hg). Fuels are divided into five classes. As SRF is used as a substitute for biomass, the determination of biomass content according to EN 15440 [243] is also recommended. The current production of SRF in Europe is 16.5 Mt. The market potential is estimated at 27–37 Mt [244]. More than half of the production is used for power plant production and the rest is used for cement kilns.

5.3.1.2 Recycling

The approach chosen depends, inter alia, on the volumes and "purity" of the streams of bio-waste to be used and on the technology readiness level (TRL). The "value" of waste is a key element of the concept of the circular economy [245]. The most important step to ensure successful recycling is the efficient separation of the individual components of MSWs. "Waste valorization" requires the separate collection of certain commodities to avoid deterioration or contamination [246]. Complications for sorting can be caused by microbiological parameters and hygienic properties (the occurrence of moulds or pesticides in food), which form the input material for composting [247].

5.3.1.3 Composting and Anaerobic Digestion

Composting is most often used in biotechnology when dealing with BDMSW. It is one of the low-cost technologies, and there are not as strict requirements for the quality of the input material as in anaerobic digestion. From an environmental point of view, compost improves soil properties, reducing doses of chemical fertilizers [248]. Compost can be used for land reclamation and filling, where it can significantly reduce the mobility of pollutants [249]. The sale of compost depends on its properties and on the raw materials available for composting. Marketing research suggests that it is preferred to use compost from yard waste rather than from the organic fraction of MSWs or with the addition of sewage sludge [250]. All EU countries use diverted green waste from MSWs for composting, but only about 5% is used as input for the anaerobic digestion (mostly kitchen waste). Compost made from MSWs must not be used in Finland, Denmark, Ireland, The Netherlands, Sweden, and Switzerland. Austria, Italy, the UK, and Hungary landfill compost from MSWs, while France, Spain, and Portugal also use these composts for agricultural purposes [251]. A transparent approach to the biodegradable component of MSWs is newly defined in regulation (EU) 2019/1009 [252]. A fertilizer product marketed in the EU may not contain compost obtained by aerobic composting of the organic part of mixed municipal waste from households separated by physical, chemical, biological, or manual means or from sewage sludge. The fertilizer product cannot contain digestate obtained by anaerobic digestion of separated organic waste from MSWs.

5.3.1.4 Other Possibilities of Using BDMSW

Research and development in the field of biorefinery of waste materials used to focus on the production of energy and fuel, but now it is focused on the production of *bulk and fine chemicals* from waste lignocellulose. Lignocellulose (the structural tissue of plants) is contained in the bio-waste of MSWs. Agricultural residues and food-based waste are more suitable for the production of chemicals because they provide homogeneous waste streams with large volumes and simpler logistics compared to bio-waste from MSWs [245]. Production of *bulk chemicals* from waste biomass can be 3.5 times more profitable than energy production [245]. The recovery of LA by fermentation from the biodegradable component of MSWs was verified by López-Gómez et al. [253], who achieved up to 93% yield of LA by introducing a pre-treatment stage of BDMSW, allowing an increase in the liberation of sugars by enzymatic hydrolysis.

An important way of using BDMSW is the production of hydrogen. Hydrogen can be produced using thermochemical processes, biological processes, water electrolysis, and radiolytic processes [254]. Thermochemical conversion has the greatest prospects, as evidenced by the opening of Ways2H and the first-of-its-kind commercial waste-to-hydrogen facility to begin supplying H_2 to customers in 2021. Ways2H is a joint venture between US-based Clean Energy Enterprises and

Tokyo-based Japan Blue Energy Corporation (JBEC). H_2 is obtained from MSWs (after removal of the inert component) and crushed into particles with a size of 0.5–3 cm. The screw conveyor brings the waste to the gasification reactor, where they are mixed with ceramic beads, which have been pre-heated to a temperature of 1000 °C. At this temperature, most organic compounds decompose to form a mixture of gases, CH_4, H_2, CO, and CO_2. Solid carbon and minerals remain in the char. The gas mixture is further fed to a reforming vessel, where steam is added, which splits methane into hydrogen, CO, and CO_2. This ensures that the amount of H_2 in the gaseous mixture is higher than 50% by volume. The char generated during gasification is burned at a temperature of 1000 °C. The resulting heat is used to heat the ceramic beads. The amount of H_2 produced from 1 ton of dried waste is 40–50 kg. The moisture of the waste contributes to a higher H_2 production. Way2H has mobile technology at its disposal (Japan), 1 ton of waste/day or stationary technology with a capacity of 50 tons d^{-1}.

Further research on the optimization of hydrogen production during the thermochemical conversion was focused on the use of CO_2 for gasification [255] or the addition of CaO as a catalyst and a sorbent of CO_2 [256]. An interesting alternative to hydrogen production is also the use of bottom ash from an MSW incineration plant. Hydrogen gas is generated through the reaction between water (at 50 °C) and essentially the metal aluminum in the bottom ash. The highest yield of 12.7 m^3 H_2 ton^{-1} of ash was obtained for the ratio S : L = 5 : 1, which, according to the theoretical calculation, corresponds to about 26% [257].

5.3.2 Sewage Sludge

Sewage sludge is the residual, semi-solid material that is produced as a by-product during sewage treatment of industrial or municipal waste water. Sludge is an aqueous suspension of solid and colloidal particles composed of organic compounds (50–60% d.m.) and inorganic compounds and minerals (40–50% d.m.). The resulting composition of the sludge is influenced not only by the nature of the inflowing waste water (inhabitants and industry) but also by sewer systems (combined sewers overflow for sewage + surface water and separate sewers).

- Organic substances form (the rest of toilet paper and plant residue, products of metabolism, and microorganisms from the sewage treatment process). Sludges contain highly complex mixtures of organic molecules, which include naturally formed compounds – cellulose, hemicellulose, proteins, peptides, lipids, and plant macromolecules with a phenol structure; fragments of lignin and lignans or aliphatic structures; and cutines and suberines [258].
- Chemical compounds of nitrogen and phosphorus (nutrients).
- Toxic substances: heavy metals and metalloids, organic micropollutants (PCBs, PAHs, PCDD/Fs, EOXs, C_{10}–C_{40} hydrocarbons, pesticides; surfactants: alkylsulfophenols, polyphenols), flame retardants (polybrominated diphenylethers), and drugs. High concentrations of PFC (perpolyfluoridated compounds) were found in sludge and compost produced using sewage treatment sludge. Attention is focused on perfluorooctane sulfonate (PFOS) and perfluorooctanoic acid (PFOA), which are potent endocrine disruptors (they decrease the immune system function and can have adverse effects on several organs, including the liver and pancreas).
- Mineral phases: quartz, feldspars, carbonates, phosphates – vivianite, apatite, Fe-oxides.
- Water (60–75%) depending on drainage technology.

5.3.2.1 Problem Parameters in Sludge

5.3.2.1.1 Contaminants of Emerging Concern (CEC)

Contaminants of emerging concern (CEC) can be concentrated in the sludge, such as pharmaceuticals and microplastics. The primary sources of CECs that enter waste water include household chemicals (detergents, cleaners, personal care products, textile fibers), pharmaceuticals, and nanoparticles. Secondary sources include rainwater, which contains microplastics released by abrasion (tires, car parts).

Nanoparticle-sized microplastics that flow in waste water to the WWTP show acute inhibition of microorganisms and enzymes and affect the course of metabolic processes, thereby reducing the efficiency of wastewater treatment. Due to their large specific surface area, which is hydrophobic, they easily capture metals, persistent organic pollutants, and pathogens on the surface of the particles [259]. The most important microplastics in sludge include synthetic fibers that are released during washing [260]. The Danish Environmental Protection Agency [261] reports up to 4.5 g kg^{-1} of microplastics in sludge dry matter (with an average value from six WWTPs). Approximately 80% of the total amount inflowing to WWTP is captured in the sludge [262]. An important source of microplastics in the sludge are residues of cosmetics (PE in liquid soap), which may contain 0.5–5% microbeads, primary microplastics [262].

5.3.2.1.2 Pharmaceuticals

The presence of antibiotics in the sludge affects the occurrence and growth of antibiotic-resistant bacteria. The antibiotics contained in the sludge in the highest concentrations are Ciprofloxacin (22 mg kg^{-1}) and levofloxacin (8.9 mg kg^{-1} quinolone antibiotic). There are also other drugs in the sludge, sorted in descending order of concentration: setraline (antidepressants), triclosan (increases resistance to bacteria – toothpaste, soap), simvastatin (lowers total cholesterol), clarithromycin, azithromycin and erythromycin (macrolide antibiotics), atorvastatin (regulates cholesterol levels), ibuprofen (anti-inflammatory), diclofenac (2.1 mg kg^{-1}, treatment of pain and inflammation), carbamazepine (neuralgia, nerve pain). The importance of the occurrence of antibiotics in sludge is constantly growing, as their consumption increases. Carbamazepine consumption in 1995 was estimated at 742 tons and in 2015 at 1214 tons, while ciprofloxacin consumption increased 7.7 times to 2318 tons over the same period [263].

Fluoroquinolones stimulate bacterial resistance and are toxic to aquatic organisms and terrestrial plants. Of the total amount flowing into WWTP, 85% is captured in the sludge. It is not biodegradable and can only be removed by advanced oxidation technologies [264]. The concentration of 100–300 μg l^{-1} of antibiotics (doxycycline, gentamicin, penicillin, nitrofurantoin, and rifampicin) fundamentally affects the decomposition of organic substances in WWTP technology, and the effect of nitrogen and phosphorus has not yet been proven. There was a significant increase in the number of antibiotic-resistant bacteria, which differed for each type of antibiotic. The largest number of bacteria were resistant to nitrofurantoin and penicillin [265]. The effect of antibiotics in sludge used on agricultural land is evidenced by studies describing their effects on arable and vegetable crops [266]. The persistence of antibiotics (10–50 days) is influenced by the ability to sorb on particles of organic matter present in soils or sludges, the physical and chemical parameters of organic compounds, and the environmental and climatic factors. Antibiotics damage soil microorganisms by changing their enzyme production, the ability to metabolize organic matter, and changes in the composition of microbial communities [267].

5.3.2.1.3 Pathogens

Pathogens such as bacteria (*Clostridia*, *Legionella*, *Yersinia*, and *Escherichia coli* O157:H7), viruses, parasites (*Giardia* and *Cryptosporidia*), and worm eggs are also found in sewage sludge [268]. Information concerning the ability to survive in sewage sludge and consequently in the soil in the case of strong pathogenic strains, such as *Escherichia coli* (EHEC) with the serotype O104:H4, is very limited [269].

The driving forces for changes in the technological line of sludge management with anaerobic stabilization within the circular economy are health protection, environmental protection, nutrient recycling, energy recycling, and water recycling. The sludge management of WWTPs will be the subject of the application of new technologies focused on the use of the energy potential of sludge as biomass. These are technologies of thermal hydrolysis, low-temperature sludge drying, pyrolysis, or gasification. These technologies, combined with anaerobic stabilization, offer comprehensive WWTP sludge treatment, fully in line with Circular Economy requirements.

5.3.2.2 Utilization

In the European Union, WWTPs produce up to 30 kg of dry sludge per capita per year [270]. The estimated annual sludge production in Europe in 2020 is 13 Mt of dry matter [271]. The approach of individual countries in the EU to sludge management is very different. In 2017, Germany burned 64.5% of the sludge produced (Figure 5.21) and Austria 53.5%. In other countries, the share of incineration is significantly lower, with Hungary 27%, France and Poland 17% [273].

The use of sludge in agriculture within the EU is currently regulated mainly by the Council Directive 86/278/EEC [274], where limits on heavy metals (Cd, Cu, Hg, Ni, Pb, and Zn) are listed. However, it is outdated and does not express the current needs of requirements for the safe application of sludge in agriculture [275]. The pros and cons of each approach to sludge recovery are listed in Table 5.9. Sludge used in agriculture can easily create a diffuse source of environmental contamination. However, some EU countries have reduced the use of sludge on agricultural land to avoid the risks associated with contaminant emergence concerns and pathogens [276].

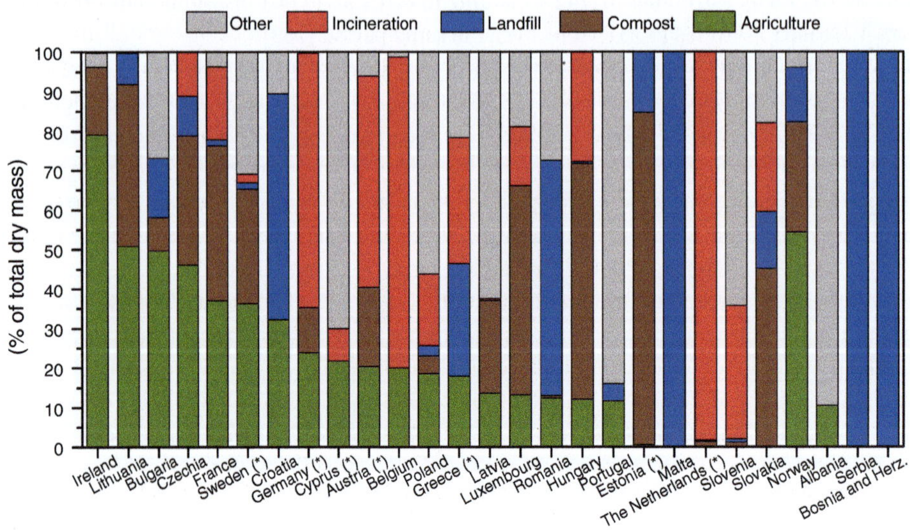

Figure 5.21 Sludge utilization in selected EU countries in 2017. (*) Data designed for 2016 instead of 2017 [272].

Table 5.9 Advantages and non-advantages for other utilization of sewage sludge.

Utilization	Advantages	Disadvantages
Farming, landscaping	Nutrients uptake	Occurrence of organic pollutants and pathogens
	Cost efficient	
Mono-incineration	Eliminated occurrence of organic micropollutants, pathogens	P recovery from ash is expensive
	Energy and heat recovery from waste	Transport is source of other pollution
	Recovery of P from ash, product with high bioavailability for plants	Drying or other sanitation process before transport
	Protection of non-renewable sources (P)	Temporary storage
	Recovery of other elements (Al, Fe) and heavy metals	
Co-incineration	Partly utilization of waste for energy recovery (10–20%)	Possible change in the properties of ash and thus its use
	Less cost intensive than mono-incineration	Drying or other sanitation process before transport
	Reduction of primary sources energy	Modification of the technology lines – fuel feed
	Reducing operational capacities due to fossil fuel cuts	
	Nutrients from slag/ash are not available	

Large WWTPs in Germany will not be allowed to use sludge as a fertilizer or as a component in compost [277]. A key factor in the revised regulation [277] is the introduction of a future obligation to recover phosphorus in all treatment plants that have a total phosphorus content of $>20\,g\,kg^{-1}$ in the sludge dry matter (Figure 5.22). There is currently a 12–15-year transition period. From 1 January 2032, there will be a ban on the application of sludge on agricultural land for WWTPs with 50 000–100 000 PE and for WWTPs >100 000 PE. In the future, it is expected that the need to obtain phosphorus will also be introduced for treatment plants with a smaller population < 50 000 PE.

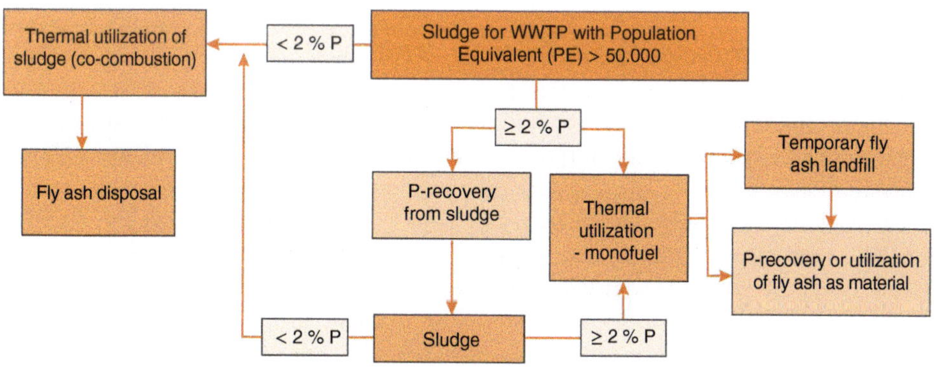

Figure 5.22 Possibilities of sludge treatment for obtaining P from WWTPs with over 50 000 PE from 2032 [278].

5.3.2.2.1 Cellulose Fibers

The monthly consumption of toilet paper in The Netherland is 1 kg month^{-1} [279]. The length of the fiber in the toilet paper is a standard parameter in the range of 1–1.2 mm. The size of the fibers affects their occurrence in waste water in the grain size class below 0.35 mm. This information was used to implement the Cellvation® process, which can be used instead of, or in conjunction with, the primary treatment of a sewage treatment plant. Up to 40% of the total suspended solids are captured when the waste water passes through the bar screen. Industrial filtration is followed by purification and sterilization at very high temperatures. The resulting product is high-quality Recell® cellulose fibers, which are widely used in various industries: construction, paper and board production, or for the production of polymer granules.

5.3.2.2.2 Possibilities of Obtaining Phosphorus – Mineral Fertilizer

Phosphorus is one of the biogenic elements. It is crucial for energy metabolism. In the form of adenosine triphosphate (ATP), it carries energy and is a key component of DNA and RNA. In an organic form, phosphorus is part of phospholipids, phosphoproteins, and nucleic acids. The daily nutritional consumption of humans for phosphorus is in the range of 1.8–2 g of P, of which approximately 1.4–1.6 g of P/PE d^{-1} (1PE = 60 g BOD_5 d^{-1}) is excreted with their excreta and dishwasher detergent to the waste water [280]. Some authors report a slightly higher throughput rate of phosphorus of 0.77 kg per person year^{-1} [281]. In 2015, 3.7 Mt of P got into the waters, which corresponds to about 19% of agricultural consumption of P, amounting to 19.51 Mt [281]. Because phosphorus is obtained from non-renewable resources (phosphate rocks), it is on the European list of critical raw materials. Critical raw materials form a priority area of the "EU Circular Economy Action Plan COM (2015)614" [282], which aims to promote their efficient use and recycling.

The sludge contains 2–4% of P, from which struvite can be produced with an estimated sales market price ranging from €188 to €763 per tonne [283], compared to the sales price per tonne of phosphate rock amounting to €220–332 [281].

Phosphorus is removed from waste water by biological accumulation in biomass (Enhanced Biological Phosphorus Removal, EBPR) or by chemical precipitation. Approximately 10% of the total amount in the inflow thus flows from the WWTP. The rest is bound by the activating sludge [284]. The concentration of P_{tot} in the inflow to the WWTP is in the range from 4 to 12 mg l^{-1} [285]. Removal of P from the waste water stream with struvite technology takes place at 80 WWTP (Japan, USA, and Europe) with an efficiency of 40–50%. Phosphorus recovery from sewage (SS) or sewage sludge ash (SSA) is much more efficient, reaching up to 90%.

5.3.2.2.3 Direct Extraction of Phosphorus from the Sludge – MAP Crystallization

The high costs, including energy consumption associated with obtaining phosphorus from ash after sludge incineration, are an incentive to find alternative methods for obtaining phosphorus by direct leaching. The highest P leaching efficiency (up to 82%) was achieved using sulfuric acid in concentrations up to 1 M, leaching time of two hours, S/L ratio (5%), and temperature of 100 °C. Fe, Mn, and Ca are also leached together with P. Ion exchange was used to remove Fe and Mn. The yield of precipitation of magnesium ammonium phosphate (MAP) = struvite mineral ($MgNH_4PO_4.6H_2O$) by the addition of ammonium and magnesium hydroxide at a molar ratio P : N : Mg = 1 : 3 : 4 reached 57% [286]. The following technologies are used in operation: REM-NUT (ion exchange + precipitation), AirPrex, PRISA (precipitation + crystallization), Ostara-Pearl reactor, P-ROC (crystallization), REPHOS, NuReSys (striper/crystallizer with the addition of $MgCl_2$, end product: Bio-Stru). A HUBER Grit Washer can be used to separate MAP crystals from the sludge suspension with a fluidized bed system with a pressure measuring system that affects the density and size of the MAP crystals.

5.3.2.3 Thermal Processes Using Sewage

Co-firing is usually the case of fuel replacement (power plant or cement factory). In the case of a municipal waste incinerator, the SS addition can be a maximum of 20%, converted to DM [287] (Figure 5.23).

5.3.2.3.1 Co-Firing

Power plants, incinerators, and cement plants can be used to obtain energy/heat, where sewage sludge (SS) can be used and co-fired. In Germany, from 2029 co-firing for sewage sludge from WWTP >100 000 PE will only be possible if the P content is <20 g kg^{-1} d.m. Mono-incineration with P-recovery or sewage sludge ash (SSA) disposal (including direct use of sewage sludge ashes as fertilizer) will be possible if the P content is >20 g kg^{-1} d.m.

Sewage sludge is considered to be a renewable energy source. It contributes to reducing greenhouse gas emissions. Combustion of SS is not considered "environmentally friendly." Dewatered sludge contains up to 75% water, which is used to remove energy. The calorific value of dry, anaerobically stabilized sewage sludge is 9–12 MJ kg^{-1}, while for food waste it reaches up to 18.5 MJ kg^{-1}.

The technology of co-firing coal with SS is used mainly in Germany and Poland [288] and is being tested in the Czech Republic. Originally, only a 5–10% addition of biomass or 5% [289] was used for large-scale coal–biomass co-firing power generation because at a higher amount of 10–20%, slagging and fouling in boilers occurred [290, 291]. Emission limits for stationary emission sources can be met even with the addition of 20–30% [288]. Studies of co-firing coal with SS show that the addition of biomass affects both the content of heavy metals in fly ash: Cr, Cu, Ni, Pb, Zn, and Fe [292] and the amount of fine particulate matter, especially particles below $PM_{2.5}$ in emissions [293].

5.3.2.3.2 Mono Incineration

There are 170 sludge incinerators in operation in the USA that use incineration technology: multiple hearths (80%), fluidized bed (15%), and electric infrared (3%). Some sludge is co-fired with MSW in combustors with suitable technology for refuse combustion. There are currently 27 mono-incineration facilities for SS in operation in Germany [287]. Mono-incineration facilities

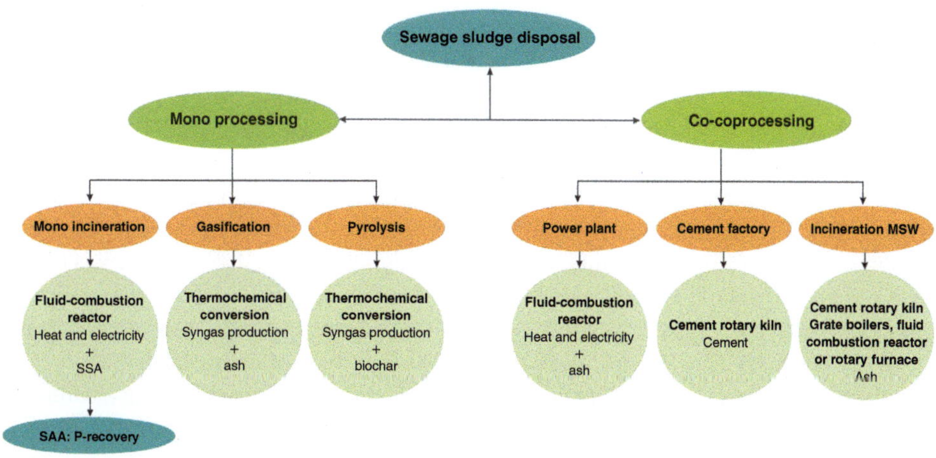

Figure 5.23 Thermal processes of the sludge utilization and disposal.

Table 5.10 Concentrations of P and other elements in sewage sludge ash [295, 296].

	P	Ca	Al	Fe	K	Mg	Cu	Zn	Pb	Ni	Cr	Cd	Hg	As	Mn	U
			$(g\ kg^{-1})$							$(mg\ kg^{-1})$						
Sewage sludge – ash	80	103	109	24	10	11	1050	1.79	137	46	56	5			1300	4.7
AVG Sewage sludge ash –11 WWTP	79	152	57.6	84.8												
11 WWTP, Germany, Min–Max	26–110	84–250	21–114	15–138												

burn SS at an operating temperature of 850–950 °C. At temperatures below 850 °C, the occurrence of odorous substances is a problem, and at temperatures above 950 °C, sintering can occur. The concentration of phosphorus in SSA is around 9%, which corresponds to the concentration contained in the mined rock [294]. However, the high concentration of Fe and Al in SSA (Table 5.10) is disadvantageous. It requires a higher consumption of sulfuric acid during its processing. A number of technological problems with the removal of Fe and Al or heavy metals have already been overcome, which is documented by the high increase in operational equipment for obtaining P from SSA (Table 5.11). The resulting products of the technology are most often phosphoric acid, dicalcium phosphate (mineral brushite), calcium hydroxyapatite and phosphate-containing ash (Table 5.11).

Ash2Phos technology (Rambol, Denmark) was chosen as an example of a successfully implemented technology, which is currently under construction (Sweden, Germany, and Denmark) with the planned commissioning in 2023–2024. After burning the sludge, the ash is dissolved in HCl (without pressure and heating). P, Fe, and Al are separated from the solution by precipitation. The solution is neutralized and heavy metals are removed. The resulting product is pure calcium phosphate, ferric chloride, and aluminum hydroxide. Calcium phosphate is used to make NPK fertilizers, while ammonium phosphate, ferric chloride and aluminum hydroxide are used to make coagulants used to treat waste water. The ash consists of 50–60% acid-insoluble silicate sand, which can be used (up to 30%) as a component to replace cement in the production of concrete. The problem may be a higher Fe content, which would limit this use.

5.3.2.3.3 *Thermochemical Conversion*

Another solution is represented by the methods of thermal conversion (pyrolysis), which is not yet industrially applied in the EU. The PYREG system (Pyreg GmbH, Germany) is being tested at three WWTPs in Germany. The dried sludge is first gasified at a temperature of 550–600 °C and then carbonized by the supply of a controlled amount of air. The resulting product (biochar) is usable as a fertilizer. The process gas released during the gasification of the input biomass is combusted at a temperature of 1250 °C in a FLOX® burner (flameless oxidation). At these temperatures, there is no condensation of organic compounds and tars. The use of pyrolysis in sludge processing is limited by the high content of heavy metals in biochar. The problem is solved by the addition of chlorinating agents (PVC, $MgCl_2$, $CaCl_2$) during pyrolysis, which significantly reduces the concentration of heavy metals by up to 86% (Zn, Mn, Cu, and Pb) in biochar and bioavailability P [308].

Municipal waste water treatment technology is a very good example for the application of circular economy principles in several levels in order to achieve the best possible material recycling – nutrient recovery (P) and energy utilization by improving conditions for anaerobic digestion, removing cellulose fibers from toilet paper, etc.

Table 5.11 Overview of technology (full scale and pilot scale) of P-recovery from SSA and from SS, main principles and final product description (Source European Sustainable Phosphorus Platform, February 2020).

Technology	Technology description	Full-scale + pilot plant		Place	Final product	Description
ECOPHOS	Wet chemical extraction + ion-exchange purification	2020/21	2	Dunkerque, France	H_3PO_4/DCP	[296]
PHOS4LIFE	SSA - leaching with sulfuric acid + solvent extraction with HCl			Full-scale planed Solothurn, Swiss	H_3PO_4, $FeCl_3$	[297]
TetraPhos, REMONDIS	SSA – leaching H_3PO_4, precipitation H_2SO_4, cation-exchange (HCl) + nanofiltration	2020	2	WFA Elverlingsen's incineration plant in Werdohl – pilot facility, Hamburg – industrial scale	H_3PO_4	[298]
AshDec (Outotec)	SSA is mixed with a Na-salt (Na_2CO_3, $NaHCO_3$ or Na_2SO_4) and heated to about 850–900 °C in a rotary kiln to modify the P-compounds to $CaNaPO_4$			Pilot plant Leoben Austria	Calcium-sodium phosphate with 80% solubility in citric acid	[299]
Ash2®Phos, Easy Mining	SSA – acid leaching (HCl or H_2SO_4), precipitation – extraction of P, Al, and Fe, lime addition P, Fe, and Al separation, neutralization – heavy metals removal		2+2	Full-scale plant: Bitterfeld-Wilfen near Berlin, Sweden. The pilot plant at the Kemira site in Helsingborg + Upsala	Calcium phosphate, iron hydroxide and aluminum hydroxide	[300]
METAWATER	SSA – alkaline leaching 50–70 °C, P is in the form of $AlPO_4$, membrane type solid–liquid-separation, dissolved P is precipitated with $Ca(OH)_2$		2+3	Full-scale plant: Gifu a Tottori (Japan)	Calcium hydroxyapatite	[301]
EUPHORE	SSA – rotary kiln (2 zone, reductive for drying and oxidation reactor) with addition KCl or $MgCl_2$ for heavy metals depletion and enhancing solubility of P.	2018	4	Oftringen and Urvier (both Switzerland), Offenheim + Mannheim (Germany)	Min. fertilizer (phosphate containing ash)	[302]

(Continued)

Table 5.11 (Continued)

Technology	Technology description	Full-scale + pilot plant		Place	Final product	Description
PHOS4Green (Glatt)	SSA is leached with H_3PO_4, addition N, K, Mg, S and trace elements	2019	1+4	Pilot plant in construction, Weimar Germany	P or NPK	[303]
MEPHREC	The dried sludge is gasified at the temperature of1450 °C; non-volatile elements are producing silico-phosphate slags.	2016/2017		Pilot plant in Nuremberg	P-slags	[304]
Electrodialysis	SSA is leached with sulfuric acids following the electrodialytic separation process (ED). Phosphorus is mobilized toward the anode and heavy metals electromigrated toward the cathode.			Pilot plant		[305]
Fertilizer industry	SSA is mixed in phosphate rock and leachate with H_3PO_4	2019		ICL Netherlands		
Kubota	Thermal treatment 1300 °C with Fe-oxide addition		4	30 full-scale furnaces in Japan	P-containing slag	[306]
TerraNova (HTC)	Hydrothermal hydrolysis carbonization process at 175 °C, 20–25 bars. Acid treatment + mechanical separation + phosphate precipitation.	2016	4	China, Duisburg	Mg/Ca-P salt	[307]

5.3.3 Cardboard

Approximately 25% of recycled paper is used for the production of cardboard and the remaining 75% is the primary source. Different types of paper and adhesives can be used for the production of corrugated board [309]. *Paper and cardboard packaging* is collected separately and sorted so that the resulting products are homogeneous and suitable for recycling. Recycling recovers cellulose fibers, which are used to make new packaging. Fibers can be used a maximum of five times to make new packaging before they become too short and damaged. Such material is eliminated from the recycling process and is used for energy production, building boards, or as a carbon source on agricultural land. The European paper industry as a whole uses 86% of renewable raw materials, of which 46% are fibers from paper recycling, 40% comes from raw wood, and about 14% is filler, calcium carbonate. More than 90% of the wood used in the production of paper and cardboard comes from the European Union. In Europe, 83% of paper and cardboard packaging (including cardboard) is recycled and another 7% is used for incineration. In the EU28, 34.6 Mt of paper and board waste was generated in 2015, of which more than 90% was recovered (including recycling) [310]. Under certain conditions, the paper is a renewable resource and the high energy potential of waste paper may be more significant than recycling [311]. Most waste paper can be used for the production of RDFs according to EN 15359 : 2011 [242] and can be classified in class III by meeting the condition of a net calorific value >15 MJ kg^{-1}. The advantage of such a fuel is the low content of ash, chlorine, and nitrogen. Waste paper can be easily compacted at low pressures [312]. In comparison with briquettes from wood waste, briquettes made from recycled paper and cardboard have lower moisture, higher specific mass, higher mechanical durability, and require a relatively high force for their rupture. In pellets or briquettes produced solely from paper, the unburned remains often prevent the entire combustion of their inner part [313]. Compared to briquettes made of wood, they have a high ash amount (newspapers 13.9% and magazines 41.0%) and a low gross calorific value: sorted office paper 13.9 MJ kg^{-1} and white wood-free books 18.0 MJ kg^{-1} [312]. There are concerns that the burning of paper briquettes represents a significant risk to human health [314] associated with the presence of additives used in printing. By burning paper briquettes, As, Hg, Cd, Pb, and Mn are released into the flue gas [315].

5.4 New Sources of Biodegradable Waste

5.4.1 Biodegradable Plastics

Today, there are about 300 types of bioplastics. Terminology concerning biodegradable plastics is rather confusing. Words "bio," "degradable," or "compostable," etc., are often used in a similar sense. However, their technical meanings are different, and the related information can influence the use, recycling, or disposal of the specific polymer [316]. The key terms are defined below.

Biodegradable plastics (BDPs) are plastics that can be split into monomers. Decomposition processes into H_2O, CO_2, and biomass take place with the help of microorganisms (bacteria and fungi) over time. The rate of decomposition depends on the chemical composition of the plastics and the environmental conditions. Although plastics may be considered biodegradable, the recycling (composting and anaerobic digestion) may not be a suitable way to process them due to the long time required to ensure their decomposition. The properties important for biodegradation depend on the chemical structure of the polymer (molecular weight, crystallinity, and physical form) and not on its source [317]. This is the reason why not all bioplastics can be considered as biodegradable [316].

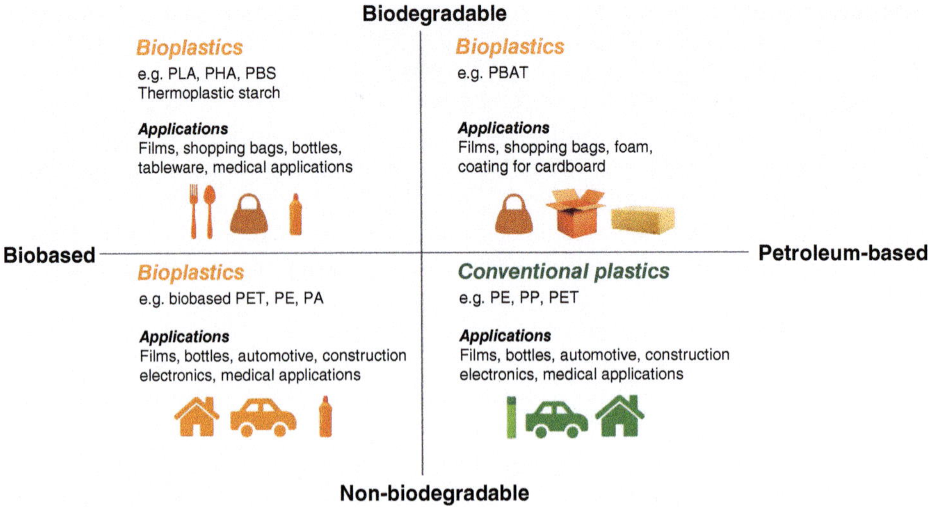

Figure 5.24 Classification of plastics [321].

According to Standard ASTM D6400-04 [318], *compostable plastics* are plastics suitable for biological decomposition in compost. They break down to water, carbon dioxide, biomass, and inorganic compounds at a rate comparable with usual composted materials (e.g. cellulose) and produce no toxic residue. Plastics classified as "compostable" must be decomposed during the standard composting process and must not in any way compromise the properties of the resulting compost.

Biodegradable plastics are divided into:

Bioplastics (bio-based polymers) are produced from plant-based polymers, such as cellulose, lignin, sugarcane, corn-starch, chitin, polylactic acid, polyhydroxyalkanoates, casein, gelatine, wheat gluten, wool, silk, animal fats, and plant oils. Bioplastics can be chemically synthesized from a bio-derived product: polylactic acid, polybutylene succinate, and polytrimethylene terephthalate [319]. Lactic acid as the raw material is produced by fermentation of sucrose or glucose and is refined to a high purity [320]. *Bioplastics include non-degradable and biodegradable materials, both important for sustainable production.*

BDPs based on fossil resources such as poly(vinyl alcohol), aliphatic polyesters (polybutylene succinate, polyglycolic acid, and polycaprolactone), and aromatic copolyesters (polybutylene succinate terephthalate) are produced by synthesis from monomers derived from fossil resources [319] (Figure 5.24).

5.4.1.1 Degradable Plastics

Degradable plastics include plastics with additives to accelerate disintegration, the most commonly used being the oxo-degrading agent d2w or EPI TDPATM. Plastics gradually break down into particles, which should then biodegrade. In a real environment, when discarded oxo-degradable plastics do not have an oxygen supply (in landfills or in the seas), the decomposition may not take place fast enough, or it is only partial. The particles can thus become microplastics with similar properties to those of microplastics from conventional plastic. These can enter the food chain of animals and humans, and their effect on human health has not yet been sufficiently studied. Due to the slow decomposition, oxo-degradable plastics are not even suitable for composting, and we cannot recycle them either, as they would decrease the quality of the recycled material.

The mechanical properties of bioplastics are worse than in conventional plastics; they have a lower tensile strength, tension, and compression [322]. These mechanical properties of bioplastics can be enhanced by chemical, enzymatic, or physical treatments. Citric acid, glyoxal, glutaraldehyde, and formaldehyde are used as crosslinkers to improve properties [323]. The production of bioplastics is about 10 times more expensive, which limits their use [324]. The solution is the production of next-generation bioplastics. The annual production of bioplastics in 2018 was approximately 2.1 Mt of the total production of plastics – 335 Mt [325]. Asia is a significant manufacturer of bioplastics. In 2017, more than 50% of the total bioplastics worldwide production was produced there. Japan has been the leader in using bioplastics in industry. Bioplastics are extensively used in the automobiles and electronics industries as well [326]. An estimate of the global demand increase is 19% annually, due to their use in packaging materials, catering products, electronics, the automotive industry, agriculture, horticulture, and the production of toys and textiles [327]. So far, a faster expansion of bioplastics has been hindered by twice the price costs of those in the production of conventional plastics and some negative properties: high vapor and oxygen permeability, poor mechanical properties, thermal instability, and lack of legislation [328].

5.4.2 Classification of Bioplastics

Bioplastics may be categorized into three classes according to their origin and process of production [326]:

- Biopolymers directly extracted from biomass, e.g., starch and cellulose-based polymers, and casein and gluten-based biopolymers.
- Biopolymers synthesized by using renewable monomers, e.g., poly-lactic acid synthesized from lactic acid.
- Biopolymers produced by microorganisms or genetically modified bacteria, e.g., polyhydroxy alkenoate.

An overview of the main types of starch-based bioplastics, polylactic acid (PLA), polyhydroxyalkanoates (PHAs), renewable polyethylene and polyvinyl chloride (PVC), cellulose, as well as protein-based polymers is given in Figure 5.25. The left column shows the chemical formulae of naturally occurring monomers and polymers that form the basis for the formation of individual bioplastic types. Their source is higher plants, or algae and cyanobacteria. The middle column contains examples of the conversion of monomers and polymers to bioplastic. The right column lists the properties, uses, and biodegradability.

5.4.2.1 Biopolymers Directly Extracted from Biomass

Starch is a natural polysaccharide present in cereal grains (corn, rice, wheat, barley, oats, sorghum), plant tubers (potatoes), roots (sweet potatoes), stems (sago palm), and legume seeds (peas and beans). Starch grains have a size of 0.5 to 175 μm and various shapes. Starch is polysaccharide composed of α-D-glucopyranosyl units, which consists of two different molecules (70–80% of amylose and 20–30% of amylopectin) [330]. Starch is a hydrophile and has poor mechanical

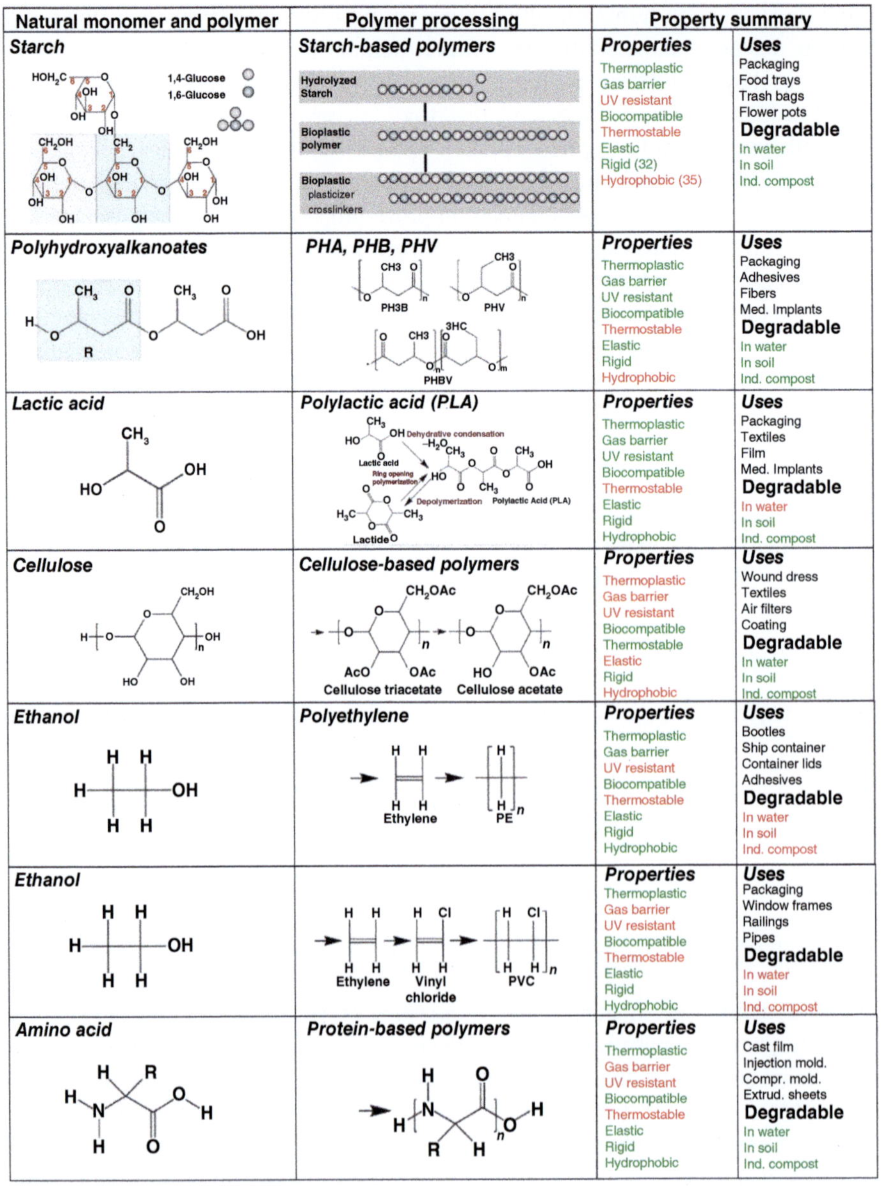

Figure 5.25 Overview of bioplastics. Starch-based polymers, polyhydroxyalkanoates (PHAs) [including polyhydroxybutyrate (PHB), polyhydroxyvalerate (PHV), poly (3-hydroxybutyrate-co-3-hydroxyvalerate) (PHBV)], polylactic acid (PLA), cellulose-based polymers, polyethylene (PE), polyvinyl chloride (PVC), and protein-based polymers [329].

properties that limit its use for the production of bioplastics, so it is often modified physically or chemically [331]. Starch serves approximately 50% for the production of the bioplastics used commercially.

Improvement of the properties of starch can be achieved by adding crosslinking agents blending with thermoplastic polymers. Various compostable natural fibers can also be added to starch in order to increase the modulus or impact toughness and reduce water uptake of the thermoplastic composites [330]. The production of starch-based bioplastics is simple, and they are widely used for packaging applications. For trade applications, the starch-based plastics are usually mixed with eco-friendly polyesters [332].

Lignocellulosic fibers are another raw material used in the production of bioplastics. The possibilities of transformation of chemicals derived from lignocellulose into monomers and their use for the synthesis of newly formed polymeric materials are reviewed by Delidovich et al. [333]. Lignocellulosic fibers are composed of cellulose, hemicellulose, and lignin, with a minor presence of pectin, water-soluble substances, and waxes [334]. The composition of the fibers depends on the type of plant, age, climate, and soil conditions. Cellulose $(C_6H_{12}O_5)_n$ is a polysaccharide composed of β-D-glucose units that are linked by β-1,4-glycosidic. Hemicellulose consists of homo- and heteropolymers containing anhydro-β-(1-4)-D-xylopyranose, glucopyranose, galactopyranose, and mannopyranose.

Lignin is a polymeric product composed of trans-coniferyl, trans-p-coumaryl and trans-sinapyl units. It forms the secondary wall of plant cells. Lignin is insoluble in water and is amorphous and hydrophobic [335]. Lignocellulosic fibers are hydrophilic. Therefore, they have poor adhesion to polymeric materials [336]. For these reasons, the surfaces of the lignocellulosic fibers are modified. Surface modifications also reduce moisture absorption, improve processability, and fiber-matrix adhesion.

Chemical processes (Figure 5.26) that improve matrix-fiber adhesion include alkalization, acetylation, silanization, benzoylation, cyanoethylation, esterification, graft copolymerization, dewaxing, isocyanates treatment, crosslinking with formaldehyde, p-phenylenediamne and phthalic anhydride, dinitrophenylation, nitration, and transesterification [337].

Pectins are a complex group of heteropolysaccharides with a variable content of methyl ester component. The simplest pectin is homogalacturonan (HG), which consists of D-galacturonic acid linked by α-(1 → 4) bonds. The carboxyl groups can be easily esterified with methanol. The ratio of esterified and non-esterified pectins determines the properties of pectin-based bioplastic [338]. A novel environmentally safe approach is the production of pectin-cellulose biofilms utilized, for example, for food packaging. They are obtained from orange waste using a solution casting method [339]. The biofilm production uses the gelling ability of pectin combined with the strength of cellulosic fibers.

Proteins are composed of alpha-amino acid chains. Proteins occur in various parts of the plant and animal organism. According to the type of plants, proteins can be classified as corn, wheat, potato, and pea. Proteins of animal origin include gelatine, casein, collagen, and keratin. Most proteins (up to 38% of the total composition) are contained in soybeans [330]. Soy protein bioplastics are used for plastics, coatings, packaging, lubricants, printing inks, and automobile parts [340]. The greater spread of soy protein bioplastics is limited by low strength and high moisture absorption. Improving the mechanical properties of soy protein-based bioplastic can be achieved by adding admixtures (lignocellulose fiber, up to 5%) and increasing mold temperature [341], or by adding 1% of formaldehyde (HCHO), which increases bending strength to 35 MPa [342]. Milk contains 2.8–3.3% total protein. The protein of cow's milk is formed by approximately 82% of casein, and the remaining 18% is serum, or whey protein [343]. Coagulation or precipitation of casein occurs at pH 4.6. Casein can be separated from milk by precipitation, deamidation, dephosphorylation, glycosylation, acylation, and enzyme-catalyzed reactions [330]. An alternative method for production of bioplastics from milk consists of substitution of formaldehyde with an acid such as vinegar or

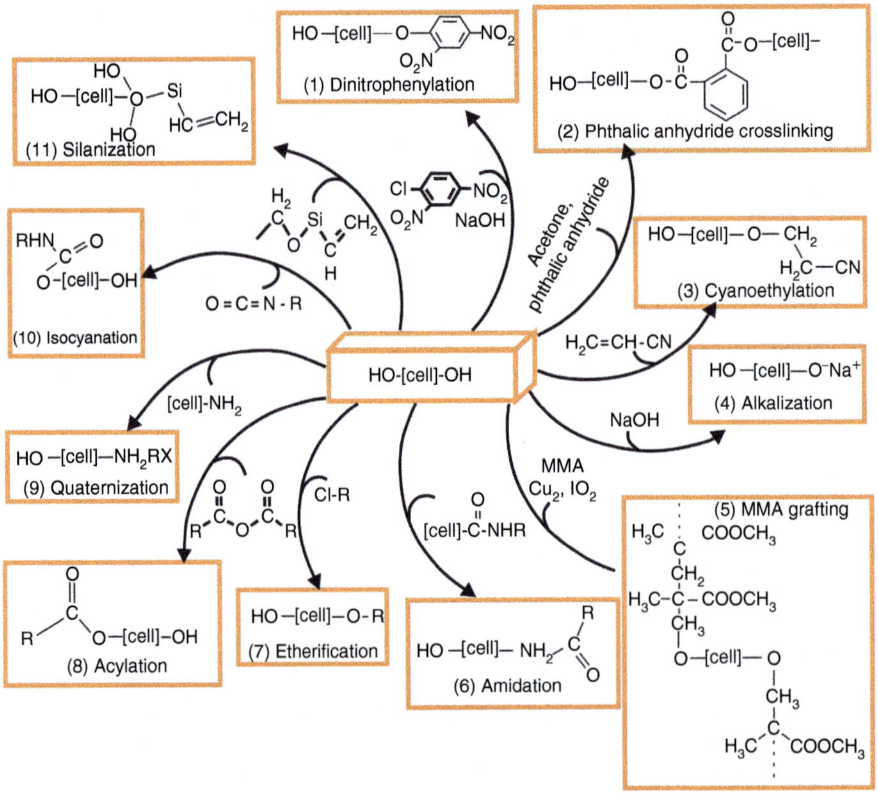

Figure 5.26 Chemical processes used for enhancement of properties of natural fiber [330].

lemon juice to perform the acidification, which results in a green procedure [344]. Casein is renewable, biodegradable, and compostable, and it allows a bioplastic to be developed with no significant aquatic toxicity.

5.4.2.2 bioPE, bioPET, bioPA

These are "traditional" types of plastics made from renewable raw materials. They decompose faster than conventional plastics, but not so fast that we can call them degradable. Some are recyclable, such as bioPE (from sugar cane) and bioPA (from castor oil) [345].

5.4.2.3 NotPLA

This is made from brown seaweed and often serves as packaging for sauces or drinks, nets for fruits and vegetables, or bags for non-food goods. Algae grow incomparably faster than standard plant production (up to 1 m day^{-1}) and do not need to be treated or fertilized in any way. In contact with the soil, they should decompose without special conditions within 4–6 weeks, without the formation of microplastics or other waste. Because they are algae, they decompose in water [345].

5.4.2.4 Biopolymers Synthesized by Using Renewable Monomers

Polylactic acid or polylactide (PLA, (–CH(CH$_3$)–CO–O–)$_n$) is a biodegradable synthetic thermoplastic polyester based on the lactic acid monomer, which is produced by fermentation of carbohydrate-rich substances, e.g., agricultural by-products. The sources of PLAs are mainly potato, corn, and sugar beet. Compared to other biodegradable polyesters, PLA is a biopolymer with the highest potential due to its wide availability and low cost. They represent approximately ¼ worldwide production of all compostable bioplastics. In some applications, PLAs can replace fossil-based polymers [346]. Purification is a critical phase in the production of PLA because it is expensive and determines the profitability of the process. Commonly used techniques are liquid extraction, ion exchange, membrane separation, electrodialysis, and reactive distillation. PLAs are mainly used for packaging, textiles, fiber, consumer goods, and automotive industries [347]. They are widely used in biomedicine, such as bioresorbable suture threads, bone fixation screws, and stent coating [348]. The biodegradability of PLA is limited, depending on molecular weight [349], time, temperature, and impurities. PLAs accumulate in soils or sewage sludge, and they are resistant to microbial attack [350]. PLAs are compostable, but only under specific conditions in industrial composting plants (high oxygen content – not less than 6%, high temperature – approximately 60–70 °C, humidity – approximately 60%, and a controlled pH value of 8.5) [351]; they are not suitable for home composting. When PLAs enter rivers and oceans, they are virtually indecomposable, like traditional plastics.

5.4.2.5 Biopolymers Produced by Microorganisms or Genetically Modified Bacteria

Poly(hydroxyalkanoate) (PHAs) are biopolyesters formed by the fermentation of sugars and lipids [352]. PHAs include P3HB (poly-(3-hydroxybutyrate)), P4HB (poly(4-hydroxybutyrate)), PHV (poly-(3-hydroxyvalerate)), and PHH (poly(3-hydroxyoctanoate)). PHAs are one of the key materials that influence the market for biodegradable polymers, because of their great potential as a substitute for conventional plastics due to their biodegradability and rubber-like properties [353]. The excellent properties of PHA include the better function as an oxygen barrier (compared with both non-biodegradable polypropylene – PP – and polyethylene terephthalate), a better water vapor barrier (than PP), and the fat and odor barrier. These properties predetermine its use for food packaging and also for medical applications, and for fine chemicals [354].

PHAs can be obtained from used cooking oils, animal by-products (fats, blood, residues from intestines), organic crop residues (peels, straw, stover, fruit pomace), and mixed domestic waste (waste bread, cheese whey, nuts), etc. Only a small amount of bacteria species used for commercial production of PHAs (*Bacillus megaterium*, *Alcaligenes latus*, *Cupriavidus necator*, and *Pseudomonas oleovorans*) is able to convert these various kinds of carbon sources into PHAs [353]. The diagram of the biosynthetic process of PHA isolation is shown in Figure 5.27.

PHAs are biodegradable by bacteria commonly found in soils – *Pseudomonas stutzeri* and *Alcaligenes faecalis* [355]. Biodegradation of PHAs under aerobic conditions takes five to six weeks.

5.4.3 BDPs Based on Fossil Resources

Poly(caprolactone) (PCL) is a synthetic fossil-based biodegradable polyester prepared by polymerization of ε-caprolactone connected with ring-opening and using a catalyst such as stannous octanoate [356]. PCLs are decomposed by lipases and esterases [148] and have biodegradation and mechanical properties that can be exactly controlled by adjusting the environmental conditions – microorganisms, enzymes, and hydrolysis [357]. The mixture of PCLs and starch has been successfully used for making trash bags in Korea. PCLs can be used as an admixture in resins to improve their processing ability and their resulting application properties. PCLs adhere well to

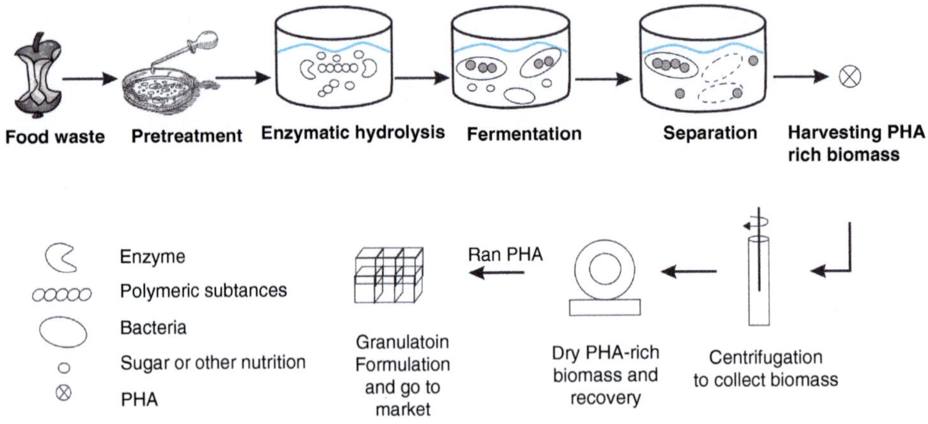

Figure 5.27 PHA biosynthesis process scheme. Pre-treatment (physical and acid treatment), enzymatic hydrolysis (converting polymeric compounds into fermentable products – e.g. lignocellulose into carbohydrate and animal fat into fatty acids), fermentation (hydrolysis of proteins and sugary compounds), biological synthesis of bioplastics, and separation [353].

many surfaces. They are easily manufactured, can be easily melted, and are non-toxic. PCLs are attracting increasing attention in the applications of medical 3D printing and tissue production due to their biodegradability [358].

Poly(butylene succinate) (PBS) is a well-known aliphatic polyester. The compounds used in the PBS synthesis are succinic acid (SA) and 1,4-butanediol (BD), which are obtained from fossil resources and are available on the market. Enzymatic catalysis is used for PBS synthesis [359].

PBSs exhibit interesting thermomechanical properties and the proven biodegradability, in combination with acceptable raw material and production costs [360]. Their main applications include environmental materials, such as compostable bags, mulching films, non-woven textiles and sheets, catering products, foams, and they are used in biomedicine for implantation [360].

Poly(vinyl alcohol) (PVA) is the most widely used water-soluble synthetic polymer for biomedical applications. PVA is the hydrolysis product of polyvinyl acetate. PVA has compatibility with many polymers and can be easily mixed with various natural materials, which extend its possible applications. The modifications of poly(vinyl alcohol) with different molecular weights and hydrolysis degrees can be further modified. This is possible using chemical procedures, such as esterification, etherification, and carbamation. These approaches provide materials suitable for a wide range of applications, such as the production of membrane fuel cells, adsorption of heavy metals and organic contaminants, separation of mixtures, chemo-sensors for detection of some molecules that are either useful or toxic, production of biomedicals, and catalysis in inorganic and organic syntheses [361]. Modified PVA with an application of graphene, hyaluronic acid, and poly(L-lactic acid) is utilized in the form of hydrogels, films, membranes, and nanoparticles. PVA foams can have plastic memory capabilities similar to those of polyurethanes.

5.4.4 Biodegradability of Plastics

The process of biodegradation is the result of biological activity. The biodegradation of polymers has three important phases:

- Biodeterioration, which represents the changes of physical and chemical properties of the polymer caused by the growth of microorganisms on or inside the surface of the polymers.

- Biofragmentation, which is the chemical splitting of polymers to oligomers and monomers by the activity of microorganisms.
- Assimilation, which means that microorganisms are supplied by the necessary carbon, energy, and nutrients from the fragmentation of polymers, and transform carbon compounds in plastic to CO_2, water, and biomass [362].

The significant factors that influence the process of plastic biodegradation in the environment are the properties of the bioplastics and environmental factors, which include temperature, moisture, pH value, and content of oxygen. Decomposition of bioplastics depending on environmental factors and different environments (compost, soils, aquatic environment) is described by Emadian et al. [317].

A number of standards (Figure 5.28) have been prepared to assess the biodegradability of plastics, which are mainly focused on the assessment of the decomposition of packaging materials during composting. Plastics in which the chemical structure changes significantly during decomposition to form CO_2, H_2O, inorganic components, and biomass without the presence of toxic residues in the compost are considered to be compostable (ISO 17088 : 2012) [363]. Plastics in which 90% conversion to CO_2 takes place during composting for a period of six months, and in the case of anaerobic digestion, min. 50% plastic within two months. are considered to be biodegradable (EN 13432 : 2000) [364].

The question is whether bioplastics are a better option for the environment than conventional plastics made from oil. It turns out, for example, that their use in healthcare brings excellent benefits. In a comprehensive life cycle assessment, some plastic products can cause less damage to the environment and humans than bioplastic products. No material is sustainable in itself. Only the behavior of society can be sustainable (Table 5.12).

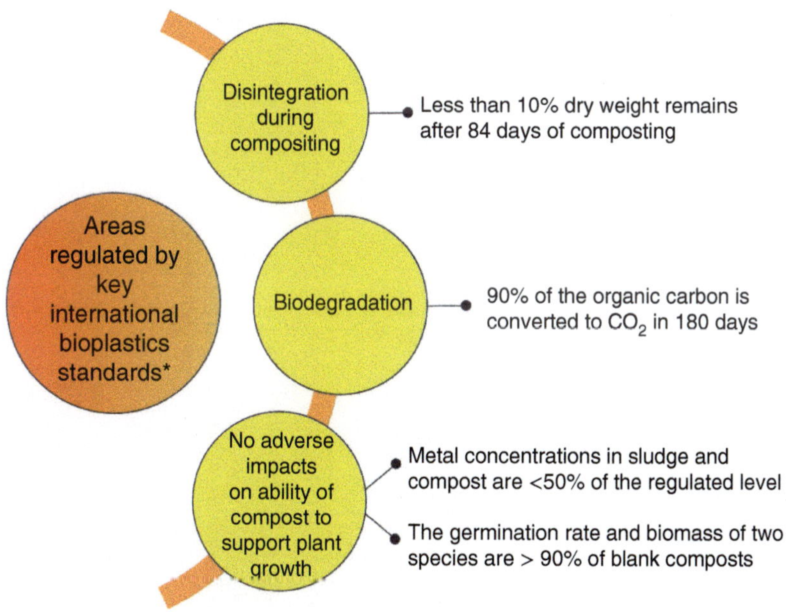

Figure 5.28 Degradation of bioplastics – assessment of bioplastic degradability based on international standards Source: International Organization for Standardization: ISO 17088, ISO 18606; European Standards: EN 13432, EN 14995; American Society for Testing and Materials: ASTM 6400, ASTM 6868 [329].

Table 5.12 Comparison of traditional plastics and bioplastics [319, 365, 366].

Traditional plastics	Bioplastics
Production	
× They are made from oil, a non-renewable resource.	+ They are made from renewable plant sources.
× A large part of traditional plastics is produced in China. They are transported by ship, which significantly contributes to the emission load.	× The production of raw materials can help development and employment in rural areas.
	+ Bioplastics are also produced from waste generated during agricultural production. Agricultural production gets another use. No extra land is occupied for growing raw materials, and no extra fertilizer is used.
	+ Biomass production can be ensured by sustainable and environmentally friendly agricultural technologies.
	× "First-generation" bioplastics are usually produced at the expense of food production, which negatively affects food availability.
	× Fossil fuels are used for agricultural production, transport and processing.
	× Renewable technologies are often more costly than technologies using oil.
Recyclability	
+ Built infrastructure for sorting plastics.	+ Only some bioplastics can be recycled.
× The production of some types of plastics is much cheaper than their recycling.	× Low awareness of the rules of proper recycling of bioplastics.
	× There are not enough industrial composting plants, and there is no system for collecting and disposing of bioplastics.
	× Recycling of unsuitable bioplastics negatively affects the quality of recycled material.

Degradability	
× Some types do not decompose for hundreds of years.	+ Some types are broken down as part of home composting.
× They produce microplastics that endanger the food chain and health.	× Some types behave similarly to traditional plastics in real conditions – they do not decompose and they can break down into microplastics.
Durability	
+ They are more durable than bioplastics.	× At present, they have a shorter service life than conventional plastics.
Use	
+ Wide.	× They are used in areas where a shorter service life can be applied – the packaging industry. It is the subject of research.
Environmental impacts	
× Products made from fossil fuels disrupt natural systems and their balance.	+ During decomposition or combustion, CO_2 produced is considered climate neutral.
× Formation of microplastics identified in all components of the environment.	+ In a short time, compostable bioplastics decompose into components that are not harmful to the environment.
× Formation of garbage stains in the oceans.	Some types of bioplastics behave similarly to conventional plastics, do not decompose, and they pollute the environment.

Explanation: × negative effect; + positive effect

References

1 Lokesh, K., Ladu, L., and Summerton, L. (2018). Bridging the gaps for a 'circular' bioeconomy: selection criteria, bio-based value chain and stakeholder mapping. *Sustainability* [Internet]. 10 (6): –1695. Available from: http://www.mdpi.com/2071-1050/10/6/1695.

2 De Corato, U. (2020). Agricultural waste recycling in horticultural intensive farming systems by on-farm composting and compost-based tea application improves soil quality and plant health: a review under the perspective of a circular economy. *Science of the Total Environment* [Internet]. 738: 139840. Available from: https://linkinghub.elsevier.com/retrieve/pii/S004896972033360X.

3 Ajila, C.M., Brar, S.K., Verma, M. et al. (2012). Bio-processing of agro-byproducts to animal feed. *Critical Reviews in Biotechnology* [Internet]. 32 (4): 382–400. Available from: http://www.tandfonline.com/doi/full/10.3109/07388551.2012.659172.

4 Iqbal, Y., Lewandowski, I., Weinreich, A. et al. (2016). *Maximising the Yield of Biomass from Residues of Agricultural Crops and Biomass from Forestry*. ECOFYS Germany GmbH: Berlin Report No.: BIENL15082.

5 Vilajeliu-Pons, A., Bañeras, L., Puig, S. et al. (2016). External resistances applied to MFC affect core microbiome and swine manure treatment efficiencies. S.J. Green (ed.). *PLoS One* [Internet]. 11 (10): e0164044. Available from: https://dx.plos.org/10.1371/journal.pone.0164044.

6 Zhao, Z., Zhang, C., Li, F. et al. (2020). Effect of compost and inorganic fertilizer on organic carbon and activities of carbon cycle enzymes in aggregates of an intensively cultivated vertisol. Paz-Ferreiro J, editor. *PLoS One* [Internet]. 15 (3): e0229644. Available from: https://dx.plos.org/10.1371/journal.pone.0229644.

7 Park, J., Cho, K.H., Ligaray, M., and Choi, M.-J. (2019). Organic matter composition of manure and its potential impact on plant growth. *Sustainability* [Internet]. 11 (8): 2346. Available from: https://www.mdpi.com/2071-1050/11/8/2346.

8 Wohde, M., Berkner, S., Junker, T. et al. (2016). Occurrence and transformation of veterinary pharmaceuticals and biocides in manure: a literature review. *Environmental Sciences Europe* [Internet]. 28 (1): 23. Available from: http://enveurope.springeropen.com/articles/10.1186/s12302-016-0091-8.

9 Chojnacka, K., Moustakas, K., and Witek-Krowiak, A. (2020). Bio-based fertilizers: a practical approach towards circular economy. *Bioresource Technology* [Internet]. 295: 122223. Available from: https://linkinghub.elsevier.com/retrieve/pii/S0960852419314531.

10 Leip, A., Ledgard, S., Uwizeye, A. et al. (2019). The value of manure - manure as co-product in life cycle assessment. *Journal of Environmental Management* [Internet]. 241: 293–304. Available from: https://linkinghub.elsevier.com/retrieve/pii/S0301479719303627.

11 Manitoba. 2015. *Properties of manure. Manitoba Agriculture*, Food and Rural Development.

12 Huang, J., Yu, Z., Gao, H. et al. (2017). Chemical structures and characteristics of animal manures and composts during composting and assessment of maturity indices. Mao J, editor. *PLoS One* [Internet]. 12 (6): e0178110. Available from: https://dx.plos.org/10.1371/journal.pone.0178110.

13 Font-Palma, C. (2019). Methods for the treatment of cattle manure – A review. C [Internet]. 5(2):27. Available from: https://www.mdpi.com/2311-5629/5/2/27.

14 Chastain, J.P., Camberato, J.J., and Skewes, P. (2001). Poultry manure production and nutrient. *Confined Animal Manure Managers Certification Program Manual.*

15 Larney, F.J., Buckley, K.E., Hao, X., and WP, M.C. (2006). Fresh, stockpiled, and composted beef cattle feedlot manure: nutrient levels and mass balance estimates in Alberta and Manitoba. *Journal of Environmental Quality* [Internet]. 35 (5): 1844–1854. Available from: http://doi.wiley.com/10.2134/jeq2005.0440.

16 Li, H., Yang, S., Sun, H., and Liu, X. (2018). Production of activated carbon from cow manure for wastewater treatment. *BioResources* [Internet]. 13 (2): 3135–3143. Available from: http://ojs.cnr.ncsu.edu/index.php/BioRes/article/view/13305.

17 Li, Y., Achinas, S., Zhao, J. et al. (2020). Co-digestion of cow and sheep manure: performance evaluation and relative microbial activity. *Renewable Energy* [Internet]. 153: 553–563. Available from: https://linkinghub.elsevier.com/retrieve/pii/S0960148120302275.

18 Moula, M., Mohamed, A., Elmorsy, A., and Farag, A. (2018). The action and interaction of different planting dates and organic fertilizers on the growth and yield of okra plants. *Japan Atomic Energy Research Institute* [Internet]. 14 (1): 1–14. Available from: http://www.sciencedomain.org/abstract/22738.

19 Augustyńska-Prejsnar, A., Ormian, M., Sokołowicz, Z. et al. (2018). Environmental impacts of pig and poultry farms. *Proceedings of Ecopole* [Internet]. 12 (1): 117–129. Available from. http://doi.org/10.2429/proc.2018.12(1)01.1.

20 Losinger, W.C. and Sampath, R.K. (2000). Economies of scale in the production of swine manure. *Arquivo Brasileiro de Medicina Veterinária e Zootecnia* [Internet]. 52 (3): 285–294. Available from: http://www.scielo.br/scielo.php?script=sci_arttext&pid=S0102-09352000000300019&lng=en&tlng=en.

21 Kyakuwaire, M., Olupot, G., Amoding, A. et al. (2019). How safe is chicken litter for land application as an organic fertilizer? A review. *International Journal of Environmental Research and Public Health* [Internet]. 16 (19): 3521. Available from: https://www.mdpi.com/1660-4601/16/19/3521.

22 Ma, J. and You, F. (2019). Superstructure optimization of thermal conversion based poultry litter valorization process. *Journal of Cleaner Production* [Internet]. 228: 1111–1121. Available from: https://linkinghub.elsevier.com/retrieve/pii/S095965261931443X.

23 Buckwell, A. and Nadeu, E. (2016). Nutrient recovery and reuse (NRR) in European agriculture. A review of the issues, opportunities, and actions. RISE Foundation, Brusel.

24 Kim, D., Olympiou, C., McCoy, C.P. et al. (2020). Time-resolved dynamics of struvite crystallization: insights from the macroscopic to molecular scale. *Chemistry – A European Journal* [Internet]. 26 (16): 3555–3563. Available from: https://onlinelibrary.wiley.com/doi/abs/10.1002/chem.201904347.

25 Mak, J. and Grieger, L. (n.d). Evaluation of air floatation and belt filter press manure separation system. [Internet]. Winnipeg: Manitoba Park; 2015 Dec p. 35. Report No.: MLMMI Report MLMMI 2015–02. Available from: https://www.manitobapork.com/images/MLMMI/2015-02/Final%20Report%202015-02%20VP%20System%20Econ_Eff.pdf.

26 Egilson, K. and Grieger, L. 2014. Manure treatment technology evaluation. Manitoba Livestock Manure Management Initiative. Report No.: MLMMI 2010–21.

27 Slevinsky, L., Plohman, G., and Racz, G. (2009). Solid/liquid separation of hog manure utilizing a settling tank. Manitoba. p. 35.

28 Zarebska, A., Romero Nieto, D., Christensen, K.V. et al. (2015). Ammonium fertilizers production from manure: a critical review. *Critical Reviews in Environmental Science and Technology* [Internet]. 45 (14): 1469–1521. Available from: http://www.tandfonline.com/doi/full/10.1080/10643389.2014.955630.

29 Vaneeckhaute, C., Lebuf, V., Michels, E. et al. (2017). Nutrient recovery from digestate: systematic technology review and product classification. *Waste and Biomass Valorization* [Internet]. 8 (1): 21–40. Available from: http://link.springer.com/10.1007/s12649-016-9642-x.

30 Frear, C., Ma, J., and Yorgey, G. (2018). *Approaches to nutrient recovery from dairy manure [Internet]* Washington State University Extension. Available from: http://pubs.cahnrs.wsu.edu/publications/wp-content/uploads/sites/2/2018/06/em112e.pdf.

31 Stolarski, M.J., Warmiński, K., Krzyżaniak, M. et al. (2020). How does extraction of biologically active substances with supercritical carbon dioxide affect lignocellulosic biomass properties? *Wood Science and Technology* [Internet]. 54 (3): 519–546. Available from: http://link.springer.com/10.1007/s00226-020-01182-5.

32 Lim, S.-F. and Matu, S.U. (2015). Utilization of agro-wastes to produce biofertilizer. *International Journal of Energy and Environmental Engineering* [Internet]. 6 (1): 31–35. Available from: http://link.springer.com/10.1007/s40095-014-0147-8.

33 Speight, J.G. (2020). Sources of water pollution. In: *Natural Water Remediation.* [Internet]. (ed. J.G. Speight), 165–198. Elsevier Available from: https://linkinghub.elsevier.com/retrieve/pii/B978012803810900005X.

34 Schatzmayr, G., Zehner, F., Täubel, M. et al. (2006). Microbiologicals for deactivating mycotoxins. *Molecular Nutrition & Food Research* [Internet]. 50 (6): 543–551. Available from: http://doi.wiley.com/10.1002/mnfr.200500181.

35 Gontard, N., Sonesson, U., Birkved, M. et al. (2018). A research challenge vision regarding management of agricultural waste in a circular bio-based economy. *Critical Reviews in Environmental Science and Technology* [Internet]. 48 (6): 614–654. Available from: https://www.tandfonline.com/doi/full/10.1080/10643389.2018.1471957.

36 Kiran, E.U., Trzcinski, A.P., Ng, W.J., and Liu, Y. (2014). Bioconversion of food waste to energy: A review. *Fuel* [Internet]. 134: 389–399. Available from: https://linkinghub.elsevier.com/retrieve/pii/S0016236114005365.

37 Cao, L., Zhang, C., Chen, H. et al. (2017). Hydrothermal liquefaction of agricultural and forestry wastes: state-of-the-art review and future prospects. *Bioresource Technology* [Internet]. 245: 1184–1193. Available from: https://linkinghub.elsevier.com/retrieve/pii/S0960852417315109.

38 Torales, A. (ed.) (2013). *Anaerobic Digestion: Types, Processes, and Environmental Impact.* New York: Nova Science Publisher 163 pp. (Environmental Science, Engineering and Technology).

39 Monlau, F., Francavilla, M., Sambusiti, C. et al. (2016). Toward a functional integration of anaerobic digestion and pyrolysis for a sustainable resource management. Comparison between solid-digestate and its derived pyrochar as soil amendment. *Applied Energy* [Internet]. 169: 652–662. Available from: https://linkinghub.elsevier.com/retrieve/pii/S0306261916302240.

40 Hans, M. and Kumar, S. (2019). Biohythane production in two-stage anaerobic digestion system. *International Journal of Hydrogen Energy* [Internet]. 44 (32): 17363–17380. Available from: https://linkinghub.elsevier.com/retrieve/pii/S0360319918331975.

41 Bolzonella, D., Fatone, F., Gottardo, M., and Frison, N. (2018). Nutrients recovery from anaerobic digestate of agro-waste: techno-economic assessment of full scale applications. *Journal of Environmental Management* [Internet]. 216: 111–119. Available from: https://linkinghub.elsevier.com/retrieve/pii/S0301479717308113.

42 Fabbri, D. and Torri, C. (2016). Linking pyrolysis and anaerobic digestion (Py-AD) for the conversion of lignocellulosic biomass. *Current Opinion in Biotechnology* [Internet]. 38: 167–173. Available from: https://linkinghub.elsevier.com/retrieve/pii/S0958166916300271.

43 Zhen, G., Lu, X., Kumar, G. et al. (2017). Microbial electrolysis cell platform for simultaneous waste biorefinery and clean electrofuels generation: current situation, challenges and future perspectives. *Progress in Energy and Combustion Science* [Internet]. 63: 119–145. Available from: https://linkinghub.elsevier.com/retrieve/pii/S0360128516301563.

44 Zeppilli, M., Lai, A., Villano, M., and Majone, M. (2016). Anion vs cation exchange membrane strongly affect mechanisms and yield of CO_2 fixation in a microbial electrolysis cell. *Chemical Engineering Journal* [Internet]. 304: 10–19. Available from: https://linkinghub.elsevier.com/retrieve/pii/S1385894716308270.

45 Villano, M., Scardala, S., Aulenta, F., and Majone, M. (2013). Carbon and nitrogen removal and enhanced methane production in a microbial electrolysis cell. *Bioresource Technology* [Internet]. 130: 366–371. Available from: https://linkinghub.elsevier.com/retrieve/pii/S0960852412017750.

46 Blasco-Gómez, R., Batlle-Vilanova, P., Villano, M. et al. (2017). On the edge of research and technological application: a critical review of electromethanogenesis. *International Journal of Molecular Sciences* [Internet]. 18 (4): 874. Available from: http://www.mdpi.com/1422-0067/18/4/874.

47 De Corato, U., Pane, C., Bruno, G.L. et al. (2015). Co-products from a biofuel production chain in crop disease management: a review. *Crop Protection* [Internet]. 68: 12–26. Available from: https://linkinghub.elsevier.com/retrieve/pii/S0261219414003378.

48 Xiu, S. and Shahbazi, A. (2015). Development of green biorefinery for biomass utilization: a review. *Trends in Renewable Energy* [Internet].: 4–15. Available from: http://futureenergysp.com/index.php/tre/article/view/8.

49 VDI 6310 BLATT 1. 2016. Classification and quality criteria of biorefineries. p. 100.

50 Kamm, B., Schönicke, P., and Ch, H. (2016). Green biorefinery – industrial implementation. *Food Chemistry* [Internet]. 197: 1341–1345. Available from: https://linkinghub.elsevier.com/retrieve/pii/S0308814615302338.

51 Ögmundarson, Ó., Herrgård, M.J., Forster, J. et al. (2020). Addressing environmental sustainability of biochemicals. *Nature Sustainability* [Internet]. 3 (3): 167–174. Available from: http://www.nature.com/articles/s41893-019-0442-8.

52 Schnyder Hand European Grassland Federation. 2010. *Grassland in a changing world: Proceedings of the 23rd General Meeting of the European Grassland Federation*, Kiel, Germany. (August 29–September 2 2010). Duderstadt: Mecke Druck und Verlag; 2010. 1094 pp.

53 Zollmann, M., Robin, A., Prabhu, M. et al. (2019). Green technology in green macroalgal biorefineries. *Phycologia* [Internet]. 58 (5): 516–534. Available from: https://www.tandfonline.com/doi/full/10.1080/00318884.2019.1640516.

54 Corona, A., Parajuli, R., Ambye-Jensen, M. et al. (2018). Environmental screening of potential biomass for green biorefinery conversion. *Journal of Cleaner Production* [Internet]. 189: 344–357. Available from: https://linkinghub.elsevier.com/retrieve/pii/S0959652618310035.

55 Santamaria-Fernandez, M., Ytting, N.K., Lübeck, M., and Uellendahl, H. (2019). Potential nutrient recovery in a green biorefinery for production of feed, fuel and fertilizer for organic farming. *Waste and Biomass Valorization* [Internet]. Available from: http://link.springer.com/10.1007/s12649-019-00842-3.

56 Federal Government. 2012. *Agency for Renewable Resources. Biorefineries roadmap as part of the German Federal Government action plans for the material and energetic utilisation of renewable raw materials.*

57 Bouxin, F.P., Clark, J.H., Fan, J., and Budarin, V. (2019). Combining steam distillation with microwave-assisted pyrolysis to maximise direct production of levoglucosenone from agricultural wastes. *Green Chemistry* [Internet]. 21 (6): 1282–1291. Available from: http://xlink.rsc.org/?DOI=C8GC02994F.

58 Takkellapati, S., Li, T., and Gonzalez, M.A. (2018). An overview of biorefinery-derived platform chemicals from a cellulose and hemicellulose biorefinery. *Clean Technologies and Environmental Policy* [Internet]. 20 (7): 1615–1630. Available from: http://link.springer.com/10.1007/s10098-018-1568-5.

59 Nguyen, H., Jamali Moghadam, M., and Moayedi, H. (2019). Agricultural wastes preparation, management, and applications in civil engineering: a review. *Journal of Material Cycles and Waste Management* [Internet]. 21 (5): 1039–1051. Available from: http://link.springer.com/10.1007/s10163-019-00872-y.

60 Sharma, G. and Singh, K. (2019). Recycling and utilization of agro-food waste ashes: syntheses of the glasses for wide-band gap semiconductor applications. *Journal of Material Cycles and Waste Management* [Internet]. 21 (4): 801–809. Available from: http://link.springer.com/10.1007/s10163-019-00839-z.

61 Akram, M.S., Cheema, M.A., Waqas, M. et al. 2020. Role of bio-fertilizers in sustainable agriculture. Preprints [Internet]; 2020030262. Available from: https://www.preprints.org/manuscript/202003.0262/v1.

62 Bhargav, S., Panda, B.P., Ali, M., and Javed, S. (2008). Solid state fermentation: an overview. *Chemical and Biochemical Engineering Quarterly* 22 (1): 49–70.

63 Abu, E.A., Onyenekwe, P.C., Ameh, D.A. et al. (2000). Cellulase (E.C.3.21.3) production from sorghum bran by *Aspergillus niger* SL 1: an assessment of pretreatment methods. In: *Proceedings of the International Conference on Biotechnology: Commercialization and Food Security*, 153–159. Abuja, Nigeria.

64 Sadh, P.K., Duhan, S., and Duhan, J.S. (2018). Agro-industrial wastes and their utilization using solid state fermentation: a review. *Bioresources and Bioprocessing* [Internet]. 5 (1): 1. Available from: https://bioresourcesbioprocessing.springeropen.com/articles/10.1186/s40643-017-0187-z.

65 Manan, M.A. and Webb, C. (2017). Design aspects of solid state fermentation as applied to microbial bioprocessing. *Journal of Applied Biology and Biotechnology* [Internet]. 4: 1. Available from: https://medcraveonline.com/JABB/design-aspects-of-solid-state-fermentation-as-applied-to-microbial-bioprocessing.html.

66 Llewellyn, D.A., Marston, T.T., Teutemacher, K.L. et al. (2010). Evaluation of low molecular weight fractions and crude enzyme preparation from a Trichoderma cellulase complex as a treatment for fibrous feeds. *Animal Feed Science and Technology* [Internet]. 160 (1–2): 39–48. Available from: https://linkinghub.elsevier.com/retrieve/pii/S0377840110002063.

67 The European Commission. (n.d). EU Forestry Explained [Internet]. Available from: https://ec.europa.eu/info/food-farming-fisheries/forestry/forestry-explained_en#euforestryexplained.

68 Whitake, S., Clotteau, M., Andonovski, V. et al. (2017). *Innovation and circular economy in the Mountain. Forest supply chain: How to close the loop? [Internet]*. Euromontana. Available from: https://www.euromontana.org/wp-content/uploads/2017/03/Innovation-and-Circular-Economy-in-the-Mountain-Forest-Supply-Chain_FINAL.pdf.

69 Mantau, U. 2012. *Energieholzverwendung in privaten Haushalten*. Marktvolumen und verwendete Holzsortimente – Abschlussbericht.

70 Sette, C.R. Jr., Hansted, A.L.S., Novaes, E. et al. (2018). Energy enhancement of the eucalyptus bark by briquette production. *Industrial Crops and Products* [Internet]. 122: 209–213. Available from: https://linkinghub.elsevier.com/retrieve/pii/S092666901830476X.

71 DAL, S., Raymundo Pavan, A.L., Augusto de Oliveira, J., and Ometto, A.R. (2015). Life cycle assessment of offset paper production in Brazil: hotspots and cleaner production alternatives. *Journal of Cleaner Production* [Internet]. 93: 222–233. Available from: https://linkinghub.elsevier.com/retrieve/pii/S0959652615000347.

72 Pásztory, Z., Mohácsiné, I.R., Gorbacheva, G., and Börcsök, Z. (2016). The utilization of tree bark. *BioResources* 11 (3): 7859–7888.

73 Sette, C.R. Jr., da Cunha, T.Q.G., Coneglian, A. et al. (2020). Does the presence of bark in the wood of fast-growing forest species significantly change the energy potential? *Bioenergy Research* [Internet]. 13 (1): 222–228. Available from: http://link.springer.com/10.1007/s12155-020-10115-w.

74 Vis M, Mantau U, Allen B, Essel R et al. CASCADES: study on the optimised cascading use of wood. [Internet]. Luxembourg: Publications Office; 2016 . Available from: http://bookshop.europa.eu/uri?target=EUB:NOTICE:ET0416305:EN:HTML.

75 Forests and the Circular Economy. 2019. *Food and Agriculture Organization of the United Nations.*

76 Dammer, L., Bowyer, C., Breitmayer, E. et al. (2016). *Mapping Study on Cascading Use of Wood Products.* Gland: World Wide Fund for Nature.

77 Stelte, W., Holm, J.K., Sanadi, A.R. et al. (2011). Fuel pellets from biomass: the importance of the pelletizing pressure and its dependency on the processing conditions. *Fuel* [Internet]. 90 (11): 3285–3290. Available from: https://linkinghub.elsevier.com/retrieve/pii/S0016236111002730.

78 Lamlom, S.H. and Savidge, R.A. (2003). A reassessment of carbon content in wood: variation within and between 41 north American species. *Biomass and Bioenergy* [Internet]. 25 (4): 381–388. Available from: https://linkinghub.elsevier.com/retrieve/pii/S0961953403000333.

79 Tumuluru, J.S., Wright, C.T., Hess, J.R., and Kenney, K.L. (2011). A review of biomass densification systems to develop uniform feedstock commodities for bioenergy application. *Biofuels, Bioproducts and Biorefining* [Internet]. 5 (6): 683–707. Available from: http://doi.wiley.com/10.1002/bbb.324.

80 McKendry, P. (2002). Energy production from biomass (part 1): overview of biomass. *Bioresource Technology* [Internet]. 83 (1): 37–46. Available from: https://linkinghub.elsevier.com/retrieve/pii/S0960852401001183.

81 Harkin, J.M. and Rowe, J.W. (1971) Bark and its possible uses. United States Forest Service, Forest Products Laboratory. p. 56. Report No.: FPL-091.

82 Isahak, W.N.R.W., Hisham, M.W.M., Yarmo, M.A., and Yun Hin, T. (2012). A review on bio-oil production from biomass by using pyrolysis method. *Renewable and Sustainable Energy Reviews* [Internet]. 16 (8): 5910–5923. Available from: https://linkinghub.elsevier.com/retrieve/pii/S1364032112003735.

83 Routa, J., Brännström, H., Anttila, P., et al. 2017. *Wood extractives of Finnish pine, spruce and birch – availability and optimal sources of compounds: a literature review.* Helsinki: Natural Resources Institute Finland (Luke). 55 pp.

84 Taflick, T., Maich, É.G., Ferreira, L.D. et al. (2015). Acacia bark residues as filler in polypropylene composites. *Polímeros* [Internet]. 25 (3): 289–295. Available from: http://www.scielo.br/scielo.php?script=sci_arttext&pid=S0104-14282015000300008&lng=en&tlng=en.

85 Migneault, S., Koubaa, A., and Perré, P. (2014). Effect of fiber origin, proportion, and chemical composition on the mechanical and physical properties of wood-plastic composites. *Journal of Wood Chemistry and Technology* [Internet]. 34 (4): 241–261. Available from: http://www.tandfonline.com/doi/abs/10.1080/02773813.2013.869604.

86 Kairytė, A., Kremensas, A., Balčiūnas, G. et al. (2020). Evaluation of self-thermally treated wood plastic composites from wood bark and rapeseed oil-based binder. *Construction and Building Materials* [Internet]. 250: 118842. Available from: https://linkinghub.elsevier.com/retrieve/pii/S0950061820308473.

87 Yazaki, Y. (2015). Utilization of flavonoid compounds from bark and wood: a review. *Natural Product Communications* [Internet]. 10 (3):1934578X1501000. Available from: http://journals.sagepub.com/doi/10.1177/1934578X1501000333.

88 Matsumae, T., Horito, M., Kurushima, N., and Yazaki, Y. (2019). Development of bark-based adhesives for plywood: utilization of flavonoid compounds from bark and wood. II. *Journal of Wood Science* [Internet]. 65 (1): 9. Available from: https://jwoodscience.springeropen.com/articles/10.1186/s10086-019-1780-x.

89 Kawada, H. (1978). Production of bark compost and its application to agriculture and forestry. *Japan Agricultural Research Quarterly* 12 (3): 173–176.

90 Basu, P. (2018). *Biomass Gasification, Pyrolysis, and Torrefaction: Practical Design and Theory*, 3e. London/San Diego: Academic Press, imprint of Elsevier. 564 pp.

91 Melissari, B. (2014). Ash related problems with high alkali biomass and its mitigation experimental evaluation. Memoria Investigaciones en Ingeniería.

92 Mudgal, D., Singh, S., and Prakash, S. (2014). Corrosion problems in incinerators and biomass-fuel-fired boilers. *International Journal of Corrosion* [Internet]. 2014: 1–14. Available from: http://www.hindawi.com/journals/ijc/2014/505306.

93 ISO 17225-2:2014. 2014. Solid biofuels — Fuel specifications and classes — Part 2: Graded wood pellets. p. 9.

94 Lanzerstorfer, C. (2015). Chemical composition and physical properties of filter fly ashes from eight grate-fired biomass combustion plants. *Journal of Environmental Sciences* [Internet]. 30: 191–197. Available from: https://linkinghub.elsevier.com/retrieve/pii/S100107421500025X.

95 Barišić, I., Netinger Grubeša, I., Dokšanović, T., and Marković, B. (2019). Feasibility of agricultural biomass fly ash usage for soil stabilisation of road works. *Materials* [Internet]. 12 (9): 1375. Available from: https://www.mdpi.com/1996-1944/12/9/1375.

96 European Pellet Council. 2018. European wood pellet consumption [Internet]. Available from: https://epc.bioenergyeurope.org/about-pellets/pellets-statistics/european-consumption.

97 ISO 17225-3:2014. 2014. Solid biofuels — Fuel specifications and classes — Part 3: Graded wood briquettes. p. 6.

98 ISO 17225-4:2014. 2014. Solid biofuels — Fuel specifications and classes — Part 4: Graded wood chips. p. 8.

99 ISO 17225-5:2014. 2014. Solid biofuels — Fuel specifications and classes — Part 5: Graded firewood. p. 10.

100 Hlavsová, A., Corsaro, A., Raclavská, H. et al. (2014). Syngas production from pyrolysis of nine composts obtained from nonhybrid and hybrid perennial grasses. *The Scientific World Journal* [Internet]. 2014: 1–11. Available from: http://www.hindawi.com/journals/tswj/2014/723092.

101 Shi, K., Yan, J., Menéndez, J.A. et al. (2020). Production of H_2-rich syngas from lignocellulosic biomass using microwave-assisted pyrolysis coupled with activated carbon enabled reforming. *Frontiers in Chemistry* [Internet]. 8: 3. Available from: https://www.frontiersin.org/article/10.3389/fchem.2020.00003/full.

102 Buffi, M., Cappelletti, A., Rizzo, A.M. et al. (2018). Combustion of fast pyrolysis bio-oil and blends in a micro gas turbine. *Biomass and Bioenergy* [Internet]. 115: 174–185. Available from: https://linkinghub.elsevier.com/retrieve/pii/S0961953418301090.

103 Guo, X., Liu, H., and Zhang, J. (2020). The role of biochar in organic waste composting and soil improvement: A review. *Waste Management* [Internet]. 102: 884–899. Available from: https://linkinghub.elsevier.com/retrieve/pii/S0956053X19307445.

104 Ruzickova, J., Koval, S., Raclavska, H. et al. (2020). A comprehensive assessment of potential hazard caused by organic compounds in biochar for agricultural use. *Journal of Hazardous Materials* [Internet].: 123644. Available from: https://linkinghub.elsevier.com/retrieve/pii/S0304389420316307.

105 Schmidt, P.H., Bucheli, T., Kammann, C. et al. 2020. EBC (2012) European biochar certificate – Guidelines for a sustainable production of biochar. Version 9.0E of 1 June 2020 [Internet]. Version 9.0E. Arbaz, Switzerland: European Biochar Foundation (EBC); 2020. 52 pp. Available from: https://www.european-biochar.org/media/doc/2/version_en_9_0.pdf.

106 Schmidt, H.-P., Hagemann, N., Draper, K., and Kammann, C. (2019). The use of biochar in animal feeding. *PeerJ* [Internet]. 7: e7373. Available from: https://peerj.com/articles/7373.

107 Huygens, D., Delgado Sancho, L., Saveyn, HGM. et al. (2019). Technical proposals for selected new fertilising materials under the Fertilising Products Regulation (Regulation (EU) 2019/1009) – Process and quality criteria, and assessment of environmental and market impacts for precipitated phosphate salts & derivates, thermal oxidation materials & derivates and pyrolysis & gasification materials, EUR 29841 EN. [Internet]. Luxembourg: Publications Office of the European Union; 2019. Available from: http://publications.europa.eu/publication/manifestation_identifier/PUB_KJNA29841ENN.

108 Wang, L., Barta-Rajnai, E., Skreiberg, Ø. et al. (2017). Impact of torrefaction on woody biomass properties. *Energy Procedia* [Internet]. 105: 1149–1154. Available from: https://linkinghub.elsevier.com/retrieve/pii/S1876610217305271.

109 Figueiró, C.G., Vital, B.R., A de CO, C. et al. (2019). Energy valorization of woody biomass by torrefaction treatment: a Brazilian experimental study. *Maderas Ciencia y Tecnologia* [Internet]. Available from: http://www.scielo.cl/scielo.php?script=sci_arttext&pid=S0718-221X2019005000302&lng=en&nrm=iso&tlng=en.

110 ISO/TS 17225-8:2016. 2016. Solid biofuels — Fuel specifications and classes — Part 8: Graded thermally treated and densified biomass fuels. p. 14.

111 Robak, K. and Balcerek, M. (2020). Current state-of-the-art in ethanol production from lignocellulosic feedstocks. *Microbiological Research* [Internet]. 240: 126534. Available from: https://linkinghub.elsevier.com/retrieve/pii/S094450132030402X.

112 Frankó, B., Galbe, M., and Wallberg, O. (2015). Influence of bark on fuel ethanol production from steam-pretreated spruce. *Biotechnology for Biofuels* [Internet]. 8 (1): 15. Available from: http://www.biotechnologyforbiofuels.com/content/8/1/15.

113 Hubert, J., Angelis, A., Aligiannis, N. et al. (2016). In vitro dermo-cosmetic evaluation of bark extracts from common temperate trees. *Planta Medica* [Internet]. 82 (15): 1351–1358. Available from: http://www.thieme-connect.de/DOI/DOI?10.1055/s-0042-110180.

114 Gustavsson, J., Cederberg, C., Sonesson, U. et al. (2011). *Global Food Losses and Food Waste: Extent, Causes and Prevention*. Rome: Food and Agriculture Organization of the United Nations. 29 pp.

115 Jörissen, J., Priefer, C., and Bräutigam, K.R. (2015). Food waste generation at household level: results of a survey among employees of two European research centers in Italy and Germany. *Sustainability* [Internet]. 7 (3): 2695–2715. Available from: http://www.mdpi.com/2071-1050/7/3/2695.

116 Volume of wasted food in the United States in 2015, by facility [Internet]. Statista. c2020. Available from: https://www.statista.com/statistics/530098/weight-of-wasted-food-by-facility-us.

117 Caldeira, C., De Laurentiis, V., Corrado, S. et al. (2019). Quantification of food waste per product group along the food supply chain in the European Union: a mass flow analysis. *Resources, Conservation and Recycling* [Internet]. 149: 479–488. Available from: https://linkinghub.elsevier.com/retrieve/pii/S0921344919302721.

118 Caldeira, C., De Laurentiis, V., and Sala, S. (2019) European Commission, Joint Research Centre. Assessment of food waste prevention actions: development of an evaluation framework to assess performance of food waste prevention actions [Internet]. . Available from: https://op.europa.eu/publication/manifestation_identifier/PUB_KJNA29901ENN.

119 European Commission. 2020. COM/2020/98 Communication from the Commission – A new circular economy action plan for a cleaner and more competitive Europe

120 Caldeira, C., Barco Cobalea, H., Serenella, S., and De Laurentiis, V. (2019). European Commission, Joint Research Centre. Review of studies on food waste accounting at Member State level [Internet]. Available from: https://op.europa.eu/publication/manifestation_identifier/PUB_KJNA29828ENN.

121 Caldeira, C., De Laurentiis, V., and Serenella, S. (2019). European Commission, Joint Research Centre. Suggestions to improve data coverage and comparability in food waste accounting studies across the EU [Internet]. Available from: https://op.europa.eu/publication/manifestation_identifier/PUB_KJNA30024ENN.

122 Caldeira, C., Caldeira, C., Sara, C., and Serenella, S. (2017). European Commission, Joint Research Centre. Food waste accounting methodologies, challenges and opportunities. [Internet]. Available from: http://publications.europa.eu/publication/manifestation_identifier/PUB_KJ1A28988ENN.

123 European Commission. (2018) Communication from the commission – A sustainable bioeconomy for Europe strengthening the connection between economy, society and the environment: updated bioeconomy strategy. Available from: https://doi.org/10.2777/792130.

124 European Parliament, Council. Directive (EU). 2018. 2018/851 of the European Parliament and of the Council of 30 May 2018 amending Directive 2008/98/EC on waste (Text with EEA relevance).

125 Beretta, C., Stoessel, F., Baier, U., and Hellweg, S. (2013). Quantifying food losses and the potential for reduction in Switzerland. *Waste Management* [Internet]. 33 (3): 764–773. Available from: https://linkinghub.elsevier.com/retrieve/pii/S0956053X12005302.

126 Corrado, S. and Sala, S. (2018). Food waste accounting along global and European food supply chains: state of the art and outlook. *Waste Management* [Internet]. 79: 120–131. Available from: https://linkinghub.elsevier.com/retrieve/pii/S0956053X18304550.

127 RedCorn, R., Fatemi, S., and Engelberth, A.S. (2018). Comparing end-use potential for industrial food-waste sources. *Engineering* [Internet]. 4 (3): 371–380. Available from: https://linkinghub.elsevier.com/retrieve/pii/S2095809917307804.

128 Dhillon, G.S., Kaur, S., and Brar, S.K. (2013). Perspective of apple processing wastes as low-cost substrates for bioproduction of high value products: A review. *Renewable and Sustainable Energy Reviews* [Internet]. 27: 789–805. Available from: https://linkinghub.elsevier.com/retrieve/pii/S1364032113004371.

129 Ravindran, R. and Jaiswal, A.K. (2016). Exploitation of food industry waste for high-value products. *Trends in Biotechnology* [Internet]. 34 (1): 58–69. Available from: https://linkinghub.elsevier.com/retrieve/pii/S0167779915002267.

130 Teigiserova, D.A., Hamelin, L., and Thomsen, M. (2020). Towards transparent valorization of food surplus, waste and loss: clarifying definitions, food waste hierarchy, and role in the circular economy. *Science of the Total Environment* [Internet]. 706: 136033. Available from: https://linkinghub.elsevier.com/retrieve/pii/S0048969719360292.

131 Sagar, N.A., Pareek, S., Sharma, S. et al. (2018). Fruit and vegetable waste: bioactive compounds, their extraction, and possible utilization: fruit and vegetable waste. *Comprehensive Reviews in Food Science and Food Safety* [Internet]. 17 (3): 512–531. Available from: http://doi.wiley.com/10.1111/1541-4337.12330.

132 Moates, G., Sweet, N., Bygrave, K., and Waldron, K. (2016). Top 20 Food waste streams – Deliverable D6.9.

133 FAO. 2020. New Food Balances data for World in 2017 [Internet]. FAOSTAT. Available from: http://www.fao.org/faostat/en/#data/FBS.

134 Mahato, N., Sharma, K., Sinha, M., and Cho, M.H. (2018). Citrus waste derived nutra–/pharmaceuticals for health benefits: current trends and future perspectives. *Journal of Functional Foods* [Internet]. 40: 307–316. Available from: https://linkinghub.elsevier.com/retrieve/pii/S1756464617306874.

135 Gupta, S.D., Kirby, W., and Pinno, B.D. (2019). Effects of stockpiling and organic matter addition on nutrient bioavailability in reclamation soils. *Soil Science Society of America Journal* [Internet]. 83 (S1) Available from: https://onlinelibrary.wiley.com/doi/abs/10.2136/sssaj2018.07.0273.

136 Plazzotta, S., Manzocco, L., and Nicoli, M.C. (2017). Fruit and vegetable waste management and the challenge of fresh-cut salad. *Trends in Food Science & Technology* [Internet]. 63: 51–59. Available from: https://linkinghub.elsevier.com/retrieve/pii/S0924224416304976.

137 Joglekar, S.N., Pathak, P.D., Mandavgane, S.A., and Kulkarni, B.D. (2019). Process of fruit peel waste biorefinery: a case study of citrus waste biorefinery, its environmental impacts and recommendations. *Environmental Science and Pollution Research* [Internet]. 26 (34): 34713–34722. Available from: http://link.springer.com/10.1007/s11356-019-04196-0.

138 Patsalou, M., Samanides, C.G., Protopapa, E. et al. (2019). A citrus peel waste biorefinery for ethanol and methane production. *Molecules* [Internet]. 24 (13): 2451. Available from: https://www.mdpi.com/1420-3049/24/13/2451.

139 Sharma, K., Mahato, N., and Lee, Y.R. (2019). Extraction, characterization and biological activity of citrus flavonoids. *Reviews in Chemical Engineering* [Internet]. 35 (2): 265–284. Available from: http://www.degruyter.com/view/j/revce.2019.35.issue-2/revce-2017-0027/revce-2017-0027.xml.

140 Sharma, K., Mahato, N., Cho, M.H., and Lee, Y.R. (2017). Converting citrus wastes into value-added products: economic and environmently friendly approaches. *Nutrition* [Internet]. 34: 29–46. Available from: https://linkinghub.elsevier.com/retrieve/pii/S0899900716302131.

141 Upadhyay, A., Sharma, H.K., and Sarkar, B.C. (2008). Characterization and dehydration kinetics of carrot pomace. Agricultural Engineering International: The CIGR Ejournal [Internet]. (X). Available from: https://cigrjournal.org/index.php/Ejounral/article/view/1015.

142 Bakshi, M. (2016). Waste to worth: vegetable wastes as animal feed. *CAB Reviews* [Internet]. 11 (012) Available from: http://www.cabi.org/cabreviews/review/20163197627.

143 Javed, A., Ahmad, A., Tahir, A. et al. (2019). Potato peel waste-its nutraceutical, industrial and biotechnological applications. *AIMS Agriculture and Food* [Internet]. 4 (3): 807–823. Available from: http://www.aimspress.com/article/10.3934/agrfood.2019.3.807.

144 Mäder, J., Rawel, H., and Kroh, L.W. (2009). Composition of phenolic compounds and glycoalkaloids α-solanine and α-chaconine during commercial potato processing. *Journal of Agricultural and Food Chemistry* [Internet]. 57 (14): 6292–6297. Available from: https://pubs.acs.org/doi/10.1021/jf901066k.

145 Sahraoui, N., Vian, M.A., El Maataoui, M. et al. (2011). Valorization of citrus by-products using microwave steam distillation (MSD). *Innovative Food Science & Emerging Technologies* [Internet]. 12 (2): 163–170. Available from: https://linkinghub.elsevier.com/retrieve/pii/S1466856411000270.

146 M'hiri, N., Ghali, R., Ben Nasr, I., and Boudhrioua, N. (2018). Effect of different drying processes on functional properties of industrial lemon byproduct. *Process Safety and Environmental Protection* [Internet]. 116: 450–460. Available from: https://linkinghub.elsevier.com/retrieve/pii/S0957582018300545.

147 Arce, A., Marchiaro, A., Martínez-Ageitos, J.M., and Soto, A. (2008). Citrus essential oil deterpenation by liquid-liquid extraction. *Canadian Journal of Chemical Engineering* [Internet]. 83 (2): 366–370. Available from: http://doi.wiley.com/10.1002/cjce.5450830226.

148 Ruiz, B. and Flotats, X. (2014). Citrus essential oils and their influence on the anaerobic digestion process: an overview. *Waste Management* [Internet]. 34 (11): 2063–2079. Available from: https://linkinghub.elsevier.com/retrieve/pii/S0956053X14002827.

149 Saeidnia, S. and Gohari, A.R. (2012). Trypanocidal monoterpenes. In: *Studies in Natural Products Chemistry*. [Internet]. (ed. A.U. Rahman), 173–190. Elsevier Available from: https://linkinghub.elsevier.com/retrieve/pii/B9780444595140000067.

150 Choi, Y.H. and Verpoorte, R. (2014). Metabolomics: what you see is what you extract: editorial. *Phytochemical Analysis* [Internet]. 25 (4): 289–290. Available from: http://doi.wiley.com/10.1002/pca.2513.

151 Bousbia, N., Vian, M.A., Ferhat, M.A. et al. (2009). A new process for extraction of essential oil from citrus peels: microwave hydrodiffusion and gravity. *Journal of Food Engineering* [Internet]. 90 (3): 409–413. Available from: https://linkinghub.elsevier.com/retrieve/pii/S0260877408003002.

152 Negro, V., Mancini, G., Ruggeri, B., and Fino, D. (2016). Citrus waste as feedstock for bio-based products recovery: review on limonene case study and energy valorization. *Bioresource Technology* [Internet]. 214: 806–815. Available from: https://linkinghub.elsevier.com/retrieve/pii/S0960852416306460.

153 Murad, H.A., Abo-Elkhair, A.G., and Azzaz, H.H. (2019). Production of xanthan gum from nontraditional substrates with perspective of the unique properties and wide industrial applications. *JSMC Microbiology*. 1: 6.

154 Mamma, D., Kourtoglou, E., and Christakopoulos, P. (2008). Fungal multienzyme production on industrial by-products of the citrus-processing industry. *Bioresource Technology* [Internet]. 99 (7): 2373–2383. Available from: https://linkinghub.elsevier.com/retrieve/pii/S0960852407004336.

155 Kennedy, J.F. and Bradshaw, I.J. (1984). Production, prosperties and applications of xanthan. *Progress in Industrial Microbiology* 19: 319–371.

156 Lopes, B.D.M., Lessa, V.L., Silva, B.M. et al. (2015). Xanthan gum: properties , production conditions, quality and economic perspective. *Journal of Food and Nutrition Research* 53 (3): 185–194.

157 Caldeira, C., Vlysidis, A., Fiore, G. et al. (2020). Sustainability of food waste biorefinery: A review on valorisation pathways, techno-economic constraints, and environmental assessment. *Bioresource Technology* [Internet]. 312: 123575. Available from: https://linkinghub.elsevier.com/retrieve/pii/S0960852420308476.

158 Ridley, B.L., O'Neill, M.A., and Mohnen, D. (2001). Pectins: structure, biosynthesis, and oligogalacturonide-related signaling. *Phytochemistry* [Internet]. 57 (6): 929–967. Available from: https://linkinghub.elsevier.com/retrieve/pii/S0031942201001133.

159 Endress, H.U. Chapter 8. (2011). Pectins: production, properties and applications. In: *Polymer Chemistry Series* [Internet]. (ed. P.A. Williams), 210–260. Cambridge: Royal Society of Chemistry Available from: http://ebook.rsc.org/?DOI=10.1039/9781849733519-00210.

160 Sakai, T., Sakamoto, T., Hallaert, J., and Vandamme, E.J. (1993). Pectin, pectinase, and protopectinase: production, properties, and applications. In: *Advances in Applied Microbiology* [Internet]. (eds. S. Neidleman and A.I. Laskin), 213–294. Elsevier Available from: https://linkinghub.elsevier.com/retrieve/pii/S0065216408705975.

161 Pilnik, W. and AGJ, V. (1970). Pectic substances and other uronides. *Biochemistry of Fruits and Their Products* 1: 53–87.

162 Fishman, M.L., Chau, H.K., Hoagland, P., and Ayyad, K. (1999). Characterization of pectin, flash-extracted from orange albedo by microwave heating, under pressure. *Carbohydrate Research* [Internet]. 323 (1–4): 126–138. Available from: https://linkinghub.elsevier.com/retrieve/pii/S000862159900244X.

163 Ueno, H., Tanaka, M., Machmudah, S. et al. (2008). Supercritical carbon dioxide extraction of valuable compounds from citrus junos seed. *Food and Bioprocess Technology* [Internet]. 1 (4): 357–363. Available from: http://link.springer.com/10.1007/s11947-007-0015-0.

164 Ptichkina, N.M., Markina, O.A., and Rumyantseva, G.N. (2008). Pectin extraction from pumpkin with the aid of microbial enzymes. *Food Hydrocolloids* [Internet]. 22 (1): 192–195. Available from: https://linkinghub.elsevier.com/retrieve/pii/S0268005X0700094X.

165 Meyer, T.S.M., Miguel, A.S.M., Fernández, D.E.R., and GMD, O. (2015). Biotechnological production of oligosaccharides — applications in the food industry. In: *Food Production and Industry* [Internet]. (ed. E. AHA). InTech Available from: http://www.intechopen.com/books/food-production-and-industry/biotechnological-production-of-oligosaccharides-applications-in-the-food-industry.

166 Jovanovic-Malinovska, R., Kuzmanova, S., and Winkelhausen, E. (2014). Oligosaccharide profile in fruits and vegetables as sources of prebiotics and functional foods. *International Journal of Food Properties* [Internet]. 17 (5): 949–965. Available from: http://www.tandfonline.com/doi/abs/10.1080/10942912.2012.680221.

167 Tuohy, K.M., Kolida, S., Lustenberger, A.M., and Gibson, G.R. (2001). The prebiotic effects of biscuits containing partially hydrolysed guar gum and fructo-oligosaccharides – a human volunteer study. *The British Journal of Nutrition* [Internet]. 86 (3): 341–348. Available from: https://www.cambridge.org/core/product/identifier/S0007114501001994/type/journal_article.

168 Rao, V.A. (2001). The prebiotic properties of oligofructose at low intake levels. *Nutrition Research* [Internet]. 21 (6): 843–848. Available from: https://linkinghub.elsevier.com/retrieve/pii/S0271531701002846.

169 Hotchkiss, A.T., Olano-Martin, E., Grace, W.E. et al. (2003). Pectic oligosaccharides as prebiotics. In: *Oligosaccharides in Food and Agriculture* [Internet]. (eds. G. Eggleston and G.L. Côté), 54–62. (ACS Symposium Series; vol. 849). Washington, DC: American Chemical Society Available from: https://pubs.acs.org/doi/abs/10.1021/bk-2003-0849.ch005.

170 Panche, A.N., Diwan, A.D., and Chandra, S.R. (2016). Flavonoids: an overview. *Journal of Nutritional Science* [Internet]. 5: e47. Available from: https://www.cambridge.org/core/product/identifier/S2048679016000410/type/journal_article.

171 Calado, J.C.P., Albertão, P.A., Oliveira, E.A. de et al. 2015. Flavonoid contents and antioxidant activity in fruit, vegetables and other types of food. AS [Internet]. 2015; 06(04): 426–435. Available from: http://www.scirp.org/journal/doi.aspx?DOI=10.4236/as.2015.64042.

172 Benavente-García, O., Castillo, J., Marin, F.R. et al. (1997). Uses and properties of *Citrus* flavonoids. *Journal of Agricultural and Food Chemistry* [Internet]. 45 (12): 4505–4515. Available from: https://pubs.acs.org/doi/10.1021/jf970373s.

173 Mandalari, G., Bennett, R.N., Bisignano, G. et al. (2006). Characterization of flavonoids and pectins from bergamot (*Citrus bergamia* risso) peel, a major byproduct of essential oil extraction. *Journal of Agricultural and Food Chemistry* [Internet]. 54 (1): 197–203. Available from: https://pubs.acs.org/doi/10.1021/jf051847n.

174 Rodríguez, R., Jiménez, A., Fernández-Bolaños, J. et al. (2006). Dietary fibre from vegetable products as source of functional ingredients. *Trends in Food Science & Technology* [Internet]. 17 (1): 3–15. Available from: https://linkinghub.elsevier.com/retrieve/pii/S092422440500244X.

175 Clemens, M.E. (ed.) (2015). *Dietary Fiber: Production Challenges, Food Sources and Health Benefits*. New York: Nova Publishers 191 pp. (Nutrition and diet research progress).

176 Yangilar, F. (2013). The application of dietary fibre in food industry: structural features, effects on health and definition, obtaining and analysis of dietary fibre: a review. *Journal of Food and Nutrition Research* 1 (3): 13–23.

177 Sharoba, A.M., Farrag, M.A., and Adb El-Salam, A.M. (2013). Utilization of some fruits and vegetables waste as a source of dietary fiber and its effect on the cake making and its quality attributes. *Journal of Agroalimentary Processes and Technologies* 19 (4): 429–444.

178 Abdel Salam, N.M. (2018). Review of ethanol production from fruit wastes. In: *Proceedings of Research World International Conference*. Rome, Italy.

179 Calabrò, P., Catalán, E., Folino, A. et al. (2018). Effect of three pretreatment techniques on the chemical composition and on the methane yields of *Opuntia ficus-indica* (prickly pear) biomass. *Waste Management & Research* [Internet]. 36 (1): 17–29. Available from: http://journals.sagepub.com/doi/10.1177/0734242X17741193.

180 Gullón, B., Yáñez, R., Alonso, J.L., and Parajó, J.C. (2008). L-lactic acid production from apple pomace by sequential hydrolysis and fermentation. *Bioresource Technology* [Internet]. 99 (2): 308–319. Available from: https://linkinghub.elsevier.com/retrieve/pii/S0960852407000119.

181 Kagan, J.J., W, P., and Smith, M.D. (1960). Citrus by-products, lactic acid production by fermentation of citrus peel juice. *Journal of Agricultural and Food Chemistry* [Internet]. 8 (3): 236–238. Available from: https://pubs.acs.org/doi/abs/10.1021/jf60109a022.

182 Show, P.L., Oladele, K.O., Siew, Q.Y. et al. (2015). Overview of citric acid production from *Aspergillus niger*. *Frontiers in Life Science* [Internet]. 8 (3): 271–283. Available from: http://www.tandfonline.com/doi/full/10.1080/21553769.2015.1033653.

183 Bodie, E.A., Bower, B., Berka, R.M., and Dunn-Coleman, N.S. (1994). Economically important organic acid and enzyme products. *Progress in Industrial Microbiology* 29: 561–602.

184 Wu, D. (2016). Recycle technology for potato peel waste processing: A review. *Procedia Environmental Sciences* [Internet]. 31: 103–107. Available from: https://linkinghub.elsevier.com/retrieve/pii/S1878029616000153.

185 Apprich, S., Tirpanalan, Ö., Hell, J. et al. (2014). Wheat bran-based biorefinery 2: valorization of products. *LWT- Food Science and Technology* [Internet]. 56 (2): 222–231. Available from: https://linkinghub.elsevier.com/retrieve/pii/S002364381300474X.

186 Mussatto, S.I. (2014). Brewer's spent grain: a valuable feedstock for industrial applications: Brewer's spent grain and its potential applications. *Journal of the Science of Food and Agriculture* [Internet]. 94 (7): 1264–1275. Available from: http://doi.wiley.com/10.1002/jsfa.6486.

187 Gebrechristos, H.Y. and Chen, W. (2018). Utilization of potato peel as eco-friendly products: a review. *Food Science & Nutrition* [Internet]. 6 (6): 1352–1356. Available from: http://doi.wiley.com/10.1002/fsn3.691.

188 Sanchez-Vazquez, S.A., Hailes, H.C., and JRG, E. (2013). Hydrophobic polymers from food waste: resources and synthesis. *Polymer Reviews* [Internet]. 53 (4): 627–694. Available from: http://www.tandfonline.com/doi/abs/10.1080/15583724.2013.834933.

189 Mushtaq, Q., Irfan, M., Tabssum, F., and Iqbal Qazi, J. (2017). Potato peels: a potential food waste for amylase production: Mushtaq et al. *Journal of Food Process Engineering* [Internet]. 40 (4): e12512. Available from: http://doi.wiley.com/10.1111/jfpe.12512.

190 Alexandri, M., Schneider, R., Papapostolou, H. et al. (2019). Restructuring the conventional sugar beet industry into a novel biorefinery: fractionation and bioconversion of sugar beet pulp into succinic acid and value-added coproducts. *ACS Sustainable Chemistry & Engineering* [Internet]. 7 (7): 6569–6579. Available from: https://pubs.acs.org/doi/10.1021/acssuschemeng.8b04874.

191 Guillard, V., Gaucel, S., Fornaciari, C. et al. (2018). The next generation of sustainable food packaging to preserve our environment in a circular economy context. *Frontiers in Nutrition* [Internet]. 5: 121. Available from: https://www.frontiersin.org/article/10.3389/fnut.2018.00121/full.

192 Swanepoel, J.C. and Goosen, N.J. (2018). Evaluation of fish protein hydrolysates in juvenile African catfish (Clarias gariepinus) diets. *Aquaculture* [Internet]. 496: 262–269. Available from: https://linkinghub.elsevier.com/retrieve/pii/S0044848618306859.

193 Ytteborg, E., Baeverfjord, G., Lock, E.J. et al. (2016). Utilization of acid hydrolysed phosphorous from herring bone by-products in feed for Atlantic salmon (Salmo salar) start-feeding fry. *Aquaculture* [Internet]. 459: 173–184. Available from: https://linkinghub.elsevier.com/retrieve/pii/S0044848616301417.

194 Malcorps, W., Kok, B., van't Land, M. et al. (2019). The sustainability conundrum of fishmeal substitution by plant ingredients in shrimp feeds. *Sustainability* [Internet]. 11 (4): 1212. Available from: https://www.mdpi.com/2071-1050/11/4/1212.

195 Uranga, J., Etxabide, A., Guerrero, P., and de la Caba, K. (2018). Development of active fish gelatin films with anthocyanins by compression molding. *Food Hydrocolloids* [Internet]. 84: 313–320. Available from: https://linkinghub.elsevier.com/retrieve/pii/S0268005X18306210.

196 Araújo, C.S., Rodrigues, A.M.C., Peixoto Joele, M.R.S. et al. (2018). Optimizing process parameters to obtain a bioplastic using proteins from fish byproducts through the response surface methodology. *Food Packaging and Shelf Life* [Internet]. 16: 23–30. Available from: https://linkinghub.elsevier.com/retrieve/pii/S2214289417301163.

197 Yaghoubzadeh, Z., Peyravii Ghadikolaii, F., Kaboosi, H. et al. (2020). Antioxidant activity and anticancer effect of bioactive peptides from rainbow trout (*Oncorhynchus mykiss*) skin hydrolysate. *International Journal of Peptide Research and Therapeutics* [Internet]. 26 (1): 625–632. Available from: http://link.springer.com/10.1007/s10989-019-09869-5.

198 Kantorek, M., Jesionek, K., Polesek-Karczewska, S. et al. (2019). Thermal utilization of meat and bone meals. Performance analysis in terms of drying process, pyrolysis and kinetics of volatiles combustion. *Fuel* [Internet]. 254: 115548. Available from: https://linkinghub.elsevier.com/retrieve/pii/S0016236119308920.

199 Conesa, J.A., Fullana, A., and Font, R. (2003). Thermal decomposition of meat and bone meal. *Journal of Analytical and Applied Pyrolysis* [Internet]. 70 (2): 619–630. Available from: https://linkinghub.elsevier.com/retrieve/pii/S0165237003000445.

200 Kalambura, S., Kričk a, T., Kiš, D. et al. (2016). Anaerobic digestion of specific biodegradable waste and final disposal. *Journal Tehnički vjesnik* [Internet]. 23: 6. Available from: http://hrcak.srce.hr/169345.

201 Animal by-products. 2020 European Commission. [Internet]. Available from: https://ec.europa.eu/food/safety/animal-by-products_en.

202 Jayathilakan, K., Sultana, K., Radhakrishna, K., and Bawa, A.S. (2012). Utilization of byproducts and waste materials from meat, poultry and fish processing industries: a review. *Journal of Food Science and Technology* [Internet]. 49 (3): 278–293. Available from: http://link.springer.com/10.1007/s13197-011-0290-7.

203 Mekonnen, T.H., Mussone, P.G., El-Thaher, N. et al. (2015). Subcritical hydrolysis and characterization of waste proteinaceous biomass for value added applications: subcritical hydrolysis and characterization of waste. *Journal of Chemical Technology and Biotechnology* [Internet]. 90 (3): 476–483. Available from: http://doi.wiley.com/10.1002/jctb.4333.

204 Bhaskar, N., Modi, V.K., Govindaraju, K. et al. (2007). Utilization of meat industry by products: protein hydrolysate from sheep visceral mass. *Bioresource Technology* [Internet]. 98 (2): 388–394. Available from: https://linkinghub.elsevier.com/retrieve/pii/S0960852405005997.

205 Selmane, D., Christophe, V., and Gholamreza, D. (2008). Extraction of proteins from slaughterhouse by-products: influence of operating conditions on functional properties. *Meat Science* [Internet]. 79 (4): 640–647. Available from: https://linkinghub.elsevier.com/retrieve/pii/S0309174007003336.

206 Chemat, F., Abert-Vian, M., Fabiano-Tixier, A.S. et al. (2019). Green extraction of natural products. Origins, current status, and future challenges. *TrAC Trends in Analytical Chemistry* [Internet].: 118, 248–163. Available from: https://linkinghub.elsevier.com/retrieve/pii/S0165993619300706.

207 Cassoni, A.C., Freixo, R., Pintado, A.I.E. et al. (2018). Novel eco-friendly method to extract keratin from hair. *ACS Sustainable Chemistry & Engineering* [Internet]. 6 (9): 12268–12274. Available from: https://pubs.acs.org/doi/10.1021/acssuschemeng.8b02680.

208 Borrajo, P., Pateiro, M., Barba, F.J. et al. (2019). Antioxidant and antimicrobial activity of peptides extracted from meat by-products: a review. *Food Analytical Methods* [Internet]. 12 (11): 2401–2415. Available from: http://link.springer.com/10.1007/s12161-019-01595-4.

209 O'Sullivan, F., O'Sullivan, J.N., Huang, J. et al. (2017). Assessment of a statistical AIF extraction method for dynamic PET studies with 15O water and 18F fluorodeoxyglucose in locally advanced breast cancer patients. *Journal of Medical Imaging* [Internet]. 5 (01): 1. Available from: https://www.spiedigitallibrary.org/journals/journal-of-medical-imaging/volume-5/issue-01/011010/Assessment-of-a-statistical-AIF-extraction-method-for-dynamic-PET/10.1117/1.JMI.5.1.011010.full.

210 Verma, L.R. and Joshi, V.K. (eds.) (2000). *Postharvest Technology of Fruits and Vegetables: Handling, Processing, Fermentation, and Waste Management*. New Delhi: Indus Publishing Company.

211 Kowalski, Z. and Krupa-Żuczek, K. (2007). A model of the meat waste management. *Polish Journal of Chemical Technology* [Internet]. 9 (4): 91–97. Available from: https://content.sciendo.com/doi/10.2478/v10026-007-0098-4.

212 Chojnacka, K., Górecka, H., Michalak, I., and Górecki, H.A. (2011). Review: valorization of keratinous materials. *Waste and Biomass Valorization* [Internet]. 2 (3): 317–321. Available from: http://link.springer.com/10.1007/s12649-011-9074-6.

213 Zoccola, M., Montarsolo, A., Mossotti, R. et al. (2015). Green hydrolysis as an emerging technology to turn wool waste into organic nitrogen fertilizer. *Waste and Biomass Valorization* [Internet]. 6 (5): 891–897. Available from: http://link.springer.com/10.1007/s12649-015-9393-0.

214 Papademas, P. and Kotsaki, P. (2019). Technological utilization of whey towards sustainable exploitation. *The Journal of Dairy Research* 7 (4): 231.

215 Božanic, R., Barukcic, I., Jakopovic, K.L., and Tratnik, L. (2014). Possibilities of whey utilisation. *Austin Journal of Nutrition and Food Sciences* 2 (7): 1036.

216 Oliveira, D.A., Benelli, P., and Amante, E.R. (2013). A literature review on adding value to solid residues: egg shells. *Journal of Cleaner Production* [Internet]. 46: 42–47. Available from: https://linkinghub.elsevier.com/retrieve/pii/S0959652612005161.

217 Cunha, T.P., Siqueira, F.B., and JNF, H. (2019). Development of sustainable eggshell waste-polyester resin composite material for using as artificial rock. *Materials Research* [Internet]. 22 (suppl 1): e20180865. Available from: http://www.scielo.br/scielo.php?script=sci_arttext&pid=S1516-14392019000700239&tlng=en.

218 Georganas, A., Giamouri, E., Pappas, A.C. et al. (2020). Bioactive compounds in food waste: A review on the transformation of food waste to animal feed. *Food* [Internet]. 9 (3): 291. Available from: https://www.mdpi.com/2304-8158/9/3/291.

219 Dou, Z., Toth, J.D., and Westendorf, M.L. (2018). Food waste for livestock feeding: feasibility, safety, and sustainability implications. *Global Food Security* [Internet]. 17: 154–161. Available from: https://linkinghub.elsevier.com/retrieve/pii/S2211912417301384.

220 Chen, T., Jin, Y., and Shen, D. (2015). A safety analysis of food waste-derived animal feeds from three typical conversion techniques in China. *Waste Management* [Internet]. 45: 42–50. Available from: https://linkinghub.elsevier.com/retrieve/pii/S0956053X1530012X.

221 Chantaro, P., Devahastin, S., and Chiewchan, N. (2008). Production of antioxidant high dietary fiber powder from carrot peels. *LWT- Food Science and Technology* [Internet]. 41 (10): 1987–1994. Available from: https://linkinghub.elsevier.com/retrieve/pii/S0023643807003969.

222 Ncobela, C.N., Kanengoni, A.T., Hlatini, V.A. et al. (2017). A review of the utility of potato by-products as a feed resource for smallholder pig production. *Animal Feed Science and Technology* [Internet]. 227: 107–117. Available from: https://linkinghub.elsevier.com/retrieve/pii/S0377840116310574.

223 Liu, D.H.F. and Lipták, B.G. (eds.) (1999). *Environmental engineers' handbook*, 2e. Boca Raton, Fla: CRC Press USA. 1431 pp.

224 Velvizhi, G., Shanthakumar, S., Das, B. et al. (2020). Biodegradable and non-biodegradable fraction of municipal solid waste for multifaceted applications through a closed loop integrated refinery platform: paving a path towards circular economy. *Science of the Total Environment* [Internet]. 731: 138049. Available from: https://linkinghub.elsevier.com/retrieve/pii/S004896972031562X.

225 Ferronato, N., Rada, E.C., Gorritty Portillo, M.A. et al. (2019). Introduction of the circular economy within developing regions: a comparative analysis of advantages and opportunities for waste valorization. *Journal of Environmental Management* [Internet]. 230: 366–378. Available from: https://linkinghub.elsevier.com/retrieve/pii/S0301479718311058.

226 Raclavská, H., Corsaro, A., Hlavsová, A. et al. (2015). The effect of moisture on the release and enrichment of heavy metals during pyrolysis of municipal solid waste. *Waste Management & Research* [Internet]. 33 (3): 267–274. Available from: http://journals.sagepub.com/doi/10.1177/0734242X14568609.

227 Chen, Y.C. and Wu, W.F. (2015). Constructing an effective prevention mechanism for MSW lifecycle using failure mode and effects analysis. *Waste Management* [Internet]. 46: 646–652. Available from: https://linkinghub.elsevier.com/retrieve/pii/S0956053X15301124.

228 Abu-Qudais, M. and Abu-Qdais, H.A. (2000). Energy content of municipal solid waste in Jordan and its potential utilization. *Energy Conversion and Management* [Internet]. 41 (9): 983–991. Available from: https://linkinghub.elsevier.com/retrieve/pii/S0196890499001557.

229 Wang, H., Wang, L., and Shahbazi, A. (2015). Life cycle assessment of fast pyrolysis of municipal solid waste in North Carolina of USA. *Journal of Cleaner Production* [Internet]. 87: 511–519. Available from: https://linkinghub.elsevier.com/retrieve/pii/S095965261400941X.

230 Laurent, A., Clavreul, J., Bernstad, A. et al. (2014). Review of LCA studies of solid waste management systems – Part II: methodological guidance for a better practice. *Waste Management* [Internet]. 34 (3): 589–606. Available from: https://linkinghub.elsevier.com/retrieve/pii/S0956053X13005710.

231 Waste atlas. 2019. Municipal Solid Waste generation (t/yr) [Internet]. Available from: http://www.atlas.d-waste.com.

232 Levaggi, L., Levaggi, R., Marchiori, C., and Trecroci, C. (2020). Waste-to-energy in the EU: the effects of plant ownership, waste mobility, and decentralization on environmental outcomes and welfare. *Sustainability* [Internet]. 12 (14): 5743. Available from: https://www.mdpi.com/2071-1050/12/14/5743.

233 Municipal waste statistics. 2020. [Internet]. Eurostat. Available from: https://ec.europa.eu/eurostat/statistics-explained/index.php/Municipal_waste_statistics.

234 Bidlingmaier, W., Sidaine, J.M., and Papadimitriou, E.K. (2004). Separate collection and biological waste treatment in the European Community. *Reviews in Environmental Science and Biotechnology* [Internet]. 3 (4): 307–320. Available from: http://link.springer.com/10.1007/s11157-004-2334-1.

235 Bölükbaş, A. and Akıncı, G. (2018). Solid waste composition and the properties of biodegradable fractions in Izmir City, Turkey: an investigation on the influencing factors. *Journal of Environmental Health Science and Engineering* [Internet]. 16 (2): 299–311. Available from: http://link.springer.com/10.1007/s40201-018-0318-2.

236 Pleissner, D. and CSK, L. (2013). Valorisation of food waste in biotechnological processes. *Sustainable Chemical Processes* [Internet]. 1 (1): 21. Available from: https://sustainablechemicalprocesses.springeropen.com/articles/10.1186/2043-7129-1-21.

237 Malinauskaite, J., Jouhara, H., Czajczyńska, D. et al. (2017). Municipal solid waste management and waste-to-energy in the context of a circular economy and energy recycling in Europe. *Energy* [Internet]. 141: 2013–2044. Available from: https://linkinghub.elsevier.com/retrieve/pii/S0360544217319862.

238 Scarlat, N., Fahl, F., and Dallemand, J.F. (2019). Status and opportunities for energy recovery from municipal solid waste in Europe. *Waste and Biomass Valorization* [Internet]. 10 (9): 2425–2444. Available from: http://link.springer.com/10.1007/s12649-018-0297-7.

239 Kamuk, B. (2013). *Waste to Energy in Low and Middle Income Countries*. Copenhagen: ISWA p. 31.

240 Koval, S., Raclavská, H., Skrobankova, H. et al. (2019). Hydrocyclone separation as a tool for reduction of the amount of heavy metals in municipal solid waste incinerator (MSWI) residues. *Acta Montanistica Slovaca* 24 (3): 188–197.

241 Hartmann, S., Koval', L., Škrobánková, H. et al. (2015). Possibilities of municipal solid waste incinerator fly ash utilisation. *Waste Management & Research* [Internet]. 33 (8): 740–747. Available from: http://journals.sagepub.com/doi/10.1177/0734242X15587545.

242 EN 15359:2011. 2011. Solid recovered fuels. Specifications and classes: BSI British Standards.

243 EN 15440:2011. 2011. Solid recovered fuels. Methods for the determination of biomass content. BSI British Standards.

244 Rada, E.C. and Ragazzi, M. (2015). RDF/SRF evolution in the MSW sector: coexistence of BMT and selective collection. *International Journal of Sustainable Development and Planning*

[Internet]. 10 (1): 109–119. Available from: http://www.witpress.com/doi/journals/SDP-V10-N1-109-119.

245 Lee, P., Sims, E., Bertham, O. et al. (2017). Towards a circular economy – waste management in the EU: study [Internet]. Available from: http://www.europarl.europa.eu/RegData/etudes/STUD/2017/581913/EPRS_STU(2017)581913_EN.pdf.

246 Ciesielczuk, T., Rosik-Dulewska, C., Poluszyńska, J. et al. (2018). Acute toxicity of experimental fertilizers made of spent coffee grounds. *Waste and Biomass Valorization* [Internet]. 9 (11): 2157–2164. Available from: http://link.springer.com/10.1007/s12649-017-9980-3.

247 Růžičková, J., Raclavská, H., Kucbel, M. et al. (2020). The potential environmental risks of the utilization of composts from household food waste. *Environmental Science and Pollution Research* [Internet]. Available from: http://link.springer.com/10.1007/s11356-020-09916-5.

248 Chelinho, S., Pereira, C., Breitenbach, P. et al. (2019). Quality standards for urban waste composts: the need for biological effect data. *Science of the Total Environment* [Internet]. 694: 133602. Available from: https://linkinghub.elsevier.com/retrieve/pii/S0048969719335272.

249 Wu, S., He, H., Inthapanya, X. et al. (2017). Role of biochar on composting of organic wastes and remediation of contaminated soils – a review. *Environmental Science and Pollution Research* [Internet]. 24 (20): 16560–16577. Available from: http://link.springer.com/10.1007/s11356-017-9168-1.

250 Eggerth, L.L., Diaz, L.F., Chang, M.T.F., and Iseppi, L. Chapter 12 (2007). Marketing of composts. In: *Waste management series* [Internet]., 325–355. Elsevier Available from: https://linkinghub.elsevier.com/retrieve/pii/S1478748207800157.

251 The World Bank (2016). *Sustainable Financing and Policy Models for Municipal Composting*. Washington, DC: The International Bank for Reconstruction and Development.

252 The European Parliament, The Council of the European Union. 2019. Regulation (EU) 2019/1009 of the European Parliament and of the Council of 5 June 2019 laying down rules on the making available on the market of EU fertilising products and amending Regulations (EC) No 1069/2009 and (EC) No 1107/2009 and repealing Regulation (EC) No 2003/2003 (Text with EEA relevance).

253 López-Gómez, J.P., Latorre-Sánchez, M., Unger, P. et al. (2019). Assessing the organic fraction of municipal solid wastes for the production of lactic acid. *Biochemical Engineering Journal* [Internet]. 150: 107251. Available from: https://linkinghub.elsevier.com/retrieve/pii/S1369703X19301780.

254 Lay, J.J., Lee, Y.J., and Noike, T. (1999). Feasibility of biological hydrogen production from organic fraction of municipal solid waste. *Water Research* [Internet]. 33 (11): 2579–2586. Available from: https://linkinghub.elsevier.com/retrieve/pii/S0043135498004837.

255 Zheng, X., Ying, Z., Wang, B., and Chen, C. (2018). Hydrogen and syngas production from municipal solid waste (MSW) gasification via reusing CO_2. *Applied Thermal Engineering* [Internet]. 144: 242–247. Available from: https://linkinghub.elsevier.com/retrieve/pii/S1359431117365535.

256 Hu, M., Guo, D., Ma, C. et al. (2015). Hydrogen-rich gas production by the gasification of wet MSW (municipal solid waste) coupled with carbon dioxide capture. *Energy* [Internet]. 90: 857–863. Available from: https://linkinghub.elsevier.com/retrieve/pii/S0360544215010178.

257 Miyake, M., Komiya, T., Saffarzadeh, A., and Shimaoka, T. (2018). Evaluating the crucial factors affecting hydrogen gas generation from municipal solid waste incineration bottom ash (MSWIBA). *Detritus* [Internet]. 2 (1): 150. Available from: https://digital.detritusjournal.com/doi/13652-111.

258 Kuchařová, J., Raclavská, H., and Raclavský, K. (2014). Organic compounds of natural origin in sludge from waste-water treatment plant. *Inzynieria Mineralna* 15: 61–66.

259 Zhang, Z. and Chen, Y. (2020). Effects of microplastics on wastewater and sewage sludge treatment and their removal: A review. *Chemical Engineering Journal* [Internet]. 382: 122955. Available from: https://linkinghub.elsevier.com/retrieve/pii/S1385894719323654.

260 Corradini, F., Meza, P., Eguiluz, R. et al. (2019). Evidence of microplastic accumulation in agricultural soils from sewage sludge disposal. *Science of the Total Environment* [Internet]. 671: 411–420. Available from: https://linkinghub.elsevier.com/retrieve/pii/S004896971931366X.

261 The Danish Environmental Protection Agency (2017). *Microplastic in Danish Wastewater Sources, Occurrences and Fate*. Copenhagen: The Danish Environmental Protection Agency. 55 pp.

262 Prata, J.C. (2018). Microplastics in wastewater: state of the knowledge on sources, fate and solutions. *Marine Pollution Bulletin* [Internet]. 129 (1): 262–265. Available from: https://linkinghub.elsevier.com/retrieve/pii/S0025326X18301346.

263 Oldenkamp, R., Beusen, A.H.W., and MAJ, H. (2019). Aquatic risks from human pharmaceuticals—modelling temporal trends of carbamazepine and ciprofloxacin at the global scale. *Environmental Research Letters* [Internet]. 14 (3): 034003. Available from: https://iopscience.iop.org/article/10.1088/1748-9326/ab0071.

264 Frade, V.M.F., Dias, M., Teixeira, A.C.S.C., and MSA, P. (2014). Environmental contamination by fluoroquinolones. *Brazilian Journal of Pharmaceutical Sciences* [Internet]. 50 (1): 41–54. Available from: http://www.scielo.br/scielo.php?script=sci_arttext&pid=S1984-82502014000100004&lng=en&tlng=en.

265 Jendrzejewska, N. and Karwowska, E. (2018). The influence of antibiotics on wastewater treatment processes and the development of antibiotic-resistant bacteria. *Water Science and Technology* [Internet]. 77 (9): 2320–2326. Available from: https://iwaponline.com/wst/article/77/9/2320/38640/The-influence-of-antibiotics-on-wastewater.

266 van de Vijver, L., Verwer, C., Smolders, G. et al. (2016). *The Cycle of Veterinary Antibiotics in the Ecosystem*. Bunnik, The Netherlands: Louis Bolk Institute. 31 pp.

267 Cycoń, M., Mrozik, A., and Piotrowska-Seget, Z. (2019). Antibiotics in the soil environment—degradation and their impact on microbial activity and diversity. *Frontiers in Microbiology* [Internet]. 10: 338. Available from: https://www.frontiersin.org/article/10.3389/fmicb.2019.00338/full.

268 Fijalkowski, K., Rorat, A., Grobelak, A., and Kacprzak, M.J. (2017). The presence of contaminations in sewage sludge – the current situation. *Journal of Environmental Management* [Internet]. 203: 1126–1136. Available from: https://linkinghub.elsevier.com/retrieve/pii/S0301479717305418.

269 Choffnes, E.R. (ed.), Institute of Medicine (U.S.), (2012). *Improving Food Safety through a One Health Approach: Workshop Summary*. Washington, DC: National Academies Press. 395 pp.

270 Van Caneghem, J., Brems, A., Lievens, P. et al. (2012). Fluidized bed waste incinerators: design, operational and environmental issues. *Progress in Energy and Combustion Science* [Internet]. 38 (4): 551–582. Available from: https://linkinghub.elsevier.com/retrieve/pii/S0360128512000160.

271 Durdević, D., Blecich, P., and Jurić, Ž. (2019). Energy recovery from sewage sludge: the case study of Croatia. *Energies* [Internet]. Available from: https://www.mdpi.com/1996-1073/12/10/1927.

272 Eurostat. 2020. Water statistics [Internet]. Available from: https://ec.europa.eu/eurostat/statistics-explained/pdfscache/1182.pdf.

273 Rosiek, K. (2020). Directions and challenges in the management of municipal sewage sludge in Poland in the context of the circular economy. *Sustainability* [Internet]. 12 (9): 3686. Available from: https://www.mdpi.com/2071-1050/12/9/3686.

274 The Council of the European Communities. 1986. Council Directive 86/278/EEC of 12 June 1986 on the protection of the environment, and in particular of the soil, when sewage sludge is used in agriculture.

275 Hudcová, H., Vymazal, J., and Rozkošný, M. (2019). Present restrictions of sewage sludge application in agriculture within the European Union. *Soil and Water Research* [Internet]. 14 (2): 104–120. Available from: https://www.agriculturejournals.cz/web/swr.htm?type=article&id=36_2018-SWR.

276 Egle, L., Rechberger, H., Krampe, J., and Zessner, M. (2016). Phosphorus recovery from municipal wastewater: an integrated comparative technological, environmental and economic assessment of P recovery technologies. *Science of the Total Environment* [Internet]. 571: 522–542. Available from: https://linkinghub.elsevier.com/retrieve/pii/S0048969716314656.

277 AbfKlärV (2017). Verordnung über die Verwertung von Klärschlamm, Klärschlammgemisch und Klärschlammkompost [Internet]. Bundesministerium der Justiz und für Verbraucherschutz. Available from: https://www.gesetze-im-internet.de/abfkl_rv_2017/BJNR346510017.html.

278 Heidecke, P. and Roskosch, A. (2019). Klärschlammverwertung in Deutschland – entwicklung und tendenzen [Internet]. Umweltwirtschaft. Available from: https://www.umweltwirtschaft.com/news/wasser-und-abwasserbehandlung/schlamm-und-reststoffbehandlung/Klaerschlamm-Klaerschlammverwertung-in-Deutschland--Entwicklung-und-Tendenzen-18794.

279 Ruiken, C.J., Breuer, G., Klaversma, E. et al. (2013). Sieving wastewater – cellulose recovery, economic and energy evaluation. *Water Research* [Internet]. 47 (1): 43–48. Available from: https://linkinghub.elsevier.com/retrieve/pii/S0043135412006112.

280 Kroiss, H., Rechberger, H., and Egle, L. (2011). Phosphorus in water quality and waste management. In: *Integrated Waste Management - Volume II* [Internet]. (ed. S. Kumar). InTech Available from: http://www.intechopen.com/books/integrated-waste-management-volume-ii/phosphorus-in-water-quality-and-waste-management.

281 Kok, D.J.D., Pande, S., van Lier, J.B. et al. (2018). Global phosphorus recovery from wastewater for agricultural reuse. *Hydrology and Earth System Sciences* [Internet]. 22 (11): 5781–5799. Available from: https://hess.copernicus.org/articles/22/5781/2018.

282 COM. (2015). 614 Communication from the Commission to the European Parliament, the Council, the European Economic and Social Committee and the Committee of the Regions – Closing the loop – An EU action plan for the Circular Economy [Internet]. Brussels: European Commission; 2015 Dec. Report No.: COM(2015) 614. Available from: https://eur-lex.europa.eu/legal-content/EN/TXT/?uri=CELEX%3A52015DC0614.

283 Molinos-Senante, M., Hernández-Sancho, F., Sala-Garrido, R., and Garrido-Baserba, M. (2011). Economic feasibility study for phosphorus recovery processes. *Ambio* [Internet]. 40 (4): 408–416. Available from: http://link.springer.com/10.1007/s13280-010-0101-9.

284 Kabbe, Ch. (2019). Global compendium on phosphorus recovery from sewage/Sludge/Ash [Internet]. Global Water Research Coalition. Available from: https://www.vesiyhdistys.fi/wp-content/uploads/2019/12/GWRCPhosphorusCompendiumFinalReport2019-March-20.pdf.

285 Bashar, R., Gungor, K., Karthikeyan, K.G., and Barak, P. (2018). Cost effectiveness of phosphorus removal processes in municipal wastewater treatment. *Chemosphere* [Internet]. 197: 280–290. Available from: https://linkinghub.elsevier.com/retrieve/pii/S0045653517321471.

286 Shiba, N.C. and Ntuli, F. (2017). Extraction and precipitation of phosphorus from sewage sludge. *Waste Management* [Internet]. 60: 191–200. Available from: https://linkinghub.elsevier.com/retrieve/pii/S0956053X16304081.

287 Wiechmann, B., Dienemann, C., Kabbe, Ch., et al. 2013. Sewage sludge management in Germany [Internet]. Dessau-Roßlau (Germany): Umweltbundesamt (UBA). 100 pp. Available from: https://www.umweltbundesamt.de/sites/default/files/medien/378/publikationen/sewage_sludge_management_in_germany.pdf.

288 Smoliński, A., Karwot, J., Bondaruk, J., and Bąk, A. (2019). The bioconversion of sewage sludge to bio-fuel: the environmental and economic benefits. *Materials* [Internet]. 12 (15): 2417. Available from: https://www.mdpi.com/1996-1944/12/15/2417.

289 Hroncová, E., Ladomerský, J., and Musil, J. (2018). Problematic issues of air protection during thermal processes related to the energetic uses of sewage sludge and other waste. Case study: co-combustion in peaking power plant. *Waste Management* [Internet]. 73: 574–580. Available from: https://linkinghub.elsevier.com/retrieve/pii/S0956053X17305974.

290 Degereji, M.U., Gubba, S.R., Ingham, D.B. et al. (2013). Predicting the slagging potential of co-fired coal with sewage sludge and wood biomass. *Fuel* [Internet]. 108: 550–556. Available from: https://linkinghub.elsevier.com/retrieve/pii/S0016236112010538.

291 Xu, Y., Yang, K., Zhou, J., and Zhao, G. (2020). Coal-biomass co-firing power generation technology: current status, challenges and policy implications. *Sustainability* [Internet]. 12 (9): 3692. Available from: https://www.mdpi.com/2071-1050/12/9/3692.

292 Barbosa, R., Lapa, N., Boavida, D. et al. (2009). Co-combustion of coal and sewage sludge: chemical and ecotoxicological properties of ashes. *Journal of Hazardous Materials* [Internet]. 170 (2–3): 902–909. Available from: https://linkinghub.elsevier.com/retrieve/pii/S0304389409007894.

293 Chang, C.C., Chen, Y.H., Chang, W.R. et al. (2019). The emissions from co-firing of biomass and torrefied biomass with coal in a chain-grate steam boiler. *Journal of the Air & Waste Management Association* [Internet]. 69 (12): 1467–1478. Available from: https://www.tandfonline.com/doi/full/10.1080/10962247.2019.1668871.

294 Abis, M., Calmano, W., and Kuchta, K. (2018). Innovative technologies for phosphorus recovery from sewage sludge ash. Detritus [Internet]. 1(0):23–29. Available from: http://doi.org/10.26403/detritus/2018.23.

295 Krüger, O., Grabner, A., and Adam, C. (2014). Complete survey of German sewage sludge ash. *Environmental Science & Technology* [Internet]. 48 (20): 11811–11818. Available from: https://pubs.acs.org/doi/10.1021/es502766x.

296 Petzet, S., Peplinski, B., and Cornel, P. (2012). On wet chemical phosphorus recovery from sewage sludge ash by acidic or alkaline leaching and an optimized combination of both. *Water Research* [Internet]. 46 (12): 3769–3780. Available from: https://linkinghub.elsevier.com/retrieve/pii/S004313541200245X.

297 Morf, L., Schlumberger, S., Adam, F., and Díaz, N.G. (2019). Urban phosphorus mining in the Canton of Zurich: phosphoric acid from sewage sludge ash. In: *Phosphorus recovery and recycling [Internet]* (eds. H. Ohtake and S. Tsuneda), 157–177. Singapore: Springer Singapore Available from: http://link.springer.com/10.1007/978-981-10-8031-9_10.

298 Lebek, M., Rak, A., and Hanßen, H. (2018). The REMONDIS TetraPhos® process at the WWTP in Hamburg (Germany). In: *Phosphorus: Polluter and Resource of the Future – Removal and Recovery from Wastewater* [Internet]. (ed. C. Schaum), 401–409. International Water Association Available from: https://iwaponline.com/ebooks/book/725/chapter/1378543/The-REMONDIS-TetraPhosR-Process-at-the-WWTP-in.

299 Hermann, L. and Schaaf, T. Outotec (AshDec®) (2019). Process for P fertilizers from sludge ash. In: *Phosphorus Recovery and Recycling* [Internet]. (eds. H. Ohtake and S. Tsuneda), 221–233. Springer Singapore: Singapore Available from: http://link.springer.com/10.1007/978-981-10-8031-9_15.

300 Kraus, F., Zamzow, M., Conzelmann, L. et al. (2019). Ökobilanzieller vergleich der P-Rückgewinnung aus dem abwasserstrom mit der düngemittelproduktion aus rohphosphaten unter einbeziehung von umweltfolgeschäden und deren Vermeidung [Internet]. Umweltbundesamt (UBA). Available from: https://www.umweltbundesamt.de/sites/default/files/medien/1410/publikationen/2019-02-19_texte_13-2019_phorwaerts.pdf.

301 Nakagawa, H. and Ohta, J. (2019). Phosphorus recovery from sewage sludge ash: A case study in Gifu, Japan. In: *Phosphorus Recovery and Recycling* [Internet]. (eds. H. Ohtake and S. Tsuneda), 149–155. Springer Singapore: Singapore Available from: http://link.springer.com/10.1007/978-981-10-8031-9_9.

302 Günther, S., Grunert, M., and Müller, S. (2018). Overview of recent advances in phosphorus recovery for fertilizer production. *Engineering in Life Sciences* [Internet]. 18 (7): 434–439. Available from: http://doi.wiley.com/10.1002/elsc.201700171.

303 PHOS4Green. (n.d) Phosphorus-recovery from ashes [Internet]. Glatt. Available from: https://www.phos4green-glatt.com/technology-79.html.

304 Herzel, H. and Deliverable, D. (2015). 4.1 Comparative review of ash processes – technical comparison on the design, operation and performances of ash processes. P-REX.

305 Guedes, P., Couto, N., Ottosen, L.M., and Ribeiro, A.B. (2014). Phosphorus recovery from sewage sludge ash through an electrodialytic process. *Waste Management* [Internet]. 34 (5): 886–892. Available from: https://linkinghub.elsevier.com/retrieve/pii/S0956053X14000737.

306 Hosho, F. (2019). Phosphorus recovery from sewage sludge by high-temperature thermochemical process (KUBOTA process). In: *Phosphorus recovery and recycling* [Internet]. (eds. H. Ohtake and S. Tsuneda), 189–199. Singapore: Springer Singapore Available from: http://link.springer.com/10.1007/978-981-10-8031-9_12.

307 Merzari, F., Langone, M., Andreottola, G., and Fiori, L. (2019). Methane production from process water of sewage sludge hydrothermal carbonization. A review. Valorising sludge through hydrothermal carbonization. *Critical Reviews in Environmental Science and Technology* [Internet]. 49 (11): 947–988. Available from: https://www.tandfonline.com/doi/full/10.1080/10643389.2018.1561104.

308 Xia, Y., Tang, Y., Shih, K., and Li, B. (2020). Enhanced phosphorus availability and heavy metal removal by chlorination during sewage sludge pyrolysis. *Journal of Hazardous Materials* [Internet]. 382: 121110. Available from: https://linkinghub.elsevier.com/retrieve/pii/S0304389419310647.

309 FEFCO (European Federation of Corrugated Board Manufacturers). 2015. Fefco, CCB (Cepi Container Board). European database for corrugated board life cycle studies.

310 Pro Carton, Barker (Truffula Ltd.) T. 2018. Comparison of car ton and plastic packaging sustainability [Internet]. Zürich: Pro Carton. p. 32. Available from: https://www.procarton.com/wp-content/uploads/2018/06/PC-Carton-Plastic-Sustainability.pdf.

311 Virtanen, Y. and Nilsson, S. (2017). *Environmental Impacts of Waste Paper Recycling*, 1e. London: Routledge. 178 pp.

312 Brožek, M. (2015). Evaluation of selected properties of briquettes from recovered paper and board. *Research in Agricultural Engineering* [Internet]. 61 (2): 66–71. Available from: http://www.agriculturejournals.cz/web/rae.htm?volume=61&firstPage=66&type=publishedArticle.

313 Malaťák J, Velebil J, and Bradna J. (2018). Specialty types of waste paper as an energetic commodity. 344.3 Kb. Available from: https://dspace.emu.ee/xmlui/handle/10492/3865.

314 Xiu, M., Stevanovic, S., Rahman, M.M. et al. (2018). Emissions of particulate matter, carbon monoxide and nitrogen oxides from the residential burning of waste paper briquettes and other fuels. *Environmental Research* [Internet]. 167: 536–543. Available from: https://linkinghub.elsevier.com/retrieve/pii/S0013935118304419.

315 Ibrahim, H.G., Ouiminga, S.K., Yonli, A. et al. (2018). Study of temperature fields and heavy metal content in the ash and flue gas produced by the combustion of briquettes coming from paper and cardboard waste. *Recycling* [Internet]. 3 (3): 32. Available from: http://www.mdpi.com/2313-4321/3/3/32.

316 Gilbert, J., Ricci, M., and Tang, J. (2015). *Biodegradable Plastics: An Overview of Compostability of Biodegradable Plastics and its Implications for the Collection and Treatment of Organic Wastes*. Vienna: ISWA – the International Solid Waste Association 32 pp.

317 Emadian, S.M., Onay, T.T., and Demirel, B. (2017). Biodegradation of bioplastics in natural environments. *Waste Management* [Internet]. 59: 526–536. Available from: https://linkinghub.elsevier.com/retrieve/pii/S0956053X1630561X.

318 ASTM International (2019). *ASTM D6400–19 Standard Specification for Labeling of Plastics Designed to Be Aerobically Composted in Municipal or Industrial Facilities*. West Conshohocken: ASTM International p. 3.

319 Song, J.H., Murphy, R.J., Narayan, R., and Davies GBH (2009). Biodegradable and compostable alternatives to conventional plastics. *Philosophical Transactions of the Royal Society B* [Internet]. 364 (1526): 2127–2139. Available from: https://royalsocietypublishing.org/doi/10.1098/rstb.2008.0289.

320 Moeller, M. and Matyjaszewski, K. (2012). *Polymer Science*. [Internet]. Burlington: Elsevier Science Available from: http://public.ebookcentral.proquest.com/choice/publicfullrecord.aspx?p=934526.

321 Zhao, X., Cornish, K., and Vodovotz, Y. (2020). Narrowing the gap for bioplastic use in food packaging: an update. *Environmental Science & Technology* [Internet]. 54 (8): 4712–4732. Available from: https://pubs.acs.org/doi/abs/10.1021/acs.est.9b03755.

322 Zhuang, H., Barth, M.M., and Cisneros-Zevallos, L. (2014). Modified atmosphere packaging for fresh fruits and vegetables. In: *Innovations in Food Packaging* [Internet]., 445–473. Elsevier Available from: https://linkinghub.elsevier.com/retrieve/pii/B9780123946010000187.

323 Marquié, C. (2001;). Chemical reactions in cottonseed protein cross-linking by formaldehyde, glutaraldehyde, and glyoxal for the formation of protein films with enhanced mechanical properties. *Journal of Agricultural and Food Chemistry* [Internet]. 49 (10): 4676–4681. Available from: https://pubs.acs.org/doi/10.1021/jf0101152.

324 Confente, I., Scarpi, D., and Russo, I. (2020). Marketing a new generation of bio-plastics products for a circular economy: the role of green self-identity, self-congruity, and perceived value. *Journal of Business Research* [Internet]. 112: 431–439. Available from: https://linkinghub.elsevier.com/retrieve/pii/S0148296319306186.

325 European Bioplastics (2020). Bioplastics market data [internet]. *European Bioplastics* Available from: https://www.european-bioplastics.org/market.

326 Ezeonu, C.S. and Folashade Temitope, D. (2018). Trends on bio-synthesis of plastics. *Advances in Biotechnology and Microbiology* [Internet]. 10 (5): 555797. Available from: https://juniperpublishers.com/aibm/AIBM.MS.ID.555797.php.

327 Ojha, S. and Kapoor, S. (2019). Bio-plastics: the suitable and sustainable alternative to polyethylene based plastics. *Acta Scientific Microbiology*. 2 (6): 145–148.

328 Arikan, E.B. and Ozsoy, H.D. (2015). A review: investigation of bioplastics. *JCEA* [Internet]. 9: 2. Available from: http://www.davidpublisher.org/index.php/Home/Article/index?id=116.html.

329 Karan, H., Funk, C., Grabert, M. et al. (2019). Green bioplastics as part of a circular bioeconomy. *Trends in Plant Science* [Internet]. 24 (3): 237–249. Available from: https://linkinghub.elsevier.com/retrieve/pii/S1360138518302723.

330 Gurunathan, T., Mohanty, S., and Nayak, S.K. (2015). A review of the recent developments in biocomposites based on natural fibres and their application perspectives. *Composites Part A: Applied Science and Manufacturing* [Internet]. 77: 1–25. Available from: https://linkinghub.elsevier.com/retrieve/pii/S1359835X15002067.

331 Davis, G. and Song, J.H. (2006). Biodegradable packaging based on raw materials from crops and their impact on waste management. *Industrial Crops and Products* [Internet]. 23 (2): 147–161. Available from: https://linkinghub.elsevier.com/retrieve/pii/S0926669005000658.

332 Marichelvam, M.K., Jawaid, M., and Asim, M. (2019). Corn and rice starch-based bio-plastics as alternative packaging materials. *Fibers* [Internet]. 7 (4): 32. Available from: https://www.mdpi.com/2079-6439/7/4/32.

333 Delidovich, I., Hausoul, P.J.C., Deng, L. et al. (2016). Alternative monomers based on lignocellulose and their use for polymer production. *Chemical Reviews* [Internet]. 116 (3): 1540–1599. Available from: https://pubs.acs.org/doi/10.1021/acs.chemrev.5b00354.

334 Bledzki, A. (1999;). Composites reinforced with cellulose based fibres. *Progress in Polymer Science* [Internet]. 24 (2): 221–274. Available from: https://linkinghub.elsevier.com/retrieve/pii/S0079670098000185.

335 Mbiada, A.A.Y., Musa, S., Richter, O. et al. (2018). Controlling surface hydrophobicity of cellulose-lignin composite coatings. *Polymers from Renewable Resources* [Internet]. 9 (2): 51–58. Available from: http://journals.sagepub.com/doi/10.1177/204124791800900201.

336 Yang, J., Ching, Y., and Chuah, C. (2019). Applications of lignocellulosic fibers and lignin in bioplastics: A review. *Polymers* [Internet]. 11 (5): 751. Available from: https://www.mdpi.com/2073-4360/11/5/751.

337 Mwaikambo, L.Y., Tucker, N., and Clark, A.J. (2007). Mechanical properties of hemp-fibre-reinforced Euphorbia composites. *Macromolecular Materials and Engineering* [Internet]. 292 (9): 993–1000. Available from: http://doi.wiley.com/10.1002/mame.200700092.

338 Malathi, A.N., Santhosh, K.S., and Udaykumar, N. (2014). Recent trends of biodegradable polymer: biodegradable films for food packaging and application of nanotechnology in biodegradable food packaging. *Current Trends in Technology and Science* 3 (2): 73–79.

339 Bátori, V., Åkesson, D., Zamani, A. et al. (2018). Anaerobic degradation of bioplastics: a review. *Waste Management* [Internet]. 80: 406–413. Available from: https://linkinghub.elsevier.com/retrieve/pii/S0956053X18305816.

340 Nishinari, K., Fang, Y., Guo, S., and Phillips, G.O. (2014). Soy proteins: a review on composition, aggregation and emulsification. *Food Hydrocolloids* [Internet]. 39: 301–318. Available from: https://linkinghub.elsevier.com/retrieve/pii/S0268005X14000319.

341 Gamero, S., Jiménez-Rosado, M., Romero, A. et al. (2019). Reinforcement of soy protein-based bioplastics through addition of lignocellulose and injection molding processing conditions. *Journal of Polymers and the Environment* [Internet]. 27 (6): 1285–1293. Available from: http://link.springer.com/10.1007/s10924-019-01430-1.

342 Yamada, M., Morimitsu, S., Hosono, E., and Yamada, T. (2020). Preparation of bioplastic using soy protein. *International Journal of Biological Macromolecules* [Internet]. 149: 1077–1083. Available from: https://linkinghub.elsevier.com/retrieve/pii/S0141813019398265.

343 Phillips, G.O. and Williams, P.A. (eds.) (2011). *Handbook of Food Proteins*. Cambridge and Philadelphia: Woodhead Publishing. 432 pp. (Woodhead Publishing Series in Food Science, Technology and Nutrition).

344 Jefferson, M.T., Rutter, C., Fraine, K. et al. (2020). Valorization of sour milk to form bioplastics: friend or foe? *Journal of Chemical Education* [Internet]. 97 (4): 1073–1076. Available from: https://pubs.acs.org/doi/10.1021/acs.jchemed.9b00754.

345 Iwata, T. (2015). Biodegradable and bio-based polymers: future prospects of eco-friendly plastics. *Angewandte Chemie, International Edition* [Internet]. 54 (11): 3210–3215. Available from: http://doi.wiley.com/10.1002/anie.201410770.

346 Castro-Aguirre, E., Iñiguez-Franco, F., Samsudin, H. et al. (2016). Poly(lactic acid) – mass production, processing, industrial applications, and end of life. *Advanced Drug Delivery Reviews* [Internet]. 107: 333–366. Available from: https://linkinghub.elsevier.com/retrieve/pii/S0169409X16300965.

347 Graupner, N., Herrmann, A.S., and Müssig, J. (2009). Natural and man-made cellulose fibre-reinforced poly(lactic acid) (PLA) composites: an overview about mechanical characteristics and application areas. *Composites Part A: Applied Science and Manufacturing* [Internet]. 40 (6–7): 810–821. Available from: https://linkinghub.elsevier.com/retrieve/pii/S1359835X09000840.

348 Casalini, T., Rossi, F., Castrovinci, A., and Perale, G. (2019). A perspective on polylactic acid-based polymers use for nanoparticles synthesis and applications. *Frontiers in Bioengineering and Biotechnology* [Internet]. 7: 259. Available from: https://www.frontiersin.org/article/10.3389/fbioe.2019.00259/full.

349 ETH, V., Rábago, K.R., Glassner, D.A., and Gruber, P.R. (2003). Applications of life cycle assessment to NatureWorks™ polylactide (PLA) production. *Polymer Degradation and Stability* [Internet]. 80 (3): 403–419. Available from: https://linkinghub.elsevier.com/retrieve/pii/S0141391002003725.

350 Hamad, K., Kaseem, M., Yang, H.W. et al. (2015;). Properties and medical applications of polylactic acid: A review. *Express Polymer Letters* [Internet]. 9 (5): 435–455. Available from: http://www.expresspolymlett.com/letolt.php?file=EPL-0005820&mi=c.

351 Gorrasi, G. and Pantani, R. (2017). Hydrolysis and biodegradation of poly(lactic acid). In: *Synthesis, Structure and Properties of Poly(Lactic Acid)* [Internet]. (eds. M.L. Di Lorenzo and R. Androsch), 119–151. (Advances in Polymer Science; vol. 279). Cham: Springer International Publishing Available from: http://link.springer.com/10.1007/12_2016_12.

352 Koller, M., Maršálek, L., de Sousa Dias, M.M., and Braunegg, G. (2017). Producing microbial polyhydroxyalkanoate (PHA) biopolyesters in a sustainable manner. *New Biotechnology* [Internet]. 37: 24–38. Available from: https://linkinghub.elsevier.com/retrieve/pii/S1871678416300115.

353 Tsang, Y.F., Kumar, V., Samadar, P. et al. (2019). Production of bioplastic through food waste valorization. *Environment International* [Internet]. 127: 625–644. Available from: https://linkinghub.elsevier.com/retrieve/pii/S0160412019301357.

354 Koch, D. and B, M. (2018). Assessing the change in environmental impact categories when replacing conventional plastic with bioplastic in chosen application fields. *Chemical Engineering Transactions* [Internet]. 70: 853–858. Available from: http://doi.org/10.3303/CET1870143.

355 Sayyed, R.Z., Wani, S.J., Alyousef, A.A. et al. (2019). Purification and kinetics of the PHB depolymerase of *Microbacterium paraoxydans* RZS6 isolated from a dumping yard. Kumar P, editor. *PLoS One* [Internet]. 14 (6): e0212324. Available from: https://dx.plos.org/10.1371/journal.pone.0212324.

356 McKeen, L.W. and Massey, L.K. (2012). *Permeability Properties of Plastics and Elastomers*, 3e. Amsterdam; Boston: Elsevier, WA. 344 pp.

357 Guarino, V., Gentile, G., Sorrentino, L., and Ambrosio, L. (2017). *Polycaprolactone: Synthesis, Properties, and Applications: Polycaprolactone: Synthesis, Properties, and Applications*. In: John Wiley & Sons, Inc., editor.*Encyclopedia of Polymer Science and Technology* [Internet]., 1–36. Hoboken: Wiley Available from: http://doi.wiley.com/10.1002/0471440264.pst658.

358 Tzeng, J.J., Hsiao, Y.T., Wu, Y.C. et al. (2018). Synthesis, characterization, and visible light curing capacity of polycaprolactone acrylate. *BioMed Research International* [Internet]. 2018: 1–8. Available from: https://www.hindawi.com/journals/bmri/2018/8719624.

359 Azim, H., Dekhterman, A., Jiang, Z., and Gross, R.A. (2006). *Candida a ntarctica* lipase B-catalyzed synthesis of poly(butylene succinate): shorter chain building blocks also work. *Biomacromolecules* [Internet]. 7 (11): 3093–3097. Available from: https://pubs.acs.org/doi/10.1021/bm060574h.

360 Gigli, M., Fabbri, M., Lotti, N. et al. (2016). Poly(butylene succinate)-based polyesters for biomedical applications: A review. *European Polymer Journal* [Internet]. 75: 431–460. Available from: https://linkinghub.elsevier.com/retrieve/pii/S0014305716300167.

361 Moulay, S. (2015). Review: poly(vinyl alcohol) functionalizations and applications. *Polymer-Plastics Technology and Engineering* [Internet]. 54 (12): 1289–1319. Available from: http://www.tandfonline.com/doi/full/10.1080/03602559.2015.1021487.

362 Lucas, N., Bienaime, C., Belloy, C. et al. (2008). Polymer biodegradation: mechanisms and estimation techniques – A review. *Chemosphere* [Internet]. 73 (4): 429–442. Available from: https://linkinghub.elsevier.com/retrieve/pii/S0045653508008333.

363 ISO 17088:2012. 2012. Specifications for compostable plastics. p. 8.

364 CEN - EN 13432. 2000. Packaging – requirements for packaging recoverable through composting and biodegradation – test scheme and evaluation criteria for the final acceptance of packaging. p. 25.

365 Geyer, R., Jambeck, J.R., and Law, K.L. (2017). Production, use, and fate of all plastics ever made. *Science Advances* [Internet]. 3 (7): e1700782. Available from: https://advances.sciencemag.org/lookup/doi/10.1126/sciadv.1700782.

366 Oever M van den, Molenveld K, Zee M van der, and Bos H. (2017). Bio-based and biodegradable plastics: facts and figures: focus on food packaging in the Netherlands [Internet]. Available from: https://doi.org/10.18174/408350.

6

Limitation for Use of Biodegradable Waste onto Soil

Biodegradable Waste Management in the Circular Economy: Challenges and Opportunities, First Edition.
Edited by Malgorzata Kacprzak, Eleonore Attard, Kari-Anne Lyng, Helena Raclavska, BalRam Singh, Eyob Tesfamariam, and Franck Vandenbulcke.

6.1

Organic, Inorganic, and Microbiological Contaminations in Biodegradable Waste

Krzysztof Fijałkowski and Anna Kwarciak-Kozłowska

Faculty of Infrastructure and Environment, Czestochowa University of Technology, Częstochowa, Poland

6.1.1 Directions for the Management of Biodegradable Waste and Biosolids

A growing population directly influences the generation of huge amounts of a variety of solid waste around the world. Accelerated industrialization and significant consumerism also led to the excessive use of natural resources [1, 2]. Research conducted by the World Bank predicts a 1.14–1.73 times increase in the per capita municipal waste generation rate in both developed and developing countries between 1995 and 2025 [3]. It is estimated that around 4 billion tons of solid waste (municipal, industrial, and hazardous waste) is produced annually worldwide, of which municipal solid waste production ranges from 1.6 to 2 billion tons [4]. In 2016, total waste generation in the EU was about 2500 million tons [5]. It is reported that the average amount of municipal waste generated per person in the European Union in 2018 was 486 kg. This amount, however, varies considerably across the Member States of the European Union and worldwide (Table 6.1.1).

Biodegradable wastes are a special type of municipal waste. The definition is that they are any kind of waste that undergoes anaerobic or aerobic degradation. It is estimated that between 120 and 140 million tons of biodegradable waste are produced in the EU each year. This is equivalent to around 300 kg of bio-waste per EU citizen per year [10].

The concept of biodegradable waste is broad and applies not only to food waste generated at the household or industrial level. They also include agro-industrial waste, garden and park waste, animal waste, bioplastic waste, and sewage sludge [11–13]. The general division of solid waste is presented in Figure 6.1.1.

Biodegradable waste is characterized by a high content of organic matter, which makes it suitable for biological recycling. Thermal treatment is also an appropriate method of neutralizing biodegradable waste due to its high energy content (11.6–15.6 TJ Gg^{-1} for solid biomass and 27.4 TJ Gg^{-1} for liquid biomass) [14]. This type of waste contains biodegradable organic fractions with a moisture content below 85–90% [15].

The limit value of organic matter in biodegradable waste share, which allows for its biological treatment, is the loss on ignition >30% [16]. Sewage sludge, often called biosolids, is characterized by a high content of biodegradable organic compounds (from 75% DS for raw sludge and 45 ÷ 55% DS for stabilized sludge) as well as a high content of nitrogen compounds (2 ÷ 7% DS) and phosphorus (1 ÷ 5% DM) [17].

Table 6.1.1 Production and utilization of biosolids and municipal waste target in the EU and selected countries [6–9].

Country/region	Municipal waste generated – kg capita^{-1} in 2018	Share of recycling and composting in 2017	Share of landfill disposal in 2017 – %	Biosolids production – million tons of dry soil year^{-1}	Biosolids utilization rate – %
Australia	—	—	—	0.36	80
Austria	579	48	7	—	—
Belgium	411	30	61	—	—
Bulgaria	423	24	75	—	—
Croatia	432	47	0	—	—
Cyprus	640	6	93	—	—
Czech Republic	351	35	49	—	—
Denmark	766	46	24	—	—
Estonia	405	23	31	—	—
European Union in total	—	—	—	9.0	40
Finland	551	41	26	—	—
France	527	41	1	—	—
Germany	615	16	82	2.3	60
Greece	497	48	26	—	—
Hong Kong, China	—			0.3	0
Hungary	381	28	20	—	—
India	—		—	2.3	80
Ireland	567	58	2	—	—
Italy	499	28	50	—	—
Japan	—	—	—	2.2	74
Latvia	407	54	1	—	—
Lithuania	464	33	54	—	—
Luxembourg	610	68	1	—	—
Malta	640	46	1	—	—
Netherlands	511	43	22	—	—
Poland	329	38	48	—	—
Portugal	508	54	1	—	—
Republic of Korea	—	—	—	1.9	6
Romania	272	34	42	—	—
Singapore	—	—	—	0.12	0
Slovakia	414	35	62	—	—
Slovenia	486	19	80	—	—
South Africa	—	—	—	1.0	80
Spain	475	58	13	—	—
Sweden	434	48	33	—	—
United Kingdom	—	14	71	1.05	85
USA	—	—	—	17.8	55

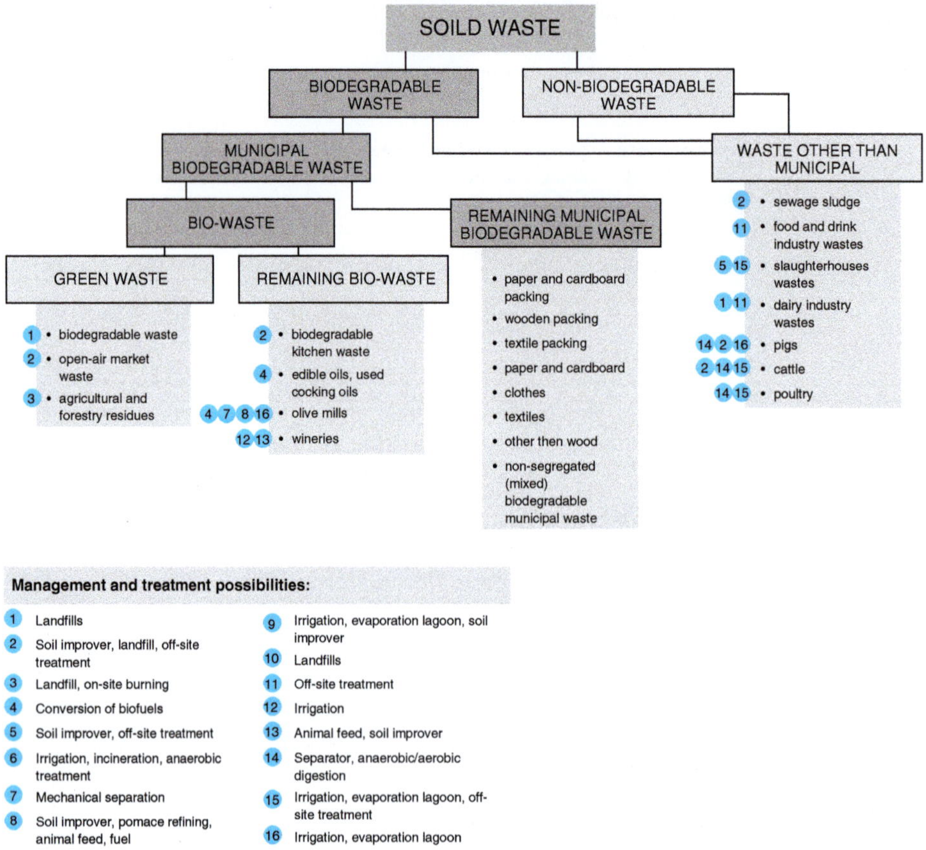

Figure 6.1.1 Division of solid waste, taking into account the directions of their processing or disposal possibilities [1, 2, 7, 11, 12].

In today's world with problems related to natural resources and the energy crisis, the importance and need to develop a new sustainable approach to managing biodegradable waste must be considered as urgent [18]. Improper disposal of these materials poses a serious threat to the quality of the environment. It can cause contamination of surface waters and groundwater as well as contribute to soil degradation [6]. If waste is not properly managed, this problem can also be a huge burden for future generations.

Biodegradable waste is of particular interest to the European Union, where efforts are being made to develop and optimize strategies for their proper use and management. Legislative regulations of the European Union in the field of biodegradable waste management were formulated in Council Directive 1999/31/EC from April 26, 1999 on the landfilling of waste [19] and in Directive 2008/98/EC of the European Parliament and of the Council from November 19, 2008 on the Waste Directive, which contains detailed provisions on recycling targets [20]. According to the directive in question, biodegradable waste is subject to an established hierarchy of handling it. In the first place, waste should be prevented, then prepared for re-use, recycled by other recovery methods, such as energy recovery, and finally neutralized, which is the least desirable solution [21]. The proposed industrial and non-industrial methods of dealing with biodegradable waste and biosolids are presented in Figure 6.1.2.

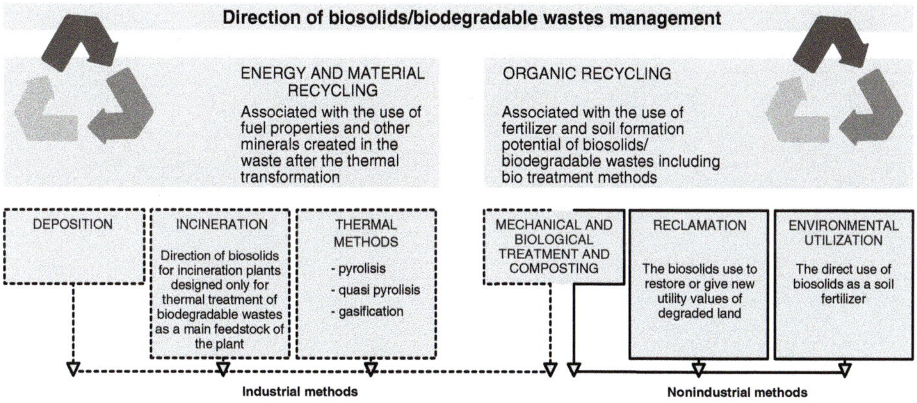

Figure 6.1.2 Direction for biosolids/biodegradable waste management [21–23].

The legal regulations of the European Union in the field of agricultural management of biodegradable waste, including sewage sludge, are mainly Council Directive 86/278/EEC of June 12, 1986 on environmental protection, in particular soil and the so-called sediment directive, which results in significant restrictions for the agricultural and natural use of sewage sludge [24]. The aim of this directive is to promote the use of biodegradable waste and sewage sludge in agriculture, while preventing its harmful effects on human health and the environment (soil and living organisms). It contains requirements for the sewage sludge or biodegradable waste quality that must be met, as well as the quality requirements of the soil on which they are to be applied. Requirements in the use of sewage sludge or biodegradable waste in agriculture are also indicated, related to the possibility of soil contamination. The limit values of heavy metals were determined in the soil where the biodegradable waste, including sewage sludge, is used, as well as the maximum annual amounts of heavy metals that can be introduced into the soil [24, 25].

The directive from year 2008 introduces regulations obliging EU member states to selectively collect biodegradable waste in order to ensure its subsequent recycling. Activities undertaken as part of the green economy are an alternative to the storage of biodegradable waste:

- ecological services for the selective collection of biodegradable waste,
- green investments generating jobs in the processing of biodegradable waste, e.g. plant composting, biorefineries, and mechanical and biological installations processing mixed municipal solid waste,
- ecological products in the form of composts (fertilizers and soil conditioners),
- heat and electricity obtained during fermentation of biodegradable waste [26, 27].

Improper disposal of biodegradable waste not only creates a serious threat to the natural environment, but also causes a loss of economic value of waste. As organic waste is, among others, a large amount of organic matter and valuable plant nutrients, recycling in agriculture seems advisable [28, 29]. The experience of countries with well-established recycling of biodegradable waste shows that the preferred recycling option is the production of compost. It is a relatively easy and cost-effective strategy to implement locally. Composting is a biological conversion of heterogeneous biodegradable waste into humus-like substances with the help of microorganisms such as

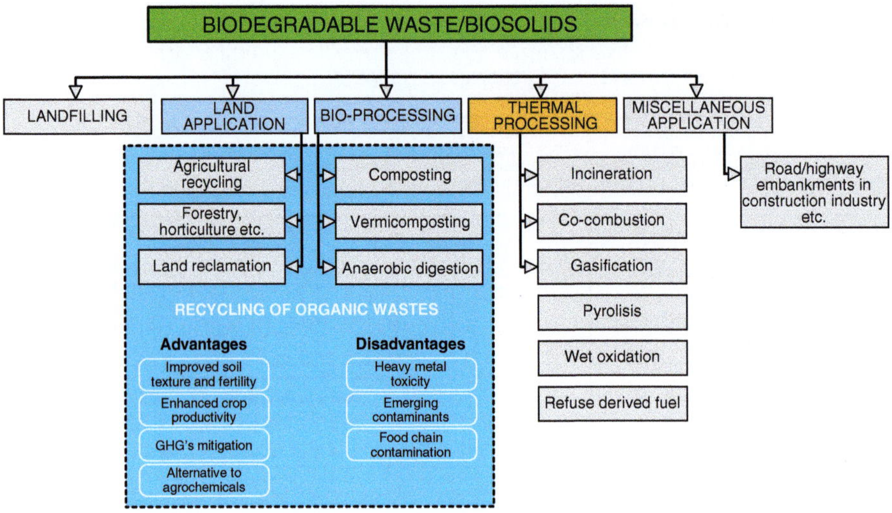

Figure 6.1.3 Distribution of biodegradable wastes considering the advantages and disadvantages of land application [2, 6, 22].

bacteria, fungi, and actinomycetes under controlled conditions of humidity, temperature, and aeration [2, 30–32]. The same is true of the vermicomposting process in which earthworms are additionally used and the resulting vermicompost is an excellent, nutrient-rich organic fertilizer or soil conditioner [33, 34].

The anaerobic stabilization of biodegradable waste consisting in its simultaneous stabilization and energy recovery is also popular [35]. The decomposition of organic matter contained in the waste takes place in four successive stages, i.e. hydrolysis, acidogenesis, acetateogenesis, and methanogenesis, which result in the production of methane and a large amount of post-fermentation material, the so-called digestate, which can be used as an additive to fertilizers [36]. The use of the above-mentioned recycling methods allows the volume of generated waste to be reduced, as a result of which we obtain an organic fertilizer that can be used in broadly understood agriculture [37]. Figure 6.1.3 presents the possibilities of handling biodegradable waste, taking into account the advantages and disadvantages of their use in land application and bioprocesses.

6.1.2 The Effect of Biodegradable Waste Addition on Soils

Degraded soils are very often deprived of organic matter and basic elements such as nitrogen, potassium, and phosphorus (NPK) necessary for the development of vegetation and the circulation of matter. Their standard source is artificial fertilizers. Growing awareness of the adverse effects of chemicals used in agriculture on human health has increased interest in organic farming, which also helps protect the environment through the rational use of natural resources [38–41]. A number of concerns related to the use of inorganic nitrogen fertilizers include, inter alia, with the problem of eutrophication, soil carbon losses, acidification and increased salinity of soils, and greenhouse gas emissions. Therefore, alternative sources of NPK are sought [42, 43]. Due to the high content of organic compounds (>50%) and macro- and microelements, biodegradable wastes such as food waste, bio-waste, manure, or sewage sludge seem to be interesting substitutes for artificial fertilizers. This waste is a valuable organic resource that can be recycled and transformed into cheap and rich fertilizer and/or soil conditioner [22, 44–46]. The usability of sewage sludge as a fertilizer is compatible with ecology and leads to the reuse of potentially hazardous waste [38, 47, 48]. The use of bio-waste compost or sewage sludge, on the one hand, provides valuable nutrients and organic matter to agricultural soil and, on the other hand, enables the reuse of wastewater treatment plant by-products and constitutes a sustainable approach to the management of large amounts of generated waste [32, 49].

Ayres et al. [50] studied natural and anthropogenic global nitrogen cycles and concluded that human activities doubled nitrogen fixation, mainly through the production of industrial fertilizers. Organic matter contained in sewage sludge, due to the relatively low content of lignin/cellulose, quickly decomposes and mineralizes, producing and increasing the concentration of nitrates in the soil [51, 52].

Pedrol et al. [53] reported that, unlike nitrogen, phosphorus or potassium fertilizers, sewage sludge/organic waste compost contributes to the preservation of the diversity and richness of plant species and significantly increases the metabolic activity of the flora (e.g. on lignite mine soils, i.e. in Spain). It was found that the added compost gradually corrected the phosphorus limitations and significantly improved plant production compared to unused soils. As a result of the degradation of organic matter, phosphorus and nitrogen are released more slowly than in the case of inorganic fertilizers, thanks to which these compounds are available for longer. Phosphorus rock reserves, a non-renewable and potentially scarce resource, have been estimated to last in the range of 120–710 years at current use rates [54]. According to, for example, the Swedish EPA, phosphorus contained in sewage sludge accounts for 15% of the needs of Swedish agriculture. Overall, it has been estimated that the amount of phosphorus in biodegradable waste is one third of chemical fertilizers imported into Sweden [55–57]. Organic and inorganic forms of phosphorus present in sewage sludge can be additionally enriched by pyrolysis of the sludge itself, which additionally produces a carbon-rich product with a porous structure (biochar) [58].

Biodegradable waste/sewage sludge, which are rich in essential nutrients for plants, may change the physico-chemical and biological properties of soils. Humic acid contained in the compost from biodegradable waste increases the cation exchange capacity (CEC) and the buffer capacity of the soil [59–61]. About half of the solid fraction of solid waste consists of organic matter, which has a significant effect on the physical properties of the soil, such as increasing water retention and movement, improving soil porosity, and soil aggregation. They also contribute to the improvement of soil properties, such as bulk density, aeration, or an increase in electrical conductivity [62]. The addition of organic matter in the form of biodegradable waste improves water infiltration and hydraulic conductivity. It is the main opportunity for carbon sequestration by adding stable organic matter and promoting plant growth (their biostimulation) during the assisted phytoremediation processes [63, 64]. The application of sewage sludge also affects the activity of important soil enzymes (arylsulfatase, phosphatase acid, and alkaline phosphatase).

On soil, added composts based on biodegradable waste may also increase the biomass of microbes [64–66] and reduce the amount of plant diseases [67]. The application of sewage sludge to soil modifies the chemical environment, community structure and functions of prokaryotes and eukaryotes.

The use of biodegradable waste can also be successfully added in soils with a seasonal deficiency of organic components. An example is the soils of the Mediterranean in summer [68]. The addition of biosolids/biodegradable waste to soils, especially degraded ones, can bring many benefits and the most important ones are presented in Figure 6.1.4.

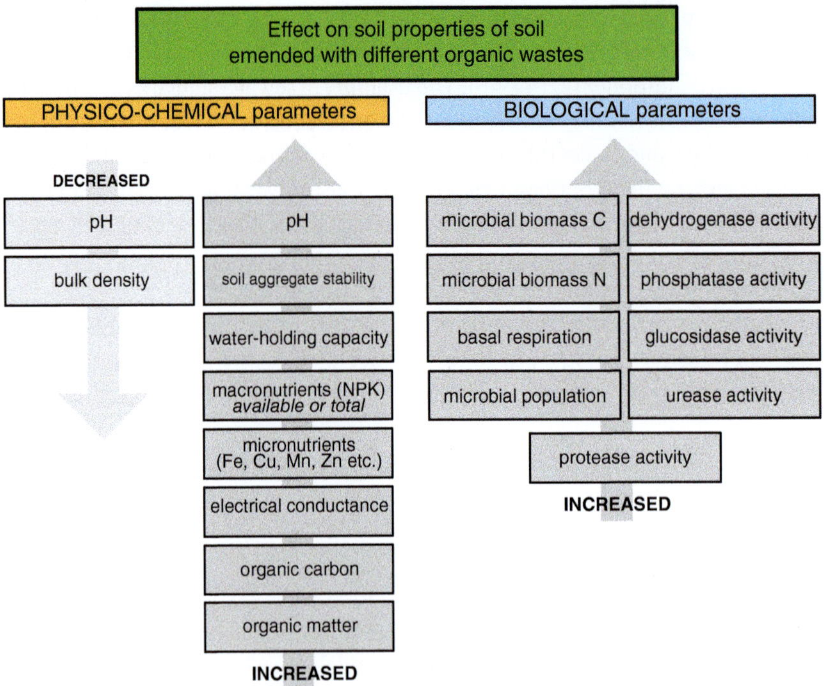

Figure 6.1.4 Effect of biosolids/biodegradable waste on degraded soils [63, 69–71].

6.1.3 Restrictions on the Use of Biodegradable Waste/Biosolids to Soil (Organic, Inorganic, and Microbiological Contaminations)

In biodegradable waste, including sewage sludge, that are rich in organic matter, macro- and micronutrients can be a potential substitute for conventional fertilizers and increase the yield of various crops [72–74]. However, many wastewater treatment plants in Europe are unable to obtain sewage sludge fit for landfill [75].

Sewage sludge as well as biodegradable waste may contain various pollutants, including heavy metals, i.e. Cd, Cu, Zn, Pb, which may be a serious limitation in the case of their use on arable land. Taking into account the concentrations of heavy metals present in sewage sludge, they can be generally summarized as follows: Zn > Cu > Cr > Ni > Pb > Cd or Zn > Cr > Pb > Cu > Ni > Cd [49, 76].

Heavy metals in concentrations higher than the permissible limits can contaminate the food chain. After exceeding the threshold of permissible limits in soil, their negative impact on the existing organisms and the environment is observed. The concentration of heavy metals in sewage sludge as well as in organic wastes generated in different countries may differ due to different technological advancement of production processes or different wastewater treatment technology (Table 6.1.2) [72, 73].

Regarding heavy metals, the limit values of their concentrations in soils where sewage sludge/biodegradable bio-waste as fertilizer are to be used have been defined in Table 6.1.3. Depending on the nature of their use (fertilization/soil fertilization and agrotechnical processing of them into compost), the permissible concentration of heavy metals is presented in Table 6.1.4 [77, 78].

Table 6.1.2 A comparison of the physico-chemical properties of biosolids of selected countries [1, 2, 7, 22].

Properties	India	China	Australia	Spain
pH	6.16–7.5	6.86–8.73	4.4–8.3	7.1–8.1
EC (ms cm^{-1})	2.28–2.7	0.667–5.01	1.6–7.9	1.2–3.9
Organic C (%)	5.52–12.6	—	—	—
Total N (%)	1.6–1.73	2.23–6.50	0.60–2.5	3–4.1
Total P (%)	0.49–1.3	1.06–2.18	0.28–0.83	2–3.6
Average K (%)	132–716.7	—	—	13 900
Total K (%)	0.8–1.26	0.46–0.62	0.18–0.45	0.24–0.47
Exchangeable K (mg kg^{-1})	208.9	593	—	—
Exchangeable Na (mg kg^{-1})	483	—	—	—
Exchangeable Ca (mg kg^{-1})	154.1	—	—	—
Fe (mg kg^{-1})	6059–14 390	0.46–2.40	13 824–18 026	31 200
Ni (mg kg^{-1})	47.17–60	52.5–202	166	<25–71
Mn (mg kg^{-1})	186.2–260	0.35–537	173	165–233
Zn (mg kg^{-1})	161–2050	0.21–1350	210–3060	560–1100
Pb (mg kg^{-1})	28.5–240	49.1–186	323	43–219
Cr (mg kg^{-1})	35.5–60	52.8–288	308	1–210
Cd (mg kg^{-1})	32.3–154.5	2.23–7.61	0.70–13.6	<0.2–3
Cu (mg kg^{-1})	186–330	0.27–975	92–1996	149–230

Table 6.1.3 Comparison of the permissible metal concentration values in relation to EU law [77, 78].

	Permissible concentrations – Mg kg^{-1} d.m.			
Heavy metals	**Light soils**	**Medium soils**	**Heavy soils**	**Directive 86/278/EWG**
Cd	1	2	3	1–3
Cu	25	50	75	50–140
Ni	20	35	50	30–75
Pb	40	60	80	50–300
Zn	80	120	180	150–300
Hg	0.8	1.2	1.5	1–1.5
Cr	50	75	100	-

The issue of the micronutrients and heavy metals concentration in composts based on biodegradable wastes/biosolids requires special attention when used in agriculture. If the concentration of heavy metals is high in the sludge itself, then its successive addition to the soil in the form of such prepared compost may cause their toxic accumulation. Inorganic pollutants are not biodegradable; therefore, they can accumulate in the soil and, consequently, can get into the food chain.

Table 6.1.4 Permissible content of heavy metals in sewage sludge intended for agricultural and non-agricultural use [77, 78].

Heavy metals	Method of natural use of sewage sludge		
	Fertilization, fertilization and land reclamation		Agrotechnical processing of sewage sludge into compost, plant preservation of land surface
	Agricultural use	Non-agricultural use	
	Concentrations of heavy metals, mg kg^{-1} d.o.m		
Pb	500	1000	1500
Cd	10	25	50
Cr	500	1000	2500
Cu	800	1200	2000
Ni	100	200	500
Hg	5	10	25
Zn	2500	3500	5000

Bioconcentration of heavy metals in soil has a phytotoxic effect on cultivated plants [74, 79]. As a consequence, a slight increase in heavy metals in the soil increases the amount of Zn, Cd, Pb, and Cu in plants. Their concentration is higher in the roots compared to other parts of the plant. This is due to the fact that the soil solution is absorbed by vascular plants, mainly through the roots. However, it cannot be ruled out that in other parts of the plant there is no high accumulation of heavy metals [2].

In addition to heavy metals like Cd, Cu, Zn, Pb, Cr, Mn, Fe, Co, Ni, metalloids such as As, Se, or non-metals and light metals such as Al are important from the point of view of toxicity. Interestingly, in some soils, e.g. with a copper deficiency, heavy metals in sewage sludge may correct the deficiency of this element [80].

In the soil, the introduced metals may move. Factors influencing their movement include: soil pH, temperature, CEC or organic matter content. Soluble organic matter has two groups of exchange sites, one of which binds Ca, Mg, Zn, Ni, Co, Mn, Cd, Pb, and Fe, and the rest bind only Cu, Pb, and Hg [79].

Increasing the amount of organic matter contributes to the binding of heavy metals in the soil, creating organometallic complexes, which consequently reduces their mobility and phytotoxicity. However, fresh manure or unripe compost, with a relatively high content of soluble organic matter, may increase the mobility of metals [2].

The most important factor controlling the movement of metals in the soil is pH, and its measurement allows you to control the bioavailability of heavy metals. The reaction affects the precipitation and dissolution of insoluble hydroxides and the formation of organic mineral complexes [79, 81]. Research conducted by Sniffer 2007 suggests that composts from sewage sludge should not be used on agricultural soils with a pH lower than 5 [82].

Moisture, structure, and soil type have a great influence on starting metals. The bioavailability of Cd is limited in calcareous soil. This is probably due to the combination of chemisorption, $CdCO_3$ precipitation [79, 83–85], and competition with Ca^{2+} ions for absorption sites on the plant root. The labile fraction of heavy metals in soil is the most toxic to plants. Therefore, it is important to reduce this fraction in the soil contaminated with sewage sludge to reduce their negative effects [79].

It was found that, apart from the composition of composts based on bio-waste or sewage sludge, the depth of compost dosing affects the impact of heavy metals contained in them on the soil. When such composts/fertilizers are applied to the soil, an initial balance is established between the soil and the liquid phase of the heavy metal. The strong bond between them means that heavy metals remain in the places where they are dosed (upper layers). It has been found that mean layers of soil enriched in heavy metals are 0–15 and 15–40 cm, approximately equivalent to the depth of incorporation of the compost [79, 80].

Composted sewage sludge is also associated with the problem of potentially toxic and toxic organic pollutants. It is particularly important due to their diverse influence on living organisms (mutagenic, carcinogenic, and endocrine disrupting effects) [79, 86, 87].

Sewage sludge contains in the range of less than 1 ppm to over 1000 ppm of over 60 000 kinds of toxic chemicals. In addition to heavy metals, there are other harmful organic pollutants, including polycyclic aromatic hydrocarbons (PAHs), polychlorinated biphenyls (PCBs), dioxins, adsorbed organic halides (AOX), pharmaceuticals, hormones, nanoparticles, and microplastics [49, 88–90].

Fourteen pathways of organic pollutants that have an impact on human health after entering soil with compost have been identified; the five most important ones are presented below:

- Lane 1: sewage sludge–soil–plant–human (from agricultural use)
- Lane 2: sewage sludge–soil–plant–human (domestic use of the sludge)
- Lane 3: sewage sludge–soil–human
- Lane 4: sewage sludge–soil–plant–animal–human
- Lane 5: sewage sludge–soil–animal–human [91]

The level of phytotoxicity of organic compounds depends on the type of soil, plant species, and the nature of sewage sludge [87]. There are also soil mixed with sewage sludges that show different solubility and absorption potential by plants and are susceptible to degradation processes in soil or volatilization or leaching processes. Potentially toxic organic compounds in sludge-based compost and biodegradable waste are usually present in very low concentrations and their behavior when mixed with soil can be divided into three broad categories:

1. Volatile compounds that are quickly released into the atmosphere,
2. Compounds that are rapidly mineralized by microorganisms,
3. Persistent compounds that are strongly adsorbed to the soil [91–94].

Degradation of many potentially toxic organic compounds contained in soil from sewage sludge is fast, but it can be intensified by adding appropriate organic waste, making it difficult for plants to absorb/uptake them [91]. The compounds of the third group – stable compounds – are very often characterized by low solubility and due to the fact that they are strongly adsorbed in the soil, they are not transferred to plant tissues [91, 95].

Recently, the so-called emerging organic (EC) pharmaceuticals (PhC) and personal care products (PCP) are of the greatest interest to researchers. Significant amounts of painkillers/anti-inflammatory drugs, antihistamines, hormones, antiseptics, antifungal drugs, hypnotics, anticancer drugs, lipid regulators, and hypertensive drugs are found in sewage sludge. Significant amounts of antibiotics, contrast agents, non-ionic surfactants, and insect repellants are also measured [49, 96]. Some of these compounds such as Ibuprofen, Diazepam, Iopromide, pesticides, and detergents (such as Triclosan) are biotransformed. Compounds such as estrogens, synthetic musks, and alkyl benzene sulphonates are resistant to biodegradation [2]. Research is still ongoing to determine the absorption potential of PPCP by crops from land where bio-based products have been used. It has been found that PPCPs are transferred to the aboveground tissues after absorption [2, 97, 98]. It is connected with the risk of the accumulation of dioxins and pharmaceuticals in soils and their subsequent translocation to various trophic levels of the food chain [99]. The more commonly used silver nanoparticles, titanium dioxide, and zinc oxide in the cosmetics, medical, or textile industries result in their appearance in sewage sludge [49]. Their presence in the soil can cause irreversible effects, such as aging. The behavior of silver nanoparticles (NC-Ag) in soil, for example, may differ due to the uniqueness of each soil, i.e. its biological, physical, and chemical properties. The transport of nanoparticles in the porous soil structure is hindered by the phenomenon of silver aggregation and dissolution [100–102]. NC-Ag oxidation in neutral soil is usually low. The resulting Ag^+ ions adsorb on the surface of the nanoparticles and migrate deep into the soil. It is not clear to what extent the silver ions and other nanoparticles are transferred into the soil solution. The issues of dependence of silver nanoparticles with soil are based on, among others on the hypothesis of Klitzke et al. [103].

According to the Klitzke hypothesis, the following dependencies must be taken into account:

- The presence of organic matter in the soil inhibits the oxidation of metallic silver, i.e. reduces the concentration of Ag ions on the surface of the nanoparticle.
- The NC-Ag sorption process on soil particles depends on the surface charge of nanoparticles [103].

Sulfidated NC-Ag reduces the solubility of Ag in soil and reduces the toxicity [100, 104]. The penetration of nanoparticles into plant cells depends on the dimensions of the pores on the cell walls, ranging from 5 to 20 nm. However, some nanomaterials are capable of influencing the chemical composition of the plant cell wall. This can result in the formation of new, larger pores and an increase in permeability. Selected nanoparticles, such as CeO_2, ZnO, or Zn, may have a toxic effect on plants exposed to their exposure, causing inhibition of germination and rooting. The harmful effects of TiO_2 on plants after the interaction with the building blocks of the cell wall include limitation of H_2O transport and reduction of transpiration, as well as inhibition of leaf growth [105]. Table 6.1.5 compares the legal restrictions from different countries for compost due to the heavy metal and organic pollutant content.

Sewage sludge used as soil additives and fertilizers should also be analyzed for the degree of contamination with pathogenic microorganisms, e.g. parasite eggs of *Ascaris lumbricoides*, *Trichuris* sp., and *Ascaris Toxocara* sp., significant amounts of fungi such as *Penicillium*, *Aspergillus*, *Trichoderma*, *Verticillium*, *Mortierelta*, *Fusarium*, *Mucor*, and *Geotrichum*, and bacteria, e.g. *Escherichia* sp., *Salmonella* sp., and *Shigella* sp. Viruses are also a large group of pathogens present in the sewage sludge produced in municipal sewage treatment plants, e.g. HIV and HCV, rotaviruses, and coronaviruses [22, 106–109].

The largest and most diverse group in sewage sludge are heterotrophic and chemautotrophic bacteria, saprophytes, and pathogens. The dominant types of microorganisms are *Gemmatimonas*, *Nitrosomonas*, *Thermomonas*, and *Truepera*, as well as *Nitrosomonas*, *Opitutus*, *Nitrospira*, and *Truepera* [49, 110]. Pathogenic viruses and bacteria keep the so-called the state of virulence, i.e. virulence that lasts for several months. Protozoa and tapeworms remain invasive for about one year, and eggs of parasitic nematodes for several years. Even very drastic physical and chemical treatments against pathogenic organisms in the wastewater are ineffective. The eggs of parasitic nematodes – human roundworm (*Asariuis lumbricoides*) and canine roundworm (*Toxocara canis*) – are the most difficult to destroy [77]. The main groups of pathogenic organisms found in sewage sludge are: intestinal bacteria, parasites, viruses, and fungi (Table 6.1.6).

Table 6.1.5 Comparison of law limits for heavy metals and organic pollutants in compost from different countries and compost produced [6, 13, 72–74].

Country	Compost produced – tones (year)	Heavy metals – mg kg $^{-1}$ dry compost								Organic pollutants			
		Cd	Pb	Zn	Cu	Cr	Cu	Ni	Hg	PAH – mg kg^{-1}	PCBs – mg kg^{-1}	AOX – mg kg^{-1}	DEHP – mg kg^{-1}
Australia	—	3	200	250	200	—	200	—	—	6	1	—	—
Austria	634 400 (2005)	0.7–1	45–120	200–500	75–150	70	70–150	25–60	0.4–0.7	6	0.2	500	—
Belgium	—	5	600	1000	100	150	100	50	—	5	0.15	250	—
Canada	—	3	150	500	100	210	100	62	0.8	—	—	—	—
Denmark	374 530 (2008)	0.8	60–120	4000	1000	—	1000	30	0.8	3	0.08		
EU – European Commission	—	0.7–1.5	100–150	200–400	100–150	—	100–150	—	—	—	—	—	—
France	2 490 000 (2005)	8	800	—	—	—	—	200	8	—	0.8	—	—
Germany	4 384 400 (2008)	2	200	400	150	150	150	50	1	—	—	—	—
India	—	5	100	1000	300	50	300	50	0.15	—	—	—	—
Luxembourg	20 677 (2005)	1.5	150	400	100	100	100	50	1	10	0.1		
Netherlands	1 603 464 (2008)	0.7–1	65–100	75–200	25–60	50	25–60	10–20	0.2–0.3	—	—	—	—
Poland	—	5	350	1500	300	—	300	—	—	—	—	—	—
Spain	610 148 (2008)	10	300	1100	450	400	450	120	7	—	—	—	—
Sweden	199 700 (2008)	3	150	500	150	150	150	50	3	—	—	—	—
UK	203 600 (2005/06)	1.5	150	400	200	100	200	50	1	—	—	—	—
USA	—	4	150	400	300	100	300	50	0.5	—	—	—	—

Table 6.1.6 The occurrence of different groups of microorganisms in sewage sludge in relation to soil biota [49].

Group of microorganism	Typical subspecies in sewage sludge	Occurrence
Viruses	Polio virus	
	Coxsackie	
	Influenza virus	
	Adenovirus	
	Astrovirus	$5.8–32.5 \times 10^3$ cfu l^{-1}
	Calicivirus	
	Coronavirus	
	Enterovirus	
	Parwovirus	>8.3 cfu 10 g DM^{-1}
	Reovirus	
	Rotavirus	
	Norwalk virus	$26–30 \times 10^4$ cfu l^{-1}
	Hepatitis A virus	
	Hepatitis E virus	
	TOTAL	
Bacteria	*Arizona hinshawii*	
	Aeromonas spp.	
	Bacillus cereus	
	Bacillus anthracis	
	Brucella spp.	
	Campylobacter jejuni	10^6 gene ml^{-1}
	Citrobacter spp.	
	Clostridium botulinum	
	Clostridium perfringens	
	Clostridium spp.	$1–1000 \times 10^3$ gene ml^{-1}
	Enterobacteriaceae	$1–50 \times 10^7$ gene ml^{-1}
	Escherichia coli	
	Klebsiella spp.	
	Legionella sp.	
	Leptospira icterohaemorrhagiae	
	Listeria monocytogenes	
	Mycobacterium tuberculosi	
	Pasteurella pseudotuberculosis	
	Proteus spp.	
	Providencia spp.	5×10^4 gene ml^{-1}
	Pseudomonas aeruginosa	
	Salmonella spp.	
	Serratia spp.	
	Shigella spp.	
	Staphylococcus aureus	
	Enterococcus spp.	
	Vibrio parahaemoliticus	
	Vibrio cholerae	
	Yersinia enterocolitica	
	TOTAL	

Table 6.1.6 (Continued)

Group of microorganism	Typical subspecies in sewage sludge	Occurrence
Fungi	*Absidia* spp.	
	Aspergillus spp.	
	Candida albicans	10^2–10^3 cfu g DW^{-1}
	Candida guillermondii	
	Candida krusei	
	Candida tropicalis	
	Cryptococcus neoformans	
	Epidermophyton spp.	
	Fusarium spp.	
	Geotrichum candidum	
	Microsporum spp.	
	Mucor spp.	
	Penicillium spp.	
	Phialophora richardsii	
	Trichoderma spp.	
	Trichosporon cutaneum	114.3–752.3 × 10^6 cfu ml^{-1}
	Trichophyton spp.	
	Verticillium spp.	
	TOTAL	
Protozoa	*Acanthomoeba*	
	Dientamoeba fragilis	
	Entamoeba hystolitica	
	Giardia lamblia	22–32 oocysts g^{-1}
	Giardia intestinalis	
	Isospora belli	
	Naeglaria fomleri	
	Palantidium coli	
	Sarcocystis spp.	
	Toxoplasma gondii	
	Cryptosporidium	12–17
	TOTAL	12–32
Helminths	*Ankylostoma duodenale*	
	Ascaris lumbricoides	
	Echinococcus granulosus	
	Echinococcus multilocularis	
	Enterobium vermicularis	
	Hymenolepsis spp.	1.5–6 eggs g^{-1}
	Necator americanus	
	Strongyloides stercoralis	
	Taenia saginata	
	Taenia solium	
	Toxocara spp.	
	Trichuris spp.	13–94
	Ascaris spp.	312–776
	TOTAL	0–2910

6.1.4 Conclusion

The use of biodegradable waste in agricultural soils can be sustainable and economical based on nutrient recycling and disposal of sewage sludge waste. However, there is also a potential risk of contaminants present in sediment, and there is growing concern that some of them may accumulate in the soil, be ingested by plants and animals grazed, and passed on to humans through the food chain. The use of biodegradable waste sludge is controlled or banned under EU legislation and others, which will help reduce their current release to the environment. In addition, the nature of the sediments changes over time and therefore stringent and periodic monitoring of products derived from biodegradable waste for use in the environment should be carried out. Overall, the potential management of waste depends on several levels, such as disposal, recovery, recycling, re-use, and prevention. It is obvious that the impact of biodegradable waste causes an increase in agricultural productivity to a certain level; however, the use of this waste in main food production and supply chains requires even more research.

References

1 Singh, R.P., Sharma, B., Sarkar, A. et al. (2014). Biological responses of agricultural soils to fly-ash amendment. In: *Reviews of Environmental Contamination and Toxicology*, vol. 232, 45–60. Cham: Springer.

2 Sharma, B., Vaish, B., Singh, U.K. et al. (2019). Recycling of organic wastes in agriculture: an environmental perspective. *International Journal of Environmental Research* 13 (2): 409–429.

3 Hoornweg, D. and Bhada-Tata, P. (2012). What a waste: a global review of solid waste management. *Urban Development Series Knowledge Papers* 15: 1–98.

4 Vaish, B., Sarkar, A., Singh, P. et al. (2016). Prospects of biomethanation in Indian urban solid waste: stepping towards a sustainable future. In: *Recycling of Solid Waste for Biofuels and Bio-Chemicals* (eds. O. Parthiba, K. Kirsten, H. Subramanian and S. Muthu), 1–29. Singapore: Springer.

5 Zeschmar-Lahl, B. (n.d.) *Wastewater Treatment Residues as Resources for Biorefinery Products and Biofuels.* https://doi.org/10.1016/B978-0-12-816204-0.00001-1.

6 Sharma, B., Sarkar, A., Singh, P., and Singh, R.P. (2017). Agricultural utilization of biosolids: a review on potential effects on soil and plant grown. *Waste Management* 64: 117–132.

7 Singh, R.P., Singh, A., and Srivastava, V. 2017. *Environmental Issues Surrounding Human Overpopulation.* IGI Global, USA.

8 Municipal waste by waste management operations, Eurostat. Retrieved from: https://ec.europa.eu/eurostat/databrowser/view/env_wasmun/default/table?lang=en (accessed December 2020).

9 Srivastava, V., Ferreira de Araujo, A.S., Vaish, B. et al. (2016). Biological response of using municipal solid waste compost in agriculture as fertilizer supplement. *Reviews in Environmental Science and Biotechnology* 15: 677–696.

10 Managing municipal solid waste – a review of achievements in 32 European countries. EEA Report No. 2/2013. 2013. Retrieved from: http://www.eea.europa.eu/publications/managing-municipalsolid-waste (accessed November 2013).

11 Wojnowska-Baryła, I., Kulikowska, D., and Bernat, K. (2020, 2088). Effect of bio-based products on waste management. *Sustainability* 12 (5).

12 Mohammed, M., Donkor, A., and Ozbay, I. (2018). Bio-drying of biodegradable waste for use as solid fuel: a sustainable approach for green waste management, agricultural waste

and residues, Anna Aladjadjiyan, IntechOpen. doi: 10.5772/intechopen.77957. Available from: https://www.intechopen.com/books/agricultural-waste-and-residues/bio-drying-of-biodegradable-waste-for-use-as-solid-fuel-a-sustainable-approach-for-green-waste-manag.

13 Kopeć, M., Chmiel, M., Gondek, K. et al. (2015). Factors influencing composting poultry waste. *Journal of Ecological Engineering* 16 (5): 93–100.

14 Intergovernmental Panel on Climate Change (IPCC) (2006). *2006 IPCC Guidelines for National Greenhouse Gas Inventories*. Prepared by the National Greenhouse Gas Inventories Programme, Eggleston H.S., Buendia L., Miwa K., Ngara T., and Tanabe K. (eds.). Japan: IGES.

15 Mata-Alvarez, J., Mace, S., and Llabres, P. (2000). Anaerobic digestion of organic solid wastes. An overview of research achievements and perspectives. *Bioresource Technology* 74 (1): 3–16.

16 Jędrczak A. 2008. Biologiczne przetwarzanie odpadów, *Wydawnictwa PWN Naukowe*.

17 Bień J, Kacprzak M, Kamizela T, et al. (2015). Komunalne osady ściekowe – zagospodarowanie energetyczne i przyrodnicze. *Wydawnictwo Politechniki Częstochowskiej*.

18 Pappu, A., Saxena, M., and Asolekar, S.R. (2007). Solid wastes generation in India and their recycling potential in building materials. *Building and Environment* 42 (6): 2311–2320.

19 European Union Directive 1999/31/WE.

20 Directive of the European Parliament and of the Council of 19 November 2008 (2008/98/CE).

21 Kwarciak-Kozłowska, A. (2019). Co-composting of sewage sludge and wetland plant material from a constructed wetland treating domestic wastewater. In: *Industrial and Municipal Sludge*, Butterworth-Heinemann. pp. 337–360.

22 Kacprzak, M., Neczaj, E., Fijałkowski, K. et al. (2017). Sewage sludge disposal strategies for sustainable development. *Environmental Research* 156: 39–46.

23 Kosobucki, P., Chmarzynski, A., and Buszewski, B. (2000). Sewage sludge composting. *Polish Journal of Environmental Studies* 9 (4): 243–248.

24 Council Directive 86/278/EEC Of 12 June 1986 on the protection of the environment, and in particular of the soil, when sewage sludge is used in agriculture.

25 Kacprzak, M. and Fijałkowski, K. (2009). Mycorrhiza and sewage sludge effect on biomass of sunflower and willow during phytoremediation of degraded terrains within zinc foundry zone. *Environment Protection Engineering* 35 (2): 181–187.

26 Przystupa, A. (2018). Biodegradable waste management as a practical example of green economy. Organizacja i Zarządzanie: kwartalnik naukowy. *Scientific Quarterly Organization and Management* 1 (41) https://doi.org/10.29119/1899-6116.2018.41.10.

27 Grobelak, A. and Jaskulak, M. (2019). Sludge multifunctions in a phytobiome – forest and plantation application including microbial aspects. In: *Industrial and Municipal Sludge* (eds. M. Narasimha, V. Prasad, P.J. de Campos Favas, et al.), 323–336. Butterworth-Heinemann.

28 Petersen, S.O., Henriksen, K., Mortensen, G.K. et al. (2003). Recycling of sewage sludge and household compost to arable land: fate and effects of organic contaminants, and impact on soil fertility. *Soil & Tillage Research* 72: 139–152.

29 Liu, T., Liu, B., and Zhang, W. (2014). Nutrients and heavy metals in biochar produced by sewagesludge pyrolysis: its application in soil amendment. *Polish Journal of Environmental Studies* 23 (1): 271–275.

30 Atalia, K.R., Buha, D.M., Bhavsar, K.A., and Shah, N.K. (2015). A review on composting of municipal solid waste. *Journal of Environmental Science, Toxicology and Food Technology* 9: 20–29.

31 Yu, H. et al. (2019). The changes in carbon, nitrogen components and humic substances during organic-inorganic aerobic co-composting. *Bioresource Technology* 271: 228–235.

32 Picariello, E., Pucci, L., Carotenuto, M. et al. (2021). Compost and sewage sludge for the improvement of soil chemical and biological quality of Mediterranean agroecosystems. *Sustainability* 13 (1): 26.

33 Garg, V.K., Suthar, S., and Yadav, A. (2011). Management of food industry waste employing vermicompost technology. *Bioresource Technology*, https://doi.org/10.1016/j.biortech.2011.11.116.

34 Manyuchi, M.M. and Phiri, A. (2013). Vermicomposting in solid waste management: a review. *International Journal of Scientific Engineering and Technology* 2 (12): 1234–1242.

35 Ferguson, R.M., Coulon, F., and Villa, R. (2016). Organic loading rate: a promising microbial management tool in anaerobic digestion. *Water Research* 100: 348–356.

36 Tambone, F., Scaglia, B., D'Imporzano, G. et al. (2010). Assessing amendment and fertilizing properties of digestates from anaerobic digestion through a comparative study with digested sludge and compost. *Chemosphere* 81 (5): 577–583.

37 Diacono, M. and Montemurro, F. (2010). Long-term effects of organic amendments on soil fertility: a review. *Agronomy for Sustainable Development* 30: 401–422.

38 Kaur, T. (2020). *Vermicomposting: An Effective Option for Recycling Organic Wastes*. Organic Agriculture. IntechOpen.

39 Reganold, J.P., Palmer, A.S., Lockhart, J.C., and Macgregor, A.N. (1993). Soil quality and financial performance on biodynamic and conventional farms in New Zealand. *Science* 260: 344–349. https://doi.org/10.1126/science.260.5106.344.

40 Mader, P., Fliessbach, A., Dubois, D. et al. (2002). Soil fertility and biodiversity in organic farming. *Science* 296: 1694–1697. https://doi.org/10.1126/science.1071148.

41 Chan, Y.C., Sinha, R.K., and Wang, W. (2011). Emission of greenhouse gases from home aerobic composting, anaerobic digestion and vermicomposting of household wastes in Brisbane (Australia). *Waste Management & Research* 29 (5): 540–548.

42 Vitousek, P.M., Mooney, H.A., Lubchenco, J., and Melillo, J.M. (2008). Human domination of Earth's ecosystems. In: *Urban Ecology* (eds. J.M. Marzluff et al.). Boston, MA: Springer.

43 University of Illinois at Urbana-Champaign. (2007. October 30). Nitrogen fertilizers deplete soil organic carbon. *ScienceDaily*. Retrieved January 5, 2021 from: http://www.sciencedaily.com/releases/2007/10/071029172809.htm.

44 Johansson, K., Perzon, M., Fröling, M. et al. (2008). Sewage sludge handling with phosphorus utilization–life cycle assessment of four alternatives. *Journal of Cleaner Production* 16 (1): 135–151.

45 Linderholm, K., Tillman, A.-M., and Mattsson, J.E. (2012). Life cycle assessment of phosphorus alternatives for Swedish agriculture. *Resources, Conservation and Recycling* 66: 27–39.

46 Pradel, M., Lippi, M., Daumer, M.L. et al. (2020). Environmental performances of production and land application of sludge-based phosphate fertilizers – a life cycle assessment case study. *Environmental Science and Pollution Research* 27: 2054–2070.

47 Calabi-Floody, M., Medina, J., Rumpel, C. et al. (2018). Smart fertilizers as a strategy for sustainable agriculture. In: *Advances in Agronomy*, vol. 147 (ed. D.L. Spark), 119–157. Academic Press.

48 Barra, P.J., Inostroza, N.G., Acuña, J.J. et al. (2016). Formulation of bacterial consortia from avocado (*Persea americana* Mill.) and their effect on growth, biomass and superoxide dismutase activity of wheat seedlings under salt stress. *Applied Soil Ecology* 102: 80–91.

49 Fijalkowski, K., Rorat, A., Grobelak, A., and Kacprzak, M.J. (2017). The presence of contaminations in sewage sludge–the current situation. *Journal of Environmental Management* 203: 1126–1136.

50 Ayres, R.U., Schlesinger, W.H., and Socolow, R.H. (1994). Human impacts on the carbon and nitrogen cycles. In: *Industrial Ecology and Global Change* (eds. R. Socolow et al.). UK: Cambridge University Press.

51 Rehman, R.A., Rizwan, M., Qayyum, M.F. et al. (2018). Efficiency of various sewage sludges and their biochars in improving selected soil properties and growth of wheat (*Triticum aestivum*). *Journal of Environmental Management* 223: 607–613.

52 Wollmann, I., Gauro, A., Müller, T., and Möller, K. (2018). Phosphorus bioavailability of sewage sludge-based recycled fertilizers. *Journal of Plant Nutrition and Soil Science* 181: 158–166.

53 Pedrol, N., Puig, C.G., Souza, P. et al. (2010). Soil fertility and spontaneous revegetation in lignite spoil banks under different amendments. *Soil and Tillage Research* 110: 134–142.

54 Günther, F. (1997). Hampered effluent accumulation process: phosphorus management and societal structure. *Ecological Economics* 21 (2): 159–174.

55 Larney, F.J., Janzen, H.H., Olson, B.M., and Olson, A.F. (2009). Erosion–productivity–soil amendment relationships for wheat over 16 years. *Soil and Tillage Research* 103: 73–83.

56 Albertsson, B., Goldstein, Pettersson, O., Ulén, B. 1997. Phosphorus – Essential, Finite and an Environmental Problem (in Swedish), Report 4730, Swedish Environmental Protection Agency, Stockholm, Sweden.

57 Björklund, A., Bjuggren, C., Dalemo, M., and Sonesson, U. (2008). Planning biodegradable waste management in Stockholm. *Journal of Industrial Ecology* 3 ((4)): 43–58.

58 Rizwan, M., Ali, S., Qayyum, M.F. et al. (2016). Mechanisms of biochar-mediated alleviation of toxicity of trace elements in plants: a critical review. *Environmental Science and Pollution Research* 23: 2230–2248.

59 Garcıa-Gil, J.C., Ceppi, S.B., Velasco, M.I. et al. (2004). Long-term effects of amendment with municipal solid waste compost on the elemental and acidic functional group composition and pH-buffer capacity of soil humic acids. *Geoderma* 121 (1): 135–142.

60 Lloret, E., Pascual, J.A., Brodie, E.L. et al. (2016). Sewage sludge addition modifies soil microbial communities and plant performance depending on the sludge stabilization process. *Applied Soil Ecology* 101: 37e46. https://doi.org/10.1016/j.apsoil.2016.01.002.

61 Kanchanapiya, P., Intaranon, N., and Tantisattayakul, T. (2020). Assessment of the economic recycling potential of a glycolysis treatment of rigid polyurethane foam waste: a case study from Thailand. *Journal of Environmental Management* 111: 638.

62 Singh, R.P. and Agrawal, M. (2008). Potential benefits and risks of land application of sewage sludge. *Waste Management* (New York, NY). 28 (2): 347–358. https://doi.org/10.1016/j.wasman.2006.12.010.

63 Alvarez-Campos, O., Evanylo, G.K., and Badzmierowski, M.J. (2018). Development and assessment of exceptional quality biosolids products for urban gardens. *Compost Science & Utilization* 26 (4): 232–243.

64 Odlare, M., Pell, M., and Svensson, K. (2008). Changes in soil chemical and microbiological properties during 4 years of application of various organic residues. *Waste Management* 28 (7): 1246–1253.

65 Leifeld, J., Siebert, S., and Kögel-Knabner, I. (2002). Biological activity and organic matter mineralization of soils amended with biowaste composts. *Journal of Plant Nutrition and Soil Science* 165 (2): 151–159.

66 Jedidi, N., Hassen, A., van Cleemput, O., and M'Hiri, A. (2004). Microbial biomass in a soil amended with different types of organic wastes. *Waste Management and Research* 22 (2): 93–99.

67 Hoitink, H.A.J. and Boehm, M.J. (1999). Biocontrol within the context of soil microbial communities: a substrate-dependent phenomenon. *Annual Review of Phytopathology* 37: 427–446.

68 Koutroubas, S.D., Antoniadis, V., Fotiadis, S., and Damalas, C.A. (2014). Growth, grain yield and nitrogen use efficiency of Mediterranean wheat in soils amended with municipal sewage sludge. *Nutrient Cycling in Agroecosystems* 100: 227–243.

69 Larney, F.J. and Angers, D.A. (2012). The role of organic amendments in soil reclamation: a review. *Canadian Journal of Soil Science* 92: 19–38.

70 Dickinson, N., Marmiroli, M., Das, B. et al. (2015). Endemic plants as browse crops in agricultural landscapes of New Zealand. *Agroecology and Sustainable Food Systems* 39: 224–242.

71 Ouimet, R., Pion, A.-P., and Hébert, M. (2015). Long-term response of forest plantation productivity and soils to a single application of municipal biosolids. *Canadian Journal of Soil Science* 95: 187–199.

72 Elmaslar-Özbas, E. and Balkaya, N. (2012). Extraction of heavy metals from compost using a mixture of Na_2EDTA and $Na_2S_2O_5$: column studies. *Journal of Soil Science and Plant Nutrition* 12 (3): 525–534.

73 Cai, Q.Y., Mo, C.H., Wu, Q.T. et al. (2007). Concentration and speciation of heavy metals in six different sewage sludge-composts. *Journal of Hazardous Materials* 147 (3): 1063–1072.

74 Sharholy, M., Ahmad, K., Mahmood, G., and Trivedi, R.C. (2008). Municipal solid waste management in Indian cities – a review. *Waste Management* 28 (2): 459–467.

75 Mininni, G., Blanch, A.R., Lucena, F., and Berselli, S. (2015). EU policy on sewage sludge utilization and perspectives on new approaches of sludge management. *Environmental Science and Pollution Research* 22 (10): 7361–7374.

76 Wilk, M. and Gworek, B. (2009). Heavy metals in sewage sludge. *Ochrona środowiska i zasobów Naturalnych* 39: 40–59.

77 Szwedziak, K. and Wozniak, A. (2005). Charakterystyka osadu pościekowego pochodzacego z oczyszczalni ścieków w Strzelacach Opolskich. *Infrastruktura i Ekologia Terenów Wiejskich* 1 (1): 147–156.

78 Mizera, A. (2003). Osady ściekowe odpadem (nie) bezpiecznym. *Ekotechnika* 3: 22–25.

79 Usman, K., Khan, S., Ghulam, S. et al. (2012). Sewage sludge: an important biological resource for sustainable agriculture and its environmental implications. *American Journal of Plant Sciences* 3 (12): 1708–1721. https://doi.org/10.4236/ajps.2012.312209.

80 European Commission 2001. Disposal and Recycling Routes of Sewage Sludge Part 3, Scientific and Technical Report, European Commission DG Environment.

81 Walker, D.J., Clemente, R., and Bernal, M.P. (2004). Contrasting effects of manure and compost on soil pH, heavy metal availability and growth of *Chenopodium album* L. in a soil contaminated by pyretic mine waste. *Chemosphere* 57: 215–135. https://doi.org/10.1016/j.chemosphere.2004.05.020.

82 Ayres, J.G., Dick, S., Semple, S., Ogden, I., & Fowler, P. (2008). Human health and the environmental impacts of using sewage sludge on forestry and for restoration of derelict land: Task 1 – Desk-based literature review of the human health impacts of spreading sewage sludge on non-agricultural land. SNIFFER. Available from: http://www.sniffer.org.uk/files/7713/4183/7997/UKLQ09_Task_1_Report_Final.pdf.

83 Zabowski, D. and Zasoki, R.J. (1987). Cadmium, copper and zinc adsorption by a forest soil in the presence of sludge leachate. *Water, Air & Soil Pollution* 36 (1): 103–113. https://doi.org/10.1007/BF00450623.

84 Alloway, B.J., Thornton, I., Smart, G.A. et al. (1988). Metal availability. In: *The Shipham Report: An Investigation into Cadmium Contamination and its Implication for Human Health,*

Science of the Total Environment, vol. 75(1) (ed. H. Morgan), 41–69. https://doi.org/10.1016/0048-9697(88)90159-3. PMID: 3222706.

85 Christensen, T. and Tjell, J.C. (1984). Interpretation of experimental results on cadmium crop intake from sewage sludge amended soil. In: *Processing and Use of Sewage Sludge* (eds. P.L. Hermite and H. Ott), 358–369. Dordrecht: Reidel.

86 Oleszczuk, P. (2010). Testing of different plants to determine influence of physico-chemical properties and contaminants content on municipal sewage sludges phytotoxicity. *Environmental Toxicology* 25: 38–47.

87 Pillai, S.D., Widmer, K.W., Dowd, S.E., and Ricke, S.C. (1996). Occurance of air borne bacteria and pathogen indicator during land application of sewage sludge. *Applied and Environmental Microbiology* 62 ((1)): 296–299.

88 Dewil, R., Baeyens, J., and Neyens, E. (2006). Reducing the heavy metal content of sewage sludge by advanced sludge treatment methods. *Environmental Engineering Science* 23: 994–999.

89 Li, S., Di, X., Wu, D., and Zhang, J. (2013). Effects of sewage sludge and nitrogen fertilizer on herbage growth and soil fertility improvement in restoration of the abandoned opencast mining areas in Shanxi, China. *Environment and Earth Science* 70: 3323–3333.

90 Mahon, A.M., O'Connell, B., Healy, M.G. et al. (2017). Microplastics in sewage sludge: effects of treatment. *Environmental Science & Technology* 51: 810–818.

91 Smith, S.R. (2009). Organic contaminants in sewage sludge (biosolids) and their significance for agricultural recycling. *Philosophical Transactions of the Royal Society A* 367: 4005–4041. https://doi.org/10.1098/rsta.2009.0154.

92 Webber, M.D., Pietz, R.I., Granato, T.C., and Svoboda, M.L. (1994). Plant uptake of PCBs and other organic contaminants from sludge-treated coal refuse. *Journal of Environmental Quality* 23: 1019–1026.

93 Duarte-Davidson, R. and Jones, K.C. (1996). Screening the environmental fate of organic contaminants in sewage sludge applied to agricultural soils: II The potential for transfers to plants and grazing animals. *Science of the Total Environment* 185: 59–70. https://doi.org/10.1016/0048-9697(96)05042-5.

94 Wilson, S.C. and Jones, K.C. (1999). Volatile organic compound losses from sewage sludge-amended soils. *Journal of Environmental Quality* 28: 1145–1153.

95 O'Connor, G.A. (1996). Organic compounds in sludge-amended soils and their potential for uptake by crop plants. *Science of the Total Environment* 185: 71–81. https://doi.org/10.1016/0048-9697(95)05043-4.

96 Wang, X., Li, C., Zhang, B. et al. (2016). Migration and risk assessment of heavy metals in sewage sludge during hydrothermal treatment combined with pyrolysis. *Bioresource Technology* 221: 560e567. https://doi.org/10.1016/j.biortech.2016.09.069.

97 Boxall, A.B.A., Johnson, P., Smith, E.J. et al. (2006). Uptake of veterinary medicines from soils into plants. *Journal of Agricultural and Food Chemistry* 54: 2288–2297.

98 Wu, C., Spongberg, A.L., Witter, J.D. et al. (2010). Uptake of pharmaceutical and personal care products by soybean plants from soils applied with biosolids and irrigated with contaminated water. *Environmental Science & Technology* 44: 6157–6161.

99 Kwarciak-Kozlowska, A. (2019) Removal of pharmaceuticals and personal care products by ozonation, advance oxidation processes, and membrane separation. In: *Pharmaceuticals and Personal Care Products: Waste Management and Treatment Technology*, Butterworth-Heinemann, pp. 151–171.

100 Włodarczyk, R. and Kwarciak-Kowlowska, A. (2019, 2019). Analysis of the sorption posibilities of silver nanoparticles of anthropogenic origin by using biochar. *Przemysl Chemiczny* 98 (1): 113–116.

101 Whitley, A.R., Levard, C., Oostveen, E. et al. (2013, 2013). Behavior of Ag nanoparticles in soil: effects of particle surface coating, aging and sewage sludge amendment. *Environmental Pollution* 182: 141–149.

102 Bone, A.J., Colman, B.P., Gondikas, A.P. et al. (2012). Biotic and abiotic interactions in aquatic microcosms determine fate and toxicity of Ag nanoparticles: Part 2 – toxicity and Ag speciation. *Environmental Science and Technology* 46: 6925–6933.

103 Klitzke, S., Metreveli, G., Peters, A. et al. (2015, 2015). The fate of silver nanoparticles in soil solution – sorption of solutes and aggregation. *Science of the Total Environment* 535: 54–60.

104 Reinsch, B.C., Levard, C., Li, Z. et al. (2012). *Environmental Science and Technology* 46: 6992–7000.

105 Jośko, I. and Oleszcuk, P. (2013). Manufactured nanomaterials: the connection between environmental fate and toxicity. *Critical Reviews in Environmental Science and Technology* 43 (23): 2581–2616.

106 Ramachandra, T.V., Bhat, S., and Shivamurthy, V. (2017). Constructed Wetlands for Tertiary Treatment of Wastewater. ENVIS Technical Report 124, Energy & Wetlands Research Group, CES, Indian Institute of Science, Bangalore 560 012.

107 Zdybel, J., Karamon, J., and Cencek, T. (2009). Występowanie jaj nicieni pasożytniczych z rodzajów Ascaris, Trichuris i Toxocara w nawozach organicznych i organiczno-mineralnych oraz osadach ściekowych. *Życie Weterynaryjne* 84: 992–996.

108 Stańczyk-Mazanek, E., Nalewajek, T., and Zabochnicka, M. (2012). Drug-resistant microorganisms in soils fertilized with sewage sludge. *Archives of Environmental Protection* https://doi.org/10.2478/v10265-012-0008-6.

109 Bibby, K. and Peccia, J. (2013). Identification of viral pathogen diversity in sewage sludge by metagenome analysis. *Environmental Science & Technology* 47: 1945–1951.

110 Zhang, P., Guo, J.-S., Shen, Y. et al. (2015). Microbial communities, extracellular proteomics and polysaccharides: a comparative investigation on biofilm and suspended sludge. *Bioresource Technology* 190: 21e28. https://doi.org/10.1016/j.biortech.2015.04.058.

6.2

Toxicity Assessment

Krzysztof Fijałkowski and Anna Kwarciak-Kozłowska

Faculty of Infrastructure and Environment, Czestochowa University of Technology, Częstochowa, Poland

6.2.1 Introduction to Toxicity of Biodegradable Wastes onto Soil

World population growth and industrialization generate huge amounts of municipal solid waste. Currently, about 3 billion inhabitants of global cities generate about 1.2 kg of solid waste per capita per day (i.e. 1.3 billion tons per year). By 2025, these amounts are likely to increase due to the estimated increase in the number of urban residents to 4.3 billion, which will result in an annual waste generation of 2.2 billion tones [1].

Due to the potential for reuse of municipal solid waste, biodegradable waste deserves special attention. The concept of biodegradable waste is broad and includes waste generated in households and food-related industries, agro-industrial waste, waste from gardens and parks, sewage sludge, or bioplastic [2–4].

Biodegradation waste is any waste that undergoes anaerobic or aerobic degradation. It is reported that every EU citizen produces around 300 kg of waste per year. This amounts to 120–140 million tons of biodegradable waste per year in the European Union alone [5].

The research for sustainable solutions in the field of biodegradable waste management is a challenge, not only for the waste management sector but also for the agricultural and industrial sectors. It is estimated that the decomposition of the organic part of municipal solid waste is responsible for around 5% of global greenhouse gas (GHG) emissions annually [6]. It aims to develop and improve a range of biodegradable waste treatment technologies, such as composting (organic fertilizer), landfill, anaerobic digestion, and thermal methods [7].

However, the implementation of some of these techniques is difficult due to high operating and investment costs as well as other problems, including environmental ones [8]. Legal regulations on the management of biodegradable waste of the European Union have been formulated in Council Directive 1999/31/EC of April 26, 1999 and in Directive 2008/98/EC of the European Parliament and of the Council of November 19, 2008. According to the directive, biodegradable waste is subject to an established hierarchy of dealing with them. In the first step, they should be prevented, then prepared for re-use, recycled (aerobic, anaerobic, reclamation of degraded soils, etc.), and then other recovery methods, e.g. energy recovery (pyrolysis, incineration), and finally disposed of (least desirable solution). With the growing demand for the recovery of natural resources and energy, it is assumed that biodegradable waste recycling is of major importance [9]. Soil quality and improvement can be restored by organic recycling of biodegradable waste in the soil [10]. Biodegradable

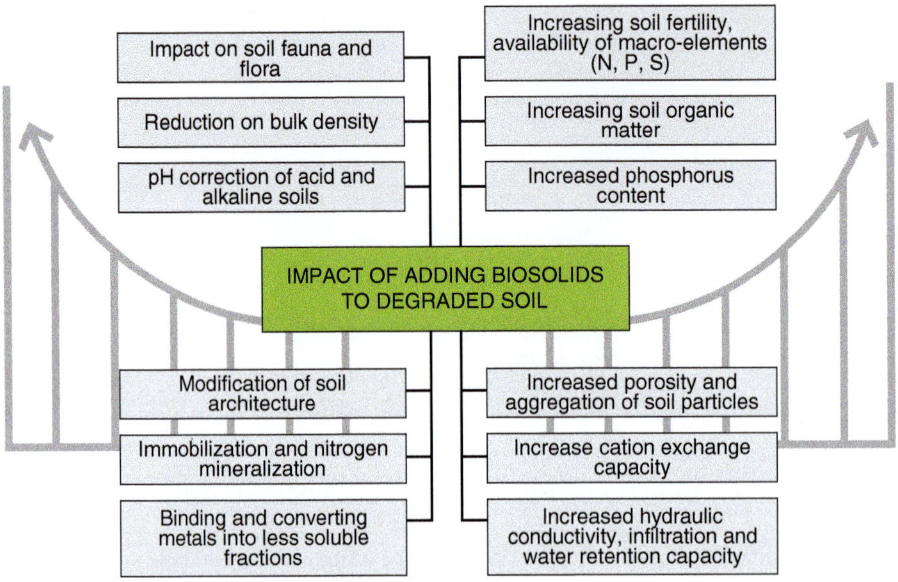

Figure 6.2.1 Impact of adding biosolids to degraded soil [20].

waste contains significant amounts of macro- and microelements and organic matter that can be used as potential organic fertilizers for plant production [11, 12].

Numerous studies that have been conducted [10, 13–15] confirm that the use of biodegradable waste improves the physical, chemical, and biological properties of soils, provides the necessary nutrients for the stimulation and development of plants, and consequently contributes to the increase in yield. They also affect the embodiment of the bulk density, aeration, or an increase in electrical conductivity [16]. It was also found that the addition of fertilizers from biodegradable waste allows for the recovery of degraded soils [17, 18]. Organic matter contained in biodegradable waste has a significant impact on the physical properties of the soil, such as increasing water retention and movement, or improving soil porosity. The addition of organic matter in the form of biodegradable waste improves water infiltration and hydraulic conductivity. It is the main opportunity for carbon sequestration by adding stable organic matter and promoting plant growth (their biostimulation) during aided phytoremediation processes [19].

The addition of biodegradable waste/sewage sludge to soils, especially degraded ones, can bring many benefits and the most important ones are presented in Figure 6.2.1.

The presence in biodegradable waste/sewage sludge of substances such as heavy metals (Cd, Cu, Ni, Pb, Zn, Hg, Cr, Se, As, Ag), organic pollutants (polycyclic aromatic hydrocarbons, adsorbed organic halides, polycyclic aromatic hydrocarbons, polychlorinated biphenyls (PCBs), dioxins, pharmaceuticals, hormones, nanoparticles, and microplastics are the subject of intense current research [21–24].

Therefore, it seems necessary to assess the toxic effects of these compounds in order to limit pollution and environmental degradation. One of the directions of modern toxicology is environmental toxicology, which deals with the mechanisms of toxic effects of various substances present in the natural environment on the human, animal, and plant organism. It covers the study of all animate nature and therefore we also call it ecotoxicology. Ecotoxicology is a young, dynamically developing branch of science that combines elements of classical toxicology with ecology, where the effects of natural and man-made pollutants on ecosystems are studied [25].

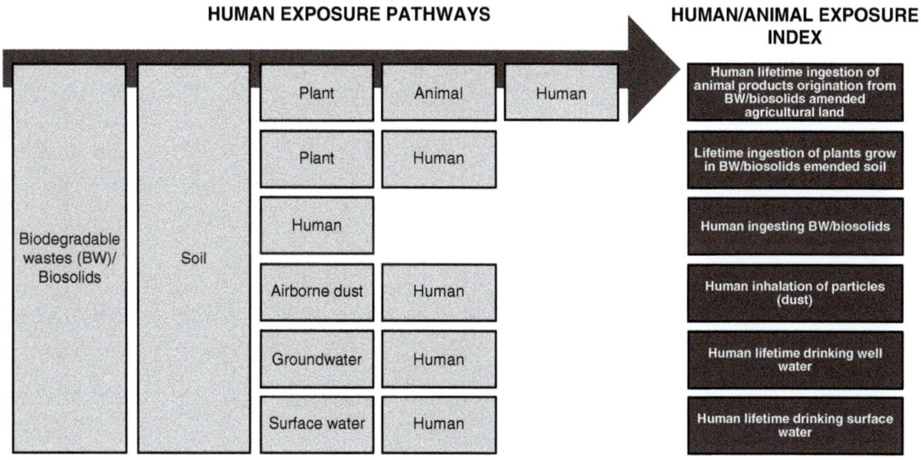

Figure 6.2.2 Human/animal exposure pathways and index of biodegradable wastes/biosolids [26].

There are six possible ways that biodegradable waste (BW) influences humans through plants, air, surface, and groundwater, and are presented in Figures 6.2.1 and 6.2.2.

6.2.2 Toxicity of Organic Compounds in Soils Treated with Biodegradable Wastes or their By-products

The presence and level of organic pollutants in the waste depends largely on the quality of the waste water, various local point sources, the physicochemical properties of individual organic compounds, and the operational parameters of the waste water treatment plant. More than 300 organic chemicals from different classes of compounds have been identified in biodegradable wastes and biosolids, with concentrations ranging from pg kg^{-1} to g kg^{-1} [27].

Some organic pollutants are known or suspected to be endocrine disruptors (EDCs), which has increased interest in their presence in waste and biosolids. The endocrine system is found in almost all animals and regulates bodily functions, including growth, development, and maturation, as well as the way various organs work. There is increasing concern about the seemingly increasing incidence of reproductive disorders and abnormal development in wildlife and reduced fertility in men, problems that can be caused by EDCs that have been anthropogenically released into the environment [28].

Organic pollutants can certainly have a negative impact on the soil/crop animal organism that receives the ground application of biodegradable waste-based fertilizers [29]. Nevertheless, the risk of contamination with compounds derived from biodegradable wastes is generally considered to be minimal due to the relatively low concentrations present and the many changes (especially biodegradation) that can occur in soils [30]. The European Union has proposed limit values for several organic pollutants (or groups of pollutants) in biodegradable wastes/biosolids, and several European countries (e.g. Sweden, Germany) have introduced several of them. Table 6.2.1 presents the most frequently detected pollutants in biodegradable waste, their potential toxicity and persistence in the soil environment, and in Table 6.2.2 he highest concentration reported pharmaceuticals and personal care products (PPCPs) detected in edible tissue of plants grown in soil amended with biosolids or manure [31]. As can be seen, the differentiation is quite large. In terms of half time, the

Table 6.2.1 Identification of risk factors persistence, bioaccumulation, and toxicity for selected contaminants present in biodegradable wastes [31–42].

Chemical class	Presistence (half-life in soil – days)	Bioaccumulation (BCF aquatic organism)	Chronic toxicity (fish mg l^{-1})
PCBs	360	25 000	0.005
PCDD/Fs	360	14 000	0.0017
PAE	50	160	11
PAH	1674	5000	0.0003–0.271
JC	1	0.06–28	0.032–202
OCB	1788–25 185	10 000 000	3.85–11.1
Brominated flame retardants			
HeptaBDE	360	18 000	–
HexaBDE	360	4800	–
PentaBDE	360	15 000	0.00069
TetraBDE	360	4600	0.003
PFOS	360	56	3
PeCB	360	1200	0.037
Anti-convulsant			
Carbamazepine	75	19	0.9
Beta-blockers			
Atenolol	75	3.2	1.1
Propanolol	30	51	0.95
Metoprolol	75	4.5	5.3
Sotalol	30	3.2	1.1
Antimicrobial			
Triclosan	120	640	0.071
Triclocarban	120	800	0.013

least dangerous organochlorine pesticides (OCBs) are JCs (chlorobenzenes). The same relationship is visible in the case of the bioconcentration factor (BCF) index. In the case of chronic toxicity, polychlorinated dibenzodioxins/polychlorinated dibenzofurans (PCDD/Fs) show the highest levels.

The authors deliberately omitted the description of heavy metals due to the fact that they are a well-known, broad, and described group of pollutants, focusing on identifying the remaining organic compounds also present in waste.

Most types of benthic organisms are bacteria, algae, and crustaceans. They play an important role in benthic communities and are a food source for larger invertebrates, juvenile fish, and water birds. Many of them directly absorb sediment particles of various origins (including biodegradable wastes) and are therefore directly exposed to sediment-related contamination along with contaminants in pore and surface waters [43]. Following the principles of healthy ecotoxicology, methods for the ecotoxicology of whole sediment and soil after their application have been developed or are

Table 6.2.2 Highest concentration reported PPCPs detected in edible tissue of plants grown in soil amended with biosolids or manure [31].

PPCP compound	Compound purpose	Plant	Highest concentration in edible tissue ng g^{-1} dw
		Biosolids-derived	
Atenolol	β-blocker	Tomato	5.57
Carbamazepine	Anticonvulsant	Collard	800
Ciprofloxacin	Antibiotic	Carrot	2.15
Diphenhydramine	Antihistamine	Tomato	21
Naproxen	Anti-inflammatory	Corn	0.8
Norfloxacin	Antibiotic	Carrot	4.5
Progesterone	Hormone	Corn	6.65
Salbutamol	Anti-inflammatory	Cabbage	47.1
Testosterone	Hormone	Tomato	5.52
Triamterene	Diuretic	Carrot	0.064
Triclocarban	Antimicrobial	Collard/lettuce/tomato	250
Triclosan	Antimicrobial	Radish	9200
Manure-derived			
Chlortetracycline	Antibiotic	Wheat	43
Gentamicin	Antibiotic	Radish	81
Monensin	Antibiotic	Radish/garlic	4
Streptomycin	Antibiotic	Lettuce	27
Sulfamethazine	Antibiotic	Corn	1200
Tylosin	Antibiotic	Onion	2
Virginiamycin	Antibiotic	Garlic	6.6

currently being developed for a wide range of organisms from diverse environments. However, the main dilemma of the assessment remains that for most of the environments assessed, we do not have data on the effects on many other species. Hence, for the purposes of this study, the following three indicators were adopted: persistence (half-life in soil – days), bioaccumulation (BCF aquatic organisms), chronic toxicity (for fish – mg l^{-1}), mainly due to the availability of data in the literature and standardization of labeling (Figures 6.2.3 and 6.2.4).

The parameter persistence shows how long it takes for the contamination to be characterized and halved by biochemical changes. This parameter indicates biodegradability and at the same time determines the degree of harmfulness of a given substance in relation to its presence in the soil environment. Soil organisms are responsible for detoxifying naturally occurring toxins such as phytotoxins and provide the soil ecosystem with a naturally occurring mechanism to detoxify man-made pollutants [44].

The BCF index is quite widely used in environmental toxicology. It expresses how much of a substance can be accumulated in the biomass, in this case fish (as a test organism) and thus get into the

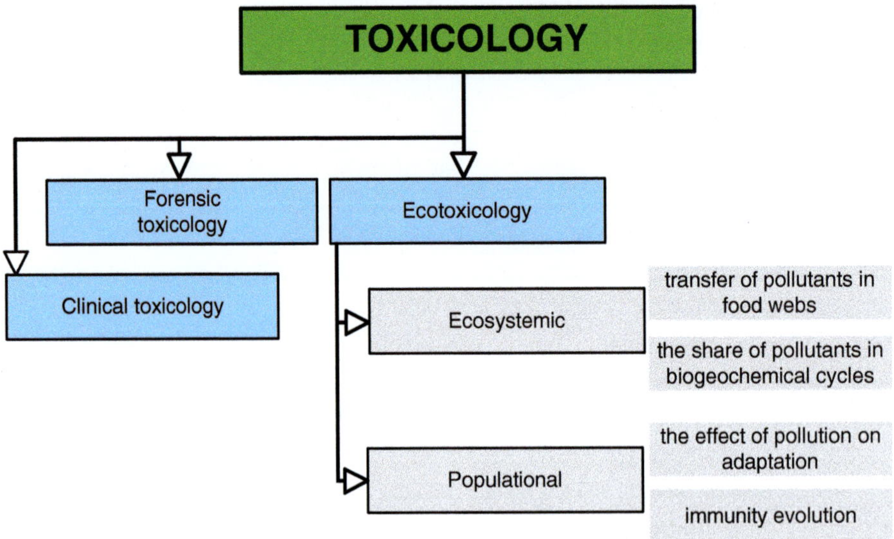

Figure 6.2.3 Classification of toxicity analysis methods [25].

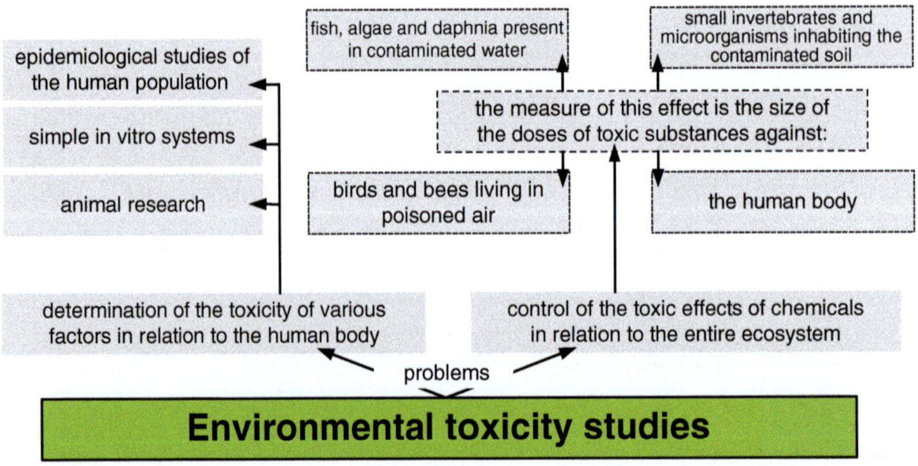

Figure 6.2.4 Types of ecotoxicological tests [25].

food chain. BCF is also used to express the concentration of substances in plant biomass, similarly to the concentration shown in Table 6.2.2. This parameter strongly indicates the toxic potential of the tested pollutant after reaching the food chain, and, in the case of pesticides, additional biomagnification is possible [45].

The last parameter in the summary is for the chronic toxicity presented to the aquatic environment where the tested organism is fish, where chronic toxicity harmful changes in the test organisms were caused by prolonged exposure to a chemical compound after repeated or continuous exposure. The tests used subtle, lower than lethal, doses (concentrations) of the substance. The observations are based on the assessment of changes in physiological activity, e.g. in food and reproductive systems, genetic disorders, and disturbances in the functioning of organs. This parameter directly indicates the amount of harmfulness of the test substance.

6.2.2.1 Phthalic Acid Esters (PAE)

Phthalic acid esters (PAE) are produced in large numbers and are mainly used as plasticizers to make plastics more flexible [29]. The vast majority of phthalates are used to plasticize polyvinyl chloride (PVC) in the manufacture of products ranging from kitchen and bathroom flooring to medical pipes, toys, footwear, electric cables, packaging, and roofing. Phthalates are used in non-PVC applications such as paints, rubber products, adhesives, cosmetics and toiletries (e.g. nail polish, perfumes), epoxy resins, adhesives, and printing inks. The widespread industrial and domestic use of phthalate-containing products causes large amounts to be washed into the sewage system [46, 47]. Total PAE concentrations in biosoils usually range from 12 to 200 mg kg^{-1} [48]. DEHP (di-2-ethylhexyl phthalate) is most often found in waste and usually accounts for 25–95% of all detected PAE [49]. Others are DnOP, BBP, DnBP, DEP, and DMP, which are commonly present in waste at low concentrations (<10 mg kg^{-1}) [50]. Under aerobic conditions in soil, PAE is readily microbiologically degraded [51] with less degradable high mass compounds and a longer half-life. Research indicates that complete degradation of PAE in soil occurs within 15 days. The concentrations of dibutyl phthalate and diethylhexyl phthalate in plastic greenhouse soils, for example, were 2.5–3 times higher than that of corresponding soils not covered by plastic greenhouses [52, 53]. In addition, Kong et al. [54] found that phthalic acid ester concentrations in film-covered soils were 74 and 208% higher than those in farmland and vegetable soils. The concentrations of diethylhexyl phthalate in 91% of biosolid samples met the limit (100 mg kg^{-1}, dry weight) proposed by the Europe Union [32].

6.2.2.2 Polycyclic Aromatic Hydrocarbons (PAH)

Polycyclic aromatic hydrocarbons (PAHs) are a complex group of organic compounds produced as preparatory products from the incomplete combustion of carbonaceous fuels such as wood, coal, diesel, fat, and tobacco. Important combustion processes include food preparation, combustion in motor vehicles, house fires, coal-fired power stations, cement plants, and ironworks, as well as volcanoes and forest fires. Thus, PAHs enter the environment mainly from air releases followed by direct aviation precipitation. They enter the sewage system from many sources, including land runoff from roads containing car exhaust particles. Due to their toxic, mutagenic, estrogenic, and carcinogenic properties, 16 PAHs have been identified by the USEPA as priority pollutants, and seven of them are considered carcinogenic [55]. The average concentration of all PAHs in biodegradable waste from industrialized countries is usually in the range of 1–100 mg kg^{-1} [56]. Individual PAHs differ significantly although the dominant compounds are usually those with 3, 4, or 5 rings [57]. The European Union proposed that the sum of 11 PAHs in biodegradable waste should not exceed 6 mg kg^{-1} in the case of land [58]. It is believed that the strong binding of PAHs to organic matter initially limits their decomposition. Following the discontinuation of fertilization, two long-term studies by Beck et al. [59] observed that 90% of the total PAHs were removed from the soil in one site and about 65% in the other. The half-life ranged from two years for naphthalene to over seven years for fluoranthenuperylene. The degradation of PAHs is greater in cultivated soils, which stimulates the decomposition of organic matter, than in non-cultivated soils such as pastures or land with trees [36].

6.2.2.3 Chlorobenzenes (JC)

Chlorobenzenes is a group of cyclic aromatic compounds in which one or more hydrogen atoms have been replaced with a chlorine atom. There are 12 different JCs: monochlorobenzene (MCB), dichlorobenzene (DCB) (three isomers), trichlorobenzene (TCB) (three isomers), tetrachlorobenzene (TCB) (three isomers), pentachlorobenzene (PeCB), and hexachlorobenzene (HCB). JCs are used as intermediates in the synthesis of pesticides and other chemicals such as engine cleaners and other solvents, and as components of dielectric fluids; 1,4-DCB is widely used as a toilet deodorant. The concentration of JC varies considerably from source to source including biodegradable waste and sewage sludge and is generally higher in industrial waste than domestic waste. The total concentration of JC in biosolids from biodegradable wastes ranges from 0.1 to 40 $mg\,kg^{-1}$ although concentrations of up to 2000 $mg\,kg^{-1}$ have been reported [55, 60]. Unfortunately, still little is known about the fate of JC after its introduction into the environment. When introduced into the soil by biosolid volatilization, it is considered to be the main mechanism of decomposition of JC. Wang and Jones [61] have shown that the half-life of JC added in processed waste to soil varied from 13 to 209 days, and this time increases with increasing chlorination of individual JCs. They also showed that only about 10% of added JC became resistant to soil decomposition by strong adsorption on soil particles, and among them CB proved to be the most durable [61].

6.2.2.4 Polychlorinated Dibenzodioxins (PCDD) and Polychlorinated Dibenzofurans (PCDF)

Dioxin is a generic term for a mixture of 219 different nearly planar tricyclic aromatic compounds belonging to the groups polychlorinated dibenzodioxins (PCDD) and polychlorinated dibenzofurans (PCDF). These compounds are not produced commercially, but are formed as undesirable co-products from the production or use of many organochlorine compounds, i.e. herbicides. Bleaching of chlorine from pulp and paper can be an important source of the release of these compounds into the environment and they are also produced in the slime of graphite electrodes from chlorine plants. In mammals, they are known to be mutagenic, carcinogenic, immunotoxic, teratogenic and hepatotoxic [62]. The concentration of dioxins in the biodegradable waste ranges from 2 to 1270 ng I-TEQ kg^{-1} from 20 to 200 ng I-TEQ kg^{-1} [63–65]. Overall, the level of PCDD/Fs in biosolids in Europe has decreased since the 1980s and this is likely a consequence of the ban on pentachlorophenol (PCP) [66]. Once incorporated into the soil and the use of processed waste, PCDDs are very persistent with a half-life in excess of 10 years. Processes such as photodegradation and volatilization have little effect on the fate of PCDDs in soils and as a result are extremely persistent [67].

6.2.2.5 Polychlorinated Biphenyls (PCBs)

CBs consist of a biphenyl ring (two benzene rings) with 10 positions where chlorine substitution can occur. PCBs are widely used as coolants and insulating liquids for transformers and capacitors, stabilizing additives in flexible PVC coatings, including electrical cables and electronic components, adhesives, wooden floor finishes, paints, and printing inks. PCBs have been shown to be persistent organic pollutants that bioaccumulate in animals and plants [62]. PCBs in biosolids are common and their concentrations range from 20 to 2000 $mg\,kg^{-1}$ [63, 68, 69]. The European Union proposed that the sum of seven of the PCB congeners (28, 52, 101, 118, 138, 153, and 180) should not exceed 800 $mg\,kg^{-1}$ [70]. In a study by Stevens et al. [63] it was found that the congeners: 28, 52, 101, 138, 153, and 180 were the most abundant in biosolids in the UK. When this waste is applied to soils, PCB is slowly lost. In a long-term study, Alcock et al. [71] and Haynes et al. [55] found that total PCB concentration decreased sharply over a 31-year period (since the last bio-waste application).

6.2.2.6 Organochlorine Pesticides (OCPs)

The term organochlorine pesticides (OCPs) refers to a wide variety of organic compounds that contain chlorine. OCPs are a diverse group of synthetic chemicals that have been used primarily against insect pests in agricultural crops and insect-transported human and pet diseases [55]. Their use has been minimized or terminated in most technologically advanced countries due to their persistence, susceptibility to biomagnification, and higher toxicity to higher organisms [72, 73]. The concentrations of OCP have decreased over the years in the environment due to restrictions on their use and, as a consequence, individual compounds are commonly present in waste in very low concentrations, and often some of them are not detectable, such as DDT and its decomposition products (DDD and DDE) in concentrations: 7.5–15.5; 15.4–47.9 up to 97.2 mg kg^{-1} [55, 69]. In the Polish biosolids, Falandysz and Strandberg [74] found that DDT and its metabolites are present in the highest concentrations, ranging from 330 to 490 mg kg^{-1}. In Greece, Katsoyiannis and Samara [69] measured DDD and DDE in Greek soils but found none of these associations. Under aerobic soil conditions, the degradation of PCBs is very slow and therefore, if present in waste, they will remain for a long time after application in soil.

6.2.2.7 Brominated Flame Retardants (BFRs)

Flame retardants incorporated into potentially flammable materials, such as plastics, rubbers, and textiles, slow down and/or inhibit the initial stage of a fire. Brominated fire retardants (BFRs) are an extremely diverse group of compounds. In 2003, the European Union introduced a ban on the use of pentaBDE and production in Europe has shifted in the direction of HBCD and TBBPA. Concentrations of PBDE in the waste have been reported in the range of 0.4 to 2600 μg kg^{-1} and 1100–1000 μg kg^{-1} for congeners 47, 99, 100, and 209, which are the most common [75–77]. After the use of biosolids from biodegradable waste, PBDEs have been shown to be persistent in soils [66]. Biodegradation is not considered to be an important degradation pathway but reports for photolysis may appear to be very relevant [78]. Nevertheless, microbial degradation of PBDEs can occur under aerobic and especially anaerobic conditions [79].

6.2.2.8 Surfactants and Related Residues

Surfactants are one of the most ubiquitous group of anthropogenic organic compounds. The world market is over 18 million tons per year, and about 40% of them are household soaps and detergents. The most common surfactant used is anionic alkyl benzene sulfonate (LAS). Much of the LAS in the waste water is adsorbed to solids and removed during sedimentation; therefore, the sludge in primary settlement tanks is generally rich in LAS concentrations ranging from 5 to 15 g l^{-1}. Biosoilds produced from waste generally have a low LAS concentration (100–500 mg k^{-1}) compared to anaerobically treated sewage sludge (5000–15 000mg kg^{-1}). When reintroduced into an aerobic environment such as soil, surfactants such as LAS are rapidly degraded [80]. Other detergent residues that have also been measured in biosolids include quaternary ammonium surfactants, and fluorescent whitening agents are also a component of most modern laundry detergents and have been measured in waste at concentrations of 5–100 mg kg^{-1}. These materials are believed to be relatively resistant to biodegradation in soil [81].

6.2.2.9 Chlorophenols (PCPs)

Chlorophenols are phenol organochlorines and are composed of 19 different compounds, with PCP being the most used in the group. They have biocidal properties and are used as disinfectants, herbicides, and fungicides. They are used as wood paint and leather preservatives, and some are important intermediates in the production of pharmaceuticals. PCP is known to be a neurotoxic immunosuppressant and is suspected to be carcinogenic [82]. Due to concerns about the toxicity of PCP and its related dioxins, their use has been severely restricted (e.g. a residential ban) in the most technologically advanced countries since the 1980s. Nevertheless, significant amounts of di-, tri-, tetra-, and penta- CPs have been measured in waste samples. The concentrations of individual compounds are often in the range 0.02–10 $mg\,kg^{-1}$ but their total concentration is in the range 10–60 $mg\,kg^{-1}$ [83]. After application to soils, chlorophenols are degraded relatively quickly in a well-documented way in microbial degradation [84].

6.2.2.10 Pharmaceuticals and Personal Care Products (PPCPs)

Pharmaceuticals and personal care products (PPCPs) in waste water are produced mainly from human feces and urine, which contain residues or metabolites, wash water containing topically applied chemicals, and residues or all of prescription drugs that have expired [85]. For example, chemicals belonging to groups such as synthetic musk are present in sludge-based biosolids (2000–500 000 $\mu g\,kg^{-1}$), antibiotics such as fluoroquinolones (1000–2500 $\mu g\,kg^{-1}$), macrolides (1–150 $g\,kg^{-1}$), and antibacterial agents, for example sulphonamides (1–200 $g\,kg^{-1}$) [85]. Anti-bulb agents such as acetaminophen (e.g. Tylenol, panadol, acetylalicylic acid (asprin), albuterol, and gemfibrozil (lowers blood fats and cholesterol) have also been found [29, 85]. There is little information available on the environmental behavior, biological activity, or ecotoxicology of these materials. There is currently no evidence that residues of pharmaceutical chemicals present in biosolids could be harmful to human health or the environment. Nevertheless, much more accurate data on concentrations, fate, and migration in soil, and their potential impact on humans and the environment will be needed in the future [86].

6.2.2.11 Natural and Synthetic Hormones

Natural and synthetic hormones are, for example b-estradiols and ethynylestradiol. Some organic contaminants are known or suspected that they are chemicals that disrupt the endocrine system (EDC). These compounds have the ability to alter or disrupt the function of the endocrine system by mimicking, antagonizing, or disrupting the biosynthesis or biodegradation of endogenous hormones. Natural steroid hormones include: progestrogens (P4), glucocorticoids, mineralocorticoids, androgens, and estrogens secreted by the endocrine systems of humans and animals. The synthetic steroids are 17β-ethinylestradiol (EE2) and mestranol (MeEE2) commonly used in oral contraceptives and hormone replacement therapy; natural steroidal estrogens are: estrone (E1), 17β-estradiol (E2), and estriol (E3) [87]. In the studies of Yang et al. [38] on samples from the runoff of soils fertilized with biosolids based on biodegradable waste, collected before application, the concentrations of two hormones (2.23 $ng\,l^{-1}$) and androstenedione (1.54 $ng\,l^{-1}$) were low. On the other hand, significantly higher concentrations of many estrogens (25.0 $ng\,l^{-1}$), androgens (216 $ng\,l^{-1}$), and progesterone (as much as 98.9 $ng\,l^{-1}$) were observed in the collected outflow samples 1, 8, and 35 days after the application of biosolids. Fent et al. [88] indicated that up to 11 900 $ng\,l^{-1}$ of P4 was found in livestock waste. It is clear that animal feces are a source of progestins in the waters adjacent to livestock farms and cattle pastures and contribute to steroid contamination of soil.

6.2.3 Conclusion

Biodegradable waste and sewage sludge by-products are characterized by a significant load of heavy metals, including other organic products. Since extremely persistent organic meanings such as dioxins and organochlorines have been effectively discontinued, contamination of biosolids with such compounds is less significant. As a result, the investigation of waste, which after processing can be improved for soil fertilization, becomes the first decrease, and the positive effects of this waste through its application to field or forest land are gaining momentum.

References

1 Hossain, M.Z., Von Fragstein, P., Von Niemsdorff, P., and Heß, J. (2017). Effect of different organic wastes on soil properties and plant growth and yield: a review. *Scientia Agriculturae Bohemica* 48 (4): 224–237.

2 Wojnowska-Baryla, I., Kulikowska, D., and Bernat, K. (2020). Effect of bio-based products on waste management. *Sustainability* 12 (5): 2088.

3 Boudjabi, S., Kribaa, M., and Chenchouni, H. (2017). Sewage sludge fertilization alleviates drought stress and improves physiological adaptation and yield performances in durum wheat (*Triticum durum*): a double-edged sword. *Journal of King Saud University – Science*: 31:1–31:9.

4 Saveyn, H. and Eder, P. (2014). *End-of-Waste Criteria for Biodegradable Waste Subjected to Biological Treatment (Compost & Digestate): Technical Proposals*. Sevilla, Spain: IPTS.

5 Agency EE. 2013. Managing municipal solid waste: a review of achievements in 32 European countries. Office for Official Publications of the European Communities.

6 UNEP U. 2011. Waste – Investing in energy and resource efficiency. Towards a green economy. 44.

7 Laner, D., Crest, M., Scharff, H. et al. (2012). A review of approaches for the long-term management of municipal solid waste landfills. *Waste Management* 32 (3): 498–512.

8 Luo, W., Chen, T., Zheng, G. et al. (2008). Effect of moisture adjustments on vertical temperature distribution during forced-aeration static-pile composting of sewage sludge. *Resources, Conservation and Recycling* 52 (4): 635–642.

9 Padmavathiamma, P.K., Li, L.Y., and Kumari, U.R. (2008). An experimental study of vermi-biowaste composting for agricultural soil improvement. *Bioresource Technology* 99 (6): 1672–1681.

10 Yu, H., Ding, W., Luo, J. et al. (2012). Long-term effect of compost and inorganic fertilizer on activities of carbon-cycle enzymes in aggregates of an intensively cultivated sandy loam. *Soil Use and Management* 28 (3): 347–360.

11 Achiba, W.B., Lakhdar, A., Gabteni, N. et al. (2010). Accumulation and fractionation of trace metals in a Tunisian calcareous soil amended with farmyard manure and municipal solid waste compost. *Journal of Hazardous Materials* 176 (1–3): 99–108.

12 Srivastava, M., Ma, L.Q., Singh, N., and Singh, S. (2005). Antioxidant responses of hyper-accumulator and sensitive fern species to arsenic. *Journal of Experimental Botany* 56 (415): 1335–1342.

13 Cavagnaro, T.R. (2014). Impacts of compost application on the formation and functioning of arbuscular mycorrhizas. *Soil Biology and Biochemistry* 78: 38–44.

14 Molina-Herrera, S. and Romanya, J. (2015). Synergistic and antagonistic interactions among organic amendments of contrasted stability, nutrient availability and soil organic matter in the regulation of C mineralisation. *European Journal of Soil Biology* 70: 118–125.

15 Ling, N., Zhu, C., Xue, C. et al. (2016). Insight into how organic amendments can shape the soil microbiome in long-term field experiments as revealed by network analysis. *Soil Biology and Biochemistry* 99: 137–149.

16 Singh, R. and Agrawal, M. (2008). Potential benefits and risks of land application of sewage sludge. *Waste Management* 28 (2): 347–358.

17 Pena, L.B., Zawoznik, M.S., Tomaro, M.L., and Gallego, S.M. (2008). Heavy metals effects on proteolytic system in sunflower leaves. *Chemosphere* 72 (5): 741–746.

18 Puga, A.P., Abreu, C.A., Melo, L.C.A. et al. (2015). Cadmium, lead, and zinc mobility and plant uptake in a mine soil amended with sugarcane straw biochar. *Environmental Science and Pollution Research* 22 (22): 17606–176014.

19 Odlare, M., Pell, M., and Svensson, K. (2008). Changes in soil chemical and microbiological properties during 4 years of application of various organic residues. *Waste Management* 28 (7): 1246–1253.

20 Sharma, B., Sarkar, A., Singh, P., and Singh, R.P. (2017). Agricultural utilization of biosolids: a review on potential effects on soil and plant grown. *Waste Management* 64: 117–132.

21 Fijalkowski, K.L. and Kwarciak-Kozlowska, A. (2020). Phytotoxicity assay to assess sewage sludge phytoremediation rate using guaiacol peroxidase activity (GPX): a comparison of four growth substrates. *Journal of Environmental Management* 263: 110413.

22 Dewil, R., Baeyens, J., and Neyens, E. (2006). Reducing the heavy metal content of sewage sludge by advanced sludge treatment methods. *Environmental Engineering Science* 23 (6): 994–999.

23 Li, S., Di, X., Wu, D., and Zhang, J. (2013). Effects of sewage sludge and nitrogen fertilizer on herbage growth and soil fertility improvement in restoration of the abandoned opencast mining areas in Shanxi, China. *Environmental Earth Sciences* 70 (7): 3323–3333.

24 Mahon, A.M., O'Connell, B., Healy, M.G. et al. (2017). Microplastics in sewage sludge: effects of treatment. *Environmental Science & Technology* 51 (2): 810–818.

25 Rejmer, P. 1997. Podstawy ekotoksykologii. Wydawn. Ekoinzynieria.

26 Clarke, R.M. and Cummins, E. (2015). Evaluation of "classic" and emerging contaminants resulting from the application of biosolids to agricultural lands: a review. *Human and Ecological Risk Assessment: An International Journal* 21 (2): 492–513.

27 Smith, S.R. (2000). Are controls on organic contaminants necessary to protect the environment when sewage sludge is used in agriculture? *Progress in Environmental Science* 2 (2): 129–146.

28 Ashby, J., Houthoff, E., Kennedy, S.J. et al. (1997). The challenge posed by endocrine-disrupting chemicals. *Environmental Health Perspectives* 105 (2): 164–169.

29 Bhandari, A. and Xia, K. (2005). Hazardous organic chemicals in biosolids recycled as soil amendments. In: *Water Pollution. The Handbook of Environmental Chemistry*, vol. 1 (eds. T.A. Kaqssim and K.J. Williamson), 217–239. Berlin, Heidelberg: Springer https://doi.org/10.1007/b11440.

30 Epstein, E. (2003). *Land Application of Sewage Sludge and Biosolids*. Boca Raton. FL, 216 pp: Lewis Publishers.

31 Prosser, R.S. and Sibley, P.K. (2015). Human health risk assessment of pharmaceuticals and personal care products in plant tissue due to biosolids and manure amendments, and wastewater irrigation. *Environment International* 75: 223–233.

32 He, L., Gielen, G., Bolan, N.S. et al. (2015). Contamination and remediation of phthalic acid esters in agricultural soils in China: a review. *Agronomy for Sustainable Development* 35 (2): 519–534.

33 Tan, W., Zhang, Y., He, X. et al. (2016). Distribution patterns of phthalic acid esters in soil particle-size fractions determine biouptake in soil-cereal crop systems. *Scientific Reports* 6 (1): 31987.

34 Staples, C.A., Adams, W.J., Parkerton, T.F. et al. (1997). Aquatic toxicity of eighteen phthalate esters. *Environmental Toxicology and Chemistry* 16 (5): 875–891.

35 Okpashi, V.E., Ogugua, V.N., Ubani, S.C. et al. (2017). Estimation of residual polycyclic aromatic hydrocarbons concentration in fish species: implication in reciprocal corollary. *Cogent Environmental Science* 3 (1): 1303979.

36 Oleszczuk, P. (2006). Persistence of polycyclic aromatic hydrocarbons (PAHs) in sewage sludge-amended soil. *Chemosphere* 65 (9): 1616–1626.

37 Bragin, G.E., Parkerton, T.F., Redman, A.D. et al. (2016). Chronic toxicity of selected polycyclic aromatic hydrocarbons to algae and crustaceans using passive dosing. *Environmental Toxicology and Chemistry* 35 (12): 2948–2957.

38 Song, Y., Wang, F., Bian, Y. et al. (2012). Chlorobenzenes and organochlorinated pesticides in vegetable soils from an industrial site, China. *Journal of Environmental Sciences* 24 (3): 362–368.

39 McKone, T.E. and Maddalena, R.L. (2007). Plant uptake of organic pollutants from soil: bioconcentration estimates based on models and experiments. *Environmental Toxicology and Chemistry* 26 (12): 2494.

40 World Health Organisation. 2004. Concise International Chemical Assessment Document 60, Chlorobenzenes other than hexaxhlorobenzene: Environmental aspects. Geneva.

41 Beckvar, N. and Lotufo, G.R. (2011). DDT and other organohalogen pesticides in aquatic organisms. In: *Environmental Contaminants in Biota: Interpreting Tissue Concentrations*, 2nd edition, ed. W. Nelson Beyer & James P. Meador, vol. 23, 47–101. Boca Raton, Florida: CRC Press.

42 Toan, V.D., Thao, V.D., and Walder, J. (2009). Residue, temporal trend and half-life time of selected organochlorine pesticides (OCPs) in surface soils from Bacninh, Vietnam. *Bulletin of Environmental Contamination and Toxicology* 82 (4): 516–521.

43 Simpson, S.L., Campana, O., and Ho, K.T. (2016). Sediment toxicity testing, Chapter 7. In: *Marine Ecotoxicology*, vol. 3, issue 2. (eds. J. Blasco, P.M. Chapman, O. Campana and M. Hampel), 199–237. Academic Press. ISBN: ISBN 9780128033715 https://doi.org/10.1016/B978-0-12-803371-5.00007-2.

44 Plants, T., Invertebrates, T., and Vertebrates, T. 2009. Soil Toxicity and Bioassessment Test Methods for Ecological Risk Assessment. Available at: https://citeseerx.ist.psu.edu/viewdoc/download?doi=10.1.1.614.8569&rep=rep1&type=pdf.

45 Kacprzak, M. and Fijałkowski, K. (2009). Mycorrhiza and sewage sludge effect on biomass of sunflower and willow during phytoremediation of degraded terrains within zinc foundry zone. *Environmental Protection Engineering* 35 (2): 181–187.

46 Marttinen, S.K., Kettunen, R.H., and Rintala, J.A. (2003). Occurrence and removal of organic pollutants in sewages and landfill leachates. *Science of the Total Environment* 301 (1–3): 1–12.

47 Palmquist, H. and Hanæus, J. (2005). Hazardous substances in separately collected grey- and blackwater from ordinary Swedish households. *Science of the Total Environment* 348 (1–3): 151–163.

48 Amir, S., Hafidi, M., Merlina, G. et al. (2005). Fate of phthalic acid esters during composting of both lagooning and activated sludges. *Process Biochemistry* 40 (6): 2183–2190.

49 Gibson, R., Wang, M.-J., Padgett, E., and Beck, A.J. (2005). Analysis of 4-nonylphenols, phthalates, and polychlorinated biphenyls in soils and biosolids. *Chemosphere* 61 (9): 1336–1344.

50 Zheng, X. and Zhou, Y. (2006). Study on phthalate esters in Beijing sewage sludges. *Journal of Beijing University of Technology* 4: 112–120.

51 Jianlong, W., Xuan, Z., and Weizhong, W. (2004). Biodegradation of phthalic acid esters (PAEs) in soil bioaugmented with acclimated activated sludge. *Process Biochemistry* 39 (12): 1837–1841.

52 Peng, X., Huang, Q., Zhang, K. et al. (2012). Distribution, behavior and fate of azole antifungals during mechanical, biological, and chemical treatments in sewage treatment plants in China. *The Science of the Total Environment* 426: 311–317.

53 Wang, W., Zhang, Y., Wang, S. et al. (2012). Distributions of phthalic esters carried by total suspended particulates in Nanjing, China. *Environmental Monitoring and Assessment* 184 (11): 6789–6798.

54 Kong, S., Ji, Y., Liu, L. et al. (2012). Diversities of phthalate esters in suburban agricultural soils and wasteland soil appeared with urbanization in China. *Environmental Pollution* 170: 161–168.

55 Haynes, R.J., Murtaza, G., and Naidu, R. (2009 [cytowane 7 styczeń 2021]. s.). Inorganic and organic constituents and contaminants of biosolids. *Advances in Agronomy* [Internet]. Elsevier;: 165–267. Available from: https://linkinghub.elsevier.com/retrieve/pii/S0065211309040048.

56 Dai, J., Xu, M., Chen, J. et al. (2007). PCDD/F, PAH and heavy metals in the sewage sludge from six wastewater treatment plants in Beijing, China. *Chemosphere* 66 (2): 353–361.

57 Oleszczuk, P. (2007). Changes of polycyclic aromatic hydrocarbons during composting of sewage sludges with chosen physico-chemical properties and PAHs content. *Chemosphere* 67 (3): 582–591.

58 Gendebien A, Davis B, and Hobson J. 2010. *Environmental, Economic, and Social Impacts of the Use of Sewage Sludge on Land: Final Report*. Part III: Project Interim Reports. Milieu Ltd., WRc, RPA.

59 Beck, A.J., Alcock, R.E., Wilson, S.C. et al. (1995). Long-term persistence of organic chemicals in sewage sludge-amended agricultural land: a soil quality perspective. *Advances in Agronomy* 55: 345–391. Available from: https://www.sciencedirect.com/science/article/abs/pii/S006521130860543X.

60 Harrison, E.Z., Oakes, S.R., Hysell, M., and Hay, A. (2006). Organic chemicals in sewage sludges. *Science of the Total Environment* 367 (2–3): 481–497.

61 Wang, M.-J., McGrath, S.P., and Jones, K.C. (1995). Chlorobenzenes in field soil with a history of multiple sewage sludge applications. *Environmental Science & Technology* 29 (2): 356–362.

62 Johnson, C. (2003). *Endocrine Disrupters in Wastewater and Sludge Treatment Processes*. In: *Environmental Practice* (eds. J.W. Birkett and J.N. Lester). Boca Raton, FL. 295 pp.: Lewis Publishers/CRC Press, 2004;6(4):325.

63 Stevens, J.L., Northcott, G.L., Stern, G.A. et al. (2003). PAHs, PCBs, PCNs, organochlorine pesticides, synthetic musks, and polychlorinated n-alkanes in UK sewage sludge: survey results and implications. *Environmental Science & Technology* 37 (3): 462–467.

64 Dudzinska, M.R. and Czerwinski, J. (2002). PCDD/F levels in sewage sludge from MWTP in South-Eastern Poland. *Organohalogen Compounds* 57: 305–308.

65 Oleszek-Kudlak, S., Grabda, M., Czaplicka, M. et al. (2005). Fate of PCDD/PCDF during mechanical–biological sludge treatment. *Chemosphere* 61 (3): 389–397.

66 Eljarrat, E., Caixach, J., and Rivera, J. (2003). A comparison of TEQ contributions from PCDDs, PCDFs and dioxin-like PCBs in sewage sludges from Catalonia, Spain. *Chemosphere* 51 (7): 595–601.

67 McLachlan, M.D., Horstmann, M., and Hinkel, M. (1996). Polychlorinated dibenzo-p-dioxins and dibenzofurans in sewage sludge: sources and fate following sludge application to land. *Science of the Total Environment* 185 (1–3): 109–123.

68 Blanchard, M., Teil, M.-J., Ollivon, D. et al. (2001). Origin and distribution of polyaromatic hydrocarbons and polychlorobiphenyls in urban effluents to wastewater treatment plants of the Paris area (France). *Water Research* 35 (15): 3679–3687.

69 Katsoyiannis, A. and Samara, C. (2004). Persistent organic pollutants (POPs) in the sewage treatment plant of Thessaloniki, northern Greece: occurrence and removal. *Water Research* 38 (11): 2685–2698.

70 CEC. 2000. Council of the European Community Working Document on Sludge. EU, Brussels, 3rd Draft.

71 Alcock, R., McGrath, S., and Jones, K. (1995). The influence of multiple sewage sludge amendments on the PCB content of an agricultural soil over time. *Environmental Toxicology and Chemistry: An International Journal* 14 (4): 553–560.

72 Angulo Lucena, R., Farouk Allam, M., Serrano Jimenez, S., and Luisa Jodral Villarejo, M. (2007). A review of environmental exposure to persistent organochlorine residuals during the last fifty years. *Current Drug Safety* 2 (2): 163–172.

73 Kolankaya, D. (2006). Organochlorine pesticide reidues and their toxic effects on the environment and organisms in Turkey. *International Journal of Environmental Analytical Chemistry* 86 (1–2): 147–160.

74 Falandysz, J. and Strandberg, B. (2004). Persistent organochlorine compounds in sludge and sediments from the Gdańsk region, Baltic Sea. *Polish Journal of Environmental Studies* 13 (2): 133–138.

75 Christensen JH, Groth BS, Vikelsøe J, and Vorkamp K. 2003. Polybrominated diphenyl ethers (PBDEs) in sewage sludge and wastewater. Method Development and Validation NERI Technical Report. (481).

76 Witters, N., Mendelsohn, R.O., Van Slycken, S. et al. (2012). Phytoremediation, a sustainable remediation technology? Conclusions from a case study. I: energy production and carbon dioxide abatement. *Biomass and Bioenergy* 39 (C): 454–469.

77 Fabrellas, B., Larrazabal, D., Martinez, M.A. et al. (2004). Presence of polybrominated diphenyl ethers in Spanish sewage sludges: important contribution of deca-BDE. *Organohalogen Compounds* 66: 3755–3760.

78 Fang, L., Huang, J., Yu, G., and Wang, L. (2008). Photochemical degradation of six polybrominated diphenyl ether congeners under ultraviolet irradiation in hexane. *Chemosphere* 71 (2): 258–267.

79 Vonderheide, A.P., Mueller-Spitz, S.R., Meija, J. et al. (2006). Rapid breakdown of brominated flame retardants by soil microorganisms. *Journal of Analytical Atomic Spectrometry* 21 (11): 1232–1239.

80 Castro, M.J., Ojeda, C., and Cirelli, A.F. (2013). *Surfactants in Agriculture. W: Green Materials for Energy, Products and Depollution*, 287–334. Springer.

81 Aasen, J., Munday, R., Samdal, I. et al. (2006). Faecal sterols and fluorescent whiteners as indicators of the source of faecal contamination. *Articles and Features* 74: 74–77.

82 USEPA. United States Environmental Protection Agency Technology Transfer Network. Air toxics web site. Pentachlorophenol' [Internet]. 2008. Available from: http://www.epa.gov/ttn/atw/hlthef/pentachl.html.

83 Bright, D. and Healey, N. (2003). Contaminant risks from biosolids land application: contemporary organic contaminant levels in digested sewage sludge from five treatment plants in greater Vancouver, British Columbia. *Environmental Pollution* 126 (1): 39–49.

84 Cho, Y.-G., Rhee, S.-K., and Lee, S.-T. (2000). Effect of soil moisture on bioremediation of chlorophenol-contaminated soil. *Biotechnology Letters* 22 (11): 915–919.

85 Jones-Lepp, T.L. and Stevens, R. (2007). Pharmaceuticals and personal care products in biosolids/sewage sludge: the interface between analytical chemistry and regulation. *Analytical and Bioanalytical Chemistry* 387 (4): 1173–1183.

86 Kinney, C.A., Furlong, E.T., Zaugg, S.D. et al. (2006). Survey of organic wastewater contaminants in biosolids destined for land application. *Environmental Science & Technology* 40 (23): 7207–7215.

87 Amoah, I.D., Singh, G., Stenström, T.A., and Reddy, P. (2017). Detection and quantification of soil-transmitted helminths in environmental samples: a review of current state-of-the-art and future perspectives. *Acta Tropica* 169: 187–201.

88 Fent, K. (2015). Progestins as endocrine disrupters in aquatic ecosystems: concentrations, effects and risk assessment. *Environment International* 84 (C): 115–130.

7

Biodegradable Waste Collection Systems

Jolanta Sobik-Szołtysek

Faculty of Infrastructure and Environment, Department of Environmental Engineering and Biotechnology, Czestochowa University of Technology, Częstochowa, Poland

Directive 2008/98/EC of the European Parliament and of the Council [1] defines bio-waste as biodegradable garden and park waste, food and kitchen waste from households, restaurants, caterers, retail outlets, and comparable waste from food processing plants. A substantial part of bio-waste is green waste such as tree and shrub branches, leaves, flowers and grass cuttings, sawdust, tree bark, fruit, and vegetables. Bio-waste does not include agricultural waste, manure, sewage sludge, forestry residues, and biodegradable waste such as natural fibers, paper, and cardboard. A synonym for the term bio-waste is the organic fraction of municipal solid waste (OFMSW). Of all municipal waste fractions, it has the greatest environmental impact [2, 3]. In addition to the concept of bio-waste, European Union law also uses the term "biodegradable waste" as defined in Council Directive 1999/31/EC [4]. It is a broader concept than bio-waste as it refers to any waste that decomposes anaerobically or aerobically, including paper and cardboard.

Ensuring efficient management of solid waste, including bio-waste, is recognized as one of the key challenges of the twenty-first century and a fundamental element of sustainable development [5]. Bio-waste is one of the primary municipal waste streams. At the same time, due to its specific properties, bio-waste has a very high potential for the circular economy. Therefore, it is critical to selectively collect bio-waste because only such a system ensures the possibility of the full utilization of its raw material and energy potential. Failure to selectively collect bio-waste has a very negative impact on the efficient management of the remaining municipal waste fractions. This is because mixed collection results in contamination of these fractions with bio-waste, which hinders or even prevents effective recovery of secondary raw materials.

7.1 Rules and Systems of Bio-waste Collection

One of the most effective methods of waste disposal is the selective collection for reuse and recycling. Nowadays, waste collection is understood to mean collecting waste, including its preliminary sorting that does not lead to a fundamental change in its nature and composition and its subsequent collection. Therefore, the process of the waste collection begins when it is dumped into the appropriate containers and ends when it is collected and emptied by collection vehicles. This process also includes transportation of waste to the place where the vehicle is emptied, e.g.

Biodegradable Waste Management in the Circular Economy: Challenges and Opportunities, First Edition.
Edited by Malgorzata Kacprzak, Eleonore Attard, Kari-Anne Lyng, Helena Raclavska, BalRam Singh, Eyob Tesfamariam, and Franck Vandenbulcke.

to treatment or disposal facilities [6, 7]. A waste collection system (WCS) can be defined as a combination of technology and human activity, with its characteristics including [8]:

- containers used for waste collection,
- the type of vehicles used to unload the waste containers,
- the method used to position and empty the containers.

All these elements are interdependent on each other and other elements of the urban environment, i.e. waste producers (citizens), urban infrastructure, and traffic. It is important, for example, to plan the locations of waste containers and the schedule for their collection to ensure that collection takes place quickly and during periods of reduced road traffic. Therefore, the way these components interact impacts on the functionality and efficiency of the system. Figure 7.1 shows a diagram of the relationships between individual elements of the WCS.

Waste containers vary greatly in their type (bags, containers, bins, barrels), the material they are made of (plastic or metal), and size. The types of vehicles used to collect or unload waste containers are also diverse. The most commonly used waste collection vehicles are generally equipped with the option of mechanical unloading of the container from the vehicle using a lifting device [10]. These vehicles often have waste compactor devices with one or more compartments dedicated to different fractions of municipal waste. For the collection of biodegradable waste characterized by a higher moisture content, specialized vehicles are increasingly used, with their design preventing the outflow of leachate from waste. The method of waste collection depends on the mechanism for emptying the container. The container can be emptied at the waste collection point (simple emptying, stationary emptying), exchanged for an empty container (exchange) or transported without emptying (with the waste inside) to its final destination. The container loading system is usually mechanized. With regard to the place of emptying (collection) of the waste container, there are two systems:

- a pick-up system (or curbside collection system), with containers standing for collection near the point of waste generation,
- a drop-off system (or bring system), with waste transferred by its producer to a central collection point and dumped into prepared containers. In this system, transport costs are lower because the collection vehicles only take over the waste from the collection center rather than separately from each property.

One of the most commonly used systems for selective collection of bio-waste is the door-to-door system. In this system, waste containers from each property are placed by the owner on the day

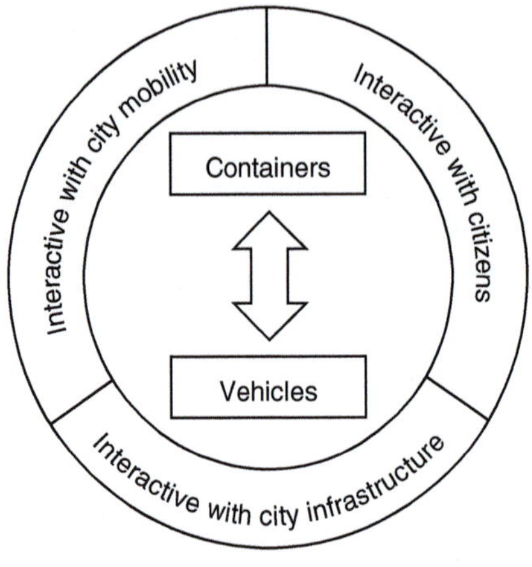

Figure 7.1 Waste collection system diagram [9].

of collection in a designated place, e.g. in front of their houses, and the waste is collected through emptying the containers [11, 12]. Waste containers can also be placed on the property and emptied there by the crew responsible for loading the waste into a vehicle [6] or transported by them to a truck, emptied, and returned to their storage location [13]. The so-called multicontainer system may also be used, with residents (waste producers) placing each fraction, including bio-waste, in special containers located so that the containers with the organic fraction are at a maximum distance of 50 m, while containers for the other fractions - at a maximum of 300 m from households [14]. Large containers for green waste can also be seasonally placed in central points of the city to serve, for example, one or several districts [12].

Municipal WCSss use containers or bags of different colors, which facilitates correct segregation of waste into particular fractions already at the collection source. In EU countries, including Poland, the following colors are used:

- blue containers: paper and cardboard,
- green containers: glass,
- yellow containers: metals and plastics, including multimaterial packaging,
- gray containers: non-segregated municipal waste.

Furthermore, biodegradable waste, with particular emphasis on bio-waste, is collected in brown containers, usually labeled "Bio." The brown color for biodegradable waste containers is generally accepted and applied in individual EU countries (Figure 7.2).

Containers for bio-waste should ensure that the quality of the waste collected is not impaired by future treatment processes. Therefore, it is suggested to equip such containers with ventilation and a grate for leachate. Figures 7.3–7.5 show examples of such solutions. Arrangement of holes on both sides of the container (Figure 7.3) ensure air circulation in the container, thus improving waste drying and reducing its weight and odors [16].

The containers can be equipped with aeration screens at the top and bottom of the container sides to allow increased air circulation within the container. The side screens and honeycomb perforations can be connected to obtain a larger ventilation area (Figure 7.4).

An innovative solution is the bio-grid in the form of a folding grid, which allows the leachate from bio-waste to fall to the bottom of the container. The leachate evaporates through the side aeration screens. The solution reduces the volume of the container by a maximum of 10% (Figure 7.5).

Some collection systems introduce an additional separation of biodegradable waste into bio-waste and green waste. In this case, two appropriately labeled separate brown containers are delivered

Figure 7.2 Sample colors of waste containers [15].

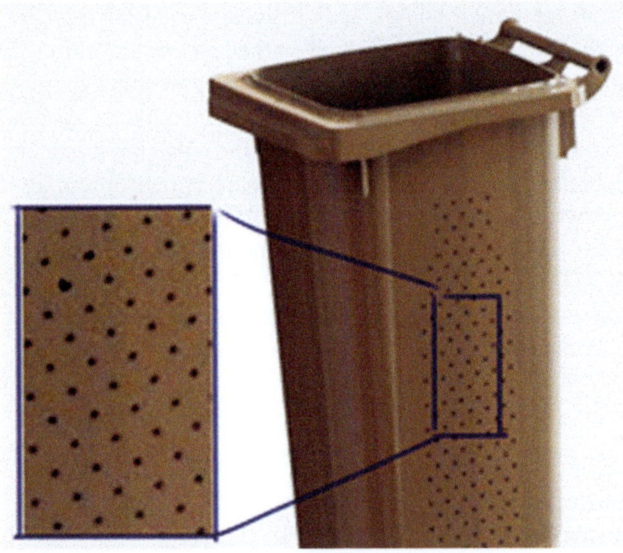

Figure 7.3 Ventilation in the bio-waste container [16]. Source: ESE World B.V.

Figure 7.4 The side ventilation in the bio-waste container [16]. Source: ESE World B.V.

to the recipients. If waste is collected in bags rather than containers, it is collected along with the packaging, i.e. the bag. Furthermore, in some countries, e.g. Poland, there is a possibility to selectively transfer green waste, free of charge, to the so-called selective municipal waste collection points organized on the territory of individual administrative units (towns, community) for their residents.

The selective collection system should be used wherever it is technically, environmentally, and economically feasible. At the same time, such a system should be organized with specialized

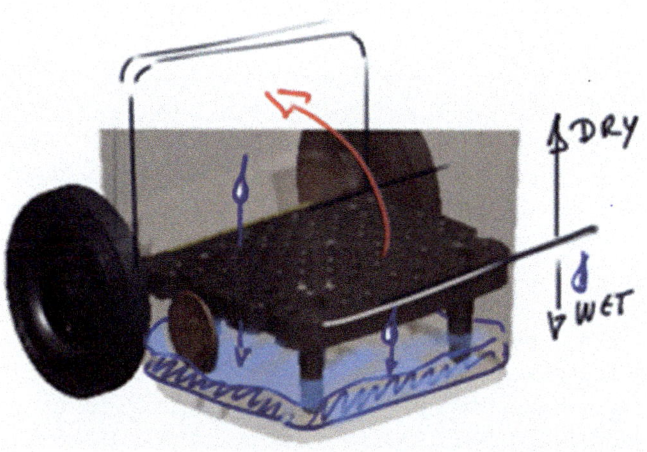

Figure 7.5 Bio-grid in the bio-waste container [16]. Source: ESE World B.V.

installations and be convenient for the users. A prerequisite for the introduction of selective waste collection of bio-waste is that the public sector bears the initial investment costs. However, if the entire waste management system is optimized, the overall economic outcome is positive both for citizens and the system [17]. The world's first selective municipal WCS started to be used in 1895 in New York, and its originator was Colonel George E. Waring, known as the "apostle of cleanliness." Over time, this idea began to be repeated around the world. In the 1970s, the first industrial-scale selective waste collection programs were introduced in the United States and Western European countries. These actions required both legal regulations and large-scale public education. They did not begin to produce measurable outcomes until 20 years later.

Proper bio-waste segregation is difficult and requires more commitment than for other segregated municipal waste, such as glass, metals, or plastics. Biodegradable waste constitutes the most problematic fraction, because its segregation and management generate higher costs, due to, for example, the necessity of a higher frequency of collection, which translates into higher waste fees. It is extremely important to maintain continuity of the collection of bio-waste throughout the year and appropriate frequency of collection from residents, following the-more-often-the-better principle. Failure to collect such waste for a longer time may cause, among other things, a nuisance to the environment (rapidly progressing decay processes, odor nuisance, etc.).

Another problem to be solved is whether green waste should be collected separately or together with kitchen waste because it is characterized by completely different parameters and the type of waste determines the installation it will be transferred to. Green waste is generated seasonally and can be collected in bulk in collection containers. The volume of kitchen waste is much lower, but it is produced on a daily basis, has a semi-liquid form, and quickly undergoes rotting and fermentation processes. Kitchen waste means vegetable and fruit scraps, peelings, egg shells, coffee and tea grounds, wilted flowers and potted plants, and food scraps, but without meat, bones, or fats. Animal products (except egg shells), pet feces, vegetable oils, ash, and soil should not be added to this type of waste. Green waste usually contains 50–60% water and more wood (lignocellulose), while kitchen waste contains no wood and its water content reaches 80%. Therefore, kitchen waste should be transferred to a biogas plant whereas green waste goes to a composting plant as it has no biogas potential. Collecting these groups of waste together significantly reduces the efficiency of the treatment systems used. Bio-waste should be collected in bulk in appropriately labeled containers, in biodegradable/compostable bags meeting standards, or paper bags. It is preferred to damp bio-waste in bulk to the container.

Satisfactory results in the selective collection of bio-waste can be achieved, among others, by introducing the PAY-as-You-Throw (PAYT) system, i.e. the users pay for as much waste as they throw away. The different tariffs reflect the amount of waste disposed of by each resident and its type. Such a system was introduced, for example, in many regions of Italy, where the frequency of bio-waste collection at individual households is measured by microchips fixed on waste containers/bags, which are scanned by the garbage truck during collection. In the regions where such a system was introduced, new fees resulted in more efficient waste prevention and segregation, raising the collection rate for segregated waste even to 90%. Similar WCSs operate in Milan and Barcelona. The widespread introduction of the PAY-as-You-Throw system would certainly contribute to the optimization of bio-waste management.

On the other hand, the door-to-door collection method makes it easier to identify the resident with the quantity and/or quality of bio-waste produced. This allows for the quality of selective collection to be enforced at the collection level. These methods can be applied to individual consumers. If communal containers are used (multifamily housing), other alternative models of bio-waste management can be used. The use of closed containers equipped with user identification technology makes it possible to link a specific person to waste generation. This type of controlled access of users to the waste container is a factor motivating their proper segregation.

7.2 Environmental Impact of the Bio-waste Collection System and Possibilities of Its Further Management

Bio-waste is one of the primary waste streams. For example, the percentage of bio-waste in the total municipal waste stream in EU countries is 34%, of which about 60% is food waste, 35% is garden waste, and the remaining 5% of bio-waste is defined as "other" [18]. Figure 7.6 illustrates the percentage of biodegradable waste in total typical municipal waste in developed and developing countries according to [19].

Systematic solutions for the collection and management of biodegradable waste are a good example of sustainable development and green economy, which includes environmentally friendly

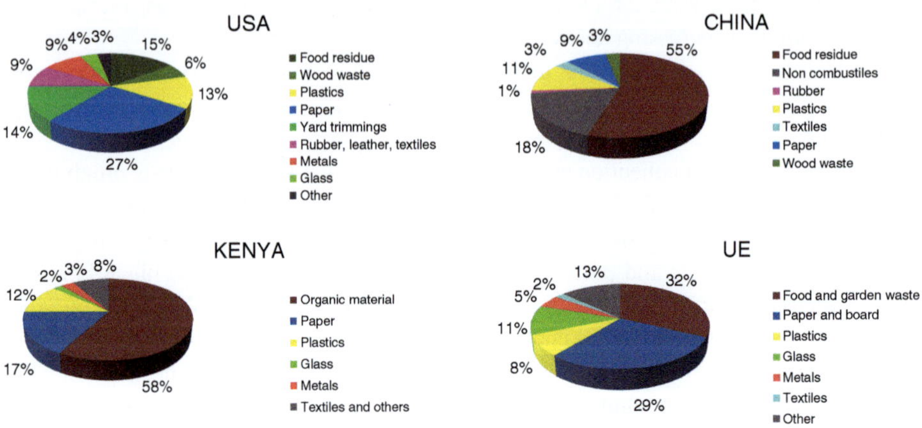

Figure 7.6 The share of biodegradable waste in the municipal waste stream [19].

technologies and renewable energy sources [20]. Management of this waste should be carried out in such a way as to exclude any danger to the environment and human health. Therefore, the effective management of bio-waste should consist in its correct selective collection in order to minimize the environmental impact and to ensure the maximum utilization of its potential as raw material and an energy source. The level of selective collection of bio-waste varies significantly between countries, both in Europe and worldwide. The efficient operation of selective WCSs requires a high level of knowledge and involvement of all citizens [21]. For example, in 2017, ca. 50% of municipal bio-waste generated in 32 European countries was collected together with residual (mixed) waste, and the other 50% was collected separately. Selective collection rates range from 80% or more in Austria and Slovenia to less than 10% in Bosnia and Herzegovina, Cyprus, Northern Macedonia, Portugal, Spain, and Turkey. Selective collection of bio-waste across the country has also been recently introduced in Malta. In some countries, selective collection of bio-waste is being introduced as a pilot project in certain cities, regions, or municipalities, while in Poland, Portugal, and Serbia, this type of collection of bio-waste is already used in the entire country [22]. In Belgium (Brussels–Capital Region and Flanders) and the Netherlands, projects are being implemented to improve and further develop already existing collection systems.

Waste management systems based on collecting waste and transporting it to landfills are obsolete and uneconomical. It is estimated that collection costs account for 40–60% of the overall solid waste management costs [23]. Selective collection of bio-waste is a prerequisite and pre-condition for the utilization of bio-waste as a resource cyclically and enables its use as valuable secondary raw materials for the production of biofuels and other value-added bio-products such as soil improvers, organic fertilizers, and biogas as an alternative energy source [21, 24, 25]. The primary objective of selective collection of the bio-fraction is therefore to minimize waste transported to landfills, increase recycling rates, and reduce greenhouse gas emissions. Bio-waste discharged to landfills is biodegraded and produces a gas composed mainly of methane, whose impact on the climate is 23 times stronger than the impact of CO_2 [26]. Methane from landfills is estimated to be the primary source of greenhouse gas emissions, accounting for ca. 3% of total EU greenhouse gas emissions [27]. US landfills are the third-largest source of methane emissions in the country, generating, for example, in 2017, 108 million metric tons of CO_2 equivalent [28]. Landfilled bio-waste also affects the quantity and quality of leachate, which poses several environmental hazards due to its pollution potential [29]. Leachate needs to be collected and treated over a long period of time and at considerable technical expense [30]. Another argument supporting the need for selective collection of bio-waste at source was provided by, for example, the research on the effect of the bio-fraction on the quality of fuels derived from municipal waste [31]. In the case of selective bio-waste collection at the level of 65–70%, the characteristics of waste fuels change by lowering the content of certain chemical substances (including heavy metals) and ash compared to fuels derived from mixed waste. Energy recovery (calorific value of the fuel) is also increased and dioxin emissions are reduced. Selective collection of bio-waste may also offer the opportunities for using it in other forms of recycling currently being explored, e.g. the production of chemicals in biorefineries.

It is difficult to indicate the best method of biodegradable waste management from the point of view of environmental protection because its environmental performance depends on many factors, including the system of collecting, storage, and quality of waste, and potential opportunities of using the obtained products (electricity, heat, gas, or compost). Therefore, for example, Life Cycle Thinking (LCT) combined with a supporting tool, i.e. Life Cycle Assessment (LCA), should be used to develop the best waste strategy [32]. The circular economy is based on the assumption that waste materials should return to the production cycle as embedded material or energy through recycling, composting, or anaerobic decomposition [33].

The literature on bio-waste treatment methods emphasizes 13 bio-waste treatment technologies, grouped into four categories [34]:

A. direct use (direct land application, direct animal feed, direct combustion),
B. biological treatment (composting, vermicomposting, black soldier fly treatment, anaerobic digestion, fermentation),
C. physico-chemical treatment (transesterification, densification),
D. thermo-chemical treatment (pyrolysis, liquefaction, gasification).

Composting and anaerobic digestion seem to be the most economically and environmentally beneficial procedures for bio-waste treatment. In the case of composting, organic matter is restored to the environment whereas the fermentation process additionally produces energy. The basic prerequisite to ensure the correct course and efficiency of these processes is a sufficiently good quality of the starting material. This can be ensured by separating bio-waste at source. Numerous scientific studies have demonstrated the differences between the composition of organic matter collected selectively and that mixed with municipal waste. It was found that, among other things, the nutrient content in bio-waste is higher for selective collection compared to mixed waste. The content of heavy metals is also lower, which supports the thesis that the migration of heavy metals from non-compostable materials to the decomposing organic matrix occurs during their mutual contact [35]. Similar results were obtained by Malamis et al. [36, 37] and Schüch et al. [31], confirming that selective collection of bio-waste is necessary to obtain high-quality compost. Thus, if biodegradable waste is processed by organic recycling to the compost of acceptable quality, the selective collection must be clean, which excludes the presence of other waste, especially plastic packaging. Most oil-based plastics are resistant to microorganisms because they do not contain the enzyme capable of degrading most artificial polymers. The enzymatic activity is further limited by the hydrophobic nature of plastics [38]. It is permissible to collect bio-waste in packaging with similar biodegradation properties and the capability of composting. However, this applies only to compostable packaging and not to biodegradable packaging. Compostable packaging is composted with bio-waste, while biodegradable but non-compostable packaging is considered to be regular plastic packaging. It was found [35] that some polyethylene bags used in selective organic waste collection processes contained large amounts of heavy metals, while compostable bags contained only small amounts of these elements. Part of the plastics that are currently labeled as compostable or biodegradable is designed to decompose only under controlled conditions in industrial bio-waste treatment plants. Under anaerobic conditions, these plastics do not degrade or degrade to a limited extent [39]. In fact, only bags made from natural materials, such as starch, are easily biodegradable in composting conditions. The strategy to use compostable bags for the collection of bio-waste varies between EU member states. In some countries, these bags are used as a means to facilitate collection and reduce contamination with conventional plastics, while in others, their use is banned [18]. The use of compostable or biodegradable plastics to manufacture products that are, for example, attached to food waste, such as fruit stickers or tea bags, can also help reduce bio-waste pollution [40]. The compostability of plastic is indicated by a certificate because the term biodegradable does not always mean ecological. It is therefore critical to develop clear and consistent marking and instructions for the use and disposal of biodegradable/compostable plastics. Figure 7.7 shows example logos for compostable materials, i.e. (a) BPI, (b) DIN CERTCO, and (c) OK compost according to [41].

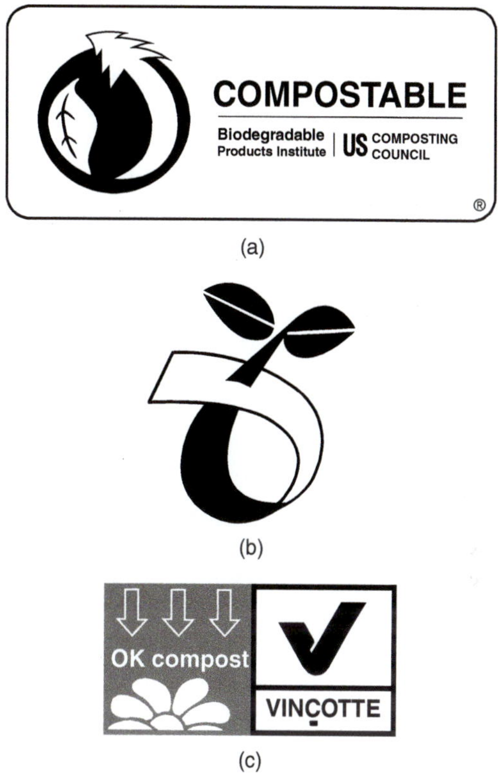

Figure 7.7 Sample markings for compostable materials [41].

Bio-waste is easily contaminated during mixed municipal waste collection (non-selective collection), with typical contaminants including heavy metals and other waste (e.g. broken glass). There is also a potential risk of contamination by persistent organic substances such as PCDD/Fs, PCBs, or PAHs, so in order to obtain compost of satisfactory quality, it is crucial to avoid the presence of such materials in bio-waste [42]. The level of bio-waste contamination depends to a large extent on the type of collection system. Experience shows that door-to-door collection systems are more likely to encourage residents to collect a separate fraction of bio-waste compared to those with containers available on the street [43]. The production of compost from contaminated bio-waste and its application to the soil can lead to contamination and accumulation of hazardous substances in the soil environment and in plants, and thus their introduction into the trophic chain. Therefore, proper control of the initial material (bio-waste) together with monitoring the quality of the compost produced is essential. Household composting, using domestic composters made of wood or plastic, is considered the most environmentally friendly method of handling domestic biodegradable waste. This method reduces the volume of bio-waste to be collected and treated, eliminates gas emissions, reduces costs associated with transportation [44], and, most importantly, ensures strict control of the stored material and increases the environmental awareness of those generating the bio-waste. One example is Slovenia, which has one of the lowest fraction of bio-waste in municipal waste generated in Europe [45]. Individuals who do not choose home composting of bio-waste are obliged to separate it for selective collection conducted by public services [43]. The possibility of widespread community composting at the local level is also considered, e.g. housing communities and small local communities, especially those in areas far from larger urban centers [46]. Proper segregation and selective collection of bio-waste to achieve the right quality of initial material is also a key factor in this case. This represents a typical example o,f closing the waste cycle. For economic reasons, mainly related to the high costs of transporting waste to processing installations, such selective collection solutions should be chosen especially by municipalities with a low population density, far away from larger industrial centers [47–49].

Information campaigns to raise awareness and training to motivate people to segregate waste are extremely important [22]. In Flanders, Belgium, more than 5000 volunteers have been trained in recent years to help residents with selective collection of bio-waste and home composting [50]. In the Spanish province of Pontevedra, a decentralized system for the separate collection and management of biodegradable waste was designed, with a combination of home composting, local composting, and small-scale composting facilities with a simultaneous promotion and training campaign for residents [51].

The characteristics of bio-waste supplied as digestion feedstock to a biogas plant are strictly dependent on the WCS and its pre-treatment. There are two basic systems in use in EU countries for the collection of this waste: collection of mixed waste and the selective collection of the different bio fractions at source. The process of anaerobic digestion of bio-waste is carried out in closed reactors, which significantly reduces the emissions of gases into the atmosphere. It is assumed that 100–200 m^3 of biogas can be obtained from one ton of bio-waste destined for biological treatment. Therefore, the combination of the possibility of recovering energy from biogas and obtaining a soil improver in the form of digestate makes this method of dealing with bio-waste in many cases the most beneficial ecologically and economically. In 2007, Europe's largest center for the use of biogas as a biofuel was opened in Lille, France. The installation processing the selectively collected organic waste produces about 4 million m^3 of biogas $year^{-1}$. The biogas is then processed into transport fuel used in the operation of buses within the urban transport network.

Composting is the dominant treatment method for separately collected bio-waste in Austria, Belgium, Cyprus, Italy, the Netherlands, Slovakia, and Spain, while anaerobic digestion methods are mostly used in Croatia, Poland, Portugal, Slovenia, Sweden, and Turkey. Among EU countries, Belgium, Cyprus, Hungary, Italy, Latvia, Poland, Portugal, Romania, Slovakia, and Spain have the possibility to treat all selectively collected municipal bio-waste. However, in order to be able to process all the municipal bio-waste (not only that selectively collected), it will be necessary to expand the technical infrastructure [18].

7.3 Examples of Bio-waste Collection Systems Used Worldwide Related to Waste Management Methods

Collecting bio-waste at source is being systematically introduced in the European Union. In some EU member states, recycling is combined with selective waste collection, e.g. in Austria, the Netherlands, Germany, Sweden, part of Belgium (Flanders), Spain (Catalonia), and northern regions of Italy, while in other countries, such as Czech Republic, Denmark, and France, composting of green waste and collection of kitchen waste together with municipal solid waste (MSW) is preferred. In all regions where selective waste collection has been introduced, this method is recognized as a very practical and effective form of waste management [52]. Under current EU regulations, from 1 January 2027, member states will be able to consider aerobically or anaerobically treated municipal bio-waste as recycled only if it is separately collected or sorted at source, i.e. in households, in a specified way. By 31 December 2023, member states have to ensure that bio-waste is separated and recycled at source or selectively collected without mixing with other types of waste.

As global statistics show that as much as one-third of the world's food is dumped to the trash, highly developed countries are paying increasing attention to reducing food waste and increasing the recovery of bio-waste in the form of kitchen and catering waste. About 88 million tons of food (173 kg per person) is wasted every year in the EU member states, which corresponds to ca. 20% of all produced food [18]. Therefore, in line with the objectives of sustainable development, the principle was adopted to reduce food waste by half by 2030 [53]. The first European city to declare Zero Waste objective was Capannori, Italy. The city has one of the highest rates of municipal waste recycling in Europe due to, for example, effective management of the bio fraction. With the system of segregation and collection used in the city, organic waste is segregated at home and collected from the residents with high frequency and then delivered to the composting plant. The fact that the city's public canteens are equipped with facilities for composting also reduces the cost of collection, transportation, and processing of waste. Free compost containers were distributed, with training on how to properly compost bio-waste. There is also a 10% discount on the waste fee for composting. Capannori's example has been followed by the Slovenian municipality of Vrhnika, which decided to become the first Zero Waste municipality in the country and achieve a level of 82% selectively collected waste by 2021. In another Italian city, Treviso, the public company Contarina implemented efficient separate waste collection at source combined with a PAYT system. Municipal waste is collected in five or six main streams, including organic waste (food leftovers) and green waste, collected in special colored bins and collected from residents from their households. Organic waste is collected with the highest frequency. For composting at source, the municipality offers discounts on waste fees of up to 50%.

In Scotland, selective collection of bio-waste increased by over 40% between 2013 and 2017. The country's government adopted a target of reducing bio-waste production by 33% by 2025. The Zero Waste Scotland fund was set up by the government and the EU to coordinate the project through legal regulations, education, training, and the implementation of effective tools to increase the recovery of bio-waste and its use to generate energy. Most bio-waste is sent for further treatment in biogas installations, increasing the percentage of renewable energy sources in the energy sector in Scotland. Lower CO_2 emissions and reduced dependence of the power industry on fossil fuels are also very important. In 2017, the Scottish biogas sector produced ca. 245 520 MWh of electricity, with a third coming from food waste. This energy is sufficient to power nearly 20 000 households. Such impressive results were achieved using appropriate legal regulations. A law introduced in Scotland in 2012 requires entrepreneurs to use the selective waste collection. However, since 2016, enterprises that generate at least 5 kg of bio-waste per week are obliged to document the fact that they sort waste in an appropriate manner. Proper bio-waste management is also facilitated by an appropriate technical base. The number of bio-waste treatment facilities enables up to 80% of the Scottish population to have access to selective bio-waste collection schemes. In 2019 the United Kingdom (of which Scotland is a part) announced that, for the first time ever, the country is using more electricity from renewable sources than from coal.

In Spain, the Catalan city of Argentona is a good example of rational bio-waste management. Residents of the city were provided with bins for the organic waste of varying sizes, which are collected using the door-to-door system, while other waste streams continue to be collected in containers placed on the streets. In the Basque Country province of Gipuzkoa, Spain, food waste and bio-waste generation has been significantly reduced due to cooperation with the local Food Bank.

In some countries, the selective collection system has been successfully applied for many years. In Germany, selective bio-waste collection was introduced in 1985. Selective collection systems vary from region to region in that country. The selective waste collection covers on-average 64% of municipal bio-waste. In Munich, such a system has been in operation for a relatively long time, since the Waste Management Corporation introduced the experimental selective bio-waste collection in 1988 and started to use it in practice in 1994. A bin system was used, with organic waste and paper residual waste collected separately. A biogas plant is also part of the bio-waste management system. The biogas plant processes ca. 22 000 tons of food and garden waste a year using dry fermentation, yielding 2600 MWh of electricity and process heat. Up to 8000 tons of compost is also obtained every year [54]. The system of waste collection from households in Frankfurt am Main is similar to the systems operating in the entire area of the former Federal Republic of Germany. Biodegradable waste is collected from brown containers placed on properties by a company indicated by the municipality. Next, they are transferred to a biological waste treatment plant where waste is processed using combined methods using anaerobic digestion and classic composting. The biogas is incinerated in the engines of the block combined heat and power plant, where both heat and electricity are generated and sold to the grid as green energy. Green waste and solid residue from the process of anaerobic digestion are subjected to the composting process. On the other hand, in the area of Frankfurt (Oder) located on the territory of the former German Democratic Republic, household (mixed) waste and biodegradable waste are collected from containers placed on private properties with the use of weighing devices on vehicles, while for bio-waste, the standard volume of a container is assumed to be 15 l per resident.

In the Swedish city of Västerås, ca. 90% of the residents segregate organic waste for the municipal biogas plant. Biodegradable waste is collected in paper bags provided by the city with printed instructions. Food scraps, garden waste, wilted flowers, potted plants, and waste paper from households are collected. Filled bags are stored in ventilated plastic containers. In blocks of flats, containers for organic waste are located in special rooms where waste is brought by residents. Waste from restaurants and catering outlets is collected in a liquid form into tanks and transported to biogas plants by liquid tankers [55].

The Czech Republic has successfully implemented the principle of producer responsibility for municipal waste management. The effective organization of recovery ensures covering the costs of handling selectively collected waste (reimbursement of costs to municipalities, sorting plants, and other entities involved in the selective collection of municipal waste). Prague has implemented a municipal waste management system typical of the Czech Republic. Selective collection of municipal waste in the city is mandatory. The selective collection system operates on several levels, with bio-waste collected at the following locations:

- selective waste collection points,
- mobile selective waste collection points,
- mobile systems for organic waste collection,
- other, mixed municipal waste in containers for mixed waste located in private properties.

Vienna uses a waste management system typical of all large Austrian cities. Municipal waste is owned by the city. Bio-waste is collected in containers and collected selectively at individual properties or in "green islands" for selective waste collection in public places. Typical containers with a capacity of not less than 110 dm^3 are mainly used for the collection of municipal waste from private properties. The selective waste collection points have a system of containers for collecting all types of municipal waste (free of charge). At some points, for example, residents have the opportunity to pick up small amounts of compost produced from the municipal waste collected in the city free of charge. Biodegradable waste from the area of Vienna is treated biologically in the anaerobic waste treatment installation (Biogas Wien) and in the Lobau composting plant. Composting has also been used as a form of management of selectively collected bio-waste in Finland, e.g. in

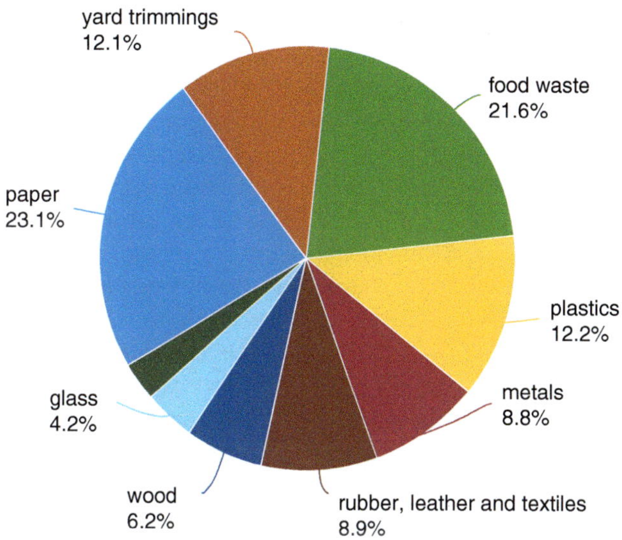

Figure 7.8 Total MSW generation in the United States by type of waste in 2018 according to [57].

the Pielinen Karelia region in North Karelia, which consists of four municipalities: Juuka, Lieksa, Nurmes, and Valtimo [56].

Bio-waste disposal is increasingly becoming a global problem. In 1960, the United States produced only 88.1 million tons of MSW. In 2014, this number rose to 258.5 million tons, and in 2018, it reached 292 million tons. Figure 7.8 shows total MSW generation in the United States by type of waste in 2018.

Most of this waste is landfilled or incinerated in incinerators. In 2018, about 12% of the 292 million tons of MSW produced in the United States was burned in waste-to-energy plants. Some 50% of the municipal waste collected was transported to landfills. This system is expensive and environmentally damaging. In recent years, more and more attention has been paid to the composting of bio-waste, which requires its selective collection. This management of bio-waste could reduce the volume of waste landfilled and incinerated in the US by at least 30%. In San Francisco, composting and recycling programs have reduced waste landfilled by 80%, with ca. 255 500 tons of organic material used for composting each year. The state of Vermont passed a universal recycling law back in 2012, banning biodegradable organic materials from landfills since 2020. To increase the amount of bio-waste used from composting, waste is collected from containers placed on the streets. The residents and businesses are encouraged to participate in composting programs including Save Money and Reduce Trash (SMART) in which residents pay less for waste collection if they throw away less waste. This represents a direct financial motivation to dispose of organic waste in composters rather than in mixed waste containers. Furthermore, organic waste is collected as regularly and frequently as the other fractions. In Portland, Oregon, organic waste is collected from residents more often than other waste. An important element of the bio-waste management system in the USA, especially kitchen waste, is the commonly used household waste grinders installed in kitchen sinks. These devices grind food leftovers very finely into several-millimeter pieces that flow down with water into the sewage system, becoming a component of municipal sewage and, consequently, of sewage sludge, formed after its treatment. This method significantly reduces the volume of kitchen waste that is non-compostable and would have to be landfilled or incinerated.

Another problem in the USA regarding bio-waste is food waste since, according to the EPA, Americans throw away nearly 40 million tons of food each year, most of which ends up in landfills. Therefore, food represents the largest component of waste in US landfills, with as much as 22% of total MSW. To reduce this waste, some states are taking measures to limit the process. According to the National Conference of State Legislatures, California, Connecticut, Massachusetts, Rhode Island, and Vermont introduced legislation to reduce the volume of food waste disposed into landfills. In California, Colorado, and Massachusetts, funding is planned for private sector composting and selective collection programs for organic fractions. In 2019, New York expanded the principles for separate collection of organic waste at businesses involved in food production and distribution. It is expected to reduce the amount of food going to landfills by nearly 100 000 tons $year^{-1}$. The United States Department of Agriculture (USDA) and the United States Environmental Protection Agency (EPA) set a goal to reduce food waste in half by 2030. According to the World Wildlife Federation, wasted food in the USA is equivalent to the greenhouse gas emissions from 37 million cars. A good example of the efficient use of selectively collected bio-waste is the Kompogas®SLO renewable energy and recycling plant in California (Figure 7.9). In this plant, local organic waste, i.e. out-of-date food, kitchen scraps, and household green waste, is transformed into renewable electricity, compost, and liquid fertilizer. This allowed for the 75% reduction in organic matter originally deposited in landfills by 2020, thus extending their life.

In China, the world's largest and most populous country, the objectives of bio-waste management are in line with global trends. Segregation of MSW at source is considered a pre-requisite for efficient recycling and ensuring high-quality end products of its treatment [59]. Figure 7.10 shows the volume of MSW collected vs. the population of China from 2012 to 2016 based on the China National Bureau of Statistics data from 2019.

MSW, including bio-waste, is divided into four main categories in China: food waste, recycled waste, landfill waste, and hazardous substances. According to Chinese researchers, the most effective method of managing municipal waste is sorting at source combined with a method of converting waste into energy. The main component of MSW is food waste with a high content of moisture and organic fraction and should therefore be used for compost production, especially in southern China, where food waste accounts for ca. 50% of MSW. To avoid the presence of contaminants that

Figure 7.9 KompogasSLO renewable energy and recycling plant in California [58]. Source: Hitachi Zosen Inova AG.

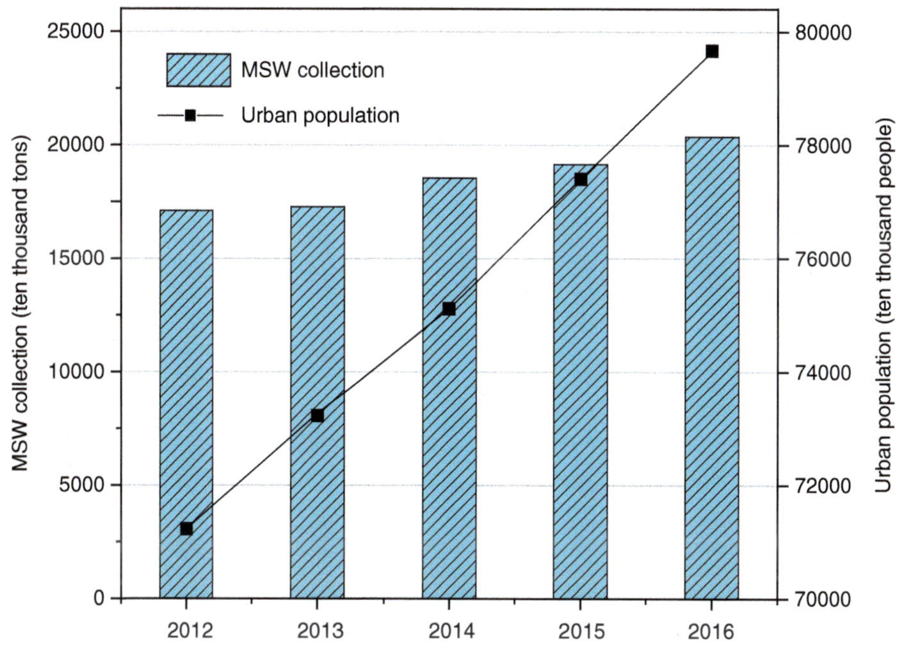

Figure 7.10 Volume of municipal solid waste (MSW) collected in China in 2012–2016 [60].

interfere with the composting process, the number of bins and the frequency of collection should be steadily increasing. Reducing the amount of food waste in the MSW stream through selective collection improves the energy value of municipal waste for energy recovery during incineration processes [60].

The amount of bio-waste generated in all Asian countries is estimated at 2.6 billion tons year^{-1}, including ca. 260 million tons year^{-1} in Japan. For example, municipal waste incineration technology has been widely adopted in Japan. Most domestic waste and ca. 60% of sewage sludge (after digestion) has been processed using this method until recently [61]. However, economic and environmental requirements have forced the country's authorities to introduce new ecological methods of bio-waste management. An example is the bio-waste recycling system used in the rural town of Ohki-cho (Fukuoka Prefecture) in western Japan. Organic waste from households is segregated at source and selectively collected together with sewage sludge by the municipality. These two waste streams are mixed and supplied together into an anaerobic digestion plant. The obtained biogas is used for energy generation and the digestate is used as slurry by the agricultural sector in local paddy fields [62].

Figure 7.11 shows the percentage of MSW incinerated for energy recovery in selected countries including Japan. The part of Scandinavia presented in the chart includes Denmark, Norway, and Sweden. The data in the chart are given for 2016 in South Korea, for 2017 in Japan, and 2018 in all other countries.

Waste management in most developing or underdeveloped countries is mainly based on uncontrolled landfilling [63] and domestic waste incineration, resulting in serious environmental and health problems for the population. Most often, this consists in unorganized landfilling of the waste outside the village or town [64]. There are virtually no municipal WCSs in operation in these regions. For example, in India, more than 90% of MSW is directly disposed of by landfilling, with the recycling rate generally not exceeding 10% [65].

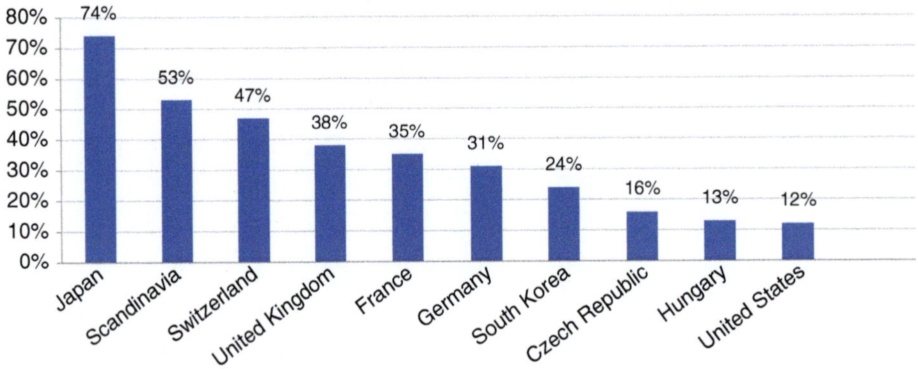

Figure 7.11 Percentage of MSW incinerated for energy recovery in selected countries of the world [57].

The organic fraction comprises 45 to over 80% of all waste generated in developing countries [66]. Most often, these countries do not have adequate facilities to process it safely. Organic waste management is one of the major problems that require special attention from developing countries [67]. This is due to the rapid and unplanned urban development, infrastructural problems, lack of financial resources, adequate technical knowledge, and legal regulations.

Organic waste represents the largest part (more than 65%) of all waste generated in Tunisia (North Africa). The three main categories of bio-waste generated here are food waste, green waste, and cattle manure. However, only 5% of them are designed for composting while there are no anaerobic digestion facilities to produce biogas [68]. At the same time, the composition of this waste varies from region to region, thus influencing the choice of the management system. The study confirmed that the technique suitable for Tunisia in terms of waste management and environmental and energy aspects is the biological treatment of biodegradable waste. Although economic analysis has shown that the returns on both composting and biogas plants would be small, socioenvironmental considerations must also be taken into account, including the reduction of the environmental risk and operational costs related to landfilling [69, 70]. The major obstacles to proper waste management in Tunisia are the lack of funding, public awareness, trained staff, reliable data, and basic technical knowledge in this field. This limits the choice of appropriate bio-waste treatment or disposal techniques that could be used under local conditions.

Difficulties in efficient waste management in some parts of the world are also exacerbated by political instability, civil wars, and military operations. For example, in the Palestinian and West Bank areas, more than 80% of waste is disposed of in open landfills and only 2–8% of municipal budgets are allocated to the management of MSW, which confirms the low priority of this sector [71]. In 2008, Sharholy et al. [63] indicated that composting should be widely used as a sustainable way of organic waste management to improve the rural economy in India. A similar course of action was indicated for African countries [72, 73]. In 2016, waste management was legally regulated in India. Biowaste is defined in the legislation as biodegradable waste from gardens and parks, food and kitchen waste from households, restaurants, caterers, and commercial establishments, and similar waste from food processing plants. This group does not include forestry or agricultural residues, manure, sewage sludge, or other biodegradable waste such as natural textiles and processed wood. There is very limited or no segregated collection of municipal waste in most cities, towns, and villages of India. This results in a mixed waste composition, which makes sustainable treatment technologies difficult or even impossible. According to the new regulations, biodegradable waste should be selectively collected at source, treated, and disposed of by composting or biomethanization. Waste collection in an area should be organized by the designated administrative bodies. As the lowest-rank authorities in the Indian government system, all local municipal units are required to prepare a solid waste management plan in accordance with the state policy and strategy in this regard [74].

In recent years, the Republic of South Africa has been facing a serious waste disposal crisis due to the high level of urbanization and industrialization of the country. The most common methods of waste disposal in this country are landfilling and incineration, which have a highly harmful impact on the environment. Every year, the amount of waste generated increases significantly. Data from the Department of Environmental Affairs show that it is increasing by ca. 11 million tons year^{-1} [75]. In 1997, 42.3 million tons of municipal organic waste was generated, while in 2014, this was already 69 million tons. In 2012, 22.9 million tons of bio-waste (agricultural, municipal, and industrial wastewater) were generated, whereas in 2014 this was already 26.2 million tons, and the upward trend continues. Due to its availability and nutrient content, this bio-waste is considered to be a suitable substrate for biohydrogen production by fermentation. It is therefore possible to produce alternative and sustainable energy fuels in South Africa that are environmentally friendly. The wide availability of biohydrogen as a clean alternative energy source can meet the country's growing energy demand. The widespread access to bio-waste for biohydrogen production should substantially improve the economics of the process. The economic analysis showed that due to the complexity of the process, i.e. low conversion efficiency, accumulation of by-products, the need for hydrogen purification, and the requirements for hydrogen storage systems, the unit price of industrial-scale biohydrogen production will be more expensive compared to energy from conventional fuels. However, in light of the adverse climate change, dwindling fossil fuel resources, and the steady increase in energy prices, biohydrogen remains the preferred alternative fuel [76].

In 2020, due to the global COVID-19 pandemic and the related health crisis, many countries temporarily modified their waste management rules on household waste management and subsequent treatment. In Belgium, for example, the door-to-door method was indicated as the preferred form of selective waste collection as it reduces the need for residents to travel further distances to dispose of their waste, thus preventing direct contact between each other. Bags with household waste, including organic waste, should be put out for collection after a seven-day quarantine to minimize the risk of the virus surviving in the waste. It is also assumed that in case of a critical shortage of manpower in the waste collection companies, waste will be collected first from places with high population density and only then from rural areas. Only when it is impossible to selectively collect residual and organic waste will all waste be collected together (as mixed waste) and transported to the incinerator. The guidelines in Finland stipulate that in the case of a delay in emptying waste containers, the waste must be carefully packed and stored in a way that does not harm the health of residents or create a risk of animals having access to the waste. However, waste from infected people should be collected separately. France continues to place a strong emphasis on maintaining selective collection, although it has been reduced in many municipalities. However, the restriction does not apply to organic and household waste. Ensuring regular collection and treatment of infectious medical waste is a priority. In Italy, which is the country most affected by the COVID-19 pandemic among countries in Europe, the separate collection has been suspended and all household waste is collected as mixed municipal waste. Waste is collected into one bag and put into another bag before it is taken out of the house to protect the waste from spillage. Bags with waste must be tied up. Municipal waste from households with COVID-19 patients staying in isolation or mandatory quarantine is classified as infectious (hazardous) medical waste and is collected and disposed of by specialized firms. In Tuscany, the local waste collection operator provides residents with a special collection kit consisting of bags, adhesive tapes, and a special container. Collection of such waste is performed in the door-to-door system every three to five days (after telephone arrangement). In Spain, local communities have the main competencies to manage waste. In Catalonia, it is assumed that selective waste collection should be maintained and waste treatment facilities should function normally. Only when the capacity of these facilities is reduced is the waste sent for treatment in alternative facilities, including facilities for the recovery of energy from waste or for controlled storage. It was also recommended that the proximity principle should be respected and that waste should be supplied to installations as close as possible to the place of generation to reduce the risk of pandemic spread through transport.

7.4 Conclusions

As one of the fractions of municipal waste, bio-waste represents a basic waste stream and, due to its characteristics, has a very high potential for the circular economy. Proper management of bio-waste is one of the main elements of sustainable development. The analysis of the bio-WCSs presented in this study confirmed that the organic fraction should be collected selectively, taking into account the objectives for its final management. Due to its high potential to be used as raw materials and for energy generation, bio-waste is a valuable secondary raw material that can be used to produce soil improvers and renewable energy during biological aerobic and anaerobic decomposition processes. Disposal of bio-waste by incineration is not justifiable from the environmental (pollution and CO_2 emissions) and economic perspectives. This waste contains a large amount of moisture and must be dried using energy-intensive processes before incineration. The use of selective collection of bio-waste reduces its landfilling, thus decreasing greenhouse gas emissions resulting from the decomposition of organic matter. Furthermore, selective waste collection eliminates the contamination of other fractions of municipal waste with this waste, which might hinder or even prevent the effective recovery of recyclable materials and/or energy.

The level of selective collection of bio-waste varies greatly, both in Europe and worldwide, ranging from 10 to nearly 90% in Zero Waste systems. The selective collection system should be used wherever it is technically, environmentally, and economically feasible. The systems used around the world are similar to each other and differ only in technical details such as the type of containers, the place and frequency of collection, and the equipment used. The importance of information and training campaigns to raise awareness and motivate residents to segregate and manage bio-waste in a sustainable manner should also be stressed, e.g. in the form of lowering fixed fees for waste collection. Efficient selective WCSs also require appropriate logistics with respect to both the frequency of waste collection and the technologies used. At the same time, investments are needed in modern and advanced bio-waste treatment facilities. Clear legal regulations are also critical to oblige waste producers to respect the principles of sustainable development, which include waste segregation and the use of waste as raw materials and for energy generation.

References

1 Directive 2008/98/EC of the European Parliament and of the Council of 19 November 2008 on waste and repealing certain Directives. *Official Journal of the European Union*, L 312/3.

2 Hoornweg, D. and Bhada-Tata, P. (2012). *What a Waste: A Global Review of Solid Waste Management*. Washington, DC: World Bank.

3 United Nations Environment Programme (UNEP) (2016). *Global Waste Management Outlook*. New York: UN.

4 Council Directive 1999. 1999/31/EC of 26 April 1999 on the landfill of waste. *Official Journal of the European Union*, L 182.

5 Wilson, D.C. (ed.) (2015). *Global Waste Management Outlook. International Solid Waste Association (ISWA) and United Nations Environmental Programme (UNEP)*. Nairobi: United Nations Environment Programme.

6 Tchobanoglous, G., Theisen, H., and Vigil, S. (1993). *Integrated Solid Waste Management: Engineering Principles and Management Issues*. New York: McGraw-Hill.

7 Bilitewski, B., Hardtle, G., and Marek, K. (1997). *Waste Management*. Berlin: Springer.

8 Bilitewski, B., Wagner, J., and Reichenbach, J. (2010). *Best Practice Municipal Waste Management*. Intecus: Federal Environmental Agency.

9 Pires, A., Martinho, G., Rodrigues, S., and Gomes, M.I. (2019). *Sustainable Solid Waste Collection and Management*. Springer.

10 Diaz, L.F., Savage, G.M., and Eggerth, L.L. (2005). *Solid Waste Management*. Paris: United Nations Environment Programme.

11 Dahlén, L. and Lagerkvist, A. (2010). Evaluation of recycling programmes in household waste collection systems. *Waste Management and Research* 28: 577–586.

12 Gonzalez-Torre, P.L., Adenso-Diaz, B., and Ruiz-Torres, A. (2003). Some comparative factors regarding recycling collection systems in regions of the USA and Europe. *Journal of Environmental Management* 69: 129–138.

13 O'Leary, P.R. (1999). *Decision Maker's Guide to Solid Waste Management*, 2e. Washington, DC: United States Environmental Protection Agency, Office of Solid Waste, RCRA Information Center.

14 Iriarte, A., Gabarrell, X., and Rieradevall, J. (2009). LCA of selective waste collection systems in dense urban areas. *Waste Management* 29: 903–914.

15 https://hej.mielec.pl/pl/11_wiadomosci/692_z_magistratu/49588_selektywna-zbi-rka-odpad-w-dlaczego-to-takie-wa-ne.html.

16 https://www.ese.com/home/produkte/garten-bio-abfall/bio-select.

17 Rodrigues, J., Oliveira, V., Lopes, P., and Dias-Ferreira, C. (2015). Door-to-door collection of food and kitchen waste in city centers under the framework of multi municipal waste management systems in Portugal: the case study of Aveiro. *Waste and Biomass Valorization* 6: 647–656.

18 Report European Environment Agency No 04/2020, 2020. *Bio-waste in Europe – Turning Challenges into Opportunities*, Luxembourg: Publications Office of the European Union, ISBN 978-92-9480-223-1, ISSN 1977-8449.

19 Jouhara, H., Czajczyńska, D., Ghazal, H. et al. (2017). Municipal waste management systems for domestic use. *Energy* 139: 485–506.

20 Przystupa, A. (2018). Biodegradable waste management as a practical example of green economy. *Organization and Management* 1 (41): 129–141.

21 Xevgenos, D., Papadaskalopoulou, C., Panaretou, V., and Moustakas, K. (2015). Success stories for recycling of MSW at municipal level – a review. *Waste and Biomass Valorization* 6 (5): 657–684.

22 Eionet. 2019. Biowaste in Europe – questionnaire to Eionet. Replies from Eionet to a Survey Carried Out in 2019, European Environment Information and Observation Network.

23 Pichtel, J. (2014). *Waste Management Practices: Municipal, Hazardous, and Industrial*, 2e. Boca Raton: Taylor and Francis Group.

24 Fricke, K., Heußner, C., Hüttner A, and Turk, T. (2018). Recycling of biowaste: experience with collection, digestion, and quality in Germany. In: *Source Separation and Recycling. The Handbook of Environmental Chemistry*, vol. 63 (eds. R. Maletz, C. Dornack and L. Ziyang), 175. Cham: Springer.

25 Lee, U., Han, J., and Wang, M. (2017). Evaluation of landfill gas emissions from municipal solid waste landfills for the life-cycle analysis of waste-to-energy pathways. *Journal of Cleaner Production* 166: 335–342.

26 www.ipcc.ch.

27 EEA. 2019. Annual European Union approximated greenhouse gas inventory for the year 2018, EEA Report No 16/2019, European Environment Agency.

28 https://uspirg.org/reports/usp/composting-america.

29 Feng, S.J., Zhao, Y., Zhang, X.L., and Bai, Z.B. (2020). Leachate leakage investigation, assessment and engineering countermeasures for tunneling underneath a MSW landfill. *Engineering Geology* 265: 105447.

30 Stegmann, R. (2005). Mechanical biological pretreatment of municipal solid waste. In: Abstracts of the Sardinia 2005, Tenth International Waste Management and Landfill Symposium, S. Margherita di Pula, Cagliari; pp. 3–7 October 2005.

31 Schüch, A., Morscheck, G., Lemke, A., and Nelles, M. (2016). Bio-waste recycling in Germany – further challenges. *Procedia Environmental Sciences* 35: 308–318.

32 http://lca.jrc.ec.europa.eu/waste.

33 Bhatia, S.K., Joo, H.S., and Yang, Y.H. (2018). Biowaste-to-bioenergy using biological methods—a mini-review. *Energy Conversion and Management* 177: 640–660.

34 Lohri, C.R., Diener, S., Zabaleta, I. et al. (2017). Treatment technologies for urban solid biowaste to create value products: a review with focus on low- and middle-income settings. *Reviews in Environmental Science and Biotechnology* 16: 81–130.

35 Huerta-Pujol, O., Gallart, M., Soliva, M. et al. (2011). Effect of collection system on mineral content of biowaste. *Resources, Conservation and Recycling* 55: 1095–1099.

36 Malamis, D., Moustakas, K., Bourka, A. et al. (2015). Compositional analysis of biowaste from study sites in Greek municipalities. *Waste and Biomass Valorization* 6: 637–646.

37 Malamis, D., Bourka, A., Stamatopoulou, E. et al. (2017). Study and assessment of segregated biowaste composting: the case study of Attica municipalities. *Journal of Environmental Management* 203: 664–669.

38 Davis, G. and Song, J.H. (2006). Biodegradable packaging based on raw materials from crops and their impact on waste management. *Industrial Crops and Products* 23: 147–161.

39 Kern, M., Turk, T., Hüttner, A., and Kojet, U. (2017). Compostable plastic bags and anaerobic digestion – field trials in four German facilities. *Müll und Abfall* 49 (2): 57–104.

40 Crippa, M., De Wilde, B., Koopmans, R. et al. (2019). *A Circular Economy for Plastics: Insights from Research and Innovation to Inform Policy and Funding Decisions.* Luxembourg: Publications Office of the European Union.

41 Kale, G., Kijchavengkul, T., Auras, R., and Rubino, M. (2007). Compostability of bioplastic packaging materials: an overview. *Macromolecular Bioscience* 7 (3): 255–277.

42 Rodrigues, L.C., Puig-Ventosa, I., López, M. et al. (2020). The impact of improper materials in biowaste on the quality of compost. *Journal of Cleaner Production* 25: 119601.

43 EC. 2015. Assessment of separate collection schemes in the 28 capitals of the EU, European Commission, Brussels. Available from: https://ec.europa.eu/environment/waste/studies/pdf/Separate%20collection_Final%20Report.Pdf.

44 Vázquez, M. and Soto, M. (2017). The efficiency of home composting programmes and compost quality. *Waste Management* 64: 39–50.

45 Žitnik, M. and Vidic, T. (2016). *Food among Waste*. Ljubljana: Statistical Office of the Republic of Slovenia.

46 Panaretou, V., Vakalis, S., Ntolka, A. et al. (2019). Assessing the alteration of physicochemical characteristics in composted organic waste in a prototype decentralized composting facility. *Environmental Science and Pollution Research* 26 (20): 20232–20247.

47 Guerrini, A., Carvalho, P., Romano, G. et al. (2017). Assessing efficiency drivers in municipal solid waste collection services through a non-parametric method. *Journal of Cleaner Production* 147: 431–441.

48 Exposito, A. and Velasco, F. (2018). Municipal solid-waste recycling market and the European 2020 horizon strategy – a regional efficiency analysis in Spain. *Journal of Cleaner Production* 172: 938–948.

49 Bartolacci, F., Del Gobbo, R., Paolini, A., and Soverchia, M. (2019). Efficiency in waste management companies: a proposal to assess scale economies. *Resources, Conservation and Recycling* 148: 124–131.

50 VLACO. 2020. Comments received from VLACO, Flemish Region of Belgium, during the Eionet consultation of this report.

51 Mato, S., Pérez-Losada, C., Martínez-Abraldes, M. and Villar, I. (2019) Towards the recycling of bio-waste: the case of Pontevedra, Spain (REVITALIZA). In: *Municipal Solid Waste Management*. El-Din Mostafa Saleh H. (ed.): IntechOpen: (5 March 2019).

52 http://ec.europa.eu/environment/waste/publications/compost_success_stories.htm.

53 EU. 2018, Directive (EU) 2018/851 of the European Parliament and of the Council of 30 May 2018 amending Directive 2008/98/EC on waste (OJ L 150, 14.6.2018, pp. 109–140).

54 Langer, G. 2017. Long-term experience with separate collection of biowaste in Munich, Presentation given at: Biowaste in the Circular Economy, Brussels.

55 http://www.waste-managementworld.com/biological-treatment.html.

56 Okkonen, L. (2008). Applying industrial ecosystem indicators: case of Pielinen Karelia, Finland. *Clean Technologies and Environmental Policy* 10: 327–339.

57 https://www.eia.gov/energyexplained/biomass/waste-to-energy.php.

58 https://kompogas-slo.com/what-we-do.

59 Zhang, D.Q., Tan, S.K., and Gersberg, R.M. (2010). Municipal solid waste management in China: status, problems and challenges. *Journal of Environmental Management* 91 (8): 1623–1633.

60 Zhu, Y., Zhang, Y., Luo, D. et al. (2020). A review of municipal solid waste in China: characteristics, compositions, influential factors and treatment technologies. *Environment, Development and Sustainability* https://doi.org/10.1007/s10668-020-00959-9.

61 Matsushita, J., Matsui, S., Tomonao, M., and Nakagawa, N. (2013). Basin management approach for stepwise water pollution control under rapid urbanization. *Journal of Japan Society of Civil Engineers* 69 (G),5: 199–206.

62 Nakamura, O. and Endo, H. 2011. Best Practice for Raw Garbage Recycling Link, Nousan-gyoson Bunka Kyokai. (in Japanese).

63 Sharholy, M., Ahmad, K., Mhamood, G., and Trivedi, R.C. (2008). Municipal solid waste management in Indian cities – a review. *Waste Management* 28: 459–467.

64 Marshall, R.E. and Farahbakhsh, K. (2013). Systems approach to integrated solid waste management in developing countries. *Waste Management* 33: 988–1003.

65 Kumar, S., Bhattacharyya, J.K., Vaidya, A.N. et al. (2009). Assessment of the status of municipal solid waste management in metro cities, state capitals, class I cities and class II towns in India: an insight. *Waste Management* 29: 883–895.

66 Al-Khatib, I.A., Monou, M., Salam, F.A. et al. (2010). Solid waste characterization, quantification and management practices in developing countries. A case study: Nablus district – Palestine. *Journal of Environmental Management* 91: 1131–1138.

67 Kaza, S., Yao, L., Bhada-Tata, P., and Frank Van Woerden, F. 2018. What a Waste 2.0: a global snapshot of solid waste management to 2050. *Urban Development*.

68 ANGED. 2014. Available from: http://www.anged.nat.tn/index.php?option=com_content&view=article&Id=146.

69 Chaher, N.E.H., Chakchouk, M., Nassour, A. et al. (2020). Potential of windrow food and green waste composting in Tunisia. *Environmental Science on Pollution Research* https://doi.org/10.1007/s11356-020-10264-7.

70 Chaher, N.E., Hemidat, S., Thabit, O. et al. (2020). Potential of sustainable concept for handling organic waste in Tunisia. *Sustainability* 12: 8167.

71 Al-Khatib, A., Arafat, H.A., Besheer, T. et al. (2007). Trends and problems of solid waste management in developing countries: a case study in seven Palestinian districts. *Waste Management* 27: 1910–1919.

72 Couth, R. and Trois, C. (2012). 2012. Sustainable waste management in Africa through CDM projects. *Waste Management* 32: 2115–2125.

73 Couth, R. and Trois, C. (2011). Waste management activities and carbon emissions in Africa. *Waste Management* 31: 131–137.

74 Ghosh, S.K. and Francesco Di Maria, F. (2018). A comparative study of issues, challenges and strategies of bio-waste management in India and Italy. *Detritus* 01: 8–17.

75 Department of Environmental Affairs (DEA). 2012. National Waste Information Baseline Report. Pretoria, South Africa; .

76 Sekoai, P.T. and Daramola, M.O. (2017). The potential of dark fermentative bio-hydrogen production from biowaste effluents in South Africa. *International Journal of Renewable Energy Research* 7 (1): 359–378.

8

Anaerobic Digestion

Anna Grosser and Ewa Neczaj

Czestochowa University of Technology (CUT), Czestochowa, Poland

8.1 Introduction

The first industrial installations using waste and sewage sludge in anaerobic digestion (AD) processes were built in Europe in the nineteenth century. However, an increase in the use of waste for biogas production has been observed since the 1990s. Currently, the AD process is focused on the production of biogas, but there are already biorefineries in which, apart from biogas, other products are generated, e.g. biomethanol and olefins. AD is primarily used for biogas production in Wastewater Treatment Plants (WWTPs), biogas plants (agricultural or waste), and in the degassing of municipal waste landfills. Theoretically, all biodegradable waste can be used as feedstock in the AD process. From the diversified waste stream, an important role is played by the organic fraction of municipal waste (OFMSW), considered to be a large potential source of renewable energy (Table 8.1) [18–20].

The most important benefits of using AD to treat biodegradable waste are [21]:

- Biogas production, which can be used for heat and energy generation, and other applications.
- Digestate can be recycled in liquid or dry form, e.g. as fertilizer.
- Disposal of different types of waste, e.g. OFMSW, food waste.
- AD system required relatively little space.

8.2 Feedstock for Anaerobic Digestion Processes

As mentioned before, any biodegradable material can be used as a feedstock for AD. Examples of such waste materials are shown in Table 8.2.

An important group of waste susceptible to AD is municipal waste. MSW is generated both in households, restaurants, and different public utility buildings. The amount and composition of this type of waste depends on a number of factors. The most important ones include: the place of their generation, society wealth and lifestyle, season of the year, urbanization degree, economic condition of the region, as well as legal regulations [24–28]. Characteristics of municipal waste also strongly depend on their collection system and pretreatment processes (Table 8.3) [30–32].

Biodegradable Waste Management in the Circular Economy: Challenges and Opportunities, First Edition.
Edited by Malgorzata Kacprzak, Eleonore Attard, Kari-Anne Lyng, Helena Raclavska, BalRam Singh, Eyob Tesfamariam, and Franck Vandenbulcke.

Table 8.1 Biogas potential of organic waste.

Feedstock	Methane yield (m^3 Mg^{-1} VS_{added})	References
OFMSW	310	[1]
OFMSW	382	[2]
Household waste	350	[3]
MSW	200	[4]
MSW	360	[5]
Fruit and vegetable wastes	420	[6]
Fruit waste	180–732	[7]
SS - OFMSW	388	[8]
SS - OFMSW	275–410	[9]
SS - OFMSW	400	[10]
OFMSW	187	[11]
OFMSW	186–222	[12]
Coffee husks	159.4	[13]
Coffee waste	126–233.7	[14]
Chicken manure	200–370	[15]
Grass silages	359–427.7	[16]
Potato peel	267	[17]
Paper (office)	368	
Sunflower (whole crop silage)	300	
Sugar beet leaves	210	
Yard waste (grass)	209	
Lemon pressing	473	

SS - OFMSW, selectively collected organic fraction of municipal solid waste.

Table 8.2 Examples of waste feedstock for AD, adopted from references [22, 23].

Substrates of preference for waste fermentation facilities	
Organic household waste like kitchen scraps	Distiller's residues
Organic scraps from communal feeding facilities	Waste from starch processing
Green residues, e.g. from public spaces (parks, cemeteries)	Residues from sugar production
Organic waste from food processing	Tea and coffee grounds
Fish-processing waste	Paper/cardboard
Dairy products	Biomass plantations for energetic purposes
Slaughterhouse wastes	Organic waste from the chemical industry
Organic waste from agriculture, e.g. straw, manure, plant residue	Residue from the cosmetics industry
Waste from fodder production	Municipal and industrial wastewater
	Sewage sludge from municipal and industrial wastewater treatment plants

Table 8.3 Characteristics of the different fractions of OFMSW [29].

	Unit	MS-OFMSW	SC-OFMSW	SS-OFMSW
TS	%	51–95	9–31	7–31
VS	% TS	29–57	70–94	63–95
TKN		1.2–3.4	2.1–3.3	1.5–3.4
P	% TS	0.05–0.22	0.3–2.6	0.4–3.9

TS, total solids, VS, volatile solids, MS–OFMSW, mechanically sorted OFMSW, SC-OFMS, source collected OFMSW, SS-OFMSW, source sorted OFMSW.

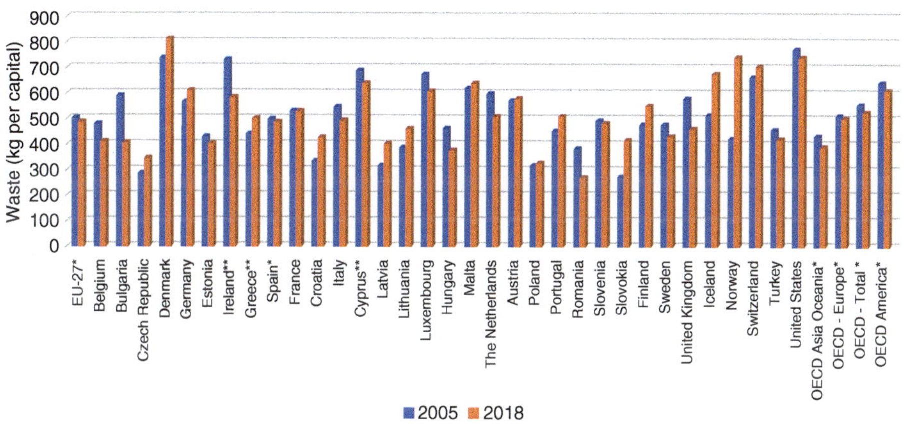

Figure 8.1 Municipal waste generated per capita in EU and selected regions in the world in 2005 and 2018 based on references [33, 34].

Figure 8.1 shows municipal waste generation in the years 2005 and 2018 by country in Europe as well as selected regions expressed in kilograms per capita. As reported by Eurostat 24% of MSW. landfilling is 27% disposed of through incineration, 31% recycled, and 17% composted.

Bio-waste constitutes 34% of municipal waste, which means its annual production at the level of 75 million Mg in Europe (EU27). In 2017 about 47.5 million Mg of municipal bio-waste was collected and disposed of through composting and/or digestion. Therefore, there is still a large mass of waste that can be used as feedstock for AD. OFMSW is characterized by a high dry matter content (from 30 to 50%), a high C : N ratio, and a low content of macro- and micronutrients. Sometimes it also contains toxic compounds, e.g. heavy metals [35]. Table 8.3 shows the characteristics of the different fractions of OFMSW.

8.3 Anaerobic Digestion – Process Chemistry

AD is a complex biochemical process of decomposition high-molecular organic substances presented in feedstock, carried out under anaerobic conditions. The process takes place in four key stages: hydrolysis, acidogenesis, acetogenesis, and methanogenesis (Figure 8.2) [38–40]. The equation summarizes the whole AD process [36]:

$$C_nH_aO_bN_Z + \left(\frac{n-a}{4} - \frac{b}{2} + \frac{3z}{4}\right) H_2O \rightarrow \left(\frac{n}{2} - \frac{a}{8} + \frac{b}{4} + \frac{3z}{8}\right) CO_2 + \left(\frac{n}{2} + \frac{a}{8} - \frac{b}{4} + \frac{3z}{8}\right) CH_4 + zNH_3$$

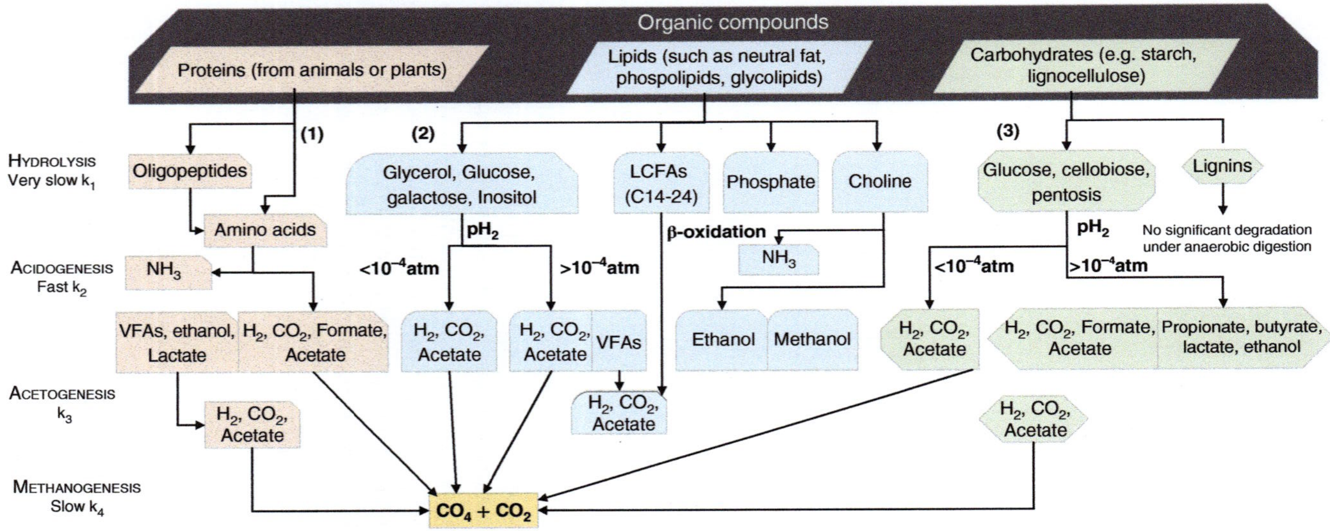

(1) – proteases from animal and plants: acid proteases (pH:1–5) and thioproteases (pH 4–8); Proteases from bacteria: metalloproteases (pH 7–8), alkaline proteases (pH:9–11)
(2) lipases, phospholipases
(3) amylases
VFAs – volatile fat acids; LCFAs – long chain fatty acids

Component	Methanogenic reaction
Lipids	$C_{50}H_{90}O_6 + 24{\cdot}5H_2O \rightarrow 34.75CH_4 + 15.25CO_2$
Carbohydrates	$C_6H_{10}O_5 + H_2O \rightarrow 3CH_4 + 3CO_2$
Proteins	$C_{16}H_{24}O_5N_4 + 14{\cdot}5H_2O \rightarrow 8.25CH_4 + 3.75CO_2 + 4NH_4^+ + 4HCO_4^-$

Figure 8.2 Anaerobic degradation of organic compounds based on references [36, 37].

Table 8.4 Composition of biogas from OFMSW [41].

Component	Content (% v/v)
CH_4	55–60 (50–75)
CO_2	35–40 (25–45)
H_2O	2 (20 °C) – 7 (40 °C)
H_2S	2 (20–20 000 ppm)
N_2	<2
O_2	<2
H_2	<1

Table 8.5 Methane content in biogas from anaerobic digestion of various feedstocks adopted [40].

Feedstock	Composition (%)
Cattle manure	50–60
Pig manure	60
Poultry waste	68
Wheat straw	78.5
Corn silage	54.5
Grass	84
Dried leaves	58
Barley straw	77
Sewage sludge	50–75

For example, assuming complete decomposition of the bio fraction of MSW, which can be described by the formula $C_{64}H_{107}O_{37}N$, the composition of the biogas would be as follows: 55% methane, 43.5% carbon dioxide, and 1.5% ammonia. Table 8.4 shows a typical composition of biogas obtained during the fermentation of the organic fraction of municipal solid waste. The methane content in biogas strongly depends on the feedstock composition (Table 8.5).

8.3.1 Factors affecting the process

8.3.1.1 Substrate specification

The biogas production during the AD process depends on process parameters, e.g. pH, temperature, organic loading rate (OLR), hydraulic retention time (HRT), feedstock pre-treatment, and inoculums (see Table 8.6) [64].

Table 8.7 shows the optimum process parameters for methanogen bacteria. It should be noted that the optimum parameters for AD strongly depends on feedstock type and must be determined for the waste group used [65].

Table 8.6 Anaerobic digestion process efficiency for different organic waste streams.

Substrate	Type of reactor (volume l)	Description of process	Organic matter removal (%)	Biogas production	Methane (%)	References
SS-OFMSW MS-OFMSW	Laboratory-scale reactors (5)	D, 55 °C, 60 d,	45.0[a)]/29.9[b)] 55.5[a)]/59.3[b)]	120[c)] 82[c)]	53.4 55.2	[42]
Bio-waste fraction from domestic wastes	Laboratory scale digestion (8.5)	37 °C, HRT = 5.7; 10 or 20 d, OLR = 15; 8.5; 4.3 kg COD m^{-3}·d	58–70[b)]	15–50[d)]	63–69	[43]
	Full-scale digestion (960–1220)[e)]	37 °C, HRT = 12 d, OLR = 8.3–8.5 kg COD m^{-3}·d	60[b)]	2234[f)]	65–70	
FW	Batch digesters (1)	Batch system, 50 ± 2 °C, 28 d	80.57[a)]	0.6[g)]	73.1	[44]
SS-OFMSW	CSTR (4.5)	55 °C, HRT = 15d, OLR = 11.4 kg VS m^{-3}·d	30[a)]	0.71[g)]	64	[45]
SS-OFMSW	Pilot-scale digestion (35)	55 °C, HRT = 15d, OLR = 2.8 kg VS m^{-3}·d	81[a)]	0.275–0.41[h)]	62	[9]
MS-OFMSW+ SS-OFMSW	CSTR (3000)	55 ± 1 °C, average HRT = 13.5d, OLR = 9.2 kg VS m^{-3}·d	–	0.23[g)]	68.7	[46]
SS-OFMSW	Laboratory-scale bioreactors of the type BioFlo 110 (9.5)	35 °C, HRT = 90 d, OLR = 2.03 kg VS m^{-3}·d	–	0.2–0.56[g)]	40–65	[47]
FVW	SCTAD (18)	35 °C, HRT = 12; 15; 20 d, 4, 6, 8% TS	58.58–75.91[a)]	0.514–0.707[g)]	50–65	[48]
FVW	SCTAD (18)	20 °C, HRT = 20 d, 4, 6% TS	54.2–66[a)]	0.386–0.625[g)]	56–58	[49]
		35 °C, HRT = 12, 15, 20 d, 4, 6, 8, 10% TS	60.3–74.4[a)]	0.183–0.705[g)]	54–65	
		55 °C, HRT = 20 d, 4, 6, 8, 10% TS	72.8–75.9[a)]	0.733–0.997[g)]	58–62	
FW	Reactors (8)	Three stages: semi-anaerobic hydrolysis, anaerobic acidogenesis, and strictly anaerobic methanogenesis 40–55 °C, HRT = 10, 12 d	–	0.199–0.223[h)]	65.6–67.4	[50]
OFMSW	Semi-continuous regime (14)	37 °C, HRT = 17 d, OLR = 0.97 kg VS m^3d	73[a)]	0.8[g)]	58	[51]

WS-OFMSW	Batch bench-scale reactor w (35)	Semi-dry, 30 °C, 60 d, 16, 13.5 11% TS,	26.1–41;8[a)]	0.273–0.314[g)]	65- 66	[52]
FW	CSTR (3)	semi-continuous mode, 37 °C, HRT = 8; 16 d, OLR = 3.7; 5.5; 7.4; 9.2; 12.9 kg VS m^{-3}·d	84.4–92.5	0.417–0.455[h)];	–	[53]
OFMSW	Laboratory-scale bioreactors (3.25)	25 °C, 100 d, TS: 45; 67.5; 112.5; 135 g l	10.11–40.43[a)]	21.7–48.5[i)]	62–72	[54]
		29 °C, 100 d, TS: 45; 67.5; 112.5; 135 g l^{-1}	22.21–58.44[a)]	42.2–141.6[i)]		
OFMSW	Pilot-scale reactor (375)	37 °C, 60 d, 10% TS	61[a)]	0.260[h)]	60	[55]
SS-OFMSW	Full scale (2200 m^3)	38.6 °C, HRT = 50–70d, OLR = 3–8 kg VS m^{-3}·d	40–45[a)]	0.40[g)]	56	[56]
Ss	CSTR (6.5)	37 °C, HRT = 10d, OLR = 2.98 kg VS m^{-3}·d	45[a)]	0.221[h)]	69	[57]
Ss	CSTR (4)	35 °C, HRT = 16, 18, 20 d; OLR = 1.56–2.092.98 kg VS m^{-3}·d	52[a)]	0.463[h)]	63	[58]
Ss	CSTR (6)	37 °C, HRT = 20d, OLR = 1.15–1.35 kg VS m^{-3}·d	50[a)]	0.300[h)]	69	[59]
KW	Acidogenic reactor (CSTR) (2.5) Methanogenic reactor (UASB) (3)	Two-phase anaerobic digestion: Stage 1 : 35–37 °C; HRT = 3.1 d; SRT = 9.3 d; TS loading rate: 16 g TS/Ld; Stage 2 : 35–37 °C; HRT = 1 d; COD loading rate: 20 gCOD dl^{-1}	86.9[b)]	0.52[j)]	–	[60]
OFW	Single-stage, CSTR (6)	HRT = 30 d; OLR = 2.4 kg VS m^{-3}·d; 35 °C	74.1[a)]	0.448[g)]	60.5	[61]
	Two-stage systems, CSTR (2 and 8)	Stage 1: HRT = 6 d; OLR = 12 kg VS m^{-3}·d; Stage 2: HRT = 24 d; OLR = 3.1 kg VS m^{-3}·d	71.6[a)]	0.450[g)]	61.0	
	Two-stage systems with recycle effluent from stage II to stage I CSTR (1.5 and 6)	Stage 1: HRT = 3 d; OLR = 14.2 kg VS m^{-3}·d Stage 2: HRT = 12 d; OLR = 2.6 kg VS m^{-3}·d	80.1[a)]	0.440[g)]	70.4	

(Continued)

Table 8.6 (Continued)

Substrate	Type of reactor (volume l)	Description of process	Organic matter removal (%)	Biogas production	Methane (%)	References
Ss	9 CSTR (20)	System I : 35 °C–35 °C-35 °C for each stage HRT = 5d	39.7[a)]	63.3[f)]		[62]
		System II: 42 °C–35 °C-35 °C for each stage HRT = 5d	43.8[a)]	84.9[f)]	–	
		System III: 35 °C–35 °C; HRT = 5/10 d	37.5[a)]	46[f)]		
		Single stage, 35 °C; HRT = 5 d	32[a)]	39.5[f)]		
SW	CSTR (1000 m^3)	38 °C, HRT = 20 d; OLR = 11 kg COD m^{-3}·d	–	0.68[k)]	–	[63]
MS	PF (270 m^3)	44 °C, HRT = 158 d; OLR = 3.5 kg COD m^{-3}·d	–	0.46[k)]	–	
	CSTR (1065 m^3)	38 °C, HRT = 158 d; OLR = 3.5 kg COD m^{-3}·d	–	0.46[k)]	–	

Substrate: SS-OFMSW, source sorted OFMSW (SS_OFMSW); MS-OFMSW, mechanically selected municipal fraction; FW, food waste; FVW, fruit and vegetables waste; WS-OFMSW, water sorted organic fraction of municipal solid waste; KW, kitchen waste; Ss, sewage sludge; OFW, oily food waste; SW, slaughterhouse waste; MS, maize silage.
Reactors: CSTR, continuously stirred tank reactors; SCTAD, semi-continuously mixed mesophilic tubular anaerobic digester; UASB, upflow anaerobic sludge bed; PF, plug flow reactor.
Process description: D, dry system, OLR, organic loading rate.
Other.

a) Based on VS.
b) Based on COD.
c) Cumulative biogas production, l.
d) l d^{-1}.
e) m^3.
f) m^3 d^{-1}.
g) Biogas yield, m^3 kg^{-1} VS.
h) Methane yield, m^3 CH_4 kg^{-1} VS.
i) Ultimate biogas production.
j) m^3 kg^{-1} TS.
k) m^3 kg^{-1} COD.

Table 8.7 The process parameters range suitable for methanogen bacteria growth and biogas production [64].

Parameter	Unit	Optimum	Extreme values
Alkalinity	mg $CaCO_3$ l^{-1}	1500–3000	1000–1500 and 3000–5000
pH	–	6.8–7.2	6.6–6.8 and 7.2–7.6
HRT	d	10–15	7–10 and 15–30
Temperature (mesophilic conditions)	°C	30–35	20–30 and 35–40
Temperature (thermophilic conditions)	°C	50–56	60–65 and 70–75
VFA	mg CH_3COOH l^{-1}	50–500	500–2000
ORP	mV	−520 do −530	−490 to −550
Biogas composition			
CH_4	% v/v	65–70	60–65 and 70–75
CO_2	% v/v	30–35	25–30 and 35–40

8.4 The Anaerobic Digestion Systems

Generally, the AD systems used in the sewage sludge treatment, wastewater treatment, agricultural wastes, and organic solid wastes are similar [22]. The most popular classification of AD systems for solid waste treatment divides them, depending on the dry matter content in the feedstock, into "dry" and "wet" (Figure 8.3). In a dry system the solid content ranges from 35 to 45%, while in

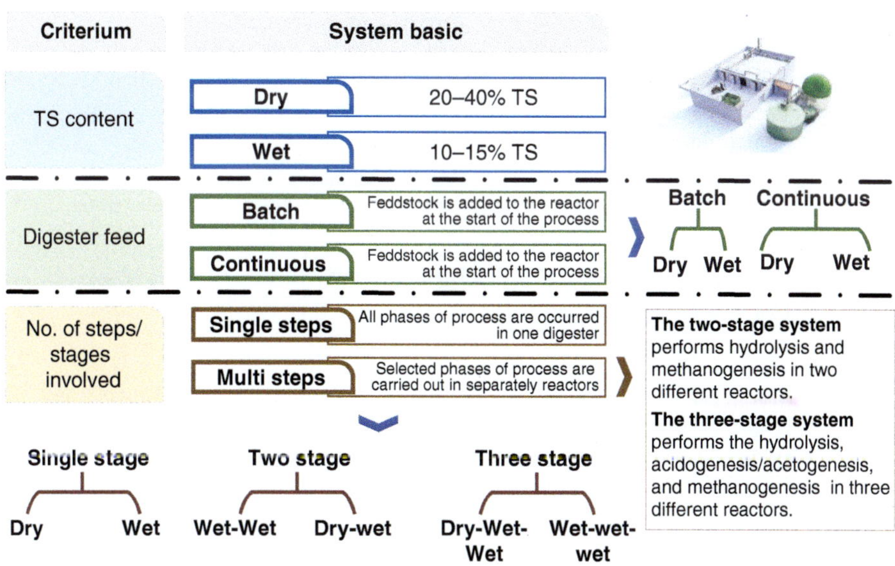

Figure 8.3 Classification of anaerobic digestion system [22, 66–68].

the second system value of this parameter it ranges between 8% and 16%, so sometimes a mixing with fresh or recalculated water is required. Some authors distinguish a third group, the so-called "semi-dry" system, in which the dry matter content is between the previously mentioned ranges [69–71]. It is possible to carry out the AD in single-stage, two-stage, and multi-stage processes. The bioreactors can be operated continuously, quasi-continuously, or in batch mode. Table 8.8 summarizes the advantages and disadvantages of selected AD systems.

Table 8.8 Pros and cons of selected AD systems [11, 72].

Criterion	Advantages	Disadvantages
Single stage dry process		
Technical aspect	Simple feedstock pre-treatment Simple bioreactor design No need to remove plastic and inert materials	Not suitable for waste streams with a dry matter content <20%
Biological aspect	High OLR Less VS losses during the pre-treatment of waste	Poor contact between microorganisms and substrates Low possibility of inhibitors dilution with water
Economic/ Environmental aspect	Low feedstock pre-treatment costs Low reactor volume Low water and heat requirement	Need for expensive specialized transport and mixing techniques
Single stage wet process		
Technical aspect	Simplified operation and mixing	Complicated waste pre-treatment Problems with sedimentation and flotation
Biological aspect	Inhibitors dilution with water	Sensitive to changes in process parameters Losses VS from the inert fraction of waste in the pre-treatment step
Economic/ Environmental aspect	Low maintenance costs	Large volumes of reactors High water and heat demand
Two stage process		
Technical aspect	Flexible operation	Complicated design
Biological aspect	High OLR Tolerance to changes in OLR and feedstock composition	Possible increase in the concentration of a toxic substance in the recirculated process water
Economic/ Environmental aspect	Higher process efficiency Low volume of bioreactors	High investment costs
Batch process		
Technical aspect	Ease of use No pre-treatment	Danger of explosion during emptying reactors
Biological aspect	Higher process efficiency than in landfills	Variable biogas production Low OLR
Economic/ Environmental aspect	Low costs Low water demand	Low process efficiency Large installation area

8.5 Commercial Technologies for AD of Biodegradable Waste

Table 8.9 shows technologies intended for AD of municipal solid waste, agriculture, and food waste, but the examples provided mainly concern the OFMSW treatment. The next table (Table 8.10) compares the methane production for various feedstock in selected technologies. As shown in Figure 8.4, most installations used biogas sources in Europe and North America (United States), although other regions progressively are deploying biogas technologies. For instance, in Ghana, biogas obtained from waste is part of a plant that integrates various methods of energy recovery (AD, pyrolysis, and solar PV). In turn, in the Indian state of Maharashtra the launch of a 4 MW installation is planned, in which biogas will be obtained from agricultural and municipal waste. In the state of Jalisco, Mexico, facility was opened that uses poultry waste as the substrate to AD. Also, in Brazil the first biogas plant based on pig manure was built [80].

Table 8.9 Examples of commercial AD technologies [73–77].

Name of technology	Number of stages	Regime of temperature	Example – throughput capacity in thousands of Mg of waste
Dranco (Organic Waste Systems, Belgium)[a)]	I	T	Tenneville (Belgium) – 39 Hotaka (Japan) – 3 Brecht II (Belgium) – 50 Alicante (Spain) – 180 Leszno (Poland) – 50
Kompogas (Kompogas, Switzerland)[a)]	I	T	Doha (Quatar) – 274 Montpellier (France) – 100 Jönköping (Sweden) – 40 Botarell (Spain) – 54 San Luis Obispo (USA)-30
Valorga (Valorga, France)[a)]	I	M or T	Amiens (France) – 85 Szanghaj (China) – 227, 5 Mons (Belgium) – 23 MSW + 35, 7 kitchen and garden waste
Biopercolat (Wehrle-Werk, Germany)[a)b)]	II	M	Baden-Württemberg (Germany) – 100
Waasa (Citec Environmental, Finland)[c)]	I	M or T	Groningen (The Netherlands) – 230 Stromossen (Finland) – 25 Kil (Sweden) – 3
ArrowBio (Arrow Ecology Ltd., Israel)[c)]	II	M	Tel Awiw (Israel) 70
BTA (Biotechnische Abfallverwer-tund GmbH, Germany)[c)]	I or II	M or T	Puławy (Poland) – 22 Verona (Italy) – 70 Dietrichsdorf (Germany) –17 Leper (Belgium) – 50 Newmarket (Canada) – 150
Entec (Environment Technology GmbH, Austria)[c)]	I	M	Lucknow (India) – 200 Kogel (Germany) – 40 Nakasorachi (Japan) – 19

Table 8.9 (Continued)

Name of technology	Number of stages	Regime of temperature	Example – throughput capacity in thousands of Mg of waste
Biostab (Ros Roca, Germany)[c]	I	T	Krosno (Poland) – 10 Salerna (Italy) – 23 Beisslingen (Germany) – 24
Linde KCA/BRV (Linde-KCA-Dresden GmbH, Germany)[a)c)]	I or II	M or T	Barcelona (Spain) – 150[c] Radenberg (Germany) – 56[c] Heppenheim (Germany) – 36[a]
Biocel (Arcadis Heidemij, The Netherlands)[d]	I	M	Lelystad (The Netherlands) – 35

a) Dry process.
b) Currently WEHRLE Umwelt GmbH.
c) Wet process.
d) Batch system, M, mesophilic conditions, T, thermophilic conditions.

Table 8.10 Comparison of the methane production for various feedstock in selected technologies [78].

	Methane yield $m^3\ Mg^{-1}$ VS		
Technology	**Kitchen waste**	**Garden waste**	**Paper and cardboard**
Valorga	634	211	353
Kompogas	460	152	256
Dranco	671	426	485
BTA	695	230	385
Biocel	498	166	277

8.6 Anaerobic Digestion Intensification Methods

Despite a high energy potential (Table 8.1), mono-digestion of some waste like OFMSW is problematic, since the AD process may have failed due to a high ratio of carbon and nitrogen in the feedstock, the high content of macroelements (sodium, potassium), deficiency of microelements (zinc, iron, molybdenum, etc.), as well as a high content of lipids ($5\,g\,l^{-1}$) [81, 82]. Furthermore, high susceptibility to biological decomposition facilitates the intense generation of VFAs whose accumulation in the digester may lead to the environment acidification, and in consequence to the process inhibition [82]. Another factor that may also affect the installation operation is negatively high OFMSW variability because its composition depends on, among other things, the number of citizens and their social condition, as well as lifestyle, geographical region, including a dominating economic activity, urbanization degree and economic condition, the season of the year or waste

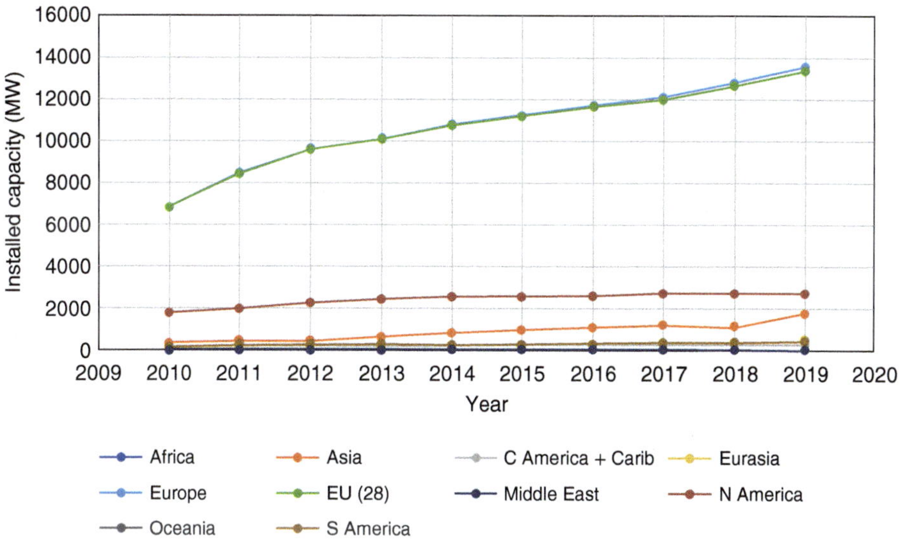

Figure 8.4 Total installed capacity of biogas installations in different regions of the world in 2010–2019 based on reference [79].

collection system, and fraction acquisition frequency [24, 26]. Therefore, much research focuses on process streamlining. Two of the most common techniques are now discussed, which are anaerobic co-digestion (AcD) and feedstock pre-treatment.

8.6.1 Pre-treatments Methods

The task of the feedstock pre-treatment is to solubilize intra- and extracellular polymers, to disintegrate fermented fractions, to increase the proper area of waste particles, as well as to change the structure of complex organic compounds (e.g. cellulose, hemicellulose), and also to change their polymerization degree, thus weakening molecular bonds between lignine and carbohydrates. Such actions are aimed at accelerating its hydrolysis, which is a phase regarded as limiting the entire process. Depending on the nature of the conditioning factor, the methods may be divided as follows: biological, chemical, thermal, as well as physical and mechanical (Figure 8.5) [85–92].

8.6.1.1 Biological Pre-treatments

Biological pre-treatment refers to using different microorganisms or enzymes to pre-treat various biodegradable waste. White-rot fungi (e.g. *Phanerochaete chrysosporium*) were found to be the most effective in the degradation of lignocellulosic materials. Brown-soft fungi (e.g. *Serpula lacrymans*), soft-rot fungi (*Paecilomyces* sp.), and bacteria (e.g. *Bacillus circulans*) may also be used for this purpose [93]. A method called enzymic hydrolysis (EH) to intensify the biodegradation of excess activated sludge was proposed by Mayhew et al. [94]. The enzymes used in EH are mainly extracted from bacteria, yeasts, and molds. The most commonly used enzymes in EH technology belong to the group of enzymes that break down proteins and polysaccharides (especially cellulose). Research has also been carried out on the enzymatic hydrolysis of waste activated sludge (WAS) and other biodegradable waste with the use of mixtures of extracted

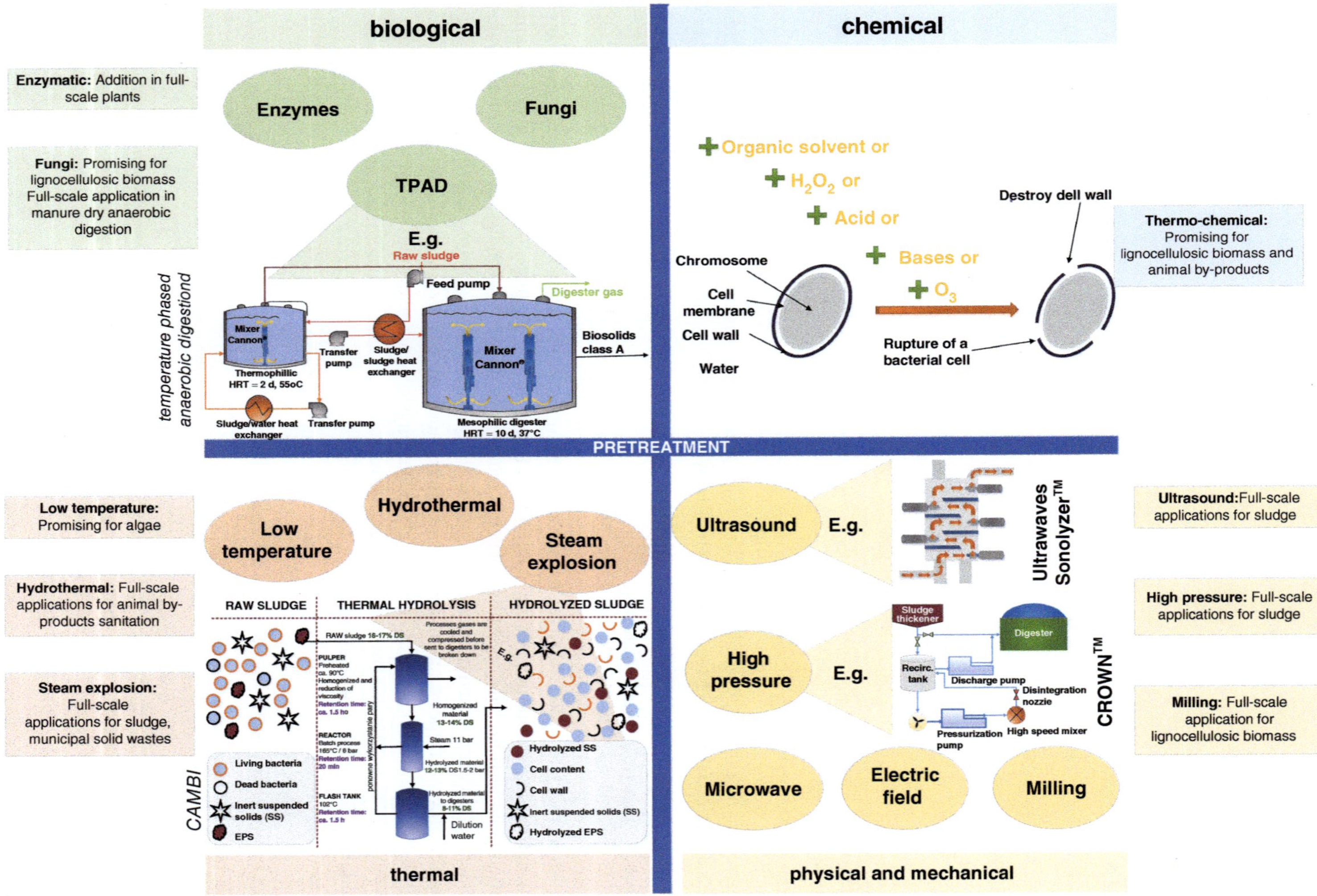

Figure 8.5 Different types of pre-treatment methods based on references [83, 84].

enzymes (proteases, lipases, and/or cellulases) with endogenous enzymes from disintegrated cells of *Aspergillus oryzae*, *Aspergillus* sp., *Streptomyces griseus*, and various strains of anaerobic bacteria [95, 96].

The main disadvantages of enzymatic pre-treatment are related to the very long retention times (several weeks) and low efficiency. On the other hand, another biological method based on two-stage fermentation, called temperature phased AD (TPAD), shows a much better efficiency. In TPAD fermentation, the first phase can be carried out under thermophilic (about 55 °C) or hyper-thermophilic (60, 70 °C) conditions. Compared to the classical technologies, the TPAD process is more stable, with a higher OLR and a higher volatile solids degradation ratio, which results in higher biogas production and better pH control [89].

8.6.1.2 Chemical Pre-treatments

Organic waste pre-treatment with alkaline solutions leads to the hydrolysis and depolymerization of carbohydrates, fats, and proteins present in the waste to compounds with a lower molecular weight, i.e. aliphatic acids, polysaccharides, and amino acids. Acidic solutions are also used for the pre-treatment of bio-waste, e.g. hydrochloric, sulfuric, and peracetic acids. However, acidic pre-treatment reducing the time required for digestion is less financially effective than alkaline pre-treatments due to the high cost. Moreover, in alkaline pre-treatments less caustic than acidic reagents are used and the process can be carried out in an ambient temperature [97, 98]. During alkaline pre-treatment disruption of the lignin structure and its exposure to enzymatic degradation take place [99]. It was found that among the chemical reagents used for the sewage sludge pre-treatment the greatest influence on their solubilization was NaOH [100]. There are a few other pre-treatment methods, e.g. hydrogen peroxide [101, 102] and ozone [103], which are still investigated as an alternative chemical method for AD enhancement.

8.6.1.3 Physical and Mechanical Pre-treatments

There are many methods that can be used to improve the physical and mechanical properties of waste, e.g. milling, griding, lysis-centrifugation, liquid shear/collision, screw press, and high pressure methods [104, 105]. This is one of the most important steps in facilitating the feeding of feedstock to digesters and also has a positive impact on the waste retention time and the efficiency of the process.

Ultrasonic (US) pre-treatment is one of the oldest methods used for disintegration of sewage sludge. US pre-treatment has an impact on the sludge structure, and increases its decomposition rate under anaerobic conditions. The produced cavitation causes cell lysis and releases proteins, carbohydrates, fats, and other intracellular compounds to supernatant water [106–108].

8.6.1.4 Thermal Pre-treatments

Thermal pre-treatment results in feedstock disintegration and hydrolysis, and has a positive impact on its biodegradability [109, 110]. The AD process can be conducted under a higher OLR with a higher biogas production. Moreover, digestate is also sterilized and free of unwanted microbes. Pre-treatment at lower temperatures is preferred because formation of recalcitrant compounds is observed at temperatures over 150 °C [111]. Thermal hydrolysis (TH) is increasingly used to intensify AD of sewage sludge, e.g. Cambi technology (see Table 8.11). TH is defined as a reaction that takes place in aqueous solutions at a specific temperature (150–180 °C) and pressure (~6 bar) [112].

Table 8.12 presents the impact of the selected pre-treatment methods on the process effectiveness. In turn, Figures 8.6 and 8.7 present selected commercial pre-treatment technologies, which are used for conditioning sewage sludge.

Table 8.11 Comparison of the Cambi technology with classic AD process [83].

Parameter	Cambi	Classic AD
Solid retention time, d	<15	>20
Volume of bioreactor, –	1/3 the volume of a conventional AD digester	1
Volatile solids, %	12	4–6
Organic loading rate, $kg\,m^{-3}\,d^{-1}$	5	2–3
pH, –	7.5–8.0	6.8–7.5
Temperature, °C	38–42	35–37
VFAs/alkalinity, –	0.1–0.5	0.1–0.5
Ammonium nitrogen, $mg\,l^{-1}$	2500–3200	600–1000
Share of methane in biogas, %	65–70	60–65
VS removal, %	70	50–55
Methane yield, $m^3\,Mg^{-1}$ VS	510	350

Table 8.12 Influence of selected methods of feedstock pre-treatment on the AD process performance.

Method	Feedstock	Pre-treatment conditions	Yield B(biogas) M(methane) $m^3\,kg^{-1}$ VS[a)]	References
CH	OFMSW	2.3 g $Ca(OH)_2$ 6 h	M : 0.15 (+172%)	[113]
B	OFMSW	Mature compost 2.5% (v/v)	M : 0.504[b)](+203%)	[114]
B		*Aspergillus awamori* 2.5%(v/v)	M : 0.413[b)](+132%)	
B		Sludge from WWTP 2.5% (v/v)	M : 0.401[b)](+125%)	
TCH		180 °C, 5Ba, 3 $g\,l^{-1}$ NaOH (N_2)	M : 0.201[b)](+13%)	
TCH		180 °C, 5Ba, 3 $g\,l^{-1}$ NaOH (O_2)	M : 0.117[b)](−34%)	
M	OFMSW	A laboratory MW Accelerated Reaction System (Mars 5®) – samples were heated from room temperature to 115, 145, and 175 °C, temperature ramp time: 40 min	115 °C: 494[c)], 439[d)] (+5%[c)], +7.6%[d)]) 145 °C: 490[c)], 414[d)] (+4.5%[c)], +1.5%[d)]) 175 °C: 385[c)], 370[d)] (−18%[c)], −9%[d)])	[115]
CH		H_2O_2 0.38 g H_2O_2 g^{-1} TS	471[c)], 494[d)] (+0.4%[c)], +14.1%[d)])	
MCH		Mars 5 + H_2O_2	496[c)] 412[d)] (+5.8%[c)], +1%[d)])	

Table 8.12 (Continued)

Method	Feedstock	Pre-treatment conditions	Yield B(biogas) M(methane) $m^3\ kg^{-1}\ VS^{a)}$	References
T	Cow manure	170 °C, 1 h	B : 0.182 (+7.8%)	[116]
	Pig manure		B : 0.385 (+13.3%)	
	Sewage sludge		B : 0.202 (+67.8%)	
	Fruit and vegetable waste		B : 0.443 (+18.5%)	
	Food waste		B : 0.781(−3.4%)	
CH	OFMSW+Sewage sludge	NaOH $4\ g\ l^{-1}$	B : 0.12 (−80%)	[8]
BCH		25 °C, pH = 7, bacterial hydrolysis $1\ g\ l^{-1}$ + NaOH $4\ g\ l^{-1}$	B : 0.87 (+45%)	
M	OFMSW+Sewage sludge	Bench sonicator, model VCX 750 Sonics (Sonics, USA) 60 min, $0.4\ W\ ml^{-1}$	22 500[e)] (+24%)	[117]
M	Grass	Speed of rotation between 200 and 1200 rpm and biomass load between 1.40 and 0.40 kg grass m^{-2}	M : 0.376 (+27%)	[118]
M	Municipal solid waste	Electromagnetic mill: 120 Hz frequency of the magnetic field rotation for 0.5, 1.0, 2.0, and 4.0 min	B : 0.238 (+25%)	[119]
M	Digested manure fibers	Applying shearing forces to reduce the particle size using heavy plates	M: up to 0,075 (+45%)	[120]
T	Late harvested hay	160–220 °C and 5–15 min	M : 0.469 (+16%)	[121]
T	Wheat straw	140–178 °C 0.5–2 h	B : 0.286 (−11%)	[122]
T	Food waste and cattle manure	121 °C 30 min	M : 0.430 (+81%)	[123]
M	Cattle manure, crude glycerin	Ultrasonic pre-treatment $520\ kJ\ kg^{-1}$ TS	M : 0.509 (280%)	[124]
M	WAS	Ultrasonic pre-treatment 5000–5035 000 $kJ\ kg^{-1}$ TS	B: up to 0.291 (up to +31%)	[125]
M	Wheat straw	Ultrasonic pre-treatment 30–60–90 min at 22 kHz and 250 W sonication	M : 0.357 (+75%)	[126]
	mung bean husk		M : 0.382 (+46%)	
	OFMSW		M : 0.331 (+71%)	
M	Fruit and vegetable waste	Ultrasonic pre-treatment $2380\ kJ\ kg^{-1}$ TS	B : 0.396 (+80%)	[127]
M	Rice straw	Microwave pre-treatment 190 °C and 4 min	M : 0.326 (+41%)	[128]
M	Dewatered sludge cake	Microwave pre-treatment. heating ramp rate of 3–11 °C/minute and final temperature of the pre-treatment of 80–160 °C	B : 0.498 (+24.7%)	[129]

(Continued)

Table 8.12 (Continued)

Method	Feedstock	Pre-treatment conditions	Yield B(biogas) M(methane) $m^3\ kg^{-1}$ VS[a)]	References
M	Pulp and paper mill sludge	Electrohydrolysis pre-treatment 15 V and 45 min	M : 0.301 (+9.8%)	[130]
M	WAS	Electrochemical pre-treatment 20 V and 40 min	B : 0.647 (+64%)	[131]
M	Sewage sludge	Pulsed electric fields: 150 and 280 $kJ\ l^{-1}$,	M : 0.420 (+20%)	[132]
CH	Maize straw	CaO (0.05–0.15 g/g TS), and 6–12 h	B : 0.290 (+37.5%)	[133]
		Peracetic acid (PAA) (0.01–0.1 g/g TS) and 6–12 h	B : 0.320 (+137.5%)	
		Both reagents	B : 0.590 (+125%)	
CH	Sewage sludge	At 8–12 pH pre-treatment (adding NaOH) for six days	B : 0.385 (+29.6%)	[134]
CH	Rice husk	3–9% (v/v) NaOH	B : 0.056 (+37.5%)	[135]
CH	manure	5–10–30 gO_3/g TS	B : 0.403 (+13%)	[136]
	Potato waste		B : 0.807 (+26.6%)	
	Mixed organic waste		B : 0.927 (+1.0%)	
CH	Barley straw	Ionic liquids: 85% of N-methylmorpholine-N-oxide at 90 °C for 3–30 h	M : 0.230 (+100%)	[137]
	Forest residue		M : 0.150 (+100%)	
B	Food waste	Two-stage anaerobic digestion was conducted in a batch mode at 39 ± 1 °C Using a 2-l glass reactor for the first stage and a 5-l glass reactor for the second stage	M : 0.392 (+19%)	[138]
B	Pulp and paper biosludge	Enzymatic pretreatment: 6 h at 37 °C and 100 rpm.	225[f)] (+26%)	[139]
B	Yard trimmings	White rot fungus; *C. subvermispora*	0.0446 (+2.3 x)	[140]
TCH	Municiapl waste	3 $g\ l^{-1}$ NaOH at 180 °C for 60 min	493[e)](+116%))	[141]
TM	Fruit and vegetable waste	CH: 0.002–0.01 g/g suspended solid Dimethyl sulphoxide UD: 5400 $kJ\ kg^{-1}$ TS	M : 0.190[f)] (+287.7)	[142]

Method: CH, chemical, B, biological, M, mechanical, T, thermal.
Other:

a) A change compared to untreated waste.
b) Microorganism's maximum specific growth rate.
c) The cumulative biogas production (whole), ml.
d) The cumulative biogas production, free liquid, ml.
e) Cumulative biogas yield.
f) $m^3\ kg^{-1}$ COD.

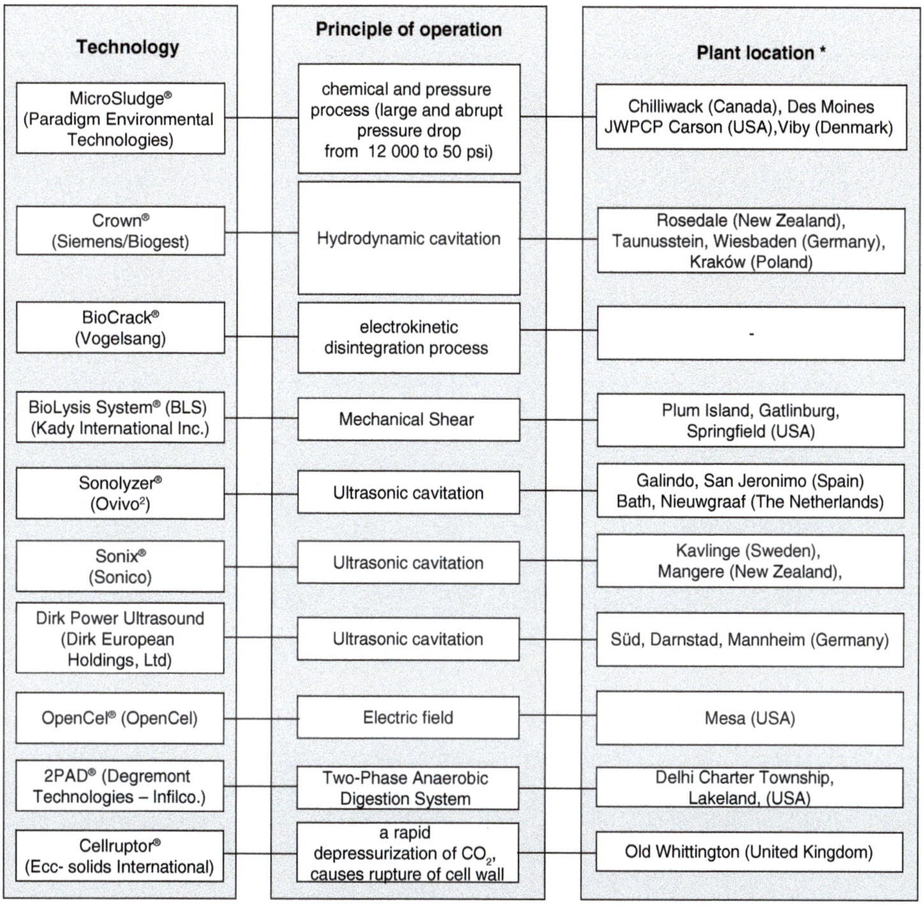

Figure 8.6 Commercial technologies for the pre-treatment of sewage sludge [83].

8.6.2 Anaerobic Co-digestion Process

Co-digestion is defined as the joint AD of a mixture of at least two types of biodegradable waste, so-called co-substrates [143]. This process is usually carried out with over 50% share of one of the co-substrates. In the sewage sludge co-digestion process the following co-substrates can be used: waste from agriculture or the agro-food industry (e.g. animal manure, waste generated in breweries, sludge from fat separators, waste from slaughterhouses, straw) and the organic fraction of municipal solid waste [144–146]. Moreover, there is also research conducted aimed at assessing the feasibility of using other organic waste as a co-substrate, such as energetic plants [147] and glycerol (a co-product generated during biodiesel production) [148]. The addition of co-substrates to digesters results in increases in the macro- and microelements content, improvement of the C/N ratio, dilution of toxic substances, an increase in the content of biodegradable fractions, and higher biogas production [149]. Generally, the addition of co-substrates increases the degree of sludge

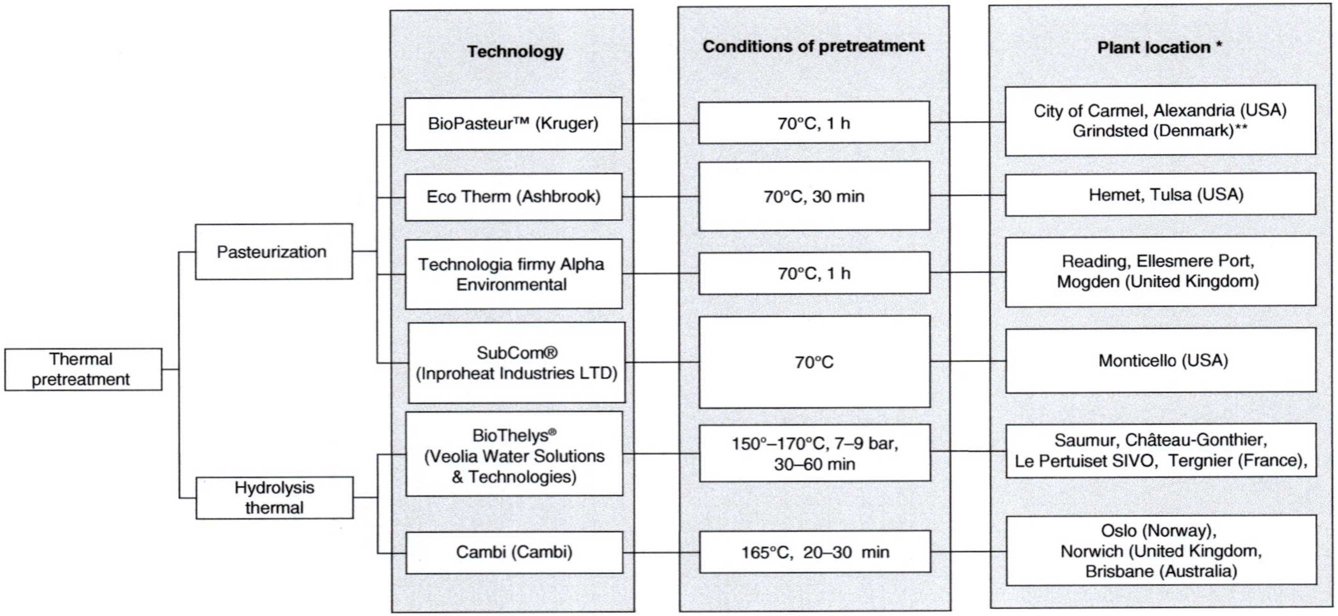

Figure 8.7 Commercial technologies for the thermal pre-treatment of sewage sludge [83].

digestion and higher biogas production. However, the co-substrate introduced to the anaerobic digester should have the following properties [150]:

- The organic substance concentration should be comparable to its content in bio-waste – an additive, in such a case, will not affect the retention time significantly.
- It should consist of an easily biodegradable organic matter with a high biogas production potential.
- It should not contain substances in concentrations that could inhibit the process.
- It should contain the number of macro- and micronutrients that allow for improvement of the specification of the main substrate.
- It should not cause any problems with clogging the line and it should enable the digesters to be supplied by means of an automatic feeder.

Table 8.13 presents the results of the co-digestion of various substrates.

8.7 Conclusion

A positive energy balance and limited environmental impact make the AD a promising management method of the organic waste. Nevertheless, some properties of bio-waste (e.g. OFMSW) contribute to the fact that it is regarded as a substrate which is problematic in anaerobic stabilization. Therefore, in the recent years, the studies have been focused on the process optimization and modification and, in consequence, its intensification. A portion of the research has concentrated on streamlining the stage regarded as limiting the entire process rate, that is hydrolysis. Apart from the most common mechanical pre-treatment of the feedstock (the size of particles is one of the factors affecting the AD result), the following is also applied: chemical, thermal, and biological methods, as well as their combinations. The data presented in the chapter indicate clearly that the solubilization of organic compounds during pre-treatment results in process streamlining, estimated biogas production, and VS removal.

The second option presented in this article also appears to be promising. The addition of the co-substrate to the anaerobic digester, most often as compared to the AD of a single substrate, allowed for the following: increasing biogas production and VS removal, as well as shortening the lag phase. Furthermore, it must be borne in mind that a correctly selected ratio of co-substrates may adjust a series of the parameters of the co-digestion mixture, thus ensuring process optimal conditions.

It is difficult to indicate the best option because each of them has its strong and weak points. Hence, the selection of a technology should depend on the properties of the waste and it should be preceded with the calculation of energy consumption, outlays, etc. Notwithstanding the process enhance method applied, it appears that the more comprehensive use of such a technology may become an effective tool in implementing the provisions included in the legal acts concerning waste (i.e. Directive 1999/31/EC, Waste Act in Polish Law), containing provisions referring to the weight reduction of municipal solid waste components subject to biological decomposition. Nonetheless, the more comprehensive industrial use of AD requires development of the system of biodegradable waste collection.

Table 8.13 The effect of introducing the co-substrate into the anaerobic digester on the efficiency of the process.

Methane yield (m^3 Mg^{-1} VS)			VS removal (%)				
Control	Co-digestion	Impact co-substrate	Control	Co-digestion	Impact co-substrate	Feedstock	References
sewage sludge plus							
143[a)]	355	+148%	NA	NA	–	Source sorted OFMSW	[151]
NA	210	–	NA	39.5	–	Source sorted bio-waste	[152]
164.5[a)]	220.6	+34%	45[a)]	48	+7%	Synthetic FMSW	[153]
310[a)]	393	+27%	NA	NA	–	Food waste	[154]
183[b)]	316	+73%	46[b)]	41	−11%	Hydromechanically separated organic fraction of municipal solid waste	[155]
176[b)]	283	+61%	49[b)]	29	−40%		
190[b)]	224	+17%[c)]	48[b)]	41[c)]	−15%		
186[b)]	166	−11%[c)]	51[b)]	39[c)]	−24%		
250[a)]	320	+28%	33[a]	46.4	+41%	Source selected OFMSW	[156]
384[a)]	641	+67%	47[a)]	59	+25%	Local butcher's shop	[157]
163[a)]	520	+219%	NA	NA	–	Grease trap waste	[158]
278[a)]	463	+67%	52[a)]	67	+29%	Grease trap sludge from meat plant	[58]
249[a)]	369	+48%	36[a)]	52	+44%	The dissolved air flotation unit of a WWTP	[159]
252[a)]	614	+144%	40[a)]	57	+43%	Fat oil and grease	[160]
294[a)]	490	+67%	52[a)]	59	+13%	Grease sludge from domestic wastewater	[161]
221[a)]	423	+91	45[a)]	60	+33	grease trap Sludge from meat industry	[57]
180[a)]	752	+318%	30[a)]	55	+83%	Polymer dewatered grease trap waste	[162]
159[a)]	473	+197%	25[a)]	50	+100%	Pre-treated grease interceptor waste	[163]
300[a)]	547	+82%	50	64.7	+29%	Grease trap sludge + OFMSW	[59]

260[a)]	300	+15%	NA	NA	–	Fruit waste	[164]
390[a)d)]	890	+128%	71	81	+14%	Organic waste from domestic refuse (swill) and	[165]
237[a)]	400	+69%	38.2[a)]	70	+83	Food waste	[166]
Other waste							
207[e)]	293	+42%	NA	NA	–	Microalgae-sludge + FOG	[167]
207[e)]	334	+61%	NA	NA	–	Pre-treated microalgae-sludge + FOG	
335[f)]	431	+29%	NA	NA	–	Food waste with pre-treated yard waste:	[168]
234[g)]	354	+51%	50.7	52.3	+3%	Dairy manure with aloe peel + vermiculite	[169]
264[h)]252.6[i)]	280.7	+5.8%[h)] +10.7%[i)]	NA	NA	–	Corn stover and wastewater from hydrothermal carbonation	[170]
166.5[j)]	323.78	+94.4%	NA	NA	–	Rice straw and food waste	[171]
409.2[k)]	580.9	+90.3%	54.96[l)]	79.37	+44%	Acorn slag waste with dairy manure	[172]
264[m)]147[n)]691[o)]	266	+1%[m)]/+81%[n)]	NA	NA	–	Whey + manure	[173]
	534	+102%[m)]/−23%[o)]	NA	NA	–	Whey + fish ensilage	
	729	+396%[n)]/+5%[o)]	NA	NA	–	Manure + fish ensilage	
	566	+114%[m)]/+285%[n)] −18%[o)]	NA	NA	–	Manure + fish ensilage + whey	
164.5[b)]275.9[p)]	396.6	+140%[b)]/+43.5%[p)]	NA	NA	–	OFMSW + fruit and vegetable waste	[174]

Table 8.13 (Continued)

Methane yield ($m^3\ Mg^{-1}$ VS)			VS removal (%)				
Control	Co-digestion	Impact co-substrate	Control	Co-digestion	Impact co-substrate	Feedstock	References
266[e] 375[q]	461	+73%[e]/+23%[q]	NA	NA	–	Olive mill solid waste and microalga	[175]
403[r] 155[m]	259	+67%[m]/−35%[r]	NA	NA	–	Spent goat straw bedding and goat cheese whey	[176]
536.6[f] 299.7[e]	469.3	+57%[e]/−12.5%[f]	NA	NA	–	Food waste + algae	[177]

NA, not available.

a) AD of sewage sludge alone.
b) AD of OFMSW alone.
c) Thermophilic AD.
d) Biogas yield.
e) AD of algae alone.
f) AD of food waste alone.
g) Reactor was feed with co-substrates and inoculum, without any exogenous additive.
h) Mono-digestion of corn stover.
i) Mono-digestion of wastewater of hydrothermal carbonation.
j) Mono-digestion of rice straw.
k) Mono-digestion of ASW.
l) COD reduction.
m) Mono-digestion of whey.
n) Mono-digestion of manure.
o) Mono-digestion of fish ensilage.
p) AD of fruit and vegetable waste alone.
q) AD of olive mill solid waste.
r) AD of spent goat straw bedding.

References

1 Rintala, J.A. and Järvinen, K.T. (1996). Full-scale mesophilic anaerobic co-digestion of municipal solid waste and sewage sludge: methane production characteristics. *Waste Management & Research* 14 (2): 163–170. Available from: http://journals.sagepub.com/doi/10.1177/0734242X9601400206.

2 Ponsá, S., Gea, T., and Sánchez, A. (2011). Anaerobic co-digestion of the organic fraction of municipal solid waste with several pure organic co-substrates. *Biosystems Engineering* 108 (4): 352–360.

3 Ferrer, I., Garfí, M., Uggetti, E. et al. (2011). Biogas production in low-cost household digesters at the Peruvian Andes. *Biomass and Bioenergy* 35 (5): 1668–1674.

4 Walker, M., Banks, C.J., and Heaven, S. (2009). Two-stage anaerobic digestion of biodegradable municipal solid waste using a rotating drum mesh filter bioreactor and anaerobic filter. *Bioresource Technology* 100 (18): 4121–4126. Available from: http://dx.doi.org/10.1016/j.biortech.2009.03.066.

5 Vogt, G.M., Liu, H.W., Kennedy, K.J. et al. (2002). Super blue box recycling (SUBBOR) enhanced two-stage anaerobic digestion process for recycling municipal solid waste: laboratory pilot studies. *Bioresource Technology* 85 (3): 291–299.

6 Bouallagui, H., Touhami, Y., Ben, C.R., and Hamdi, M. (2005). Bioreactor performance in anaerobic digestion of fruit and vegetable wastes. *Process Biochemistry* 40 (3–4): 989–995.

7 Gunaseelan, V.N. (2004). Biochemical methane potential of fruits and vegetable solid waste feedstocks. *Biomass and Bioenergy* 26 (4): 389–399.

8 Del Borghi, A., Converti, A., Palazzi, E., and Del Borghi, M. (1999). Hydrolysis and thermophilic anaerobic digestion of sewage sludge and organic fraction of municipal solid waste. *Bioprocess Engineering* 20 (6): 553–560.

9 Davidsson, Å., Gruvberger, C., Christensen, T.H. et al. (2007). Methane yield in source-sorted organic fraction of municipal solid waste. *Waste Management* 27 (3): 406–414.

10 Li, Y., Park, S.Y., and Zhu, J. (2011). Solid-state anaerobic digestion for methane production from organic waste. *Renewable and Sustainable Energy Reviews* 15 (1): 821–826. Available from: http://dx.doi.org/10.1016/j.rser.2010.07.042.

11 Guendouz, J., Buffière, P., Cacho, J. et al. (2010). Dry anaerobic digestion in batch mode: design and operation of a laboratory-scale, completely mixed reactor. *Waste Management* 30 (10): 1768–1771. Available from: http://dx.doi.org/10.1016/j.wasman.2009.12.024.

12 Owens, J.M. and Chynoweth, D.P. (1993). Biochemical methane potential of municipal solid waste (MSW) components. *Water Science and Technology* 27 (2): 1–14.

13 Beyene, A., Kassahun, Y., Addis, T. et al. (2012). The impact of traditional coffee processing on river water quality in Ethiopia and the urgency of adopting sound environmental practices. *Environmental Monitoring and Assessment* 184 (11): 7053–7063.

14 Lahcen, S.A.L., Ahmed, S.I., Joute, Y., and El Bari, H. (2018). Characterization and evaluation of methanogenic potential of coffee waste. *Journal of Materials and Environmental Science* 9 (2): 655–661.

15 Dróżdż, D., Wystalska, K., Malińska, K. et al. (2020). Management of poultry manure in Poland – current state and future perspectives. *Journal of Environmental Management* 264: 110327.

16 Himanshu, H., Murphy, J.D., Grant, J., and O'Kiely, P. (2018). Antagonistic effects on biogas and methane output when co-digesting cattle and pig slurries with grass silage in in vitro

batch anaerobic digestion. *Biomass and Bioenergy* 109: 190–198. Available from: https://doi.org/10.1016/j.biombioe.2017.12.027.

17 Dhar, H., Kumar, S., and Kumar, R. (2017). A review on organic waste to energy systems in India. *Bioresource Technology* 245: 1229–1237. Available from: https://doi.org/10.1016/j.biortech.2017.08.159.

18 Grosser, A., Worwag, M., Neczaj, E., and Grobelak, A. (2013). Semi-continuous anaerobic co-digestion of mixed sewage sludge and waste fats of vegetable origin. *Rocznik Ochrona Srodowiska* 15: 2108–2101, 2125.

19 Romero-Güiza, M.S., Vila, J., Mata-Alvarez, J. et al. (2016). The role of additives on anaerobic digestion: a review. *Renewable and Sustainable Energy Reviews* 58: 1486–1499.

20 Mata-Alvarez, J., Dosta, J., and Peces, M. (2014). A critical review on anaerobic co-digestion achievements between 2010 and 2013. *Renewable and Sustainable Energy Reviews* 36: 412–427. Available from: http://dx.doi.org/10.1016/j.rser.2014.04.039.

21 Grosser, A., Worwag, M., Neczaj, E., Kamizela, T. 2013 Co-digestion of organic fraction of municipal solid waste with different organic wastes: A review. In: *Environmental Engineering IV – Proceedings of the Conference on Environmental Engineering IV*. pp. 231–242. Available from: http://www.scopus.com/inward/record.url?eid=2-s2.0-84880151950&partnerID=tZOtx3y1.

22 Technologien, N. and Wirtschaftsberatung, B. (2000). *Anaerobic Methods of Waste Treatment*. Tech Information W2e.

23 Scarlat, N., Dallemand, J., and Fahl, F. (2018). Biogas: developments and perspectives in Europe. *Renewable Energy* 129: 457–472. Available from: https://doi.org/10.1016/j.renene.2018.03.006.

24 den, B.E., Jedrczak, A., Kowalski, Z. et al. (2010). A review of municipal solid waste composition and quantities in Poland. *Waste Management* 30 (3): 369–377.

25 Hanc, A., Novak, P., Dvorak, M. et al. (2011). Composition and parameters of household bio-waste in four seasons. *Waste Management* 31 (7): 1450–1460.

26 Reddy, P.J. (2011). *Municipal Solid Waste Management*, vol. 9, 2012. The Netherlands: CRC Press.

27 Kumar, A. and Samadder, S.R. (2017). A review on technological options of waste to energy for effective management of municipal solid waste. *Waste Management* 69: 407–422. Available from: https://doi.org/10.1016/j.wasman.2017.08.046.

28 Grobelak, A., Czerwińska, K., and Murtaś, A. (2019). General considerations on sludge disposal, industrial and municipal sludge. In: *Industrial and Municipal Sludge* (eds. M.N.V. Prasad, P.J. de Campos Favas, M. Vithanage and S.V. Mohan), 135–153. Elsevier.

29 Cecchi, F., Traverso, P., Pavan, P., and Bolzonella, D. (2002). Characteristics of the OFMSW and behavior of the anaerobic digestion process. In: *Biomethanization of the Organic Fraction of Municipal Solid Wastes* (ed. J. Mata-Alvarez). Iwa Publishing.

30 Hartmann, H. and Ahring, B.K. (2006). Strategies for the anaerobic digestion of the organic fraction of municipal solid waste: an overview. *Water Science and Technology* 53 (8): 7–22.

31 Chand Malav, L., Yadav, K.K., Gupta, N. et al. (2020). A review on municipal solid waste as a renewable source for waste-to-energy project in India: current practices, challenges, and future opportunities. *Journal of Cleaner Production* 277: 123227. Available from: https://doi.org/10.1016/j.jclepro.2020.123227.

32 Wiśniowska, E., Grobelak, A., Kokot, P., and Kacprzak, M. (2019). Sludge legislation-comparison between different countries. In: *Industrial and Municipal Sludge* (eds. M.N.V. Prasad, P.J. de Campos Favas, M. Vithanage and S.V. Mohan), 201–224. Elsevier.

33 OECD.stat. Municipal Waste, Generation and Treatment. Available from: https://stats.oecd.org/Index.aspx?DataSetCode=MUNW.

34 Eurostat. Municipal Waste Generated, in Selected Years, 1995–2018 (kg per capita). Available from: https://ec.europa.eu/eurostat/statistics-explained/index.php?title=File:Municipal_waste_generated,_in_selected_years,_1995-2018_(kg_per_capita).png#filelinks.

35 Hartmann, H., Angelidaki, I., and Ahring, B.K. (2002). Co-digestion of the organic fraction of municipal waste with other waste types. In: *Biomethanization of the Organic Fraction of Municipal Solid Wastes* (ed. J. (e.) Mata-Alvarez). IWA Publishing.

36 Gallert, C. and Winter, J. (2005). Bacterial metabolism in wastewater treatment systems. *Environmental Biotechnology*: 1–48. Available from: http://dx.doi.org/10.1002/3527604286.ch1.

37 Wang, P., Wang, H., Qiu, Y. et al. (2018). Microbial characteristics in anaerobic digestion process of food waste for methane production – a review. *Bioresource Technology* 248: 29–36. Available from: https://doi.org/10.1016/j.biortech.2017.06.152.

38 Zhang, L., Loh, K.C., and Zhang, J. (2019). *Enhanced biogas production from anaerobic digestion of solid organic wastes: Current status and prospects*, vol. 5, Bioresource Technology Reports, 280–296. Elsevier. Available from: https://doi.org/10.1016/j.biteb.2018.07.005.

39 Tabatabaei, M., Aghbashlo, M., Valijanian, E. et al. (2020). A comprehensive review on recent biological innovations to improve biogas production, Part 1: upstream strategies. *Renewable Energy* 146: 1204–1220. Available from: https://doi.org/10.1016/j.renene.2019.07.037.

40 Anukam, A., Mohammadi, A., Naqvi, M., and Granström, K. (2019). A review of the chemistry of anaerobic digestion: methods of accelerating and optimizing process efficiency. *Processes* 7 (8): 504.

41 Edelmann, W. (2003). Products, impacts and economy of anaerobic digestion of OFMSW. In: *Biomethanization of the Organic Fraction of Municipal Solid Wastes* (ed. J. (e.) Mata-Alvarez), 265–301. Iwa Publishing.

42 Forster-Carneiro, T., Pérez, M., and Romero, L.I. (2008). Anaerobic digestion of municipal solid wastes: dry thermophilic performance. *Bioresource Technology* 99 (17): 8180–8184.

43 Gallert, C., Henning, A., and Winter, J. (2003). Scale-up of anaerobic digestion of the biowaste fraction from domestic wastes. *Water Research* 37 (6): 1433–1441.

44 Zhang, R., El-mashad, H.M., Hartman, K. et al. (2007). Characterization of food waste as feedstock for anaerobic digestion. *Bioresource Technology* 98: 929–935.

45 Angelidaki, I., Cui, J., Chen, X., and Kaparaju, P. (2006). Operational strategies for thermophilic anaerobic digestion of organic fraction of municipal solid waste in continously stirred tank reactors. *Environmental Technology* 27 (8): 855–861.

46 Bolzonella, D., Innocenti, L., Pavan, P. et al. (2003). Semi-dry thermophilic anaerobic digestion of the organic fraction of municipal solid waste: focusing on the start-up phase. *Bioresource Technology* 86 (2): 123–129.

47 Maroun, R. and El Fadel, M. (2007). Start-up of anaerobic digestion of source-sorted organic municipal solid waste in the absence of classical inocula. *Environmental Science and Technology* 41 (19): 6808–6814.

48 Bouallagui, H., Ben Cheikh, R., Marouani, L., and Hamdi, M. (2003). Mesophilic biogas production from fruit and vegetable waste in a tubular digester. *Bioresource Technology* 86 (1): 85–89.

49 Bouallagui, H., Haouari, O., Touhami, Y. et al. (2004). Effect of temperature on the performance of an anaerobic tubular reactor treating fruit and vegetable waste. *Process Biochemistry* 39 (12): 2143–2148.

50 Kim, J.K., Oh, B.R., Chun, Y.N., and Kim, S.W. (2006). Effects of temperature and hydraulic retention time on anaerobic digestion of food waste. *Journal of Bioscience and Bioengineering* 102 (4): 328–332.

51 Fernández, A., Sánchez, A., and Font, X. (2005). Anaerobic co-digestion of a simulated organic fraction of municipal solid wastes and fats of animal and vegetable origin. *Biochemical Engineering Journal* 26 (1): 22–28.

52 Dong, L., Zhenhong, Y., and Yongming, S. (2010). Semi-dry mesophilic anaerobic digestion of water sorted organic fraction of municipal solid waste (WS-OFMSW). *Bioresource Technology* 101 (8): 2722–2728. Available from: http://dx.doi.org/10.1016/j.biortech.2009.12.007.

53 Nagao, N., Tajima, N., Kawai, M. et al. (2012). Maximum organic loading rate for the single-stage wet anaerobic digestion of food waste. *Bioresource Technology* 118: 210–218. Available from: http://dx.doi.org/10.1016/j.biortech.2012.05.045.

54 Rao, M.S. and Singh, S.P. (2004). Bioenergy conversion studies of organic fraction of MSW: kinetic studies and gas yield-organic loading relationships for process optimisation. *Bioresource Technology* 95 (2): 173–185.

55 Nguyen, P.H.L., Kuruparan, P., and Visvanathan, C. (2007). Anaerobic digestion of municipal solid waste as a treatment prior to landfill. *Bioresource Technology* 98 (2): 380–387.

56 Bolzonella, D., Pavan, P., Mace, S., and Cecchi, F. (2006). Dry anaerobic digestion of differently sorted organic municipal solid waste: a full-scale experience. *Water Science and Technology* 53 (8): 23–32.

57 Grosser, A. and Neczaj, E. (2018). Sewage sludge and fat rich materials co-digestion – performance and energy potential. *Journal of Cleaner Production* 198: 1076–1089. Available from: https://linkinghub.elsevier.com/retrieve/pii/S0959652618321097.

58 Luostarinen, S., Luste, S., and Sillanpää, M. (2009). Increased biogas production at wastewater treatment plants through co-digestion of sewage sludge with grease trap sludge from a meat processing plant. *Bioresource Technology* 100 (1): 79–85. Available from: http://linkinghub.elsevier.com/retrieve/pii/S096085240800549X.

59 Grosser, A., Neczaj, E., Singh, B.R. et al. (2017). Anaerobic digestion of sewage sludge with grease trap sludge and municipal solid waste as co-substrates. *Environmental Research* 155: 249–260. Available from: http://linkinghub.elsevier.com/retrieve/pii/S0013935117302402.

60 Zhang, B., Zhang, L.-L., Zhang, S.-C. et al. (2005). The influence of pH on hydrolysis and acidogenesis kitchen wastes. *Environmental Technology* 26 (3): 329–340.

61 Wu, L.J., Kobayashi, T., Li, Y.Y., and Xu, K.Q. (2015). Comparison of single-stage and temperature-phased two-stage anaerobic digestion of oily food waste. *Energy Conversion and Management* 106: 1174–1182. Available from: http://dx.doi.org/10.1016/j.enconman.2015.10.059.

62 Salsali, H., Parker, W., and Sattar, S. (2005). Influence of staged operation of mesophilic anaerobic digestion on microbial reduction. *Proceedings of the Water Environment Federation* 2005 (11): 4571–4586. Available from: https://accesswater.org/publications/-291781/influence-of-staged-operation-of-mesophilic-anaerobic-digestion-on-microbial-reduction.

63 Theuerl, S., Klang, J., Heiermann, M., and De Vrieze, J. (2018). Marker microbiome clusters are determined by operational parameters and specific key taxa combinations in anaerobic digestion. *Bioresource Technology* 263: 128–135. Available from: https://doi.org/10.1016/j.biortech.2018.04.111.

64 Gerardi, M.H. (2003). *The Microbiology of Anaerobic Digesters*. Wiley.

65 Meegoda, J.N., Li, B., Patel, K., and Wang, L.B. (2018). A review of the processes, parameters, and optimization of anaerobic digestion. *International Journal of Environmental Research and Public Health* 15 (10): 2224.

66 Van, D.P., Fujiwara, T., Tho, B.L. et al. (2020). A review of anaerobic digestion systems for biodegradable waste: configurations, operating parameters, and current trends. *Environmental Engineering Research* 25 (1): 1–17.

67 Kothari, R., Pandey, A.K., Kumar, S. et al. (2014). Different aspects of dry anaerobic digestion for bio-energy: an overview. *Renewable and Sustainable Energy Reviews* 39: 174–195. Available from: http://dx.doi.org/10.1016/j.rser.2014.07.011.

68 Dry – solid-state anaerobic digestion AD biogas plant. Available from: https://commons.wikimedia.org/wiki/File:Dry_-_solid-state_anaerobic_digestion_AD_biogas_plant.jpg.

69 Yaashikaa, P.R., Kumar, P.S., Saravanan, A. et al. (2020). Bioconversion of municipal solid waste into bio-based products: a review on valorisation and sustainable approach for circular bioeconomy. *Science of the Total Environment* 748: 141312. Available from: https://doi.org/10.1016/j.scitotenv.2020.141312.

70 Curry, N. and Pillay, P. (2012). Biogas prediction and design of a food waste to energy system for the urban environment. *Renewable Energy* 41: 200–209. Available from: http://dx.doi.org/10.1016/j.renene.2011.10.019.

71 Khalid, A., Arshad, M., Anjum, M. et al. (2011). The anaerobic digestion of solid organic waste. *Waste Management* 31 (8): 1737–1744. Available from: http://dx.doi.org/10.1016/j.wasman.2011.03.021.

72 Zeshan, K.O.P. and Visvanathan, C. (2012). Effect of C/N ratio and ammonia-N accumulation in a pilot-scale thermophilic dry anaerobic digester. *Bioresource Technology* 113: 294–302. Available from: http://dx.doi.org/10.1016/j.biortech.2012.02.028.

73 Jenkins, B.M., Williams, R.B. et al. (2008). Current Anaerobic Digestion Technologies Used for Treatment of Municipal Organic Solid Waste. 2008 (March).

74 OWS. Biogas Plants. Available from: https://www.ows.be/biogas-plants/references/brecht-i-1992-and-brecht-ii-2000.

75 Kompogas® Plants. Available from: https://www.hz-inova.com/cms/en/?page_id=1870#slo.

76 Valorga International References. Available from: http://www.valorgainternational.fr/en/pag8-OUR-REFERENCES.html.

77 Plants designed according to the BTA® Process. Available from: http://www.bta-international.de/en/referenzen/reference-list.html.

78 Chavez-Vazquez, M. and Bagley D.M. 2002. Evaluation of the performance of different anaerobic digestion technologies for solid waste treatment. In: *An International Perspective on Environmental Engineering: joint 2002 CSCE/ASCE International Conference on Environmental Engineering*; . pp. 1–15. Available from: http://library1.nida.ac.th/termpaper6/sd/2554/19755.pdf.

79 IRENA (2020). *Renewable Capacity Statistics 2020*, 66. International Renewable Energy Agency. Available from: https://irena.org/-/media/Files/IRENA/Agency/Publication/2020/Mar/IRENA_RE_Capacity_Statistics_2020.pdf.

80 REN21. 2020. Renewables 2020 global status report. Available from: https://abdn.pure.elsevier.com/en/en/researchoutput/ren21(5d1212f6-d863-45f7-8979-5f68a61e380e).html.

81 Kim, M., Chowdhury, M.M.I., Nakhla, G., and Keleman, M. (2017). Synergism of co-digestion of food wastes with municipal wastewater treatment biosolids. *Waste Management* 61: 473–483. Available from: http://dx.doi.org/10.1016/j.wasman.2016.10.010.

82 Iacovidou, E., Ohandja, D.-G., and Voulvoulis, N. (2012). Food waste co-digestion with sewage sludge–realising its potential in the UK. *Journal of Environmental Management* 112: 267–274.

83 Neczaj, E. 2016. The methods used to intensifity a biogas production from sewage sludge. *Wydawnictwo Politechniki Częstochowskiej*; . Monografie – Politechnika Częstochowska. Available from: https://books.google.pl/books?id=LMo8ngAACAAJ.

84 Carrere, H., Antonopoulou, G., Affes, R. et al. (2016). Review of feedstock pretreatment strategies for improved anaerobic digestion: from lab-scale research to full-scale application. *Bioresource Technology* 199: 386–397. Available from: http://dx.doi.org/10.1016/j.biortech.2015.09.007.

85 Kumar, B., Bhardwaj, N., Agrawal, K. et al. (2020). Current perspective on pretreatment technologies using lignocellulosic biomass: an emerging biorefinery concept. *Fuel Processing Technology* 199: 106244.

86 Pramanik, S.K., Suja, F.B., Zain, S.M., and Pramanik, B.K. (2019). The anaerobic digestion process of biogas production from food waste: prospects and constraints. *Bioresource Technology Reports* 8: 100310. Available from: https://doi.org/10.1016/j.biteb.2019.100310.

87 Carlsson, M., Lagerkvist, A., and Morgan-Sagastume, F. (2012). The effects of substrate pre-treatment on anaerobic digestion systems: a review. *Waste Management* 32 (9): 1634–1650. Available from: http://dx.doi.org/10.1016/j.wasman.2012.04.016.

88 Mata-Alvarez, J., Macé, S., and Llabres, P. (2000). Anaerobic digestion of organic solid wastes. An overview of research achievements and perspectives. *Bioresource Technology* 74: 3–16.

89 Carrère, H., Dumas, C., Battimelli, A. et al. (2010). Pretreatment methods to improve sludge anaerobic degradability: a review. *Journal of Hazardous Materials* 183 (1–3): 1–15.

90 Elliott, A. and Mahmood, T. (2007). Pretreatment technologies for advancing anaerobic digestion of pulp and paper biotreatment residues. *Water Research* 41 (19): 4273–4286.

91 Grosser, A. and Celary, P. (2019). Biogas (methane production) and energy recovery from different sludges. In: *Industrial and Municipal Sludge* (eds. M.N.V. Prasad, P.J. de Campos Favas, M. Vithanage and S.V. Mohan), 705–740. Elsevier.

92 Grosser, A. and Neczaj, E. 2017. Pretreatment methods as a means of boosting methane production from sewage sludge and its mixtures with grease trap sludge. In: Kaźmierczak, B., Kutyłowska, M., Piekarska, K., Jouhara, H., and Danielewicz, J. (eds.), *E3S Web Conference.* 22:00058. Available from: http://www.e3s-conferences.org/10.1051/e3sconf/20172200058.

93 Zhang, J., Zhou, H., Liu, D., and Zhao, X. (2019). Pretreatment of lignocellulosic biomass for efficient enzymatic saccharification of cellulose. In: *Lignocellulosic Biomass to Liquid Biofuels* (eds. A. Yousuf, F. Sannino and D. Pirozzi), 17–65. Academic Press. Available from: http://dx.doi.org/10.1016/B978-0-12-815936-1.00002-2.

94 Mayhew, M., Le, M.S., Brade, C.E., and Harrison, D. (2003). The united utitlities "enzymic hydrolysis process" – validation of phased digestion at full scale to enhance pathogen removal. *Proceedings of the Water Environment Federation* 2003 (1): 1000–1013.

95 Yang, Q., Luo, K., ming, L.X. et al. (2010). Enhanced efficiency of biological excess sludge hydrolysis under anaerobic digestion by additional enzymes. *Bioresource Technology* 101 (9): 2924–2930. Available from: http://dx.doi.org/10.1016/j.biortech.2009.11.012.

96 Ojewumi, M., Obielue, B., Emetere, M. et al. (2018). Alkaline pre-treatment and enzymatic hydrolysis of waste papers to fermentable sugar. *Journal of Ecological Engineering* 19 (1): 211–217. Available from: http://www.journalssystem.com/jeeng/alkaline-pre-treatment-and-enzymatic-hydrolysis-of-waste-papers-to-fermentable-sugar,79404,0,2.html.

97 Li, H., Li, C., Liu, W., and Zou, S. (2012). Optimized alkaline pretreatment of sludge before anaerobic digestion. *Bioresource Technology* 123: 189–194. Available from: http://dx.doi.org/10.1016/j.biortech.2012.08.017.

98 Kamizela, T., Kowalczyk, M., and Zawieja, I. (2020). The use of chemical methods and magnetic field in conditioning and dewatering of digested sewage sludge. *Water* (Switzerland). 12 (6): 1–13.

99 Chen, Y., Stevens, M.A., Zhu, Y. et al. 2015. Understanding of alkaline pretreatment parameters for corn stover enzymatic saccharification. *Fuel Production from Non-Food Biomass Corn Stover*. 131–153.

100 Li, C., Li, H., and Zhang, Y. (2015). Alkaline treatment of high-solids sludge and its application to anaerobic digestion. *Water Science and Technology* 71 (1): 67–74.

101 Gundupalli, M.P. and Bhattacharyya, D. (2017). Recovery of reducing sugar from food waste: optimization of pretreatment parameters using response surface. *Methodology*: 161–172.

102 Zawieja, I. and Wolny, L. (2018). Effect of hydrogen peroxide on nitrogen forms in sewage sludge subjected to the anaerobic stabilization. *Desalination and Water Treatment* 117: 301–308.

103 Ariunbaatar, J., Panico, A., Frunzo, L. et al. (2014). Enhanced anaerobic digestion of food waste by thermal and ozonation pretreatment methods. *Journal of Environmental Management* 146: 142–149. Available from: http://dx.doi.org/10.1016/j.jenvman.2014.07.042.

104 Kumar, A.K. and Sharma, S. (2017). Recent updates on different methods of pretreatment of lignocellulosic feedstocks: a review. *Bioresource Bioprocess* 4 (1).

105 Nah, I.W., Kang, Y.W., Hwang, K.Y., and Song, W.K. (2000). Mechanical pretreatment of waste activated sludge for anaerobic digestion process. *Water Research* 34 (8): 2362–2368.

106 Li, X., Guo, S., Peng, Y. et al. (2018). Anaerobic digestion using ultrasound as pretreatment approach: changes in waste activated sludge, anaerobic digestion performances and digestive microbial populations. *Biochemical Engineering Journal* 139: 139–145. Available from: http://dx.doi.org/10.1016/j.bej.2017.11.009.

107 Zawieja, I., Włodarczyk, R., and Kowalczyk, M. (2019). Biogas generation from sonicated excess sludge. *Water* (Switzerland) 11 (10): 2127.

108 Zawieja, I. (2020). Effect of physical and chemical disintegration of excess sludge on the susceptibility to biodegradation. *Desalination and Water Treatment* 186: 341–349.

109 Barber, W.P.F. (2016). Thermal hydrolysis for sewage treatment: a critical review. *Water Research* 104: 53–71. Available from: http://dx.doi.org/10.1016/j.watres.2016.07.069.

110 Neczaj, E. and Grosser, A. (2019). Biogas production by thermal hydrolysis and thermophilic anaerobic digestion of waste-activated sludge. In: *Industrial and Municipal Sludge* (eds. M.N.V. Prasad, P.J. de Campos Favas, M. Vithanage and S.V. Mohan), 741–781. Elsevier.

111 Zhang, D., Feng, Y., Huang, H. et al. (2020). Recalcitrant dissolved organic nitrogen formation in thermal hydrolysis pretreatment of municipal sludge. *Environment International* 138: 105629. Available from: https://doi.org/10.1016/j.envint.2020.105629.

112 Kepp, U., Machenbach, I., Weisz, N., and Solhelm, O.E. (2000). Enhanced stabilisation of sewage sludge through thermal hydrolysis – three years of experience with full scale plant BT – disposal and utilisation of sewage sludge: treatment methods and application modalities, October 13, 1999–October 15, 1999. *Water Science and Technology* 42 (9): 89–96.

113 Llore, C.E. and Lo, M. (2008) Effect of Alkaline Pretreatment on Anaerobic Digestion of Solid Wastes. 28:2229–2234.

114 Fdez.-Güelfo, L.A., Álvarez-Gallego, C., Sales Márquez, D., and Romero García, L.I. (2011). The effect of different pretreatments on biomethanation kinetics of industrial Organic Fraction of Municipal Solid Wastes (OFMSW). *Chemical Engineering Journal* 171 (2): 411–417.

115 Shahriari, H., Warith, M., Hamoda, M., and Kennedy, K.J. (2012). Anaerobic digestion of organic fraction of municipal solid waste combining two pretreatment modalities, high temperature microwave and hydrogen peroxide. *Waste Management* 32 (1): 41–52. Available from: http://dx.doi.org/10.1016/j.wasman.2011.08.012.

116 Qiao, W., Yan, X., Ye, J. et al. (2011). Evaluation of biogas production from different biomass wastes with/without hydrothermal pretreatment. *Renewable Energy* 36 (12): 3313–3318. Available from: http://dx.doi.org/10.1016/j.renene.2011.05.002.

117 Cesaro, A., Naddeo, V., Amodio, V., and Belgiorno, V. (2012). Enhanced biogas production from anaerobic codigestion of solid waste by sonolysis. *Ultrasonics Sonochemistry* 19 (3): 596–600. Available from: http://dx.doi.org/10.1016/j.ultsonch.2011.09.002.

118 Tsapekos, P., Kougias, P.G., and Angelidaki, I. (2018). Mechanical pretreatment for increased biogas production from lignocellulosic biomass; predicting the methane yield from structural plant components. *Waste Management* 78: 903–910. Available from: https://doi.org/10.1016/j.wasman.2018.07.017.

119 Litti, Y., Kovalev, D., Kovalev, A. et al. (2018). Increasing the efficiency of organic waste conversion into biogas by mechanical pretreatment in an electromagnetic mill. *Journal of Physics Conference Series* 1111 (1): 012013.

120 Tsapekos, P., Kougias, P.G., Frison, A. et al. (2016). Improving methane production from digested manure biofibers by mechanical and thermal alkaline pretreatment. *Bioresource Technology* 216: 545–552. Available from: http://dx.doi.org/10.1016/j.biortech.2016.05.117.

121 Bauer, A., Lizasoain, J., Theuretzbacher, F. et al. (2014). Steam explosion pretreatment for enhancing biogas production of late harvested hay. *Bioresource Technology* 166: 403–410. Available from: http://dx.doi.org/10.1016/j.biortech.2014.05.025.

122 Theuretzbacher, F., Lizasoain, J., Lefever, C. et al. (2015). Steam explosion pretreatment of wheat straw to improve methane yields: investigation of the degradation kinetics of structural compounds during anaerobic digestion. *Bioresource Technology* 179: 299–305. Available from: http://dx.doi.org/10.1016/j.biortech.2014.12.008.

123 Arelli, V., Begum, S., Anupoju, G.R. et al. (2018). Dry anaerobic co-digestion of food waste and cattle manure: impact of total solids, substrate ratio and thermal pre treatment on methane yield and quality of biomanure. *Bioresource Technology* 253: 273–280. Available from: https://doi.org/10.1016/j.biortech.2018.01.050.

124 Ormaechea, P., Castrillón, L., Suárez-Peña, B. et al. (2018). Enhancement of biogas production from cattle manure pretreated and/or co-digested at pilot-plant scale. Characterization by SEM. *Renewable Energy* 126: 897–904.

125 Lizama, A.C., Figueiras, C.C., Herrera, R.R. et al. (2017). Effects of ultrasonic pretreatment on the solubilization and kinetic study of biogas production from anaerobic digestion of waste activated sludge. *International Biodeterioration & Biodegradation* 123: 1–9.

126 Prajapati, K.K., Pareek, N., and Vivekanand, V. (2018). Pretreatment and multi-feed anaerobic co-digestion of agro-industrial residual biomass for improved biomethanation and kinetic analysis. *Front Energy Research* 6. Available from: https://www.frontiersin.org/article/10.3389/fenrg.2018.00111/full.

127 Zeynali, R., Khojastehpour, M., and Ebrahimi-Nik, M. (2017). Effect of ultrasonic pretreatment on biogas yield and specific energy in anaerobic digestion of fruit and vegetable wholesale market wastes. *Sustainable Environment Research* 27 (6): 259–264. Available from: https://doi.org/10.1016/j.serj.2017.07.001.

128 Kainthola, J., Shariq, M., Kalamdhad, A.S., and Goud, V.V. (2019). Enhanced methane potential of rice straw with microwave assisted pretreatment and its kinetic analysis. *Journal of Environmental Management* 232: 188–196. Available from: https://doi.org/10.1016/j.jenvman.2018.11.052.

129 Hosseini Koupaie, E. and Eskicioglu, C. (2016). Conventional heating vs. microwave sludge pretreatment comparison under identical heating/cooling profiles for thermophilic advanced anaerobic digestion. *Waste Management* 53: 182–195. Available from: http://dx.doi.org/10.1016/j.wasman.2016.04.014.

130 Veluchamy, C., Raju, V.W., and Kalamdhad, A.S. (2017). Prerequisite – an electrohydrolysis pretreatment for anaerobic digestion of lignocellulose waste material. *Bioresource Technology* 235: 274–280. Available from: http://dx.doi.org/10.1016/j.biortech.2017.03.137.

131 Yu, B., Xu, J., Yuan, H. et al. (2014). Enhancement of anaerobic digestion of waste activated sludge by electrochemical pretreatment. *Fuel* 130: 279–285. Available from: http://dx.doi.org/10.1016/j.fuel.2014.04.031.

132 Kopplow, O., Barjenbruch, M., and Heinz, V. (2004). Sludge pre-treatment with pulsed electric fields. *Water Science and Technology* 49 (10): 123–129.

133 Ramos-Suárez, J.L., Gómez, D., Regueiro, L. et al. (2017). Alkaline and oxidative pretreatments for the anaerobic digestion of cow manure and maize straw: factors influencing the process and preliminary economic viability of an industrial application. *Bioresource Technology* 241: 10–20. Available from: http://dx.doi.org/10.1016/j.biortech.2017.05.054.

134 Wang, T., Xu, B., Zhang, X. et al. (2018). Enhanced biogas production and dewaterability from sewage sludge with alkaline pretreatment at mesophilic and thermophilic temperatures. *Water, Air, and Soil Pollution* 229 (3): 1–10.

135 Syafrudin, Nugraha, W.D., Agnesia, S.S. et al. (2018). Enhancement of biogas production from rice husk by NaOH and enzyme pretreatment. *E3S Web Conference* 31: 2–7.

136 Somers, M.H., Azman, S., Sigurnjak, I. et al. (2018). Effect of digestate disintegration on anaerobic digestion of organic waste. *Bioresource Technology* 268: 568–756. Available from: https://doi.org/10.1016/j.biortech.2018.08.036.

137 Kabir, M.M., Niklasson, C., Taherzadeh, M.J., and Horváth, I.S. (2014). Biogas production from lignocelluloses by N-methylmorpholine-N-oxide (NMMO) pretreatment: effects of recovery and reuse of NMMO. *Bioresource Technology* 161: 446–450. Available from: http://dx.doi.org/10.1016/j.biortech.2014.03.107.

138 De Gioannis, G., Muntoni, A., Polettini, A. et al. (2017). Energy recovery from one- and two-stage anaerobic digestion of food waste. *Waste Management* 68: 595–602. Available from: http://dx.doi.org/10.1016/j.wasman.2017.06.013.

139 Bonilla, S., Choolaei, Z., Meyer, T. et al. (2018). Evaluating the effect of enzymatic pretreatment on the anaerobic digestibility of pulp and paper biosludge. *Biotechnology Reports* 17: 77–85.

140 Zhao, J., Ge, X., Vasco-Correa, J., and Li, Y. (2014). Fungal pretreatment of unsterilized yard trimmings for enhanced methane production by solid-state anaerobic digestion. *Bioresource Technology* 158: 248–252. Available from: http://dx.doi.org/10.1016/j.biortech.2014.02.029.

141 Bala, R., Gautam, V., and Mondal, M.K. (2019). Improved biogas yield from organic fraction of municipal solid waste as preliminary step for fuel cell technology and hydrogen generation. *International Journal of Hydrogen Energy* 44: 164–173. Available from: https://doi.org/10.1016/j.ijhydene.2018.02.072.

142 Shanthi, M., Rajesh Banu, J., and Sivashanmugam, P. (2019). Synergistic effect of combined pretreatment in solubilizing fruits and vegetable residue for biogas production: hydrolysis, energy assessment. *Fuel* 250: 194–202. Available from: https://doi.org/10.1016/j.fuel.2019.04.001.

143 Rabii, A., Aldin, S., Dahman, Y., and Elbeshbishy, E. (2019). A review on anaerobic co-digestion with a focus on the microbial populations and the effect of multi-stage digester configuration. *Energies* 12 (6): 1106.

144 Alqaralleh, R., Kennedy, K., Delatolla, R., and Sartaj, M. (2017). Biogas recovery from hyper-thermophilic anaerobic co-digestion of thickened waste activated sludge, organic fraction of municipal solid waste and fat, oil and grease. *Journal of Bioremediation & Biodegradation* 8 (5): 1000408.

145 Panpong, K., Srisuwan, G., O-Thong, S., and Kongjan, P. (2014). Anaerobic co-digestion of canned seafood wastewater with glycerol waste for enhanced biogas production. *Energy Procedia* 52: 328–336. Available from: http://dx.doi.org/10.1016/j.egypro.2014.07.084.

146 Sosnowski, P. (2003). Anaerobic co-digestion of sewage sludge and organic fraction of municipal solid wastes. *Advances in Environmental Research* 7: 609–616.

147 Schievano, A., D'Imporzano, G., and Adani, F. (2009). Substituting energy crops with organic wastes and agro-industrial residues for biogas production. *Journal of Environmental Management* 90 (8): 2537–2541. Available from: http://dx.doi.org/10.1016/j.jenvman.2009.01.013.

148 Fountoulakis, M.S. and Manios, T. (2009). Enhanced methane and hydrogen production from municipal solid waste and agro-industrial by-products co-digested with crude glycerol. *Bioresource Technology* 100 (12): 3043–3047. Available from: http://dx.doi.org/10.1016/j.biortech.2009.01.016.

149 Mata-alvarez, J., Dosta, J., Romero-güiza, M.S. et al. (2014). A critical review on anaerobic co-digestion achievements between 2010 and 2013. *Renewable and Sustainable Energy Reviews* 36: 412–427. Available from: http://dx.doi.org/10.1016/j.rser.2014.04.039.

150 Nayono, S. 2009. Anaerobic digestion of organic solid waste for energy production. MSc Dissertation.

151 Bolzonella, D., Battistoni, P., Susini, C., and Cecchi, F. (2006). Anaerobic codigestion of waste activated sludge and OFMSW: the experiences of Viareggio and Treviso plants (Italy). *Water Science and Technology* 53 (8): 203–211.

152 Cavinato, C., Bolzonella, D., Pavan, P. et al. (2013). Mesophilic and thermophilic anaerobic co-digestion of waste activated sludge and source sorted biowaste in pilot- and full-scale reactors. *Renewable Energy* 55: 260–265. Available from: http://dx.doi.org/10.1016/j.renene.2012.12.044.

153 Nielfa, A., Cano, R., and Fdz-Polanco, M. (2015). Theoretical methane production generated by the co-digestion of organic fraction municipal solid waste and biological sludge. *Biotechnology Reports* 5 (1): 14–21. Available from: http://dx.doi.org/10.1016/j.btre.2014.10.005.

154 Koch, K., Plabst, M., Schmidt, A. et al. (2016). Co-digestion of food waste in a municipal wastewater treatment plant: comparison of batch tests and full-scale experiences. *Waste Management* 47: 28–33. Available from: http://dx.doi.org/10.1016/j.wasman.2015.04.022.

155 Borowski, S. (2015). Co-digestion of the hydromechanically separated organic fraction of municipal solid waste with sewage sludge. *Journal of Environmental Management* 147: 87–94. Available from: http://dx.doi.org/10.1016/j.jenvman.2014.09.013.

156 Cabbai, V., De Bortoli, N., and Goi, D. (2016). Pilot plant experience on anaerobic codigestion of source selected OFMSW and sewage sludge. *Waste Management* 49: 47–54. Available from: http://dx.doi.org/10.1016/j.wasman.2015.12.014.

157 Martínez, E.J., Gil, M.V., Fernandez, C. et al. (2016). Anaerobic codigestion of sludge: addition of butcher's fat waste as a cosubstrate for increasing biogas production. *PLoS One* 11 (4): e0153139. Available from: http://www.ncbi.nlm.nih.gov/pubmed/27071074.

158 Yalcinkaya, S. and Malina, J.F. (2015). Anaerobic co-digestion of municipal wastewater sludge and un-dewatered grease trap waste for assessing direct feed of grease trap waste in municipal digesters. *International Biodeterioration & Biodegradation* 104: 490–497. Available from: http://dx.doi.org/10.1016/j.ibiod.2015.08.007.

159 Silvestre, G., Rodríguez-Abalde, A., Fernández, B. et al. (2011). Biomass adaptation over anaerobic co-digestion of sewage sludge and trapped grease waste. *Bioresource Technology* 102 (13): 6830–6836. Available from: http://dx.doi.org/10.1016/j.biortech.2011.04.019.

160 Wan, C., Zhou, Q., Fu, G., and Li, Y. (2011). Semi-continuous anaerobic co-digestion of thickened waste activated sludge and fat, oil and grease. *Waste Management* 31 (8): 1752–1758. Available from: http://dx.doi.org/10.1016/j.wasman.2011.03.025.

161 Noutsopoulos, C., Mamais, D., Antoniou, K. et al. (2013). Anaerobic co-digestion of grease sludge and sewage sludge: the effect of organic loading and grease sludge content. *Bioresource Technology* 131: 452–459. Available from: http://dx.doi.org/10.1016/j.biortech.2012.12.193.

162 Wang, L., Aziz, T.N., and de los Reyes, F.L. (2013). Determining the limits of anaerobic co-digestion of thickened waste activated sludge with grease interceptor waste. *Water Research* 47 (11): 3835–3844. Available from: http://dx.doi.org/10.1016/j.watres.2013.04.003.

163 Kabouris, J.C., Tezel, U., Pavlostathis, S.G. et al. (2009). Methane recovery from the anaerobic codigestion of municipal sludge and FOG. *Bioresource Technology* 100 (15): 3701–3705. Available from: http://dx.doi.org/10.1016/j.biortech.2009.02.024.

164 Fonoll, X., Astals, S., Dosta, J., and Mata-Alvarez, J. (2015). Anaerobic co-digestion of sewage sludge and fruit wastes: evaluation of the transitory states when the co-substrate is changed. *Chemical Engineering Journal* 262: 1268–1274. Available from: http://dx.doi.org/10.1016/j.cej.2014.10.045.

165 Zupančič, G.D., Uranjek-Ževart, N., and Roš, M. (2008). Full-scale anaerobic co-digestion of organic waste and municipal sludge. *Biomass and Bioenergy* 32 (2): 162–167.

166 Dai, X., Duan, N., Dong, B., and Dai, L. (2013). High-solids anaerobic co-digestion of sewage sludge and food waste in comparison with mono digestions: stability and performance. *Waste Management* 33 (2): 308–316. Available from: http://dx.doi.org/10.1016/j.wasman.2012.10.018.

167 Solé-bundó, M., Garfí, M., and Ferrer, I. (2020). Pretreatment and co-digestion of microalgae, sludge and fat oil and grease (FOG) from microalgae-based wastewater treatment plants. *Bioresource Technology* 298: 122563. Available from: https://doi.org/10.1016/j.biortech.2019.122563.

168 Panigrahi, S., Sharma, H.B., and Dubey, B.K. (2020). Anaerobic co-digestion of food waste with pretreated yard waste: a comparative study of methane production, kinetic modeling and energy balance. *Journal of Cleaner Production* 243: 118480. Available from: https://doi.org/10.1016/j.jclepro.2019.118480.

169 Xu, H., Yun, S., Wang, C. et al. (2020). Improving performance and phosphorus content of anaerobic co-digestion of dairy manure with aloe peel waste using vermiculite. *Bioresource Technology* 301: 122753. Available from: https://doi.org/10.1016/j.biortech.2020.122753.

170 Wang, F., Yi, W., Zhang, D. et al. (2020). Anaerobic co-digestion of corn Stover and wastewater from hydrothermal carbonation. *Bioresource Technology* 315: 123788. Available from: https://doi.org/10.1016/j.biortech.2020.123788.

171 Kainthola, J., Kalamdhad, A.S., and Goud, V.V. (2020). Optimization of process parameters for accelerated methane yield from anaerobic co-digestion of rice straw and food waste. *Renewable Energy* 149: 1352–1359. Available from: https://doi.org/10.1016/j.renene.2019.10.124.

172 Wang, Z., Yun, S., Xu, H. et al. (2019). Mesophilic anaerobic co-digestion of acorn slag waste with dairy manure in a batch digester: focusing on mixing ratios and bio-based carbon accelerants. *Bioresource Technology* 286: 121394. Available from: https://doi.org/10.1016/j.biortech.2019.121394.

173 Vivekanand, V., Girma, D., Eijsink, V.G.H., and Horn, S.J. (2018). Synergistic effects of anaerobic co-digestion of whey, manure and fish ensilage. *Bioresource Technology* 249: 35–41. Available from: https://doi.org/10.1016/j.biortech.2017.09.169.

174 Pavi, S., Kramer, L.E., Gomes, L.P. et al. (2017). Biogas production from co-digestion of organic fraction of municipal solid waste and fruit and vegetable waste. *Bioresource Technology* 228: 362–367. Available from: http://dx.doi.org/10.1016/j.biortech.2017.01.003.

175 de la Lama, D. and Borja, R. (2019). Anaerobic co-digestion of olive mill solid waste and microalga *Scenedesmus quadricauda*: effect of different carbon to nitrogen ratios on process performance and kinetics. *Journal of Applied Phycology* 31: 3583–3591.

176 Puntano, N.F., Mancilla-leyt, J.M., and Fern, M.J. (2020). Batch mesophilic anaerobic co-digestion of spent goat batch mesophilic anaerobic co-digestion of spent goat straw bedding and goat cheese whey: comparison with the mono-digestion of the two sole substrates. *Journal of Environmental Management* 280: 111733.

177 Zhang, L., Li, F., Kuroki, A. et al. (2020). Methane yield enhancement of mesophilic and thermophilic anaerobic co-digestion of algal biomass and food waste using algal biochar: semi-continuous operation and microbial community analysis. *Bioresource Technology* 302: 122892. Available at: https://doi.org/10.1016/j.biortech.2020.122892.

9

Types of Composting

Biodegradable Waste Management in the Circular Economy: Challenges and Opportunities, First Edition.
Edited by Malgorzata Kacprzak, Eleonore Attard, Kari-Anne Lyng, Helena Raclavska, BalRam Singh, Eyob Tesfamariam, and Franck Vandenbulcke.

9.1

Composting

Shahriar Shams[1] and Beston F. Nore[2]

[1] *Civil Engineering (CE) Programme Area, Faculty of Engineering, Universiti Teknologi Brunei (UTB), Gadong, Brunei Darussalam*
[2] *Food Science Technology (FST) Programme Area, School of Applied Sciences and Mathematics, Universiti Teknologi Brunei (UTB), Gadong, Brunei Darussalam*

9.1.1 Introduction

In the era of rapid development, urbanization and increased population, solid waste management has been evolving continuously with the aim to provide waste reduction at source, with an emphasis on recycling, composting, and resource recovery. Today, the major challenge is shifting toward zero waste management. Hence, there is a persistent need for diverting waste from going to the landfill and converting waste to resources by composting. The handling and/or process management of municipal solid waste (MSW) is a burning concern globally, both for developed and developing countries. The prevailing MSW management is inadequate to sustain the rapid increase in waste generation and possess a substantial hazard to the environment and inhabitants. In many countries, the agri-food waste generated from human activities comprises 50–60% of total MSW. Therefore, composting can play an important role and feasible option by utilizing biodegradable or organic waste as part of waste management strategies. Generally speaking, the composting act has been recognized as an environmentally sound method of treating waste due to less pollution [1]. Cities and towns that generate thousands of tons of organic waste can reduce waste disposal problems by composting. Organic waste comprising vegetables can be used for composting to produce fertilizers rich in nutrients [2]. Composting, a key process in the waste hierarchy, is one of the most favorable options to recycle wastes into useful fertilizers in a more sustainable manner to reduce the bulk of biodegradable waste [3]. Composting is an excellent way to restore soil and nutrients within the land [4]. Decomposition of organic matter has been used as a means of waste recycling back into the land to improve soil structure, composition, and ecological fertility.

9.1.1.1 History and Concept of Composting

Composting for agriculture (agri-compost) by humans has been practised in rural areas for centuries and even millennium. Farmers typically re-use agricultural and some animal waste on their fields to boost the productivity of the land. Agri-compost was already known to Mesopotamians, Romans, Greeks, and Islamic dynasty as recorded with numerous citations in religious source books on how to cultivate efficiently and boost soil fertility. The Akkadian dynasty practised the use of manure in composting in their agricultural activities for the Mesopotamian farm lands, an exercise that was recorded in clay tablets [5]. In Europe about 170 years ago, because of overexploitation of the soil as a result of farming practices, the soil nutrient content was depleted, which led to a substantial decrease in harvesting productivity and prompted farmers to explore better and more sustainable farming. Eventually, basic composting practises were invented as a long-term solution to the problem of low productivity [6]. An increase in agriculture reproduction terminology has been placed on the three R concepts (3Rs): *Reduce, Reuse, and Recycle.* The overall composting objectives provide the means to accomplish all 3Rs. Through the composting process the waste volume sent to the landfill has reduced substantially and the organic matter is recycled into a useful soil amendment, instead of dumping. The composting process is a way of harnessing the natural process of decomposition of wasted matter and speeding up the decay of waste. Global demand to handle vast-scale waste management processes, and at the same time to increase production of huge amounts of agri-food for an increasingly populated earth, has become a challenge to balance these activities in an environmentally friendly manner. To save our planet with sustainable living and technical/industrial development has led to a revitalized interest in the development of large-scale, commercial, and municipal composting systems and also the invention of various composting/recycling tools for small-scale, urban mini-farmers and backyard composting. The overall aim of any composting act is to enhance the decomposition time of crude organic matter by complex fermentation processes governed by various organisms. Raw organic wasted agri-material polymers are digested and re-digested by many tiny organisms from microorganisms to earthworms, leading to components that are gradually metabolized (degraded) and also catabolized (re-synthesized). Farmers and urban gardeners often use the terms organic matter, compost, and humus as homologous and apparent identities, but there are critical important differences in the aims of composting that need to be discussed.

9.1.2 Composting Process

The selection of composting method or its process depends on, for example, the availability of organic materials, its characteristics, labour, and cost, which are a few of various available options. Composting also allows opportunities to explore new ideas and research related to waste management. A controlled composting process aims to eliminate odor, emissions, and reduction of decomposition time, which are new dimensions to improve the composting process.

The composting assembly consists of an incubator for the biological decomposition process, which converts various organic wastes and/or by-products (leaves, fruit, vegetables, and food wastes) into soil humus-like substances under the right conditions of oxygen, temperature, and moisture, assisted by worms and microorganisms (Figure 9.1.1).

The fundamental difference between an open dump or piled waste and composting is controlled decomposition with reduction of odor and nuisance for composting, while in the former case there is a rotten smell and potential hazard threat to human health by attracting various pathogen carriers, such as insects, rodents, and pathogenic microorganisms.

Composting can be carried out in both aerobic and anaerobic conditions. Aerobic composting is based on the principle that works in above-ground composting structures. It usually takes place in a free-standing pile or in a container providing air ventilation, such as an open-sided bin or a tumbler with aeration holes or cavities. As long as there is a plentiful supply of aeration availability, aerobic

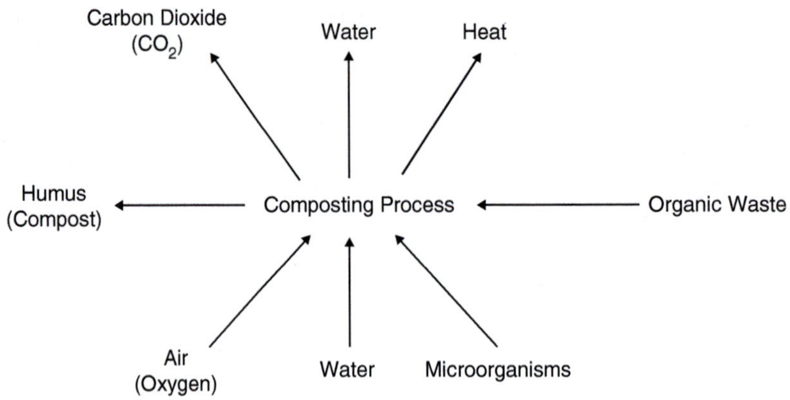

Figure 9.1.1 The major composting processes.

Table 9.1.1 Difference between aerobic and anaerobic processes.

Process	Aerobic	Anaerobic
Smell	Odorless	Strong smelly odors
Duration	Require less time to decompose as the oxygen and heat speed up the process, usually six weeks	Require more time to decompose the organic matter, usually six months
Oxygen	Fuelled by oxygen	Fuelled by bacteria and moisture; no oxygen required
Turning	Weekly turning required	No turning required
Space	Need smaller space	Need larger space
Heat	Produces heat as the bacteria consume energy	Only small amount of heat
Uses	The compost, manure, and compost tea from this method produce useful nutrients for plants	Cattle manure is decomposed using this method to produce methane for combustion purposes

decomposers operate quicker and more effectively compared to anaerobic decomposers, and provide a finished compost process on a faster timetable. Anaerobic organisms work under much lower oxygen tension and therefore most of the anaerobic techniques take place deep underground in pits or trenches. The anaerobic organisms operate at a slower pace but it is difficult to track the progress without digging into the holes for checking. This method often excludes smelly gas as a by-product of their exertions, since anaerobic composting produces methane gas. Table 9.1.1 shows a summary of the major differences between aerobic and anaerobic composting.

9.1.3 Key Factors Affecting Composting

9.1.3.1 Oxygen

Composting is an aerobic process; therefore, the presence of oxygen is important during the composting process. An adequate level of oxygen is necessary for efficient composting to minimize the composting period and to ensure stable end products. An inadequate supply of oxygen will lead to an anaerobic process and as a result the compost pile may start to smell, while an excess flow of oxygen may result in undesirable cooling down of the composting process. The compost pile needs to contain some pore spaces to allow the flow of oxygen. This can be achieved by mixing small particles with larger particles, which will create the required pore spaces to ensure an adequate flow of oxygen. Turning the pile as the broken-down material starts to settle and the process slows down also fluffs up the pile and introduces air to keep the oxidation process moving.

9.1.3.2 Moisture Content

This is one of the most important factors influencing the composting process. Ideally, the moisture content should be between 50 and 60% of the compost weight [7]. Excess moisture (over 60%) generally replaces air (oxygen) from compost piles and creates an aerobic condition, but moisture content below 12% can cause metabolic activity for microorganisms to cease. The exothermic process in the composting process may lead to a decrease in moisture content and therefore it is necessary to add water to maintain the moisture content level. Hence, during the heat of a dry summer, water must be added to the compost pile, while in wet or rainy seasons the compost pile must be covered to avoid excess moisture content.

9.1.3.3 Temperature

Temperature is an important parameter for monitoring the composting process [8]. The compost process typically takes place best between the optimum temperatures of 40 and 65 °C [9]. This is the temperature range within which the microbes involved in the process function best and reproduce to keep the process moving as well as killing pathogens significantly. The reactions in the composting process release heat, resulting in an increase in temperature within the compost piles. Since microbial activity during the composting process is significantly influenced by temperature changes, it is essential to maintain the temperature of the process within the optimal range for the microorganisms.

9.1.3.4 pH

Organic materials with a wide range of pH values can be composted, but the more desirable pH range for composting is between 5.5 and 8.5. pH influences the microbial activities during composting [10, 11]. At the beginning of the composting process, acid-producing bacteria become more active and convert complex materials to organic acid intermediates, hence reducing the pH level. These organic acids are soon utilized by succeeding microorganisms, resulting in an increase in the pH value. Sometimes lime is used to increase the pH level at the beginning of the composting process, particularly if there is a sharp drop of pH level with some composting materials and the process appears to be ineffective. However, proper attention has to be paid when adding lime, because in many situations it may have adverse effects at a later stage of the process.

9.1.3.5 Nutrients

The presence of nitrogen, phosphorus, potassium, and some other trace elements helps the composting process. However, an appropriate carbon-nitrogen ratio (C/N) is essential for a successful nutrient balance for most of the microorganisms in the composting process. Due to the high microbial activity, an initial value of the C/N ratio between 25 and 40 : 1 is recommended [12]. Bulking agents could be used to adjust the C/N ratios [11].

9.1.3.6 Feed Materials Size

Feed materials having smaller particles size usually increase the rate of decomposition due to more surface area being exposed for microorganisms to populate. Consequently, smaller particles will be broken down more quickly to a useful compost. When adding coarse materials to the pile, such as woody branches, corn cobs, or tough plant stalks, it is helpful to chip, grind, or cut up these materials for faster composting. However, particles that are too fine may restrict air circulation, leading to an anaerobic condition. The particle size distribution is important for the final compost in order to maintain its water holding capacity [13].

9.1.4 Types of Composting

9.1.4.1 Onsite Composting

Using onsite composting gives the opportunity to utilize wood waste and keeps most of the organic material in its original ecological system. This method is suitable for yard trimmings and small quantities of food scraps. However, it is not suitable for animal products and large quantities of food scraps. Onsite composting is suitable when leaves and grass cuttings are kept in a pile and left to decompose naturally. As a result, the nutrients from the leaves and grass will go back to the soil and thus is called grass cycling. One of the advantages of using onsite composting is that the climate and seasons do not present a major challenge. Minor adjustments can be made when an undesirable weather condition approaches, such as during rainy seasons. Also, the process is quite straightforward and takes very little time or equipment. Furthermore, the leaf pile can be used as mulch around trees and shrubs to retain moisture.

9.1.4.2 Passive Windrow Composting

Passive windrow composting is the simplest form of composting in which compost piles remain static while air is fed into them through perforated pipes buried in the piles. There is no need to turn the piles upside down apart from occasional turning to check their porosity. Usually the piles are small enough to allow passive air movement, otherwise anaerobic zones will form. A passive windrow system comprising a porous base of wood chips can successfully compost agricultural waste in the winter [14]. The basic construction of a passive windrow system contains a perforated pipe in the bottom, formed as a loop, with a non-perforated pipe connected to the blower. The installation of the loop is covered with a porous base-layer, followed by the composting mass, which is then covered with a biofilter. The perforated pipes having a length of 10 cm are perpendicular to the length of the windrow and are usually 1 m apart in the porous base. Piles are covered with the finished compost of 15–20 cm. The composting cycle usually takes 50–80 days to obtain. Thermophilic temperatures are achieved after the piles heat up. Meanwhile, the oxygen levels are highest at the bottom of the pile and decrease gradually as they move toward the top of the pile. The composting process lasts for 21 days with forced aeration and then may be left to mature in heaps without forced aeration. A common problem with the passive pile is that the particle size needs to

Table 9.1.2 Pros and cons of passive windrow composting.

Pros	Cons
• No turning required on a weekly basis, hence less labor required • Less odor • Easy to achieve high temperatures • Requires less area in comparison to turned windrow composting	• Too much dryness in the compost pile due to air flow • Requires electricity for forced aeration • Requires use of finished compost or mulch for each pile • Requires thorough mixing for decomposition • More water needed to maintain moisture as compared to traditional compost pile

be less than 3–4 cm, and thus an initial step of shredding is often necessary. The pros and cons of passive windrow composting are given in Table 9.1.2.

9.1.4.3 Turned Windrow Composting

Turned windrow composting, also known as aerated composting, consists of rows of long piles called "windrows," which is suitable for large volumes of composts. Windrow composting involves organic waste collected from a community, restaurants, food processing industries, etc. The windrows are aerated either manually or by mechanically turning the piles. The ideal pile height is between 1 and 2 m to produce heat and maintain enough air flow and temperatures and have a width of 4–5 m [15]. The advantage of windrow composting is that diverse wastes such as poultry wastes and yard trimmings can be composted using this method.

9.1.4.3.1 Requirements for Turned Windrow Composting

- Windrow composting is often laborious as it requires periodical turning with various material mixtures.
- Windrows are required to be covered by plastic to prevent loss of moisture, particularly during warm and arid climates.
- Windrows should be shaped in such a way that excess water runs off the top rather than being absorbed by the pile.
- Windrow composting can work in a wider range of cool-weather conditions, even cold climates. Often the outside of the pile might freeze, but in its core a windrow can reach 60 °C.
- The bottom of the windrow should have a leachate collection system to avoid groundwater being contaminated.
- Windrow composting requires laboratory testing for heavy metals and bacteria as a large-scale operation procedure and might be subjected to regulatory enforcement.

9.1.4.4 Vermicomposting Product of the Composting Process Using Various Species of Worms

Vermicomposting consists of using earthworms such as red wigglers or white worms for composting organic residues, which include food wastes, bedding materials, gardening waste, papers, etc. The products resulting from vermicomposting act as excellent soil conditioners due to their composition of water-soluble nutrients rich in organic contents. It is a recycling technique that is detoxifying, odorless, and non-thermophilic and is suitable for the environment [16, 17].

The main objectives of vermicomposting are to process decomposition of biological materials as quickly and efficiently as possible. Earthworms can consume practically all kinds of organic matter equivalent to their own body weight. The secretion from earthworms, known as casts, contains eightfold as many microorganisms as their microbial feed. These microorganisms are sustainable for proper plant growth. Worm casts also contain several times more nitrogen, phosphorus, and potassium levels than normal soil. These minerals are needed for healthy plant growth and in addition the large numbers of beneficial soil microorganisms in worm casts have a synergic positive effect on the overall soil quality. Humic acid in worm casts has an additional benefit in keeping the soil at a perfect pH balance, and the casts also contain plant growth factors similar to those found in seaweed.

A proper moist compost heap dimension of 2.4 m wide, 1.2 m deep, and 0.6 m high can support a worm colony of more than 50 000. The movement of earthworms in the soil boost bacterial growth and spread and also allow better aeration suitable for breaking down organic matter under aerobic conditions. Thus, heap turning is not required, since earthworms do the mixing and aeration conditioning. The ideal environmental settings for worm colonization are a shallow pit with ample humidity, as they cannot tolerate high temperatures. As vermicomposting processes are vulnerable to high temperatures, ideally they should be kept at less than 35 °C. Therefore, it is less likely to kill pathogens, which are usually killed during traditional thermophilic composting where the temperature exceeds 70 °C. Apart from excess temperature, inadequate ventilation or excess moisture leads to sour odors in the bin, and can kill the worms. The advantage of vermicomposting is that it is a short duration process with no requirement for frequent turning of composting materials and has an adequate surface area available for a homogenous resultant product.

It is good to bear in mind that vermicomposting requires larger spaces, because worms are large surface feeders and will not thrive in compost-soil material more than a meter in depth and are more vulnerable to weather fluctuations, such as temperature, freezing conditions, and water-saturation. It requires more resources in terms of time, cost, and labor to start up and maintain worm propagation optimally. Vermicomposting is ideal for a wide-range cool climate, but is not suitable in very cold or hot climate zones. The difference between vermicomposting and traditional composting is explained in Table 9.1.3.

9.1.4.5 Barrel Composting

Barrel composting has gained popularity as a low-cost sustainable waste management option in developing countries. Household organic waste is kept in the barrel for a period of one month for the composting process under aerobic conditions. The barrel is perforated along its cylindrical surface and a perforated pipe is inserted in the barrel from top to bottom. The top of the barrel is covered to protect it from sunlight and rainwater. After aerobic digestion is complete, the end product is taken out for the further maturation stage from the side door of the barrel located at the bottom. Rahman and Al-Muyeed [18] proposed an empirical equation for aerobic digestion for selected parameters that influence the composting process under natural environmental conditions suitable for tropical countries.

The empirical equation is given below:

$$Y = A\,x_1{}^p\,x_2{}^q\,x_3{}^r$$

where Y = volume loss (fraction percentage), function of moisture content, temperature, and retention time. The variable parameters are A, p, q, and r. These can be adjusted according to local

Table 9.1.3 Difference between traditional composting and vermicomposting.

Vermicomposting	Traditional composting
Pathogens present in the food waste, manure, or any relevant feed are not destroyed due to temperatures ranging between 10 and 30 °C as required for worms in the compost piles.	Suitable for killing of pathogens due to heat generated up to 65 °C during aerobic breakdown of organic matter.
Worm compost is activated by mesophilic microbes that require more moderate temperatures.	Compost piles can get hot due to the presence of thermophilic (or "heat-loving") microbes that can thrive in those temperatures.
Much faster process may take as little as 8–12 weeks.	Slow process usually takes six to nine months.
Vermicomposting bins are relatively shallow, turning is not required, and worms provide the necessary oxygen for aeration.	Requires good amount of turning of compost piles as the depth of the compost piles makes it difficult for aeration.
More available surface area as composting can process with a limited depth with worms assisting in composting.	Less space is required for a given volume of compost due to increased depth of piles.
Comparatively expensive due to the controlled environment, use of worms, \$300–1200 per cubic yard.	Less expensive due to simple requirements such as compost bins, layering of browns and greens, \$30 for a single yard.

environmental conditions. *A* varies from 0.19 to 0.20 for hot and humid tropical climates, while

x_1 = percentage of moisture content (%), x_2 = reaction temperature (°C), and

x_3 = retention time (days)

Problem Case on Barrel Composting

Problem Case: A barrel containing 250 L has been chosen to install at a barrel composting plant at the Sungai Paku Landfill site at Tutong district. The population of the nearby located village is 500. The plant will operate throughout the year. The average temperature measured inside the waste is 43 °C. The composition of the solid waste in a 100 kg sample is given in Box 9.1.1.

The targeted loss of volume is 54.19% and the average density of the waste is 450 kg m^{-3}. The waste generation rate is 1.42 kg/capita/day. Design the barrel composting plant by calculating the number of barrels required. For the tropical condition

$$y = 0.198\, x_1^{0.943} x_2^{-0.754} x_3^{0.022}$$

Solution: In the composting process the non-combustible materials must be screened for digestion of the waste. Therefore, in this 100 kg sample, cardboard (8 kg), plastics (5 kg), and wood (5 kg) must be screened out.

Weight of the wet mass = (100 − 8 − 5 − 5) = 82 kg and weight of the dry mass (54.4 − 7.5 − 4.9 − 4) = 38 kg. Therefore, moisture content = (82 − 38)/82 × 100 = 54%. Total percentage of the non-combustible material = 18%

Using the following equation,

$$y = 0.198\, x_1^{0.943} x_2^{-0.754} x_3^{0.022}$$

$$0.5419 = 0.198\,(54)^{0.943}\,(43)^{-0.754}\, x_3^{0.022}$$

i. The retention time is calculated as $x_3 = 40$ days
ii. The total waste generated for the composting plant
$= (1.42\ \text{kg/capita/day})\ (500\ \text{people})\ (40\ \text{days}) = 28\,400\ \text{kg}$
iii. The compostable waste is = 82% of the total waste = $28\,400 \times 0.82 = 23\,288$ kg
iv. The volume of the compostable waste, $V = M/\rho = 23\,288/450 = 51.751\ \text{m}^3$
$= 51.751 \times 1000 = 51\,751$ L

The required number of barrels = $51\,751/250 = 207$ for the composting plant.

Box 9.1.1

Component	Wet mass, kg	Dry mass, kg
Food wastes	45	18
Paper	22	15
Cardboard	8	7.5
Plastics	5	4.9
Garden trimming	15	5
Wood	5	4
Total	**100**	**54.4**

9.1.4.6 Self-Turning Composting

A novel composting scheme that combines gravity mixing and vertical composting together is known as self-turning reactor (STR) technology [19, 20]. Gravity mixing consumes no external energy and is able to mix or turn common compost materials within a few moments in both a continuous and batch manner. The concept is to compost organic waste in a stack of STR units, which allows material to flow vertically downward by gravity, as shown in Figure 9.1.2.

A mixture of organic waste and compost material enters the system from the top. The mixture is then subjected to a number of turning cycles, which results in a compost product from the bottom of the stack. Each STR unit offers two functions: (i) composting the input material in a batch manner and (ii) aerating and restructuring, releasing excess water. These two functions are performed in the reactor and self-turning unit, respectively. The compost mixture is elevated to charge its potential energy and is then put into the topmost STR unit. The material will be retained in the reactor component for some days for composting and will be released to a self-turning unit, within a few seconds, before sending it to the next STR unit for the next cycle. The cycle will be repeated until well degraded organic biomass is obtained as a compost product. This STR unit can be extended vertically to increase the number of turning cycles and can be expanded horizontally to increase the overall treatment capacity in parallel. This design concept ensures that the extendibility of the system scale can be as large as demanded. The STR units are connected vertically, which utilizes land space effectively. The self-turning unit utilizes gravitational energy in mixing and turning materials, which has a major advantage over mechanical turning processes in terms of energy and time consumption. The system is designed using a simple mechanism, which also has the advantage of reducing maintenance and operational costs.

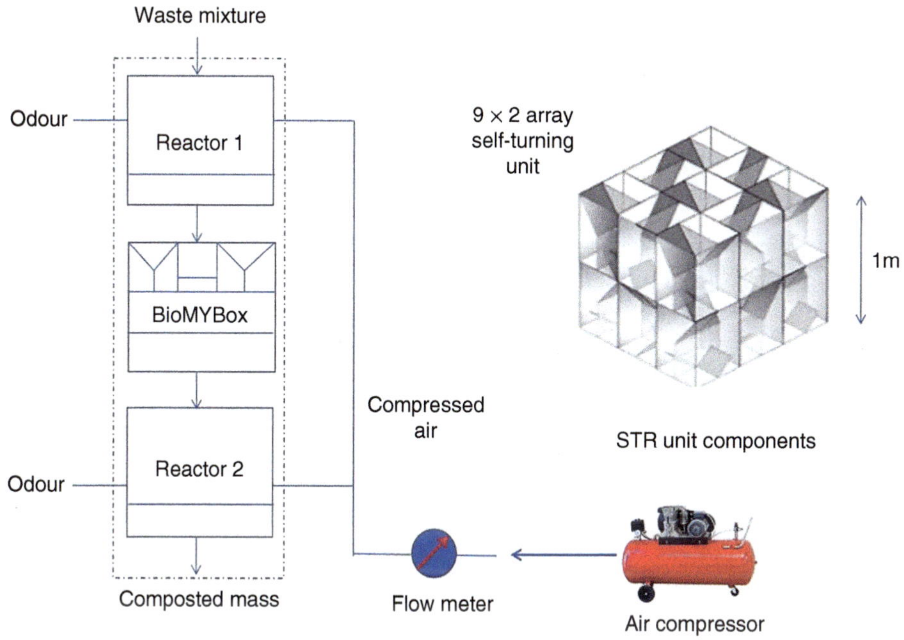

Figure 9.1.2 STR process. Source: Adopted from [20].

9.1.4.7 Sandwich or Sheet Composting

A compost sandwich is composed of multiple layers of organic materials (from green to brown dry leaves) directly on a hard-ground bed that allows a good water-retaining capability of any rainfall. The nutritionally poor land will be transformed into a fertile loamy soil. The sandwich compost ingredients are all found in the traditional compost pile, although sandwich composts are larger in size. The creative assembly of compost materials is a key toward successful achievement, with flattened cardboard, newspaper, brown material (rich in carbon, dry leaves, shredded bark, wood shavings, hay), green materials (rich in nitrogen, grass clippings, vegetable peelings, seedless weeds), water, and topsoil. Sheet composting is a variation of sandwich composting, where the compost ingredients are under a sheet barrier or heavy-gauge plastic with some random holes to enrich moisture conditioning, with manure, paper, carton/box/cardboard packages, and so on. There is switching between browns and greens, layering all of the materials onto the bed (Figure 9.1.3). The compost sandwich is covered with a sheet and the edges are secured with bricks, stakes, or whatever is available. After this, the composting is left alone for at least several months. With this type of composting, there is no need to dig or till it into the pre-existing soil during the decomposition process or after.

9.1.4.7.1 Advantages of Sandwich or Sheet Composting

- Excellent for re-cycling. The conversion of waste ingredients into fertile soil, i.e. any organic food waste, old newspaper, carboard boxes, kitchen waste, etc.
- No digging required. The materials in sheet composting are thrown onto a layered heap directly on the ground, which is a classical form without any digging. The only small hole digging required is the actual act of seeding/plantation.
- No weeding. During the composting process, the composting land does not require weeding, since the top soil covering the composting layers is clear of weeds.

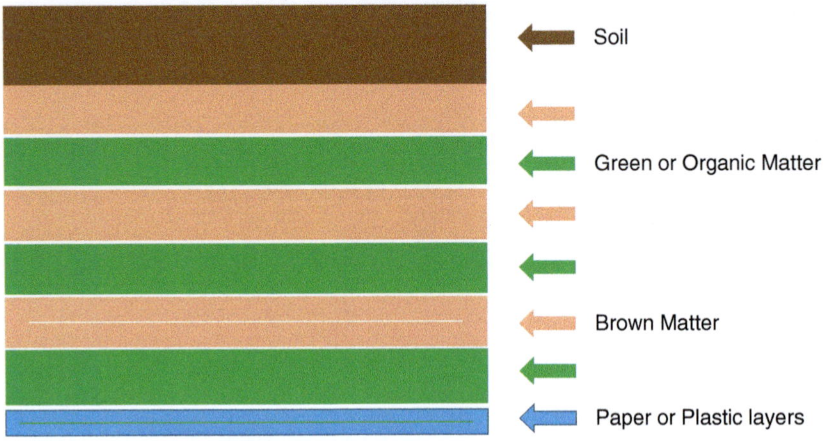

Figure 9.1.3 Schematic illustration of sandwich or sheet composting. Paper or cardboard material is shown in blue, organic nitrogen-rich vegetation (grass cutting, kitchen waste, manure, fruits, etc.) in green, carbon materials (wood shaving, sawdust, small branches, shredded papers) in light brown, and, finally, a 7.5 cm layer of earlier compost or soil in dark brown.

- No external fertilizers. When the composting process is ended, the sheet heap decomposes into a nutrient-rich growing medium. This counts as a pure organic process, free from chemical fertilizers.
- Utilization for non-productive land. Any hard, stony, or poor soil, where growing is difficult, could be converted to a productive agricultural land. The layered composting can be laid out on any piece of ground.

9.1.4.7.2 Building Requirements

- **A suitable location.** A suitable position of land can be used for composting ground with regards to the type of growing choices. For that only two main criteria need to be considered for the location spot – the water supply and natural daylight.
- **Collect the material.** Here, we describe a list of materials that should be present in the compost sandwich pile:
  - Dried-brown matter (source of carbon). Rich sources of carbon are dried leaves, straw, wood chips, sawdust, dried grass, cardboard, non-colored paper, small branches, etc.
  - Green-wet matter (source of nitrogen and moisture). Any scrap (from fruits, vegetables, rice, grains, and bred), lawn clippings, weeds, plant cuttings, and old flowers, but not any animal manure.
- **Exclude the following matter.** It is an absolute requirement to exclude the following waste materials in the compost, for a variety of reasons:
  - Inorganic matter: plastic, polyethene, polystyrene, etc.
  - Other food: fat, fish, bones, and any dairy products.
  - Pet manures: dog, cat, or any other pet droppings.
  - Coal ashes: as it contains a large amount of sulfur.
  - Colored printed papers: as they contain many harmful or toxic chemicals.
  - Diseased plants: due to the risk of spreading disease, avoid diseased plants. Burning is the most suitable disposal possibility.

Table 9.1.4 Difference between various types of composting.

Composting types	Onsite	Vermicomposting	Barrel composting	Self-turning composting	Turned window	Passive window	Sandwich composting
Climate condition (°C)	A	13–25	A	A	A	A	A
Duration (weeks)	104	12–16	3–6	12–24	12–24	12–24	16–32
Cost (scale 1–5)[a)]	1	2	3	1	3	4	1
Setting up resources (scale 1–5)[a)]	1	2	4	1	5	3	1
Applicability to urban areas	Yes	Yes	Yes	Yes	No	Yes	Yes

a) 1 being favorable and 5 being non-favorable; A stands for adaptable.

- **Assembling the materials.** Once you have the materials assembled you should roughly sort them into two categories, browns and greens in a sandwich/sheet manner (Figure 9.1.3). Basically, the brown material allows aeration and ventilation of the heap, in addition to acting as a carbon source. The green material supplies all the needed nutrients, such as nitrogen and potassium. The more green-mass you have, the faster the heap will decompose. The illustration gives an example of the sandwich/sheet composting system.

To summarize, based on various criteria such as climatic conditions, duration, cost, setting up resources, and application to the urban area, self-turning composting and sandwich composting are the most promising types of composting, followed by vermicomposting, as shown in Table 9.1.4.

9.1.5 Case Studies from Developed and Developing Countries

Many developing countries like Bangladesh, India, Pakistan, and Ethiopia are implementing sustainable composting projects, which not only creates employment opportunities for urban poor but also contributes to the reduction in poverty and environmental sustainability, two important goals of Millennium Development Goals (MDGs). Therefore, emphasis is given to mobilizing the urban poor through such composting projects in order to generate livelihoods. However, the challenges exist when it comes to segregation of waste due to confusion on what needs to be segregated. Besides, people's attitude and reluctance to handling waste and negligence toward waste pickers in the society are some of the other challenges in developing countries. The successful composting project like Waste Concern in Bangladesh is a good example of how cooperation among waste collectors and waste producers can assist in managing solid waste while creating employment opportunities for the urban poor. Many NGOs in developing countries are managing waste with particular emphasis on composting. Focus is now given to many coastal cities where the population is increasing.

An effective composting process may be achieved in many coastal cities by utilizing fish waste from the wet/fish markets and drift seaweed rich in potassium to convert organic matter to soil conditioner due to large amounts of minerals essential for plant growth, such as nitrogen, phosphorus, and calcium [21]. Some countries such as the United Kingdom have a municipal initiative on a curbside collection of food and garden wastes for recycling [22]. The recycling methods they use include commercial in-vessel composting. Countries that do not have these initiatives can have the option of recycling their food and/or garden waste through backyard composting. Many developed countries are encouraging home or backyard composting to minimize the amount of biodegradable organic waste that goes from the total household waste to the landfill. The future vision for home composting is to provide a versatile, low-cost approach for waste handling and to facilitate a sustainable life-style toward recycling habit for citizens and urban populated areas. However, it still requires active engagement of a higher proportion of home owners in urban regions to decrease the environmental impact on soil, air, and underground water in the landfill. This could be achieved by establishment of regional associations and awareness teams in public organizations in order to encourage home composting at the municipal level.

9.1.6 Cutting Edge Technologies for Composting

9.1.6.1 Smart Composting System

The Smart Composting System (SCS) is a new concept evolving with advancement of sensors and Information Communication Technology (ICT) that can be applied to make composting efficient and automatic without relying on conventional manual handling or use of labor. It can remove the hard work of turning windrow piles, introducing aeration into the composting piles, and eliminates intensive monitoring. In this method, drums or disposal bins are fitted with wireless sensors that record temperature and moisture content, and provides an automatic alert to the user system (hand-held mobile/computer) and triggers rotation of the bin/drum or container for proper mixing with air and spraying of water to maintain the minimum temperature and moisture levels suitable for composting based on the framework of the Smart Waste System developed by various researchers [23–25].

9.1.6.2 Additives for Enhancing Compost Products

Normally, composting is a slow process and the maturation is typically a time-consuming practice [26], but additives like jaggery and polyethylene glycol can be applied to accelerate the composting operation time with superior quality. Onsite composting and vermicomposting are the most sustainable strategies to biodegrade organic wastes into precursor compounds, which are important as soil conditioners or potting media. However, the major drawback of these processes is the emission of greenhouse gases, the release of odorous molecules, and the existence of potentially toxic compounds. These various types of additives, such as organic, inorganic, or biological additives, can be used to reduce gas emissions and mobility of heavy metal ions.

Biological enhancers refer to microorganisms such as *Alcaligenes*, *Bacillus*, *Clostridium*, *Enterococcus*, and *Lactobacillus*, which act as inoculum for the composting pile [27]. These microorganisms are either kept and maintained during the thermophilic phase or sold commercially. As additives, effective strains of isolated microorganisms with the highest bio-degrative efficacy are commonly sold commercially [28].

A suitable quantity of organic additives such as grass clippings, straw, wooden chips, and chopped wood pallets should be selected carefully, taking into consideration the carbon nitrogen (C/N) ratio of the initial waste-pile mixtures to ensure organic matter degradation and prevent N leaching during the composting process [29]. Recently, biochar, a highly aromatic pyrolysis product [30], has

Table 9.1.5 Types of additives with advantages and disadvantages.

Additives	Advantages	Disadvantages
Biological	Degradation of complex compounds Increase in microbial activity Reduction in odor Reduction in GHG emission	Increase in composting time
Organic	Increase in aeration Reduction in odor Reduction in GHG emission Increase in microbial activity	Increase in composting time
Mineral	Buffer of pH Metal adsorption Nutrient retention	Presence of heavy metals

Source: Adopted from Barthod et al. [35].

gained enormous attention as it has a high capability as an organic additive to stabilize the process of composting [31].

On the other hand, mineral neutralizing additives such as fly ash, Ca-bentonite, phosphate rock, lime, or zeolite [32, 33] are favorable to reduce heavy metal mobility. Furthermore, organic and mineral additive mixtures in the form of lime and biochar have been shown to be effective in reducing the heavy metal content in composts [34].

The potential advantages and disadvantages of biological, organic, and mineral additives are shown in Table 9.1.5. A careful trade-off principle must be found between these additives in the process of selecting items. A combination of worms and additives seems to increase plant available nutrient contents, and in the meantime allow a decrease in N leaching, heavy metal mobility, and composting time.

9.1.6.3 Advanced Composting Technologies for Small-Scale Communities

The small-scale composting systems for MSW are pleasant to handle and reduce the collection labor time, lower transportation costs, and eliminate soring and environmental hazard contaminations. It is particularly suitable for those communities that have no access to a central composting facility. The microbial and physico-chemical environment in composting can be affected by the diversity of the microbial population, temperature, bulking agent, aeration, and chemical properties of raw materials such as the C/N ratio and moisture content.

Therefore, the interaction mix among biological, chemical, and physical factors are critical to the best quality yield of the composting process. Thus, a viable system of control and process optimization is required for reproducibility and maintaining good composting quality. One example of modern composting technologies has been designed with six compartment composting reactors ($50 \times 20 \times 25$ cm), as shown in Figure 9.1.4.

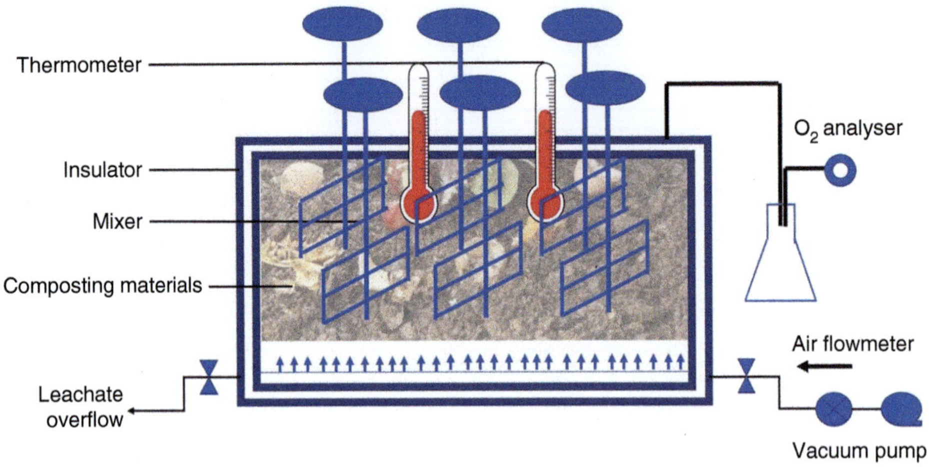

Figure 9.1.4 A six-compartment modern composting technology design. Source: Adopted from [36].

Six mixers were dedicated for each reactor to obtain the best result. An inlet was connected to supply air to the reactor and the flow was controlled by a vacuum pump. A proper plate sparger was installed at the bottom of the reactor for optimal air distribution in the system. The exhaust gas was monitored by a gas monitoring system and then discharged into a flask containing a 1 M H_2SO_4 solution to absorb volatile odor compounds, including NH_3. To maintain a stable temperature in the reactor, heat insulating layers were applied on the reactor walls, covering the reactor entirely. Each composting batch run was at least 30 days. The breakdown process of organic matter was achieved by microbial activities and was also accompanied by heat release (due to exothermic reactions of microbial metabolism pathways) to the composting material. However, the temperature declined as the composting progressed, since the amount of degradable matter decreased. As soon as the content of the organic matter diminishes following the degradation process, the temperature in the reactor will drop, reaching normal ambient temperatures.

9.1.7 Conclusion

Composting is a multidisciplinary concept aimed for producing a sustainable waste managing system, mainly to minimize exposure of waste-dirt risks and also to produce cheap organic fertilizers for backyard, rural farmers, and urban agri-businesses. Composting is a natural act to improve the environment and promote recycling of waste generated from diverse human activities. Determination of the composting need in any populated area requires the best economical durability with a minimum input of human labor. Designing the best ratio of a C/N compost requires mixing various types of waste originating from gardening, farming, organic waste of food, or waste from agri-food industries. On the other hand, water management in harvesting rain water and recycling wastewater is an additional concern to incorporate in the overall composting processes. The MSW governors should provide loans for small communities to set up composting facilities and encourage organic farming, creating a favorable trade-off market to promote the use of composts instead of artificial/chemical soil fertilizers. This will lead to the creation of job opportunities leading to a sustainable composting technology, an alternative to landfill disposal for organic waste.

Acknowledgments

Special gratitude goes to Professor Dayang Hjh Zhorah binti Haji Sulaiman (VC) and Professor Mohamed Hasnain Isa (AVC for Research), in creating a research-friendly environment at UTB. The authors are also grateful to UTB for maintaining a UTB-internal fund for basic research.

Conflict of Interest

The authors declare no competing financial interest in this work.

References

1 Tweib, S.A., Rahman, R.A., and Kalil, M.S. (2011). A Literature Review on the Composting. In: *2011 International Conference on Environment and Industrial Innovation IPCBEE*, vol. 12, 124–127. Singapore: IACSIT Press.

2 Varma, V.S. and Kalamdhad, A.S. (2014). Bio-conversion of organic waste using composting technologies – a review. *Int. J. Environ. Technol. Manage.* 17 (6): 483–507.

3 Atalia, K.R., Buha, D.M., Bhavsar, K.A., and Shah, N.K. (2015). A review on composting of municipal solid waste. *IOSR J. Environ. Sci., Toxicol. Food Technol. (IOSR-JESTFT)*, 9 (5): 20–29.

4 Argun, Y.A., Tunc, A.K., Calisir, U., and Kilinc, N. (2017). Composting as waste management method. *J. Int. Environ. Appl. Sci.* 12: 244–255.

5 Smith, M., Friend, D., and Johnson, H. (2007). Composting for the Homeowner. *Composting for the Homeowner*. University of Illinois, Available at https://web.extension.illinois.edu/homecompost/history.cfm.

6 Blum, B. (1992). Composting and the roots of sustainable agriculture. *Agric. Hist.* 66 (2): 171–188. *JSTOR*.

7 Petric, I., Helic, A., and Avdic, E.A. (2012). Evolution of process parameters and determination of kinetics for co-composting of organic fraction of municipal solid waste with poultry manure. *Bioresour. Technol.* 117: 107–116.

8 Chen, Z., Zhang, S., Wen, Q., and Zheng, J. (2015). Effect of aeration rate on composting of penicillin mycelial dreg. *J. Environ. Sci.* 37: 172–178.

9 Rich, N. and Bharti, A. (2015). Assessment of different types of in-vessel composters and its effect on stabilization of MSW compost. *Int. Res. J. Eng. Technol.* 2 (3): 1–6.

10 Chan, M., Selvam, A., and Wong, J.W.C. (2016). Reducing nitrogen loss and salinity during "struvite" food waste composting by zeolite amendment. *Bioresour. Technol.* 200: 838–844.

11 Zhang, L. and Sun, X. (2016). Influence of bulking agents on physical, chemical, and microbiological properties during the two-stage composting of green waste. *Waste Manag.* 48: 115–126.

12 Petric, I., Avdihodzic, E., and Ibric, N. (2015). Numerical simulation of composting process for mixture of organic fraction of municipal solid waste and poultry manure. *Ecol. Eng.* 75: 242–249.

13 Zhang, L. and Sun, X. (2014). Changes in physical, chemical, and microbiological properties during the two-stage co-composting of green waste with spent mushroom compost and biochar. *Bioresour. Technol.* 171: 274–284.

14 Lynch, N.J. and Cherry, R.S. (1996). Winter composting using the passively aerated windrow system. *Compost Sci. Util.* 4 (3): 44–52.

15 Xu, D., Zhao, S., Xiong, Y. et al. (2015). Biological, physicochemical, and spectral properties of aerated compost extracts: influence of aeration quantity. *Commun. Soil Sci. Plant Anal.* 46 (18): 2295–2310.

16 Goswami, L., Pratihar, S., Dasgupta, S. et al. (2016). Exploring metal detoxification and accumulation potential during vermicomposting of tea factory coal ash: sequential extraction and fluorescence probe analysis. *Sci. Rep.* 6: 30402. Available at: http://dx.doi.org/10.1038/srep30402.

17 Pathma, J. and Sakthivel, N. (2012). Microbial diversity of vermicompost bacteria that exhibit useful agricultural traits and waste management potential. *SpringerPlus* 1: 1–19.

18 Rahman, M.H. and Al-Muyeed, A. (2010). *Solid and Hazardous Waste Management, ITN, BUET, Centre for Water Supply and Waste Management*. Dhaka, Bangladesh: BUET.

19 Bongochgetsakul, N. (2005). Integrated analytical system for optimizing design of large-scale composting and development of serial self-turning reactor system, published PhD thesis, Tokyo: University of Tokyo.

20 Sungsomboon, P.-y., Chaisomphob, T., Bongochgetsakul, N., and Ishida, T. (2013). Pilot-scale tests of an innovative "serial self-turning reactor" composting technology in Thailand. *Waste Manag. Res.* 31 (2): 212–222.

21 Illera-Vivesa, M., Labandeirab, S.S., Britoc, L.M. et al. (2015). Evaluation of compost from seaweed and fish waste as a fertilizer forhorticultural use. *Sci. Hortic.* 186: 101–107.

22 DEFRA (2018). Statistics on waste managed by local authorities in England in 2018/19. *National Statistics*.

23 Mahajan, S.A., Kokane, A., Shewale, A. et al. (2017). Smart waste management system using IoT. *Int. J. Adv. Eng. Res. Sci.* 4 (4).

24 Yerraboina, S., Kumar, N.M., Parimala, K.S., and Jyothi, N.A. (2018). Monitoring the smart garbage bin filling status: an IoT application towards waste management. *Technology* 9 (6): 373–381.

25 Yusof, N.M., Jidin, A.Z., and Rahim, M.I. (2017). Smart garbage monitoring system for waste management. In: *MATEC Web of Conferences, EDP Sciences*, vol. 97, 01098.

26 Lim, S., Lee, L.H., and Wu, T.Y. (2016). Sustainability of using composting and vermicomposting technologies for organic solid waste biotransformation, recent overview, greenhouse gases emissions and economic analysis. *J. Clean. Prod.* 111: 262–278.

27 Jurado, M.M., Suarez-Estrella, F., Lopez, M.J. et al. (2015). Enhanced turnover of organic matter fractions by microbial stimulation during lignocellulosic waste composting. *Bioresour. Technol.* 186: 15–24. Available at: https://doi.org/10.1016/j.biortech.2015.03.059.

28 Manu, M.K., Kumar, R., and Garg, A. (2017). Performance assessment of improved composting system for food waste with varying aeration and use of microbial inoculum. *Bioresour. Technol.* 234: 167–177. Available at: https://doi.org/10.1016/j.biortech.2017.03.023.

29 Doublet, J., Francou, C., Poitrenaud, M., and Houot, S. (2011). Influence of bulking agents on organic matter evolution during sewage sludge composting; consequences on compost organic matter stability and N availability. *Bioresour. Technol.* 102: 1298–1307. Available at: https://doi.org/10.1016/j.biortech.2010.08.065.

30 Lehmann, J. and Joseph, S. (eds.) (2015). *Biochar for Environmental Management: Science, Technology and Implementation*. Taylor and Francis.

31 Waqas, M., Nizami, A.S., Aburiazaiza, A.S. et al. (2018). Optimization of food waste compost with the use of biochar. *J. Environ. Manage.* 216: 70–81.

32 Luo, W.H., Yuan, J., Luo, Y.M. et al. (2014). Effects of mixing and covering with mature compost on gaseous emissions during composting. *Chemosphere* 117: 14–19. Available at: https://doi.org/10.1016/j.chemosphere.2014.05.043.

33 Wang, Q., Li, R., Cai, H. et al. (2016). Improving pig manure composting efficiency employing Ca-bentonite. *Ecol. Eng.* 87: 157–161. Available at: https://doi.org/10.1016/j.ecoleng.2015.11.032.

34 Awasthi, M.K., Wang, Q., Huang, H. et al. (2016). Effect of biochar amendment on greenhouse gas emission and bio-availability of heavy metals during sewage sludge co-composting. *J. Clean. Prod.* 135: 829–835. Available at: https://doi.org/10.1016/j.jclepro.2016.07.008.

35 Barthod, J., Rumpel, C., and Dignac, M.-F. (2018). Composting with additives to improve organic amendments. A review. *Agron. Sustain. Dev.*, Springer Verlag/EDP Sciences/INRA 38 (2): 17. Available at: https://doi.org/10.1007/s13593-018-0491-9.hal-01731152.

36 Zhang, B., Lye, L., Kazemi, K., Lin, W. (2013). Development of Advanced Technologies for Municipal Organic Waste Treatment in Small Communities in Mewfoundland and Labrador, The Harris Centre Memorial University.

9.2

Vermicomposting

Agnieszka Rorat and Franck Vandenbulcke

Laboratoire de Génie Civil et géo-Environnement, Univ. Lille, Institut Mines-Télécom, Univ. Artois, Junia, Lille, France

9.2.1 Introduction

Vermicomposting is an aerobic bioprocess in which earthworms are employed as natural bioreactors in order to transform organic residues into a valuable product named vermicompost [1]. Those small invertebrates cooperate with microorganisms, which perform the main part of the decomposition of organic matter. Consequently, vermicomposting may be considered as an interesting solution for the treatment of organic wastes (e.g. municipal solid wastes, sewage sludge, etc.) [2] according to the principles of the circular economy. Moreover, the vermicompost, which is considered as one of the best amendments due to its composition, can be widely used in sustainable agriculture and replace or reduce the use of fertilizers [3, 4]. Even though the role of earthworms is well recognized, and that earthworms belonging to the *Eisenia* genus are often employed in composting of kitchen and garden wastes, their potential for treatment of organic wastes on an industrial scale is relatively new, as well as their ability to improve the compost quality, for example by degrading organic contaminants. The scientific interest in the topic has increased lately and special attention has been made to the possible application of vermicompost in agriculture and phytoremediation, which can be estimated through the important number of articles published in the last decade (Figure 9.2.1).

9.2.2 Earthworms and Their Role in the Process

The earthworms, known as "soil engineers," are detritivorous dwelling organisms that play a major role in soil ecosystems. They are oligochetae annelids with metameric segmentation and are hermaphrodites, reproducing by cross- or self-fertilization. Earthworms employed in the process of vermicomposting are mostly epigeic species, considered to be litter dwellers and transformers living near the soil surface. The epigeic species used in vermicomposting are characterized by their ability to support changes in physico-chemical conditions and often show a strong resistance to pollutants. A few species are being applied in the process of vermicomposting on an industrial scale, showing a good capacity to survive in the hostile (changing) conditions and the ability to convert the organic wastes rapidly and successfully (Figure 9.2.2). *Eisenia fetida* is the most commonly used species, also called the "Californian Red Worm," characterized by high reproduction

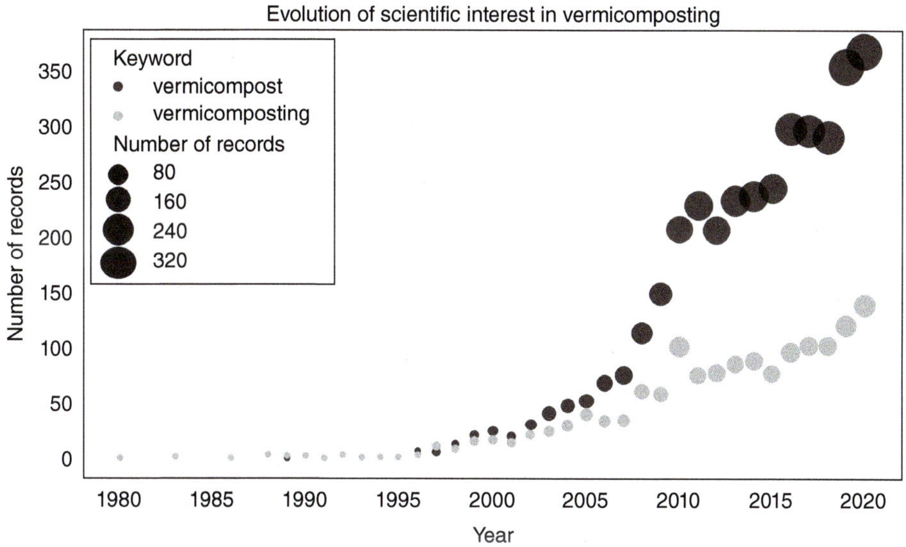

Figure 9.2.1 Evolution of scientific interest in vermicomposting and vermicompost according to the data collected from the *Scopus* database.

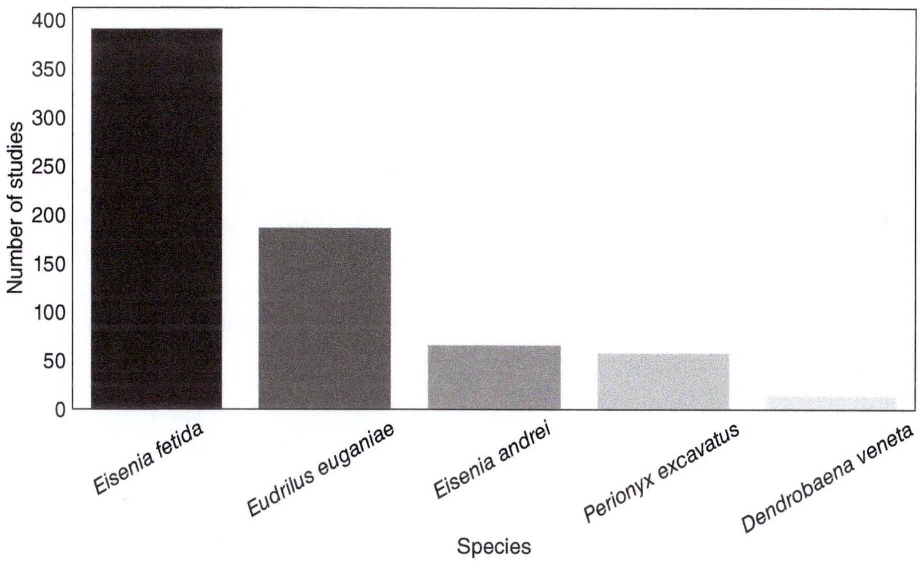

Figure 9.2.2 Comparison of the popularity of selected earthworm species in the vermicomposting process across the articles concerning vermicomposting (according to results obtained using the Scopus database).

rates, a short life cycle, a low body weight, a short maturation time, and well-developed defense mechanisms, which allow them to cope with many different types of organic waste [1]. *Eisenia fetida* and *E. andrei* earthworms are closely related and were originally described as different morphotypes of *E. fetida* according to the differences in body pigmentation [5]. Consequently, some early studies refer to those two species as *E. fetida* or *E. andrei*, not taking into account the distinction between them. Actually, based on barcoding studies (sequencing of the Cytochrome Oxidase 1 mitochondrial gene), we know that *E. fetida* and *E. andrei* are two distinct species [6].

However, recently it was found that those two species are capable of asymmetrical hybridization and fertile hybrids were found and described [7]. Interestingly, the hybrids are very efficient cadmium bio-accumulators with very well-developed mechanisms of Cd detoxification [8].

Earthworms can be used in monocultures, although sometimes multiple earthworm species are applied. For instance, Hussain et al. [9] used the consortium of *E. fetida, Eudrilus eugeniae*, and *Perionyx excavates*, which produced a compost of better quality compared to the product obtained by vermicomposting performed by monocultures of those species.

According to Dominguez et al. [10], two main phases of contribution of earthworms in the decomposition of organic matter have been identified and confirmed by multiple scientific studies:

- The active phase, in which earthworms ingest and process the organic matter,
- The maturation-like phase, in which microorganisms perform the process of decomposition of already pre-processed material [11].

Thus, the success of vermicomposting depends on the mutual relationship between the earthworms and microbes [12]. The influence of worms on the substratum is twofold. Directly, earthworms influence it by degrading, homogenizing, and shredding and, indirectly, mostly by increasing the surface area for microorganisms but also impacting the microbial community composition in various ways [2]. For instance, it has been shown that *E. fetida* reduced the microbial biomass and activity but increased the microbial diversity in the experiment on composting of vegetable wastes (cabbage, lettuce, and potato peels) [13]. Authors claim that this could be related to the composition of the microbial community present in the earthworm's digestive tract, which could consequently lead to the development of the analogous microbial community in the vermicompost. This likely hypothesis could be tested using genomic tools to describe the microbial communities hosted in the digestive tract of worms.

Lately, the role of an earthworm's mucus has also been highlighted. Mucus, which can be defined as a nutrient-rich extracellular polymeric substance, is excreted by those invertebrates mainly to facilitate their movements [14]. Moreover, as it contains a mixture of carbohydrates and protein-like substances, it can also have an impact on soil structure and soil functions, e.g. it may serve as a nutrient source for plants and has a priming effect on stimulation of microbial activity [15, 16]. Similarly, it was shown that the mucus of epigeic earthworms could participate and accelerate the mineralization and humification during vermicomposting [15].

9.2.3 Parameters That Affect the Vermicomposting Process

Although composting earthworms are generally resistant and adapt rapidly to changing environmental conditions, some parameters should be controlled and optimized in order to ensure the most rapid and effective transformation of organic wastes (Figure 9.2.3). The vermicomposting system needs to be well aerated in order to provide enough oxygen during the process and the moisture should be kept relatively high. Moisture content is often in the range between 60 and 80% [17], although some sources indicate a range between 40 and 55% [18]. An acceptable pH for earthworms and microorganisms ranges from 5 to 9 with an optimum around 7. Note that absorption of heavy metals by earthworms (bioavailability) can be influenced by pH [19, 20]. The optimal temperature for earthworm's metabolic activity and reproduction ranges from 12 to 28 °C, while the preferred C/N ratio is 30 : 1 [21]. If those parameters cannot be found in the organic wastes, bulking agents can be added and mixed in order to provide bulk density, moisture control, pH, and the C/N ratio [22]. Moreover, those substances can reduce the harmful effects of toxic compounds,

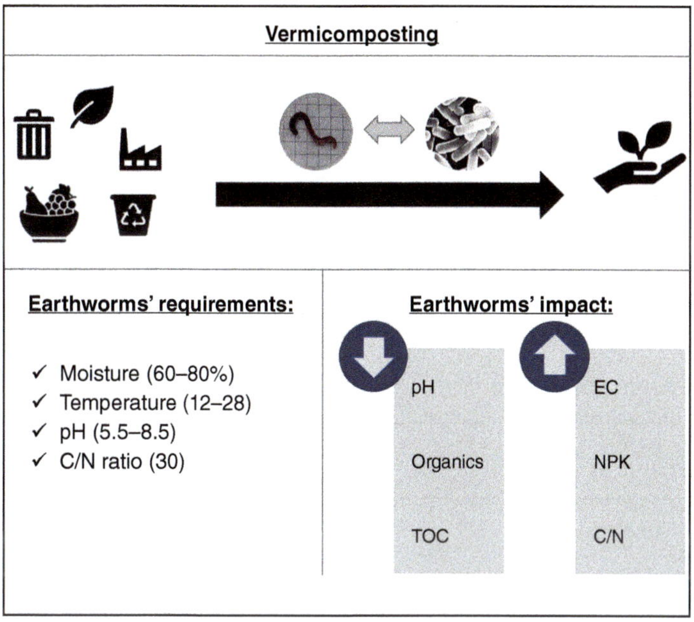

Figure 9.2.3 Earthworm requirements and actions during the vermicomposting process (TOC, total organic carbon; NPK, nitrogen, phosphorus, potassium).

for example when sewage sludge is used as a main substrate [2]. The most used bulking agents are cow dung, poultry waste, sawdust, sugarcane trash, woodchips, tree leaves, etc. They can also be organic wastes. If the added substances have a direct or indirect effect on composting parameters and improve/enhance the process, they can be named additives [23]. For example, it has been shown that a biochar amendment significantly increased the reproduction rate of *E. fetida* earthworms during integrated composting/vermicomposting of sewage sludge [24]. In this case, this carbon-rich by-product coming from the thermal degradation of organic wastes under pyrolysis created favorable conditions for degradation of biowastes. Among other products, it allows an increase in microbial growth, reproduction of earthworms, aeration, porosity, nutrient retention, etc. [25]. However, we cannot say that all biochar creates favorable conditions since biochars can be very different in origin and composition. Another parameter that affects the vermicomposting process is the stocking density of earthworms used in the process, which should be assessed separately for different species. It is crucial to choose an optimal worm density that allows a product of quality to be obtained in a short time, but with a minimum influence on their growth and reproduction [2].

9.2.4 Impact of Vermicomposting on Physicochemical Parameters of the Substrate and Stabilization of the Final Product

The cooperation of earthworms and microorganisms leads to important changes in the physicochemical parameters of the substrate, from color and odor to the content of heavy metals and some biological parameters. Generally, the pH of the composted mixture decreases due to nitrogen and phosphorus mineralization as well as the production of CO_2 related to the metabolic activity of earthworms and microorganisms [2, 26, 27]. A few authors reported the increase of electrical conductivity (EC, reflects the salinity of an organic material) in relation to an increase in mineral salt and/or the release of soluble salts due to degradation of organics [27, 28]. The metabolic activity of earthworms and microorganisms leads to the decrease in total organic carbon (TOC). Indeed, as shown before, the mutual biological action of earthworms and microorganisms present in the growing medium causes C loss in the form of CO_2 from the substrate during degradation and mineralization of organic matter. Also, the increase in earthworms' biomass, due to conversion of some part of the organic fraction of the substrate, can cause a decrease in C from the substrate [2, 28]. Multiple studies have shown that the concentrations of various micro- and macronutrients (NPK) increase after vermicomposting, as well as the total concentration of humic acids and plant growth hormones (e.g. auxins, gibberellins, cytokines) [17]. It is also noteworthy that earthworms may have an impact on the final concentration and/or bioavailability of certain pollutants, for example heavy metals [2] and nanoparticles [29]. Nevertheless, the evaluation of toxicity of the obtained product can be performed if the process is applied to a substrate that can potentially contain important quantities of dangerous compounds, e.g. sewage sludge [2].

9.2.5 Characteristics of Vermicompost

In general, vermicomposts are considered to be products of quality that can be applied into soil as organic amendments and replace the chemical fertilizer. Different properties of the vermicompost make it an interesting soil amendment. Firstly, several parameters, namely porosity, water-holding capacity, and available nutrients, are higher than in other organic fertilizers, while the C/N ratio is lower [30]. Moreover, the participation of earthworms in the process of composting assures a better structure and texture and a lower content of heavy metals with more developed and diverse microbial communities. In summary, the product is rich in nutrients, odorless, light and like-peat, exhibits high porosity, aeration, drainage, high water retention capacity, modified pH, low electrical conductivity, high cation exchange capacity, and high microbial activity [31]. Of course, the properties of the final product vary depending on the used raw materials [30], but, on the whole, vermicomposts are a source of macro- and micronutrients, vitamins, enzymes, antibiotics, growth hormones, and immobilized microflora [30]. Table 9.2.1 presents examples of the nutrient composition of vermicompost obtained from various substrates.

Therefore, differences between the chemical composition of vermicompost and raw materials (substrates) may be important [35]. Lately, a few studies have proposed mathematical models in order to predict the nutrient recovery from solid waste under different vermicomposting treatments. For instance, Hosseinzadeh et al. applied successfully multiple linear regression (MLR) and artificial neural network (ANN) models to simulate and predict total nitrogen (TN) and total phosphorus (TP) recovery in the vermicomposting process [36]. ANN and response surface methodology (RSM) models were also applied in order to study the changes in physico-chemical and maturity parameters on the proportions of cattle dung and floral wastes for better optimization of ratios of those two substrates [37]. Authors have shown that the ANN model was more adapted than RSM to provide appropriate proportions of floral and cattle dung waste for the optimized vermicomposting process along with a high rate of earthworm biomass generation. In another study, Faverial et al. used the Bayesian network modeling method in order to assess the impact of the raw materials and the procedure of composting on the product quality. While the concentration and the quality of raw materials were recognized as the principal factors affecting the organic matter and nutrient contents as well as the stability of the compost, vermicomposting did not improve those parameters compared to classic composting [38].

Table 9.2.1 The composition of several vermicomposts obtained from various substrates.

	Kitchen wastes	Dyeing sludge/Cattle dung	Household solid wastes	Biological sludge	Chemical sludge
pH	–	–	8.2 ± 0.4	8 ± 0.5	7.7 ± 1.2
C:N ratio		15.46 ± 0.57	16.5 ± 3	14.5 ± 2	15 ± 6
TC (%)	–	–	–	–	–
OC (%)	10.30	37.12 ± 0.11	–	–	–
N (%)	0.85	2.40 ± 1.20	2.2 ± 0.4	2.6 ± 0.5	2.2 ± 0.3
P (%)	0.15	1.49 ± 0.81	0.72 ± 0.3	0.5 ± 0.3	0.6 ± 0.2
K (%)	–	1.90 ± 0.08	–	–	–
Na (%)		1.41 ± 0.38	–	–	–
Ca (%)	1.96	–	–	–	–
Mg (%)	0.80	–	–	–	–
References	[32]	[33]	[34]	[34]	[34]

In the future, it may be very interesting to apply the methods of machine learning in order to model and predict the quality of the obtained vermicompost, taking into account not only the parameters of the raw materials but also those related to the earthworms, e.g. density, species, weight, etc. Even though nowadays a lot of data exist, some dedicated experiments may be of great interest in order to feed the databases and perform the most effective calculations.

9.2.6 Environmental Application of Vermicompost

Vermicompost is a compost-based organic amendment that can be a supplement to the soil in order to improve and enhance plant growth. Many different advantages of using this type of amendment have been described, including increasing flower and fruit set and yield, controlling diseases and pests, increasing beneficial microbial biomass, improving the formation of roots, and increasing the tolerance to different stress factors [39].

The recent meta-analysis by Blouin et al. [40] confirmed the basis of 68 scientific articles published between 1997 and 2016 that the effects of vermicomposts on plant growth are remarkable (Table 9.2.2). Considering the observations on the increase in plant root (57%), shoot (78%), and total biomass values (13%), the authors confirmed that this organic amendment can be useful in organic farming. The authors indicated a few factors that may affect those effects. The optimal proportion of vermicompost in growing media was estimated at 20–60% of the growing medium volume. While different substrates were taken into account in the quantitative analysis, the most effective original material used for vermicomposting was cattle manure with an increase of 27% in yield, 113% in shoot biomass, and up to 88% in root biomass, followed by pig manure, food waste, and paper waste.

The methods used in order to enhance post-transplant growth of urban trees and prevent their mortality are often considered as unsustainable as the massive use of inorganic fertilizers is a common practice. Thus, eco-friendly substitutes are of great interest and vermicompost application has been proposed lately as an amendment for three tree species (fast-growing *Betula platyphylla* and *Larix kaempferi* and slow-growing *Chamaecyparis obtusa*) [41]. The authors have shown that the

Table 9.2.2 The influence of vermicomposting on shoot, root, and total biomass and plant yield (according to a quantitative review [40]).

Parameter	Overall increase (%)
Shoot biomass	78
Root biomass	57
Total biomass	13
Plant yield	26

vermicompost treatment is beneficial and could be considered as a solution for growing urban tree stocks for sustainable afforestation programs.

The composts and vermicomposts were also applied as components in soil-less horticultural substrates [42]. Moreover, many studies have shown that those organic medias can be applied as biosorbents for the removal of pesticides, PAHs, herbicides, and some inorganic pollutants from aqueous media and from soils [43], which opens the discussion of their potential use in remediation of degraded and/or polluted soils [35, 44]. According to Sharma and Garg [17] the vermicompost participate in the bioremediation due to strong incorporation of humic substances with toxic compounds, which accelerates their degradation as well as the presence of the rich microbial communities. Moreover, as mentioned earlier, earthworm mucus can promote soil decontamination and defense mechanisms may lead to adsorption of some substances including inorganic contaminants, e.g. Cd via binding to metallothioneins [6]. However, in the latter case, decontamination would be effective after removal of earthworms from the compost before they die, which can release toxic compounds back into the composter.

9.2.7 Conclusions

Vermicomposting is an eco-friendly technology that can be applied in waste management strategies. It is considered to be a low-cost and low-emission process, in which many different types of organic waste can be transformed into the valuable product named vermicompost.

Acknowledgements

This work was supported by an internal research grant from the Czestochowa University of Technology. The research leading to these results has received funding from the EnviSafeBioC project – contract No. PPI/APM/2018/1/00029/U/001. The project was financed by the Polish National Agency for Academic Exchange (NAWA).

References

1 Rorat, A. and Vandenbulcke, F. Earthworms converting domestic and food industry wastes into biofertilizer. In: *Industrial and Municipal Sludge* (eds. M.N.V. Prasad, P.J. de Campos Favas, M. Vithanage and S.V. Mohan), 83–106. Butterworth-Heinemann.

2 Suleiman, H., Rorat, A., Grobelak, A. et al. (2017). Determination of the performance of vermicomposting process applied to sewage sludge by monitoring of the compost quality and immune responses in three earthworm species: *Eisenia fetida*, *Eisenia andrei* and *Dendrobaena veneta*. *Bioresource Technology* 241: 103–112.
3 Włóka, D., Placek, A., Smol, M. et al. (2019). The efficiency and economic aspects of phytoremediation technology using *Phalaris arundinacea* L. and *Brassica napus* L. combined with compost and nano SiO_2 fertilization for the removal of PAH's from soil. *Journal of Environmental Management* 234: 311–319.
4 Włóka, D., Placek, A., Rorat, A. et al. (2017). The evaluation of polycyclic aromatic hydrocarbons (PAHs) biodegradation kinetics in soil amended with organic fertilizers and bulking agents. *Ecotoxicology and Environmental Safety* 145: 161–168.
5 Domínguez, J. (2018). Earthworms and vermicomposting. In: *Earthworms – The Ecological Engineers of Soil*. Sajal Ray, IntechOpen. Available from: http://dx.doi.org/10.5772/intechopen.76088.
6 Rorat, A., Vandenbulcke, F., Gałuszka, A. et al. (2017). Protective role of metallothionein during regeneration in *Eisenia andrei* exposed to cadmium. *Comparative Biochemistry and Physiology Part C: Toxicology and Pharmacology* 203: 39–50.
7 Plytycz, B., Bigaj, J., Osikowski, A. et al. (2018). The existence of fertile hybrids of closely related model earthworm species, *Eisenia andrei* and *E. fetida*. *PLoS One* 13 (1): e0191711. Available from: https://dx.plos.org/10.1371/journal.pone.0191711.
8 Jaskulak, M., Rorat, A., Kurianska-Piatek, L. et al. (2021). Species-specific Cd-detoxification mechanisms in lumbricid earthworms Eisenia andrei, *Eisenia fetida* and their hybrids. *Ecotoxicology and Environmental Safety* 208: 111425.
9 Hussain, N., Das, S., Goswami, L. et al. (2018). Intensification of vermitechnology for kitchen vegetable waste and paddy straw employing earthworm consortium: assessment of maturity time, microbial community structure, and economic benefit. *Journal of Cleaner Production* 182: 414–426.
10 Domínguez, J., Aira, M., Kolbe, A.R. et al. (2019). Changes in the composition and function of bacterial communities during vermicomposting may explain beneficial properties of vermicompost. *Scientific Reports* 9 (1): 1–11. Available from: https://www.nature.com/articles/s41598-019-46018-w.
11 Domínguez, J., Aira, M., and Gómez-Brandón, M. (2010). Vermicomposting: earthworms enhance the work of microbes. In: *Microbes at Work* (eds. H. Insam, I. Franke-Whittle and M. Goberna), 93–114. Heidelberg: Springer-Verlag, Berlin. Available from: https://doi.org/10.1007/978-3-642-04043-6_5.
12 Rorat, A., Suleiman, H., Grobelak, A. et al. (2016). Interactions between sewage sludge-amended soil and earthworms – comparison between *Eisenia fetida* and *Eisenia andrei* composting species. *Environmental Science and Pollution Research* 23 (4): 3026–3035. Available from: https://link.springer.com/article/10.1007/s11356-015-5635-8.
13 Huang, K., Li, F., Wei, Y. et al. (2013). Changes of bacterial and fungal community compositions during vermicomposting of vegetable wastes by *Eisenia foetida*. *Bioresource Technology* 150: 235–241.
14 Guhra, T., Stolze, K., and Totsche, K.U. 2019. The role of extracellular polymeric earthworm mucus for the formation of organo-mineral associations in soil. Vol. 21, Geophysical Research Abstracts.
15 Huang, K. and Xia, H. (2018). Role of earthworms' mucus in vermicomposting system: biodegradation tests based on humification and microbial activity. *Science of the Total Environment* 610–611: 703–708.

16 Marichal, R., Mathieu, J., Couteaux, M.M. et al. (2011). Earthworm and microbe response to litter and soils of tropical forest plantations with contrasting C:N:P stoichiometric ratios. *Soil Biology and Biochemistry* 43 (7): 1528–1535.

17 Sharma, K. and Garg, V.K. (2019). Vermicomposting of waste: a zero-waste approach for waste management. In: *Sustainable Resource Recovery and Zero Waste Approaches* (eds. M.J. Taherzadeh, K. Bolton, J. Wong and A. Pandey), 133–164. Elsevier.

18 Lim, S.L., Lee, L.H., and Wu, T.Y. (2016). Sustainability of using composting and vermicomposting technologies for organic solid waste biotransformation: recent overview, greenhouse gases emissions and economic analysis. *Journal of Cleaner Production* 111: 262–278.

19 Kaushik, P. and Garg, V.K. (2004). Dynamics of biological and chemical parameters during vermicomposting of solid textile mill sludge mixed with cow dung and agricultural residues. *Bioresource Technology* 94 (2): 203–209.

20 Bhat, S.A., Singh, J., and Vig, A.P. (2018). Earthworms as organic waste managers and biofertilizer producers. *Waste and Biomass Valorization* 9: 1073–1086.

21 Singh, R.P., Singh, P., Araujo, A.S.F. et al. (2011). Management of urban solid waste: vermicomposting a sustainable option. *Resources, Conservation and Recycling* 55 (7): 719–729.

22 Richa Gupta, M.B. (2013). Implementation of bulking agents in composting: a review. *Journal of Bioremediation & Biodegradation* 4 (7): 1–3. Available from: https://www.omicsonline.org/implementation-of-bulking-agents-in-composting-a-review-2155-6199.1000205.php?aid=19176.

23 Barthod, J., Rumpel, C., and Dignac, M.-F. (2018). Composting with additives to improve organic amendments. A review. *Agronomy for Sustainable Development* 38 (2): 17. Available from: https://hal.archives-ouvertes.fr/hal-01731152.

24 Malińska, K., Golańska, M., Caceres, R. et al. (2017). Biochar amendment for integrated composting and vermicomposting of sewage sludge – the effect of biochar on the activity of *Eisenia fetida* and the obtained vermicompost. *Bioresource Technology* 225: 206–214.

25 Khan, M.B., Cui, X., Jilani, G. et al. (2020). New insight into the impact of biochar during vermi-stabilization of divergent biowastes: literature synthesis and research pursuits. *Chemosphere* 238: 124679.

26 Nath, S. and Singh, K. (2016). Analysis of different nutrient status of liquid bio-fertilizer of different combinations of buffalo dung with gram bran and water hyacinth through vermicomposting by *Eisenia fetida*. *Environment, Development and Sustainability* 18 (3): 645–656. Available from: https://link.springer.com/article/10.1007/s10668-015-9666-6.

27 Kumar Badhwar, V., Singh, S., and Singh, B. (2020). Biotransformation of paper mill sludge and tea waste with cow dung using vermicomposting. *Bioresource Technology* 318: 124097.

28 Yadav, A. and Garg, V.K. (2019). Biotransformation of bakery industry sludge into valuable product using vermicomposting. *Bioresource Technology* 274: 512–517.

29 Courtois, P., Rorat, A., Lemiere, S. et al. (2019). Ecotoxicology of silver nanoparticles and their derivatives introduced in soil with or without sewage sludge: a review of effects on microorganisms, plants and animals. *Environmental Pollution* 253.

30 Lim, S.L., Wu, T.Y., Lim, P.N., and Shak, K.P.Y. (2015). The use of vermicompost in organic farming: overview, effects on soil and economics. *Journal of the Science of Food and Agriculture* 95 (6): 1143–1156.

31 A'ali, R., Jafarpour, M., Kazemi, E., and Pessarakli, M. (2017). Effects of raw materials on vermicompost qualities. *Journal of Plant Nutrition* 40 (11): 1635–1643.

32 Bhat, M.R. and Limaye, S. (2012). Nutrient status and plant growth promoting potential of prepared vermicompost. *International Journal of Environmental Science and Technology* 3 (1): 312–321.

33 Ahmad Bhat, S., Singh, J., and Pal Vig, A. (2013). Vermiremediation of dyeing sludge from textile mill with the help of exotic earthworm *Eisenia fetida* Savigny. *Environmental Science and Pollution Research International* 20 (9): 5975–5982. Available from: https://link.springer.com/article/10.1007/s11356-013-1612-2.

34 Amouei, A.I., Yousefi, Z., and Khosravi, T. (2017). Comparison of vermicompost characteristics produced from sewage sludge of wood and paper industry and household solid wastes. *Journal of Environmental Health Science and Engineering* 15 (1): 1–6.

35 Pereira, M.D.G., Cardoso De Souza Neta, L., Fontes, M.P.F. et al. (2014). An overview of the environmental applicability of vermicompost: from wastewater treatment to the development of sensitive analytical methods. *Scientific World Journal* 2014: 1–14.

36 Hosseinzadeh, A., Baziar, M., Alidadi, H. et al. (2020). Application of artificial neural network and multiple linear regression in modeling nutrient recovery in vermicompost under different conditions. *Bioresource Technology* 303: 122926.

37 Sharma, D., Kumar Pandey, A., Durg Yadav, K., and Kumar, S. (2021). Response surface methodology and artificial neural network modelling for enhancing maturity parameters during vermicomposting of floral waste. *Bioresource Technology* 324: 124672. Available from: https://linkinghub.elsevier.com/retrieve/pii/S0960852421000109.

38 Faverial, J., Cornet, D., Paul, J., and Sierra, J. (2016). Multivariate analysis of the determinants of the end-product quality of manure-based composts and vermicomposts using bayesian network modelling. *PLoS One* 11 (6): e0157884. Available from: https://dx.plos.org/10.1371/journal.pone.0157884.

39 Stewart-Wade, S.M. (2020). Efficacy of organic amendments used in containerized plant production: Part 1 – compost-based amendments. *Scientia Horticulturae* 266: 108856.

40 Blouin, M., Barrere, J., Meyer, N. et al. (2019). Vermicompost significantly affects plant growth. A meta-analysis. *Agronomy for Sustainable Development* 39 (4, 15): 1. Available from: https://doi.org/10.1007/s13593-019-0579-x.

41 Dao, H.T.T., Seo, J.M., Hernandez, J.O. et al. (2020). Effective placement methods of vermicompost application in urban tree species: implications for sustainable urban afforestation. *Sustainability* 12 (14): 5822. Available from: https://www.mdpi.com/2071-1050/12/14/5822.

42 Paradelo, R., Basanta, R., and Barral, M.T. (2019). Water-holding capacity and plant growth in compost-based substrates modified with polyacrylamide, guar gum or bentonite. *Scientia Horticulturae*. (Amsterdam). 243: 344–349.

43 Paradelo, R., Vecino, X., Moldes, A.B., and Barral, M.T. (2019). Potential use of composts and vermicomposts as low-cost adsorbents for dye removal: an overlooked application. *Environmental Science and Pollution Research* 26: 21085–21097. Available from: https://doi.org/10.1007/s11356-019-05462-x.

44 Huang, M., Zhu, Y., Li, Z. et al. (2016). Compost as a soil amendment to remediate heavy metal-contaminated agricultural soil: mechanisms, efficacy, problems, and strategies. *Water, Air, and Soil Pollution* 227: 1–18. Available from: https://link-springer-com.ressources-electroniques.univ-lille.fr/article/10.1007/s11270-016-3068-8.

10

Biorefineries

Anna Grosser and Ewa Neczaj

Czestochowa University of Technology (CUT), Czestochowa, Poland

10.1 Introduction

For the economic use of biomass and production of chemicals, biofuels, and energy (heat and electricity) integrated processing methods are required. They include the separation of the so-called precursors (i.e. specific groups of compounds), most often by physical (physico-chemical) methods. The intermediates undergo biochemical, chemical, and thermal transformations. In the 1990s of the twentieth century the term biorefineries appeared, to define plants dealing with processing different origins of biomass. According to the most frequently cited definition proposed by the National Renewable Energy Laboratory (NREL USA), "a biorefinery is a facility that integrates biomass conversion processes and equipment to produce fuels, power, and value-added chemicals from biomass. Biorefinery is analogous to today's petroleum refinery, which produces multiple fuels and products from petroleum." The most common definitions for biorefineries are presented in Table 10.1. All definitions include the upgrading of biomass, showing which technical level depends on the type of end products.

Although the concept of a biorefinery is analogous to that of petroleum refineries, the differences are fundamental (Figure 10.1). During the petrochemical refinery process fossil recourses are transformed into energy (e.g. fuel) and other products like chemicals, plastic, etc., while in a biorefinery a wide range of bio-based products like bioenergy, biochemicals, biomaterials, food, and feed are produced from the biomass feedstock.

10.2 Concept and Classification of Waste Biorefineries

Very often a biorefinery is associated with a plant transforming biomass from agriculture. However, it should be emphasized that in fact biorefineries can use a huge range of materials as a feedstock, including many groups of biodegradable waste (Figure 10.2). Regardless of the feedstock type, various mechanical, biochemical, chemical, and thermochemical technological processes must be involved in order to transform it into useful bioproducts. Their type and technical level depend on the properties of the raw material and the final product to be obtained.

Biodegradable Waste Management in the Circular Economy: Challenges and Opportunities, First Edition.
Edited by Malgorzata Kacprzak, Eleonore Attard, Kari-Anne Lyng, Helena Raclavska, BalRam Singh, Eyob Tesfamariam, and Franck Vandenbulcke.

Table 10.1 Definition of a biorefinery [1].

Source	Definition
NREL	A biorefinery is a facility that integrates biomass conversion processes and equipment to produce fuels, power, and value-added chemicals from biomass. A biorefinery is analogous to today's petroleum refinery, which produces multiple fuels and products from petroleum.
US Department of Energy (DOE)	A biorefinery is an overall concept of a processing plant where biomass feedstocks are converted into a spectrum of valuable products.
International Energy Agency (IEA)	The sustainable processing of biomass into a spectrum of bio-based products (food, feed, chemicals, materials) and bioenergy (biofuels, power and/or heat).
Oak Ridge National Laboratory (US)	A facility that processes and converts biomass into value-added products, ranging from biomaterials to biofuels such as ethanol or important feedstocks for the production of chemicals.

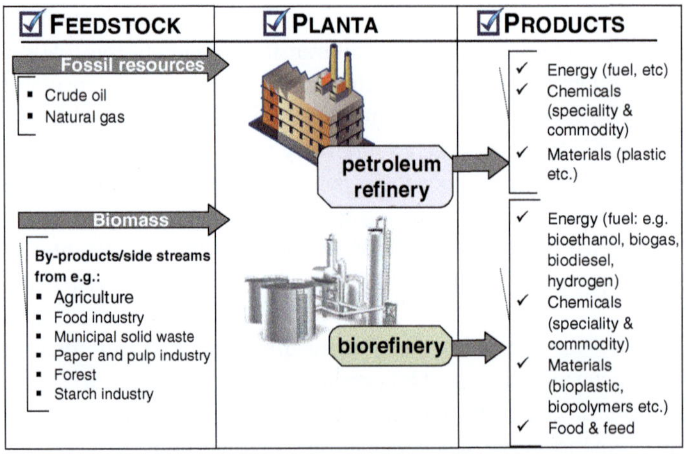

Figure 10.1 Comparison of a petroleum refinery and a biorefinery [2, 3].

There are many classifications of biorefineries (Figure 10.3). One of the simplest classifications based on conversion technologies to obtain various bioproducts distinguishes phases I, II, and III biorefineries. In phase I biorefineries feedstock is converted into one product via one process, e.g. biodiesel production from sunflowers. In phase II biorefineries it is possible to produce multiple products from one feedstock using different technical processes, e.g. production ethanol, high fructose syrup, corn oil, and starch using grain as the feedstock. The most advanced biorefinery is the phase III biorefinery, where a mixture of biomass is used to produce multiple products using a combination of different technologies, e.g. whole crop biorefineries and the lignocellulosic feedstock biorefinery (LCFB) where cellulosic biomass and wastes are used.

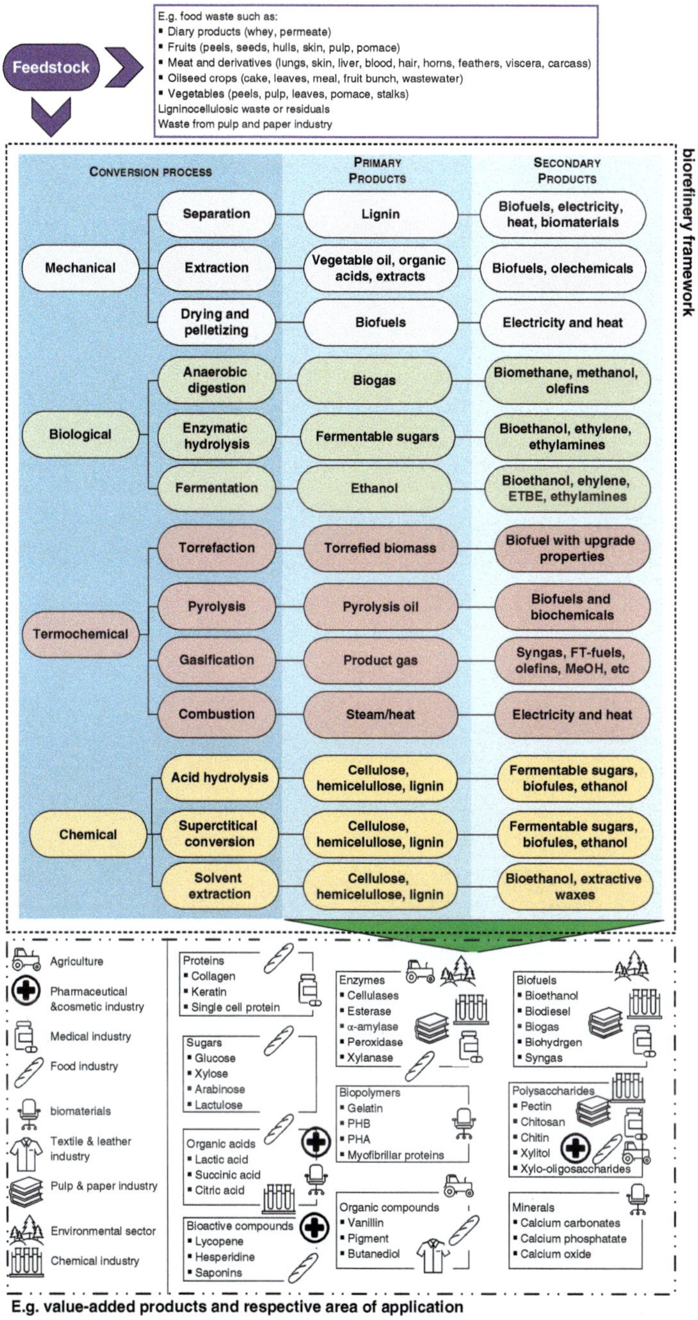

Figure 10.2 Visualization of a waste biorefinery with main conversion processes as well as their valued-added products and representative area of application [1–9].

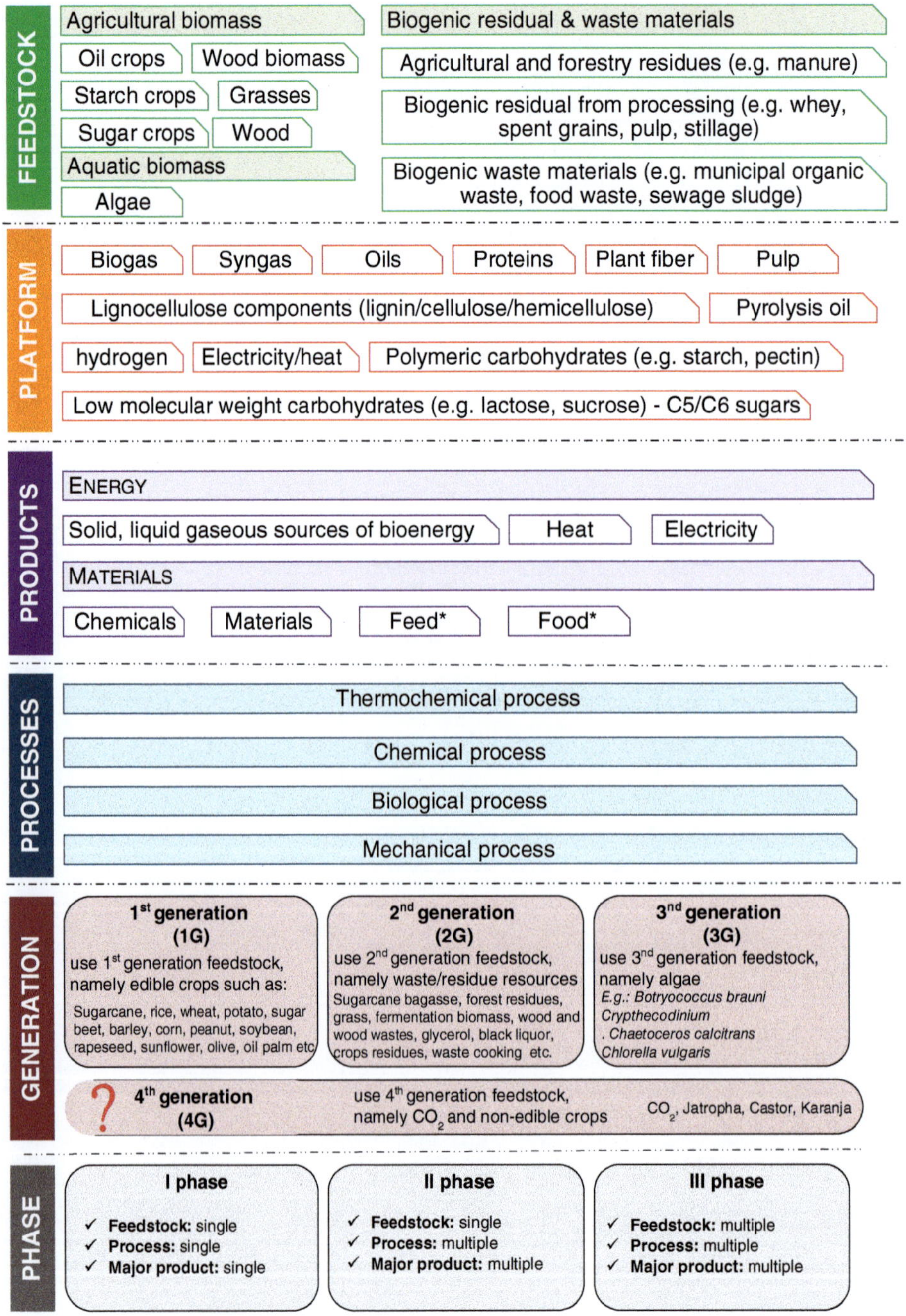

Figure 10.3 Examples for possible biorefinery classification [2, 10–15] * as a co-product; in a rectangle, an element of biorefinery classification.

Table 10.2 Classification of biorefineries based on their concepts [13, 16].

Concept	Type of feedstock	Predominant technology	TRL
Green biorefineries (GBR)	Wet biomass: green grasses and green crops, such as lucerne and clover	Pre-treatment, pressing, fractionation, separation, digestion	9
Whole crop biorefineries (WCBR)	Whole crop (including straw) cereals such as rye, wheat, and maize	Dry or wet milling, biochemical conversion	7–8
Ligno cellulosic feedstock biorefineries (LCFBR)	Lignocellulosic-rich biomass: e.g. straw, chaff, reed, miscanthus, wood	Pre-treatment, chemical and enzymatic hydrolysis, fermentation, separation	7–9
Two platform concept biorefineries (TPCBR)	All types of biomass	Combination of sugar platform (biochemical conversion) and syngas platform (thermochemical conversion)	6–8
Thermochemical biorefineries (TCBR)	All types of biomass	Thermochemical conversion: torrefaction, pyrolysis, gasification, HTU, product separation, catalytic synthesis	5–7
Marine biorefineries (MBR)	Aquatic biomass: microalgea and macroalgae (seaweed)	Cell disruption, product extraction and separation	5–6

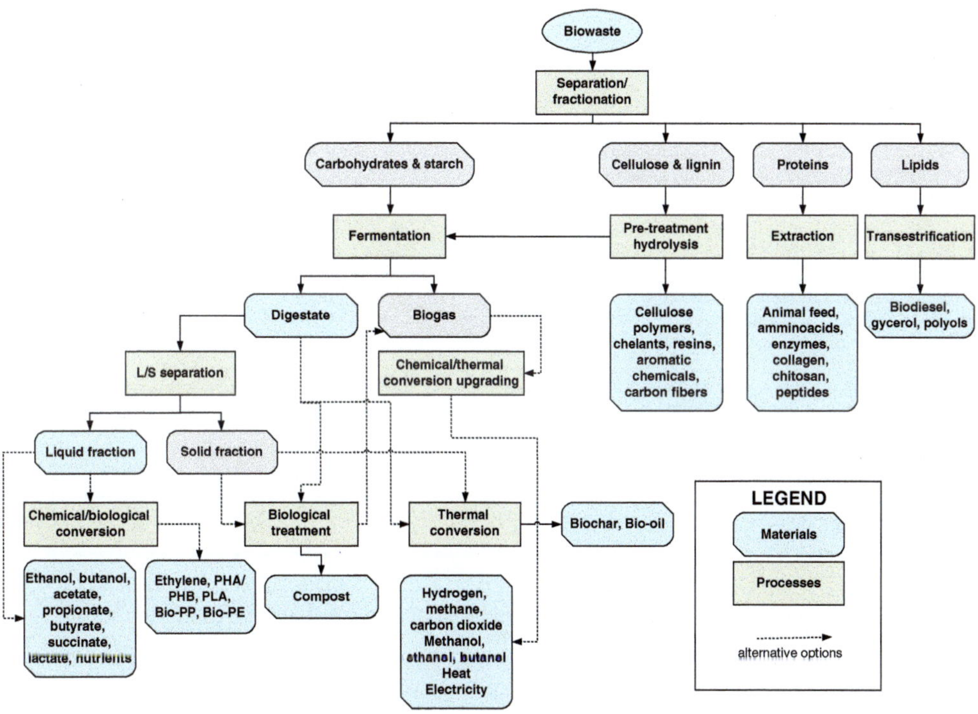

Figure 10.4 Conception of a multiplatform anaerobic biorefinery producing biofuels and biomolecules from biowaste [17].

In an IEE report is shown the classification of biorefineries based on their concepts (Table 10.2) [13]. Some of the biorefinery concepts are under development and show different Technology Readiness Levels (TRL). Generally, the main goal is to use the widest possible group of raw materials, including those from the waste sector, to produce the widest possible range of products.

Alibardi et al. [17] showed the concept of a multiplatform anaerobic biorefinery where biowaste can be used to produce biofuels and biomolecules (Figure 10.4). Their development and implementation brought waste management closer to the concept of sustainable development. In order to use the potential of waste fully as a raw material for biorefineries, many barriers, including social, legislative, economic, and technical, must be overcome.

The concept of a zero waste biorefinery producing biofuels and biomolecules from algal biomass seems very promising (Figure 10.5). This concept allows for the simultaneous solution of two problems: wastewater treatment and the management of algae biomass for the production of energy and various materials.

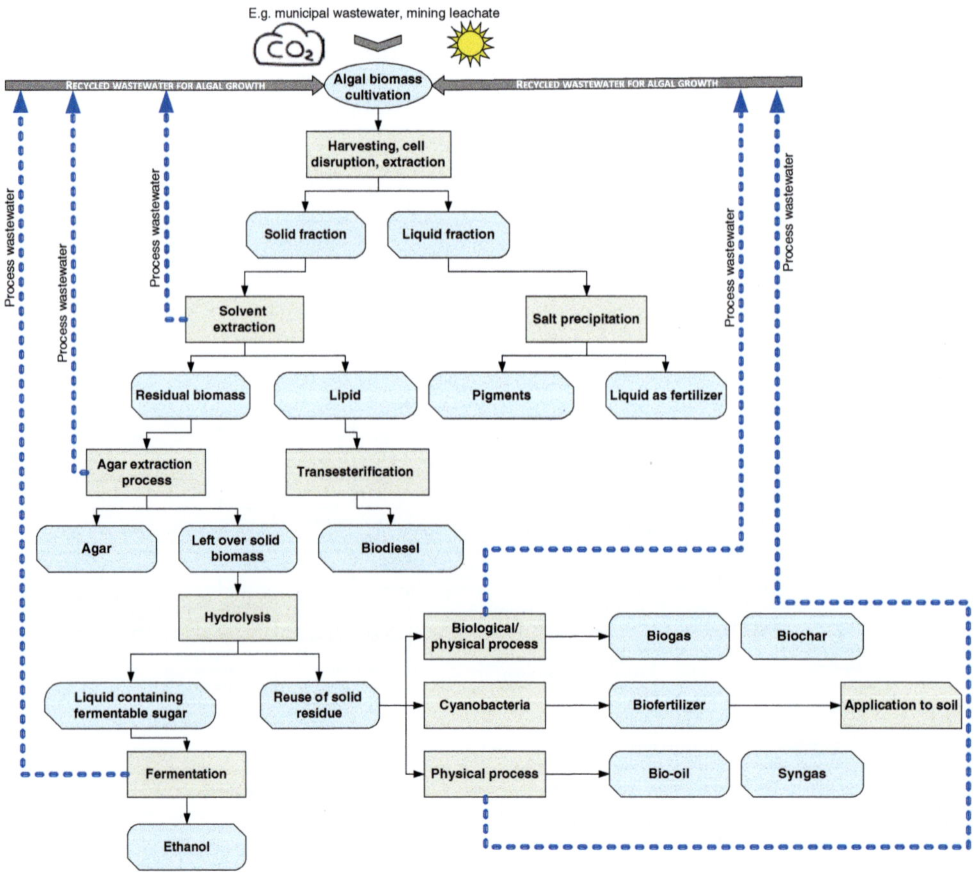

Figure 10.5 Conception of a zero waste biorefinery producing biofuels and biomolecules from algal biomass [18–22].

10.3 Source of Feedstock for Biorefineries

The term "feedstock" refers to raw materials to feed or supply the biorefinery process. The main feedstock in biorefineries is biomass. The renewable energy directive (2009/28/EC) gives the following definition of biomass: "Biomass means the biodegradable fraction of products, wastes and residues from biological origin from agriculture (including vegetable and animal substances), forestry and related industries including fisheries and aquaculture, as well as the biodegradable fraction of industrial and municipal waste." Figure 10.6 shows how the feedstock market for biorefineries has changed in recent years. Although these data are difficult to compare, mainly due to the different classification of feedstock, it is clearly visible that waste is starting to be used in biorefineries (Figure 10.6a). The number of scientific publications in this area has also increased. Among the 149 publications analyzed from 2017, 18% concern the use of industrial and domestic waste in biorefineries. Currently, there are also numerous research and development projects financed by the EU in this area (Table 10.3).

Agricultural resources including: sugar/starch-based feedstock, oil/fat-based feedstock, and secondary residues and by/co-products of industry utilizing agricultural products are the feedstock source used by most biorefineries in all EU countries (Table 10.4) [24]. Unfortunately, only 13 use municipal and industrial waste to produce energy and materials.

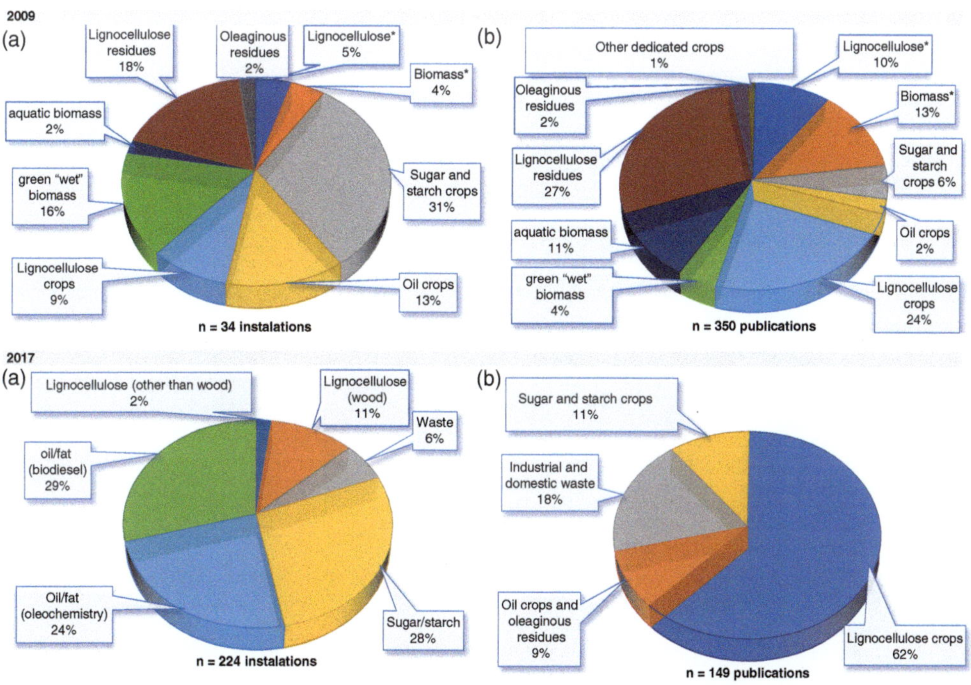

Figure 10.6 Feedstock used in Europe 2009 and 2017 in (a) existing or planed biorefineries; (b) indicated as raw material in systematic literature reviews, where * shows the lack of further specification [23].

Table 10.3 Examples of project funded by EU to support LCFBR [24–26].

Project name	Feedstock	Period	Total cost, €
BIOSKOH http://bioskoh.eu	Wheat straw and other biomass	2016–2021	21.568.195
EUCALIVA http://eucaliva.eu	Eucalyptus waste	2017–2021	1 795 009.88
GRACE http://www.grace-bbi.eu	Non-food industrial crops	2017–2022	12 324 632.86
IFERMENTER https://ifermenter.eu	Forestry sugar residual streams	2018–2022	3 997 825
LIGNIOX http://www.ligniox.eu	Lignin streams originating from variable raw materials and processes	2017–2021	4 338 374.88
LIGNOFLAG http://www.lignoflag-project.eu	Lignocellulosic feedstock	2017–2022	24.738.840
SSUCHY https://www.ssuchy.eu	Plant fibers	2017–2021	4 457 194.75
SWEETWOODS https://sweetwoods.eu	Hardwood	2018–2022	20 959 745
UNRAVEL http://unravel-bbi.eu	Agricultural and forestry residues	2018–2022	3 603 545
WOODZYMES https://www.woodzymes.eu	Forest biomass (hardwood and softwood)	2018–2021	3 253 874

Table 10.4 Category of feedstock used in biorefineries in the EU [27].

Origin	Feedstock categories	Number of facilities
Agriculture	Sugar/starch-based feedstock	216
Agriculture	Oil/fat-based feedstock	275
Marine	Oil/fat-based feedstock	34
Agriculture	Agricultural residues (in the field)	76
Agriculture	Secondary residues and by/co-products of industry utilizing agricultural products	111
Agriculture	Intermediate products derived from agriculture-based feedstock	23
Agriculture	Vegetable fibers	67
Agriculture	Other agricultural products	13
Forestry	Wood	77
Forestry	Forestry intermediate products or secondary residues and by/co-products of industry utilizing forestry products	124
Grasses and SRC	Grasses and SRC (short-rotation coppice), including derivatives	57
Waste	Waste	13

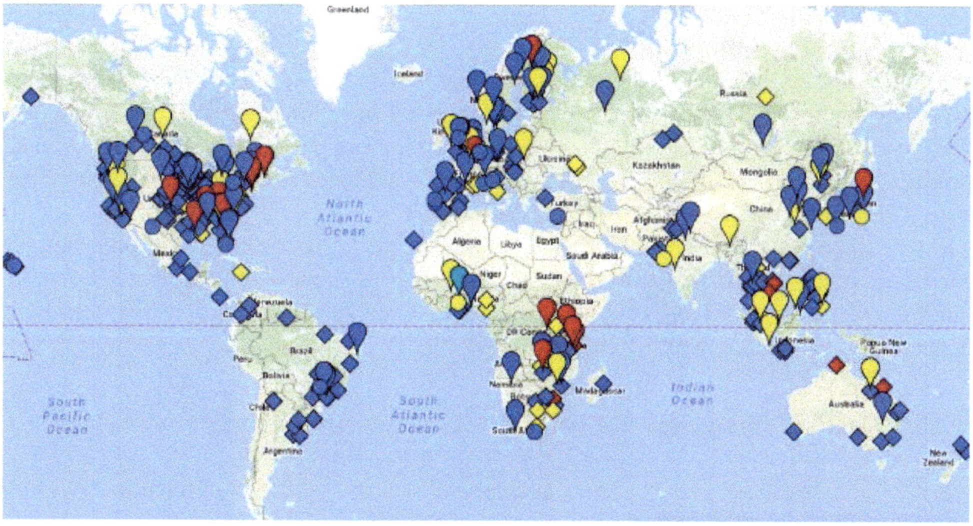

Figure 10.7 Biorefineries in the world (blue mark – operational plants; yellow mark – development plants; red mark – plants whose development has been suspended) [28].

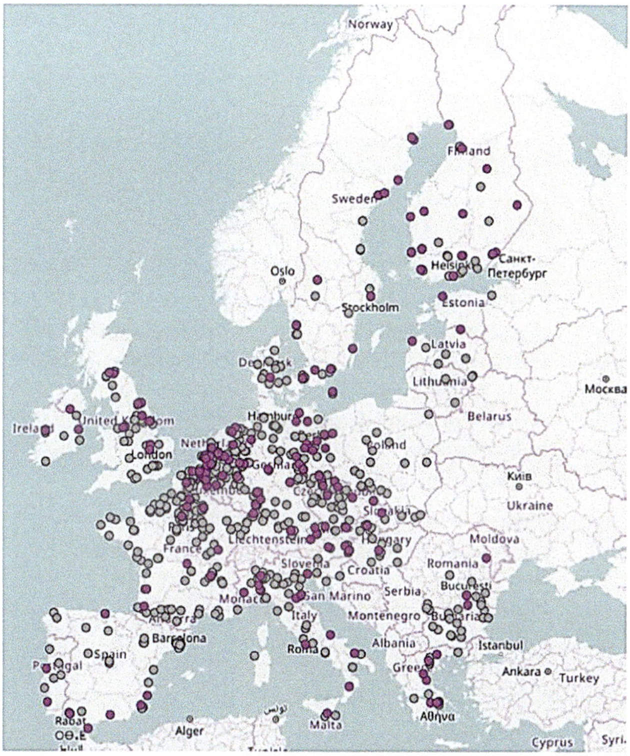

Figure 10.8 Biorefineries in Europe [27].

10.4 Biorefineries in the World

Figure 10.7 shows the location of biorefineries in the world, taking into account the existing plants, development plants, and plants that are planned for balding. Most of the plants are located in the US and Europe [25]. In 2018, there were 803 biorefineries operating in Europe, of which 507 produced chemicals, 363 liquid biofuels, while in 141 bio-based composites and fibers were generated (Figure 10.8) [26].

Table 10.5 shows examples of biorefineries around the world in which waste from various industries and municipal waste are used [29–31]. Examples of plants show that it is possible to produce from waste not only biogas but also other bioproducts such as renewable hydrocarbons, e.g. biodiesel and bioethanol.

Table 10.5 Examples of biorefineries in the world [29–31].

Company	Feedstock	Description	Location
Mercurius	Agricultural residue, woody biomass, municipal solid waste (MSW)	***Conversion technology***: Hybrid (Mercurius's REACH Technology) ***Primary product:*** Renewable hydrocarbons (jet fuel and diesel blend stocks) ***Biofuel capacity (m^3 yr^{-1}):*** 95 ***Scale:*** Pilot	Ferndale, Washington, USA
Fulcrum	MSW	***Conversion technology:*** Thermochemical, gasification ***Primary product***: Renewable hydrocarbons (synthetic crude oil) ***Biofuel capacity (m^3 yr^{-1}):*** 39740 ***Scale:*** Pioneer	McCarran, Nevada, USA
Emerald Biofuels	Inedible corn oil, food processing waste, animal fats, and yellow/brown greases	***Conversion technology:*** Thermochemical – hydrogenated esters and fatty acids ***Primary product:*** Renewable carbons (diesel) ***Biofuel capacity (m^3 yr^{-1}):*** 310400 ***Scale***: Pioneer	Plaquemine, Louisiana, USA
Inbicon IBUS	Straw	***Conversion technology***: Hydrolysis, fermentation ***Platform:*** C6/C5 sugars and lignin ***Products:*** Bioethanol, animal feed (molasses with hemicellulose sugars), lignin, electricity, and heat ***Scale:*** Pilot	Fredericia, Denmark
UPM Biofuels (UPM Lappeenranta Biorefinery)	Tall oil	***Conversion technology:*** Pre-treatment, hydro-treatment ***Platform:*** oil ***Products***: Biodiesel ***Scale:*** Commercial	Lappeenranta, France

Table 10.5 (Continued)

Company	Feedstock	Description	Location
Biodiesel Amsterdam, Noba, Orgaworld and Rotie (Green Mills Project)	Fryer fat, other organic waste	***Conversion technology***: Esterification, fermentation ***Platform:*** Starch, C5/C6 sugars ***Products:*** Biodiesel, bioethanol, biogas, fertilizers ***Capacity (yr):*** 25 million m^3 biogas, 113 million liters of biodiesel, and 5 million liters of bioethanol ***Scale:*** Commercial	Port of Amsterdam, Netherlands
Highmark renewables	Wheat, manure, slaughtering waste	***Conversion technology:*** Hydrolysis, fermentation, combustion ***Platform:*** C6 sugars, biogas ***Products:*** Bioethanol, animal feed, fertilizer, electricity, and heat ***Scale:*** Commercial	Vegreville, Alberta, Canada
Borregaard (Sarpsborg Biorefinery)	Lignocellulosic crops or residues	***Conversion technology:*** Cooking, fermentation, separation ***Platform:*** C5/C6 sugars, lignin, biogas ***Products:*** Bioethanol, lignin, cellulose, vanillin, biogas ***Scale:*** Commercial	Sarpsborg, Norway

10.5 Economic Aspect

The central importance for the economic analysis and assessment of the biorefinery concept has the following factors: (i) availability and feedstock prices – aspect also includes the transportation and logistics of raw materials and auxiliary and operating materials, in particular enzymes solvents and catalysts as well as the costs related to feedstock pre-treatment (even up to 75% of the total generation costs may originate from only the feedstock price); (ii) type of conversion technology/biorefinery technologies; (iii) scale of production (size of the biorefinery); (iv) kind of primary and secondary products (Table 10.6) [33–35]. It is worthy to note that the geographic locations of the waste biorefineries may also impact the net operational cost as well as the annual capital cost. It is associated with national/regional biofuel policies, funding policies (subsidies), the market for bio-based products, and the availability of feedstocks [36].

The following cost category should be included in the economic evaluation of biorefineries: capital expenditure (CAPEX), operational expenditures related to placed investments (OPEX such as: write-offs, imputed interest, maintenance, taxes, insurance, administration), material and energy stream costs (raw material supply, auxiliary and operating materials, energy supply, disposal supply, transport cost, material products, heat, electricity), labor costs, overheads, and an overall evaluation (operating results, net present value) (Table 10.7) [13].

Table 10.6 Cost of production of different bioproducts for a plant size of 0.17 MM m^3 per year [32].

	Ethanol				Butanol	Biodiesel
	Corn	Sugarcane	Corn stover[a)]	Corn stover[b)]	Corn	Soybean
Feedstock cost ($ m^{-3})	237.8[c)] 190.2[d)]	245.7	134.7	150.6	649.9[e)]	583.8
Total production cost ($M yr^{-1})	74	58	64	58	88	121
Total investment ($M)	131	88	183	241	276	23
Production cost ($ m^{-3})	404.2	340.8	391	348.7	517.8	673.6
Energy density (MJ m^{-3})	21 274				27 826	33 321
Production cost with energy equivalent to gasoline ($ m^{-3})	615.5	515.1	594.4	528.3	602.3	655.1 723.8[f)]

a) Biochemical conversion.
b) Thermochemical conversion.
c) Dry mill.
d) Wet mill.
e) Shelled corn.
f) Equivalent to diesel.

Table 10.7 Cost of production of various products in biorefineries based on reference [13].

		Case study			
		#1	#2	#3	#4
Process		Lignocellulosic biomass conversion	Fermentation	Fermentation	Lignocellulosic biomass conversion
Concept		Lignocellulose biorefinery	Sugar biorefinery	Waste biorefinery	Lignocellulose biorefinery
Raw material (feedstock)		Corn stover 1535 TJ/a 764 kt/a	Sugar cane 1 015 938 kg/a	Food waste 83 000 t/a	Round wood 1 549 200 t
Products		Ethanol 4400 TJ/a Electricity 387 TJ/a	PHB 46200 kg/a	PLA 10624 t/a Lipids 12 118 t/a Animal feed 64 657 t/a	Pulp 600 000 t Lignin 70 427 t Heat 1 478 632 GJ
Auxiliaries		Heat 3273 TJ Chemical inputs 82 727 t/a	Energy 309 007 MJ Chemical inputs 407 668 kg/a	Electricity 10 439 GJ Chemical inputs 22 438 t/a	Energy 780 000 GJ Chemical inputs 139 453 t
Cost	Investment	422 Mio. €	606 673 €	116.5 Mio. US$	11 Mio €
	Feedstock	48 Mio. €	345 419 €/a	16.2 Mio. US$	1.5 Mio €
	Operating	26 Mio. €	–	–	–
	Labour	3 Mio. €	–	–	–

Table 10.7 (Continued)

	Case study			
	#1	**#2**	**#3**	**#4**
Annual costs	127 Mio. €	862 080 €	79.5 Mio. US$	207 Mio €
Specific costs	0.61 € l^{-1} EtOH	19 € kg^{-1} PHB	7.49 US$ kg^{-1} PLA	–
Revenues	140.7 Mio. €	172 788 €	PLA: 55.4 Mio. US$ Lipids: 6.1 Mio. US$ Animal feed: 29.1 Mio. US$ Food waste treatment 6.4 Mio. US$	–
Specific revenues	~ 0.68 € l^{-1} EtOH	3.74 € kg^{-1} PHB	9.13 US$ kg^{-1} PLA	633 € t^{-1} Reference flow (600 000 t pulp)
Conversion rates (efficiencies)	Corn stover to EtOH 0.35 MJEtOH/MJ By-products to EtOH 0.46 MJEtOH/MJ	Sugar cane to PHB 22 kg kg^{-1}	–	Pulp to lignin 8.5 t t^{-1} lignin Black liquor to lignin 10.8 t t^{-1} lignin
Share of cost	Write-offs 11% Imputed interest 20% Maintenance 3% Insurance 3% Raw material supply 39% Auxiliary and operating material 21% Other 3%	Write-offs 4% Imputed interest 4% Maintenance7% Insurance 1% Raw material supply 40% Auxiliary and operating material 31% Energy supply 1% Overheads 8% Other 4%	Write-offs 7% Imputed interest 7% Maintenance 9% Insurance 1% Administration 1% Raw material supply 20% Auxiliary and operating material 1% Energy supply 11% Transport cost 3% Water supply costs 3% Labour cost 11% Other 16% Overheads 10%	Aux. materials 12% Aux. energy3% Feedstock75% Personal 4% Maintenance 6%

Case study # 1: two-platform (C5 and C6 sugars, lignin) biorefinery to produce bioethanol, electricity, and heat from corn stover.
Case study # 2: two-platform (C5 and C6 sugars, biogas) biorefinery to produce the biopolymer polyhydroxybutyrate (PHB), electricity, and heat from sugar beet or sugar cane.
Case study # 3: three-platform (C6 sugar, animal feed, lipids) biorefinery to produce the biopolymer PLA, animal feed, and lipids from food waste.
Case study # 4: three-platform (pulp, lignin, energy) biorefinery to produce pulp, lignin, and energy from wood chips.

10.6 Conclusion and Future Perspective

A strengths, weaknesses, opportunities, and threats (SWOT) analysis of waste biorefineries is presented in Table 10.8.

The key to the development of a waste biorefinery is to intensify the research and development (R&D) focus on inter alia enhance conversion processes, improve the biodegradable waste collection system, and develop new strategies for homogenization and stabilization of waste. However, the activities undertaken in this area should also relate to such issues as: promoting products and processes in order to gain acceptance of future consumers; integration of processes with the existing infrastructure; exchange of experience; activities to characterize better the available waste (create a base of the feedstock resource) and the potential products derived from it, which will allow the development of different integration strategies, such as processing and product markets and managing the feedstock supply chain (Figure 10.9). Moreover, a biorefinery design consistent with the concept of sustainable development should take into account the following aspects: ecological, social, technical, and economic (Figure 10.10).

Table 10.8 SWOT analysis for waste biorefinery based on [7, 13, 33, 37–39].

Strengths	Weaknesses
• Transformation waste into value-added products and energy; • Minimizing raw material requirements; • Positive political framework and public opinion for bio-based products and renewable energies; • Reduced market risks due to wide spectrum of products; • Experience in the conversion of biomass in different type of biorefineries, which allow to tackle both technical and non-technical problems into waste biorefinery; • Reduced GHG emissions.	• Broad unclassified and undefined area; • Limitations in the availability of waste biorefinery technologies; furthermore, sometimes their concepts are unclear; • Involvement of stakeholders of different sectors are necessary (e.g. energy, chemical, farming, forestry, and food processing); • High investment and economic risk; • Feedstock impurities; • Profitability is strongly dependent on current market prices of the products generated; • Social acceptance; • Consumer acceptance of the products; • Low amount of validated plants; • Further improvement of feedstock sustainable collection, sorting, and the sustainable provision of raw materials is essential; • Basic research and applied research is not sufficient in some areas of the decomposition and conversion of biomass.
Opportunities	**Threats**
• A significant contribution to sustainable development; • Resource efficiency; • A strengthening of industrial and energy sectors; • Development of multipurpose biorefineries; • Opening new markets.	• Limited availability of feedstock (associated with, for example, climate change, policies, logistics); • Price competition on the sales market; • Economic change and drop in fossil fuel prices; • Fast implementation of other renewable energy technologies feeding the market requests; • Uncertainty on target products and markets; • Competition distortions due to globally differing sustainability standards; • (High) investment capital for pilot and demo installation difficult to fund; • Fluctuating (long-term) governmental policies.

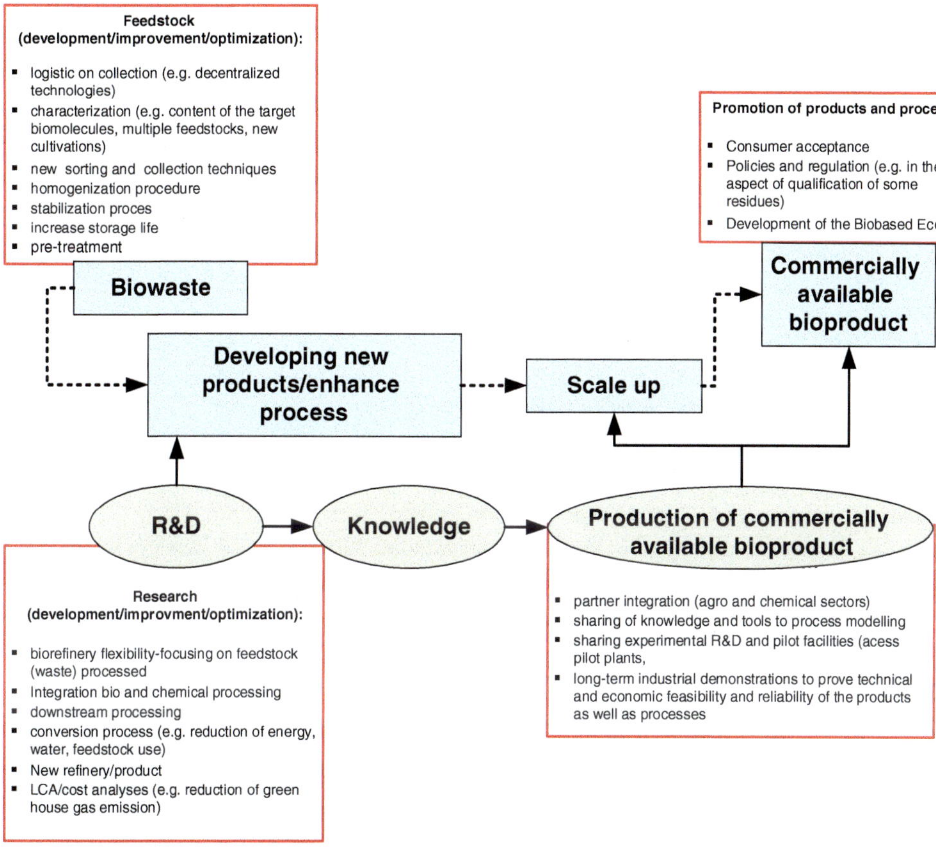

Figure 10.9 R&D and innovation needs of the waste biorefinery [40].

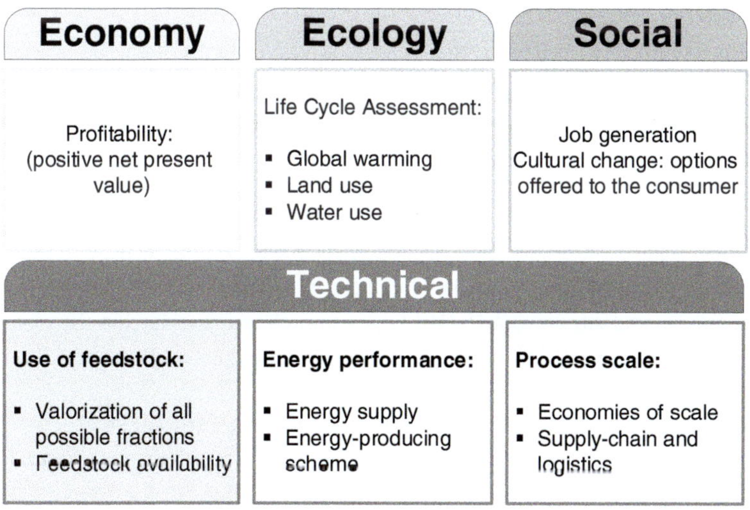

Figure 10.10 Important aspects that need to be considered during the design of sustainable biorefineries [11].

References

1 Management Systems in Production Engineering (2016). Biorefineries – New Green Strategy for Development of Smart and Innovative Industry.

2 Clark, J.H. and Deswarte, F.E.I. (2008). The biorefinery concept – an integrated approach. In: *Introduction to Chemicals from Biomass* (ed. J. Clark and F. Deswarte), 1–20. Wiley http://onlinelibrary.wiley.com/doi/10.1002/9780470697474.ch1/summary.

3 Ferreira, J.A., Agnihotri, S., and Taherzadeh, M.J. (2019). Waste biorefinery, Chapter 3. In: *Sustain Resource Recovery and Zero Waste Approaches* (ed. M.J. Taherzadeh, K. Bolton, J. Wong and A. Pandey), 35–52. Elsevier.

4 Venkata Mohan, S., Nikhil, G.N., Chiranjeevi, P. et al. (2016). Waste biorefinery models towards sustainable circular bioeconomy: critical review and future perspectives. *Bioresource Technology* 215: 2–12. http://dx.doi.org/10.1016/j.biortech.2016.03.130.

5 Caldeira, C., Vlysidis, A., Fiore, G. et al. (2020). Sustainability of food waste biorefinery: a review on valorisation pathways, techno-economic constraints, and environmental assessment. *Bioresource Technology* 312: 123575. https://doi.org/10.1016/j.biortech.2020.123575.

6 Kumar, P.S., Saravanan, A., Jayasree, R., and Jeevanantham, S. (2020). Food industry waste biorefineries. In: *Refining Biomass Residues for Sustainable Energy and Bioproducts* (ed. R.P. Kumar, E. Gnansounou, J.K. Raman and G. Baskar), 407–426. Elsevier Inc. http://dx.doi.org/10.1016/B978-0-12-818996-2.00018-1.

7 Ubando, A.T., Felix, C.B., and Chen, W.H. (2020). Biorefineries in circular bioeconomy: a comprehensive review. *Bioresource Technology* 299: 122585. https://doi.org/10.1016/j.biortech.2019.122585.

8 Atsonios, K., Nesiadis, A., Detsios, N. et al. (2020). Review on dynamic process modeling of gasification based biorefineries and bio-based heat & power plants. *Fuel Processing Technology* 197: 106188. https://doi.org/10.1016/j.fuproc.2019.106188.

9 Dahiya, S., Kumar, A.N., Shanthi Sravan, J. et al. (2018). Food waste biorefinery: sustainable strategy for circular bioeconomy. *Bioresource Technology* 248: 2–12. https://doi.org/10.1016/j.biortech.2017.07.176.

10 German Federal Government (2012). Biorefineries Roadmap as a part of the German Federal Government action plans for the material and energetic utilisation of renewable raw materials, 108. https://www.bmbf.de/upload_filestore/pub/Roadmap_Biorefineries_eng.pdf%0A https:/www.bmbf.de/pub/Roadmap_Biorefineries_eng.pdf%0A https://www.ieabioenergy.task42-biorefineries.com/upload_mm/b/0/0/38ff080b-03f1-4328-8c47-2226f10717c9_Borefinery Roadmap Ger.

11 Serna-Loaiza, S., Miltner, A., Miltner, M., and Friedl, A. (2019). A review on the feedstocks for the sustainable production of bioactive compounds in biorefineries. *Sustainability* 11 (23): 6765.

12 Gutiérrez, C.D.B., Serna, D.L.R., and Alzate, C.A.C. (2017). A comprehensive review on the implementation of the biorefinery concept in biodiesel production plants. *Biofuel Research Journal* 4 (3): 691–703.

13 Lindorfer, J., Lettner, M., Hesser, F. et al. (2019). Technical, economic and environmental assessment of biorefinery concepts. *IEA Bioenergy Task* 42: 55. https://www.ieabioenergy.com/wp-content/uploads/2019/07/TEE_assessment_report_final_20190704-1.pdf.

14 Kumar, B., Bhardwaj, N., Agrawal, K. et al. (2020). Current perspective on pretreatment technologies using lignocellulosic biomass: an emerging biorefinery concept. *Fuel Processing Technology* 199: 106244.

15 van Ree, R. and van Zeeland, A. (2014). IEA Bioenergy Task 42 Biorefining Sustainable and synergetic processing of biomass into marketable food & feed ingredients, chemicals, materials and energy (fuels, power, heat). Hilde Holleman. http://www.IEA-Bioenergy.Task42-Biorefineries.com.

16 Xuan, T.D., Sakanishi, K., Nakagoshi, N., and Fujimoto, S. (2012). Biorefinery: concepts, current status, and development trends. *International Journal of Biomass Renewables* 1 (2): 1–8.

17 Alibardi, L., Astrup, T.F., Asunis, F. et al. (2020). Organic waste biorefineries: looking towards implementation. *Waste Management* 114: 274–286. https://doi.org/10.1016/j.wasman.2020.07.010.

18 Parvathy Eswari, A., Anu Alias Meena, R., Yukesh Kannah, R. et al. (2020). Bioconversion of marine waste biomass for biofuel and value-added products recovery. In: *Refining Biomass Residues for Sustainable Energy and Bioproducts* (ed. R.P. Kumar, E. Gnansounou, J.K. Raman and G. Baskar), 481–507. Elsevier http://dx.doi.org/10.1016/B978-0-12-818996-2.00022-3.

19 Masri, M.A., Jurkowski, W., Shaigani, P. et al. (2018). A waste-free, microbial oil centered cyclic bio-refinery approach based on flexible macroalgae biomass. *Applied Energy* 224: 1–12. https://doi.org/10.1016/j.apenergy.2018.04.089.

20 Bhatia, L., Bachheti, R.K., Garlapati, V.K., and Chandel, A.K. (2020). Third-generation biorefineries: a sustainable platform for food, clean energy, and nutraceuticals production. *Biomass Conversion and Biorefinery* 1–16.

21 Koutra, E., Tsafrakidou, P., Sakarika, M., and Kornaros, M. (2019). Microalgal biorefinery. In: *Microalgae Cultivation for Biofuels Production*, 163–185. Amsterdam, The Netherlands: Elsevier.

22 Rajak, R.C., Jacob, S., and Kim, B.S. (2020). A holistic zero waste biorefinery approach for macroalgal biomass utilization: a review. *Science of the Total Environment* 716: 137067. https://doi.org/10.1016/j.scitotenv.2020.137067.

23 Wenger, J. and Stern, T. (2019). Reflection on the research on and implementation of biorefinery systems – a systematic literature review with a focus on feedstock. *Biofuels, Bioproducts and Biorefining* 13 (5): 1347–1364.

24 Hassan, S.S., Williams, G.A., and Jaiswal, A.K. (2019). Lignocellulosic biorefineries in Europe: current state and prospects. *Trends in Biotechnology* 37 (3): 231–234. http://dx.doi.org/10.1016/j.tibtech.2018.07.002.

25 Ahorsu, R., Medina, F., and Constantí, M. (2018). Significance and challenges of biomass as a suitable feedstock for bioenergy and biochemical production: a review. *Energies* 11 (12): 3366.

26 Hassan, S.S., Williams, G.A., and Jaiswal, A.K. (2019). Moving towards the second generation of lignocellulosic biorefineries in the EU: drivers, challenges, and opportunities. *Renewable and Sustainable Energy Reviews* 101: 590–599.

27 Parisi, C. (2018). *Research Brief: Biorefineries Distribution in the EU*, 1–8. European Commission – Joint Research Center.

28 Nguyen, Q., Bowyer, J., Howe, J. et al. (2017). Production of sustainable bioplastic. *Biofuels/Biorefinery Development Report Card* 265896: 1–15. http://www.annualreviews.org/doi/10.1146/annurev-environ-101813-013253.

29 Castilla-Archilla, J., O'Flaherty, V., and Lens, P.N.L. (2019). Biorefineries: industrial innovation and tendencies. In: *Biorefinery* (ed. J.-R. Bastidas-Oyanedel and J.E. Schmidt), 3–35. Cham: Springer.

30 Gírio, F., Marques, S., Pinto, F. et al. (2017). Biorefineries: targeting energy, high value products and waste valorisation. In: *Lecture Notes in Energy*, vol. 57 (ed. M. Rabaçal, A.F. Ferreira, C.A.M. Silva and M. (e.) Costa), 1–6. Cham: Springer International Publishing.

31 Alzate, C.A.C., Botero, J.M., and Marulanda, V.A. (2018). *Biorefineries: Design and Analysis*. CRC Press.

32 Maity, S.K. (2015). Opportunities, recent trends and challenges of integrated biorefinery: part II. *Renewable and Sustainable Energy Reviews* 43: 1446–1466.

33 Brunnhofer, M., Gabriella, N., Schöggl, J.P. et al. (2020). The biorefinery transition in the European pulp and paper industry – a three-phase Delphi study including a SWOT-AHP analysis. *Forest Policy and Economics* 110: 101882. https://doi.org/10.1016/j.forpol.2019.02.006.

34 Teigiserova, D.A., Hamelin, L., and Thomsen, M. (2019). Review of high-value food waste and food residues biorefineries with focus on unavoidable wastes from processing. *Resources, Conservation and Recycling* 149: 413–426. https://doi.org/10.1016/j.resconrec.2019.05.003.

35 Dhamodharan, K., Ahlawat, S., Kaushal, M., and Rajendran, K. (2020). Economics and cost analysis of waste biorefineries. In: *Refining Biomass Residues for Sustainable Energy and Bioproducts* (ed. R.P. Kumar, E. Gnansounou, J.K. Raman and G. Baskar), 545–565. Elsevier http://dx.doi.org/10.1016/B978-0-12-818996-2.00025-9.

36 Nizami, A.S., Rehan, M., Waqas, M. et al. (2017). Waste biorefineries: enabling circular economies in developing countries. *Bioresource Technology* 241: 1101–1117.

37 Maity, S.K. (2015). Opportunities, recent trends and challenges of integrated biorefinery: part I. *Renewable and Sustainable Energy Reviews* 43: 1427–1445.

38 Senthil Kumar, P. and Yaashikaa, P.R. (2020). Sources and operations of waste biorefineries. In: *Refining Biomass Residues for Sustainable Energy and Bioproducts* (ed. R.P. Kumar, E. Gnansounou, J.K. Raman and G. Baskar), 111–133. Elsevier http://dx.doi.org/10.1016/B978-0-12-818996-2.00005-3.

39 de Jong, E. and Jungmeier, G. (2015). Biorefinery concepts in comparison to petrochemical refineries. In: *Industrial Biorefineries and White Biotechnology* (ed. A. Pandey, R. Hofer, M. Taherzadeh, et al.), 3–33. Elsevier.

40 Fava, F., Totaro, G., Diels, L. et al. (2015). Biowaste biorefinery in Europe: opportunities and research and development needs. *New Biotechnology* 32 (1): 100–108.

11

Impact of Processing Technology on the Chemical Contaminants Occurrence in End Products

Eyob Habte Tesfamariam[1], Taruvinga Badza[1], and Craig Cogger[1,2]

[1] *Department of Plant and Soil Sciences, Natural and Agricultural Sciences, University of Pretoria, Pretoria, South Africa*
[2] *(Emeritus) Department of Crop and Soil Sciences, Washington State University, Puyallup, WA, USA*

11.1 Introduction

Globally, an unprecedented rise and accumulation of wastes are encountered, attributable to increasing human activities [1]. World population growth has seen the rise in urbanization and industrialization, leading to increasing wastewater volumes discharged into wastewater treatment plants (WWTPs) globally. Population growth holds an increased demand for pharmaceutical products and chemicals in the health, beauty, and farming industries. Recently, it was reported that 100–200 thousand tons of antibiotics are consumed across the world annually [2]. Most of these contaminants are considered "unregulated" [3] compared to additional contaminant groups, such as the heavy metals regulated for end products targeted for recycling specifically in agricultural lands. From 80 to 90% of antibiotics consumed end up being excreted by humans and animals as primary compounds into the environment [4].

Such contaminants recently became a cause for concern, attributable to their potential human and environmental health challenges [5]. In addition to their potential to cause health challenges to people and other animal groups, they introduce endocrine disruptors into the aquatic environments. This was noticed in several aquatic organisms, such as fish and frogs, while in the terrestrial spaces, mice and rats were also observed as affected. Another challenge is that they are ineffectively removed by the existing conventional wastewater treatment technologies [3]. Initial conventional WWTP processes (primary and secondary treatments) fail to completely remove the emerging contaminants [6]. Several contaminants could not be digested and could not be used as a carbon source by microbes, which may hinder microorganism activities, generating their bioaccumulation of food restraints during the treatment process [7]. This made them more problematic, and hence there is a need for continuous research on sustainable practices of managing such contaminants and/or materials containing these contaminants.

11.2 Biodegradable Waste Streams (Sources) and Their Associated Chemical Contaminant Occurrence

This chapter focuses on "emerging" contaminants [8], which include pharmaceutical and personal care products (PPCPs) [9] and endocrine-disrupting compounds (EDCs) [10] (Table 11.1).

Biodegradable Waste Management in the Circular Economy: Challenges and Opportunities, First Edition.
Edited by Malgorzata Kacprzak, Eleonore Attard, Kari-Anne Lyng, Helena Raclavska, BalRam Singh, Eyob Tesfamariam, and Franck Vandenbulcke.

Although various sources of these contaminants exist, most studies focused on wastewaters and wastewater treatment plants (WWTPs) [9, 10], through which such contaminants are transferred into various environmental spaces.

Variations in contaminant concentrations are reportedly driven by factors such as sources of waste or season of the year [15, 22]. Some research identified variations originating from animal classes [23]. For example, monensin is consistently detected from wastewater and groundwater associated with cattle facilities [24], while ractopamine and lincomycin were commonly observed from swine facilities. Industry types [25] or the point along or within the waste stream [23] from which the wastes could have been collected are additional factors determining the compound concentration variations. The waste degree and status, such as raw (untreated), primary, or secondary treated waste [11], liquid/effluent or dried biosolids [14, 19], and/or WWTP in conjunction with wastewater treatment process/system applied [11, 22], were identified as the additional pollutant variation source concentrations. Antibiotics were higher in municipal, livestock, and pharmaceutical WWTP, in contrast with hospital WWTPs with a higher proportion of caffeine and carbamazepine than antibiotic proportions [25].

The concentration of some compounds, such as diuron, gemfibrozil, and bisphenol A, increased after primary treatment, while others like triclosan and triclocarban did not indicate changes [19]. This behavior of compounds was observed in both aqueous and solid phases of the digested sludges, predominantly observed in solid sludge compared to liquid. In additional studies, sulfamethoxazole, ketoprofen, and carbamazepine increased in the effluent, following biological sludge treatment processes, as compared to influent concentrations [18, 22]. The proportion of pharmaceutical compounds of caffeine and carbamazepine from pharmaceutical and hospital WWTPs were higher in effluent than their concentrations in the influent materials [25], an indication of how detected concentrations differ with the sources of wastes. The observed scenarios, where concentrations of chemical compounds are higher in their effluents or biosolids than in influents, were attributed to the generation of secondary metabolites of such compounds during treatment [26]. It is reported that carbamazepine's glucuronides concurrent with supplementary conjugated metabolites are converted into secondary metabolites during treatment; therefore, higher concentrations were observed in effluents than in influents [27].

11.3 Impact of Existing Processing Technologies on the Fate of Chemical Contaminants in End Products

Since most of these contaminants are new/emerging, a lack of published health standards exists to provide guidelines on treating and removal of these contaminants [10, 28]. Limited knowledge exists on the impact of various commonly existing wastewater treatment technologies in removing these pollutants [22]. It is under these auspices that various potential treatment technologies [28] were and are still under study in attempting to comprehend how differently they perform on the removal of contaminants during wastes and wastewater treatment. Technologies ranging from conventional WWTPs to more advanced technologies were studied for different contaminant removals in various operating conditions. Wide ranges of removal magnitudes were recorded for diverse global contaminants. Fundamentally, it was identified that several factors govern the degree of contaminant removals, including process-specific factors, contaminant functional groups, physicochemical properties, charges, shape, size, and hydrophobicity [29, 30].

Table 11.1 Occurrence and concentrations of some compounds detected in certain wastewaters.

Contaminants type/class	Contaminants/pollutants	Compound concentration (ng l)	Type of waste	Reference(s)
Pharmaceutical and personal care products				
Antibiotics	Amoxicillin	64–3000	Manure, swine slurry, slaughter wastes	[11–13]
	Azithromycin	37–1300		[11, 14]
	Chloramphenicol	14.5–1050		[12]
	Chlortetracycline	1472–1986		[11]
	Ciprofloxacin	22–1100		[14–17]
	Clarithromycin	130–7000		[11, 14]
	Clindamycin	3.41–4.24		[11]
	Enoxacin	15–25		[16]
	Enrofloxacin	3.5–270		[14, 18]
	Erythromycin	50–480		[11, 16, 17]
	Lemefloxacin	45		[16]
	Lincomycin	4.9–510		[11, 14, 18]
	Minocycline	<100		[14]
	Norfloxacin	<5.5–3700		[12, 14, 18]
	Ofloxacin	6.4–7780		[12, 14]
	Oxytetracycline	<5.8–4200		[11, 12, 18]
	Roxithromycin	0.6–547		[12, 14]
	Penicillin	3.1–27		[14]
	Sarafloxacin	25–250		[16, 18]
	Sulfamethazine	44–360		[11, 14, 18]
	Sulfamethoxazole	<5–3400		[11, 12, 14–18]
	Tetracycline	3.5–3600		[11, 12, 14, 18]
	Trimethoprim	59–1600	Anaerobic digested effluent sludge (20–30 SRT days)	[11, 12, 15–17, 19]

(Continued)

Table 11.1 (Continued)

Contaminants type/class	Contaminants/pollutants	Compound concentration (ng l)	Type of waste	Reference(s)
	Tylosin	21–720	Sewage waste/biosolids/municipal sludge	[14, 18]
	Vancomycin	2500		[13]
	Paracetamol	305–1200	Anaerobic digested effluent sludge (20–30 SRT days)	[15, 19]
	Primidone	180–200	Anaerobic digested effluent sludge (20–30 SRT days)	[15, 17, 19]
	Clozapine	0.5	Anaerobic digested effluent sludge (20–30 SRT days)	[19]
Antifungal/ antimicrobial	Miconazole	1.6–25		[14]
	Metronidazole	35.2		[15]
	Thiabendazole	6.2–140		[14]
	Triclocarban	3.1–10 500	Anaerobic digested effluent sludge (20–30 SRT days)	[14, 19]
NSAIDs	Acetaminophen	16–130 000		[13, 14, 17]
	Codeine	80–3300		[14]
	Diclofenac	250–2668	Anaerobic digested effluent sludge (20–30 SRT days)	[13, 15, 17, 19]
	Fluoxetine	0.3	Anaerobic digested effluent sludge (20–30 SRT days)	[19]
	Ibuprofen	16–25 000	Anaerobic digested effluent sludge (20–30 SRT days)	[14, 19]
	Indomethacin	400–800		[13, 15]
	Ketoprofen	1000–2000		[13, 15]
	Naproxen	0.3	Anaerobic digested effluent sludge (20–30 SRT days)	[19]
	Salicylic acid	750		[15]

Beta-blockers	Atenolol	200–750	Anaerobic digested effluent sludge (20–30 SRT days)	[13, 19]
	Metoprolol	377–780		[15]
	Propranolol	90.1		[15]
Anticonvulsants	Carbamazepine	4.7–12 000	Anaerobic digested effluent sludge (20–30 SRT days)	[13, 15, 19]
Lipids regulators	Gemfibrozil	1600	Anaerobic digested effluent sludge (20–30 SRT days)	[19]
Hormones	Estrone	1.04	Thermophilic and mesophilic anaerobic digested liquid sludge spiked with organic micropollutants	[20]
	17β-estradiol	30	Thermophilic and mesophilic anaerobic digested liquid sludge spiked with organic micropollutants	[20]
	17-α-ethinylestradiol	140–620	Thermophilic and mesophilic anaerobic digested liquid sludge spiked with organic micropollutants	[17, 20]
Endocrine disrupting compounds	4-nonylphenol	0.1–13.5	Wastewater effluent from activated sludge treatment plants ($n = 9$)	[21]
	4-tert-octylphenol	0.1–1.25	Wastewater effluent from activated sludge treatment plants ($n = 9$)	[21]
	4-tert-octylphenol-polyethoxylates	0.1–5	Wastewater effluent from activated sludge treatment plants ($n = 9$)	[21]
	4-nonylphenolpoly-ethoxylates	1.13–27	Wastewater effluent from activated sludge treatment plants ($n = 9$)	[21]
	4-nonylphenolethoxy-carboxylic acids	50–1000	Wastewater effluent from activated sludge treatment plants ($n = 9$)	[21]
	Ethylenediaminetetra-acetic acid	30–1200	Wastewater effluent from activated sludge treatment plants ($n = 9$)	[21]
	Bisphenol A	0.1–30	Activated and anaerobic digested effluent sludge (20–30 SRT days)	[19, 21]
	Triclosan	28.5–12 500	Activated and anaerobic digested effluent sludge (20–30 SRT days)	[14, 17, 19, 21]
Herbicides	Diuron	0.2		[19]

11.3.1 Contaminant Removal through Conventional Wastewater Treatment Plant Processes

11.3.1.1 Removal Through Activated Sludge Treatment

Activated sludge treatment (AST) is a combination of primary and secondary treatment processes [7]. This process is the major conventional technology for contaminant removal globally [30]. Efficiencies of the removal of various contaminants differ [31]. Contaminant removal in AST occurs through biological transformation or biodegradation [32, 33], although abiotic degradation, volatilization, and adsorption [31] and sorption on biological flocs also play a role [17]. During these treatment processes, adjusting WWTP operational conditions with chemical and/or physical properties of compounds could improve removal efficiency [31, 33].

The solid retention time (SRT) is one of the WWTP operational conditions correlated with the removal of certain contaminants. Research indicates an effective SRT of 10–15 days for contaminant removal [34]. The maximum removal of chemicals, such as androgens, estrogens, nonylphenolics, and metals at the SRT condition, was observed in 27 days [32]. The observed contrasting operating conditions on the studied contaminants indicates heterogeneity and diversity in the chemistry of contaminants and how these properties govern their removal during wastewater treatment, rendering their removal difficult [32]. This, therefore, explains the reasons behind a wide variation in removal magnitude observed by various global studies, as presented in Table 11.2.

Enhanced treatment systems, such as membrane bioreactors (MBR), full-scale membrane bioreactors (full-scale MBR), GSBR, a horizontal or upflow anaerobic sludge bed, and the anoxic/oxic (A/O) system improved removal efficiency of the conventional WWTP systems, although sometimes no significant difference between some of these and conventional systems was established [11]. Flocculation using aluminum sulfate [2] increased the efficiency of tetracycline removal from >50% to between 83 and 100% (Table 11.2). Additional compounds were not removed as efficiently in the same flocculation system [2].

The removal of hormones and EDCs (73–100%) is indicated as effective in AST in conjunction with additional systems. The lowest removal efficiency has been reported [42] for the removal of phytoestrogens in the activated sludge system with primary and secondary treatments. A full-scale membrane reactor performance was 95% and higher in removing estrogen, testosterone, androsterone.

Activated sludge treatment (AST) inadequately performed in the removal of many antibiotic compounds, but more interesting are the groups such as amides (propyphenazone, carbamazepine, and crotamiton) [17, 35], trimethoprim, and sulfamethoxazole [26, 36], which recorded negative minimum removal percentages to a maximum of 87%. Negative removal percentages indicate an increase in those compounds during treatment, and due to the regeneration of some secondary metabolites, such as with carbamazepine, its glucoronides and other conjugated metabolites are converted by enzymes during wastewater treatment into their parent compound [6, 43]. This indicates the inefficiency of the treatment system for such compounds.

For the anti-inflammatory/analgesics class, the treatment process performed the degradation of the compounds exceptionally well, with 90–98% removal of ibuprofen, paracetamol, naproxen, and aspirin. These high removals could be attributed largely to high microbial degradation, while >60% of ketoprofen and diclofenac were removed [26, 35]. Triclosan is mainly found in several household consumables, such as soaps, toothpaste, and numerous personal care products, attributed to its antimicrobial active component [44]. Its removal in a full-scale MBR system was 68% (Table 11.2). Some studies reported high removal values above 90% in AST [17]. Such high values are attributed to its hydrophobicity and high sorption constant, which enhance it to be strongly sorbed to sludge flocs [44].

Table 11.2 Contaminant compounds removal efficiency through activated treatment systems.

Contaminant type and class	Compounds	Removal efficiency (%)	Operating conditions	Reference(s)
Pharmaceutical and personal care products				
Antibiotics	Triclosan	68	Full-scale MBR	[26]
	Thymol	95	Primary + secondary treatment, SRT = 4–8 days, HRT = 7–9 hours	[35]
	Erythromycin	<−20–70	Activated biological treatment + membrane filter	[17]
	Trimethoprim	−140–62	Full-scale MBR, denitrification and nitrification, SRT ~ 10–12 days, HRT ~ 15 hours	[26, 36]
	Sulfamethoxazole	−138–87	Full-scale MBR, denitrification and nitrification, SRT ~ 10–12 days, HRT ~ 15 hours	[26, 36]
Analgesics	Ibuprofen	98	Full-scale MBR	[26]
	Paracetamol	99	Full-scale MBR	[26]
	Diclofenac	60	Full-scale MBR	[26]
	Naproxen	98	Full-scale MBR	[26]
	Ketoprofen	<50	Primary + secondary treatment, SRT = 4–8 days, HRT = 7–9 hours	[35]
	Fenoprofen	65–95	Primary + secondary treatment, SRT = 4–8 days, HRT = 7–9 hours	[35]
	Aspirin	>90	Primary + secondary treatment, SRT = 4–8 days, HRT = 7–9 hours	[35]
Amide	Diethyltoluamide	10–97	Primary + secondary treatment, SRT = 4–8 days, HRT = 7–9 hours	[35]
	Propyphenazone	−700-85	Primary + secondary treatment, SRT = 4–8 days, HRT = 7–9 hours	[35]
	Carbamazepine	−200–79	Primary + secondary treatment, SRT = 4–8 days, HRT = 7–9 hours	[35]
	Crotamiton	−50–65	Primary + secondary treatment, SRT = 4–8 days, HRT = 7–9 hours	[35]

(Continued)

Table 11.2 (Continued)

Contaminant type and class	Compounds	Removal efficiency (%)	Operating conditions	Reference(s)
	Naphthalene	94	GSBR, aerobic, SRT = 17, HRT = 24	[37]
	Acenaphthene	83	GSBR, aerobic, SRT = 17, HRT = 24	[37]
	Naphthylene	82	GSBR, aerobic, SRT = 17, HRT = 24	[37]
	Tetracyclines, Sulfamethoxazole, Ampicillin	62.5–80.9	Upflow anaerobic sludge bed (UASB) - SRT = 6 months, HRT = 12 hours, Temperature = 35 °C with anoxic-oxic tank - SRT = 15 days, HRT = 12 hours	[2]
Antibiotics	SD, SMX, SM2, SDM, SCP, TMP, TC, OTC, CTC, DOXY, CIP, ENR, OFL, ERY, ROX, TYL, AMP, CLX	80.7	Upflow anaerobic sludge bed (UASB) - SRT = 6 months, HRT = 12 hours, Temperature = 35 °C	[2]
		43.4	Anoxic-oxic (A/O)- SRT = 15 days, HRT = 12 hours.	[2]
		28.6	Secondary clarifier - HRT =12.8 hours	[2]
		52.4	Flocculation - HRT of 2 hours, Aluminum sulfate at 2.5–5 g l dosages	[2]
	Enrofloxacin	73.9	Upflow anaerobic sludge bed (UASB) - SRT = 180, HRT = 12 hours, Temperature = 35 °C	[2]
	SCP, TMP, SDM, DOXY, CTC, CIP, ENR, OFL, ETM, AMP, CLX	85	Upflow anaerobic sludge bed (UASB) - SRT = 180, HRT = 12 hours, Temperature = 35 °C	[2]
	SCP, DOXY, CTC, ETM	77	Anoxic-oxic (A/O) - SRT = 15 days, HRT = 12 hours	[2]
	OTC,	83.7	Flocculation - HRT of 2 hours, Aluminum sulfate at 2.5–5 g l dosages	[2]
	DOXY	91	Flocculation - HRT of 2 hours, Aluminum sulfate at 2.5–5 g l dosages	[2]
	ROX, CLX	100	Flocculation - HRT of 2 hours, Aluminum sulfate at 2.5–5 g l dosages	[2]

Antidepressant	Amitriptyline	30	Full-scale MBR	[26]
Alkylphenol hormones	4-nonylphenols	70	Full-scale activated sludge, SRT = 1.7–10 days, HRT = 1.2–14 hours	[38]
	4-tert-octylphenol	<50	Full-scale activated sludge, SRT = 1.7–10 days, HRT = 1.2–14 hours	[38]
Beta blocker	Atenolol	92	Full-scale MBR	[26]
Lipid regulator	Bezafibrate	−38–100	A combination of different technologies for each of the five sampled AST processes	[33]
	Gemfibrozil	90	Full-scale MBR	[26]
Plasticizers	Bisphenol - A	<50	Full-scale activated sludge, SRT = 1.7–10 days, HRT = 1.2–14 hours	[38]
	Diethyl phthalate	59–79	20% higher removal (above 59%) was from membrane bioreactors	[39]
Hormones and endocrine disrupting compounds				
Natural estrogens	Estrone	>95	Full-scale membrane bioreactor (MBR)	[26]
	Estriol	92	UV/H_2O_2 oxidation at pH = 3, UV wavelength = 253.7 nm, UV fluence rate = 0.069 mW cm^{-2}, H_2O_2 dosage = 0.5 mol l^{-1}	[40]
	17β-estradiol	>98	SRT = 11–13 days	[41]
	17α-ethinylestradiol	90	SRT = 11–13 days	[41]
Natural androgens	Testosterone	96	Full-scale activated sludge, SRT = 1.7–10 days, HRT = 1.2–14 hours	[38]
	Androsterone	100	Full-scale MBR	[26]
Artificial synthetic estrogens/androgens				
	Phytoestrogens	73–88	Activated sludge with primary and secondary clarifiers	[42]

11.3.2 Contaminants Removal by Constructed Wetlands (CWs) and Other Treatment Systems

Natural and constructed wetlands (CWs) are used for water and wastewater treatments. Some could be unplanted while others are planted, typically to *Typha angustifolia* and/or *Phragmites Australis*. Their operating and maintenance costs are low. They do not need external sources of energy for their operations, rendering them a good low-tech option. They require more land than other wastewater treatment systems, which is a drawback where land is scarce.

Under wetland systems, removal of contaminants occurs through complex physical, chemical, and microbial interactions [45, 46], including biodegradation, sorption, sedimentation, microbial, and plant uptake and phytovolatilization [46]. Attributable to the complex nature of these removal processes under this technology, it is reportedly challenging to determine the primary removal mechanism for a group(s) of contaminants [46].

11.3.2.1 Antibiotic Removal

The performance of CWs on antibiotic removal was extensively studied (Table 11.3). However, due to variations of operating parameters of wetlands and contaminant properties, a wide range of removal efficiency has been reported (Table 11.3). A wetland microcosm study [46] assessed the removal of four pharmaceutical compounds (carbamazepine, diclofenac, ibuprofen, and naproxen). The study indicated the efficient removal of ibuprofen (80%) and naproxen (91%) in a planted wetland. The removal of two other compounds was lower (60% and below) in the unplanted wetland. In the same study, carbamazepine and diclofenac were inadequately removed (55% and below) regardless of the CW being planted or not. These compounds are associated with higher hydrophobicity; the low removal efficiency is attributed to this characteristic. Despite high removal efficiencies recorded for trimethoprim, acetaminophen, sulfamethoxazole, among others, lincomycin was not removed from either the planted or unplanted CW [48].

11.3.2.2 Removal of Hormones and EDCs by CWs

Hormone and EDC removal vary in CWs. Over 99% of Bisphenol A was removed in a hybrid CW planted with *P. Australis* [51]. In a comparison of various wastewater treatment processes for the removal of EDCs, the concentrations of E2, E3, and NP were remarkably reduced under CW [54]. The study also indicates that removal efficiency is season dependent.

11.3.3 Removal Through Advanced Oxidation Processes (AOPs)

Electrochemical oxidation, ionizing radiation, ozonation, Fenton and Fenton-like processes, photocatalytic oxidation, sulfate-radical oxidation, and ultrasound are some of the AOPs involved in contaminant removal from both water and wastewaters [55–57]. In brief, AOPs work through the generation of highly active radical oxidizers, attacking organic pollutants. Predominantly, catalysts are added, enhancing and driving the removal processes [57, 58]. In Fenton and Fenton-like reactions, hydroxy (OH) is generated through activating hydrogen peroxide (H_2O_2) [59]. Semiconductors are employed for light photon absorption in the photocatalysts processes. Electrons (e^-) and holes (h^+) are generated when these electrons are excited; they then react with electron acceptors leading to generation of O_2^- and $\cdot OH$ [60]. During ozonation, ozone is decomposed to generate radical oxidizers (O_2^- and $\cdot OH$) [61], while sulfate-based AOPs developed where sulfate radicals (SO_4^-) are generated. This chapter focuses on the efficiencies of each of these contaminant removals from wastes and little on the details of each of these AOP working principles.

Table 11.3 Contaminant compounds removal efficiency through constructed wetlands (CWs) and additional selected treatment systems.

Treatment technology	Pollutant	Removal efficiency (%)	Operating conditions	References
Constructed wetland (unplanted)	Tetracycline	38	Laboratory experimentation: swine wastewater treated.100 ug l of chemical was initially added	[47]
	Enrofloxacin	57		
	Carbamazepine	29	Horizontal subsurface flow design, sand filters. HRT = 4 days, Depth = 0.6 m,	[46]
	Ibuprofen	60	Removal was mainly by sorption	
	Diclofenac	41–47		
	Naproxen	52		
	Doxycycline	79		[48]
	Amoxicillin	6		
	Clarithromycin	32		
	Trimethoprim	95		
	Sulfamethoxazole	73		
Constructed wetland (planted)	Tetracycline	75–94	Laboratory experimentation: *Phragmites Australis* planted. Duration = 7 days, 100 ug l of chemical added initially, Removal was mainly by the plants through adsorption on roots, plant uptake, and degradation	[45, 47]
	Enrofloxacin	94–98		
	Ceftiofur	56		
		95–100	Microcosm experimentation, *Phragmites Australis* planted.	[49]
			Kept under greenhouse conditions Depth = 0.3 m, HRT = 20 weeks	
	Carbamazepine	28	Horizontal subsurface flow design, sand filters. HRT = four days, Depth = 0.6 m, three-year matured *Typha angustifolia*,	[46]
			Removal was mainly by sorption	

(Continued)

Table 11.3 (Continued)

Treatment technology	Pollutant	Removal efficiency (%)	Operating conditions	References
	Naproxen	91		
	Ibuprofen	71–80		
	Diclofenac	48–55		
		86–98	Secondary wastewater into a pond (depth = 1 m, HRT = 4 days) then into a surface flow CW of depth = 0.2, HRT = 8.5 days, *Typha angustifolia* and *Phragmites Australis*	[50]
	Triclosan	74–93		
	Oxybenzone	42–94		
	Tonalide	62–97		
	Tributyl phosphate	26–80		
	Diazinon	10–82		
	Terbutrin	52–73		
	Doxycycline	63–75	*Typha angustifolia* planted	[48]
		62–71	*Phragmites Australis* planted	
	Amoxicillin	39–45	*Typha angustifolia* planted	
		21–27	*Phragmites Australis* planted	
	Ampicillin	29–32	*Typha angustifolia*. No removal *in Phragmites Australis* planted CW	
	Clarithromycin	22–39	*Typha angustifolia* planted	[48]
		11–31	*Phragmites Australis* planted	
	Erythromycin	64	*Phragmites Australis* No removal observed in *Typha angustifolia* planted CW	
	Lincomycin	Not removed	Not removed in both plant species	
	Trimethoprim	65–92	*Typha angustifolia* planted	
		82–99	*Phragmites Australis* planted	
	Sulfamethoxazole	65–80	*Typha angustifolia* planted	
		59–87	*Phragmites Australis* planted	

	Acetaminophen	99	Hybrid CW: wastewater pumped through three units: vertical (VF) subsurface to horizontal (HF) subsurface to free water surface flow. VF & HF were planted with *Phragmites australis*, depth = 2.3, while a mixture of plant species planted in free surface flow, depth = 0.3 m, HRT = 5.1 days	[51]
	Bisphenol A	>99		
Aerobic/ Anaerobic digestion	Estrone	<1	Upflow anaerobic packed bed biofilm reactor (UAnPBBR)	[52]
	1 7β-estradiol	<1		
	17α-ethynylestradiol	2		
	Caffeine	11		
	Estrone	99	Aerobic trickling filter biofilm reactor (TF)	
	1 7β-estradiol	99		
	17α-ethynylestradiol	20		
	Caffeine	92		
	Naproxen	99	Aerobic membrane bioreactor (AeMBR), MLSS = 4.87 g l, HRT = 24 hours, SRT = 50 days, Temperature = 22 °C, Dissolved oxygen = 2.5 mg l	[53]
	Amitriptyline	80		
	4-tert-octyphenol	>95		
	Atrazine	7		
	Ketoprofen	95		
	Bisphenol A	95		
	Naproxen	53	Anaerobic membrane bioreactor (AnMBR), MLSS = 15.42 g l, HRT = 48 hours, SRT = 250 days, Temperature = 35 °C	
	Amitriptyline	99		
	4-tert-octyphenol	>95		
	Atrazine	35		
	Ketoprofen	24		
	Bisphenol A	30		

Various AOP performances on pollutant removal efficiencies are reviewed and presented in Table 11.4. Several technologies are highly efficient although their efficacy still varies among compounds. Treatment technology operation conditions and associated catalysts are major drivers attributed to specific magnitudes of contaminants removal. The removal of contaminants by AOPs presented in the literature indicated a wide variation based on treatment technology and compound type.

11.3.3.1 Antibiotics Removal

Antibiotic removal from pharmaceutical wastewater using AOPs indicated a wide range of removal efficiencies [2]. Among the 18 antibiotics studied in four AOPs (UV, ozonation, Fenton, and Fenton/UV), the removal magnitude ranged from as low as <30–100%. UV was not as effective as the other technologies. In a study involving UV irradiation alone, the removal efficiency for the antibiotic sulfamethazine was 68.6% [63]. This was 26–30% lower than the UV operated with iron catalysts and/or H_2O_2.

The other three technologies (Fenton, ozonation, and Fenton/UV) removed 99% of tetracycline and oxytetracycline. They also performed well in removing additional antibiotics (Table 11.4) [2]. A heterogeneous Fenton oxidation process could remove >90% of the tetracycline from the wastewater [62]. In the same study, the H_2O_2 alone removed only 20% of tetracycline. Most technologies presented in Table 11.4 removed (with >70%) a wide range of antibiotics from wastes.

11.3.3.2 Hormones and Endocrine Disrupting Compounds (EDCs) Removal

Using UV/H_2O_2 oxidation, regarded as one of the AOPs, on activated sludge had a 92% efficiency on estriol hormone (E2) removal under the stated operating conditions presented in Table 11.3. This was the highest efficiency observed as driven by a pH of 3. As the pH increased to levels 5 and 7, the removal efficiency dropped to 89%. Other hormone removals were also assessed; E1, EE2, E3, and BPA had 97, 95, 94 and 89% removal, respectively. Their removal efficiency dropped significantly when exposing the process to higher (5 and 7) pH levels [40]. Of importance was also the combination of UV and H_2O_2 to enhance the process to outperform the removal efficiencies under UV irradiation or H_2O_2 oxidation individually. The reported technologies (Table 11.4) employed to degrade/remove EDCs and/or hormones, indicated a significant removal of these compounds with an efficiency of 98% when conditions were optimized. It is not clear, though, how effective a single set of conditions would be in a wide range of compounds.

11.4 Conclusion

The presence of pharmaceuticals, personal care products, and EDCs in the environment pose potential health and ecosystem hazards. Several factors govern the removal of these contaminants, ranging from technology operation conditions and technology type to contaminant properties, rendering it challenging to choose an option in treating wastewaters. Such wastewaters contain multiple emerging contaminants and would not require a single treatment option for their removal. Therefore, one technology may best remove some contaminants but is not effective for another group. There is no one-size-fits-all scenario concerning technologies required for contaminant removal from various wastes. AOPs are a promising option; however, cost and complexity are limitations.

Since no treatment process can be deemed the best in removing all contaminant groups, measures towards mitigating their environmental entry are critical. Research on sustainable technologies limiting compounds originating from using the afore-mentioned products is crucial.

Table 11.4 Contaminant compounds removal efficiency through advanced oxidation processes (AOPs).

Contaminant type and class	Compounds	Removal efficiency (%)	Technological process used	Operating conditions	References
Pharmaceutical and personal care products					
Antibiotics	Tetracycline	>90	Heterogeneous Fenton Oxidation process: $Fe^0/CeO_2 + H_2O_2$	Initial TC = 100 mg l^{-1}, catalyst = 0.1 g l^{-1}, H_2O_2 = 100 mmol l^{-1}, pH_0 = 5.8, T = 26 °C)	[62]
	Tetracycline,	65.7–97.2	UV/H_2O_2	[Fe^0] 0.3 mM, [H_2O_2] 0.3 mM and UV 6 W, Time 30 minutes, pH = 7	[63]
	Oxytetracycline, Sulfamethazine	85–99	$UV/Fe^0/H_2O_2$		
	Sulfadiazine	98.8	$UV/Fe^0/H_2O_2$	[Fe^0] 0.3 mM, [H_2O_2] 0.3 mM and UV 6 W, Time 60 minutes, pH = 7	
	Sulfamethoxazole	100	Fenton Oxidation: Fe@MesoC catalyst	pH = 4.0; initial SMX = 20 mg l^{-1}; catalyst = 0.2 g l^{-1}; H_2O_2 = 3 mM; temperature = 25 °C	[64]
	Sulfamethazine	<10	Fe^0/H_2O_2	[Fe^0] 0.3 mM, [H_2O_2] 0.3 mM and UV 6 W, Time 60 minutes, pH = 7	[63]
		<20	pre-Fe^0/H_2O_2		[63]
		68.6	UV irradiation		[63]
		94.3	UV/H_2O_2		[63]
		99.4	$UV/Fe^0/H_2O_2$		[63]
		100	UV/pre-Fe^0/H_2O_2		[63]
	Metacycline	86.4	Microwave-assisted Fenton reaction (MAFR)	$CuCo_2O_4$ catalyst =10 mg, H_2O_2 = 500 ul^{-1}, 50 mg l^{-1} of MTC, 90 T°C, Time = 4 minutes, Energy of MAFR was 17.3 kJ mol^{-1}	[65]
	Lincomycin	94	Fenton-like reaction: GFe0.5/H_2O_2	GFe0.5 = 0.01 g l^{-1}, LCM = 20 mg l^{-1}, H_2O_2 = 1 mM, T = 25 °C, pH = 5.8, Time = 90 minutes	[66]
	Metronidazole	84.3	Heterogeneous Fenton-like process	0.1 g l^{-1} $FeNi_3/SiO_2$, pH =7, H_2O_2–150 mg l^{-1} and metronidazole = 20 mg l^{-1}	[67]
	TC, OTC	~99	Ozonation, Fenton, Fenton/UV	500 ml wastewater, At 25 ± 2 °C with continuous stirring at 300 rpm for 2 hours, pH = 7.0	[2]

Table 11.4 (Continued)

Contaminant type and class	Compounds	Removal efficiency (%)	Technological process used	Operating conditions	References
	TC, OTC	~65	UV irradiation	500 ml wastewater, At 25 ± 2 °C with continuous stirring at 300 rpm, UV dose of 165 mJ cm^{-2} for 2 hours, pH = 7.0	[2]
	SD, SMX, SM2, SDM, SCP, TMP, TC, OTC, CTC, DOXY, CIP, ENR, OFL, ERY, ROX, TYL, AMP, CLX	<30–100	UV irradiation	500 ml wastewater, At 25 ± 2 °C with continuous stirring at 300 rpm for 2 hours, pH = 7.0	[2]
		75–100	Ozonation	500 ml wastewater, at 25 ± 2 °C with continuous stirring at 300 rpm for 2 hours, pH = 7.0, Ozone concentration of 50 mg l	[2]
		70–100	Fenton	500 ml wastewater, At 25 ± 2 °C with continuous stirring at 300 rpm, $FeSO_4.7H_2O$, H_2O_2 (30% w/w) solution at Fe^{2+}: H_2O_2 molar ratio of 1 : 10, pH = 3 for 2 hours	[2]
		80–99	AOP - Fenton/UV	500 ml wastewater, Continuous stirring at 300 rpm, Temperature = 25 ± 2 °C, $FeSO_4.7H_2O$, H_2O_2 (30% w/w) solution at Fe^{2+}: H_2O_2 molar ratio of 1 : 10, pH = 3 for 2 hours, 165 mJ cm^{-2} UV exposure	[2]
Phthalate esters-EDCs	Diethyl phthalate	>98	UV/peroxymonosulfate (PMS) oxidation	254 nm UV/PMS, pH = 7, $[PMS]_0$ = 10 mM, $[DEP]_0$ = 0.4 mM, $[MeOH]_0$ = 100 mM, $[TBA]_0$ = 30 mM, Time = 10 minutes	[68]
	Di-(2-ethylhexyl) phthalate	90	UV-activated peroxymonosulfate (PMS) oxidation	Temperature = 20 °C, pH = 7, and [PMS]: [DEHP] = 100 : 1, 10 mg l^{-1} DEHP, Time = 20 minutes	[69]
Herbicides	2,4-dichloropheno-xiacetic acid (2,4-D)	100	Elecro-chemical Advanced oxidation Process	[2,4-D]: 20 mg l^{-1}, pH = 3, Time = 20 minutes, Applied voltage: 4 V, Volume: 100 ml, [Na2SO4]: 50 mM	[70]
Phenols	4-nitrophenol	99	Elecrocatalyst Advanced Oxidation Process	Co0.5@C + PMS (30 mg, 0.5 g l), Co_3O_4@C anode, SnO_2/CC cathode, Time = 60 minutes	[71]

References

1 Gavrilescu, M., Demnerová, K., Aamand, J. et al. (2015). Emerging pollutants in the environment: present and future challenges in biomonitoring, ecological risks and bioremediation. *New Biotechnology* 32 (1): 147–156.

2 Hou, J., Chen, Z., Gao, J. et al. (2019). Simultaneous removal of antibiotics and antibiotic resistance genes from pharmaceutical wastewater using the combinations of up-flow anaerobic sludge bed, anoxic-oxic tank, and advanced oxidation technologies. *Water Research* 159: 511–520.

3 Ávila, C., Matamoros, V., Reyes-Contreras, C. et al. (2014). Attenuation of emerging organic contaminants in a hybrid constructed wetland system under different hydraulic loading rates and their associated toxicological effects in wastewater. *Science of the Total Environment* 470–471: 1272–1280.

4 Yan, C., Yang, Y., Zhou, J. et al. (2013). Antibiotics in the surface water of the Yangtze Estuary: occurrence, distribution and risk assessment. *Environmental Pollution* 175: 22–29.

5 Pal, A., He, Y., Jekel, M. et al. (2014). Emerging contaminants of public health significance as water quality indicator compounds in the urban water cycle. *Environment International* 71: 46–62.

6 Tran, N.H., Reinhard, M., and Gin, K.Y.H. (2018). Occurrence and fate of emerging contaminants in municipal wastewater treatment plants from different geographical regions-a review. *Water Research* 133: 182–207.

7 Zaied, B.K., Rashid, M., Nasrullah, M. et al. (2020). A comprehensive review on contaminants removal from pharmaceutical wastewater by electrocoagulation process. *Science of the Total Environment* 726: 138095.

8 Wilkinson, J., Hooda, P.S., Barker, J. et al. (2017). Occurrence, fate and transformation of emerging contaminants in water: an overarching review of the field. *Environmental Pollution* 231: 954–970.

9 García, J., García-Galán, M.J., Day, J.W. et al. (2020). A review of emerging organic contaminants (EOCs), antibiotic resistant bacteria (ARB), and antibiotic resistance genes (ARGs) in the environment: increasing removal with wetlands and reducing environmental impacts. *Bioresource Technology* 307: 123228.

10 Gogoi, A., Mazumder, P., Tyagi, V.K. et al. (2018). Occurrence and fate of emerging contaminants in water environment: a review. *Groundwater for Sustainable Development* 6: 169–180.

11 Tran, N.H., Chen, H., Reinhard, M. et al. (2016). Occurrence and removal of multiple classes of antibiotics and antimicrobial agents in biological wastewater treatment processes. *Water Research* 104: 461–472.

12 Minh, T.B., Leung, H.W., Loi, I.H. et al. (2009). Antibiotics in the Hong Kong metropolitan area: ubiquitous distribution and fate in Victoria Harbour. *Marine Pollution Bulletin* 58 (7): 1052–1062.

13 Matsuo, H., Sakamoto, H., Arizono, K. et al. (2011). Behavior of pharmaceuticals in waste water treatment plant in Japan. *Bulletin of Environmental Contamination and Toxicology* 87 (1): 31–35.

14 Guerra, P., Kim, M., Shah, A. et al. (2014). Occurrence and fate of antibiotic, analgesic/anti-inflammatory, and antifungal compounds in five wastewater treatment processes. *Science of the Total Environment* 473–474: 235–243.

15 Papageorgiou, M., Kosma, C., and Lambropoulou, D. (2016). Seasonal occurrence, removal, mass loading and environmental risk assessment of 55 pharmaceuticals and personal care

products in a municipal wastewater treatment plant in central Greece. *Science of the Total Environment* 543: 547–569.

16 Dinh, Q.T., Moreau-Guigon, E., Labadie, P. et al. (2017). Occurrence of antibiotics in rural catchments. *Chemosphere* 168: 483–490.

17 Yang, X., Flowers, R.C., Weinberg, H.S. et al. (2011). Occurrence and removal of pharmaceuticals and personal care products (PPCPs) in an advanced wastewater reclamation plant. *Water Research* 45 (16): 5218–5228.

18 Karthikeyan, K.G. and Meyer, M.T. (2006). Occurrence of antibiotics in wastewater treatment facilities in Wisconsin, USA. *Science of the Total Environment* 361 (1): 196–207.

19 Yang, S., Hai, F.I., Price, W.E. et al. (2016). Occurrence of trace organic contaminants in wastewater sludge and their removals by anaerobic digestion. *Bioresource Technology* 210: 153–159.

20 Gonzalez-Gil, L., Papa, M., Feretti, D. et al. (2016). Is anaerobic digestion effective for the removal of organic micropollutants and biological activities from sewage sludge? *Water Research* 102: 211–220.

21 Barber, L.B., Loyo-Rosales, J.E., Rice, C.P. et al. (2015). Endocrine disrupting alkylphenolic chemicals and other contaminants in wastewater treatment plant effluents, urban streams, and fish in the Great Lakes and Upper Mississippi River Regions. *Science of the Total Environment* 517: 195–206.

22 Di Marcantonio, C., Chiavola, A., Dossi, S. et al. (2020). Occurrence, seasonal variations and removal of organic micropollutants in 76 wastewater treatment plants. *Process Safety and Environmental Protection* 141: 6172.

23 Wei, R., Ge, F., Huang, S. et al. (2011). Occurrence of veterinary antibiotics in animal wastewater and surface water around farms in Jiangsu Province, China. *Chemosphere* 82 (10): 1408–1414.

24 Bartelt-Hunt, S., Snow, D.D., Damon-Powell, T. et al. (2011). Occurrence of steroid hormones and antibiotics in shallow groundwater impacted by livestock waste control facilities. *Journal of Contaminant Hydrology* 123 (3): 94–103.

25 Sim, W.-J., Lee, J.-W., Lee, E.-S. et al. (2011). Occurrence and distribution of pharmaceuticals in wastewater from households, livestock farms, hospitals and pharmaceutical manufactures. *Chemosphere* 82 (2): 179–186.

26 Phan, H.V., Hai, F.I., McDonald, J.A. et al. (2015). Nutrient and trace organic contaminant removal from wastewater of a resort town: comparison between a pilot and a full scale membrane bioreactor. *International Biodeterioration and Biodegradation* 102: 40–48.

27 Bueno, M.J.M., Gomez, M.J., Herrera, S. et al. (2012). Occurrence and persistence of organic emerging contaminants and priority pollutants in five sewage treatment plants of Spain: two years pilot survey monitoring. *Environmental Pollution* 164: 267–273.

28 Bolong, N., Ismail, A.F., Salim, M.R. et al. (2009). A review of the effects of emerging contaminants in wastewater and options for their removal. *Desalination* 239 (1): 229–246.

29 Heo, J., Kim, S., Her, N. et al. (2020). Chapter 5 – Removal of contaminants of emerging concern by FO, RO, and UF membranes in water and wastewater. In: *Contaminants of Emerging Concern in Water and Wastewater* (eds. A.J. Hernández-Maldonado and L. Blaney), 139–176. Butterworth-Heinemann.

30 Liu, Z.-H., Kanjo, Y., and Mizutani, S. (2009). Removal mechanisms for endocrine disrupting compounds (EDCs) in wastewater treatment – physical means, biodegradation, and chemical advanced oxidation: a review. *Science of the Total Environment* 407 (2): 731–748.

31 Shreve, M.J. and Brennan, R.A. (2019). Trace organic contaminant removal in six full-scale integrated fixed-film activated sludge (IFAS) systems treating municipal wastewater. *Water Research* 151: 318–331.

32 Petrie, B., McAdam, E.J., Lester, J.N. et al. (2014). Assessing potential modifications to the activated sludge process to improve simultaneous removal of a diverse range of micropollutants. *Water Research* 62: 180–192.

33 Li, Y., Niu, X., Yao, C. et al. (2019). Distribution, removal, and risk assessment of pharmaceuticals and their metabolites in five sewage plants. *International Journal of Environmental Research and Public Health* 16: 4729.

34 Falås, P., Longrée, P., la Cour Jansen, J. et al. (2013). Micropollutant removal by attached and suspended growth in a hybrid biofilm-activated sludge process. *Water Research* 47 (13): 4498–4506.

35 Nakada, N., Tanishima, T., Shinohara, H. et al. (2006). Pharmaceutical chemicals and endocrine disrupters in municipal wastewater in Tokyo and their removal during activated sludge treatment. *Water Research* 40 (17): 3297–3303.

36 Göbel, A., McArdell, C.S., Joss, A. et al. (2007). Fate of sulfonamides, macrolides, and trimethoprim in different wastewater treatment technologies. *Science of the Total Environment* 372 (2): 361–371.

37 Ofman, P., Struk-Sokołowska, J., Skoczko, I. et al. (2020). Alternated biodegradation of naphthalene (NAP), acenaphthylene (ACY) and acenaphthene (ACE) in an aerobic granular sludge reactor (GSBR). *Journal of Hazardous Materials* 383: 121184.

38 Drewes, J.E., Hemming, J., Ladenburger, S.J. et al. (2005). An assessment of endocrine disrupting activity changes during wastewater treatment through the use of bioassays and chemical measurements. *Water Environment Research* 77 (1): 12–23.

39 Cases, V., Alonso, V., Argandoña, V. et al. (2011). Endocrine disrupting compounds: a comparison of removal between conventional activated sludge and membrane bioreactors. *Desalination* 272 (1): 240–245.

40 Zhang, A. and Li, Y. (2014). Removal of phenolic endocrine disrupting compounds from waste activated sludge using UV, H_2O_2, and UV/H_2O_2 oxidation processes: effects of reaction conditions and sludge matrix. *Science of the Total Environment* 493: 307–323.

41 Andersen, H., Siegrist, H., Halling-Sørensen, B. et al. (2003). Fate of estrogens in a municipal sewage treatment plant. *Environmental Science & Technology* 37 (18): 4021–4026.

42 Laganà, A., Bacaloni, A., De Leva, I. et al. (2004). Analytical methodologies for determining the occurrence of endocrine disrupting chemicals in sewage treatment plants and natural waters. *Analytica Chimica Acta* 501 (1): 79–88.

43 Kasprzyk-Hordern, B., Dinsdale, R.M., and Guwy, A.J. (2009). The removal of pharmaceuticals, personal care products, endocrine disruptors and illicit drugs during wastewater treatment and its impact on the quality of receiving waters. *Water Research* 43 (2): 363–380.

44 Singer, H., Müller, S., Tixier, C. et al. (2002). Occurrence and fate of a widely used biocide in the aquatic environment: field measurements in wastewater treatment plants, surface waters, and lake sediments. *Environmental Science & Technology* 36 (23): 4998–5004.

45 Carvalho, P.N., Araújo, J.L., Mucha, A.P. et al. (2013). Potential of constructed wetlands microcosms for the removal of veterinary pharmaceuticals from livestock wastewater. *Bioresource Technology* 134: 412–416.

46 Zhang, D.Q., Tan, S.K., Gersberg, R.M. et al. (2011). Removal of pharmaceutical compounds in tropical constructed wetlands. *Ecological Engineering* 37 (3): 460–464.

47 Carvalho, P.N., Basto, M.C.P., and Almeida, C.M.R. (2012). Potential of Phragmites Australis for the removal of veterinary pharmaceuticals from aquatic media. *Bioresource Technology* 116: 497–501.

48 Hijosa-Valsero, M., Fink, G., Schlüsener, M.P. et al. (2011). Removal of antibiotics from urban wastewater by constructed wetland optimization. *Chemosphere* 83 (5): 713–719.

49 Santos, F., Almeida, C.M.R., Ribeiro, I. et al. (2019). Potential of constructed wetland for the removal of antibiotics and antibiotic resistant bacteria from livestock wastewater. *Ecological Engineering* 129: 45–53.

50 Matamoros, V. and Salvadó, V. (2012). Evaluation of the seasonal performance of a water reclamation pond-constructed wetland system for removing emerging contaminants. *Chemosphere* 86 (2): 111–117.

51 Ávila, C., Bayona, J.M., Martín, I. et al. (2015). Emerging organic contaminant removal in a full-scale hybrid constructed wetland system for wastewater treatment and reuse. *Ecological Engineering* 80: 108–116.

52 Lin, S., Rong, K., Lamichhane, K.M. et al. (2020). Anaerobic-aerobic biofilm-based digestion of chemical contaminants of emerging concern (CEC) and pathogen indicator organisms in synthetic wastewater. *Bioresource Technology* 299: 122554.

53 Liu, W., Song, X., Huda, N. et al. (2020). Comparison between aerobic and anaerobic membrane bioreactors for trace organic contaminant removal in wastewater treatment. *Environmental Technology & Innovation* 17: 100564.

54 Qiang, Z., Dong, H., Zhu, B. et al. (2013). A comparison of various rural wastewater treatment processes for the removal of endocrine-disrupting chemicals (EDCs). *Chemosphere* 92 (8): 986–992.

55 Dewil, R., Mantzavinos, D., Poulios, I. et al. (2017). New perspectives for advanced oxidation processes. *Journal of Environmental Management* 195: 93–99.

56 Wang, J. and Zhuan, R. (2020). Degradation of antibiotics by advanced oxidation processes: an overview. *Science of the Total Environment* 701: 135023.

57 Ghernaout, D., Elboughdiri, N., and Ghareba, S. (2020). Fenton technology for wastewater treatment: dares and trends. *Open Access Library Journal* 7: 1–26.

58 Niu, L., Wei, T., Li, Q. et al. (2020). Ce-based catalysts used in advanced oxidation processes for organic wastewater treatment: a review. *Journal of Environmental Sciences* 96: 109–116.

59 He, J., Yang, X., Men, B. et al. (2016). Interfacial mechanisms of heterogeneous fenton reactions catalyzed by iron-based materials: a review. *Journal of Environmental Sciences* 39: 97–109.

60 Singh, P., Sharma, K., Hasija, V. et al. (2019). Systematic review on applicability of magnetic iron oxides–integrated photocatalysts for degradation of organic pollutants in water. *Materials Today Chemistry* 14: 100186.

61 Faria, P.C.C., Órfão, J.J.M., and Pereira, M.F.R. (2008). A novel ceria–activated carbon composite for the catalytic ozonation of carboxylic acids. *Catalysis Communications* 9 (11): 2121–2126.

62 Zhang, N., Chen, J., Fang, Z. et al. (2019). Ceria accelerated nanoscale zerovalent iron assisted heterogenous Fenton oxidation of tetracycline. *Chemical Engineering Journal* 369: 588–599.

63 Pan, Y., Zhang, Y., Zhou, M. et al. (2019). Enhanced removal of antibiotics from secondary wastewater effluents by novel UV/pre-magnetized Fe^0/H_2O_2 process. *Water Research* 153: 144–159.

64 Tang, J. and Wang, J. (2018). Fenton-like degradation of sulfamethoxazole using Fe-based magnetic nanoparticles embedded into mesoporous carbon hybrid as an efficient catalyst. *Chemical Engineering Journal* 351: 1085–1094.

65 Qi, Y., Mei, Y., Li, J. et al. (2019). Highly efficient microwave-assisted fenton degradation of metacycline using pine-needle-like $CuCO_2O_4$ nanocatalyst. *Chemical Engineering Journal* 373: 1158–1167.

66 Ouyang, Q., Kou, F., Tsang, P.E. et al. (2019). Green synthesis of Fe-based material using tea polyphenols and its application as a heterogeneous Fenton-like catalyst for the degradation of lincomycin. *Journal of Cleaner Production* 232: 1492–1498.

67 Nasseh, N., Taghavi, L., Barikbin, B. et al. (2019). FeNi3/SiO2 magnetic nanocomposite as an efficient and recyclable heterogeneous Fenton-like catalyst for the oxidation of metronidazole in neutral environments: adsorption and degradation studies. *Composites Part B: Engineering* 166: 328–340.

68 Lei, Y., Lu, J., Zhu, M. et al. (2020). Radical chemistry of diethyl phthalate oxidation via UV/Peroxymonosulfate process: roles of primary and secondary radicals. *Chemical Engineering Journal* 379: 122339.

69 Huang, J., Li, X., Ma, M. et al. (2017). Removal of di-(2-ethylhexyl) phthalate from aqueous solution by UV/peroxymonosulfate: influencing factors and reaction pathways. *Chemical Engineering Journal* 314: 182–191.

70 Yang, W., Zhou, M., Oturan, N. et al. (2019). Enhanced activation of hydrogen peroxide using nitrogen doped graphene for effective removal of herbicide 2,4-D from water by iron-free electrochemical advanced oxidation. *Electrochimica Acta* 297: 582–592.

71 Zhu, M., Zhang, L., Liu, S. et al. (2020). Degradation of 4-nitrophenol by electrocatalysis and advanced oxidation processes using Co_3O_4@C anode coupled with simultaneous CO_2 reduction via SnO_2/CC cathode. *Chinese Chemical Letters* 31 (7): 1961–1965.

12

Effect of End Products on the Soil Microbial Communities

Biodegradable Waste Management in the Circular Economy: Challenges and Opportunities, First Edition.
Edited by Malgorzata Kacprzak, Eleonore Attard, Kari-Anne Lyng, Helena Raclavska, BalRam Singh, Eyob Tesfamariam, and Franck Vandenbulcke.

12.1

Impact of End Products on Microbial Communities with Special Reference to Pathogens

Daquan Sun[1], Eléonore Attard[2], and Rémy Guyoneaud[2]

[1] *Soil & Water Research Infrastructure, Biology Centre CAS, Na Sádkách, České Budějovice, Czech Republic*
[2] *Universite de Pau et des Pays de l'Adour, E2S UPPA, CNRS, IPREM, Pau, France*

12.1.1 Introduction

The recovery of waste and the production of an end product after treatment (composting, vermicomposting, anaerobic digestion (AD) with co-production of energy, for example) is a widespread technique for waste re-use. Land application of end products (biosolids, manure, dairy slurry, digestates, etc.) provides valuable nutrients to soil while offering a solution to recycle products that would have been landfilled. It participates in the circular economy of the waste sector through their use as organic materials and nutrients for plant production [1]. This report from the European Commission points out the risks in relation to this use, in particular in relation to metals (lead, mercury, and copper as examples) and to organic compounds such as nonylphenols, endocrine disruptors, PCBs, and PFAs. Nothing is given in this report about the risk from microorganisms.

In composts or digestates from AD, different microorganisms can be found depending on the treatment, i.e. thermic treatments or anaerobiosis. These microorganisms will be spread with the land application of the end product (Figure 12.1.1). The risks are considered depending on the concentration in the end product (Coliforms for example) and do not consider what is happening after the product is spread, apart from a crop production point of view. On the one hand, these end products may be sources of contaminants (pathogens) but they can also disrupt the proper functioning of biogeochemical cycles, including those of carbon and of course nitrogen, and few studies have dealt with these aspects until now (for recent surveys see references [2–7].

It is evident that in the first place the pathogenic risk is the one to be considered on soils that are intended for plant production. All of these end products carry pathogens to the soil that can affect human and animal health. The challenge is to assess and control the microbiological risk in the end products and in the environment, i.e. the soil, where pathogens can survive and grow. *Escherichia coli*, *Salmonella*, *Shigella*, *Clostridium*, and *Mycobacterium tuberculosis* are the major microbial pathogens encountered in sewage sludges and manures. However, a high diversity of others pathogens, eventually opportunistic, can also be recorded, such as *Streptococcus*, *Yersinia enterocolitica*, *Listeria monocytogenes*, *Campylobacter jejuni*, *Legionella pneumophila*, *Pseudomonas aeruginosa*, *Staphylococcus aureus*, and *Vibrio cholera* [8–11]. In activated sludges, a metagenomic study has revealed that up to 2.3% of the total bacterial community can be composed of human bacterial pathogens [12].

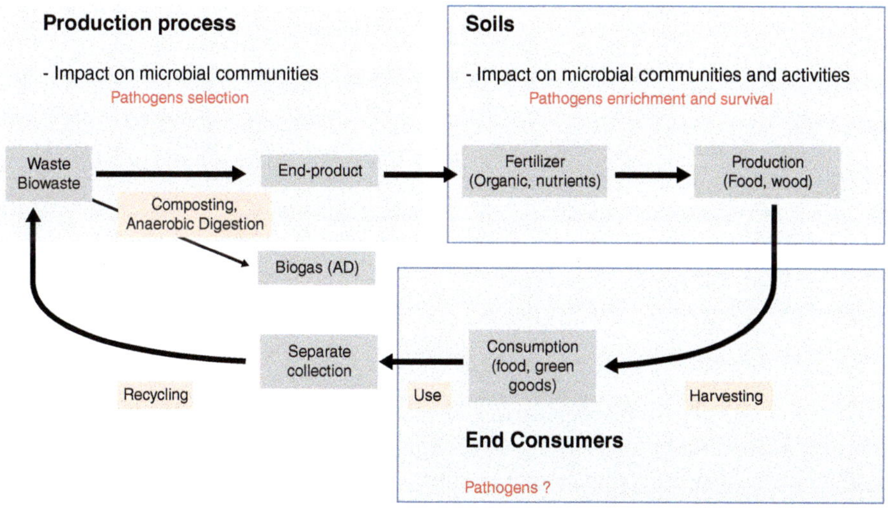

Figure 12.1.1 Risks with pathogens in the circular economy view of organic wastes.

To reduce health hazards, the EU directive 86/278/EEC on the use of sewage sludge in agriculture determines the basic requirements within the territory of Europe. Briefly, for microorganisms in sewage sludge, this directive stipulates that conventional treatments should achieve a reduction of 99.9% of *E. coli* with less than 1×10^3 CFU.g^{-1} (dry weight), less than 3×10^3 spores of *Clostridium perfringens* per gram (dry weight), and no *Salmonella spp.* detectable in 50 g (wet weight). Concerning compost and digestate, the EU Fertilizing Products Regulation (2019/1009) stipulates that five samples should contain less than 1×10^3 *E. coli* per gram or mL and no *Salmonella* should be detectable in 25 g. In contrast, for manure and dairy slurry, land spreading is authorized without any treatment and no regulation concerning microbiological parameters is required.

12.1.2 Pathogen Survival in End Products

12.1.2.1 Survival Depending on the Identity of Microorganisms

The probability to add bacterial pathogens able to survive in soil will firstly depend on their ability to survive and/or maintain in end products. The most important factors affecting survival in these end products are temperature, pH, water content, and UV light (sunlight exposure). This is true for the vast majority of bacteria, but for those producing spores, such as *Clostridium*, resistance to environmental factors will be stronger and their survival will be extended. Survival can also differ among non-spore-forming bacteria. Using a metagenomics survey, Ju et al. [13] have shown that among 83 bacterial pathogens, three of them (*Collinsella aerofaciens*, *Streptococcus salivarius*, and *Gordonia bronchialis*) are still detected after AD. Similarly, Goberna et al. [14] have observed that AD completely removed *E. coli* and *Salmonella* but not *Listeria*, which persisted at a significant level (1.7×10^4 CFU.g^{-1}). In another study performed after WWTP treatment and composting, survival of the four bacteria targeted was significantly different (*Salmonella* spp., *C. jejuni*, *E. coli*, and *C. perfringens*) [15]. Overall, these differences in survival among bacterial pathogens highlight the difficulty in finding suitable organisms to indicate contamination of end products. It should also be mentioned that there are possible technical issues of monitoring bacterial pathogens in end products. Indeed, bacteria can be present and viable but in a non-culturable state (called a VBNC state). This implies that some pathogens can be overlooked whereas there are present and able to re-grow when conditions are favorable again. This was confirmed by a qPCR survey [16], showing that after mesophilic and thermophilic digestion, *E. coli* was respectively one and five orders more abundant with a measurement by qPCR than by a standard culturing method. Survival can also be affected by the indigenous compost microorganisms (regardless of the physico-chemical parameters of the compost). As shown by Sidhu et al. [17], regrowth of *Salmonella* was more likely to occur in sterilized rather than non-sterilized compost.

12.1.2.2 Survival Depending on the Treatments Applied to the End Products

As mentioned before, the most important factors affecting bacterial survival in end products are temperature, pH, water content, and UV light (sunlight exposure). All these parameters varied due to the different treatments applied to end products. In this section, we will focus on the most commonly used treatments, i.e. AD and composting (excluding home composting).

12.1.2.2.1 Composting

Composting is widely used to produce organic fertilizer from waste and to reduce pathogen concentration. The inactivation of pathogens is mainly due to exposure to high temperatures during the thermophilic phase of the composting process (>55 °C) (see Chapter 9.1 for more details), but pathogens also have to compete with compost growing microorganisms. This led to unfavorable conditions for pathogens, which have to compete for nutrients and to resist antimicrobial compounds and NH_3 produced by the compost growing microorganisms [18]. Overall, composting of all type of waste is an effective method to reduce pathogens (except for spore-forming *Clostridium*). For instance, Deportes et al. [19] have observed a huge decrease of fecal coliforms from 10^8 to 10^2 CFU.g^{-1} dw and no detection of *Salmonella* spp. after 20 days of municipal solid waste composting. Similarly, in different livestock manures, composting has been shown to be an effective sanitation method where pathogen content can decrease until an undetectable level is reached [20–23]. This is the case for thermophilic composting; if not properly applied, pathogens may persist in insufficiently self-heated areas and re-grow at the end of the process. In addition, sanitation efficiencies may depend on the composting techniques used. Pereiro Neto et al. [24] have shown that aerated piles were more efficient than windrows to sanitize sewage sludge. However, contradictory results have also been reported with the disappearance of pathogens in windrow composting [25]. Actually, several parameters can affect the reduction of pathogens, such as the nature of the initial substrate, the water content [26], and the maturity of the compost [27]. For instance, using an *in vitro* experiment, Paniel et al. [27] have inoculated three bacterial pathogens (*L. monocytogenes*, *Salmonella infantis*, and *Enterococcus faecalis*) in composts collected at different stages of the composting process. Whereas in mature composts these bacteria did not survive, in not yet stabilized compost the bacteria were able to persist. This persistence in immature composts is particularly risky due to the potential regrowth of the bacterial pathogens to hazardous levels (e.g. reference [20]).

Another composting process is vermicomposting with the addition of earthworms (see Chapter 9.2 for more details on this process). This process is considered as an efficient way to reduce bacterial pathogens [28–33]. For instance, Aira et al. [32] and Rodríguez-Canché et al. [33] have shown a substantial decrease of bacterial pathogens in sludge and manure submitted to vermicomposting. However, few studies have focused on this topic and contradictory results are reported, as in the study by Monroy et al. [29]. These authors have observed an increase in total coliform numbers in vermireactors fed with high doses of pig slurry.

Mechanisms underlying the decrease of bacterial pathogen are not well known, but this is generally attributed to the direct and indirect effects of the addition of earthworms. Direct effects are due to mechanical disruption of bacteria through enzymatic digestion and antimicrobial substances produced by earthworms. This explains why different authors observed different efficiencies of sanitation depending on the species of earthworm used [34]. Indirect effects are thought to rely on stimulation of other microbial species, leading to competition and antagonism. Overall, several authors suggested that vermicomposting could be seen as a complementary method to mitigate drawbacks of composting, notably the problems of survival and regrowth of pathogens due to the lack of a homogeneous composting process.

12.1.2.2.2 Anaerobic Digestion

AD is a process through which organic substrates are degraded by a microbial consortium in the absence of oxygen. Four microbial communities perform successively hydrolysis, acidogenesis, acetogenesis, and methanogenesis (for more details see Chapter 8).

Several studies have reported inactivation of the cultivable microbial community, as reviewed by Jiang et al. [35] and Franke-Whittle and Insam [36]. A recent metagenomics survey, which has the advantage of taking into account cultivable and non-cultivable microorganisms, has confirmed AD efficiency; in this study Ju et al. [13] have observed a decrease of 70–95% of the pathogen abundance in sewage sludge treated by AD. However, different factors can affect AD efficiency in decreasing the pathogen load. As mentioned above in this chapter, the identity of bacteria can be important, with spore-forming bacteria able to survive many treatments including AD [37–39]. Temperature is another major factor. Indeed, AD can be run at mesophilic (25–37 °C) or thermophilic temperatures (55–65 °C). Overall thermophilic AD is more efficient in reducing pathogen concentrations but is also far less stable than mesophilic AD. In thermophilic conditions, pathogen inactivation is mainly due to exposure to high temperatures. In mesophilic conditions a cluster of different factors are involved, such as pH, substrate competition between pathogens and autochthonous microorganisms, and concentration of free ammonia and volatile fatty acids. This leads to a significant decrease with, for instance, in 10 full-scale biogas plants run in mesophilic conditions, a reduction of 2 log of Enterobacteriaceae (6.85×10^3 to 1.82×10^1) and 3 log of fecal coliforms (1.82×10^4 to 2.45×10^1) [40]. Behind these averaged values, efficiencies differed between the biogas plants, due to different pH values and ammonia concentrations.

AD can also have "side effects" on pathogens. We have discussed in a recent paper review [41] that, during AD, silver nanoparticles undergo chemical reactions (mainly sulfidation), leading to considerably less toxic effects for soil microorganisms once applied to agricultural lands. This suggests that the kind of treatments applied to the end products (with or without anaerobic conditions) may have an indirect impact on the ability of the soil autochthonous microflora to compete with the pathogens supplied.

12.1.3 Pathogen Survival in Soils

The direct consequence of pathogen survival in end products is a possible soil contamination through land spreading. Different factors influence the survival and the ability to regrow of bacterial pathogens; the main factors are now described.

12.1.3.1 Pathogen Survival Depending on the Identity of the Bacterial Pathogen and Competition with Autochthonous Microorganisms

Several studies have been conducted during the last few decades to assess the survival abilities between different bacterial species. They were performed using soil microcosms amended with end products (sludge, manure, etc.) inoculated with bacterial pathogens. For instance, Franz et al. [42] and Semenov et al. [43] found that *E. coli* and *Salmonella* declined progressively in soils amended with manure, but the *Salmonella* strains survived significantly longer. Goberna et al. [14] found a complete decay of *E. coli* and *Salmonella* but not for *Listeria*, with a reduction of only one order of magnitude within three months. Interestingly, in this study, the authors inoculated manure and digestate in γ-sterilized soils and non-sterilized soils to assess the importance of the autochthonous soil microorganisms. In γ-sterilized soils, without competition from other microorganisms, pathogens were more abundant than in the native non-sterilized soils. This is consistent with other studies reporting a slower decline of *E. coli* in sterilized soil amended with manure [44, 45]. This demonstrates that besides the intrinsic nature of the pathogens to survive in soils, competition is a key factor determining the survival rate. Underlying mechanisms were clarified by the study of Van Elsas et al. [46]. Using an assembly experiment and dilution to provide an extinction approach, they experimentally manipulated soil microbial diversity and then inoculated an *E. coli* strain. The diversity of the autochthonous soil microorganisms had a clear impact on the survival of *E. coli*; in soils with the weakest diversity, survival was high, whereas in soils with the highest diversity, survival was low. The authors stated that this negative correlation was explained by different efficiencies to compete for resources between *E. coli* and the autochthonous soil microorganisms.

12.1.3.2 Pathogen Survival Depending on Soil Abiotic Parameters

Soil texture is an important factor driving the pathogen survival rate, mainly due to impacts on soil moisture, the nutrient retention capacity, and the presence of anaerobic conditions. Clayey soils with a finer texture generally support a higher pathogen survival rate than sandy soils with a coarser texture [47, 48]. For instance, after the supply of biosolids to soils, Lang and Smith [47] observed that a decrease of 90% of the *E. coli* populations is achieved, which is faster in a sandy loam than a silty clay soil (100 vs. 200 days respectively). In addition, when moisture increased, the same level of pathogen reduction is achieved within only 20 days. However, this is not a general trend and some studies have shown no impact on moisture or pathogen survival in very dry soils [44, 49]. The effects of moisture seem to be different depending on the type of pathogen. For instance, *Salmonella enterica* persisted much longer than *E. coli* 0157:H7 when submitted to moisture stress [50].

Generally, neutral pH conditions allowed better survival rates of enteropathogen bacteria [42, 51], but due to their presence in the gastrointestinal tract hosts, these bacteria withstand and survive acidic pH conditions [42]. On the contrary, alkaline conditions led to a faster decline of *E. coli*, *Salmonella typhimurium*, and *E. faecalis* after lime addition [52, 53].

Temperatures affect the survival of pathogens with a more intense and faster decline with increasing mean temperatures. Several studies have been performed to assess the importance of this factor on the die-off of several pathogens (e.g. references [43] and [54]), but temperature fluctuations have also to be taken into account. Semenov et al. [43] have submitted soil microcosms inoculated with manure to different temperature oscillations. They have shown that *E. coli* O157:H7 and *S. typhimurium* declined more intensely at larger than at smaller amplitudes of temperature oscillations.

12.1.3.3 Pathogen Survival Depending on Land-Use Management

With organic matter of end products supplied to soils, nutrients become available for the autochthonous microflora and can improve growth efficiency and competitiveness. For instance, in a previous study [55], the authors have observed a significant increase in the potential microbial respiration until five weeks after supply of AD sludge to soils. Even functions linked to the N cycle, denitrification and nitrification, are increased within the next few days after sludge applications in two different soils (unpublished data). Based on this, Ellis et al. [56] hypothesized that with the same quantity of pathogen inoculated, decline of the pathogen will depend on the sludge: soil ratio. They did not observe a clear proportional relationship but they observed a greater attenuation of both *E. coli* and *Salmonella* strains in microcosms containing 25% of sludge than in the microcosms containing 0, 50, and 75% of sludge. Similarly, *E. coli* declined more rapidly in soils amended with biosolids than in unamended soils [44, 57]. Overall, this shows that application of an end product can increase the antagonistic interactions with autochthonous microflora and improve the decay of pathogens supplied to the soil. However, the application method, subsurface deposition or broadcast application was not identified as a parameter affecting the rate of bacteria die-off [58].

Overall, through its impacts on soil biotic and abiotic parameters, land-use management can be an important factor involved in the pathogen survival in the environment. Semenov et al. [59] monitored survival of *E. coli* O157:H7 in 18 manure-amended soils under organic or conventional management. Decline of this pathogen followed significantly a different pattern depending on the land-use management. However, a clear and unique trend for different pathogens and different soils is difficult to observe. For Franz et al. [42], *E. coli* O157:H7 declined faster in organic than in conventional soils and this was positively correlated with total carbon and nitrogen contents. However, in the same soils, two *Salmonella* strains declined with the same rate whatever the land-use and without any correlation with abiotic parameters.

12.1.3.4 Concluding Remarks and Need for Future Research

Use of end products to agronomic purposes has many advantages but it cannot be considered as an innocuous practice. This comes with different risks for animal and human health and also for the equilibrium of the soil ecosystem. Pathogen supplied via end products can survive and even grow in soil and eventually reach aquatic habitats. A major concern relies on spore-forming pathogens that can resist treatments applied to sanitize end products. This can lead to contamination of cattle, food, and water aimed for human consumption. Several studies have been performed to analyze the survival of the pathogen in the end product, but it is still unclear how end product characteristics affect pathogen survival. For instance, mesophilic AD sanitizes end products that are likely to be due to actions of volatile fatty acids and NH_3, but this mechanism and the importance of each compound would require better attention. Addition of end products to soil means that various compounds and allogenous microorganisms are supplied to the soil environment, where crucial functional microbial communities can be directly or indirectly affected. To minimize these risks, different practices are carried out, with among the most significant being practices to reduce pathogens in animal feces, proper manure storage, and applications and optimized treatments of end products (compost and AD). An abundant bibliography is available on the impact of end products on global microorganism communities, but mainly microbial communities involved in broad functions such as carbon respiration have been targeted, and knowledge is scarce on the effects of end products on microbial communities involved in other crucial soil functions, such as nitrification and denitrification. In addition, the vast majority of studies have been performed on short time scales, with laboratory experiments. In the field, end products are applied on a regular basis in a complex and changing environment. Therefore, long-term field trials with several applications over the years would be helpful to understand better the impact on the soil microbial communities and the survival and regrowth of pathogens. Recently, interesting studies have shown the importance of soil microbial diversity to prevent the survival of allogenous pathogens. However, generally, the pathogen was supplied directly into the soil and not with the end product. Given that organic matter and other compounds can stimulate or simply affect autochthonous microbial communities, survival of a pathogen would be better understood using more realistic conditions, with pathogens brought via an end product supply to the soil.

12.1.4 Modification of Bio-Wastes for Alteration of Soil Microbial Communities

The first question here is why should extra resources be allocated to the modification of bio-wastes. The answer could be straightforward: while most bio-wastes are valuable for nutrient retrieval and C sequestration, they are wastes containing a mixture of contaminates that pose environmental threats by direct use within the concept of the circular economy.

With various inputs in urban areas, municipal sewage sludge encompasses heavy metals, polycyclic aromatic hydrocarbons (PAHs), antibiotics, pathogens, and viruses [60], which is not eco-friendly when it finds its ways directly into the environment. For example, antibiotics in sewage outlets accumulate antibiotic-resistant bacterial communities, which are risky if they disperse into the environment.

A recent attractive research spot is about the transport of microplastics in the environment [61]. Evidence shows that microplastics could be integrated into the food web adjacent to oceans where plastics are not well handled. The microplastics could enter soils with bio-wastes to make composts as a soil amendment. They are likely to migrate between the soil microorganisms and fauna, and possibly be assimilated by fauna tissues and ultimately by plants via food chains.

To close the loop of circular economy (Figure 12.1.2), a pyrolysis-based technology provides a way for bio-waste management with potential benefits for energy generation and nutrient reuse [62]. The output of pyrolysis of bio-waste includes pyrogenic carbon (PC), a stable C consortium against microbial decomposition, could increase C pools in soils. It is important and interesting to discuss how a modified PC interacts with soil microbial communities and alters their functions.

12.1.4.1 The Main Properties of Pyrogenic Carbon for Modification

Pyrolysis breaks down organic substances and rearranges their elemental compositions under a zero or limited oxygen environment [63]. During the formation of the final solid substrate – PC – a large proportion of N, S, O, H, and Cl in gases are ready to leave the pyrolysis system [64], whereas most C, P, K, and metals are retained by the PC, depending on the degree of pyrolysis, the retention time, and the highest temperature.

Unfortunately, there are no consistent positive plant responses, which are sometimes null or negative when PC is applied into soils with or in the presence of inorganic fertilizers [65]. The first reason is that not all nutrients concentrated in PC during pyrolysis are directly available to plants and microorganisms. For example, sodium and potassium are readily used by plants, whereas phosphorus binds with metals and is difficult to utilize [66]. This explains why positive effects on the

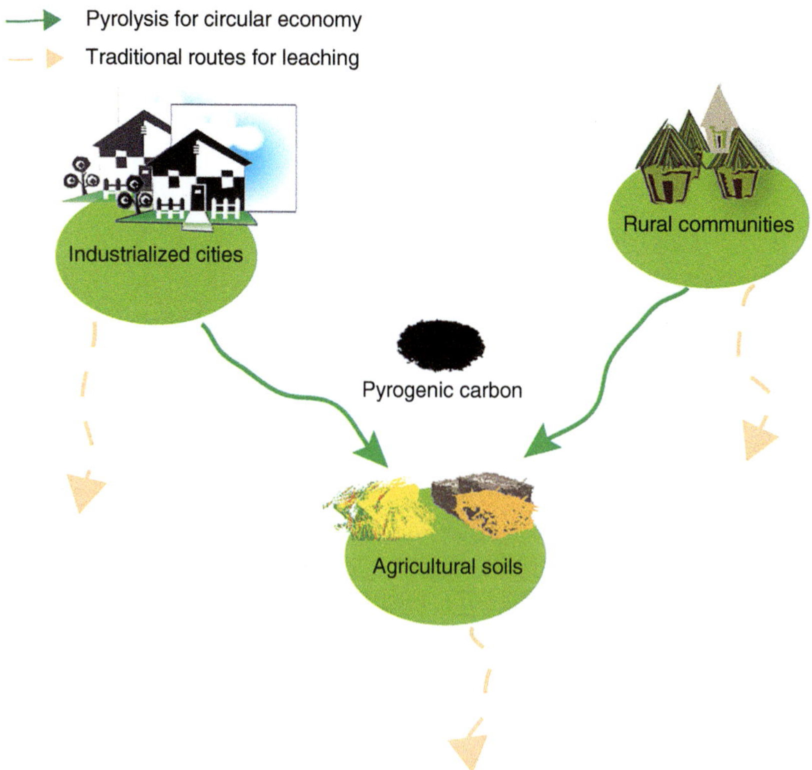

Figure 12.1.2 Illustration of the use of the pyrolysis technique in the circular economy for both nutrient retrieval, waste management, and C sequestration (Source: Courtesy of Daquan Sun). Otherwise, the direct circle of bio-wastes triggers environmental issues.

crop yield are mostly achieved when PC is applied into sandy soils, possibly owing to the fast release of K and other minerals to the barren soils.

The absorption capacity of PC could be likened to clays, which leads to a competition in acquiring the nutrients with plants, partly explaining the negative effects on the crop yield. For instance, the addition of PC had little effect on plant yield in the prairie soils in Canada [67], possibly owing to similar capacities to nutrient adsorption between PC and the soils. Similarly, many experiments combined PC with inorganic fertilizers to increase the total amount of available nutrients, which, however, did not achieve the expected positive results [68]. Considering it in a positive way, PC could reduce the leaching of nutrients because of its inherent absorption trait (Fe and Mn oxidants and functional groups), which triggers many attempts to use PC for waste water treatment.

The natural properties of PC include its high alkalinity [69], which further contributes to the immobilization of nutrients. This is another mechanism that gives plants less opportunities to acquire elements in the presence of high concentrations of PC in soils. While there are claims to use PC for combating soil acidification, in fact not all soils need to be adjusted due to their pH. For example, with the benign purpose of alleviating soil acidity, addition of PC attenuates potato scraps, possibly owing to the alteration of soil pH (Figure 12.1.3).

A high dose of PC does not increase plant yield, which could thus be credited to unavailable nutrients, competition of nutrients with plants, and immobilization of elements by a high pH. Compared with the raw bio-materials, PC has developed many functional groups, and many of them are hydrophobic [70]. As a result, PC may retain less water than the original materials, such as wheat

Figure 12.1.3 Pyrogenic carbon addition was found to attenuate potato scraps, possibly owing to the increase of soil pH. Arrows indicate pyrogenic carbon particles. This is an example showing that not all acidic soils need to be adjusted in all circumstances (Source: Courtesy of Krištůfek Václav).

straws vs. its PC, despite its higher surface area. The hydrophobicity is a detrimental factor, not only for water retention but also for the transition of water-soluble nutrients to microorganisms and plant roots.

12.1.4.2 Biomodification of Pyrogenic Carbon

12.1.4.2.1 Aging Pyrogenic Carbon by Composting

Aerobic composting is a well-documented and widely used process of degradation of organic matter by microbial activities during a relative short course. It is also a process of humification with the formations of humic substances. There is evidence that humic substances can bind with PC as colloidals [71], implying synergistic effects if they interact well.

Interestingly, studies have shown that biochar as a bulk agent or an additive could accelerate the organic matter degradation in compost, reduce the composting time [72], and enhance the humification level [73], probably because of labile C sources and the leveled-up sodium and potassium in fresh PC nurturing composting microorganisms. Because of its porous structures and hydrophobicity, PC as a bulking agent could increase the aeration to boost microbial decomposition when effective ventilation procedures are limited (Figure 12.1.4).

When labile C in PC is depleted by microbial metabolization, the mineralization rates become lower, leading to a decrease in the loss of ammonia. It is also assumed that the reduced loss of nitrogen during composting is attributed to the ammonia absorbed by PC [74] and the altered abundance of denitrifying bacteria [75].

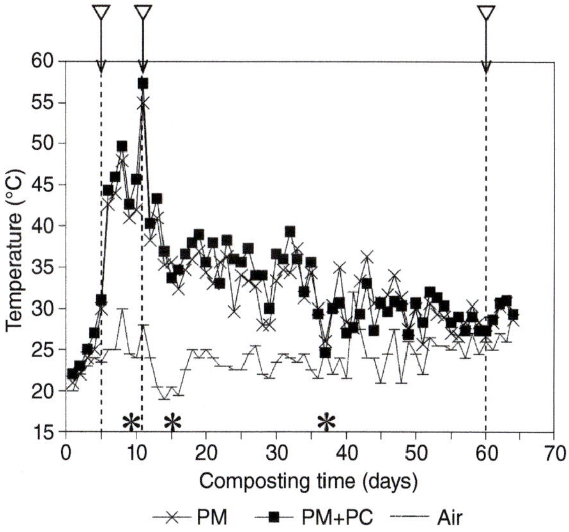

Figure 12.1.4 Pyrogenic carbon (PC) addition at 2% of dry weight increases the composting temperature compared with no pyrogenic carbon addition (PM). Asteroids show manually a turnover of composting piles for ventilation. The composting regime constitutes poultry manure and rice straws (Source: Courtesy of Daquan Sun).

The presence of PC stimulated microbial activities in the initial stage of the composting, followed by a sharp temperature increase, which further changes the microbial communities. It should be noted that comparing microbial communities at a specific time between composting regimes with and without PC can be biased, because PC may alter the composting process. It is therefore necessary to compare general or functional microbial communities at the same composting stages according to other composting indicative assays, such as temperature, the C/N ratio, and pH.

Anaerobic composting is a relatively slow process for charging PC with organic substances, if less heavy metal or other detrimental elements are present in the system.

It is interesting that most humic substances are hydrophilic, which could alleviate the hydrophobicity of PC if they are well combined. It should also be understood that composting with PC is also a natural inoculation of PC with microbial communities developed during composting. If the beneficial microbial communities such as Actinomycetes reside in PC pores at the final stage of composting, they are likely to exert their functions in soils along with the composted PC as a soil amendment.

12.1.4.2.2 Aging Pyrogenic Carbon by Vermicomposting

Vermicomposting is a process where various species of worms and earthworms decompose bio-wastes such as food wastes and sewage sludge to a small to large extent. The end product, worm castings, contain a high level of saturable nutrients such as micronutrients, which could be used as a nutrient-rich organic fertilizer for gardens or organic farming.

PC is hydrophobic, with less capacity for holding water-soluble nutrients and core microbial communities (CMCs) (Figure 12.1.5a), which may explain why earthworms are less likely to be nurtured, thus leading to their avoidance [76]. There is, however, evidence showing a reduced avoidance of earthworms to pre-moistened PC [77]. PC could be aged through the guts of earthworms no matter whether they are digested intentionally or unintentionally or blended with worm castings by their burrowing activities (Figure 12.1.5b). While there is no direct evidence showing that earthworms deliberately swallow PC as their food, they are potential food as colloidals mixed with bacteria and fungal mycelium.

The toxicity of heavy metals in PC could be reduced, if there is any, by producing a good ratio of PC and substrates to earthworms during vermicomposting [78]. As a result, it is expected that this would produce a well "nutrient-charged" PC as a soil amendment.

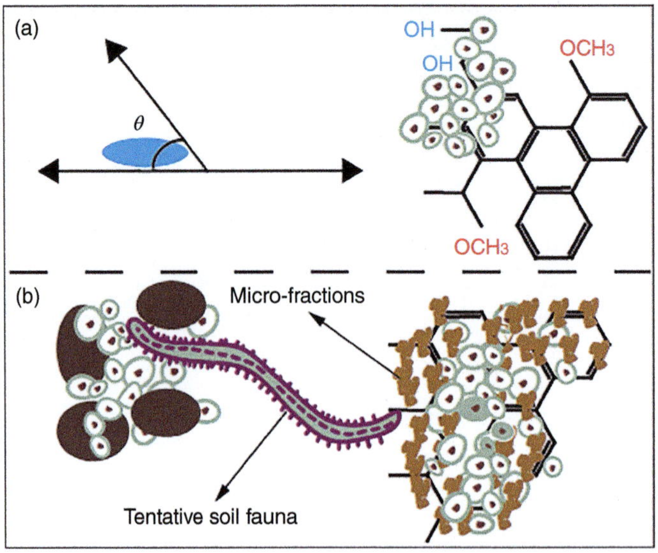

Figure 12.1.5 Bio-oxidation by conversion of hydrophobic functional groups such as CH_3O^- to hydrophilic OH^-, resulting in the extension angle of water (Θ) to decrease hydrophobicity (a); pyrogenic carbon may have a "refugee" effect on bacteria by protecting the inoculants from grazers, but earthworms could devour a colloidal as food (b).

12.1.4.2.3 Pyrogenic Carbon as a Carrier of Functional Microbial Inoculants

A bourgeoning direction of research is to use PC as a carrier for functional microbial inoculants to increase the value of PC, implying another way to modify PC for its use in soils. For example, the use of PC and microbial inoculants increases spinach growth (Figure 12.1.6).

The viability of the inoculants suffers from the exogenous physical–chemical environment (i.e. low nutrient conditions and predation by protozoa and nematodes) and the strong competence of indigenous bacterial communities [79]. As a result, their cell numbers decline rapidly and are unable to exert functions as expected.

PC can be produced from locally available bio-wastes and is pre-sterilized during the pyrolysis process, facilitating inoculation as the last step in the production process. There are abundant raw

Figure 12.1.6 Pyrogenic carbon with inoculants (left), pyrogenic carbon (middle), and with no application (right); microbial inoculants act as a function for solubilizing potassium and excreting IAA.

materials for its production including agricultural and forestry residues, compost, and manures, many of which are considered as waste products and require disposal fees.

Organic and inorganic nutrients retained on fresh PC can be used for inoculants along with other supplemental nutrients. However, culturing or enriching inoculants in PC, depending on the original nutrients, could not last long because of their limited content. Luckily that they do not die but go into dormancy until revived by suitable conditions. Both single or CMCs are candidates for PC inoculation.

However, the pore structure, the hydrophobicity, and absorption capacities of fresh PC could be further modified to improve the survival and functions of inoculants. The pore sizes of PO can either be altered during pyrolysis with different temperatures and retention times, or could be prepared into smaller sizes such as nano sizes. The hydrophobicity will delay the formation of bio-films on PC by expelling water. Some bacterial strains excrete bio-surfactants [80], which could increase the water retention capacity of PC. The adsorption capacity of PC can be enhanced by adding MgO or other metal oxidants [81].

12.1.4.3 Link Soil Microbial Functions to the Modified Pyrogenic Carbon

The diversity of soil microbial communities depends on soil heterogeneity [82]. This means that the changeable niches such as variable C sources can consistently affect the microbial community structures. For instance, the whole microbial inoculants separated from soils recover soon when inoculated back to sterile origin soils; however, microbial richness decreased during culturing with PC as carriers (Figure 12.1.7).

Microbial diversity does not directly correspond to their functions. For example, a high general microbial diversity does not guarantee a high crop yield; a high N-fixing bacterial community does not ensure a higher N-fixing capacity. It is therefore necessary to discuss in a specific scenario when altering soil microbial functions by modified PC as a soil amendment, where both the general or special microbial communities play diverged functions for the ecosystem. It should be understood that the aforementioned processes of aging or charging PC with nutrients or microbial inoculants will virtually alter the included microbial functional communities.

12.1.4.3.1 Conventional Agriculture

Conventional agriculture, as a widely used practice in agriculture, provides a good opportunity for introducing PC into soils with the purpose of C sequestration and soil conditioning. Since inorganic

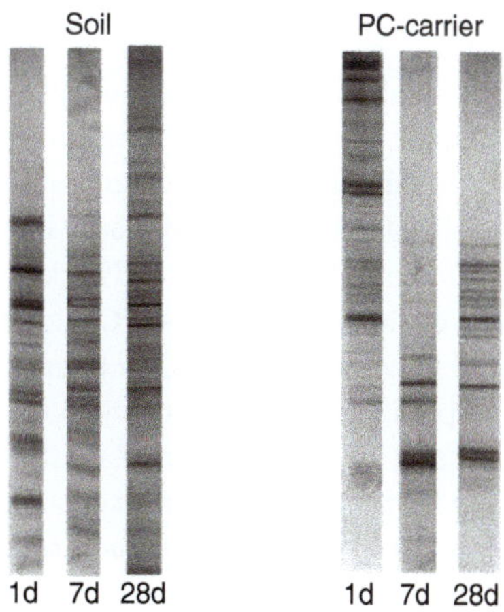

Figure 12.1.7 Denaturing gradient gel electrophoresis (DGGE) analysis of 16S rDNA gene fragments of whole soil microbial communities cultured by soil and pyrogenic carbon (PC). A recovery of bacterial community occurs in soil but not in the PC carrier.

Figure 12.1.8 A phosphorus-enriched pyrogenic carbon stimulates canola growth in prairie soils in Canada (Source: Courtesy of Raju). The right plot shows a P-enriched pyrogenic carbon and the left plot shows one without a P-enriched pyrogenic carbon.

fertilizer is usually used at a high dose for conventional agriculture, PC with relatively less available nutrients to the soil cannot trigger profound benefits for plant growth [65].

Both organic and inorganic nutrients in PC stimulate the general bacterial or fungal biomass, and enhance their roles in nutrient cycling, where it competes for nutrients with crops. It is also likely that the high pH of PC will further immobilize nutrients such as Ca, Fe, and Mo [83]. This is another mechanism where plants will have less opportunities to acquire these elements, even though there are high concentrations of them in soils.

A great deal of phosphorus (P) retained on PC cannot directly be used by plants as a P fertilizer [66]. Arbuscular mycorrhizal fungi are capable of assisting plants in acquiring unavailable P species, if a successful symbiosis is formed. If the P status in soils is not too high or too low, it is beneficial for arbuscular mycorrhiza fungi (AMF) symbiosis with plants. Not all plants are able to form a symbiosis with AMF; for example, most canola cultivars are not suitable candidate hosts for AMF [84]. Interestingly, a P-enriched PC was found to increase canola yield with reduced application of P fertilizers (Figure 12.1.8).

Attention has been given to the gross microbial biomass instead of successional functional groups since conventional agriculture is shaping microbial communities toward a direction away from natural successions. The pore structure of PC is likely to provide habitats that may be beneficial for microbial activities. Microbial inoculants with specific functions could be combined with PC. However, it would not be economical to use inoculants on a large scale since their functions could be curbed by indigenous microbial communities. If it is feasible, using manure, composted PC, and beneficial microorganisms could have synergistic beneficial effects on soils.

Another precaution should be considered for a high dose of PC in conventional agricultural soils. As it could be claimed that PC reduces the leaching of nutrients [85], it could also infer that plants need to compete with PC in the acquisition of nutrients. Also, like clays, PC absorbs herbicides and pesticides, which is likely to lead to dilution effects on the chemicals for target organisms or grasses. This is likely to increase the costs to farmers.

As PC can absorb organic substances, it would probably enhance the accumulation of root exudates. If a specific composition of root exudates accumulates, it favors some microbial groups. When the detrimental strains are dominant or at a sufficient level to inject plants, plant disease occurs

Figure 12.1.9 Pyrogenic carbon inoculated of UW4 with functions of excreting ACC relieves the symptoms of dwarfing and wilted leaves of corn under saline stress (2.3 NaCl g kg^{-1}).

and could be attenuated due to the presence of PC. Instead of continuous cropping, crop rotation is necessary when soils have received a high dose of PC.

12.1.4.3.2 Greenhouses

One of the common issues in nursery greenhouses is the development of plant-borne diseases. The accumulated spores of plant diseases incur the following development of vegetation Comprehensive measures have been taken to achieve a result that ensures the low occurrence of diseases and desirable qualities of products. Measures such as sterilization by high temperature or r-radiation have to be used to remove all pathogens along with beneficial microbial communities. This opens up opportunities for using PC combined with antagonistic inoculants against plant root pathogens in the future. It is also possible to introduce functional microbial communities such as inoculants excreting hormones such as IAA or ACC (Figure 12.1.9).

12.1.4.3.3 No-Tillage or Reduced Tillage

No-tillage or reduced tillage becomes trendy because it provides beneficial effects such as reduced soil loss by wind erosion. Leaving stubbles and mulches prevent water evaporation and less disturbance of tillage reduces microbial-induced mineralization. Machinery is needed to ensure seeds and fertilizers are successfully added into soils without tilling the top soil layers.

Modified PC can be applied occasionally by not disturbing soils too frequently. The porous structures of PC are likely to reduce soil compaction by long-term residence. If possible, combining PC with other organic amendments such as green manure or manures or compost could induce multiple beneficial effects.

Biodisturbance such as "burrowing" by earthworms could lead to the translocation of PC to deeper soil layers. No-tillage and residue retention, as well as residues of herbicides, alter fauna diversity and abundance [86]. If PC is well mixed with decayed residues, second consumers such as mites, insects, or nematodes and collembolan could devour them altogether. This may lead to "gut aging" of PC for modification of its properties. Polysaccharides in earthworm castings help to stabilize soil aggregates [87]. Colloidals formed by PC, fungi, bacteria, and polysaccharides could increase the stability of soils. The formation of colloidals has also been suggested in order to decrease the loss of phosphorus. In the presence of PC, some soil food web constituents, such as bacterivores and fungivores, could have a different fate since their physical environment is altered in no-tillage or reduced tillage.

12.1.4.3.4 Plant Restoration

The use of PC for plant restoration in disturbed soils such as post-mining sites does not seem to be economical. However, it provides a good opportunity for the disposal of biowastes with a high dose of heavy metals in post-mining sites, because the above-ground biomass acquired there is usually not for human and livestock consumption.

There is a growing body of evidence showing CMCs associated with a given host plant or a specific type of soil can lead to a positive plant–soil relationship [88, 89]. Interestingly, inoculation with late-successional soil communities can increase the performance of plant species in early soils [90], especially for some demanding plant species. This suggests that positive plant soil feedback between early-successional plants and their soil organisms may be restored by microbial inoculation with late-successional soils. The tentative effects on litter decomposition are driven by core microbial communities (CMCs) at large soil fractions, but root–microbe interactions are driven by CMCs in microfractions (Figure 12.1.10). However, it is not clear how the structure of microbial inoculation steers the plant community development toward different target states, and how to ensure their function in the undeveloped soils.

Disparity in functions of CMCs on plant species over temporal and spatial scales may be one of the reasons for the limited attempts for their use in soil restoration. This may also explain why current

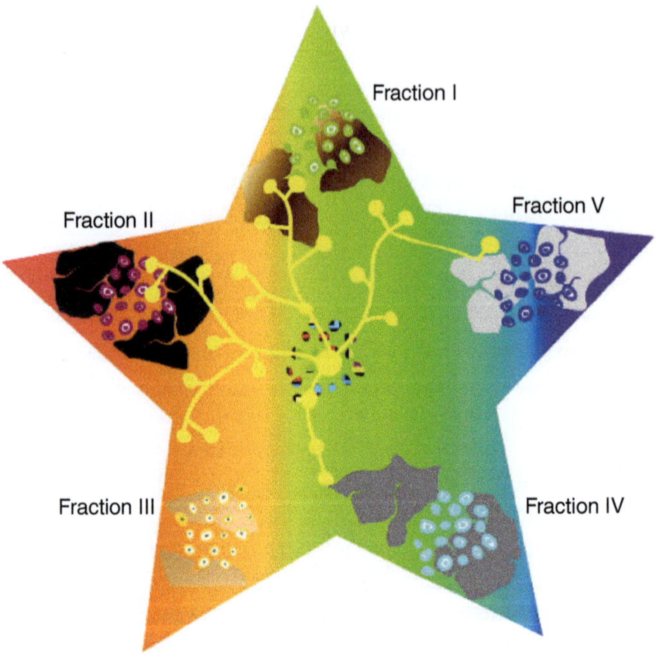

Figure 12.1.10 A schematic illustration of selected processes that CMCs mediate above- and below-ground plant–soil interactions, which are dependent on their residence on fine particulate organic matter (fPOM).

studies only focus on using PC to carry single plant growth promoting bacteria (PGPB) rather than CMCs. A combined investigation of CMCs associated with demanding plant species thus appears crucial to investigate the composition of CMCs and their linkages to soil fractions in succession soils, and to clarify their fate and functions after being coated with PC for soil restoration during their transplantations.

12.1.5 Conclusion

The thermal pyrolysis is a potential technique that links waste management, energy and nutrient reuse, soil restoration, and C sequestration in the scheme of a circular economy. One of the main merits of using PC is for C sequestration in soils, but the potential negative effects in agriculture should be treated with caution, given the competition of nutrients with plants, an unnecessary increase in soil pH, and the use of defective herbicides or pesticides in conventional agriculture. Although the resistance of PC to microbial decomposition, implying its persistent effects on the physical–chemical properties and therefore microbial communities of soils, is known, there is little consensus about the threshold that a specific soil could accept PC without steering soil functions in a negative direction. It is therefore necessary to understand to what extent the physical, chemical, and biological processes could alter the properties of PC.

Aging of fresh PC through interacting with organic substances, soil biota and other properties is a promising and seems to be a required measure to alleviate the potential downstream negative effects. Inoculating PC with composting microbial communities or beneficial microorganisms (a single strain, a collection of synergistic strains, or a core microbial community) is a way to change soil microbial communities after soil application. Their co-existence further alters the whole soil microbial functions regulating soil functions for nutrient recycling, development of plant-borne diseases, and soil restoration. Scientific advancements continue to contribute to our in-depth understanding of the roles of PC in the environment. The mitigation of negative effects from modified bio-wastes to soils is a key step for closing the loop of the circular economy, with the ultimate goal of C sequestration.

References

1 WOOD (2019). Digestate and compost as fertilisers: risk assessment and risk management options. European Commission, Report No. 40039CL003i3.
2 Azeem, M., Hale, L., Montgomery, J. et al. (2020). Biochar and compost effects on soil microbial communities and nitrogen induced respiration in turfgrass soils. *PLoS One* 15 (11): e0242209.
3 Gryta, A., Frąc, M., and Oszust, K. (2020). Genetic and metabolic diversity of soil microbiome in response to exogenous organic matter amendments. *Agronomy* 10 (4): 546.
4 Möller, K. Effects of anaerobic digestion on soil carbon and nitrogen turnover, N emissions, and soil biological activity. A review. *Agronomy for Sustainable Development* 35: 1021–1041.
5 Sadet-Bourgeteau, S. (2018). Lasting effect of repeated application of organic waste products on microbial communities in arable soils. *Applied Soil Ecology* 125: 10.
6 Oszust, K. (2019). Characterization of microbial functional and genetic diversity as a novel strategy of biowaste ecotoxicological evaluation. *International Journal of Environmental Science and Technology* 16: 14.

7 García-Sánchez, M. (2015). Effect of digestate and fly ash applications on soil functional properties and microbial communities. *European Journal of Soil Biology*: 12.

8 Dudley, D.J., Guentzel, M.N., Ibarra, M.J. et al. (1980). Enumeration of potentially pathogenic bacteria from sewage sludges. *Applied and Environmental Microbiology* 39 (1): 118–126.

9 Ye, L. and Zhang, T. (2011). Pathogenic bacteria in sewage treatment plants as revealed by 454 pyrosequencing. *Environmental Science & Technology* 45 (17): 7173–7179.

10 Sidhu, J. and Toze, S. (2009). Human pathogens and their indicators in biosolids: a literature review. *Environment International* 35 (1): 187–201.

11 Manyi-Loh, C., Mamphweli, S., Meyer, E. et al. (2016). An overview of the control of bacterial pathogens in cattle manure. *International Journal of Environmental Research and Public Health* 13 (9): 843.

12 Cai, L. and Zhang, T. (2013). Detecting human bacterial pathogens in wastewater treatment plants by a high-throughput shotgun sequencing technique. *Environmental Science & Technology* 47 (10): 5433–5441.

13 Ju, F., Li, B., Ma, L. et al. (2016). Antibiotic resistance genes and human bacterial pathogens: co-occurrence, removal, and enrichment in municipal sewage sludge digesters. *Water Research* 91: 1–10.

14 Goberna, M., Podmirseg, S.M., Waldhuber, S. et al. (2011). Pathogenic bacteria and mineral N in soils following the land spreading of biogas digestates and fresh manure. *Applied Soil Ecology* 49: 18–25.

15 Wéry, N., Lhoutellier, C., Ducray, F. et al. (2008). Behaviour of pathogenic and indicator bacteria during urban wastewater treatment and sludge composting, as revealed by quantitative PCR. *Water Research* 42 (1–2): 53–62.

16 Higgins, M.J., Chen, Y.-C., Murthy, S.N. et al. (2007). Reactivation and growth of non-culturable indicator bacteria in anaerobically digested biosolids after centrifuge dewatering. *Water Research* 41 (3): 665–673.

17 Sidhu, J. (2001). The role of indigenous microorganisms in suppression of salmonella regrowth in composted biosolids. *Water Research* 35 (4): 913–920.

18 Bernal, M.P., Sommer, S.G., Chadwick, D. et al. (2017). Current approaches and future trends in compost quality criteria for agronomic, environmental, and human health benefits. In: *Advances in Agronomy*, 143–233. Elsevier Available from: https://linkinghub.elsevier.com/retrieve/pii/S0065211317300196.

19 Deportes, B.-G. and Zmirou, B. (1998). Microbial disinfection capacity of municipal solid waste (MSW) composting. *Journal of Applied Microbiology* 85 (2): 238–246.

20 Ros, M., García, C., and Hernández, T. (2006). A full-scale study of treatment of pig slurry by composting: kinetic changes in chemical and microbial properties. *Waste Management* 26 (10): 1108–1118.

21 Vinnerås, B. (2007). Comparison of composting, storage and urea treatment for sanitising of faecal matter and manure. *Bioresource Technology* 98 (17): 3317–3321.

22 Mc Carthy, G., Lawlor, P.G., Coffey, L. et al. (2011). An assessment of pathogen removal during composting of the separated solid fraction of pig manure. *Bioresource Technology* 102 (19): 9059–9067.

23 Millner, P., Ingram, D., Mulbry, W., and Arikan, O.A. (2014). Pathogen reduction in minimally managed composting of bovine manure. *Waste Management* 34 (11): 1992–1999.

24 Pereira Neto, J.T., Stentiford, E.I., and Mara, D.D. (1986). Comparative survival of pathogenic indicators in windrow and static pile. In: *Compost: Production, Quality and Use* (eds. M. De

Bertoldi, M.P. Ferranti, P. l'Hermite and F. Zucconi), 276–295. London and New York: Elsevier Applied Science.

25 Hay, J. (1996). Pathogen destruction and biosolids composting. *Biocycle*: 67–76.

26 Turner, C. (2002). The thermal inactivation of E. coli in straw and pig manure. *Bioresource Technology* 83 (2).

27 Paniel, N., Rousseaux, S., Gourland, P. et al. (2010). Assessment of survival of *Listeria monocytogenes, Salmonella infantis* and *Enterococcus faecalis* artificially inoculated into experimental waste or compost. *Journal of Applied Microbiology* 108 (5): 1797–1809.

28 Elving, J., Ottoson, J.R., Vinnerås, B., and Albihn, A. (2010). Growth potential of faecal bacteria in simulated psychrophilic/mesophilic zones during composting of organic waste: *Salmonella* regrowth in compost. *Journal of Applied Microbiology* 108 (6): 1974–1981.

29 Monroy, F., Aira, M., and Domínguez, J. (2009). Reduction of total coliform numbers during vermicomposting is caused by short-term direct effects of earthworms on microorganisms and depends on the dose of application of pig slurry. *Science of the Total Environment* 407 (20): 5411–5416.

30 Edwards, C.A.S.S. (2011). Human pathogen reduction during vermicomposting. In: *Vermiculture Technology* (eds. C.A. Edwards, N.Q. Arancon and R. Sherman), 249–261. Florida: Taylor and Francis Group.

31 Soobhany, N., Mohee, R., and Garg, V.K. (2017). Inactivation of bacterial pathogenic load in compost against vermicompost of organic solid waste aiming to achieve sanitation goals: a review. *Waste Management* 64: 51–62.

32 Aira, M., Gómez-Brandón, M., González-Porto, P., and Domínguez, J. (2011). Selective reduction of the pathogenic load of cow manure in an industrial-scale continuous-feeding vermireactor. *Bioresource Technology* 102 (20): 9633–9637.

33 Rodríguez-Canché, L.G., Cardoso Vigueros, L., Maldonado-Montiel, T., and Martínez-Sanmiguel, M. (2010). Pathogen reduction in septic tank sludge through vermicomposting using Eisenia fetida. *Bioresource Technology* 101 (10): 3548–3553.

34 Monroy, F., Aira, M., and Domínguez, J. (2008). Changes in density of nematodes, protozoa and total coliforms after transit through the gut of four epigeic earthworms (Oligochaeta). *Applied Soil Ecology* 39 (2): 127–132.

35 Jiang, Y. (2020). Inactivation of pathogens in anaerobic digestion systems for converting biowastes to bioenergy: a review. *Renewable and Sustainable Energy Reviews* 120: 109654.

36 Franke-Whittle, I.H. and Insam, H. (2013). Treatment alternatives of slaughterhouse wastes, and their effect on the inactivation of different pathogens: a review. *Critical Reviews in Microbiology* 39 (2): 139–151.

37 Sahlstro, L. (2003). A review of survival of pathogenic bacteria in organic waste used in biogas plants. *Bioresource Technology* 87: 6.

38 Bagge, E., Sahlström, L., and Albihn, A. (2005). The effect of hygienic treatment on the microbial flora of biowaste at biogas plants. *Water Research* 39 (20): 4879–4886.

39 Elmerdahl Olsen, J. and Errebo Larsen, H. (1987). Bacterial decimation times in anaerobic digestions of animal slurries. *Biological Wastes* 21 (3): 153–168.

40 Orzi, V., Scaglia, B., Lonati, S. et al. (2015). The role of biological processes in reducing both odor impact and pathogen content during mesophilic anaerobic digestion. *Science of the Total Environment* 526: 116–126.

41 Courtois, P., Rorat, A., Lemiere, S. et al. (2019). Ecotoxicology of silver nanoparticles and their derivatives introduced in soil with or without sewage sludge: a review of effects on microorganisms, plants and animals. *Environmental Pollution* 253: 578–598.

42 Franz, E., van Diepeningen, A.D., de Vos, O.J., and van Bruggen, A.H.C. (2005). Effects of cattle feeding regimen and soil management type on the fate of *Escherichia coli* O157:H7 and *Salmonella enterica* serovar typhimurium in manure, manure-amended soil, and lettuce. *Applied and Environmental Microbiology* 71 (10): 6165–6174.

43 Semenov, A.V., van Bruggen, A.H.C., van Overbeek, L. et al. (2007). Influence of temperature fluctuations on *Escherichia coli* O157:H7 and *Salmonella enterica* serovar typhimurium in cow manure. *FEMS Microbiology Ecology* 60: 10.

44 Jiang, X., Morgan, J., and Doyle, M.P. (2002). Fate of *Escherichia coli* O157:H7 in manure-amended soil. *Applied and Environmental Microbiology* 68 (5): 2605–2609.

45 Xing, J., Sun, S., Wang, H. et al. Response of soil native microbial community to *Escherichia coli* O157:H7 invasion. *Environmental Pollution* 261: 114225.

46 van Elsas, J.D., Chiurazzi, M., Mallon, C.A. et al. (2012). Microbial diversity determines the invasion of soil by a bacterial pathogen. *Proceedings of the National Academy of Sciences of the United States of America* 109 (4): 1159–1164.

47 Lang, N.L. and Smith, S.R. (2007). Influence of soil type, moisture content and biosolids application on the fate of *Escherichia coli* in agricultural soil under controlled laboratory conditions: *E. coli* decay in sludge-amended soil. *Journal of Applied Microbiology* 103 (6): 2122–2131.

48 Brennan, F.P., Moynihan, E., Griffiths, B.S. et al. (2014). Clay mineral type effect on bacterial enteropathogen survival in soil. *Science of the Total Environment* 468–469: 302–305.

49 Berry, E.D. and Miller, D.N. (2005). Cattle feedlot soil moisture and manure content: II. Impact on *Escherichia coli* O157. *Journal of Environmental Quality* 34 (2): 656–663.

50 Ongeng, D., Muyanja, C., Geeraerd, A.H. et al. (2011). Survival of *Escherichia coli* O157:H7 and *Salmonella enterica* serovar typhimurium in manure and manure-amended soil under tropical climatic conditions in Sub-Saharan Africa: fate of enteric bacteria in the tropics. *Journal of Applied Microbiology* 110 (4): 1007–1022.

51 van Elsas, J.D., Semenov, A.V., Costa, R., and Trevors, J.T. (2011). Survival of Escherichia coli in the environment: fundamental and public health aspects. *The ISME Journal* 5 (2): 173–183.

52 Avery, L.M., Williams, A.P., Killham, K., and Jones, D.L. (2009). Heat and lime-treatment as effective control methods for *E. coli* O157:H7 in organic wastes. *Bioresource Technology* 100 (10): 2692–2698.

53 Nyberg, K.A., Vinnerås, B., Lewerin, S.S. et al. (2011). Treatment with $Ca(OH)_2$ for inactivation of *Salmonella typhimurium* and *Enterococcus faecalis* in soil contaminated with infected horse manure: treatment of *Salmonella* in soil and horse manure. *Journal of Applied Microbiology* 110 (6): 1515–1523.

54 Arrus, K.M., Holley, R.A., Ominski, K.H. et al. (2006). Influence of temperature on Salmonella survival in hog manure slurry and seasonal temperature profiles in farm manure storage reservoirs. *Livestock Science* 102 (3): 226–236.

55 Courtois, P., Rorat, A., Lemiere, S. et al. (2020). Medium-term effects of Ag supplied directly or via sewage sludge to an agricultural soil on *Eisenia fetida* earthworm and soil microbial communities. *Chemosphere*: 128761.

56 Ellis, S., Tyrrel, S., O'Leary, E. et al. (2018). Proportion of sewage sludge to soil influences the survival of *Salmonella* Dublin and *Escherichia coli*. *Clean: Soil, Air, Water* 46 (4): 1800042.

57 Schwarz, K.R., Sidhu, J.P.S., Pritchard, D.L. et al. (2014). Decay of enteric microorganisms in biosolids-amended soil under wheat (*Triticum aestivum*) cultivation. *Water Research* 59: 185–197.

58 Saunders, O., Harrison, J., Fortuna, A.M. et al. (2012). Effect of anaerobic digestion and application method on the presence and survivability of *E. coli* and fecal coliforms in dairy waste applied to soil. *Water, Air, and Soil Pollution* 223 (3): 1055–1063.

59 Semenov, A.V., Franz, E., Van Overbeek, L. et al. (2008). Estimating the stability of *Escherichia coli* O157:H7 survival in manure-amended soils with different management histories: stability of *Escherichia coli* O157:H7 survival. *Environmental Microbiology* 10 (6): 1450–1459.

60 Pepper, I.L., Brooks, J.P., and Gerba, C.P. Land application of organic residuals: municipal biosolids and animal manures. In: *Environmental and Pollution Science*, 419–434. Academic Press.

61 Singh, A.P. and Devi, A.S. (2019). Microplastics and single use plastics: a curse of over consumerism. *International Journal of Advanced Scientific Research and Management* 4: 384–388.

62 Raheem, A., Sikarwar, V.S., He, J. et al. (2018). Opportunities and challenges in sustainable treatment and resource reuse of sewage sludge: a review. *Chemical Engineering Journal* 337: 616–641.

63 Lehmann, J. and Joseph, S. (eds.) (2015). *Biochar for Environmental Management: Science, Technology and Implementation*. Routledge.

64 Demirbas, A. (2004). Effects of temperature and particle size on bio-char yield from pyrolysis of agricultural residues. *Journal of Analytical and Applied Pyrolysis* 72: 243–248.

65 Biederman, L.A. and Harpole, W.S. (2013). Biochar and its effects on plant productivity and nutrient cycling: a meta-analysis. *GCB Bioenergy* 5: 202–214.

66 Sun, D., Hale, L., Kar, G. et al. (2018). Phosphorus recovery and reuse by pyrolysis: applications for agriculture and environment. *Chemosphere* 194: 682–691.

67 Ahmed, H.P. and Schoenau, J.J. (2015). Effects of biochar on yield, nutrient recovery, and soil properties in a canola (*Brassica napus* L)-wheat (*Triticum aestivum* L) rotation grown under controlled environmental conditions. *Bioenergy Research* 8: 1183–1196.

68 Schulz, H. and Glaser, B. (2012). Effects of biochar compared to organic and inorganic fertilizers on soil quality and plant growth in a greenhouse experiment. *Journal of Plant Nutrition and Soil Science* 175: 410–422.

69 Oginni, O., Yakaboylu, G.A., Singh, K. et al. Phosphorus adsorption behaviors of MgO modified biochars derived from waste woody biomass resources. *Journal of Environmental Chemical Engineering* 8: 103723.

70 Kinney, T.J., Masiello, C.A., Dugan, B. et al. (2012). Hydrologic properties of biochars produced at different temperatures. *Biomass and Bioenergy* 41: 34–43.

71 Yang, W., Shang, J., Sharma, P. et al. (2019). Colloidal stability and aggregation kinetics of biochar colloids: effects of pyrolysis temperature, cation type, and humic acid concentrations. *Science of the Total Environment* 658: 1306–1315.

72 Sánchez-García, M., Alburquerque, J.A., Sánchez-Monedero, M.A. et al. (2015). Biochar accelerates organic matter degradation and enhances N mineralisation during composting of poultry manure without a relevant impact on gas emissions. *Bioresource Technology* 192: 272–279.

73 Dias, B.O., Silva, C.A., Higashikawa, F.S. et al. (2010). Use of biochar as bulking agent for the composting of poultry manure: effect on organic matter degradation and humification. *Bioresource Technology* 101: 1239–1246.

74 Hua, L., Wu, W., Liu, Y. et al. (2009). Reduction of nitrogen loss and Cu and Zn mobility during sludge composting with bamboo charcoal amendment. *Environmental Science and Pollution Research* 16: 1–9.

75 Steiner, C., Das, K.C., Melear, N., and Lakly, D. (2010). Reducing nitrogen loss during poultry litter composting using biochar. *Journal of Environmental Quality* 39: 1236–1242.

76 Weyers, S.L. and Spokas, K.A. (2011). Impact of biochar on earthworm populations: a review. *Applied and Environmental Soil Science* 2011: 1–12.

77 Li, D., Hockaday, W.C., Masiello, C.A., and Alvarez, P.J. (2011). Earthworm avoidance of biochar can be mitigated by wetting. *Soil Biology and Biochemistry* 43: 1732–1737.

78 Malińska, K., Golańska, M., Caceres, R. et al. (2017). Biochar amendment for integrated composting and vermicomposting of sewage sludge – the effect of biochar on the activity of *Eisenia fetida* and the obtained vermicompost. *Bioresource Technology* 225: 206–214.

79 Covarrubias, S.A., de-Bashan, L.E., Moreno, M., and Bashan, Y. (2012). Alginate beads provide a beneficial physical barrier against native microorganisms in wastewater treated with immobilized bacteria and microalgae. *Applied Microbiology and Biotechnology* 93: 2669–2680.

80 Gautam, K.K. and Tyagi, V.K. (2006). Microbial surfactants: a review. *Journal of Oleo Science* 55: 155–166.

81 Li, R., Wang, J.J., Zhou, B. et al. (2017). Simultaneous capture removal of phosphate, ammonium and organic substances by MgO impregnated biochar and its potential use in swine wastewater treatment. *Journal of Cleaner Production* 147: 96–107.

82 Ekschmitt, K. and Griffiths, B.S. (1998). Soil biodiversity and its implications for ecosystem functioning in a heterogeneous and variable environment. *Applied Soil Ecology* 10: 201–215.

83 Wagner, A. and Kaupenjohann, M. (2014). Suitability of biochars (pyro-and hydrochars) for metal immobilization on former sewage-field soils. *European Journal of Soil Science* 65: 139–148.

84 Owen, K.J., Clewett, T.G., and Thompson, J.P. (2010). Pre-cropping with canola decreased *Pratylenchus thornei* populations, arbuscular mycorrhizal fungi, and yield of wheat. *Crop & Pasture Science* 61: 399–410.

85 Laird, D., Fleming, P., Wang, B. et al. (2010). Biochar impact on nutrient leaching from a Midwestern agricultural soil. *Geoderma* 158: 436–442.

86 Melman, D.A., Kelly, C., Schneekloth, J. et al. (2019). Tillage and residue management drive rapid changes in soil macrofauna communities and soil properties in a semiarid cropping system of Eastern Colorado. *Applied Soil Ecology* 143: 98–106.

87 Zhang, H. and Schrader, S. (1993). Earthworm effects on selected physical and chemical properties of soil aggregates. *Biology and Fertility of Soils* 15: 229–234.

88 Lemanceau, P., Blouin, M., Muller, D., and Moënne-Loccoz, Y. (2017). Let the core microbiota be functional. *Trends in Plant Science* 22: 583–595.

89 Wubs, E.J., Van der Putten, W.H., Bosch, M., and Bezemer, T.M. (2016). Soil inoculation steers restoration of terrestrial ecosystems. *Nature Plants* 2: 1–5.

90 Frouz, J., Toyota, A., Mudrák, O. et al. (2016). Effects of soil substrate quality, microbial diversity and community composition on the plant community during primary succession. *Soil Biology and Biochemistry* 99: 75–84.

13

The Use of End Products on Soil

Biodegradable Waste Management in the Circular Economy: Challenges and Opportunities, First Edition.
Edited by Malgorzata Kacprzak, Eleonore Attard, Kari-Anne Lyng, Helena Raclavska, BalRam Singh, Eyob Tesfamariam, and Franck Vandenbulcke.

13.1

The Use of Biodegradable End Products on Soil – Impact on Plant Productivity

BalRam Singh[1], Anna Grobelak[2], Aneta Kowalska[2], and Åsgeir Almås[1]

[1]*Faculty of Environmental Sciences and Natural Resource Management, Norwegian University of Life Sciences, Ås, Norway*
[2]*Faculty of Infrastructure and Environment, Czestochowa University of Technology, Częstochowa, Poland*

13.1.1 Introduction

Production of biodegradable waste production is increasing because of a rising population, agricultural intensification, and the need for improved food production [1]. Currently, three billion residents of world cities generate about 1.3 billion Mg of waste per year, which may increase to 2.2 billion Mg per year by 2025 [2]. The reuse of wastes for agricultural purposes to improve soil properties and increase crop yield is a good solution for minimizing these problems. Nowadays, with the increasing demand to conserve natural resources and energy, recycling of wastes assumes major importance [3].

Such wastes can include a large variety of wastes such as animal manure, fish sludge, compost, biogas digestate, waste water sludge, garden waste, slaughter waste, and wood processing waste. Application of these wastes to soils can serve a dual purpose, i.e. (i) for disposal of solid waste from municipalities and agricultural operations and (ii) to improve chemical and physical soil properties, which in turn promotes improved crop performance. Furthermore, sustainable management strategies for disposal and recycling are required to reduce costs and negative environmental outcomes of landfill deposition [4]. For some of these wastes, especially animal manure and bio-solids (treated or stabilized sewage sludge (SS)), application to agricultural land is widely practiced and has shown positive effects on soil fertility and plant biomass [5, 6], along with improvements in soil chemical, physical, biological, and microbial properties [7, 8]. Bio-wastes are usually rich in organic matter (OM) and can contain essential plant nutrients such as nitrogen (N), phosphorus, sulfur, and potassium, but may also contain elevated concentrations of heavy metals, organic contaminants, and pathogens [9], which need to be taken into consideration while applying such wastes in agricultural soils. Because of the potential risks in using biosolids for agriculture, the use of biosolids to rebuild degraded land has become increasingly popular [10], as their application may promote topsoil development and enhance the re-establishment of vegetation, especially in degraded environments [11].

Biowaste (compost, manure, or various organic wastes) additions to soil have long been considered important in maintaining the quality of both natural and managed soils, principally because of their capacity to provide nutrients to the soil's living organisms and through their influence on the soil physical properties [12]. Several studies have reported that organic biowastes can improve soil physical, chemical, and biological properties, providing essential plant nutrients to stimulate plant growth and yield [13–15].

Although multiple investigations have documented the impact of biowastes in agricultural systems, their impacts in soil–plant systems associated with mine site reclamation and contaminated soils has received less attention. Some of these soils represent a destroyed soil–plant ecosystem. Restoration of these sites by adding biodegradable wastes provides a unique opportunity to utilize these wastes for reclamation of such soils by improving soil fertility, thus leading to increased productivity. The use of proper fertilization is also aimed at keeping the soil in good condition and preventing its depletion. Environmentally safe management of organic fertilizers is an important element in building a sustainable waste management system, aiding animal production, and maintaining the biological balance of the environment. Currently, in extensive agricultural crops, exogenous organic matter (EOM) can be an excellent source of soil organic matter (SOM), which is constantly being lowered as a result of cultivation treatments [16].

The objective of this chapter is to provide an overview of the impacts of biowaste application on soil quality and crop productivity in contaminated and degraded soils by looking into changes in soil properties and crop yield caused by the application of biowastes to soils.

13.1.2 Biodegradable Bio-wastes and Nutrient Management

Biodegradable bio-wastes consist of garden and park waste, food and kitchen waste from households, restaurants, caterers, and retail premises, and those from food processing plants.

Bio-waste treatment technologies are the processes that convert discarded bio-waste into new products with potentially some value. These technologies are grouped into four main categories: (i) direct use, (ii) biological treatment, (iii) physico-chemical treatment, and (iv) thermo-chemical treatment. Bio-wastes produced through these technologies can be used for animal food, soil amendments, and energy recovery. In the context of this paper, bio-wastes as soil amendments to improve soil quality and crop productivity will be the focus.

Agricultural bio-wastes, a source of carbon and plant nutrients, can be processed into different types of soil amendments with benefit for both crops and soils. However, with increasingly intense agricultural practices, soils are progressively vulnerable, especially in the tropics. Rapid carbon turnover (three to five times faster than in temperate regions) and extraction, decreasing nutrient retention and water storage capacity, and decreasing erosion resistance are highlighting the need for carbon and plant nutrient replenishment. This can be achieved by recycling organic waste into agriculture [17].

Waste management to organic resource utilization and to recover nutrients is the key way to have sustainable and eco-friendly agricultural production. Food demands of an ever-increasing population are on the increase and hence sustainable intensification is one of the ways to reduce the environmental impact of food production. Since humans and animals consume nutrients from crops and produce nutrient-rich waste, this waste from humans can fulfill 22% of the demand for phosphorus (P) [18]. However, the value of nutrient recovery from these wastes is very low, and they also contain heavy metals, pathogenic microorganisms, and odors [19]. Recycling nutrients is emerging as an economically sustainable method to replace synthetic fertilizers and reduce the emission of greenhouse gases (GHG) and eutrophication of waterways.

The traditional agricultural sources of EOM are mainly organic fertilizers such as manure and compost. Currently, other EOM additives are becoming more and more popular, such as sewage sludge, by-products from the food industry, composted fractions of municipal and industrial waste, digestate waste, or bone meal of animal origin [16].

Composting is useful in recycling nutrients and managing organic wastes and converting organic wastes into a marketable product. Composting is one of the best known recycling processes for the formation of conditioner from organic wastes and could help to improve the texture of light sandy soil, increase water retention, and enlarge system availability (the NPK (nitrogen, phosphorus and potassium) typical percentages in compost are N, 1.2%; P, 0.7%; K, 1.2%) and uptake of nutrients [18]. In the process of composting municipal sewage sludge, a decrease in the toxicity of heavy metals was noted. The analysis of the three-stage sequential chemical extraction (by the European Community Bureau of Reference, usually called the BCR method) confirmed that metals, after the composting of sewage sludge, passed mainly to forms II and III, i.e. practically inaccessible to the soil ecosystem under normal conditions of use [20].

Vermicomposting is an improved way of composting in which earthworms produce useful manure by reducing the harmful effects of waste. Vermicomposting is a combinable process in which earthworms and other microbes produce useful manures (vermicompost) and reduce the harmful effects of waste while maintaining the nutritional level [21]. The vermicompost main features include a higher surface area, a low value of the carbon to nitrogen ratio, i.e. C : N, maximum nutrients availability, a water-holding capacity, and increased porosity [22, 23].

In addition to soil conditioners, valuable composts from mechanical–biological waste treatment can also be obtained. Other organic waste, including sewage sludge from treatment plants in dairies, is also valuable from the point of fertilization. The sewage sludge from dairy sewage treatment contains significant amounts of fertilizing elements, which makes them attractive as an organic fertilizer or components for compost production. For sludge of dairy origin, the following average values of nutrients were recorded: K 3.1, Na 5.3, Ca 9.3, Mg 1.8, N 2.3, P 4.5 g/kg, and organic matter (OM) 75% of dry matter. Sewage sludge from a dairy wastewater treatment plant, compared to municipal sludge, has much lower contents of heavy metals and Ca [24]. Placek et al. [25] also successfully used sewage sludge from food processing plants in the process of remediation of soils contaminated with heavy metals (Figure 13.1.1).

A relatively new material is digestate from biogas plants. This biosolid is characterized by a high bioavailability of nutrients for plants. Studies have confirmed that up to 80% of the total nitrogen content is in the ammonium form, while for manure it is 10–15%. This form of nitrogen is easily absorbed by plants, quickly builds up into plant organic compounds, and due to exchangeable sorption in the soil, this form is less susceptible to leaching into groundwater than nitrates, reducing the problem of soil eutrophication [16].

13.1.3 Effect on Properties and Quality of Soils

13.1.3.1 Physical Properties

Application of bio-wastes leads to increased soil organic carbon, which in turn stimulates hydraulic conductivity by improving aggregate stability and porosity of agricultural soils [26]. According to Blanchet et al. [27], with the addition of manure, SOC increased significantly (by 6.2%) compared to a mineral fertilizer application. Similarly, Scotti et al. [28] found that with the incorporation of municipal solid waste (MSW) and manure, after one year of study, SOC increased by 36% and 25%, respectively, compared to the control plot.

Figure 13.1.1 Assisted phytoremediation of heavy metal contaminated sites (left control, right sewage sludge application); effect after four years of the study. Source: Adopted from Placek et al. [25].

In Germany, during a long-term investigation, Emmerling et al. [29] observed that all parameters related to soil quality were improved by the bio-waste compost application. Mamedov et al. [30] evaluated the effects of three different types of waste, with chemical amendments on aggregate stability of three types of soil before and after subjecting the soils to six rainstorms. They found that the addition of organic wastes remarkably improved macroporosity and the stability of macroaggregate. Organic matter contributes to the formation of soil aggregates, which in turn promotes the formation of pores in the soil. Consequently, such soils show a higher moisture retention capacity and gas exchange, root development, hydraulic conductivity, as well as reduced bulk density (BD). Porosity also positively influences the nutrient cycle and microbial activity [31]. Hemmat et al. [32] reported that the long-term incorporation of sewage sludge, municipal solid waste compost (MSWC), and farmyard manure into soil increased SOC but decreased soil bulk density (BD) and the lower BD values were associated with higher SOC (Table 13.1.1).

The use of organic fertilizers significantly improves the sorption capacity of soils and contributes to the reduction of their acidification. Inorganic fertilizers increase the acidification of agricultural soils. With a strongly acidic pH of the soil, a strongly acidic humus is formed, which looses the properties of soil particles sticking together, which results in the loss of the soil lumpy structure. Thus, the water and air conditions in the soil deteriorate and the microbiological activity in the soil is inhibited [16]. Studies have found that an increase in soil pH is often the main factor influencing the accumulation and decomposition of organic matter in soil after EOM application. Few cases have seen a drop in soil pH following the introduction of EOM, but these have been very few experiences [33]. One of the most important effects of the formation and stabilization of soil aggregates is the reduction of the erosion of these soils and the decrease of surface runoff. The more complex organic particles, such as lignins and humic substances, will improve the soil structure [16].

Table 13.1.1 Soil physical properties as affected by long-term application of different biowastes and inorganic fertilizers.

Treatment	SOC (g kg^{-1})	Bulk density (Mg m^{-3})
25 Mg SS ha^{-1}	16.2d	1.32ab
100 Mg SS ha^{-1}	28.0a	1.03c
25 Mg MSWC ha^{-1}	14.5d	1.33ab
100 Mg MSWC ha^{-1}	22.2b	1.03c
25 Mg FYM ha^{-1}	14.9d	1.32ab
100 Mg 100 ha^{-1}	22.4b	1.06c
Inorganic fertilizer	10.1c	1.37a
Control	8.9c	1.34ab

Within columns, means followed by different letters differ at $P < 0.05$ as tested by LSD.
SS: sewage sludge, MSWC: municipal solid waste compost, FYM: farmyard manure, SOC: soil organic carbon, BD: bulk density.
Source: Modified and adapted from Hemmat et al. [32].

13.1.3.2 Chemical Properties

Biowastes are good sources of plant nutrients and influence chemical properties of soil. In a 50-year long-term study, Blanchet et al. [27] found that manure application improved chemical properties of soil and provided a significant amount of P and K. Scotti et al. [28] found that in the soil treated with MSW and manure, total nitrogen was increased by 60 and 40%, respectively, compared to the untreated soil after one year of study. They also observed that with the addition of manure, available P content was 36% higher than in the untreated soil. Nest et al. [34] reported that after a long-term (40–50 years) experiment with manure application, extractable P was increased two to four times as compared to a mineral P fertilizer application. The authors proved the application of manure to be more effective in increasing soil pH and P availability than the mineral fertilizer.

Meena et al. [35] conducted a field experiment in India and examined the combined use of MSW compost, gypsum-enriched compost, rice straw compost, and chemical fertilizers on biological and chemical properties of saline soil in a mustard–pearl millet cropping system. They found that the MSW application with mineral fertilizer increases noticeably the availability of N, P, and K in soil over the control plot after a mustard and pearl millet harvest (Table 13.1.2).

Biosolids application increased plant N concentrations in ryegrass and rapeseed and increased N uptake into the seeds of sorghum, increasing seed quality and yield (Figure 13.1.2). Biosolids application did result in lower N leaching compared with urea, irrespective of plant species, and N leaching was unaffected by mixing the biosolids with sawdust [36] (Figure 13.1.2).

Moreover, biosolids have a wide range of C/N ratios. If the biosolid contains too much N to C, then N is released into the soil in ammonium form and can be quickly oxidized to nitrate by the microorganisms and then leached into the groundwater. Conversely, C is mineralized and the N in the soil is immobilized in the microbial biomass [31]. Unlike mineral fertilizers, the content of nutrients in biosolids is sometimes unbalanced.

Table 13.1.2 Effects of organic amendments and mineral fertilizers on nutrient availability (kg ha^{-1}) after harvest of mustard and pearl millet crops.

	After mustard harvest			After pearl millet harvest		
Treatment	**N**	**P**	**K**	**N**	**P**	**K**
T1	84d	11d	176f	85	11d	171d
T2	136c	32c	251e	145b	33c	255c
T3	138c	31c	255 cd	152ab	37b	258c
T4	146c	33bc	266bc	161ab	38b	269b
LSD (P = 0.05)	15.0	3.0	12.3	19.9	2.5	10.2

T1, Control; T2, 100% recommended dose of fertilizer (RDF); T3, rice-straw compost (RSC) at 14 t ha^{-1}; T4, municipal solid waste compost (MSWC) at16 t ha^{-1}. Within columns, means followed by the same letter are not significantly different at $P = 0.05$ according to Duncan's multiple range test for separation of means.
Source: Modified and adapted from Meena et al. [35].

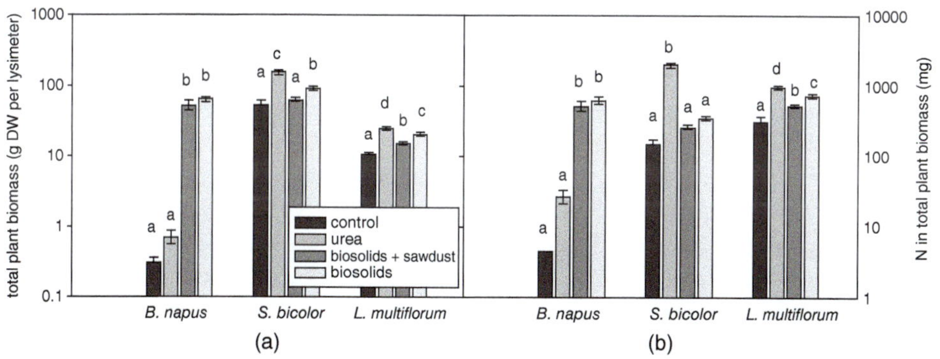

Figure 13.1.2 Total plant biomass (including seed and root biomass) at the end of the experiment (a) and total N recovered from plant biomass (b) from sorghum (*Sorghum bicolor*), rapeseed (*Brassica napus*), and ryegrass (*Lolium multiflorum*). Different levels of significance $p \leq$ £0.05 are represented by lowercase letters. Source: Adopted from Esperschuetz et al. [36].

13.1.3.3 Biological Properties

Bio-wastes influence remarkably the biological properties of soil. Accordingly, Meena et al. [35] observed that a combined use of MSW and chemical fertilizers increased microbial activities. Similarly, Scotti et al. [28] reported that phosphomonoesterase (+260% and +219%), β-glucosidase (+180% and +74%), and fluorescein diacetate hydrolase (+160% and +101%) enzymes were increased for MSW and manure, respectively, whereas urease enzyme was increased only in plots treated with MSW compost. With the addition of sewage sludge (SS), soil biological parameters such as microbial communities and enzymatic activities (dehydrogenase, phosphatase, protease, and cellulase) were stimulated, as reported by Zoghlami et al. [37]. Meena et al. [35] also demonstrated that municipal solid waste compost (MSWC) at 8 t ha^{-1} +50% with recommended dose fertilizer (RDF) gave significant improvement in dehydrogenase, alkaline phosphatase, and urease activity as well as soil fertility in terms of available nutrients over use of 100% RDF alone after the second year of applications of mustard and pearl millet. The use of biosolids has a complex effect on the condition of microorganisms, including the increase in the number and metabolic activity of microorganisms, and causes changes in the their structure. Microorganisms are a very sensitive and early indicator of the condition of the soil environment [16]. In the microbiological tests of soils after the use of biosolids (digestate from biogas plants, compost, and bone meal), stimulation of soil microorganism communities and increased genetic and functional diversity were found, but this effect was periodic. The long-term use of this waste in agriculture seems to be safe for the functioning of genetic diversity of indigenous microorganisms living in the soil [38].

13.1.4 Productivity in Wastes and End Products Amended Soils

As stated above, biowastes not only affect soil properties and soil quality but also contribute to growth and development of plants and thus improve agricultural productivity. The positive effects of waste on plant growth were also proved by Tampio et al. [39]. The authors evaluated the agronomic characteristics of five urban waste types and registered by 5–30% higher ryegrass yields. Debiase et al. [40] reported that incorporated organic wastes gave 32% higher yields than an unfertilized field. Moreover, the authors confirmed that SS application ensured a 12% higher yield compared to MSW addition of a wheat field. These amendments not only increase crop yields, but also minimize the risk of nitrogen leaching from the soil. Meena et al. [35] reported higher grain and straw yield of mustard and pearl millet with integrated use of organic amendments and 50% of a recommended dose (RDF) compared to 100% RDF alone (Table 13.1.3). Grain yields of crops were significantly affected by application of RSC and MSWC along with 50% RDF. Treatment receiving RSC +50% RDF increased grain yield by 10% and 28% for mustard and pearl millet, respectively, relative to 100% RDF alone.

Studies conducted on post-fermentation waste in European countries, analyzing their impact on the yield of cultivated plants, led to the conclusion that their fertilization value was between manure and inorganic fertilizers, and often even equaled the latter [31].

High rates of biosolids application to low-fertility soil resulted in a significant growth response of rapeseed, sorghum, and ryegrass compared with control treatments. Rapeseed produced negligible biomass in the control and urea treatments due to a lack of nutrients available to maintain the growth of this species [41]. Placek et al. [42] found that sewage sludge, municipal compost, and lake chalk application to post-Zn mine soil improved the soil quality for healthy plant growth and development. Giant Miscanthus was characterized by an extensive root system. Moreover, higher root biomass production of Giant Miscanthus growing on uncontaminated soil (lignite mine soil) was recorded. Smaller differences between root growth of Scots Pine cultivated in smelter-polluted soil and in post-mine soil were observed.

In the study of the effect of natural fertilizers in the organic matter balancing calculation in soil, it was found that the highest calculated soil reproduction ratio according to Eich and Kundler [43, 44] was recorded for the cultivation of catch crops (+0.7) and subsequent fertilization with manure (+0.35) and liquid manure (+0.28) in the cultivation of cereals, while the highest soil degradation rates were recorded for the cultivation of potatoes and maize alone. The highest increase in the level of organic matter (by 21%) was recorded after fertilization with manure (Table 13.1.4) [43]. The method of Eich and Kundler, in a simple and quick way, allows estimates to be made on whether organic matter is degraded in each crop rotation and tillage system or whether it increases (reproduction). The values of these factors for various plants and the different soil types correspond

Table 13.1.3 Yield (Mg ha^{-1}) of mustard and pearl millet as influenced by MSWC, RSC and mineral fertilizers.

	Grain yield	
Treatment	**Mustard**	**Pearl millet**
T1	1.73*c*	2.11*b*
T2	2.10*b*	2.34*ab*
T3	2.17*ab*	2.35*ab*
T4	2.20*ab*	2.53*a*
T5	2.27*ab*	2.59*a*
T6	2.50*a*	2.70*a*
LSD (P = 0.05)	0.34	0.35

T1, Control; T2, 100% recommended dose of fertilizer (RDF); T3, rice-straw compost (RSC) at 14 t ha^{-1}; T4, municipal solid waste compost (MSWC) at 16 t ha^{-1}; T5, RSC at 7 t ha^{-1} + 50% RDF; T6, MSWC at 8 t ha^{-1} + 50% RDF. Within columns, means followed by the same letter is not significantly different at $P = 0.05$ according to Duncan's multiple range test for separation of means.
Source: Modified and adapted from Meena et al. [35].

Table 13.1.4 Balance of soil organic matter (OM) for a farm (breeding cows and pigs).

Plant or fertilizer	**Reproduction (+)/degradation rate (–) (Eich – Kundler, after [43])**	**The mass of the organic substance (T)**
Winter wheat		
Winter barley	−0.53	−2.45
Spring barley	−0.53	−2.14
Potatoes	−1.4	−0.8
Maize	−1.15	−2.92
Postharvest	+0.7	5.43
Manure	+0.35	21.90
Slurry	+0.28	0.90
Straw	+0.21	10.50

Source: Changed and adapted from Kuczuk and Wacław [43].

to the amount of organic matter in Mg ha^{-1} with which the soil will be enriched (+, soil reproduction factor) or depleted (–, soil degradation factor) as a result of the one-year cultivation of the plant concerned. The research confirmed the importance of organic fertilization in the balancing of SOM. Plants (cereals, potatoes, maize) cultivated as the main crop have a degrading effect on the soil. The use of organic fertilizers balances the organic matter and reduces the risk of soil depletion.

In the studies conducted in Poland and the Czech Republic, where post-fermentation waste from biogas plants was used, the yield of maize and wheat was recorded at a level like the equivalent mineral fertilization, or even higher. The biogas digestate used (pH 7.5, C 40%) (residue from the production of fries, pulp, corn; anaerobic fermentation) was rich in easily digestible for plants micro and macro elements (N 4.4–6.9%, P 0.7–2.8%). Similar results were obtained for animal bone meal (pH 6.1, C 40%, N 8.4%, P 6.42%), except for crops on clay soil. On the other hand, the composts used (agricultural from manure; pH 8.4, C 24%, N 2.6%, P 3.35%) and industrial (sewage sludge and other waste; pH 8.3, C 23%, N 1.7%, P 0.39%) gave slightly lower results of plant yield in the first year of experimentation, which is probably related to the insufficient amount of nitrogen available to plants due to the slow mineralization of organic matter [45].

13.1.5 Summary and Conclusions

The use of biodegradable end products on soil is becoming an important issue nowadays. The main factors driving this trend of biosolids usage are as follows:

- Decreasing soil productivity
- Soil erosion, depletion, and drought
- Increasing GHG emissions
- Increased social aspect concerning the soil protection
- Required increase in food production areas due to growing human populations
- Proven impact on plant productivity
- Increasing number of farms without livestock
- Research for new soil cultivation systems providing soil resistance to degradation and sudden climate changes
- The 3R (reuse, recycle, revenue) strategy and similar law regulations and legislation about the biosolids' fate, reuse, and treatment
- Growing public interest and awareness of ecological crops cultivation

SOM is a fundamental parameter determining the physical, chemical, and biological functions of soil, including fertility, sorption capacity, and water capacity. Organic matter in soils reduces their susceptibility to water and wind erosion and leaching of valuable plant nutrients.

The use of organic fertilizers has the greatest impact on the positive balance of SOM, while the cultivation of plants with the use of inorganic fertilizers, even after post-harvest plant cultivation, does not allow a positive balance to be achieved. Additionally, more and more frequent agricultural monocultures exacerbate the problem of soil impoverishment. The most important anthropogenic factors contributing to changes in the content of organic matter in the soil are agricultural land use, intensity of agricultural crops, agricultural technology, and the application of good agricultural practice, including crop rotation and fertilization methods. The decrease in organic matter content is a clear indicator of soil degradation and decreased fertility.

Organic matter is, in a way, a store of carbon dioxide, but it can also be a source of it, depending on the use of the soil. Maintaining adequate SOM is also important because of the importance of soil in CO_2 sequestration, which reduces the greenhouse effect.

The source of EOM in the soil can be traditional organic fertilizers and composts, but also sewage sludge, digestate from biogas plants, and municipal, industrial, and animal waste. Currently, the amount of waste produced or separated from other waste streams is increasing every year. Biosolids can contain hazardous substances such as organic compounds, heavy metals, and other trace elements, and therefore they should be used in accordance with legal regulations. Scientists also point out that biosolids can provide many valuable micronutrients that cannot be found in nutrient-specific inorganic fertilizers. Increasing the standard of living of the population leads to an increase in the amount of organic waste, which in a way constitutes a potential fertilizer. The safe use of biosolids is an issue that still requires research. This is due to the fact that (i) not all pollutants are on the list of pollutants routinely tested, (ii) the synergistic and antagonistic effects occurring are difficult to assess, (iii) organic pollutants on soil particles may be biodegradable or easily accessible, and (iv) an unsustainable chemical composition can be a threat. The most valuable information on the use of biosolids is provided by long-term field tests, which also consider soil resistance (to the applied additives) and reproducibility of soil characteristics (return to the original state).

Acknowledgements

This work was supported by an internal research grant of Czestochowa University of Technology. The research leading to these results has received funding from the EnviSafeBioC project – Contract No. PPI/APM/2018/1/00029/U/001. The project is financed by the Polish National Agency for Academic Exchange (NAWA).

References

1 Río, A.V., Morales, N., Isanta, E. et al. (2011). Thermal pre-treatment of aerobic granular sludge: impact on anaerobic biodegradability. *Water Research* 45 (18): 6011–6020.

2 Hossain, M.Z., Von Fragstein, P., Von Niemsdorff, P., and Heß, J. (2017). Effect of different organic wastes on soil propertie s and plant growth and yield: a review. *Scientia Agriculturae Bohemica* 48 (4): 224–237.

3 Padmavathiamma, P.K., Li, L.Y., and Kumari, U.R. (2008). An experimental study of vermi-biowaste composting for agricultural soil improvement. *Bioresource Technology* 99: 1672–1681. https://doi.org/10.1016/j.biortech.2007.04.028.

4 Amajirionwu, M., Connaughton, N., McCann, B. et al. (2008). Indicators for managing biosolids in Ireland. *Journal of Environmental Management* 88 (4): 1361–1372.

5 Mok, H.F., Majumder, R., Laidlaw, W.S. et al. (2013). Native Australian species are effective in extracting multiple heavy metals from biosolids. *International Journal of Phytoremediation* 15 (7): 615–632.

6 Tejada, M., Hernandez, M.T., and Garcia, C. (2009). Soil restoration using composted plant residues: effects on soil properties. *Soil and Tillage Research* 102 (1): 109–117.

7 Cytryn, E., Kautsky, L., Ofek, M. et al. (2011). Short-term structure and functional changes in bacterial community composition following amendment with biosolids compost. *Applied Soil Ecology* 48 (2): 160–167.

8 Li, Y.M., Chaney, R.L., Siebielec, G., and Kerschner, B.A. (2000). Response of four turfgrass cultivars to limestone and biosolids – compost amendment of a zinc and cadmium contaminated soil at Palmerton, Pennsylvania. *Journal of Environmental Quality* 29 (5): 1440–1447.

9 Bolan, N., Kunhikrishnan, A., Thangarajan, R. et al. (2014). Remediation of heavy metal(loid)s contaminated soils – to mobilize or to immobilize? *Journal of Hazardous Materials* 266: 141–166.

10 Dere, A.L., Stehouwer, R.C., Aboukila, E., and McDonald, K.E. (2012). Nutrient leaching and soil retention in mined land reclaimed with stabilized manure. *Journal of Environmental Quality* 41 (6): 2001–2008.

11 Lu, Q., He, Z.L., and Stoffella, P.J. (2012). Land application of biosolids in the USA: a review. *Applied and Environmental Soil Science* 2012: 201462. https://doi.org/10.1155/2012/201462.

12 McGeehan, S.L. (2012). Impact of waste materials and organic amendments on soil properties and vegetative performance. *Applied and Environmental Soil Science* 2012: 907831. https://doi .org/10.1155/2012/907831.

13 Ling, N., Zhu, C., Xue, C. et al. (2016). Insight into how organic amendments can shape the soil microbiome in long-term field experiments as revealed by network analysis. *Soil Biology and Biochemistry* 99: 137–149. https://doi.org/10.1016/j.soilbio.2016.05.005.

14 Liu, M., Hu, F., Chen, X. et al. (2009). Organic amendments with reduced chemical fertilizer promote soil microbial development and nutrient availability in a subtropical paddy field: the influence of quantity, type and application time of organic amendments. *Applied Soil Ecology* 42: 166–175. https://doi.org/10.1016/j.apsoil.2009.03.006.

15 Xie, H., Li, J., Zhu, P. et al. (2014). Long-term manure amendments enhance neutral sugar accumulation in bulk soil and particulate organic matter in a Mollisol. *Soil Biology and Biochemistry* 78: 45–53. https://doi.org/10.1016/j.soilbio.2014.07.009.

16 Maly, S. and Siebielec, G. (2015). Badania egzogennej materii organicznej w celu bezpiecznego stosowania do gleby. Centralny Instytut Badań i Nadzoru w Rolnictwie, Brno (Republika Czeska), ss. 151 [in Polish].

17 Smith, J.U., Fischer, A., Hallett, P.D. et al. (2015). Sustainable use of organic resources for bioenergy, food and water provision in rural sub-Saharan Africa. *Renewable and Sustainable Energy Reviews* 50: 903–917.

18 Ahmed, M., Ahmad, S., Qadir, G. et al. (2019). Innovative processes and technologies for nutrient recovery from wastes: a comprehensive review. *Sustainability* 11 (18): 4938.

19 Mehta, C.M., Khunjar, W.O., Nguyen, V. et al. (2015). Technologies to recover nutrients from waste streams: a critical review. *Critical Reviews in Environmental Science and Technology* 45 (4): 385–427.

20 Ignatowicz, K. (2017). The impact of sewage sludge treatment on the content of selected heavy metals and their fractions. *Environmental Research* 156: 19–22.

21 Rorat, A., Suleiman, H., Grobelak, A. et al. (2016). Interactions between sewage sludge-amended soil and earthworms – comparison between *Eisenia fetida* and *Eisenia andrei* composting species. *Environmental Science and Pollution Research* 23 (4): 3026–3035.

22 Lim, S.L., Wu, T.Y., Lim, P.N., and Shak, K.P.Y. (2015). The use of vermicompost in organic farming: overview, effects on soil and economics. *Journal of the Science of Food and Agriculture* 95 (6): 1143–1156.

23 Lim, S.L., Lee, L.H., and Wu, T.Y. (2016). Sustainability of using composting and vermicomposting technologies for organic solid waste biotransformation: recent overview, greenhouse gases emissions and economic analysis. *Journal of Cleaner Production* 111: 262–278.

24 Dąbrowski, W. (2011). Effectiveness of constructed wetlands for dairy wastewater treatment. *Ecological Chemistry and Engineering A* 18 (2): 175–181.

25 Placek, A., Singh, B.R., Grobelak, A. et al. (2018). Phytosequestration of carbon in Miscanthus× giganteus and *Pinus sylvestris* L. in degraded zinc smelter and post-mining soils. Phytosequestration of carbon in Miscanthus× giganteus and *Pinus sylvestris* L. in degraded zinc smelter and post-mining soils. *Pedosphere* 28 (3): 555–560.

26 Eibisch, N., Durner, W., Bechtold, M. et al. (2015). Does water repellency of pyrochars and hydrochars counter their positive effects on soil hydraulic properties? *Geoderma* 245–246: 31–39. https://doi.org/10.1016/j. geoderma.2015.01.009.

27 Blanchet, G., Gavazov, K., Bragazza, L., and Sinaj, S. (2016). Responses of soil properties and crop yields to different inorganic and organic amendments in a Swiss conventional farming system. *Agriculture, Ecosystems and Environment* 230: 116–126. https://doi.org/10.1016/j.agee.2016.05.032.

28 Scotti, R., Pane, C., Spaccini, R. et al. (2016). On-farm compost: a useful tool to improve soil quality under intensive farming systems. *Applied Soil Ecology* 107: 13–23. https://doi.org/10.1016/j.apsoil.2016.05.004.

29 Emmerling, C., Udelhoven, T., and Schneider, R. (2010). Long-lasting impact of biowaste-compost application in agriculture on soil-quality parameters in three different crop-rotation systems. *Journal of Plant Nutrition and Soil Science* 173: 391–398. https://doi.org/10.1002/jpln.200900348.

30 Mamedov, A.I., Bar-Yosef, B., Levkovich, I. et al. (2014). Amending soil with sludge, manure, humic acid, orthophosphate and phytic acid: effects on aggregate stability. *Soil Research* 52: 317–326. https://doi.org/10.1071/SR13334.

31 Nkoa, R. (2014). Agricultural benefits and environmental risks of soil fertilization with anaerobic digestates: a review. *Agronomy for Sustainable Development* 34 (2): 473–492.

32 Hemmat, A., Aghilinategh, N., Rezainejad, Y., and Sadeghi, M. (2010). Long-term impacts of municipal solid waste compost, sewage sludge and farmyard manure application on organic carbon, bulk density and consistency limits of a calcareous soil in Central Iran. *Soil and Tillage Research* 108 (1–2): 43–50. https://doi.org/10.1016/j.still.2010.03.007.

33 Hargreaves, J.C., Adl, M.S., and Warman, P.R. (2008). A review of the use of composted municipal solid waste in agriculture. *Agriculture, Ecosystems & Environment* 123 (1–3): 1–14.

34 Nest, T.V., Ruysschaert, G., Vandecasteele, B. et al. (2016). The long- term use of farmyard manure and compost: effects on P availability, orthophosphate sorption strength and P leaching. *Agriculture, Ecosystems and Environment* 216: 23–33. https://doi.org/10.1016/j.agee.2015.09.009.

35 Meena, M.D., Joshi, P.K., Narjary, B. et al. (2016). Effects of municipal solid waste compost, rice-straw compost and mineral fertilisers on biological and chemical properties of a saline soil and yields in a mustard-pearl millet cropping system. *Soil Research* 54 (8): 958–969. https://doi.org/10.1071/SR15342.

36 Esperschuetz, J., Anderson, C., Bulman, S. et al. (2016a). Production of biomass crops using biowastes on low-fertility soil: 1. Influence of biowastes on plant and soil quality. *Journal of Environmental Quality* 45 (6): 1960–1969.

37 Zoghlami, R.I., Hamdi, H., Mokni-Tlili, S. et al. (2016). Changes in light-textured soil parameters following two successive annual amendments with urban sewage sludge. *Ecological Engineering* 95: 604–611.

38 Frąc, M., Oszust, K., and Lipiec, J. (2012). Community level physiological profiles (CLPP), characterization and microbial activity of soil amended with dairy sewage sludge. *Sensors* 12 (3): 3253–3268.

39 Tampio, E., Marttinen, S., and Rintala, J. (2016). Liquid fertilizer products from anaerobic digestion of food waste: mass, nutrient and energy balance of four digestate liquid treatment systems. *Journal of Cleaner Production* 125: 22–32.

40 Debiase, G., Montemurro, F., Fiore, A. et al. (2016). Organic amendment and minimum tillage in winter wheat grown in Mediterranean conditions: effects on yield performance, soil fertility and environmental impact. *European Journal of Agronomy* 75: 149–157.

41 Esperschuetz, J., Bulman, S., Anderson, C. et al. (2016b). Production of biomass crops using biowastes on low-fertility soil: 2. Effect of biowastes on nitrogen transformation processes. *Journal of Environmental Quality* 45 (6): 1970–1978.

42 Placek, A., Grobelak, A., Hiller, J. et al. (2017). The role of organic and inorganic amendments in carbon sequestration and immobilization of heavy metals in degraded soils. *Journal of Sustainable Development of Energy, Water and Environment Systems* 5 (4): 509–517. Available from: https://doi.org/10.13044/j.sdewes.d5.0166.

43 Kuczuk, A. and Wacław, S. (2014). Fertilization and farm fertilizing balance as an agri-environmental indicator. *Journal of Research and Applications in Agricultural Engineering* 59 (2): 50–56.

44 Kundler, P., Eich, D., and Liste, J. (1981). Do more than replacing only: increased supply of organic substance to soil-significant requirement for higher increase in plant production. *Neue Bauernzeitung* 22: 8–9.

45 Meller, E., Niedźwiecki, E., Rogalska, P. et al. (2015). Fertiliser value and trace element content of composts produced from different wastes. *Journal of Ecological Engineering* 16 (4): 154–160.

13.2

Biodegradable End Products for Bioremediation of Degraded Areas

Aneta Kowalska[1], Anna Grobelak[1], Åsgeir Almås[2], and BalRam Singh[2]

[1] *Faculty of Infrastructure and Environment, Czestochowa University of Technology, Częstochowa, Poland*
[2] *Faculty of Environmental Sciences and Natural Resource Management, Norwegian University of Life Sciences, Ås, Norway*

13.2.1 Introduction

The problem of degraded soils is receiving more attention as the areas of degraded soils are ever increasing due to anthropogenic activity. There is, therefore, an urgent need to implement all possible actions directed to improve their quality and protect areas from further degradation. Rapid expansion and mismanagement of soil sources contribute to more land degradation, resulting in a shrinkage of the areas useful for agriculture purposes. Soil degradation constitutes the problem not only for the agricultural sector but also leads to leaching of contaminants into deeper horizons of soils, causing groundwater contamination with toxic pollutants. Hence, this problem requires active cooperation of politicians and scientists to develop and implement methods of effective improvement of degraded areas [1]. Bioremediation of degraded soils seems to constitute an effective and economic method for soil reclamation as compared to physical and chemical methods. Bioremediation involves the activity of living organisms able to accumulate or degrade toxic substances present in degraded soils. Such processes may be conducted by microorganisms or plants (phytoremediation) [1, 2]. Bioremediation uses the ability of selected living organisms to grow under stress conditions such as in toxic compounds. In bioremediation, the enzymes are synthetized by microorganisms [3, 4]. Nevertheless, in many cases, the degradation of soil is so advanced that proper growth and development of living organisms is strongly limited or even impossible [5]. It is caused not only by high concentrations of contaminants (e.g. heavy metals) but also, in many areas, it is combined with a low content or absolute lack of organic matter and nutrients [6]. In such cases, the addition of biodegradable wastes to soils may be helpful in supporting growth of key organisms that carry out the remediation process.

The ever-increasing production of biowastes worldwide necessitates recycling of these wastes due to their high content of organic matter. Biowastes involve the wastes that are susceptible to biological degradation and may be used as a medium for organisms living in the soil [7]. Such methods have been widely used in agriculture since the dawn of civilization because farmers used organic agricultural waste for soil fertilization. At present, it is desirable to use the biodegradable end products fully in order to advance sustainable development in a circular economy. The use of biodegradable end products for bioremediation purposes requires meeting the specific regulations for their composition and doses. Application of biodegradable end products also contributes to carbon sequestration, increasing the soil organic carbon stock in the soil and in a further perspective mitigating climate changes [8]. In this chapter the aim is to bring current knowledge on the impact of biodegradable biowaste end products on remediation of degraded soils in order to improve their quality and productivity.

13.2.2 Degraded Soils and Magnitude

Soil degradation worldwide, beside its agriculture and ecological problems, also affects food security. In developing countries, the problem with soil degradation is most serious [9]. Soil degradation may occur in many forms including, for instance, soil desertification in dry climates and low productivity soils characterized by poor root development in more humid climates [10, 11]. Degraded soils also lead to the development of hardpans and compaction, leading to limited oxygen diffusion and water infiltration and seriously decreasing the moisture storage capacity of the soil [11].

Soil degradation may occur in different kinds including, for example, soil erosion, nutrient and organic matter depletion, acidification, contamination, and salinization. Soil erosion is a major and serious problem closely connected with degraded soils with a lack of vegetation cover, poor soil structure, soil compaction, and sloppy lands. Such a phenomenon is the effect of soil exposition due to various deleterious weather conditions [12].

Soil degradation has been assigned the definition of "global pandemic" by DeLong et al. [13], as it does not constitute a problem only in specific region, but it is rather a global problem. Intensive farming coupled with soil mismanagement and non-judicious use of fertilizers have contributed to the loss of fertile soil, resulting in a loss of productivity by 50% caused by soil erosion and desertification, which affects yields and causes ahunger problem [14].

Global estimates of total degraded areas range widely (Table 13.2.1) [15]. Estimations of soil degradation have been conducted by several attempts, such as the Global Assessment of Soil Degradation (GLASOD). This method has been widely used and is considered to be the complete tool providing consistent information about continental soil degradation. Other methods, such as the FAO's Global Assessment of Lands Degradation and Improvement project (GLADA) and TerraSTAT, are also used. The TerraSTAT method shows degraded land areas for country data, in comparison to the actual degraded area in GLASOD [15].

13.2.3 Bioremediation and Technologies

Many methods have been implemented to improve soil quality of degraded soils, but some of them (e.g. chemical: catalysis, ion exchange, flushing, and thermal treatment) require large financial outlays and may, on a large scale, contribute to further degradation via application of additional chemicals [16]. One of the economic and effective methods is bioremediation, which is widely used and recommended by researchers and ecologists [17]. Bioremediation refers to the method of soil rehabilitation by use of biological agents, such as microorganisms, plants, or enzymes [18, 19]. In this method, both catabolic and anabolic reactions may be used. Bioremediation is more cost-effective in comparison to conventional methods, especially due to the possibility of carrying it out in situ and ex situ, thus reducing the cost of the reclamation process significantly [17, 20]. Although this process has many advantages, it also suffers from many limitations. The proper growth and development of the organisms used depends on the external conditions and also on the kind and concentration of the contaminants. Selection of appropriate organisms for bioremediation is essential for an effective reclamation [21].

Table 13.2.1 Continental and global estimated soil degradation (million ha) [15].

Area	GLASOD	GLADA	FAO TerraSTAT
Europe	158	65	403
Asia	453	912	2501
Africa	321	660	1222
North America	140	469	796
South America	139	398	851
Australia and Pacific	6	236	368
Total	1216	2740	6140

Bioremediation may be based on different reactions and techniques. One of them is bioaugmentation, which refers to the application of selected microorganisms into degraded soils, which can survive in polluted areas and decomposed contaminants. Microbes are introduced to soils in an immobilized form as carriers (e.g. biocarbon, nanosilica), or may be introduced directly to the targeted area. Application of carriers in this technique allows an assessment to be made of higher microbial survival and more effective bioaugmentation [22]. Biostimulation refers to all constructive and supporting techniques that aim to provide optimal growth and proliferation conditions for native microorganisms in order to mobilize them in decomposition processes [23]. These treatments may include, for instance, adjustment of the pH value (e.g. liming) or bioventilation. Bioremediation technologies also provide the possibility of using the potential plants in the phytoremediation process. This technology focuses on the use of plants tolerant to adverse conditions and that can decompose or over-accumulate contaminants in their above-ground tissues.

A common problem for each bioremediation technology for degraded areas is the lack of nutrients that are necessary for metabolic reactions and growth of microorganisms and plants. To meet this challenge, use of organic end products in a circular economy has received greater attention in recent years [24]. This treatment increases soil nutritive potential and accelerates metabolic processes leading to decomposition of pollutants and accelerating soil quality improvement. Sewage sludge, as an end product, is produced worldwide in large amounts and has been widely used in soil reclamation processes [7]. Similarly, compost and vermicompost produced under aerobic fermentation have also gained popularity in assisted bioremediation technologies. Both end products show a high potential as soil conditioners due to their high organic matter content, which is a key element for depleted soils [25]. Since the final treatment of sewage sludges and composts consist of the major cost, their end use as soil conditioners in assisted bioremediation may lead to cost cutting to some extent. In recent years, more and more attention has been paid on the biochar application into a soil due to its strong absorption ability [26]. Besides the end products mentioned above, other wastes, e.g. biowastes from the wood processing industry, such as bioash, due to its alkaline properties (useful in raising soil pH) [27, 28], and animal manure, due to their high organic matter content, may also be used in bioremediation [24].

13.2.4 Improvement in Properties of Degraded Soils

In degraded soils, application of organic end products is known to alter their productivity potential. Many studies have shown a positive impact of end product application during the bioremediation [24, 28]. As mentioned above, end products rich in organic matter (e.g. sewage sludge, composts, vermicompost) increase soil fertility. Application of fertilizers based on lime to stabilize bio-waste into acidic grasslands resulted in improved soil fertility, leading to higher pasture production [29]. Seyedalikhani et al. [30] tested potential usage of sewage sludge to improve soil fertility of degraded stony silt loam soil aiming for essential oil production. They observed increased soil productivity in terms of increased biomass up to 86%. Organic matter in end products mobilizes microbial activity, increases soil respiration and soil enzyme activities, improves soil organic mineralization, and leads to higher soil productivity [31–33]. Therefore, the positive influence of organic end products on soil bioremediation has been noticed in higher soil microbial biomass [34] and diversity of soil microbial communities [35].

Assisted bioremediation with end products alters soil properties. Many studies have showed that the end products, depending on their source and composition, may influence soil pH either by lowering [36] or by increasing the pH value [37, 38]. Therefore, an end product application in degraded soil results in improved total organic carbon (TOC), N, P, K, Fe content, C/N ratio, and cation-exchange capacity [33, 36]. Bioremediation of degraded soil with an assisted application of end products may result in improved electron conductivity, dissolved organic carbon, and humic substances [39]. It has been observed that the impact of end products on soil bulk density (BD) may lead to either decreased or increased BD [40, 41].

The application of end products rich in organic matter for bioremediation of lands contaminated with metal trace elements and other contaminants influenced immobilization of pollutants in the organic structure, making them unavailable for living organisms and lowering soil toxicity. It was observed that the use of end products, such as sewage sludge in assisted phytoremediation, resulted in higher concentrations of metals in above-ground tissue, indicating progressive soil restoration [42]. Similarly, in soil contaminated with petroleum hydrocarbons and spiked soil amended with compost, biochar, and consortia increased hydrocarbon removal up to 85% [43]. A similar effect was also observed in the soil contaminated with polychlorinated bisphenols (PCBs), where significantly higher degradation of PCBs under bioremediation assisted by green garden waste biochar was observed [44]. Higher bioremediation efficiency and lower phytotoxicity of soil have been observed in areas degraded by the sulfur mining industry threated with post-flotation lime and sewage sludge [1].

13.2.5 Limitations and Negative Impacts on Soil Quality

The application of end products to a soil is strictly regulated by national laws due to the presence of undesirable elements in them. Before their soil application, end products need to meet a specific regulation, especially for organic matter content, concentration of heavy metals and other contaminants, and sanitary safety included in European standards and regulations on the use of organic waste in the soil (e.g. EU Directive 99/31/EC; Council Directive 86/278/EEC) [45–47]. The limitations for heavy metal concentration in sewage sludge aimed for agricultural usage are separately regulated by local law [47].

Despite those limitations, application of end products to degraded soils could lead to their degradation. Therefore, end products must be carefully used and monitored for soil restoration purposes. As stated previously, it was found that the application of end products to a degraded soil negatively influenced the groundwater through leaching of toxic heavy elements [48]. The end products obtained from municipal wastes may contain dangerous pathogens due to an improper hygienization process, thus constituting additional environmental risks for degraded soil [49]. It is therefore essential to conduct hygienization properly in order to avoid increased soil contamination and toxicity. Moreover, some end products may also contain toxic compounds and their soil application may contribute to the intensification of soil degradation [50].

13.2.6 Productivity of Improved Degraded Soils

Soil degradation is considered not only from ecological aspects but also, more often, from economic aspects. The impact assessment included in the proposal for an EU Soil Thematic Strategy [51] estimated that the cost of soil degradation due to soil erosion may vary from €0.7 to €14.0 billion, based on estimations made of 13 of the largest EU Member States. The impact assessment also expected the annual costs of the on-site effects of soil erosion to be about €40–860 million [52]. This estimation only confirms the necessity of soils restoration. According to the United Nations Convention to Combat Desertification, soil erosion and agriculture and industrial activities are professed to be among the main processes causing land degradation.

In the degraded soils, restorative biosolids are the main important soil improvers. Because of developments in industrial pre-treatment programs over the years, biosolids have generally metal concentrations much lower than regulations require. In comparison to many other organic soil amendments, biosolids are greatly characterized and often are readily available at low cost for use in degraded lands. In soil remediation, it is important to include a mixture of N-rich materials with C-rich materials to reduce the potential for N leaching while providing adequate organic matter. In general, a bulk amendment having a C : N ratio between 20 : 1 and 40 : 1 is suggested [53].

Degraded soil when amended with biosolids increases its productivity. In the research carried out by Grobelak [54], it was shown that after sewage sludge application under field conditions many basic soil parameters were significantly increased (Table 13.2.2). This study was conducted on a heavy metals contaminated area where industrial sewage sludge from food treatment was applied at the rate of 45 Mg ha^{-1}.

Brownfield soils are often characterized by low values of pH, sorption capacity (CEC), moisture, and carbon and biogenic elements. On the other hand, biosolids such as sewage sludge are characterized by a high content of organic matter and a relatively large amount of nitrogen and phosphorus. Grass grows very well on soils fertilized with biosolids. They have access to a high amount of nutrients and, due to their intensive growth and a very extensive and dense root system,

Table 13.2.2 Changes in soil parameters after sewage sludge application under field conditions [54].

Soil property	Control soil	Soil + sewage sludge (SS)	Control soil	Soil + sewage sludge (SS)
			Results afte· 5 months of the field study	
pH in H_2O	4.86 ± 0.5	6.55 ± 0.7	5.31 ± 0.04	6.8 ± 0.02
pH in 1 M KCl	4.63 ± 0.3	6.24 ± 0.8	5.2 ± 0.01	6.7 ± 0.08
CEC [cmol(+) kg^{-1} dm]	3.52 ± 0.13	8.02 ± 0.1	3.6 ± 0.2	6.1 ± 0.6
Humic acids [%]	08. ± 0.4	0.9. ± 0.3	0.9 ± 0.09	1.01 ± 0.08
C total [g kg^{-1} dm]	14.76 ± 2.3	29.27 ± 2.07	14.39 ± 0.32	24.5 ± 0.35
N total [g kg^{-1} dm]	0.703 ± 4.2	2.950 ± 0.9	0.86 ± 2.5	1.765 ± 0.25
P bioavailable [mg kg^{-1} dm]	51.7 ± 0.63	128.5 ± 0.73	71.4 ± 0.03	120.42 ± 0.01

Figure 13.2.1 Examples of grass biomass grown on heavy metal contaminated soil (Pb 1200 mg kg^{-1}, Zn 700 mg kg^{-1}, Cd 14 mg kg^{-1}); Fa, *Festuca arundinacea* Schreb.; Ac, *Agrostis capillaris* L.; Dg, *Dactylis glomerata* L.; left pot with SS application, right- control [54].

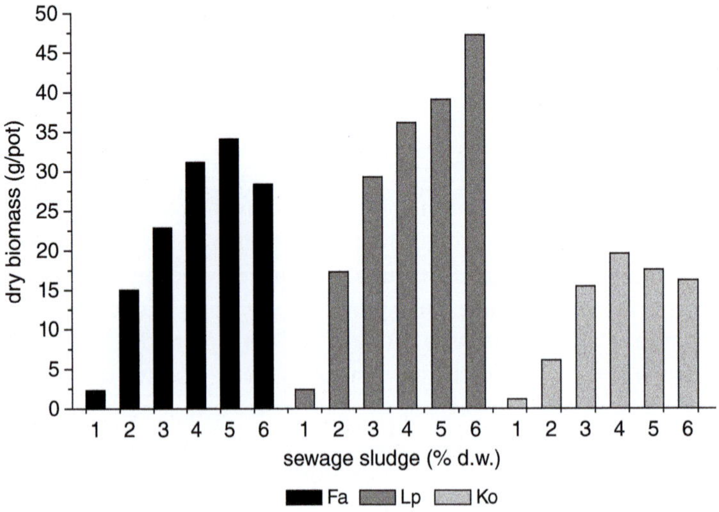

Figure 13.2.2 Grass biomass yield in heavy metal contaminated soil treated with increasing doses of sewage sludge (pot experimental study); Fa, *Festuca arundinacea* Schreb.; Lp, *Lolium perenne* L.; Ko, *Festuca ovina* L. [54].

they limit the leaching of nutrients, which consequently limits soil ground water contamination (Figure 13.2.1) [54].

Obtaining higher yields of plant biomass largely depends on the amount of additional biosolids. In a pot experiment, where the percentage of sewage sludge in the soil was increased from 1 to 6% on a dry matter basis (DM), the plant biomass yield of all grass species was increased progressively, with the effect more visible for *Lolium perenne* (Figure 13.2.2) [54].

13.2.7 Summary and Conclusions

Land degradation will persist as a key worldwide problem for the twenty-first century because of its antagonistic impact on agronomic productivity, the environment, and its result on food security and the quality of life. Moreover, the declining soil quality leads to diminishing economic growth in countries where the economy is primarily based on agricultural production.

The use of biodegradable wastes and their end products on degraded soil have shown to remediate them by improving their soil quality in terms of physical, chemical, and biological properties, on one hand, and the cost-effective waste disposal and utilization, on the other. This may also allow replacement of costly commercial fertilizers, thus making agricultural production and the soil remediation process more cost effective.

Due to declining soil quality resulting largely from human-induced degradation, soil scientists must work closely with social scientists, economists, and policy makers to develop a public policy. The awareness of the threats of land degradation in society and political leadership will also increase. We must also quantify and create awareness that degradation results in loss of intergenerational equity and the value of bequests. We need to acknowledge that local actions have a global impact. As reported in the literature, bioremediation of soils with the addition of biosolids has economic justification, especially in terms of local aspects.

Acknowledgements

This work was supported by an internal research grant of Czestochowa University of Technology. The research leading to these results has received funding from the EnviSafeBioC project – contract No. PPI/APM/2018/1/00029/U/001. The project is financed by the Polish National Agency for Academic Exchange (NAWA).

References

1 Joniec, J., Oleszczuk, P., Jezierska-Tys, S., and Kwiatkowska, E. (2019). Effect of reclamation treatments on microbial activity and phytotoxicity of soil degraded by the Sulphur mining industry. *Environmental Pollution* 252: 1429–1438. https://doi.org/10.1016/j.envpol.2019.06.066.

2 Grobelak, A., Placek, A., Grosser, A. et al. (2017). Effects of single sewage sludge application on soil phytoremediation. *Journal of Cleaner Production* 155: 189–197.

3 Chibuike, G.U. and Obiora, S.C. (2014). Heavy metal polluted soils: effect on plants and bioremediation. *Methods. Applied and Environmental Soil Science* 2014: 1–12.

4 Gałązka, A. (2015). Zanieczyszczenia gleb substancjami ropopochodnymi z uwzględnieniem biologicznych metod ich oczyszczania. *Kosmos* 64 (1): 145–164.

5 Khan, M.A., Khan, S., Khan, A., and Alam, M. (2017). Soil contamination with cadmium, consequences and remediation using organic amendments. *Science of the Total Environment* 601–602: 1591–1605. https://doi.org/10.1016/j.scitotenv.2017.06.030.

6 Świątek, J. and Grobelak, A. (2018). Organic soil amendments in assisted phytoremediation of heavy metals contaminated soil - basic Phytotoxicity markers. *Engineering and Protection of Environment* 21 (4): 335–346. https://doi.org/10.17512/ios.2018.4.2.

7 Kacprzak, M., Neczaj, E., Fijałkowski, K. et al. (2017). Sewage sludge disposal strategies for sustainable development. *Environmental Research* 156: 39–46. https://doi.org/10.1016/j.envres.2017.03.010.

8 Demessie, A., Singh, B.R., and Lal, R. (2017). Soil carbon sequestration: Ethiopia. In: *Encyclopedia of Soil Science*, 3e (ed. R. Lal et al.), 2066–2072. Taylor & Francis https://doi.org/10.1081/e-ess3-120053587.

9 FAO (2011). The state of the world's land and water resources for food and agriculture (SOLAW) – managing systems at risk. Food and Agriculture Organization of the United Nations, Rome and Earthscan, London.

10 Tebebu, T.Y., Bayabil, H.K., Stoof, C.R. et al. (2017). Characterization of degraded soils in the humid. *Land Degradation and Development* 1901: 1891–1901. https://doi.org/10.1002/ldr.2687.

11 Elhaja, M.E., Ibrahim, I.S., Adam, H.E., and Csaplovics, E. (2014). Soil aggregate stability and wind erodible fraction in a semi-arid environment of White Nile state, Sudan. *Proceedings of SPIE 9260, Land Surface Remote Sensing II* 926017: https://doi.org/10.1117/12.2069182.

12 Balaguer-Puig, M., Marqués-Mateu, Á., Lerma, J.L., and Ibáñez-Asensio, S. (2017). Quantifying small-magnitude soil erosion: geomorphic change detection at plot scale. *Land Degradation & Development* 29 (3): 825–834. https://doi.org/10.1002/ldr.2826.

13 DeLong, C., Cruse, R., and Wiener, J. (2015). The soil degradation paradox: compromising our resources when we need them the most. *Sustainability (Switzerland)* 7 (1): 866–879. https://doi .org/10.3390/su7010866.

14 Eswaran, H., Lal, R., and Reich, P. (2019). Land degradation: an overview. In: *Response to Land Degradation* (ed. E.M. Bridges, I.D. Hannam, L.R. Oldeman, et al.), 20–35. https://doi.org/10 .1201/9780429187957-4.

15 Gibbs, H.K. and Salmon, J.M. (2015). Mapping the world's degraded lands. *Applied Geography* 57: 12–21. https://doi.org/10.1016/j.apgeog.2014.11.024.

16 Zhang, B., Guo, Y., Huo, J. et al. (2019). Combining chemical oxidation and bioremediation for petroleum polluted soil remediation by BC-nZVI activated persulfate. *Chemical Engineering Journal* 382: 123055.

17 Azubuike, C.C., Chikere, C.B., and Okpokwasili, G.C. (2016). Bioremediation techniques–classification based on site of application: principles, advantages, limitations and prospects. *World Journal of Microbiology and Biotechnology* 32 (11): 180. https://doi.org/10.1007/s11274-016-2137-x.

18 Mishra, A., Mishra, S.P., Arshi, A. et al. (2019). Plant-microbe interactions for bioremediation and phytoremediation of environmental pollutants and agro-ecosystem development. In: *Bioremediation of Industrial Waste for Environmental Safety* (ed. R. Bharagava and G. Saxena), 415–436. Singapore: Springer https://doi.org/10.1007/978-981-13-3426-9_17.

19 Verma, M.L., Thankur, M., Randhawa, J.S. et al. (2020). Biotechnological applications of fungal enzymes with special reference to bioremediation. In: *Environmental Biotechnology, Environmental Chemistry for a Sustainable World*, vol. 45, vol. 2 (ed. K. Gothandam, S. Ranjan, N. Dasgupta and E. Lichtfouse), 221–247. Cham: Springer https://doi.org/10.1007/978-3-030-38196-7_10.

20 Quintella, C.M., Mata, A.M.T., and Lima, L.C.P. (2019). Overview of bioremediation with technology assessment and emphasis on fungal bioremediation of oil contaminated soils. *Journal of Environmental Management* 241 (March 2018): 156–166. https://doi.org/10.1016/j.jenvman.2019 .04.019.

21 Guerra, A.B., Oliveira, J.S., Silva-Portela, R.C.B. et al. (2018). Metagenome enrichment approach used for selection of oil-degrading bacteria consortia for drill cutting residue bioremediation. *Environmental Pollution* 235: 869–880. https://doi.org/10.1016/j.envpol.2018.01.014.

22 Kowalska, A. and Grobelak, A. (2018). Immobilisation of selected bacteria for remediation on various media. *Engineering and Protection of Environment* 21 (4): 461–472. https://doi.org/10 .17512/ios.2018.4.11.

23 Chen, X. and Achal, V. (2019). Biostimulation of carbonate precipitation process in soil for copper immobilization. *Journal of Hazardous Materials* 368: 705–713. https://doi.org/10.1016/j .jhazmat.2019.01.108.

24 Grobelak, A., Świątek, J., Murtaś, A., and Jaskulak, M. (2019). Cadmium-induced oxidative stress in plants, cadmium toxicity, and tolerance in plants: from physiology to remediation.

In: *Cadmium Toxicity and Tolerance in Plants* (ed. M. Hasanuzzaman, M.N.V. Prasad and M. Fujita), 213–231. Academic Press https://doi.org/10.1016/b978-0-12-814864-8.00009-7.

25 Cooper, J., Greenberg, I., Ludwig, B. et al. (2020). Effect of biochar and compost on soil properties and organic matter in aggregate size fractions under field conditions. *Agriculture, Ecosystems & Environment* 295: 106882. https://doi.org/10.1016/j.agee.2020.106882.

26 Kong, L., Gao, Y., Zhou, Q. et al. (2018). Biochar accelerates PAHs biodegradation in petroleum-polluted soil by biostimulation strategy. *Journal of Hazardous Materials* 343: 276–284. https://doi.org/10.1016/j.jhazmat.2017.09.040.

27 Cruz-Paredes, C., Wallander, H., Kjøller, R., and Rousk, J. (2017). Using community trait-distributions to assign microbial responses to pH changes and cd in forest soils treated with wood ash. *Soil Biology and Biochemistry* 112: 153–164. https://doi.org/10.1016/j.soilbio.2017.05.004.

28 Qin, J., Hovmand, M.F., Ekelund, F. et al. (2017). Wood ash application increases pH but does not harm the soil mesofauna. *Environmental Pollution* 224: 581–589. https://doi.org/10.1016/j.envpol.2017.02.041.

29 Mosquera-Losada, M.R., Amador-García, A., Rigueiro-Rodríguez, A., and Ferreiro-Domínguez, N. (2019). Circular economy: using lime stabilized bio-waste based fertilisers to improve soil fertility in acidic grasslands. *Catena* 179: 119–128. https://doi.org/10.1016/j.catena.2019.04.008.

30 Seyedalikhani, S., Esperschuetz, J., Dickinson, N.M. et al. (2020). Biowastes promote essential oil production on degraded soils. *Industrial Crops and Products* 145: 112108. https://doi.org/10.1016/j.indcrop.2020.112108.

31 Mekki, A., Aloui, F., and Sayadi, S. (2017). Influence of biowastes compost amendment on soil organic carbon storage under arid climate. *Journal of the Air and Waste Management Association* 67: 867–877. https://doi.org/10.1080/10962247.2017.1374311.

32 Leifeld, J., Siebert, S., and Kögel-Knabner, I. (2002). Biological activity and organic matter mineralization of soils amended with biowaste composts. *Journal of Plant Nutrition and Soil Science* 165 (2): 151. https://doi.org/10.1002/1522-2624(200204)165:2<151::aid-jpln151>3.0.co;2-t.

33 Hamdi, H., Hechmi, S., Khelil, M.N. et al. (2019). Repetitive land application of urban sewage sludge: effect of amendment rates and soil texture on fertility and degradation parameters. *CATENA* 172: 11–20. https://doi.org/10.1016/j.catena.2018.08.015.

34 Tanaka, H., Kyaw, K.M., Toyota, K., and Motobayashi, T. (2005). Influence of application of rice straw, farmyard manure, and municipal biowastes on nitrogen fixation, soil microbial biomass N, and mineral N in a model paddy microcosm. *Biology and Fertility of Soils* 42 (6): 501–505. https://doi.org/10.1007/s00374-005-0043-8.

35 Montiel-Rozas, M.M., Domínguez, M.T., Madejón, E. et al. (2018). Long-term effects of organic amendments on bacterial and fungal communities in a degraded Mediterranean soil. *Geoderma* 332: 20–28. https://doi.org/10.1016/j.geoderma.2018.06.022.

36 Angin, I., Aslantas, R., Gunes, A. et al. (2016). Effects of sewage sludge amendment on some soil properties, growth, yield and nutrient content of raspberry (*Rubus idaeus* L.). *Erwerbs-Obstbau* 59 (2): 93–99. https://doi.org/10.1007/s10341-016-0303-9.

37 Mohamed, B., Olivier, G., François, G. et al. (2018). Sewage sludge as a soil amendment in a Larix decidua plantation: effects on tree growth and floristic diversity. *Science of the Total Environment* 621: 291–301. https://doi.org/10.1016/j.scitotenv.2017.11.283.

38 Shaheen, S.M., Balbaa, A.A., Khatab, A.M. et al. (2019). Biowastes alone and combined with sulfur affect the phytoavailability of Cu and Zn to barnyard grass and sorghum in a fluvial alkaline soil under dry and wet conditions. *Journal of Environmental Management* 234: 440–447. https://doi.org/10.1016/j.jenvman.2018.12.106.

39 Fang, S., Tsang, D.C.W., Zhou, F. et al. (2016a). Stabilization of cationic and anionic metal species in contaminated soils using sludge-derived biochar. *Chemosphere* 149: 263–271. https://doi.org/10.1016/j.chemosphere.2016.01.060.

40 Andreev, N., Ronteltap, M., Lens, P.N.L. et al. (2016). Lacto-fermented mix of faeces and bio-waste supplemented by biochar improves the growth and yield of corn (*Zea mays* L.). *Agriculture, Ecosystems & Environment* 232: 263–272. https://doi.org/10.1016/j.agee.2016.08.012.

41 Aggelides, S.M. and Londra, P.A. (2000). Effects of compost produced from town wastes and sewage sludge on the physical properties of a loamy and a clay soil. *Bioresource Technology* 71 (3): 253–259. https://doi.org/10.1016/S0960-8524(99)00074-7.

42 Kacprzak, M., Grobelak, A., Grosser, A., and Prasad, M.N.V. (2014). Efficacy of biosolids in assisted phytostabilization of metalliferous acidic sandy soils with five grass species. *International Journal of Phytoremediation* 16: 593–608. https://doi.org/10.1080/15226514.2013.798625.

43 Hussain, F., Hussain, I., Khan, A.H.A. et al. (2018). Combined application of biochar, compost, and bacterial consortia with Italian ryegrass enhanced phytoremediation of petroleum hydrocarbon contaminated soil. *Environmental and Experimental Botany* 153: 80–88. https://doi.org/10.1016/j.envexpbot.2018.05.012.

44 Hayat, A., Hussain, I., Soja, G. et al. (2019). Organic and chemical amendments positively modulate the bacterial proliferation for effective rhizoremediation of PCBs-contaminated soil. *Ecological Engineering* 138: 412–419. https://doi.org/10.1016/j.ecoleng.2019.07.038.

45 Mandal, S., Kunhikrishnan, A., Bolan, N.S. et al. (2016). Application of biochar produced from biowaste materials for environmental protection and sustainable agriculture production. In: *Environmental Materials and Waste: Resource Recovery and Pollution Prevention* (ed. M.N.V. Prasad and K. Shih), 73–90. London: Academic Press https://doi.org/10.1016/B978-0-12-803837-6.00004-4.

46 Hudcová, H., Vymazal, J., and Rozkošný, M. (2019). Present restrictions of sewage sludge application in agriculture within the European Union. *Soil and Water Research* (2): 104–120. https://doi.org/10.17221/36/2018-swr.

47 Mininni, G., Blanch, A.R., Lucena, F., and Berselli, S. (2015). EU policy on sewage sludge utilization and perspectives on new approaches of sludge management. *Environmental Science and Pollution Research* 22 (10): 7361–7374. https://doi.org/10.1007/s11356-014-3132-0.

48 Campos, T., Chaer, G., dos Paulo, Santos, L. et al. (2019). Leaching of heavy metals in soils conditioned with biosolids from sewage sludge. *Floresta e Ambiente* 26 (spe1): e20180399. Epub October 3, 2019. https://doi.org/10.1590/2179-8087.039918.

49 Zhen, G., Lu, X., Kato, H. et al. (2017). Overview of pretreatment strategies for enhancing sewage sludge disintegration and subsequent anaerobic digestion: current advances, full-scale application and future perspectives. *Renewable and Sustainable Energy Reviews* 69 (March 2016): 559–577. https://doi.org/10.1016/j.rser.2016.11.187.

50 Fijalkowski, K., Rorat, A., Grobelak, A., and Kacprzak, M.J. (2017). The presence of contaminations in sewage sludge – the current situation. *Journal of Environmental Management* 203: 1126–1136. https://doi.org/10.1016/j.jenvman.2017.05.068.

51 Commission of the European Communities (2006). Communication from the Commission to the Council, the European Parliament, the European Economic and Social Committee and the Committee of the Regions, Thematic Strategy for Soil Protection, Brussels, 22.9.2006. COM(2006)231 final ([SEC(2006)620], [SEC(2006)1165]).

52 Panagos, P., Standardi, G., Borrelli, P. et al. (2018). Cost of agricultural productivity loss due to soil erosion in the European Union: from direct cost evaluation approaches to the use of macroeconomic models. *Land Degradation & Development* 29 (3): 471–484.

53 Haering, K.C., Daniels, W.L., and Feagley, S.E. (2000). Reclaiming mined land with biosolids, manures and papermill sludge. In: *Reclamation of Drastically Disturbed Lands* (ed. R.I. Barnhisel, R.G. Darmody and W.L. Daniels), 615–644. Madison, WI: American Society of Agronomy-Crop Science Society of America-Soil Science Society of America. Monograph #41. 1082 pp.

54 Grobelak, A. (2012). Chemophytostalization of soils contaminated with C, Zn and Pb, PhD Thesis, Czestochowa University of Technology [in Polish].

13.3

The Use of Biodegradable End Products for Reclamation of Metal Contaminated Soils

Åsgeir Almås[1], *Aneta Kowalska*[2], *Anna Grobelak*[2], *and BalRam Singh*[1]

[1] *Faculty of Environment and Natural Resources Managements, Norwegian University of Life Sciences, Ås, Norway*
[2] *Faculty of Infrastructure and Environment, Czestochowa University of Technology, Częstochowa, Poland*

13.3.1 Introduction

Reclamation of contaminated sites is carried out by various soil management techniques: biological, physical, and chemical. The technologies include immobilization, isolation, toxicity reduction, physical separation, and extraction [1, 2]. Many of these approaches have been used in full-scale [3]. The choice of stabilization or decontamination techniques also depends on site characteristics such as soil type, especially clay content, soil organic material, soil pH, the nature and magnitude of pollution, and current land use. Washing and physico-chemical extraction are labor and cost-intensive processes and logically are less attractive alternatives compared to other technologies. In addition to the costs, soil removal and washing degrade soil structure, increase oxidation of soil organic matter, and increase the risk of soil loss by erosion [4]. Physical processes include soil replacement, soil isolation, soil vitrification, and electro-kinetic remediation, while chemical ones typically include immobilization, encapsulation, and soil washing. The biological processes may include phytoremediation and microbial assisted phytoremediation [1], but also make use of biodegradable wastes to stabilize metals in order to reduce their availability to plants and other soil living organisms. There are also examples where various methods and materials are used in combination to stabilize and limit the bioavailability of metals and metalloids in contaminated soils [2, 5–7].

Biodegradable bio-wastes consist of garden and park waste, food and kitchen waste from households, restaurants, caterers and retail premises, and those from food processing plants. Bio-waste treatment technologies are the processes that convert discarded bio-waste into new end products with potentially commercial value into agriculture and other commercial production systems.

The main objective of the chapter is to provide an overview and discuss some biological reclamation options suitable for remediating and stabilizing heavy metal contaminated land, with a special focus on metal binding properties of biodegradable waste in such soils.

13.3.2 Soil Contamination in Europe

Heavy metals fill an important part in our society as essential nutrients for plants, animals, and humans [8]. However, in elevated concentrations, they are hazardous, as they may disrupt soil living organisms [9, 10], have fatal impacts on biochemical transformations, plants and even humans [11–14]. Even if they are naturally present in the environment, they can become up concentrated and contaminate largely through mining and processing of metal ores for their industrial applications [5, 15, 16]. Mineral oil and trace metals represent the largest source of soil contaminants (60%), and the cost of managing contaminated sites is estimated to amount to approximately 6 billion euros [17]. Generally, the expansion of transport, infrastructure, and progressive urbanization effects increases soil contamination, and thereby also heavy metals [18].

Despite the emissions of heavy metals, metalloid emissions have reduced, with Hg emissions reducing by 93% in comparison to 1990 [17]. Toth et al. [19]] have presented a study on the legacy of heavy metals and metalloid emissions recovered from European soil. The study showed that a large part of the samples contained total concentrations above the assessment threshold, indicating large areas where precautionary measures are needed [19]. The data implies that 6.25% of the cultivated lands required assessment or remediation actions, which in fact equals 137 000 km^2. According to the Joint Research Centre (JRC), heavy metals are the main contaminating compounds affecting soil and groundwater in Europe (34.8%) [20]. A review published by Hou et al. [21] has identified not only typical metal contaminating sources (ranging from lithogenic origin, transportation, atmospheric deposition, wastewater and runoff from industrial and mining facilities, fertilizer application, livestock manure, and sewage sludge) but also the need to pinpoint the sources and geochemical qualities of the sites as well as integrating GIS data in multivariate analysis and assess the temporal trends.

Despite the recognition of the importance of soil quality in Europe, and the need for its protection, an agreement on the best strategy has not yet been achieved. The EU "Soil Thematic Strategy" has been prepared and classifies the key soil threats in the EU [22]. This includes erosion, contamination, and loss of soil biodiversity, etc. Law-based activities driven to protect soils are mainly derived from other legislation acts. The EU-Regulation 2019/1009 [23] on fertilizing products set a threshold value for concentration of contaminants to limit soil pollution. Facing the need for soil protection, the European Union Commission in December 2019 set out a European Green Deal. This summoned the EU Biodiversity Strategy for the 2030 meeting, where the topic was to "Bringing nature back into our lives". The concept was adopted by the EU Commission in 2020. The program includes the EU Nature Restoration Plan which protects soil from the sealing and urbanization. Therefore, the Commission is preparing a proposal for a legal act regarding EU nature restoration targets in 2021 aimed at restoring degraded land, especially those with a high potential for carbon sequestration.

The adoption of the law documents for European countries clearly defines soil protection strategies to combat biodiversity deterioration, provide safety and quality of food production, and protect water quality. Hazardous concentrations of metals and metalloids in soils may increase the risk of biomagnification in soil-living species, affecting their biological processes and survival [24].

13.3.3 Immobilization of Metals and Metalloids by Inorganic and Organic Amendments

Stabilization of hazardous metal contaminated soils is possible by reducing the fraction of labile metals to subtoxic concentrations by specific adsorption or by (co)precipitation following application of inorganic or organic additives. The geochemical mechanisms in the process of metal sorption are principally the same, but the density and stability of the functional groups in action may vary whether they are inorganic or organic wastes (such as biodegradable wastes) or amendments are applied.

Successful stabilization of metal contaminated sites has been demonstrated in several studies, for instance by making use of synthetic zeolites, hydroxyapatite [25, 26], and phosphate-induced products or inorganic Mn-, Fe-, or Al-based additives [5, 27–32]. The pH-variable charges on oxide minerals are deprotonated when pH gets sufficiently high, reducing or making metal cations non-available to organisms. Almås et al. [5] made use of an alkaline Fe-rich slag (pH >11) in combination with either an $FeSO_4$ or an acidic organic soil (moderately contaminated) to optimize the pH for a combined sorption of Pb and Sb. The Pb is increasingly immobilized as pH increases, whereas immobilization of Sb is more efficient at the acidic pH range. The slag and organic soil treatment was almost as effective in reducing the Pb leaching as the slag and $FeSO_4$ combination, whereas for Sb, although leaching of the non-treated soil was already below the EU guideline of $150\,\mu g\,l^{-1}$ [33], the slag and organic treatment reduced the Sb leaching only by about 35–40% compared to the non-treated soil. The slag and $FeSO_4$ combination reduced the Sb leaching by >95%, but the leaching increased slightly with time. This demonstrates the complexity in remediating a site contaminated with, for example, metal cation and metalloids, where the speciation of both the target elements and the immobilizing properties of the additives make a difference [27, 34]. Due to the alkalinity of some of the minerals used, the immobilizing effect on metal cations can be a combination of high pH precipitation of metals or as a result of permanent or pH-dependent surface charges. Other inorganic additives (e.g. phosphates) are very effective metal immobilizers and the resulting precipitates are generally very stable [35, 36]. However, such additives add little extra value to the soil and their soil-living organisms besides reducing the toxic effects. The sources may also be somewhat expensive to obtain. The metal association to oxides are typically of inner-sphere complexation where the ions largely form covalent bounding with the electron donating ligand after dehydration (strong bindings) or are in the outer sphere where the ion maintains its hydration cape and the binding is electrostatic. The latter is a weaker bond [37]. The long-term effect of applying inorganic sorbents to contaminated soils has been demonstrated in studies of, for example, contaminated shooting range soils [29]. Okkenhaug et al. [29] showed no remobilization of Pb and Sb during a four-year field study where an Fe-based additive together with limestone was applied. This is in line with other similar studies where the long-term effect of inorganic additives has been applied [38].

Making use of biodegradable bio-waste in the reclamation of metal degraded soils makes sense as the functional groups (dominated by carboxylic and phenolic acid groups) deprotonate and bind metal cations, at the same time as providing energy to respiring soil living fauna. The initial metal immobilizing effect may decline, however, particularly in situations where easy degradable bio-waste, such as manure and compost, have been used. Depending on the quality of the bio-waste material, the lignin fraction will only slowly or incompletely be degraded and stored as stable humic material [39], providing binding sites for metal binding [40], while much of the other organic structures eventually will be respired and lost as CO_2 and H_2O. A problematic effect of adding an organic matter source to soil is the potential increase in dissolved organic carbon (DOC) in pore water. McBride and Martinez [32] demonstrated the issue of DOC-induced mobilization of, for example, Cu.DOC is well known to mobilize heavy metals, e.g. Cu and Pb [41]. DOC leaching is dependent on several soil factors (type of soil organic matter, pH, and quality of metal hydroxides) and is worth monitoring, but it is probably most severe shortly after mixing in. The overall effect from mixing in organic waste seems to be positive from a remediation perspective, as discussed in this review. The metal binding to humic substances is complex as the stoichiometry and acid properties of the donating ligands is very heterogenous. The metal binding can take place at specific single- or multielectron donating sites (intrinsic binding) and through electrostatic forces. In general, the metal binding can be described by a combination of intrinsic metal binding and electrostatic forces in action [42]. In principle, the intrinsic (covalent) binding is described by the dissociation reaction of the functional groups $R^ZH = R^{Z-1} + H^+$ and the association reaction $R^Z + M^z = RM^{Z+z-}$, where Z is the humic charge and z is the charge of the combining ion, R is the humic molecule, and M is the metal in action. In the WHAM/Model VI [42], the electrostatic forces are described by the electric interaction factor, w, the charge of the combining ion, z, and the charge of the humic molecule, Z. Thus, the metal binding is characterized by a combination of the electrostatic and covalent forces in action: $K = K_{int}K_{elect}$, where K_{int} represents the covalent bindings and K_{elect} represents the electrostatic potentials at the binding sites. The electron donating ligands are typically carboxylic acids (–COO^-), phenolic acids (OH^-), and some groups containing N and S. The carboxylic groups are the least discriminating and most abundant cation-binding groups, whereas the s-containing groups are the least abundant and most discriminating cation-binding groups [40]. The isolation and treatment of metal degraded soil using humic acids (HAs) is an example of applying organic matter sources, where functional groups have been shown to immobilize trace metals [43].

13.3.4 Nutrient Supply Through Biowaste Use

Although metal retention by biodegradable bio-wastes may be geochemically less effective compared to inorganic alternatives, such amendments provide several additional positive features that facilitate revegetation and soil microbial activity [44–46]. Organic material provides energy to soil microbial fauna that fuel microbial mediated redox reactions in soil, such as microbial liberation of organically bound N, P, K, and S, etc. and their following transformations. One of the overall reclamation goals is to revegetate metal-degraded land by stabilizing exposed surface soils to prevent erosion and to reduce potential leaching of dissolved metal contaminants [4]. Madejon et al. [4] presented a study where contaminated soils had been treated, making use of various organic residues (i.e. composted biosolid and sugar beet lime). They found that in addition to an increase in soil pH and total organic carbon (TOC), the easily extractable metal fraction was reduced. As a result, the treatment had stimulated regrowth with improved nutritional status (N, P, K, S, etc.) of the vegetation. The result of providing nutrients to revegetate a contaminated site was shown in a long-term field study published by Palmroth et al. [47]. Their study investigated the remediation effect of a combined hydrocarbon and heavy metal contaminated site, making use of, for example, compost and additional fertilizers. The results showed a significant and positive effect of adding nutrients to the contaminated land. The compost increased the regrowth and reduced the indication of toxicity, whereas the addition of NPK (nitrogen, phosphorus and potassium) facilitated the revegetation significantly. Organic matter in the compost probably stabilized soil metals and no uptake was found.

13.3.5 Recycling of Biowaste

Another important component of this strategy is that it allows the disposal of bio-waste into agriculture and other commercial production systems, making them a valuable resource and avoiding environmental contamination of soils and ground waters by their land-filling action. The use of municipal waste [28, 45], household or wood residues, etc. [48], is attractive from a bioeconomy perspective as such organic waste material can be used to recycle nutrients through the soil ecosystem [49, 50] while functional groups adsorb the labile metal fractions. A secondary effect is the sequestration of organic C in soil, reducing its potential release of CO_2, CH_4, and possibly also N_2O if organic residues alternatively are deposited on landfills or are burned, etc. Sequestered organic matter also improves the water holding capacity and the development of improved soil structure and in consequence improve soil fertility and quality. Good soil structure increases the water infiltration capacity and from that reduces the risk of surface soil erosion [39]. All of this promotes revegetation and hence facilitates reclamation of contaminated soil.

13.3.6 Persistence and Stabilization of Bio-waste

Although stabilization of metals contaminates sites, like a mine site, making use of biodegradable wastes have shown to be very effective in improving several soil parameters and stimulating soil life and vegetation cover [51]. The long-term metal immobilization potential is an issue as compost and other biodegradable organic materials in soil decompose as they provide energy and nutrients to soil fauna. The issue of metal-induced DOC-leaching has already been addressed [32]. Schmidt et al. [52] have shown that the persistence of soil organic matter deriving from different sources (plant material, microbial origin, different biological sources) varies substantially. Their persistence in soil ranges from a few years to a time between 50 to 100 years in maximum. The fire-derived organic matter, however, has been shown to persist for several hundred years and the "Terra Preta" phenomenon is a well-known example [53]. Biochar does not strictly fall into the definition of biodegradable bio-waste as the pyrolysis stabilizes the feed stock, making it nearly unavailable for further biological decomposition [52]. We discuss, however, the application of biochar and phytoremediation technologies in the context of reclamation of metal contaminated soils making use of bio-waste material, as a combination of several technologies seems reasonable [6]. The effect of treating soil with biochar has been shown to stimulate plant growth in degraded and low fertile soils, and the effect depends on several factors [54], but strongest in coarse soils low in soil organic matter. That is mostly because biochar increases the water holding capacity of such soils [55]. The quality is to a large extent determined by the feed stock and pyrolysis temperature [56]. From a heavy metal soil remediation perspective, the use of biochar is highly interesting as it provides long-lasting organic structures in soil positive for plant growth [57], and depending on the method of pyrolysis, functional acid groups remain on the structure capable of binding metals. Hence, biochar does increase cation exchange capacity (CEC), but it also increases the alkalinity in soil [58, 59]. The effect from the treated contaminated soils with biochar has been thoroughly studied [56, 60, 61]. Biochar can be applied together with or pre-treated with compost, manure, and other organic residues to improve the nutritional status of the degraded soil.

13.3.7 Biowaste in Combination with Inorganic Material

Some of the problems connected to accurate assessments of using biodegradable waste in addition to its limited long-term effect are associated with the heterogeneity and incomplete definition of composition, organic material quality, and functional groups taking part in metal immobilization. Even the biochemistry of biochar varies, despite homogenization through pyrolysis and the temperature used for this process. This is related to feedstock quality, pyrolysis temperature, and process. Moreover, a fraction of the organic material supplied will decompose or increase the concentration of dissolved organic matter in pore water, and the metals associated to these dissolved organic complexes may leach [32, 41]. This effect may be stronger if the pH increases [62]. However, a combination of organic and inorganic additives or pyrolysis residues may prove to be the better option, as such a combination will combine the application of energy and nutrients and prolong the duration of immobilization. An elegant example of the latter has been demonstrated by Hamid et al. [7] and colleagues, who tested the immobilization effect of various additives, where a "green stabilizing agent" (GSA), composed of manure, lime, and sepiolite or zeolite in combination, reduced the bioaccumulation of Cd and Pb very effectively. The study demonstrated that a combination of organic and inorganic additives was, in fact, a better approach in reclaiming metal contaminated soils as it provides initial fertilization through the use of quickly accessible nutrients in manure or compost, etc., permanent binding sites, and generally an increase in soil pH. Several other studies have demonstrated the benefits of combining additional surfactants to immobilize heavy metals and to stimulate soil-living organisms and plants [2, 63, 64]. However, care should be taken in planning and during remediation. McBride and Martinez [32] demonstrated a substantial reduction in Mo, As, Cd, Pb, Cr, and Cu solubility when non-crystalline Fe and Al hydroxides were mixed in a contaminated soil, Mn-oxide in situations where DOC increased, and raised mobility of Pb, As and Cd. They concluded that the overall effects are metal-specific and also dependent on pH change. Thus, are pre-tests and post-monitoring essential to make sure pH is stable and that the mobilizing effect of DOC is minimized or at least monitored.

It has been found that the establishment of a permanent vegetation cover is the ultimate goal in the reclamation of contaminated soils, as it limits the spread of contaminated soil by dust or erosion. It is well known that some plants are tolerant to high metal concentrations and others are even accumulating metals and metalloids.

13.3.8 Vegetation Cover and Phytostabilization

Phytoremediation makes use of plants to remediate and revegetate contaminated sites. It is considered environmentally friendly, attractive, esthetically pleasing, non-invasive, energy efficient, and cost-effective technology to potentially clean up sites with low-to-moderate levels of heavy metals. Phytoremediation, based on different uptake mechanisms, is categorized into phytostabilization, photoevaporation, and phytoextraction.

Phytostabilization decreases the bioavailability and mobility of heavy metals in soils due to the stabilization effects of plants [65]. This process does not reduce the concentration of heavy metals present in contaminated soil but prohibits their off-site movement and restricts heavy metals in the vadose zone of plants through accumulation by roots or precipitation within the rhizosphere [66]. Metal excluder plants accumulate high levels of heavy metals from soil into their roots with limited transport to their aerial parts [67]. These plants have a small potential for metal extraction but are highly efficient for phytostabilization purposes [67]. Willows (*Salix* spp.) are considered ideal plants for both phytoextraction and phytostabilization owing to their elevated tolerance to heavy metals, high territorial expansion, and intense evapotranspiration [65]. Organic and inorganic amendments (as discussed above) may facilitate plant-induced immobilization of heavy metals in soil [68].

During phytoevaporation or volatilization, heavy metals are taken up from the soil and converted into less toxic vapors, which are then released into the atmosphere through the transpiration process of the plants. In phytovolatilization, metals are assimilated into volatile organic compounds that are then released into the atmosphere as biomolecules [69]. Some species, such as *Arabidopsis thaliana* and *Brassica juncea*, are capable of taking up metal(loid)s and transferring them to

Table 13.3.1 Examples of hyperaccumulator plants accumulating high levels of heavy metal(loid)s in their above ground parts.

Plant species	Heavy metals	Concentration in plant aerial parts (mg kg^{-1})	Threshold level for hyper-accumulator	Type of remediation	Fold higher than threshold level	References
Thlaspi caerulescens	Cd	5 000	100	Phytoextraction	50	[71]
Solanum photeinocarpum	Cd	158	100	Phytoextraction	1.6	[72]
Phyllanthus serpentinus	Ni	38 100	1 000	Phytoextraction	38.1	[73]
Thlaspi caerulescens		16 200	1 000	Phytoextraction	16.2	[71]
Helianthus annuus	Pb	5 600	1 000	Phytoextraction	5.6	[71]
Eleocharis acicularis	As	1 470	1 000	Phytoextraction	1.5	[74]
Potentilla griffithii	Zn	11 400	10 000	Phytoextraction	1.1	[75]
Eleocharis acicularis	Cu	20 200	1 000	Phytoextraction	20.2	[74]

Source: Extracted from Khalid et al. [1].

gaseous species inside the plant, followed by their release into the atmosphere [70]. However, the practical and ethical application of heavy metal(loid) phytovolatilization for soil remediation is highly questionable, as once the metal has been volatilized, no one has control over its fate in the atmosphere.

Phytoextraction involves the clean-up of heavy metal(loid)s from soil by means of plant uptake. During phytoextraction, heavy metals are transferred from the soil to plant biomass. Plant biomass is comparatively very easy to recycle, dispose of, treat, or oxidize compared to soil. Phytoextraction guarantees a permanent removal of metal(loid)s from the contaminated sites. However, phytoextraction is suitable for those sites that are polluted by low-to-moderate levels of metals, because most plant species are not able to sustain heavily polluted sites. Plant species that can be effectively used for phytoextraction must have (i) high metal-accumulation capability in their above-ground parts, (ii) tolerance to high concentration of metals, (iii) the ability to grow fast with high biomass, and (iv) a profuse root system. Some hyper-accumulator plants that can accumulate high levels of these metals (many folds higher than the minimum threshold level of hyper-accumulators) are shown in Table 13.3.1. These plant families mainly included Brassicaceae, Asteraceae, Violaceae, Fabaceae, Caryophyllaceae, Euphorbiaceae, Lamiaceae, Flacourtiaceae, and Poaceae.

The advantages of phytoremediation technologies in general, compared to other classical remediation methods, are (i) highly economical, (ii) less disruptive to the soil and environment, (iii) no need of disposal sites, (iv) high public acceptance, (v) no excavation or transport of contaminated media, and (vi) feasible for multimetal contaminated sites [76]. The disadvantages of phytoextraction are as follows: (i) depends on the growing conditions essential for plants and microorganisms, (ii) large-scale tasks need considerable experience and expertise in using agricultural equipment, (iii) depends on plant tolerance to metal(loid)s, and (iv) a relatively long time period required to completely remediate the sites.

References

1 Khalid, S., Shahid, M., Niazi, N.K. et al. (2017). A comparison of technologies for remediation of heavy metal contaminated soils. *Journal of Geochemical Exploration* 182: 247–268.

2 Mao, X.H., Jiang, R., Xiao, W., and Yu, J.G. (2015). Use of surfactants for the remediation of contaminated soils: a review. *Journal of Hazardous Materials* 285: 419–435.

3 Mulligan, C.N., Yong, R.N., and Gibbs, B.F. (2001). Remediation technologies for metal-contaminated soils and groundwater: an evaluation. *Engineering Geology* 60 (1–4): 193–207.

4 Madejon, E., de Mora, A.P., Felipe, E. et al. (2006). Soil amendments reduce trace element solubility in a contaminated soil and allow regrowth of natural vegetation. *Environmental Pollution* 139 (1): 40–52.

5 Almås, Å.R., Pironin, E., and Okkenhaug, G. (2019). The partitioning of Sb in contaminated soils after being immobilization by Fe-based amendments is more dynamic compared to Pb. *Applied Geochemistry* 108: 8.

6 Souza, L.R.R., Pomarolli, L.C., and da Veiga, M. (2020). From classic methodologies to application of nanomaterials for soil remediation: an integrated view of methods for decontamination of toxic metal(oid)s. *Environmental Science and Pollution Research* 27 (10): 10205–10227.

7 Hamid, Y., Tang, L., Wang, X.Z. et al. (2018). Immobilization of cadmium and lead in contaminated paddy field using inorganic and organic additives. *Scientific Reports* 8: 10.

8 Almås, Å.R. and Singh, B.R. (2016). Trace metal contamination. In: *Encyclopedia of Soil Science* (ed. R. Lal), 2364–2368. Boca Raton: Taylor and Francis Group.

9 Almås, Å.R., Bakken, L.R., and Mulder, J. (2004). Changes in tolerance of soil microbial communities in Zn and Cd contaminated soils. *Soil Biology and Biochemistry* 36 (5): 805–813.

10 Almås, Å.R., Mulder, J., and Bakken, L.R. (2005). The trace metal exposure of soil bacteria depends on their position in the soil matrix. *Environmental Science and Technology* 39 (16): 5927–5932.

11 Duruibe, J.O., Ogwuegbu, M.O.C., and Egwurugwu, J.N. (2007). Heavy metal pollution and human biotoxic effects. *International Journal of Physical Sciences* 2 (5): 112–118.

12 Giller, K.E., Witter, E., and McGrath, S.P. (1998). Toxicity of heavy metals to microorganisms and microbial processes in agricultural soils: a review. *Soil Biology and Biochemistry* 30 (10–11): 1389–1414.

13 Nagajyoti, P.C., Lee, K.D., and Sreekanth, T.V.M. (2010). Heavy metals, occurrence and toxicity for plants: a review. *Environmental Chemistry Letters* 8 (3): 199–216.

14 Almås, Å.R., Lombnæs, P., Sogn, T.A., and Mulder, J. (2006). Speciation of Cd and Zn in contaminated soils determined by DGT-DIFS, and WHAM Model/ VI in relation to uptake by spinach and ryegrass. *Chemosphere* 62 (10): 1647–1655.

15 Okkenhaug, G., Zhu, Y.G., Luo, L. et al. (2011). Distribution, speciation and availability of antimony (Sb) in soils and terrestrial plants from an active Sb mining area. *Environmental Pollution* 159 (10): 2427–2434.

16 Almås, Å.R. and Manoko, M.L.K. (2012). Trace element concentrations in soil, sediments, and waters in the vicinity of Geita Gold Mines and North Mara Gold Mines in Northwest Tanzania. *Soil and Sediment Contamination* 21 (2): 135–159.

17 Panagos, P., Van Liedekerke, M., Yigini, Y., and Montanarella, L. (2013). Contaminated sites in Europe: review of the current situation based on data collected through a European network. *Journal of Environmental and Public Health* 2013: 11.

18 EEA (2020). National emissions reported to the Convention on Long-range Transboundary Air Pollution (LRTAP Convention) (accessed 17 November 2020).

19 Toth, G., Hermann, T., Da Silva, M.R., and Montanarella, L. (2016). Heavy metals in agricultural soils of the European Union with implications for food safety. *Environment International* 88: 299–309.

20 JRC (2020). Eionet NRC Soil data collection on contaminated sites. https://www.eea.europa.eu/data-and-maps/data/external/eionet-nrc-soil-data-collection-1 (17 November 2011).

21 Hou, D.Y., O'Connor, D., Nathanail, P. et al. (2017). Integrated GIS and multivariate statistical analysis for regional scale assessment of heavy metal soil contamination: a critical review. *Environmental Pollution* 231: 1188–1200.

22 EU (2006). Communication from the Commission to the Council, the European Parliament, the European Economic and Social Committee and the Committee of the Regions – Thematic Strategy for Soil Protection, COM(2006)0231. https://eur-lex.europa.eu/legal-content/EN/TXT/?uri=CELEX:52006DC0231 (accessed 17 November 2020), EU 2006).

23 EU (2019). Regulation (EU) 2019/1009 of the European Parliament and of the Council of 5 June 2019 laying down rules on the making available on the market of EU fertilising products and amending Regulations (EC) No. 1069/2009 and (EC) No. 1107/2009 and repealing Regulation (EC) No. 2003/2003. https://eur-lex.europa.eu/legal-content/EN/TXT/?uri=CELEX%3A32019R1009 (accessed 17 November 2020).

24 Abdu, N., Abdullahi, A.A., and Abdulkadir, A. (2017). Heavy metals and soil microbes. *Environmental Chemistry Letters* 15 (1): 65–84.

25 Chen, X.B., Wright, J.V., Conca, J.L., and Peurrung, L.M. (1997). Evaluation of heavy metal remediation using mineral apatite. *Water, Air, and Soil Pollution* 98 (1–2): 57–78.

26 Boisson, J., Ruttens, A., Mench, M., and Vangronsveld, J. (1999). Evaluation of hydroxyapatite as a metal immobilizing soil additive for the remediation of polluted soils. Part 1. Influence of hydroxyapatite on metal exchangeability in soil, plant growth and plant metal accumulation. *Environmental Pollution* 104 (2): 225–233.

27 Boisson, J., Mench, M., Vangronsveld, J. et al. (1999). Immobilization of trace metals and arsenic by different soil additives: evaluation by means of chemical extractions. *Communications in Soil Science and Plant Analysis* 30 (3–4): 365–387.

28 Grobelak, A., Placek, A., Grosser, A. et al. (2017). Effects of single sewage sludge application on soil phytoremediation. *Journal of Cleaner Production* 155: 189–197.

29 Okkenhaug, G., Gebhardt, K.A.G., Amstaetter, K. et al. (2016). Antimony (Sb) and lead (Pb) in contaminated shooting range soils: Sb and Pb mobility and immobilization by iron based sorbents, a field study. *Journal of Hazardous Materials* 307: 336–343.

30 Zhao, F.J., Ma, Y.B., Zhu, Y.G. et al. (2015). Soil contamination in China: current status and mitigation strategies. *Environmental Science and Technology* 49 (2): 750–759.

31 Mench, M., Vangronsveld, J., Lepp, N.W., and Edwards, R. (1999). Physico-chemical aspects and efficiency of trace element immobilization by soil amendments. In: *Metal-Contaminated Soils: In Situ Inactivation and Phytorestoration* (eds. J. Vangronsveld and S.D. Cunningham), 151–182. Springer-Verlag.

32 McBride, M.B. and Martinez, C.E. (2000). Copper phytotoxicity in a contaminated soil: remediation tests with adsorptive materials. *Environmental Science and Technology* 34 (20): 4386–4391.

33 EU (1999). Council Directive 199/31/EC of 26 April 1999 on the Landfill of Waste. Brussels: OJ L 182.

34 Guo, G.L., Zhou, Q.X., and Ma, L.Q. (2006). Availability and assessment of fixing additives for the in situ remediation of heavy metal contaminated soils: a review. *Environmental Monitoring and Assessment* 116 (1–3): 513–528.

35 Wang, Y.S., Dai, J.G., Wang, L. et al. (2018). Influence of lead on stabilization/solidification by ordinary Portland cement and magnesium phosphate cement. *Chemosphere* 190: 90–96.

36 Zeng, G.M., Wan, J., Huang, D.L. et al. (2017). Precipitation, adsorption and rhizosphere effect: the mechanisms for phosphate-induced Pb immobilization in soils – a review. *Journal of Hazardous Materials* 339: 354–367.

37 Stumm, W. and Morgan, J.J. (1996). Aquatic chemistry. In: *Chemical Equilibria and Rates in Natural Waters* (eds. J.J. Morgan, J.L. Schnoor and A. Zehnder), 1–22. New York: Wiley Interscience.

38 Ettler, V., Tomasova, Z., Komarek, M. et al. (2015). The pH-dependent long-term stability of an amorphous manganese oxide in smelter-polluted soils: implication for chemical stabilization of metals and metalloids. *Journal of Hazardous Materials* 286: 386–394.

39 Blume, H.P., Brümmer, G.W., Fleige, H. et al. (2016). *Scheffer/Schachtschabel Soil Science*, 1st English edition. Heidelberg: Springer.

40 Tipping, E. (2002). *Cation Binding by Humic Substances*, 1–434. Cambridge: Cambridge University Press.

41 Tipping, E., Rieuwerts, J., Pan, G. et al. (2003). The solid solution partitioning of heavy metals (Cu, Zn, Cd, Pb) in upland soils of England and Wales. *Environmental Pollution* 125 (2): 213–225.

42 Tipping, E. (1998). Humic ion-binding Model VI: an improved description of the interactions of protons and metal ions with humic substances. *Aquatic Geochemistry* 4: 3–48.

43 Clemente, R. and Bernal, M.P. (2006). Fractionation of heavy metals and distribution of organic carbon in two contaminated soils amended with humic acids. *Chemosphere* 64 (8): 1264–1273.

44 Garcia-Carmona, M., Romero-Freire, A., Aragon, M.S. et al. (2017). Evaluation of remediation techniques in soils affected by residual contamination with heavy metals and arsenic. *Journal of Environmental Management* 191: 228–236.

45 Smith, S.R. (2009). A critical review of the bioavailability and impacts of heavy metals in municipal solid waste composts compared to sewage sludge. *Environment International* 35 (1): 142–156.

46 Almås, Å.R., Afanou, A.K., and Krogstad, T. (2014). Impact of lignosulfonate on solution chemistry and phospholipid fatty acid composition in soils. *Pedosphere* 24 (3): 308–321.

47 Palmroth, M.R.T., Koskinen, P.E.P., Pichtel, J. et al. (2006). Field-scale assessment of phytotreatment of soil contaminated with weathered hydrocarbons and heavy metals. *Journal of Soils and Sediments* 6 (3): 128–136.

48 Kiikkila, O., Pennanen, T., Perkiomaki, J. et al. (2002). Organic material as a copper immobilising agent: a microcosm study on remediation. *Basic and Applied Ecology* 3 (3): 245–253.

49 Fiksel, J. and Lal, R. (2018). Transforming waste into resources for the Indian economy. *Environment and Development* 26: 123–128.

50 van der Kooij, S., van Vliet, B.J.M., Stomph, T.J. et al. (2020). Phosphorus recovered from human excreta: a socio-ecological-technical approach to phosphorus recycling. *Resources, Conservation and Recycling* 157: 11.

51 Touceda-Gonzalez, M., Alvarez-Lopez, V., Prieto-Fernandez, A. et al. (2017). Aided phytostabilisation reduces metal toxicity, improves soil fertility and enhances microbial activity in Cu-rich mine tailings. *Journal of Environmental Management* 186: 301–313.

52 Schmidt, M.W.I., Torn, M.S., Abiven, S. et al. (2011). Persistence of soil organic matter as an ecosystem property. *Nature* 478 (7367): 49–56.

53 Lehmann, J. (2007). A handful of carbon. *Nature* 447 (7141): 143–144.

54 Jeffery, S., Verheijen, F.G.A., van der Velde, M., and Bastos, A.C. (2011). A quantitative review of the effects of biochar application to soils on crop productivity using meta-analysis. *Agriculture, Ecosystems and Environment* 144 (1): 175–187.

55 Fellet, G., Marchiol, L., Delle Vedove, G., and Peressotti, A. (2011). Application of biochar on mine tailings: effects and perspectives for land reclamation. *Chemosphere* 83 (9): 1262–1267.

56 Beesley, L., Moreno-Jimenez, E., Gomez-Eyles, J.L. et al. (2011). A review of biochars' potential role in the remediation, revegetation and restoration of contaminated soils. *Environmental Pollution* 159 (12): 3269–3282.

57 Keiluweit, M., Nico, P.S., Johnson, M.G., and Kleber, M. (2010). Dynamic molecular structure of plant biomass-derived black carbon (Biochar). *Environmental Science and Technology* 44 (4): 1247–1253.

58 Atkinson, C.J., Fitzgerald, J.D., and Hipps, N.A. (2010). Potential mechanisms for achieving agricultural benefits from biochar application to temperate soils: a review. *Plant and Soil* 337 (1–2): 1–18.

59 Martinsen, V., Alling, V., Nurida, N.L. et al. (2015). pH effects of the addition of three biochars to acidic Indonesian mineral soils. *Soil Science & Plant Nutrition* 61 (5): 821–834.

60 Bian, R.J., Joseph, S., Cui, L.Q. et al. (2014). A three-year experiment confirms continuous immobilization of cadmium and lead in contaminated paddy field with biochar amendment. *Journal of Hazardous Materials* 272: 121–128.

61 Teodoro, M., Trakal, L., Gallagher, B.N. et al. (2020). Application of co-composted biochar significantly improved plant-growth relevant physical/chemical properties of a metal contaminated soil. *Chemosphere* 242: 10.

62 Lofts, S., Woof, C., Tipping, E. et al. (2001). Modelling pH buffering and aluminium solubility in European forest soils. *European Journal of Soil Science* 52: 189–204.

63 Mohamed, B.A., Ellis, N., Kim, C.S., and Bi, X.T. (2017). The role of tailored biochar in increasing plant growth, and reducing bioavailability, phytotoxicity, and uptake of heavy metals in contaminated soil. *Environmental Pollution* 230: 329–338.

64 Karer, J., Zehetner, F., Dunst, G. et al. (2018). Immobilisation of metals in a contaminated soil with biochar-compost mixtures and inorganic additives: 2-year greenhouse and field experiments. *Environmental Science and Pollution Research* 25 (3): 2506–2516.

65 Bart, S., Motelica-Heino, M., Miard, F. et al. (2016). Phytostabilization of As, Sb and Pb by two willow species (S-viminalis and S-purpurea) on former mine technosols. *Catena* 136: 44–52.

66 Bolan, N.S., Park, J.H., Robinson, B. et al. (2011). Phytostabilization. A green approach to contaminant containment. In: *Advances in Agronomy*, vol. 112 (ed. D.L. Sparks), 145–204. San Diego: Elsevier Academic Press Inc.

67 Ali, H., Khan, E., and Sajad, M.A. (2013). Phytoremediation of heavy metals – concepts and applications. *Chemosphere* 91 (7): 869–881.

68 Parra, A., Zornoza, R., Conesa, E. et al. (2016). Evaluation of the suitability of three Mediterranean shrub species for phytostabilization of pyritic mine soils. *Catena* 136: 59–65.

69 Marques, A., Rangel, A., and Castro, P.M.L. (2009). Remediation of heavy metal contaminated soils: phytoremediation as a potentially promising clean-up technology. *Critical Reviews in Environmental Science and Technology* 39 (8): 622–654.

70 Verbruggen, N., Hermans, C., and Schat, H. (2009). Molecular mechanisms of metal hyperaccumulation in plants. *The New Phytologist* 181 (4): 759–776.

71 Zhang, F.Q., Zhang, H.X., Xia, Y. et al. (2011). Exogenous application of salicylic acid alleviates cadmium toxicity and reduces hydrogen peroxide accumulation in root apoplasts of *Phaseolus aureus* and *Vicia sativa*. *Plant Cell Reports* 30 (8): 1475–1483.

72 Chaney, R.L., Broadhurst, C.L., and Centofani, T. (2010). Phytoremediation of soil trace elements. In: *Trace Elements in Soil* (ed. P.S. Hooda), 311–352. Blackwell Publishing Ltd.

73 Sakakibara, M., Ohmori, Y., Nguyen, T.H.H. et al. (2011). Phytoremediation of heavy metal-contaminated water and sediment by *Eleocharis acicularis*. *Clean: Soil, Air, Water* 39 (8): 735–741.

74 Hu, P.J., Qiu, R.L., Senthilkumar, P. et al. (2009). Tolerance, accumulation and distribution of zinc and cadmium in hyperaccumulator *Potentilla griffithii*. *Environmental and Experimental Botany* 66 (2): 317–325.

75 Mahar, A., Wang, P., Ali, A. et al. (2016). Challenges and opportunities in the phytoremediation of heavy metals contaminated soils: a review. *Ecotoxicology and Environmental Safety* 126: 111–121.

76 Koptsik, G.N. (2014). Problems and prospects concerning the phytoremediation of heavy metal polluted soils: a review. *Eurasian Soil Science* 47 (9): 923–939.

13.4

Biodegradable Waste: Ecotoxicological/Environmental Assessment

Courtois Pauline, Beauchamp Axel, and Vandenbulcke Franck

Univ. Lille, Institut Mines-Télécom, Univ. Artois, Junia, ULR 4515 – LGCgE – Laboratoire de Génie Civil et géo-Environnement, Lille, France

13.4.1 Introduction

The spreading of organic matter on agricultural plots is a very old practice. The use of manure and decomposing food/green waste to fertilize the soil is a technique encountered in works dating back to antiquity [1]. Today, the increase in the world population and industrialization generates the manufacture of billions of tons of organic waste in the European Union (EU) each year. Organic wastes are very diverse: i.e., sewage sludge, compost from kitchen and garden wastes, animal manures, papermill/wood sludge, and waste from various industries (food and drinks preparation industries, like dairies or edible oil or chemical/pharmaceutical companies) [2]. The animal manure alone represented 1 500 000 000 tons in 2009 [3] and the sewage sludge production alone represented 305 000 000 tons in 2016 in the EU [4]. These large quantities of biodegradable wastes, the pollution caused by the storage of these substances in landfills [5], and the cost of landfilling and incineration of wastes [6] mean that the authorities are increasingly encouraging agricultural spreading. Several European directives are driving the reuse of biosolids, such as the Sewage Sludge Directive [7], the Urban Wastewater Treatment Directive [8], the Protection of Waters Against Pollution by Nitrates Directive [9], the Landfill of Waste Directive [10], the Waste Directive [11], and the Renewable Energy Directive [12]. In France, for example, the majority of organic matter spread in soils is animal manure (77.5%), 10% is industrial sludge, and the rest correspond to other types of wastes [13].

All these products are rich in organic matter and a source of minerals of interest, such as nitrogen, phosphorus, potassium, etc. They can therefore amend/fertilize agricultural soils. Many studies demonstrate their benefits on yield, the amount of organic matter in soil, and the water holding capacity [14–16]. However, these substances also contain many contaminants, such as metals, organic contaminants, antibiotics, or microplastics. For example, in sewage sludge, many contaminants are metals (Cd, Cu, Ni, Pb, Zn, Hg, Cr, Se, As, Ag) as well as organic compounds (adsorbable organic halogen compounds, benzo[a]pyrene, chlorobenzene, dibutylphthalate, di-2-(ethylhexyl)phthalate, diethylphthalate, di-n-butylphthalate, endocrine discruptors, ethylene diaminetetra-acetic acid, hexachlorocyclohexane, linear alkylbenzene sulphonates, nonylphenole, nonylphenole ethoxylates, polynuclear aromatic hydrocarbons, polybrominated biphenyls, polybrominated diphenyl ether, chlorinated paraffins, chlorinated biphenyle, polychlorinated dibenzo-p-dioxins and –furans, pentachlorophenol, persistent organic pollutants, tributyltin, bis-tributyltin

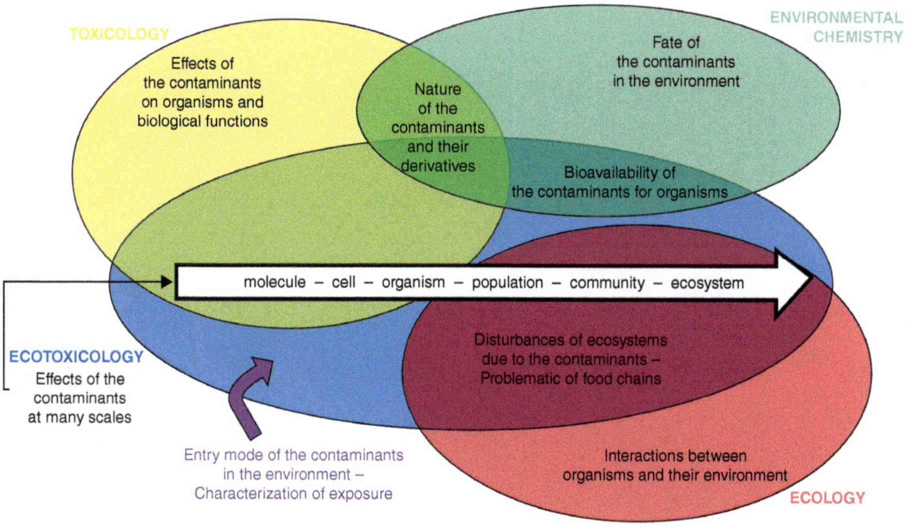

Figure 13.4.1 Area of expertise in ecotoxicology.

oxide, volatile organic compounds, tetracyclines, sulfonamides, fluoroquinolones) [17]. In animal manures, metals are also found in large quantities (especially Cu and Zn, but in a lesser extent Cd, As, Pb, Ni, Cr, Hg), and many veterinary medicines are found (mainly antibiotics, but also antiparasitic compounds and dewormers) as well as pesticides [3, 18]. In composts, pieces of glass and plastic can also be found [19]. These contaminants are then diffused in the terrestrial ecosystem due to spreading.

The evaluation of the effects of these compounds has therefore become a necessity in order to limit pollution and degradation of the environment. The term ecotoxicology is derived from ecology and toxicology words. The term was coined by Truhaut in the 1970s and means "the study of the harmful effects of chemicals upon ecosystems" [20]. The study of the toxicity of pollutants continues, but the way of entry of the pollutant in the environment, the behavior and the biodegradation or transformation of the pollutant, its bioaccumulation by organisms, and the detoxifying systems/capabilities of organisms are new parameters being studied in ecotoxicology [21]. The Figure 13.4.1, inspired by the chapter Ecotoxicology of Environmental Pollutants [22], shows the complexity of ecotoxicology, which straddles ecology, chemistry, and toxicology.

The ecotoxicological evaluation of biodegradable wastes that lands on the soil goes through many stages, based on the framework for ecological risk assessment [23]:

- Monitoring and identifying the composition of these fertilizers;
- Formulation of the problem (contaminants) and characterization of the exposure;
- Characterization of the effects on microorganisms, plants, and animals in the laboratory with ecotoxicity testing and bioassays;
- More realistic laboratory and in-situ studies and characterization of the risk: calculation of the probability of negative ecological effects in-situ.

All these studies make it possible to adapt the legislation in order to reduce risks as much as possible. In addition, there are already some laws concerning the spreading of waste on the soil but they concern especially sewage sludge, the spreading of other wastes being less controlled (Table 13.4.1)

Table 13.4.1 European regulations on the composition of biosolids that can be applied.

Law	Waste	Quantities (tons dry matter per hectare)		Heavy metals ($mg\,kg^{-1}$)		Organic compounds ($mg\,kg^{-1}$)		Nitrates (kg per year per hectare)
European Directive 86/278/EEC adapted for each country, area or type of soil	Sewage sludge	Example of Belgium: it depends on areas and type of land (arable or pasture): between 2 tons every 2 years and 12 tons every 3 years	Example of Denmark: 10 tons per year	Example of severe country: Austria: Cd: 10 Cr: 500 Cu: 500 Hg: 10 Ni: 100 Pb: 500 Zn: 2000 As: 20 Mo: 20 Co: 100	Example of lax country: Spain: Cd: 40 Cr: 1750 Cu: 1750 Hg: 25 Ni: 400 Pb: 1200 Zn: 4000 As: no control Mo: no control Co: no control	Example of Austria: dioxins and furans: 50 PCBs: 0.2 AOX: 500 PAH: 6	Example of France: PCBs: 0.8 PAH: 1.5–4	
European directive 91/676/ECC	Animal manure							170–210

(European Directive 86/278/ECC; European Directive 91/676/EEC). The frequency of analysis is highly variable, according to the country in the European Union; for example for metals, the frequency goes from every three months (Denmark) to every six years (Poland) [24].

13.4.2 Ecotoxicological Standardized Assays

To date, there are more than 30 standardized tests from the ISO (International Organization for Standardization) and 10 standardized tests from the OECD (Organization for Economic Co-operation and Development) to assess the quality of a soil, also called ecotoxicity testing or bioassays for some.

Previously there had been interest in the physico-chemical properties of soils, such as texture, aggregation, humidity, porosity and apparent density of the soil, total C and N, mineral nutrients, matter organic, or the cation exchange capacity [25]. This is why several standardized tests try to assess the effects of pollutants on the physico-chemical properties of soils (Table 13.4.2). These tests mainly aimed to evaluate the inhibition of ionic exchanges and the dispersion and leaching of the pollutant in the soil, such as, for example, the test OECD 312:2004, which describes how to measure the leaching potential of a pollutant in a soil column.

However, assessing the negative effects on organisms living in this ecosystem is now a priority. Soil is one of the most complex ecosystems that shelters many and very diverse organisms in a complex matrix: microorganisms, plants, and animals that live in or in close contact with the soil. The key to the functioning of this ecosystem is precisely this great biodiversity: organisms perform biological functions called ecosystem services, benefiting the human population among others, allowing organic matter to become available and ensuring the formation of soils, the continuity of the cycle of nutrients and primary production, but also by regulating gas flow and water absorption [26]. For this reason, a large majority of standardized tests consist in evaluating the negative effects of one or more pollutants on the various soil organisms. Initially, the tests have been developed to assess the toxic effects of a pollutant on a test organism in artificial soil recommended by ISO and the OECD: the artificial soil OECD test, as described in OECD no. 207. OECD soil is a mixture of sand, kaolin, and peat, to which, depending on the test used, it is sometimes allowed to add a little organic matter. This substrate is then artificially contaminated (spiked) with the studied pollutant. The use of a common soil for all studies allows great repeatability and comparison of results and thus greater reliability of toxicity data. However, this soil is obviously very different from natural soils and the behavior of pollutants and organisms are modified there. This is why the use of standardized natural soils has quickly spread. These standardized soils are natural soils from a specific location that all laboratories can obtain. Such soils exist like the LUFA or the RefeSol soils, for example, which allow bioassays with more realism to be achieved while still offering great repeatability and comparability of results between different laboratories. These soils arrive dried and are then rehydrated and artificially contaminated. However, the use of non-standard field soils is increasing. Their use preserves the microflora naturally present in the soil by avoiding the drying step. They also allow for greater realism since the contaminant can be studied in the type of soil in which it usually lands. Finally, these tests on natural soils make it possible to evaluate the results on many soils that have variable characteristics. In order to preserve this comparability of results as much as possible, the soils used are carefully characterized (granulometry, pH, maximal water retention capacity, C : N ratio, organic matter concentration, potential cation exchange capacity, metal concentrations, etc.) during the studies carried out. ISO standards guide researchers in the way to best describe these soils (as ISO 15175:2018, ISO 10390:2005, ISO 10694:1995, ISO

Table 13.4.2 ISO standardized tests to assess soil quality using physico-chemical tests.

No.	Release year	Title	Principle	Measure of
ISO 13536	1995	Soil quality – Determination of the potential cation exchange capacity and exchangeable cations using barium chloride solution buffered at pH = 8.1	Measurements in field soils	Potential cation exchange capacity and content of exchangeable Na, K, Ca, Mg
ISO 11271	2002	Soil quality – Determination of redox potential – Field method	Measurements in field soils	Soil redox potential
OECD no. 312	2004	Leaching in soil columns	Adding of pollutant or aged residues of it in a soil column with artificial rain to provoke a leaching process	Leaching potential of the compound and its transformation products
OECD no. 318	2017	Dispersion stability of nanomaterials in simulated environmental media	Measurement of dispersion of a nanomaterial compound in simulated environmental media	Ability to conserve a colloidal dispersion
ISO 23470	2018	Soil quality – Determination of effective cation exchange capacity (CEC) and exchangeable cations using a hexamminecobalt (III) chloride solution	Measurement on air-dried soil according to ISO 11464	Cation exchange capacity and content of exchangeable cations (Al, Ca, Fe, K, Mg, Mn, Na)
ISO 11260	2018	Soil quality – Determination of effective cation exchange capacity and base saturation level using barium chloride solution	Measurement on air-dried soil according to ISO 11464	Cation exchange capacity and content of exchangeable cations (Na, K, Ca, Mg)

11047:1998). The use of soils sampled in situ offers the possibility of artificially contaminating the soil, of comparing the toxicity/behavior of a contaminant in field soils exhibiting different pedologic characteristics, or to study the aging of historic contaminations in natural soils, etc.

Certain tests describe the laboratory conditions to assess the effects of a given pollutant on one or more organisms (microbial, plant, or animal) (Tables 13.4.3–13.4.5). A limited number of test species is recommended. The required conditions are precise and in general describe the temperature, the brightness, the photoperiod, the type of container to be used, and the duration of exposure. The concentrations to be used generally depend on the type of pollutant studied. The concentrations without effect and with 100% mortality vary greatly from one pollutant to another.

The activities and biodiversity of microorganisms are much studied since microorganisms play a key role in ecosystem services by participating in soil fertility and stability by participating in biogeochemical cycles and they also degrade contaminants. Therefore, the measure of the degradation of organic contaminants by microorganisms has been realized for almost 40 years using the test OECD 304A (1981), such as the analysis of the biodegradability of antibiotics excreted by farm animals and by soil microorganisms [27]. Then tests like ISO 14240-1:1997 and 14240-2:1997 were developed to measure the microbial biomass in soils. More recently, important activities in the soil have been evaluated using ISO 15685:2012, which assesses the potential for nitrification, or ISO 20130:2018 and ISO/TS 22939:2019, which measure the activity of many enzymes of microbial origin.

Plants are also being studied because they provide organic matter both to the soil during their degradation and to all terrestrial animals. An impact on cultivated species could also have a direct impact on humans. The animals tested come from breeding in general but may also be animals taken in situ in certain situations. The direct effects on life traits have been monitored for a long time, such as germination, growth, and reproduction using ISO 22030:2005 or OECD no. 208 (2006) tests. The genotoxicity of substances on plants has also been studied as with ISO 29200:2013. Finally, the bioavailability of the pollutant and its bioaccumulation by plants is another parameter that has been strongly studied, as with the ISO 16198:2015 test.

Finally, invertebrates living in close contact with the soil are an integral part of the ecotoxicological assessment. As with plants, life traits in animals are studied as a priority with tests OECD no. 207 (1984) for earthworms, ISO 20963:2005 for beetles, ISO 11267:2014 for collembola, OECD no. 220 (2016) for enchytraeids, OECD no. 226 (2016) for mites, or ISO 10872:2020 for nematodes. Similar tests exist for various taxa, but they are not intended to replace one another; they aim to assess the impact on different ecological niches present in the soil ecosystem. Avoidance is also an observed parameter since it translates a disturbance of the animals facing the introduction of a pollutant. It is observed, for example, in earthworms (ISO 17512-1: 2008 or OECD no. 222 of 2016) or in springtails (ISO 17512-2). Finally, the bioavailability and bioaccumulation are evaluated via tests like that in OECD no.317 (2010) in oligochaetes.

Some tests are intended to test the effects of certain families of pollutants and cannot be generalized, for example the ISO 16751:2020, which is a test intended to assess the availability of non-polar organic compounds only. When this is the case, the tests specify the pollutants for which the test is unsuitable. For example, the ISO 20963:2005 test, which observes the survival of the larvae of *Oxythyrea funesta* in the face of cuticular and food exposure, specifies that this test does not make it possible to assess the toxicity of volatile substances. In addition, many guides exist to help researchers choose the tests that are best suited to their subject of study, such as ISO 17616:2019, which makes it possible to choose the best bioassay to use to best characterize the studied soil.

Other ISO tests are evaluation methods through field studying or sampling (Tables 13.4.3 and 13.4.5), such as the ISO 18311:2016 test for microbial and animal organisms or the ISO 23611-1:2018

Table 13.4.3 ISO standardized tests to assess soil quality with tests based on microorganisms.

No.	Release year	Title	Principle	Measure of
OECD no. 304A	1981	Inherent biodegradability in soil	Evaluate of the mineralization rate of a compound labeled with ^{14}C in soil	Release of $^{14}CO_2$ from microorganisms
ISO 14240-1	1997	Soil quality – Determination of soil microbial biomass – Part 2: Substrate-induced respiration method	In unknown soils	Active aerobic, heterotrophic microbial biomass
ISO 14240-2	1997	Soil quality – Determination of soil microbial biomass – Part 2: Fumigation-extraction method	Measurement of total extractable organic biomass from freshly killed microorganisms in unknown soils	Microbial biomass
OECD no. 217	2000	Soil microorganisms: Carbon transformation test	Exposure of natural microorganism community from soil to a pollutant	Carbon dioxide released or oxygen consumed
OECD no. 216	2000	Soil microorganisms: Nitrogen transformation test	Exposure of natural microorganism community from soil to a pollutant	Quantity of nitrates (nitrogen transformation activity)
OECD no. 308	2002	Aerobic and anaerobic transformation in soil	Evaluate of the mineralization rate of a compound labeled with ^{14}C in column of loam soil	Release of $^{14}CO_2$ from microorganisms
ISO 16072	2002	Soil quality – Laboratory methods for determination of microbial soil respiration	Compare microbial respiration of unknown soils with/without addition of a substrate to evaluate the global activity and the microbial biomass	O_2 uptake and CO_2 release
ISO/TS 10832	2009	Soil quality – Effects of pollutants on mycorrhizal fungi – Spore germination test	Direct acute toxicity test on spores of mycorrhizal fungus *Glomus mosseae*	Germination

Table 13.4.3 (Continued)

No.	Release year	Title	Principle	Measure of
ISO 15685	2012	Soil quality – Determination of potential nitrification and inhibition of nitrification – Rapid test by ammonium oxidation	Compare nitrification activity in unknown soils	Ammonium oxidation and inhibition of nitrification
ISO 18311	2016	Soil quality – Method for testing effects of soil contaminants on the feeding activity of soil dwelling organisms – Bait-lamina test	Compare activity of organisms in field soils or in soils sampled as in ISO 23611-6	Feeding activity of soil organisms (microbial and zootique)
ISO 18187	2016	Soil quality – Contact test for solid samples using the dehydrogenase activity of *Arthrobacter globiformis*	Compare activity of a soil bacteria	Inhibition of dehydrogenase activity of *Arthrobacter globiformis*
ISO 17601	2016	Soil quality – Estimation of abundance of selected microbial gene sequences by quantitative PCR from DNA directly extracted from soil	Measure of abundance of selected microbial gene sequences from soil DNA extract from unknown soil by qPCR	Abundance of selected microbial gene sequences
ISO 20130	2018	Soil quality – Measurement of enzyme activity patterns in soil samples using colorimetric substrates in micro-well plates	Compare enzyme activities in unknown soils	Arylamidase, arylsulfatase, β-galactosidase, α-glucosidase, β-glucosidase, N-acetyl-glucosaminidase, acid, alkaline and global phosphatases, urease activities

(*continued*)

Table 13.4.3 (Continued)

No.	Release year	Title	Principle	Measure of
ISO/TS 22939	2019	Soil quality – Measurement of enzyme activity patterns in soil samples using fluorogenic substrates in micro-well plates	Compare enzyme activities in unknown soils	Arylsulfatase, α – glucosidase, β-glucosidase, Cellubisidase, β-xylosidase, phosphodiesterase, chitinase, phosphomonoesterase, leucine-aminopeptidase, alanine-aminopeptidase activities
ISO 23753-1	2019	Soil quality – Determination of dehydrogenases activity in soils – Part 1: Method using triphenyltetrazolium chloride (TTC)	Measurement of activity of deshydrogenases in unknown soils	Dehydrogenases activity
ISO 23753-2	2019	Soil quality – Determination of dehydrogenases activity in soils – Part 2: Method using iodotetrazolium chloride (INT)	Measurement of activity of deshydrogenases in unknown soils	Dehydrogenases activity

test for invertebrates. They pretend to visualize the real impact on the environment of the pollutant present and allow monthly or annual monitoring to be carried out in situ in order to assess, for example, the benefits of a bioremediation technique in application. The ISO 23611-2:2006, ISO 23611-2:2007, ISO 23611-1:2018, and ISO 23611-3:2019 tests allow the sampling of springtails and mites, nematodes, earthworms, and enchytraeids respectively. Once harvested, biodiversity can be measured and based on the functional traits of the species found, allowing an assessment to be made of the impact of the contaminant(s) present *in situ*.

All these tests provide important information on the behavior of pollutants and their toxicity on selected organisms, i.e. test species. They are often used and are reliable to evaluate acute exposure in controlled/laboratory conditions. However, informative from a toxicological point of view, these tests do not give a reliable idea of what is really going on in the environment. In this context, (i) the toxicity of a complex mixture like a biodegradable waste can be evaluated using standardized tests, (ii) these tests are useful to assess the short-term effects of the addition of complex mixtures like biodegradable wastes to soil, but are much less appropriate to assess long-term and chronic effects following aging, partial degradation, biotransformation or repeated applications, and (iii)

Table 13.4.4 ISO standardized tests to assess the quality of soils with tests on plants.

No.	Release year	Title	Principle	Measure of
ISO 22030	2005	Soil quality – Biological methods – Chronic toxicity in higher plants	Chronic direct exposure of *Brassica rapa* and *Avena sativa* in artificial, standard natural or unknown soils	Growth and reproduction
ISO 17126	2005	Soil quality – Determination of the effects of pollutants on soil flora – Screening test for emergence of lettuce seedlings (*Lactuca sativa* L.)	Direct exposure of *Lactuca sativa* seeds in standard natural or unknown soils	Emergence
OECD no. 208	2006	Terrestrial plant test: Seedling emergence and seedling growth test	Exposure of plants in treated soil	Emergence and early growth
ISO 11269-1	2012	Soil quality – Determination of the effects of pollutants on soil flora – Part 1: Method for the measurement of inhibition of root growth	Direct exposure of plants in standard natural or unknown soil	Root elongation
ISO 11269-2	2012	Soil quality – Determination of the effects of pollutants on soil flora – Part 2: Effects of contaminated soil on the emergence and early growth of higher plants	Direct exposure of plants in standard natural or unknown soil	Emergence and early growth
ISO 29200	2013	Soil quality – Assessment of genotoxic effects on higher plants – *Vicia faba* micronucleus test	Direct exposure of *Vicia faba* plant in unknown soil or in the water extract of it	Chromosome breakage and dysfunction of the mitotic spindle (genotoxic effects)
ISO 16198	2015	Soil quality – Plant-based test to assess the environmental bioavailability of trace elements to plants	Estimate the bioavailability of trace elements to plants	Concentration in shoots and roots, net uptake flux in plants

(*continued*)

Table 13.4.4 (Continued)

No.	Release year	Title	Principle	Measure of
ISO 18763	2016	Soil quality – Determination of the toxic effects of pollutants on germination and early growth of higher plants	Direct exposure of plants in unknown soil	Germination and early growth
ISO 21479	2019	Soil quality – Determination of the effects of pollutants on soil flora - Leaf fatty acid composition of plants used to assess soil quality	Compare the quality of soils by analyzing the leaves of plant species cultivated in a field of *Lactuva sativa* grown under controlled conditions in unknown soil (comparison with data in control soils)	Fatty acid composition of leaves

they do not allow the most hazardous compounds to be followed specifically or to identify emerging contaminants among all the contaminants contained in these mixtures.

13.4.3 New Ecotoxicological Approaches

In this context, it appears necessary to carry out more realistic and thus complex experiments in order to evaluate the effect of mixtures and to study emerging or poorly understood pollutants.

To illustrate this point, the example of silver metal will be investigated. Ag has been incorporated more and more into everyday consumer products for two decades now in the form of nanoparticles (AgNPs) due to its strong biocidal properties. The life cycle of these products generates a lot of releases into the environment and in particular into waste water. These Ag nanoparticles (AgNPs) are then trapped in the sewage sludge, which will be spread on agricultural soils without the Ag content being controlled. AgNPs have been extensively studied via various bioassays. *In vitro* results showed that AgNPs can seriously affect various organisms. AgNPs can cause a decrease of microbial activities like respiration and nitrification [28–31]. AgNPs can reduce the plant growth and induce oxidative stress in several plants [32–38]. Also, AgNPs can affect reproduction [39–41] and sometimes even survival of animals like nematodes (*Caenorhabditis elegans*) and annelids (*Eisenia fetida*) [42, 43]. To be satisfied with simplified studies on test organisms on the contaminant released would thus lead to a very pessimistic assessment of the release of AgNPs into the environment. However, a few studies in the field have shown that AgNPs are transformed into nanoparticulate silver sulfide (Ag_2S-NPs) during the passage of waste water through sewers and mainly during the treatment of sludge by anaerobic digestion. Some bioassays on Ag_2S were then carried out showing the lower toxicity of these Ag_2S-NPs compared to AgNPs [41, 44–47], but

Table 13.4.5 ISO standardized tests to assess soil quality based on animal testing.

No.	Release year	Title	Principle	Measure of
OECD no. 207	1984	Earthworm, acute toxicity tests	48 h exposure of *Eisenia fetida* via filter paper or artificial soil at five different concentrations	Survival
ISO 20963	2005	Soil quality – Effects of pollutants on insect larvae (*Oxythyrea funesta*) – Determination of acute toxicity	Exposure through cuticular and food absorption	Survival
ISO 23611-2	2006	Soil quality – Sampling of soil invertebrates – Part 2: Sampling and extraction of micro-arthropods (Collembola and Acarina)	Method of sampling, extracting and preserving collembolans and mites from field soils	Use them as bioindicators
ISO 23611-4	2007 (new version under development)	Soil quality – Sampling of soil invertebrates – Part 4: Sampling, extraction and identification of soil-inhabiting nematodes	Method of sampling and handling nematodes from field soils	Use them as bioindicators
ISO 17512-1	2008	Soil quality – Avoidance test for determining the quality of soils and effects of chemicals on behavior – Part 1: Test with earthworms (*Eisenia fetida* and *Eisenia andrei*)	Direct exposure of earthworms in unknown soil	Avoidance behavior
OECD n°317	2010	Bioaccumulation in terrestrial Oligochaetes	Direct exposure of enchytraeids or earthworms in artificial or unknown soil: uptake then elimination phase	Bioaccumulation factor, uptake rate constant, elimination rate constant

(*continued*)

Table 13.4.5 (Continued)

No.	Release year	Title	Principle	Measure of
ISO 17512-2	2011	Soil quality – Avoidance test for determining the quality of soils and effects of chemicals on behavior – Part 2: Test with collembolans (*Folsomia candida*)	Direct exposure of springtails to assess the bioavailability of contaminant in unknown soil	Avoidance behavior
ISO 11268-1	2012	Soil quality – Effects of pollutants on earthworms – Part 1: Determination of acute toxicity to *Eisenia fetida/Eisenia andrei*	Chronic cutaneous and food exposure on artificial soil, standard natural, or unknown soil	Survival
ISO 11268-2	2012	Soil quality – Effects of pollutants on earthworms – Part 2: Determination of effects on reproduction of *Eisenia fetida/Eisenia andrei*	Chronic cutaneous and food exposure on artificial soil, standard natural, or unknown soil	Reproduction
ISO 11267	2014	Soil quality – Inhibition of reproduction of Collembola (*Folsomia candida*) by soil contaminants	Dermal and alimentary exposure of *Folsomia candida* in artificial, standard natural, or unknown soil	Reproduction
ISO 11268-3	2014	Soil quality – Effects of pollutants on earthworms – Part 3: Guidance on the determination of effects in field situations	Description of field techniques for assessing the effects of substances on earthworms	
ISO 16387	2014	Soil quality – Effects of contaminants on Enchytraeidae (*Enchytraeus sp.*) - Determination of effects on reproduction	Chronic cutaneous and food exposure on unknown soil	Reproduction

Table 13.4.5 (Continued)

No.	Release year	Title	Principle	Measure of
OECD no. 220	2016	Enchytraeid reproduction test	10 weeks exposure of *Enchytraeus albidus* in artificial soil at different concentrations	Reproduction and survival
OECD no. 232	2016	Collembolan reproduction test in soil	4 weeks exposure of *Folsomia candida* or *Folsomia fimetaria* at several concentrations in artificial soil	Survival and reproduction
OECD no. 226	2016	Predatory mite (*Hypoaspis* (Geolaelaps) *aculeifer*) reproduction test in soil	2 weeks exposure of *Hypoaspis aculeifer* females at several concentrations in artificial soil	Survival and reproduction
OECD no. 222	2016	Earthworm reproduction test (*Eisenia fetida/Eisenia andrei*)	8 weeks exposure of *Eisenia fetida* or *Eisenia Andrei* in artificial soil at different concentrations	Behavior, morphology, biomass, and reproduction
ISO 23611-1	2018	Soil quality – Sampling of soil invertebrates – Part 1: Hand-sorting and extraction of earthworms	Method of sampling and handling earthworms from field soils	Use them as bioindicators
ISO 15952	2018	Soil quality – Effects of pollutants on juvenile land snails (Helicidae) – Determination of the effects on growth by soil contamination	Cutaneous and dietary exposure on artificial, standard natural, or unknown soil	Growth and survival
ISO 21285	2019	Soil quality – Inhibition of reproduction of the soil mite (*Hypoaspis aculeifer*) by soil contaminants	Chronic food exposure of *Hypoaspis aculeifer* in artificial, standard natural, or unknown soil	Reproduction

(continued)

Table 13.4.5 (Continued)

No.	Release year	Title	Principle	Measure of
ISO 23611-3	2019	Soil quality – Sampling of soil invertebrates – Part 3: Sampling and extraction of enchytraeids	Method of sampling, handling, and extracting enchytraeids from field soils	Use them as bioindicators
ISO 10872	2020	Water and soil quality – Determination of the toxic effect of sediment and soil samples on growth, fertility, and reproduction of *Caenorhabditis elegans* (Nematoda)	Exposure of *Caenorhabditis elegans* in unknown soil	Growth, fertility, and reproduction
ISO 23266	2020	Soil quality – Test for measuring the inhibition of reproduction in oribatid mites (*Oppia nitens*) exposed to contaminants in soil	Chronic cutaneous and food exposure of oribat mite *Oppia nitens* in artificial, standard natural, or unknown soil	Reproduction
ISO/AWI 16387	Under development	Soil quality – Effects of contaminants on Enchytraeidae (*Enchytraeus* sp.) – Determination of effects on reproduction		
ISO/AWI 16387	Under development	Soil quality – Effects of contaminants on Enchytraeidae (*Enchytraeus* sp.) – Determination of effects on reproduction		
ISO/CD 24032	Under development	Soil quality – In situ caging of snails to assess bioaccumulation of contaminants		
ISO/AWI 11267	Under development	Soil quality – Inhibition of reproduction of Collembola (*Folsomia candida*) by soil contaminants		

can still reduce survival and reproduction of earthworms, for example, at high concentrations [48]. The extrapolation of the results stays difficult due to a significant lack of realism compared to the actual mode of exposure in the environment, as explained in the review by Courtois et al. [50]. This is why a few rare, complex studies are starting to emerge attempting to expose natural microflora, plants, and/or animals to realistic concentrations of AgNPs transformed into Ag_2S-NPs in sewage sludge, which is applied to freshly harvested agricultural soil. These studies were able to show that the main risk of the supply of Ag_2S in realistic concentrations via sewage sludge spreading in soil seemed to be mainly the bioaccumulation in plants and animals, which is very species-dependent [46, 48, 49, 51–53]. The risk of transfer to higher trophic levels remains to be explored. The impact on microorganisms remains even more unclear because the results depend even more on the type of soil and therefore on the microbial community exposed to Ag_2S. In some soils, the microbial community seems to adapt and/or acclimatize to the presence of the contaminant and regulate its activities [44, 51, 54], but in other types of soils, there may be reductions in potential respiration or nitrification [55–59]. It is necessary to carry out more other studies, always more realistic, if one wishes to establish a precise environmental risk assessment of the presence of Ag_2S, derived from AgNPs, in sewage sludge spread on agricultural soils. This example perfectly illustrates (i) the need to take into account the transformations that contaminants undergo during their life cycle because these transformations can profoundly modify their toxicity and (ii) the difficulty of going beyond the assessment of acute toxicity and documenting the environmental risk, i.e. the transfer of a contaminant into the food web, the impact on soil functions and therefore on the ecosystem services provided by the soil. An environmental assessment is particularly necessary in the case of contaminants provided by mixtures as complex as biodegradable wastes, some being produced in very large quantities and largely redeposited in the natural environment.

13.4.4 Conclusions

To conclude, standardized tests are often used and are reliable to evaluate acute exposure in controlled/laboratory conditions. Informative from a toxicological point of view, these tests do not give a reliable idea of what is really going on in the environment. It is therefore necessary to detach from these tests to carry out complex experiments mimicking the full course of the pollutant before and after its application in the soil with, for example, the treatments undergone before the application of the spreadable waste (composting, vermicomposting, aerobic digestion, anaerobic digestion, mixing with other organic wastes, etc.). Moreover, experimental application of a biodegradable waste amendment to the soil must be as realistic as possible (quantity, degree of contamination) and thus in accordance with the procedures recommended by national and international organizations.

References

1 Zimring, C.A. and Rathje, W.L. (eds.) (2012). *Encyclopedia of Consumption and Waste: The Social Science of Garbage*. SAGE Publications 1225 pp.

2 European Commission (2001). Survey of wastes spread on land. https://ec.europa.eu/environment/waste/studies/compost/landspreading.pdf.

3 Bloem, E., Albihn, A., Elving, J. et al. (2017). Contamination of organic nutrient sources with potentially toxic elements, antibiotics and pathogen microorganisms in relation to P fertilizer

potential and treatment options for the production of sustainable fertilizers: a review. *Science of the Total Environment* 607–608: 225–242. https://doi.org/10.1016/j.scitotenv.2017.06.274.

4 Eurostat (2020). Sewage sludge production and disposal (data for 2017). https://appsso.eurostat.ec.europa.eu/nui/submitViewTableAction.do.

5 Njoku, P.O., Edokpayi, J.N., and Odiyo, J.O. (2019). Health and environmental risks of residents living close to a landfill: a case study of Thohoyandou landfill, Limpopo Province, South Africa. *International Journal of Environmental Research and Public Health* 16 https://doi.org/10.3390/ijerph16122125.

6 Kacprzak, M., Neczaj, E., Fijałkowski, K. et al. (2017). Sewage sludge disposal strategies for sustainable development. *Environmental Research* 156: 39–46. https://doi.org/10.1016/j.envres.2017.03.010.

7 Directive 86/278/ECC (1986). Directive no. 86/278 du 6 December 1986 relative à la protection de l'environnement et notamment des sols, lors de l'utilisation des boues d'épuration en agriculture | AIDA. https://aida.ineris.fr/consultation_document/1081.

8 Directive 91/271/EEC (1991). Council Directive 91/271/EEC of 21 May 1991 concerning urban waste-water treatment. https://eur-lex.europa.eu/legal-content/EN/TXT/HTML/?uri=CELEX:31991L0271&from=FR.

9 Directive 91/676/EEC 1991). Council Directive 91/676/EEC of 12 December 1991 concerning the protection of waters against pollution caused by nitrates from agricultural sources. Off. J. 375 31121991 P 0001–0008 Chapter 15, Vol. 10 P 0192. https://eur-lex.europa.eu/legal-content/EN/TXT/HTML/?uri=CELEX:31991L0676&from=FR.

10 Directive 1999/31/EC (1999). Council Directive 1999/31/EC of 26 April 1999 on the landfill of waste. Off. J. 182 16071999 P 0001–0019. https://eur-lex.europa.eu/legal-content/EN/TXT/HTML/?uri=CELEX:31999L0031&from=FR.

11 Directive 2008/98/EC (2008). Directive 2008/98/EC of the European Parliament and of the Council. https://eur-lex.europa.eu/legal-content/EN/TXT/HTML/?uri=CELEX:32008L0098&from=EN.

12 Directive 2009/28/EC (2009). Directive 2009/28/EC of the European Parliament and of the Council on the Promotion of the Use of Energy from Renewable Sources. https://eur-lex.europa.eu/legal-content/EN/TXT/HTML/?uri=CELEX:32009L0028&from=FR.

13 CGEDD report n° 009801-01, CGAAER n° 14074 (2015). Les épandages sur terres agricoles des matières fertilisantes d'origine résiduaire. Mission prospective sur les modalités d'encadrement et de suivi réglementaire. https://agriculture.gouv.fr/sites/minagri/files/cgaaer_14074_2015_rapport.pdf.

14 Cele, E.N. and Maboeta, M. (2015). A greenhouse trial to investigate the ameliorative properties of biosolids and plants on physicochemical conditions of iron ore tailings: implications for an iron ore mine site remediation. *Journal of Environmental Management* 165: 167–174. https://doi.org/10.1016/j.jenvman.2015.09.029.

15 Liu, X.-J.A., Fike, J.H., Galbraith, J.M. et al. (2015). Effects of harvest frequency and biosolids application on switchgrass yield, feedstock quality, and theoretical ethanol yield. *GCB Bioenergy* 7: 112–121. https://doi.org/10.1111/gcbb.12124.

16 Shu, W., Price, G.W., Sharifi, M., and Cade-Menun, B.J. (2016). Impact of annual and single application of alkaline treated biosolids on soil extractable phosphorus and total phosphorus. https://www.agr.gc.ca/eng/abstract/publication?id=37025000010735.

17 Prasad, M.N.V., de Campos Favas, P.J., Vithanage, M., Mohan, S.V. (2019). Industrial and Municipal Sludge – Emerging Concerns and Scope for Resource Recovery.

https://www-elsevier-com.ressources-electroniques.univ-lille.fr/books/industrial-and-municipal-sludge/vara-prasad/978-0-12-815907-1.

18 McKinnon, K., Serikstad, G.L., Eggen, T., 2014. Contaminants in manure – a problem for organic farming? Presented at the ISOFAR, Turkey.

19 Cattle, S.R., Robinson, C., and Whatmuff, M. (2020). The character and distribution of physical contaminants found in soil previously treated with mixed waste organic outputs and garden waste compost. *Waste Management* 101: 94–105. https://doi.org/10.1016/j.wasman.2019.09.043.

20 Truhaut, R. (1977). Ecotoxicology: objectives, principles and perspectives. *Ecotoxicology and Environmental Safety* 1: 151–173. https://doi.org/10.1016/0147-6513(77)90033-1.

21 Walker, C.H., Greig-Smith, P.W., Crossland, N.O., and Brown, R. (1991). Ecotoxicology. In: *Animals and Alternatives in Toxicology* (eds. M. Balls, J. Bridges and J. Southee), 223–251. London. Frame: Palgrave.

22 Bláha, L. and Hofman, J. (2020). Ecotoxicology of environmental pollutants. In: *Advanced Nano-Bio Technologies for Water and Soil Treatment, Applied Environmental Science and Engineering for a Sustainable Future* (eds. J. Filip, T. Cajthaml, P. Najmanová, et al.), 549–572. Cham: Springer International Publishing https://doi.org/10.1007/978-3-030-29840-1_27.

23 US EPA (1992). Framework for Ecological Risk Assessment. US EPA. https://www.epa.gov/risk/framework-ecological-risk-assessment.

24 European Commission, DG Environment (2001). Disposal and recycling routes for sewage sludge - Part 2 – Regulatory report 65.

25 Cardoso, E.J.B.N., Vasconcellos, R.L.F., Bini, D. et al. (2013). Soil health: looking for suitable indicators. What should be considered to assess the effects of use and management on soil health? *Science in Agriculture* 70: 274–289. https://doi.org/10.1590/S0103-90162013000400009.

26 Lavelle, P., Decaëns, T., Aubert, M. et al. (2006). Soil invertebrates and ecosystem services. *European Journal of Soil Biology ICSZ* 42: S3–S15. https://doi.org/10.1016/j.ejsobi.2006.10.002.

27 Vaclavik, E., Halling-Sørensen, B., and Ingerslev, F. (2004). Evaluation of manometric respiration tests to assess the effects of veterinary antibiotics in soil. *Chemosphere* 56: 667–676. https://doi.org/10.1016/j.chemosphere.2004.02.018.

28 Beddow, J., Stolpe, B., Cole, P. et al. (2014). Effects of engineered silver nanoparticles on the growth and activity of ecologically important microbes: impact of silver nanoparticles on environmental microbes. *Environmental Microbiology Reports* 6: 448–458. https://doi.org/10.1111/1758-2229.12147.

29 Choi, O. and Hu, Z. (2008). Size dependent and reactive oxygen species related nanosilver toxicity to nitrifying bacteria. *Environmental Science & Technology* 42: 4583–4588. https://doi.org/10.1021/es703238h.

30 Choi, O., Deng, K.K., Kim, N.-J. et al. (2008). The inhibitory effects of silver nanoparticles, silver ions, and silver chloride colloids on microbial growth. *Water Research* 42: 3066–3074. https://doi.org/10.1016/j.watres.2008.02.021.

31 Michels, C., Perazzoli, S., and Soares, M. (2017). Inhibition of an enriched culture of ammonia oxidizing bacteria by two different nanoparticles: silver and magnetite. *Science of the Total Environment* 586: 995–1002. https://doi.org/10.1016/j.scitotenv.2017.02.080.

32 Barbasz, A., Kreczmer, B., and Oćwieja, M. (2016). Effects of exposure of callus cells of two wheat varieties to silver nanoparticles and silver salt ($AgNO_3$). *Acta Physiologiae Plantarum* 38: 76. https://doi.org/10.1007/s11738-016-2092-z.

33 Cvjetko, P., Milošić, A., Domijan, A.-M. et al. (2017). Toxicity of silver ions and differently coated silver nanoparticles in *Allium cepa* roots. *Ecotoxicology and Environmental Safety* 137: 18–28. https://doi.org/10.1016/j.ecoenv.2016.11.009.

34 Geisler-Lee, J., Wang, Q., Yao, Y. et al. (2012). Phytotoxicity, accumulation and transport of silver nanoparticles by *Arabidopsis thaliana*. *Nanotoxicology* 7: 323–337. https://doi.org/10.3109/17435390.2012.658094.

35 Jiravova, J., Tomankova, K.B., Harvanova, M. et al. (2016). The effect of silver nanoparticles and silver ions on mammalian and plant cells in vitro. *Food and Chemical Toxicology* 96: 50–61. https://doi.org/10.1016/j.fct.2016.07.015.

36 Pradas del Real, A.E., Vidal, V., Carrière, M. et al. (2017). Silver nanoparticles and wheat roots: a complex interplay. *Environmental Science & Technology* 51: 5774–5782. https://doi.org/10.1021/acs.est.7b00422.

37 Song, U., Jun, H., Waldman, B. et al. (2013). Functional analyses of nanoparticle toxicity: a comparative study of the effects of TiO_2 and Ag on tomatoes (*Lycopersicon esculentum*). *Ecotoxicology and Environmental Safety* 93: 60–67. https://doi.org/10.1016/j.ecoenv.2013.03.033.

38 Stampoulis, D., Sinha, S.K., and White, J.C. (2009). Assay-dependent Phytotoxicity of nanoparticles to plants. *Environmental Science & Technology* 43: 9473–9479. https://doi.org/10.1021/es901695c.

39 Roh, J., Sim, S.J., Yi, J. et al. (2009). Ecotoxicity of silver nanoparticles on the soil nematode *Caenorhabditis elegans* using functional Ecotoxicogenomics. *Environmental Science & Technology* 43: 3933–3940. https://doi.org/10.1021/es803477u.

40 Schultz, C.L., Wamucho, A., Tsyusko, O.V. et al. (2016). Multigenerational exposure to silver ions and silver nanoparticles reveals heightened sensitivity and epigenetic memory in *Caenorhabditis elegans*. *Proceedings of the Royal Society B: Biological Sciences* 283: 20152911. https://doi.org/10.1098/rspb.2015.2911.

41 Starnes, D.L., Unrine, J.M., Starnes, C.P. et al. (2015). Impact of sulfidation on the bioavailability and toxicity of silver nanoparticles to *Caenorhabditis elegans*. *Environmental Pollution* 196: 239–246. https://doi.org/10.1016/j.envpol.2014.10.009.

42 Garcia-Velasco, N., Gandariasbeitia, M., Irizar, A., and Soto, M. (2016). Uptake route and resulting toxicity of silver nanoparticles in *Eisenia fetida* earthworm exposed through standard OECD tests. *Ecotoxicology* 25: 1543–1555. https://doi.org/10.1007/s10646-016-1710-2.

43 Kim, S.W., Nam, S.-H., and An, Y.-J. (2012). Interaction of silver nanoparticles with biological surfaces of *Caenorhabditis elegans*. *Ecotoxicology and Environmental Safety* 77: 64–70. https://doi.org/10.1016/j.ecoenv.2011.10.023.

44 Kraas, M., Schlich, K., Knopf, B. et al. (2017). Long-term effects of sulfidized silver nanoparticles in sewage sludge on soil microflora. *Environmental Toxicology and Chemistry* 36: 3305–3313. https://doi.org/10.1002/etc.3904.

45 Levard, C., Hotze, E.M., Colman, B.P. et al. (2013). Sulfidation of silver nanoparticles: natural antidote to their toxicity. *Environmental Science & Technology* 47: 13440–13448. https://doi.org/10.1021/es403527n.

46 Pradas del Real, A.E., Castillo-Michel, H., Kaegi, R. et al. (2016). Fate of ag-NPs in sewage sludge after application on agricultural soils. *Environmental Science & Technology* 50: 1759–1768. https://doi.org/10.1021/acs.est.5b04550.

47 Schlich, K., Hoppe, M., Kraas, M. et al. (2018). Long-term effects of three different silver sulfide nanomaterials, silver nitrate and bulk silver sulfide on soil microorganisms and plants. *Environmental Pollution* 242: 1850–1859. https://doi.org/10.1016/j.envpol.2018.07.082.

48 Lahive, E., Matzke, M., Durenkamp, M. et al. (2017). Sewage sludge treated with metal nanomaterials inhibits earthworm reproduction more strongly than sludge treated with metal metals in bulk/salt forms. *Environmental Science Nano* 4: 78–88. https://doi.org/10.1039/C6EN00280C.

49 Colman, B.P., Arnaout, C.L., Anciaux, S. et al. (2013). Low concentrations of silver nanoparticles in biosolids cause adverse ecosystem responses under realistic field scenario. *PLoS One* 8: e57189. https://doi.org/10.1371/journal.pone.0057189.

50 Courtois, P., Rorat, A., Lemiere, S. et al. (2019). Ecotoxicology of silver nanoparticles and their derivatives introduced in soil with or without sewage sludge: a review of effects on microorganisms, plants and animals. *Environmental Pollution* 253: 578–598. https://doi.org/10.1016/j.envpol.2019.07.053.

51 Courtois, P., Rorat, A., Lemiere, S. et al. (2020). Medium-term effects of ag supplied directly or via sewage sludge to an agricultural soil on *Eisenia fetida* earthworm and soil microbial communities. *Chemosphere* 269: 128761. https://doi.org/10.1016/j.chemosphere.2020.128761.

52 Kampe, S., Kaegi, R., Schlich, K. et al. (2018). Silver nanoparticles in sewage sludge: bioavailability of sulfidized silver to the terrestrial isopod *Porcellio scaber*: silver nanoparticles in sewage sludge. *Environmental Toxicology and Chemistry* 37: 1606–1613. https://doi.org/10.1002/etc.4102.

53 Velicogna, J.R., Schwertfeger, D.M., Jesmer, A.H. et al. (2017). The bioaccumulation of silver in *Eisenia andrei* exposed to silver nanoparticles and silver nitrate in soil. *NanoImpact* 6: 11–18. https://doi.org/10.1016/j.impact.2017.03.001.

54 Judy, J.D., Kirby, J.K., Creamer, C. et al. (2015). Effects of silver sulfide nanomaterials on mycorrhizal colonization of tomato plants and soil microbial communities in biosolid-amended soil. *Environmental Pollution* 206: 256–263. https://doi.org/10.1016/j.envpol.2015.07.002.

55 Durenkamp, M., Pawlett, M., Ritz, K. et al. (2016). Nanoparticles within WWTP sludges have minimal impact on leachate quality and soil microbial community structure and function. *Environmental Pollution* 211: 399–405. https://doi.org/10.1016/j.envpol.2015.12.063.

56 Samarajeewa, A.D., Velicogna, J.R., Schwertfeger, D.M. et al. (2019). Effect of silver nanoparticle contaminated biosolids on the soil microbial community. *NanoImpact* 14: 100157. https://doi.org/10.1016/j.impact.2019.100157.

57 Schlich, K. and Hund-Rinke, K. (2015). Influence of soil properties on the effect of silver nanomaterials on microbial activity in five soils. *Environmental Pollution* 196: 321–330. https://doi.org/10.1016/j.envpol.2014.10.021.

58 Schlich, K., Hoppe, M., Kraas, M. et al. (2017). Ecotoxicity and fate of a silver nanomaterial in an outdoor lysimeter study. *Ecotoxicology* 26: 738–751. https://doi.org/10.1007/s10646-017-1805-4.

59 Schlich, K., Hoppe, M., Kraas, M. et al. (2018). Long-term effects of three different silver sulfide nanomaterials, silver nitrate and bulk silver sulfide on soil microorganisms and plants. *Environmental Pollution* 242, 1850–1859. doi: 10.1016/j.envpol.2018.07.082.

14

Restoration, Sequestration and Modeling of Carbon in Degraded Soils

Kennedy Were[1], BalRam Singh[2], Eleanor Milne[3], and George Ayaga[4]

[1] Kenya Agricultural and Livestock Research Organization, Food Crops Research Centre – Kabete, Nairobi, Kenya
[2] Faculty of Environmental Sciences and Natural Resource Management, Norwegian University of Life Sciences, Ås, Norway
[3] The Natural Resource Ecology Laboratory, Colorado State University, Fort Collins, CO, USA
[4] Kenya Agricultural and Livestock Research Organization, Food Crops Research Centre – Alupe, Busia, Kenya

14.1 Introduction

14.1.1 Soils and Soil Degradation

The world's soil resources are finite, non-renewable and difficult to restore when degraded; yet they anchor all forms of life, ecosystem functions, and human well-being. Soils deliver multiple ecosystem services, such as the provision of food, fodder, fuel, fibre, and habitat for flora and fauna; the support of nutrient cycling, primary production, soil formation, and biological diversity; and the regulation of diseases, pests, climate, erosion, and hydrological processes, including water purification, infiltration, percolation, drainage, stream flow, and water storage [1]. Despite their significance, the world's soils have not been widely appreciated, leading to deterioration of their structure, quality, and properties through changes in land use and generations of unsustainable land management practices associated with agriculture, grazing, and forest management. This has compromised the capacity of the soils to produce ecosystem goods and services, as well as plant materials of good quality and quantity, especially in the tropics and subtropics [2]. Some estimates indicate that soil degradation lowered soil-based ecosystem services by 60% between 1950 and 2010 [3] and is costing us between USD 6.3 and 10.6 trillion annually [4]. This does not auger well for food and nutrition security, ecological resilience and stability, viable livelihoods, and a land degradation-neutral world envisaged by the sustainable development goals (SDGs). Thus, the world's soils need to be protected from the major physical, chemical, and biological degradation threats and processes. Prevention of soil degradation is far more cost-effective than restoring and rehabilitating degraded soils, and the cost of inaction is too high [1].

According to Lal [2, 3], physical soil degradation processes and threats are characterized by a decline in soil structure and tilth, often leading to surface sealing and crusting, compaction and hard setting, erosion, reduced infiltration capacity, and desertification. Important among the chemical soil degradation processes are acidification, leaching, salinization, nutrient depletion, reduced cation exchange capacity (CEC), contamination, and toxicities, while biological degradation processes manifest through the loss of soil biodiversity (e.g. earthworms), a decline in soil organic matter (SOM), and an increase in soil-borne pathogens (Figure 14.1). An interplay of these degradation processes affects other ecosystem components and functions, and the attendant socioecological

Biodegradable Waste Management in the Circular Economy: Challenges and Opportunities, First Edition.
Edited by Malgorzata Kacprzak, Eleonore Attard, Kari-Anne Lyng, Helena Raclavska, BalRam Singh, Eyob Tesfamariam, and Franck Vandenbulcke.

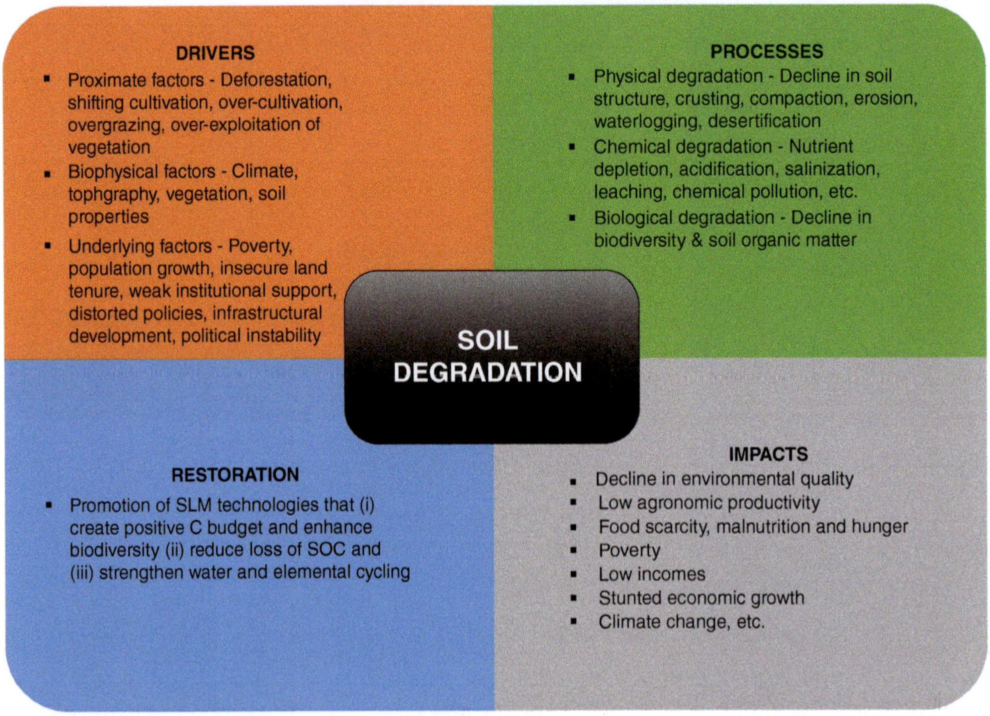

Figure 14.1 Drivers, processes, impacts, and strategies for restoring soil degradation.

responses that reinforce or suppress the driving forces lead to new conditions or a downward spiral of decline in environmental quality. For instance, the culmination might be disruptions in the provision of ecosystem services (e.g. elemental cycling, soil formation, C sequestration, flood regulation, and water purification) and a decline in net biome productivity [3].

Even though soil degradation is a natural process, it is aggravated by complex, and little understood, interactions among the biophysical, socioeconomic, and political factors that operate within specific spatial and temporal domains [5]. Proximate factors of soil degradation, which operate at the local levels, comprise over-cultivation (extractive farming), overgrazing, deforestation (agricultural expansion), shifting cultivation and overexploitation of vegetation [6–8] (Figure 14.1). The proximate factors are underpinned by fundamental sociopolitical and economic structures and processes (e.g. population pressure, poverty, land tenure, marketing, income, institutional support, policies, and political stability) and biophysical factors (e.g. terrain, vegetation, intrinsic soil properties, and climate), which operate diffusely at regional and global levels, and influence the kind and effectiveness of degradation processes. For instance, population growth can trigger several processes, such as deforestation, diminishing landholdings, increasing pressure on agricultural land, and the population shift from high potential agricultural areas to marginal lands and food shortages, which then fuel different soil degradation processes [6].

The nature, extent, and severity of soil degradation at the regional, national, and global scales are, however, variable and controversial owing to the paucity of data obtained through systematic assessment [1]. Most of the available estimates are based on qualitative expert judgments or proxy land surface parameters retrieved from satellite data (e.g. normalized difference vegetation index – NDVI). The Global Assessment of Soil Degradation (GLASOD) project commissioned by

Table 14.1 Extent of the status of human-induced soil degradation in the world.

	Region (10^6 ha)						
Degradation process	**Asia**	**Africa**	**South and Central America**	**North America**	**Europe**	**Austral-asia**	**World**
Water erosion	440	227	169	60	115	83	1094
Wind erosion	222	187	47	35	42	16	548
Nutrient depletion	14	45	72	–	3	–	135
Salinization	53	15	4	–	4	1	76
Compaction	10	18	4	1	33	2	68
Others (pollution, water-logging, acidification, and subsidence of organic soils)	8	2	10	–	22	1	43
Total	747	494	306	96	219	103	1964

Note: The other degradation processes include: pollution, water-logging, acidification and subsidence of organic soils.
Source: Adapted from Bouma and Batjes [9].

the United Nations Environment Programme (UNEP) estimated that by 1990, 15% (about 1.96 billion ha) of the world's total land surface had been degraded, out of which 38% was agricultural land, 21% was permanent pastures (rangeland), and 18% was forest and woodland (Table 14.1) [9, 10]. Recent studies have also indicated that more than 50% of all fertile, food-producing soils globally are now classified as degraded, many of them severely degraded, with 25–40 billion tonnes being lost annually [11, 12], and that by 2050, less than 10% of the Earth's surface will have escaped substantial human impact [13]. Although chemical degradation (especially nutrient depletion) and physical degradation (especially water and wind erosion) were reported as the most prevalent processes, they often result in biological degradation (especially depletion of SOM), transforming soils into net sources of greenhouse gases (GHGs) with positive feedback to global warming [3, 14, 15]. For instance, the SOM content tends to be lower in soils with severe nutrient depletion because of the concomitant reduction in plant production. This implies that investment in sustainable land management (SLM) practices to build, restore and enhance carbon (C) in degraded soils is vital for achieving land degradation neutrality (LDN)[1] and securing global environmental benefits. Literature abounds with evidence of SLM benefits, including *inter alia* climate change mitigation through enhancement of C sequestration in the soil and plants, and reduction of GHG emissions. According to ELD [4] about USD 480 billion can also be generated by enhancing SOC stocks, considering that SOC is the largest C pool in the terrestrial biosphere.

In the subsequent sections, we highlight the linkages between soil degradation and changes in SOC stocks, SLM technologies for restoring, building, and enhancing C in degraded soils, and the impact of enhanced SOC stocks on the resilience of the overall ecosystem. We also discuss how modeling tools can be leveraged to assist in assessing the C footprints and GHGs within SLM projects, using western Kenya as a case study.

1 The UNCCD defines LDN as a state where the amount and quality of land resources necessary to support ecosystem functions and services and enhance food security remain stable or increase within specified temporal and spatial scales and ecosystems.

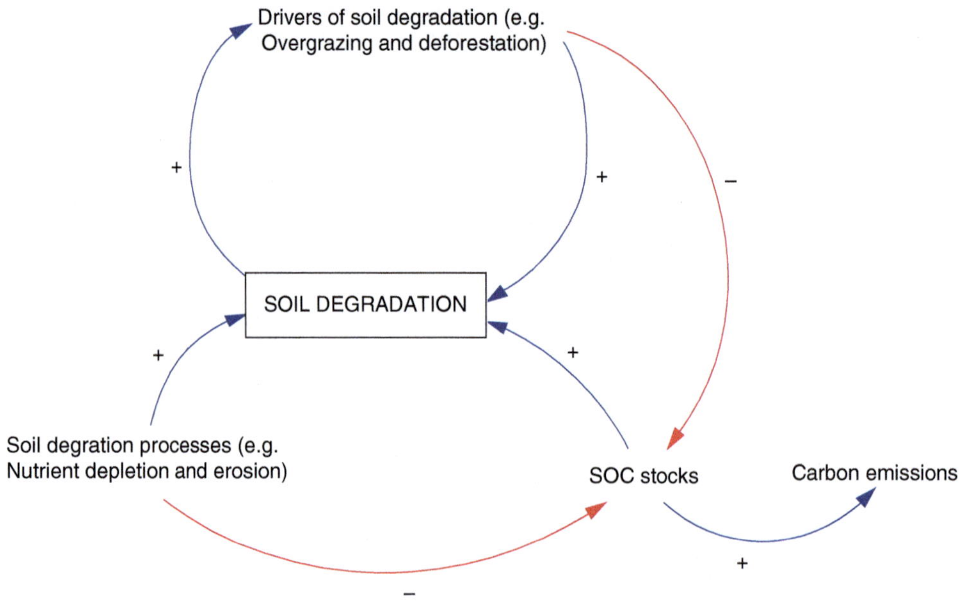

Figure 14.2 Interactions and feedbacks between soil degradation, the process and drivers of degradation, SOC, stocks and C emissions. Each arrow indicates a causal relationship, which can be large or small, immediate or delayed, and an increasing (+) or decreasing (–) effect.

14.2 Soil Degradation and SOC Stocks

As mentioned above, nearly a quarter of the Earth's surface has been degraded. Such a dramatic change coupled with the soil degradation processes are also reflected in the changes in SOC stocks and fluxes. For example, Olson et al. [16] and Chappell et al. [17] stated that accelerated erosion, which is the most widespread soil degradation process, changes the dynamics of SOC stocks in agricultural land units by detaching and preferentially transporting the fine, C- and nutrient-rich sediment, stimulating the mineralization of organic matter (OM) in the eroded sediment through oxidation and release of carbon dioxide (CO_2). Soil nutrient depletion also leads to a decrease in net primary productivity (NPP) and C input to the soil, while acidification affects the plant root exudates, exposing carbonates and releasing C [18]. Chotte et al. [13] added that apart from the soil degradation processes, most of their drivers also perturb the C cycle. For example, (over)-cultivation alters soil aeration, which accelerates microbial decomposition and oxidation of SOM to CO_2, whereas overgrazing, overexploitation of vegetation, and deforestation (clear-cutting) result in soil exposure, enhancing soil temperatures, and SOM decay. Taken together, the degradation processes and their drivers aggravate the depletion of the SOC pool on-site and emission of CO_2 off-site, with negative implications for global warming and climate change (Figure 14.2). The immediate socioeconomic consequence of this is a reduction in agronomic productivity, translating to food scarcity, low incomes, poverty, and stunted economic growth.

14.3 Restoration and Sequestration of C in Degraded Soils

There is an intricate linkage between the restoration of degraded soils and C sequestration. Soil C sequestration in this context refers to the uptake and conversion of atmospheric CO_2 to OM

Figure 14.3 Severe gully erosion in Baringo, Kenya. Source: © Duncan Kebut.

through photosynthesis by plants and the subsequent transfer of the OM into the soil reservoir for long-term storage [3, 14]. Having lost a huge fraction of the antecedent SOC pool by erosion, decomposition, and leaching, reversing the trends of soil degradation (Figure 14.3) through the adoption of appropriate SLM practices offers the highest potential for C sequestration [19, 20]. Such restorative practices can generate the surest and cost-effective winning set of solutions to improve human well-being, ecosystem services, and the resilience of the overall system, meaning reduced vulnerability to the impact of environmental change [21]. Lal [14] suggested that strategies, technologies, innovations and management practices for the restoration and re-carbonization of degraded soils should strive to create a positive ecosystem C budget through:

(a) Increasing the soil application of biomass C and enhancing the transformation to SOM;
(b) Minimizing the losses of SOC, water, and nutrients;
(c) Enhancing biodiversity and the mean residence time of SOC.

14.3.1 SLM Technologies for Restoring and Sequestering C

Scientists agree that SLM is the primary instrument for fulfilling SDG 15.3 of restoring degraded soils and achieving land degradation neutrality by 2030 [13]. Managing landscapes sustainably is about using land resources (i.e. soil, water, vegetation, and biodiversity) to maximize economic and social benefits, while simultaneously maintaining, enhancing, and ensuring the long-term productive potential and ecological functions of these resources. It means less CO_2 emissions and more removals from the atmosphere for storage in the terrestrial reservoir of C. A suite of proven SLM practices and approaches can be applied to restore, conserve, and enhance C in degraded soils at the local scale, including inter alia conservation agriculture (CA), agroforestry (AF), mulching, crop residue retention, integrated nutrient management (INM), controlled grazing at optimal stocking rates, organic manuring, green manuring, cover crops, improved plant varieties with greater root mass, crop rotations, and soil and water conservation (SWC) structures (e.g., terraces, retention ditches, and cut-off drains) (Figure 14.4). Notably, each of these SLM technologies has its own trade-offs and there is no one-size-fits-all option for the myriad soil degradation processes occurring

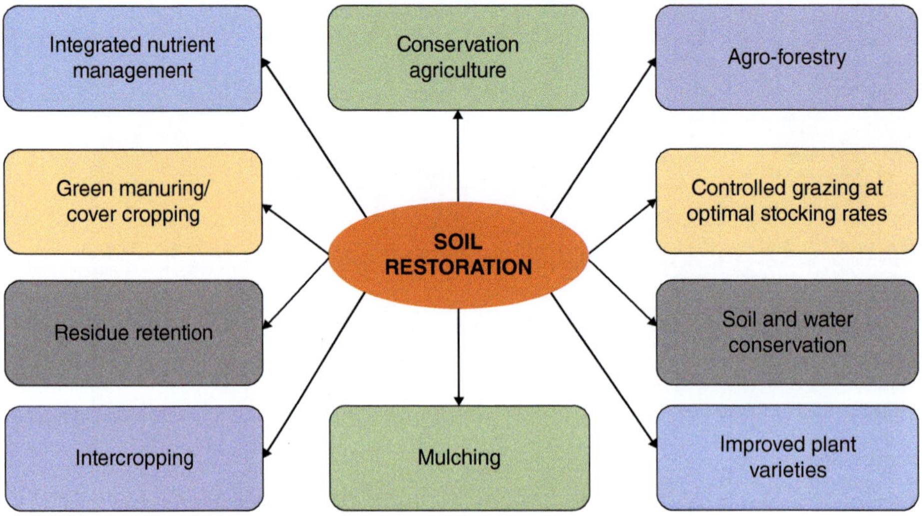

Figure 14.4 Multiple pathways to restoring, conserving, and enhancing C in degraded soils.

in diverse contexts. Thus, countries should select options that create multiple wins, maximize return on investment, and are technically feasible and socially acceptable.

To start with, CA technology is characterized by minimum mechanical soil disturbance, maintenance of at least 30% permanent organic soil cover, and diversified cropping through intercropping and crop rotations [3, 12, 22]. Through these three fundamental principles, CA offers multiple environmental benefits if properly implemented. For example, minimum soil disturbance protects soil from breaking down and releasing C to the atmosphere, avoids soil compaction, reduces erosion, lowers atmospheric CO_2 emissions through reduced fossil fuel consumption, saves time, money, and energy due to reduced tillage, improves water infiltration into the soil, and enhances efficiency in the use of inputs (e.g. fertilizer) through direct placements [23]. In addition, permanent organic soil cover (Figure 14.5) suppresses weeds, reduces exposure to erosion, improves SOM and nutrient status, and moderates soil surface temperatures and evaporation, which preserves soil moisture, increases the soil water content, and acts as insurance against drought. Further, diversified cropping controls weeds, insects, pests, and diseases through crop rotations, promotes good soil structure, fosters above- and below-ground biodiversity and natural biological processes, which contribute to nutrient cycling and improve nutrient- and water-use efficiency, reduce the risk of total crop failure in the case of drought and disease outbreak, and allow maximum utilization of the available nutrients. According to Lal [24], under CA, about 0.5–1.0 t C ha^{-1} yr^{-1} can be sequestered in the humid temperate areas, 0.2–0.5 t C ha^{-1} yr^{-1} in the humid tropics, and 0.1–0.2 t C ha^{-1} yr^{-1} in the semi-arid zones. The estimates of other studies are in the range of 0.05–0.36 t C ha^{-1} yr^{-1} [25], 0.57 ± 0.14 t C ha^{-1} yr^{-1} [26], 0.33 t C ha^{-1} yr^{-1} [27], and 0.37 t C ha^{-1} yr^{-1} [28]. Differences in input of biomass C, soil moisture, temperature regimes and susceptibility to erosion are the major factors explaining variability in SOC sequestration under CA [29].

INM is another important SLM technology to overcome deficiencies in nutrients supplied by soils, sequester C, and increase SOM content. It combines the best of organic and conventional practices in a productive, sustainable, and environmentally sound way [30]. In other words, INM maximizes the use of organic resources and improved germplasm, minimizes nutrient losses, allows timely and judicious use of inorganic fertilizers based on need and economic viability, and maintains and enhances beneficial soil organisms and biological processes [31, 32]. For

Figure 14.5 Cowpea established under maize stalk residues (left) and maize intercropped with *Mucuna pruriens* as a dense cover crop (right) in a CA system in western Kenya.

successful implementation, the technology should be adapted to the local conditions and needs. A typical INM practice is intercropping and rotating improved maize with a legume and applying both organic and inorganic fertilizers (Figure 14.6) in conjunction with water harvesting, or an SWC technology, such as residue retention [33, 34]. By boosting the soil physical, chemical, and biological characteristics (e.g. the enhanced supply of N and other essential nutrients in the soil, reduced soil water and nutrient losses, as well as improved soil porosity and water infiltration), INM strategies stimulate biomass production, which determines the amount of C stored in the plant biomass and the amount returned to the soil. The World Bank [28] reported that the average effect of fertilizer use on soil C was an additional 0.26 t C ha^{-1} yr^{-1}, whereas Oladele and Braimoh [31] found a slightly higher C sequestration rate of 0.63 t C ha^{-1} yr^{-1} after reviewing different fertilizer experiments in Africa. Regarding organic manuring, Kapkiyai et al. [35] found that after 18 years of field experimentation in Kenya, C was sequestered at the rate of 0.17 ± 0.07 t C ha^{-1} yr^{-1}, while Kundu et al. [36] observed a sequestration rate of 0.56 ± 0.02 t C ha^{-1} yr^{-1} after 30 years of field experimentation in India.

Furthermore, rotation of green manure cover crops (e.g. lablab bean – *Dolichos lablab, Desmodium, Canavalia*, and velvet bean – *Mucuna pruriens*) with cereals (e.g. maize – *Zea mays* L.) is also a promising SLM option for C sequestration in degraded soils (Figure 14.7). The idea is to build soil fertility with plants having the capacity to fix nitrogen (N) from the atmosphere through root associations with the N-fixing bacteria (i.e. biological nitrogen fixation – BNF). When the green manure crop is terminated, the C in its biomass is transferred to the soil for storage. Besides, the green manures also feed the micro-biome and keep the soil covered, while controlling soil erosion, weeds, and pests. Singh et al. [37] reported that after 12 years of *Sesbania* green manuring in India, C was sequestered at the rate of 0.09 ± 0.03 t C ha^{-1} yr^{-1}, while Sisti et al. [38] observed that after 13 years of a wheat/soybean– maize cropping system under zero tillage with vetch (*Vicia* spp.) of green manuring in southern Brazil, C was sequestered at the rate of 0.66 ± 0.26 t C ha^{-1} yr^{-1}. Poeplau and Don [39] also conducted a meta-analysis to assess C sequestration in agricultural soils through cover cropping and found that the time since the introduction of cover crops in crop rotations was linearly correlated with changes in SOC stocks with an annual change rate of 0.32 ± 0.08 t C ha^{-1} yr^{-1} in a mean soil depth of 22 cm and during the observed period of up to 54 years. By enhancing residue input and OM in the soil, legume-based cropping systems improve soil biodiversity and aggregate stability and reduce erosion. Agro-ecosystems with high biodiversity also absorb and sequester more C. This can be ascribed to the critical roles in key ecosystem functions

Figure 14.6 Farmers applying inorganic fertilizer (left) and farmyard manure (right) on their farms in western Kenya.

Figure 14.7 A field under total soil cover with *Mucuna pruriens* in western Kenya. *M. pruriens* is planted in rotation with either sole maize or a maize-bean intercrop. This allows the soil to be re-nourished before planting the next crop.

that soil biota performs, such as biomass decomposition, nutrient cycling, moderating atmospheric CO_2, improving aggregation, and creating disease suppressive soils [3].

Regenerative land uses that favor conservation and improvement of soil, water, and air quality also have the potential for soil C sequestration and restoration. For instance, introducing agro-forestry (association of trees and shrubs with crops or pastures) can result in the accumulation of C in the biomass of above-ground and below-ground growth, as well as in the deposition of

SOM. It can also increase biodiversity, stabilize the soil, improve water infiltration, and diversify the farmer's yields. Vagen et al. [25] reported high potentials for C sequestration after the establishment of improved (tree) fallow systems with the attainable rates ranging from 0.07 to 1.37 t C ha^{-1} yr^{-1}. Lorenz and Lal [40] also established that about 0.06 t C ha^{-1} yr^{-1} was sequestered in the topsoil (0–15 cm) of a 25-year-old agro-forestry system consisting of Cacao (*Theobroma cacao*) and Salm wood (*Cordia alliodora*) in Ghana. Moreover, Luedeling et al. [41] noted that agro-forestry systems, such as parklands, live fences, and home gardens in Africa accumulated C at the rate of 0.2–0.8 t C ha^{-1} yr^{-1}, while rotational woodlots sequestered at the rate of 2.2–5.8 t C ha^{-1} yr^{-1}. An analysis of C sequestration potentials revealed that the East African agro-forestry systems above 25 years old gained C at a mean rate of 0.2 t C ha^{-1} yr^{-1} [42]. Desmessie et al. [43] reported that the SOC stock under the chronosequence of 12–50 years of AF and farmland uses in Ethiopia varied from 28.2 to 98.9 Mg ha^{-1} or 12 to 43% of the stock under the natural forest (NF). The SOC was less by 6.2 Mg ha^{-1} yr^{-1} for AF of 12 years and by 0.9 Mg ha^{-1} yr^{-1} for AF for 50 years as compared with NF. The variations in C sequestration rates can be explained by the differences in agro-forestry systems (e.g. type, age, tree species, and management), soils (e.g. type, texture, depth, the severity of degradation, and management) and climates [5].

Moreover, improved grazing land and pasture management can contribute positively toward optimal C sequestration and restoration of degraded soils. This can be achieved through several SLM activities; for example, introducing improved grass species with higher productivity and C allocation to deeper roots, exclosures, restoring degraded grazing lands through exclosures, improving the soil *fauna*, fertilization, irrigation, legume integration, and rotational, seasonal, or controlled grazing regime with optimal stocking rates, and adjusting the timing of grazing [44, 45]. Poor management of the stocking rates can reduce the plant cover and biomass productivity through overgrazing and trampling, leading to a decline in soil structure and an increase in erosion hazards. The World Bank [28] indicated that attainable rates of soil C sequestration through improved management was about 0.8 t C ha^{-1} yr^{-1} on average in Africa. Conant et al. [46] synthesized data from various studies and confirmed that C sequestration rates were positive for most improved grazing and pasture management practices, with an average of 0.47 t C ha^{-1} yr^{-1} across all studies. A review of land management and precipitation effects on C sequestration in the rangelands by Derner and Schuman [47] also echoed and amplified these results.

In addition, improvements in water management through micro-irrigation technologies, such as drip irrigation can be effective in providing sinks of C in the soil and reducing GHG emissions, especially in the semi-arid regions. These technologies ensure that the right amount of water is placed close to the growing plant at the right time and that fewer drops of water are used per crop. The agro-ecological merits of microirrigation technologies are varied, including the enhancement of biomass production [22], which boosts the amount of organic materials returned to the soil and builds up SOC pools both in the short and long runs. Gebeyehu and Soromessa [48] noted that the mean SOC stock was higher by 2.85 t C ha^{-1} (3.44%) in the surface soils (0–30 cm depth) of irrigated compared to the rain-fed farming systems in northwest Ethiopia. The irrigated farming system sequestered C at the rate of 0.41 t C ha^{-1} yr^{-1}.

Besides improvements in water management technologies, SOM content can also be enhanced through the retention of crop residues (Figure 14.5), as opposed to burning. For example, Raji and Ogunwole [49] observed sequestration at the rate of 0.24 t C ha^{-1} yr^{-1} in the 0–15 cm depth of soils after incorporation of crop residues in a study conducted in Nigeria. Similarly, Namirembe et al. [42] reported a mean C sequestration rate of 0.8 ± 0.9 t C ha^{-1} yr^{-1} upon the retention of residues in the East African croplands for 10–25 years.

Figure 14.8 A retention ditch with a fodder embankment for catching and retaining the incoming runoff until it infiltrates into the soil.

Finally, establishing SWC structures, such as terraces, cut-off drains, and retention ditches in degraded lands (Figure 14.8) can help in maintaining pedological processes, reducing the loss of soil and water, strengthening the cycling of C and other vital elements (e.g. N, P, and S), and, ultimately, building resilient soils. Studies on the C sequestration rates of SWC measures are few, but a mean rate of 0.49 t C ha^{-1} yr^{-1} was calculated for terraces established in the East African croplands for a period of less than 10 years [42].

The excavated soil is placed in a mound on the upper side of the ditch to form an embankment, where a fodder crop is established to stabilize the mound and prevent filling the ditch with soil. Photo credit: Authors.

14.3.2 Impact of SOC Restoration and Sequestration

Restoration and re-carbonization of degraded soils have a direct positive impact on soil quality, plant health, and whole ecosystems. This is because SOC is a major determinant of the soil's physical, chemical, and biological properties, which in turn influence its capacity to perform ecological functions and support ecosystem services. For example, SOC supplies nutrients, enhances CEC, nurtures biodiversity, increases aggregation and stability of the soil structure, purifies water, filters pollutants of agro-chemicals, and immobilizes metals. As such, increases in C and OM in the soil potentially advance many of its functions and flow of ecosystem benefits, including inter alia nutrient retention and availability, infiltration capacity, plant available water capacity, the resistance of soils against erosion, and nutrient use efficiency. Besides, building C also increases the productivity of degraded soils, with positive consequences on agricultural yields, food and nutrition security, human well-being, and local economies. Higher productivity translates to more C in biomass, more C in the soils, and climate change mitigation. Altogether, these enhance agricultural sustainability and the resilience of (agro-) ecosystems and society [21, 50, 51]. Lal [20] approximated that an increase in the C pool of degraded agricultural soils by 1 ton can increase the yields of wheat, maize, and cowpea by 20–40 kg ha^{-1}, 10–20 kg ha^{-1}, and 0.5–1 kg ha^{-1} respectively. Further, he estimated that C sequestration in the world's degraded soils can offset emissions by 0.2–0.4 Gt C yr^{-1}.

14.3.3 Modeling the C Benefits of SLM Interventions

As mentioned in Section 14.3.1, SLM interventions have the capacity to deliver multiple environmental benefits, including climate change mitigation by reducing GHG emissions and sequestering C in the soil. The most direct means of estimating and tracking SOC stock changes and GHG emissions attributed to various SLM activities is to measure the sequential changes in SOC over time. However, such measurements are complicated by the spatial and temporal variability of SOC and understanding the effects of SLM activities can be quite complex. Carbon-balance models and tools provide a means of unraveling this complexity, as well as estimating information where comprehensive field measurements are lacking due to logistical, cost, and time constraints [52]. Further, the models and tools allow for the exploration of hypothesis and future scenarios and thus can be useful in planning and long-term predictions. This makes SOC modeling one of the most essential approaches for determining the overall impact of SLM activities and scenarios on C stocks and GHG balance.

There are many land-based models and tools that have been used successfully to simulate the evolution and dynamics of SOC under different environments and management practices. Interested readers are referred to Chotte et al. [13] and Milne et al. [53] for a detailed presentation of the existing SOC models and tools. In brief, they range from simple models and tools that treat SOC as a single pool to complex ones that consider multiple SOC pools. There are also models and tools that are dedicated to analysing SOC alone and those that quantify it alongside other ecosystem metrics. The C tools are more simplified, while the models are more detailed. That is, the tools evaluate SOC using simpler statistical and empirical approaches, often allowing the use of embedded default datasets in the context of more comprehensive C accounting or socioeconomic analyses. By their design, C tools are more suitable when low to moderate levels of certainty are required. Examples are EX-ACT, Carbon Benefits Project (CBP), and Cool Farm Tool. In contrast, C models represent more explicitly the processes impacting SOC (e.g. decomposition) and can yield results that are of relatively high certainty. However, not all degradation processes are represented due to limitations of data, understanding the linkages to SOC processes, or computation. The classic and widely used process-based C models are RothC and DAYCENT, which can also be linked to a geographical information system (GIS). For example, Molina et al. [54] and Yokozawa et al. [55] used the RothC model to determine the rates of SOC changes following a shift from conventional to conservation tillage in Mexico, and after application of organic manure was applied in Japanese arable soils respectively. It must be stated, though, that there is no model or tool that can meet all needs; hence, selection of the best-fit option depends mostly on its current capacity and the availability of data for use.

14.3.3.1 Estimating the C Benefits of SLM Interventions Using the CBP Modelling Tools: A Case Study of Western Kenya

Agriculture in western Kenya suffers from low productivity due to the degraded landscape. The soils are predominantly acidic and low in fertility, with the yields of food staples, such as maize and common bean, averaging 1 t ha^{-1} and 5 t ha^{-1}, respectively. Consequently, the smallholder farmers have encroached into the adjacent Kakamega and Nandi forests in the quest for supplementary agro-ecosystem services and farm incomes. Degradation of the agro-resource base is attributed to

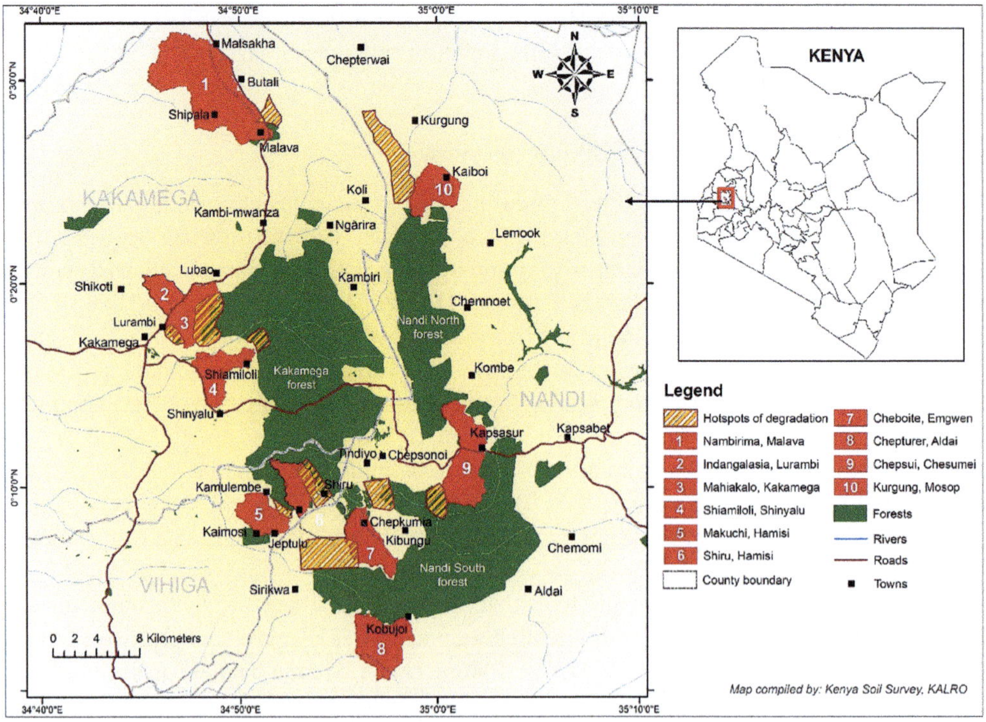

Figure 14.9 Location of the project implementation sites in western Kenya.

unsustainable practices carried out under conventional farming; for example, continuous cultivation with low nutrient inputs, mono-cropping, removal of crop residues, and full tillage. Some of these practices enhance climate change by intensifying the emissions of GHGs, such as CO_2, into the atmosphere. Therefore, widespread adoption of SLM technologies is being promoted by a five-year land management project funded by the Global Environment Facility (GEF) to reduce the adverse impacts of conventional farming and restore the degraded lands in 10 micro-catchments located within a 5-km radius from the edge of the Kakamega and Nandi forests in Kakamega, Vihiga, and Nandi Counties (Figure 14.9).

The project used a combination of field data and CBP tools for *ex-ante* assessment of the potential C stock changes and GHG emissions attributed to the various SLM activities being implemented in one of the 10 microcatchments; i.e. the Makuchi micro-catchment in Vihiga County (see no. 5 on Figure 14.9) from 2017 to 2022. Essentially, three scenarios were developed and analyzed; namely, the *initial land use*, *baseline*, and *project* scenarios. *Initial land use* refers to the status of land use at year 0, the *baseline* (business-as-usual) scenario represents a future where the *status quo* remained without SLM interventions, and the *project* scenario depicts a future where SLM activities were implemented. Thereafter, the three land use categories (i.e. forestland, annual crops, and agro-forestry) for each scenario were described (Table 14.2).

Table 14.3 shows the evolution and estimates of C stocks after five years for the two scenarios. It is evident that the various SLM activities being implemented by the land management project in western Kenya have potential C and GHG benefits. For example, the results indicate that a total of about 3807 t C can be sequestered in the landscape at the rate of 0.55 t C ha^{-1} yr^{-1}. Out of this, forests and agro-forestry practices (i.e. alley cropping, woodlots, windbreaks, and riparian buffers)

Table 14.2 Description of the scenarios and land use categories used for the CBP modeling.

Scenario land use	Initial land use	Baseline (*without-project*)	Project (*with-project*)
Forestland	• 96 ha of forest land • 3 ha is cleared per year • 3000 m^3 of fuel wood is gathered per year	• 81 ha of forest land • 3 ha is cleared per year • 2025 m^3 of fuel wood is gathered per year	• 96 ha of forest land • No clearing • 1800 m^3 of fuel wood is gathered per year
Annual crops	• 1150 ha of cropland • Maize-bean intercrop • Full tillage • 125 kg of N fertilizer applied per ha (18% N) • Crop residue is collected • Improved crop varieties	• 1157 ha of cropland • Maize-bean intercrop • Full tillage • 125 kg of N fertilizer applied per ha (18% N) • Crop residue is collected • Improved crop varieties	• 1132 ha of cropland • Maize-bean intercrop • Reduced tillage • 187 kg of N fertilizer applied per ha (10% N) • Crop residue is retained • Improved crop varieties • Organic manuring • Cover cropping • Liming
Agro-forestry	• 181 ha of agro-forestry • Woodlots, alley cropping, and windbreaks with diverse tree species	• 189 ha of agro-forestry • Woodlots, alley cropping, and windbreaks with diverse tree species	• 199 ha of agro-forestry • Woodlots, alley cropping, and windbreaks and riparian buffers with diverse tree species

can sequester about 1250 t C in woody biomass at the rate of 0.17 t C ha^{-1} yr^{-1}. Similarly, avoided deforestation (forest preservation) in the project scenario can contribute about 1267 t C in woody biomass at the rate of 0.18 t C ha^{-1} yr^{-1}. Further, SLM activities, including optimal fertilizer use, organic manuring, reduced tillage, cover cropping, residue retention, intercropping, liming, and use of improved seeds can collectively add about 1378 t C at the rate of 0.19 t C ha^{-1} yr^{-1} to the soil C stocks. This amount can offset the emission of nitrous oxide (N_2O) equivalent to about 319 t CO_2 from the use of N fertilizer, ensuring that the SLM activities still yield a net C benefit.

The foregoing case study demonstrates the utility of integrating the CBP tools to determine the incremental C benefits resulting from SLM project interventions. Wang et al. [56] also confirmed this in a study that evaluated the C impact of planting wolfberry (*Lycium barbarum* L.) on the saline soils of Jingtai oasis in Gansu, China. Overall, the CBP modeling approach can greatly reduce sampling efforts, but should be backed with periodic field measurements to validate the results.

14.4 Summary and Conclusions

Soil degradation is a major sustainable development challenge, the main symptoms of which are erosion and nutrient depletion. It impacts agronomic productivity, food and nutrition security, biodiversity, climate change, and economic well-being. This chapter has shed light on the degradation of the world's soil resources, the impact on C stocks, SLM interventions to restore and sequester C in degraded soils, and the applicable modeling tools for assessing the C footprints of SLM activities. We have noted that soil degradation is a natural process, but exacerbated by complex interactions between the biophysical, socioeconomic and political factors operating in specific spatial

Table 14.3 Summary report following the UNFCCC common reporting guidelines.

GHG source and sink categories	Baseline scenario (emissions and removals)		Project scenario (emissions and removals)		GHG balance/Carbon benefits (2017–2022)			
• Agriculture	CO_2	N_2O	CO_2	N_2O	Total t CO_2 e	t CO_2 e ha^{-1} yr^{-1}	Total t C	t C ha^{-1} $yr.^{-1}$
	tons CO_2 equivalent (t CO_2 e)							
A. Agricultural soils		4052		4372	319	0.04	87	0.01
• Land use change and forestry								
A. Forest and other woody biomass	−18 310		−22 884		−4574	−0.64	1250	−0.17
B. Forest and grassland conversion	4636		0		−4636	−0.65	1267	−0.18
C. CO_2 emissions and removals from soil	224		−4818		−5042	−0.71	1378	−0.19
Total	**−13 450**	**4052**	**−22 701**	**4372**	**−13 932**	**−2.00**	**3807**	**−0.55**

Notes:

- GHG balance/C benefits is project minus baseline scenarios.
- Signs for removals are (−) and for emissions (+).
- Forest and other woody biomass include biomass growth and losses from fuel wood gathering.
- Forest and grassland conversions include emissions from deforestation.
- UNFCCC is the United Nations framework convention on climate change; GHG is greenhouse gas; C is carbon; CO_2 is carbon dioxide; CH_4 is methane; N_2O is nitrous oxide.
- 44/12 is the factor used for converting C to CO_2 equivalent.

and temporal contexts. Several SLM options (e.g. CA, agro-forestry and INM) exist that countries can consider to restore and increase C storage in their degraded soils. These should be objectively selected and scaled out, taking into account the different biophysical and socioeconomic settings. The SLM options also provide manifold global benefits, some of these being GHG, C, ecosystem services, and biodiversity benefits. However, estimating and tracking these benefits can be quite complex and difficult. Thus, several C models and tools, such as the CBP, RothC, and DAYCENT, have been developed to help SLM project designers, managers, and evaluators to unravel the complexity, as well as estimate and report the C impact of project interventions, even in data-scarce environments. Countries can advance the adoption of best SLM practices to restore degraded soils by (i) formulating policies that foster multisectoral and multistakeholder partnerships, (ii) integrating land-based approaches to climate change mitigation and adaptation into national strategies, (iii) promoting awareness and sharing experiences of the multiple benefits associated with SLM, (iv) ensuring full and integrated accounting of SOC in all sectors, and (v) monitoring the three indicators of land degradation neutrality, including the trends in land cover, land productivity, and terrestrial C stocks.

References

1 Lorenz, K., Lal, R., and Ehle, K. (2019). Soil organic carbon stock as an indicator for monitoring land and soil degradation in relation to United Nations' Sustainable Development Goals. *Land Degradation and Development* 30: 824–838.

2 Lal, R. (1997). Degradation and resilience of soils. *Philosophical Transactions of the Royal Society of London B* 352: 997–1010.

3 Lal, R. (2015). Restoring soil quality to mitigate soil degradation. *Sustainability* 7: 5875–5895.

4 ELD (2015). The value of land: prosperous lands and positive rewards through sustainable land management. *The Economics of Land Degradation*, 170 pp. http://www.eld-initiative.org (accessed 13 July 2020).

5 Were, K.O., Singh, B.R., and Ayaga, G. (2020). Land use changes and sustainable land management practices for soil carbon sequestration in sub-Saharan African agro-ecosystems. In: *Climate Impacts on Agricultural and Natural Resources Sustainability in Africa* (eds. B.R. Singh et al.), 41–59. Cham: Springer.

6 Muchena, F.N., Onduru, D.D., Gachini, G.N. et al. (2005). Turning the tides of soil degradation in Africa: capturing the reality and exploring opportunities. *Land Use Policy* 22: 23–31.

7 Nkonya, E., Mirzabaev, A., and von Braun, J. (eds.) (2016). *Economics of Land Degradation and Improvement – A Global Assessment for Sustainable Development*. Springer International Publishing AG Switzerland.

8 Tully, K., Sullivan, C., Weil, R. et al. (2015). The state of soil degradation in sub-Saharan Africa: baselines, trajectories, and solutions. *Sustainability* 7: 6523–6552.

9 Bouma, J. and Batjes, N.H. (2000). Trends of worldwide soil degradation. *Hohenheimer Umwelttagung* 32: 33–43.

10 Oldeman, L.R. (1994). The global extent of soil degradation. In: *Soil Resilience and Sustainable Land Use* (eds. D.J. Greenland and I. Szabolcs), 99–118. Wallingford, UK: CAB International.

11 UNCCD (2015). Desertification, land degradation and drought (DLDD): some global facts and figures. United Nations Conventions to Combat Desertification. http://www.unccd.int/Lists/SiteDocumentLibrary/WDCD/DLDD%20Facts.pdf (accessed 13 July 2020).

12 FAO and ITPS (2015). *Status of the World's Soil Resources – Main Report*. Rome, Italy: Food and Agriculture Organization of the United Nations and Intergovernmental Technical Panel on Soils.

13 Chotte, J.L., Aynekulu, E., Cowie, A. et al. (2019). Realising the carbon benefits of sustainable land management practices: guidelines for estimation of soil organic carbon in the context of land degradation neutrality planning and monitoring. A report of the Science-Policy Interface. United Nations Convention to Combat Desertification (UNCCD), Bonn, Germany.

14 Lal, R. (2014). Climate strategic soil management. *Challenges* 5: 43–74.

15 Lal, R. (2014). Soil conservation and ecosystem services. *International Soil and Water Conservation Research* 2 (3): 36–47. *Journal of Soil and Water Conservation* 70(3):55A–62A.

16 Olson, K.R., Al-Kaisi, M., Lal, R. et al. (2016). Impact of soil erosion on soil organic carbon stocks. *Journal of Soil and Water Conservation* 71 (3): 61–67.

17 Chappell, A., Webb, N.P., Leys, J.F. et al. (2019). Minimising soil organic carbon erosion by wind is critical for land degradation neutrality. *Environmental Science and Policy* 93: 43–52.

18 Ai, M., Sun, Y., Yan, B., and Wei, Y. (2018). A summary of the impact of land degradation on soil carbon sequestration. *Materials Science and Engineering A* 394: 052028.

19 FAO (2013). Climate-smart agriculture sourcebook. http://www.fao.org/docrep/018/i3325e/i3325e.pdf (accessed 28 July 2020).

20 Lal, R. (2004). Soil carbon sequestration to mitigate climate change. *Geoderma* 123: 1–22.

21 Scherr, S.J. and Shames, S. (2009). Mitigating climate change through food and land use: world-watch report No. 179. Worldwatch Institute, Washington. http://www.worldwatch.org/system/files/179%20Land%20Use.pdf (accessed July 10, 2020).

22 Aune, J.B. and Coulibaly, A. (2015). Micro-dosing of mineral fertilizer and conservation of agriculture in sustainable agricultural intensification in sub-Saharan Africa. In: *Sustainable Intensification to Advance Food Security and Enhance Climate Resilience in Africa* (eds. R. Lal, B.R. Singh, D.L. Mwaseba, et al.), 223–234. Berlin: Springer.

23 Farooq, M. and Siddique, K.H.M. (eds.) (2015). *Conservation Agriculture*. Switzerland: Springer International Publishing.

24 Lal, R. (1999). Soil management and restoration for C sequestration to mitigate the greenhouse effect. *Progress in Environmental Science* 1: 307–326.

25 Vagen, T.G., Lal, R., and Singh, B.R. (2005). Soil carbon sequestration in sub-Saharan Africa: a review. *Land Degradation and Development* 16: 53–71.

26 West, T.O. and Post, W.M. (2002). SOC sequestration rates by tillage and crop rotation. A global data analysis. *Soil Science Society of America Journal* 66: 1930–1946.

27 Franzluebbers, A.J. and Follet, R.F. (2005). Greenhouse gas contributions and mitigation potential in agricultural regions of North America: introduction. *Soil and Tillage Research* 83: 1–8.

28 The World Bank (2012). Carbon sequestration in agricultural soils. Report No. 67395-GLB.

29 Lal, R. (2015b). Sequestering carbon and increasing productivity by conservation agriculture. *Journal of Soil and Water Conservation* 70 (3): 55A–62A. https://doi.org/10.2489/jswc.70.3.55A.

30 Agriculture for Impact (2014). No ordinary matter: conserving, restoring and enhancing Africa's soils. A Montpellier Panel Report.

31 Oladele, O. and Braimoh, A. (2014). Potential of agricultural land management activities for increased soil carbon sequestration in Africa – a review. *Applied Ecology and Environmental Research* 12 (3): 741–751.

32 Sanginga, N. and Woomer, P.L. (2009). *Integrated Soil Fertility Management in Africa: Principles, Practices and Development Process*, 263. Tropical Soil Biology and Fertility Institute of the Centre for Tropical Agriculture.

33 Lal, R. (2009). Soil degradation as a reason for inadequate human nutrition. *Food Security* 1: 45–57.

34 Zingore, S., Mutegi, J., Agesa, B. et al. (2015). Soil degradation in sub-Saharan Africa and crop production options for soil rehabilitation. *Better Crops with Plant Food* 99 (1): 24–26.

35 Kapkiyai, J.J., Karanja, N.K., Qureshi, J.N. et al. (1999). Soil organic matter and nutrient dynamics in a Kenyan nitisol under long-term fertilizer and organic input management. *Soil Biology and Biochemistry* 31: 1773–1782.

36 Kundu, S., Bhattacharyya, R., Prakasha, V. et al. (2007). Carbon sequestration and relationship between carbon addition and storage under rain-fed soybean-wheat rotation in a sandy loam soil of the Indian Himalayas. *Soil and Tillage Research* 92: 87–95.

37 Singh, G., Jalota, S.K., and Singh, Y. (2007). Manuring and residue management effects on physical properties of a soil under the rice-wheat system in Punjab, India. *Soil and Tillage Research* 94: 229–238.

38 Sisti, C.P.J., dos Santos, H.P., Kohhann, R. et al. (2004). Change in carbon and nitrogen stocks in soil under 13 years of conventional or zero tillage in southern Brazil. *Soil and Tillage Research* 76: 39–58.

39 Poeplau, C. and Don, A. (2015). Carbon sequestration in agricultural soils via cultivation of cover crops – a meta-analysis. *Agriculture, Ecosystems and Environment* 200: 33–41.

40 Lorenz, K. and Lal, R. (2014). Soil organic carbon sequestration in agroforestry systems. A review. *Agronomy for Sustainable Development* 34 (2): 443–454.

41 Luedeling, E., Sileshi, G., Beedy, T., and Diez, J. (2011). Carbon sequestration potential of agro-forestry in Africa. In: *Carbon Sequestration Potential of Agro-Forestry Systems: Opportunities and Challenges* (eds. B.M. Kumar and P.H. Nair), 61–84. Dordrecht: Springer.

42 Namirembe, S., Piikki, K., Sommer, R. et al. (2020). Soil organic carbon in agricultural systems of six countries in East Africa – a literature review of status and carbon sequestration potential. *South African Journal of Plant and Soil* 37: 35–49. https://doi.org/10.1080/02571862.2019.1640296.

43 Desmessie, A., Singh, B.R., and Lal, R. (2013). Soil carbon and nitrogen stocks under chronosequences of farm and traditional agroforestry land uses in Gambo District Southern Ethiopia. *Nutrient Cycling in Agroecosystems* 95: 365–375.

44 Smith, P., Martino, D., Cai, Z. et al. (2007). Agriculture. In: *Climate Change 2007: Mitigation. Contribution of Working Group III to the Fourth Assessment Report of the Intergovernmental Panel on Climate Change* (eds. B. Metz, O.R. Davidson, P.R. Bosch, et al.). Cambridge, United Kingdom and New York, NY, USA: Cambridge University Press.

45 UNFCCC (2008). Challenges and opportunities for mitigation in the agricultural sector. Technical Paper GE-08-64351.

46 Conant, R.T., Cerri, C.E.P., Osborne, B.B., and Paustian, K. (2017). Grassland management impacts on soil carbon stocks: a new synthesis. *Ecological Applications* 27 (2): 662–668.

47 Derner, J.D. and Schuman, G.E. (2007). Carbon sequestration and rangelands: a synthesis of land management and precipitation effects. *Journal of Soil and Water Conservation* 62 (2): 77–85.

48 Gebeyehu, G. and Soromessa, T. (2018). Status of soil organic carbon and nitrogen stocks in Koga Watershed Area, Northwest Ethiopia. *Agriculture and Food Security* 7: 9.

49 Raji, B.A. and Ogunwole, J.O. (2006). Potential of soil carbon sequestration unver various land use in the sub-humid and semi-arid savannah of Nigeria: Lessons from long-term experiments. *International Journal of Soil Science* 1 (1): 33–43.

50 Bernazzani, P., Bradley, B.A., and Opperman, J.J. (2012). Integrating climate change into habitat conservation plans under the U.S. Endangered Species Act. *Environmental Management* 49: 1103–1114.

51 Milder, J.C., Majanen, T., and Scherr, S.J. (2011). Performance and potential of conservation agriculture for climate change adaptation and mitigation in sub-Saharan Africa. Ecoagriculture Discussion Paper No. 6. EcoAgriculture Partners, Washington. http://ecoagriculture.org/wp-content/uploads/2015/08/PerformancePotentialofConservationAg.pdf (accessed 10 July 2020).

52 Milne, E., Neufeldt, H., Rosenstock, T. et al. (2013). Methods for the quantification of GHG emissions at the landscape level for developing countries in smallholder contexts. *Environmental Research Letters* 8: 015019. https://doi.org/10.1088/1748-9326/8/1/015019.

53 Milne, E., Neufeldt, H., Smalligan, M. et al. (2012). Methods for the quantification of emissions at the landscape level for developing countries in smallholder contexts. CCAFS Report No. 9. Copenhagen, Denmark: CGIAR Research Program on Climate Change, Agriculture and Food Security (CCAFS). www.ccafs.cgiar.org.

54 Molina, L.G., Pérez, E.C.M., and Pérez, A.B. (2017). Simulation of soil organic carbon changes in Vertisols under conservation tillage using the RothC model. *Science in Agriculture* 74 (3): 235–241.

55 Yokozawa, M., Shirato, Y., Sakamoto, T. et al. (2010). Use of the RothC model to estimate the carbon sequestration potential of organic matter application in Japanese arable soils. *Soil Science & Plant Nutrition* 56 (1): 168–176. https://doi.org/10.1111/j.1747-0765.2009.00422.x.

56 Wang, Y., Zhao, C., Ma, Q. et al. (2015). Carbon benefits of wolfberry plantation on secondary saline land in Jingtai Oasis, Gansu – a case study on application of the CBP model. *Journal of Environmental Management* 157: 303–310.

15

Impact of Treatment of Biodegradable Waste on Nutrient Recovery

Eyob Habte Tesfamariam[1], Craig Cogger[1,2], and John Ngoni Zvimba[3,4]

[1] The Department of Plant and Soil Sciences, University of Pretoria, Pretoria, South Africa
[2] (Emeritus) Department of Crop and Soil Sciences, Washington State University, Puyallup, WA, USA
[3] Water Research Commission, Pretoria, South Africa
[4] Department of Physics, Geography and Environment Science, Great Zimbabwe University, Masvingo, Zimbabwe

15.1 Introduction

The global population is expected to rise, reaching 8.6 billion by 2030 and 9.8 billion by 2050 [1]. This population increase will put immense pressure, already strained, on natural resources, partially attributable to significant new demands and challenges associated with waste disposal [2]. Existing unsustainable patterns of consumption and production have already caused significant environmental degradation while reinforcing social inequity and poverty [3]. The United Nations developed an agenda on "Sustainable Developmental Goals (SDGs)" to address these challenges, with several countries striving to achieve their mandates. This agenda anticipates addressing 17 core areas of public concern. Of these areas, 15 relate directly or indirectly to waste recycling, with an emphasis on food security, health, and the environment [4].

Recycling of bio-waste, with suitable quality for agricultural lands, improves soil physical properties by lowering the soil bulk density, increasing porosity [5], improving aggregate stability, saturated hydraulic conductivity [6], plant available water, and the infiltration rate [7, 8]. Applying bio-waste improves soil chemical properties by increasing soil organic matter [5, 9] and several associated essential chemical properties, promoting crop growth. A bio-waste application creates a conducive soil environment for microbial activity [9], providing carbon and energy, and may also enhance the biodiversity and size of the soil microbial population [10]. The role of bio-wastes in improving essential soil parameters (soil physical, chemical, and biological properties) is critical for crop production under the ever-changing environment.

Non-hazardous biowaste, such as biosolids [11], poultry litter [12], plant biomasses, and agri-food residues [13], are rich in mineral macronutrients (nitrogen, phosphorus, potassium, calcium, sulfur, and magnesium) and micronutrients (iron, boron, chlorine, manganese, zinc, copper, molybdenum, and nickel). Bio-wastes are therefore considered as potential sources of nutrients for crop production as a replacement and/or supplement to commercial inorganic fertilizers [14]. Various bio-wastes hold varying nutrient proportions with varying fertilizer values [15].

Human excreta is the largest source of phosphorus in urban areas, with humans typically excreting 1.6–1.7 g of phosphorus daily, mostly found in urine. In developing countries, the organic fraction of municipal solid waste may become an even larger sink. Within biological waste

Biodegradable Waste Management in the Circular Economy: Challenges and Opportunities, First Edition.
Edited by Malgorzata Kacprzak, Eleonore Attard, Kari-Anne Lyng, Helena Raclavska, BalRam Singh, Eyob Tesfamariam, and Franck Vandenbulcke.

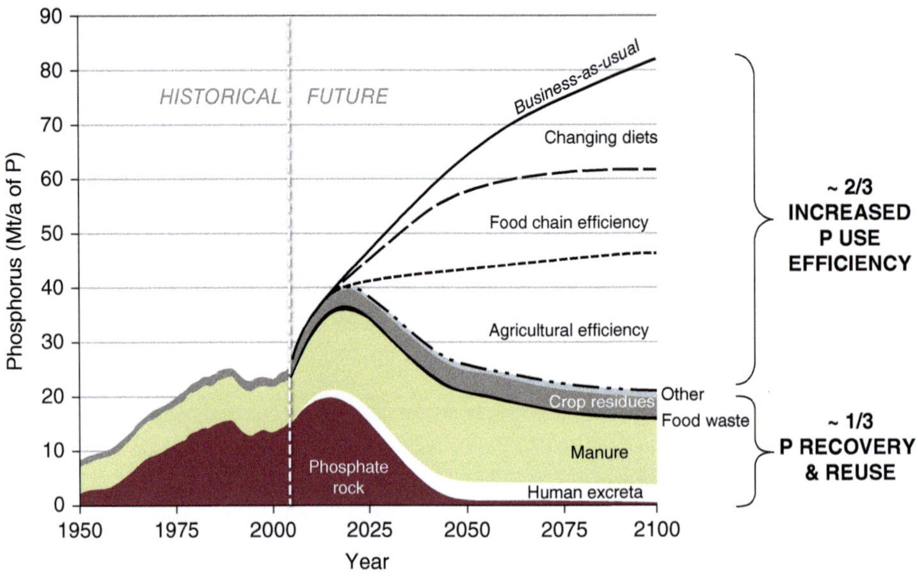

Figure 15.1 Illustration of future demands for phosphorus [17].

water treatment plants, up to 90% of the phosphorus load may be incorporated in sewage sludge. Considering that natural phosphorus reserves are declining and expected to fall short of demand by 2033, sewage waste has the potential to be a major additional phosphorus source. Some argue that this source of phosphorus can simultaneously alleviate the challenges facing developing countries concerning sanitation, offering a "low-cost source of phosphorus in the form of struvite", a magnesium ammonium phosphate-based fertilizer.

Attaining a sustainable phosphorus (P) cycle involves an integrated systems approach, including the efficient use, re-use, and recovery of phosphorus in waste streams. To achieve this, it has been estimated that the global business-as-usual demand will have to decrease by 70%, enabling the remaining demand to be met by recovery from all waste streams [16]. Figure 15.1 illustrates how future demands for phosphorus can be met through efficiency in use, changes in demand, and recovery from waste streams.

Various types of non-hazardous biowastes exist, with various nutrient contents and availability profiles. In the discussion below, non-hazardous bio-wastes are classified into the following three categories: agricultural biodegradable wastes (manure, crop, and forest residues), municipal biodegradable wastes (municipal wastes, sewage sludge), and industrial biodegradable wastes (waste from the food industry and slaughterhouses).

15.2 Nutrient Composition of Biodegradable Municipal Solid Waste

Biodegradable municipal solid waste includes food wastes (from residential houses, restaurants, and schools), garden waste (lawn trimmings, tree leaves, twigs, and branches), and waste from groceries (vegetable and fruit wastes). Proportions of the various municipal waste components vary from country to country depending on the lifestyle, economic situation, waste management regulations, industrial structure [18], scale of the study (urban vs. family house), and season [19].

For instance, of the total waste matter, non-citrus fruits and vegetables (44%) followed by citrus fruits (14%) dominated the urban solid waste from Prague, Czech Republic [19]. This contrasted with the family grass-dominated waste from the same city (29%), followed by leaves (28%), while fruit and vegetables accounted for 7%. The study also indicated that the composition of urban waste remained consistent among seasons, unlike family waste. The nutrient composition of these various solid municipal wastes is now discussed, with emphasis provided on the three key major crop nutrients: nitrogen, phosphorus, and potassium.

15.2.1 Nitrogen

The nitrogen content of urban bio-waste from Prague, Czech Republic, ranged between 1.4 and 1.8% and did not show variation between seasons [19]. The nitrogen content of family bio-waste from the same city was also within the same range (1.2–1.6%) but varied significantly between seasons. Additional studies from Denmark reported nitrogen concentrations of 1.6–1.9% for household waste comprising food waste and small flower volumes [20]. Nitrogen contents are reported to be twice higher (3.7%) in household waste from Greenland [21] than from Poland (6.0%), which are double those of Greenland for household food waste, mainly composed of fruit and vegetable residues, meat, bread, tea leaves, cooled rice, and potatoes [22].

The nitrogen content of food waste, comprising a combination of seafood, meat, bread, pasta, potatoes, vegetable, and condiments collected from restaurants near the University of South Carolina, USA, was 3.2% [23], while a higher nitrogen content value of 5.3% for food waste from a restaurant serving rice, vegetables, meats, and seafood in Taiwan was established [24]. Conversely, a much lower nitrogen content (1.6–2.4%) of food waste from a community center was reported [25].

Additional studies investigated the nutrient composition of individual household bio-wastes, such as the nitrogen content of raw vegetables (Chinese cabbage, potatoes, cabbage, tomatoes, and onion), as 2.9% [26]. The total Kjeldahl nitrogen (TKN) content of various household bio-wastes are: lettuce (4.5%), spinach (4.2%), rice (4.5%), chickpeas (8.2%), beans (8.7%), bread (5.3%), potato (4.9%), orange (3.3%), lemon (2.1%), banana (1.7%), and coffee (1.9%) [27]. The nitrogen content of non-marketable broccoli with its leaves and tomatoes from a horticultural industry were 3.7 and 2.3% respectively [28].

The nitrogen content of garden waste from Aarhus, comprising mostly leaves, flowers, twigs, and stems was 0.3–0.8% (Arhus) [29].This varied with the season and between various components of garden waste. For instance, the nitrogen content of wood (0.2%) was significantly lower than leaves, flowers, and twigs (0.6%).

15.2.2 Phosphorus

The phosphorus content of urban bio-waste is lower than the phosphorous content of household and restaurant wastes. The phosphorous content of urban bio-waste in Prague, Czech Republic, ranged between 0.1 and 0.2% [19]. This is at the low end of the range of phosphorous content of family bio-waste from the same city (0.1–0.4%). Household bio-waste phosphorus concentrations of 0.26–0.29% were reported in Denmark [20]. This contrasts with the phosphorus content of household waste from Greenland (0.73%), which was more than twice higher [21].

The phosphorus content of food waste collected from restaurants near the University of South Carolina, comprising a combination of seafood, meat, bread, pasta, potatoes, vegetables, and condiments was 0.42% [23].This is more than twice higher than that reported (0.1–0.3%) for community food waste [25]. The phosphorus content of non-marketable broccoli with its leaves and tomatoes from a horticultural industry was 0.64 and 0.62% respectively [28].

The phosphorus content of garden waste from Aarhus was mainly comprised of leaves, flowers, twigs, and stems and indicated 0.06–0.17% [29]. This varied with the season. The study also reported that the phosphorus content of wood (0.04%) was much lower than that of leaves, flowers, and twigs (0.12%).

15.2.3 Potassium

The potassium content of urban bio-waste ranged between 1.1 and 1.8% [19]. The potassium content of family bio-waste (1.0–2.1%) in the same city was within the same range as the urban waste, though it was slightly higher on the upper end of the range. Household bio-waste potassium concentrations of 1.5–1.8% were reported in Denmark [20]. Additional studies [20] also reported potassium concentrations (1.3%) of household bio-waste within the ranges [19], despite variations in geographical location.

The potassium content of food waste from restaurants near the University of South Carolina, comprising a combination of seafood, meat, bread, pasta, potatoes, vegetable, and condiments was 0.5% [23]. In contrast, the potassium content of non-marketable broccoli with its leaves and tomatoes from a horticultural industry were 4.3 and 2.9% respectively [28].

The potassium content of garden waste from Arhus (Denmark), mainly comprising leaves, flowers, twigs, and stems, was 0.5–1.4% [29]. The study indicated that the potassium content of the garden waste varied among seasons and between the various components of garden waste, such as the content of wood (0.3%) and leaves, flowers, and twigs (1.2%) [29].

15.3 Nutrient Composition of Municipal Waste Water

Sewage sludge has the second-highest concentration of phosphorus [30] that can be used as a source of phosphorus in agricultural lands, but consideration must be taken of the fact that existing natural phosphorus deposits are expected to last for only the next 30–60 years [31]. In this chapter, the term municipal waste water includes untreated domestic waste water from residential areas and airplane toilets. The nutrient composition of domestic waste water influent varies spatially and temporarily [32]. This variation in the nutrient composition is discussed in the subsequent section.

15.3.1 Nitrogen

A typical untreated domestic waste water comprises 20–85 $mg\,l^{-1}$ of nitrogen, of which 8–35 $mg\,l^{-1}$ is organic and 12–50 $mg\,l^{-1}$ is inorganic [33]. The nitrogen content of domestic waste water from a residential condominium in the Municipality of Florianopolis in Southern Brazil falls within this range: $58 \pm 19\,mg\,l^{-1}$ nitrogen of which $49 \pm 16\,mg\,l^{-1}$ was NH_3–N, $0.1 \pm 0.3\,mg\,l^{-1}$ NO_2–N and 0.02 ± 0.1 NO_3–N [34]. This contrasts with the nitrogen content of liquid effluent originating from airplane toilets stationed at the Guarulhos International airport in Sao Paulo (Brazil) (4215 $mg\,l^{-1}$), which was 50-fold higher. This variation is mainly attributable to the dilution effect of untreated domestic waste water, where an enormous volume of water is used to flush the toilet and to clean house utensils and clothes, further diluting the nutrient concentration. The dilution effect could be verified by previous studies reporting NH_3–N concentrations ranging between $1990 \pm 420\,mg\,l^{-1}$

(women's urine average) to 4140 ± 870 mg l^{-1} (men's urine average) from a urine-diverting flush toilet and waterless urinals [35]. The difference in NH_3–N concentration between urinals for males and females is attributable to the dilution effect from leaky valves, allowing some flushing to enter the urinal collection tank for females [35]. This agrees with studies reporting that 80% of the total N in municipal waste water originates from urine [36, 37]. Similar findings indicate that 69% of the nitrogen in domestic waste water making it through to waste water treatment plants reports urine despite urine accounting for only 1% of domestic waste water [38]. In contrast, human feces have a much lower amount of total nitrogen, 4.0 ± 0.4% (dry wt), than urine [39].

15.3.2 Phosphorus

Typical untreated domestic waste water comprises 4–15 mg l^{-1} of phosphorus, of which 1–5 mg l^{-1} is organic and 3–10 mg l^{-1} is inorganic [33]. Phosphorus concentrations of 6.7 ± 1.9 mg l^{-1} of total P, of which 4.1 ± 1.9 mg l^{-1} was PO_4-P, are within the ranges despite varying origins (the USA vs. Brazil) [33, 34]. A large portion (40%) [36] to 45% [37, 38], of the phosphorus in domestic waste water originated from urine, despite urine accounting for only 1% of domestic waste water. High concentrations of phosphorus therefore characterize urine [35, 40]. Phosphorus concentrations of 106 ± 17 mg l^{-1} (women's urine average) to 242 ± 23 mg l^{-1} (men's urine average) total P are reported from urine-diverting flush toilets [35]. This is 10- to 20-fold higher than the phosphorus concentration in domestic waste water. The phosphorus concentration of liquid effluent from airplane toilets (430 mg l^{-1} PO_4–P) was significantly higher [40]. In contrast, the phosphorus composition of human feces, 1.1 ± 0.2% P_2O_5, is exceptionally low [39].

15.3.3 Potassium

Urine accounts for 60% of the potassium load in domestic waste water, making it through to waste-water treatment plants despite accounting for only 1% of the domestic waste water [38]. The potassium content of urine from urine-diverting flush toilet and waterless urinals at Eawag's principal building (Dubendorf, Switzerland) was 854 ± 143 mg l^{-1} (urine average of females) and 1470 ± 130 mg l^{-1} (urine average of males), with potassium concentration of human feces reported at 2.8 ± 0.2% K_2O [39].

15.4 Nutrient Composition of Agricultural Biodegradable Waste

Agricultural biodegradable wastes include manures of various domestic animals, plant residues, and green materials. The nutrient composition of various agricultural biodegradable wastes differs considerably among sources and duration of exposure to the atmosphere, among other factors. The nitrogen, phosphorus, and potassium concentrations of various agricultural biodegradable wastes are now discussed.

15.4.1 Nitrogen

Poultry manure is reported to have a higher nitrogen content than other animal manures. The nitrogen content of animal manure ranges from as low as 0.65 ± 0.03% for cow dung mixed with urine [41] to as high as 7.1 ± 0.7% for poultry [42]. Poultry manure is rich in nitrogen compared to other animal manures. The nutrient composition of poultry manure could vary depending on the diet fed to the birds and the amount of feed and water mixed with the manure [43]. Reports on the nitrogen content of poultry manure differ from as low as 1% [44] to as high as 7.1 ± 0.7% [42]. Literature reports the following poultry nitrogen concentration values: 3.9 ± 0.7–7.1 ± 0.7% [42], 1–5.5% [44], 2.21 ± 0.04% [26], 5.3% [45], 2.3% [28], and 4.3% [46].

Like poultry, the nitrogen concentration of cattle and swine manure could vary depending on the diet fed to the cattle and the amount of feed and water mixed with the manure. A wide range of cattle manure nitrogen content was reported in the broad literature, ranging from as low as 0.65 ± 0.03% [41] to as high as 5.74 ± 0.01% [47]. The low nitrogen concentrations of 0.65 ± 0.03% were collected from cowsheds where the dung was mixed with urine. The nitrogen content reports of cattle manure from other studies include 1–2.8% [44], 1.4% [48], 1.6 ± 0.1–1.8 ± 0.04% [49], and 1.94–2.90% [50].

The nitrogen content of swine manure varies from as low as 0.20 ± 0.04% from a piggery waste water influent [51] to as high as 4.3% [50]. Reports from other studies regarding the nitrogen concentration of swine manure, such as that of [52] 0.5%, [53] 1.7%, [54] 2.2–2.3 $g\,l^{-1}$, [55] 2.7%, [45] 2.9%, and [44] 3.8%, fall within this range.

The nitrogen content of crop residues ranges from 0.6 to 4.0% [56]. Usually, legume crops have higher nitrogen concentrations than cereals. Nitrogen content of the most common cereals, such as wheat straw (0.56%) and maize stalk (0.93 ± 0.03%) [26], are low compared with legumes such as sesbania leaves (4.0%), cowpea stem (1.1%), and cowpea leaves (2.0%) [56]. Other residues, such as rice husk, also have a considerable nitrogen content (1.2%) [24].

15.4.2 Phosphorus

The phosphorus content of poultry manure is higher than in cattle and swine manures. Animal manure phosphorus concentrations could vary from as low as 0.03 ± 0.01% [42] to as high as 2.6% [45] for poultry manure. Other studies on poultry phosphorus content, such as those of [42] 1.5 ± 0.1% and [28] 1.5%, fall within this range. Cattle manure phosphorus content ranges of 0.4% [50] to 1.5% [48] are reported in the literature, which is lower than in poultry. Similar to cattle manure, reported swine manure phosphorus contents are 0.05 ± 0.02% [51], 0.65–0.84 $mg\,l^{-1}$ [54], 0.87% [45], and 1.17% [53], which are also lower than poultry.

15.4.3 Potassium

Poultry manure has a higher potassium content than in cattle and swine manure. The mean potassium content of animal manure ranges between 0.5% for swine manure [45] to 3.8% for laying hens [28]. The reported potassium contents of poultry are 2.2% [45], 2.0 ± 0.2–2.7 ± 0.1% [42], and 3.8% [28], which are higher than those reported for swine and cattle manure. Swine potassium contents of 0.55% [45], 1.38–1.41 $g\,l^{-1}$ [54], and 1.58% [53] were reported. The reported potassium contents of cattle manure of 0.63 ± 0.04% [41] and 1.7% [48] were slightly higher than that of swine manure. Legume residues such as cowpea stem have a higher potassium content (2.5%) than cereal residues, such as rice (1.38%) and maize (1.31%) [56].

15.5 The Nutrient Composition of Industrial Biodegradable Waste (Food Industry, Slaughterhouse)

Industrial biodegradable waste includes wastes from the food industry, slaughterhouses, leather production industries, etc. The nitrogen contents of some industrial biodegradable waste from slaughterhouses (10.1 ± 0.8%) [57] are much higher than in plant residues and animal manure. The reported high nitrogen contents from the slaughterhouse comprised sheep and goat stomach and rumen. The stomach contents from the sheep and goat slaughterhouse had lower nitrogen contents of 3.6 ± 0.2% compared to their stomach and rumen.

The nitrogen contents of 8.3 ± 0.8% were reported from a pig slaughterhouse [22]. High nitrogen contents of 10.2% were reported from leather manufacturing bio-waste [41]. The nitrogen content of food processing industries, such as sugar beet pulp (1.6%) [58] and olive mill waste (1.36%) [59], was significantly lower than that of slaughterhouse bio-waste. Substantial amounts of phosphorus were also reported from the sheep and goat slaughterhouse.

15.6 Impact of Treatment of Biodegradable Waste on Nutrient Recovery from Bio-waste

Several authors reported that waste processing technologies affect the composition of the waste by-product, hence affecting the nutrient content and its availability for crop uptake [60]. In this review, it was reported that compared to secondary sludge, primary sludge has more lipids and fiber but less protein (nitrogen) and phosphorus [60]. A study also indicated that the mean Kjeldahl N of aerobically digested sludge (7.2 ± 0.8%) was significantly higher than that of anaerobically digested sludge (4.2 ± 1.1%) and raw sludge (3.0 ± 0.5%) [61]. The study also indicated that the phosphorus content did not significantly vary between the aerobic stabilization (2.8 ± 0.8%) and the anaerobic stabilization (2.4 ± 0.2%), but both were higher than no biological stabilization (0.70 ± 0.01%).

Thermal hydrolyses pre-treatment is reported to have increased the total Kjeldahl nitrogen (TKN) and phosphorus content of biosolids [62]. In their study, Wang et al. [62] discovered that the concentration of TKN (5.2 ± 1.3%) and phosphorus (TP) (3.4 ± 0.6%) of class A biosolid that had undergone thermal hydrolysis pre-treatment followed by anaerobic digestion was much higher than lime stabilized class B biosolid of the same origin (TKN 3.9 ± 0.6%; TP 1.2 ± 0.1%). In contrast, the concentration of potassium in the class A biosolid (thermal hydrolysis pre-treated anaerobically digested – THP-AD) (0.085 ± 0.002%) was almost half that of the lime stabilized class B sludge (0.146 ± 0.004%). It was reasoned that the low concentration of potassium in the THP-AD is attributable to the increase in the mobility and dissolution of potassium [62]. The lower concentration of total nitrogen in the lime stabilized class B biosolid is also attributed to the dilution effect because of the increase in the solids and the increase in alkalinity attributable to the added lime, which enhanced NH_4^+ to NH_3 transformation, facilitating volatilization losses [62].

The role of anaerobic digestion on the solubilization of nutrients, such as nitrogen, phosphorus, potassium, calcium, and magnesium, was investigated [45]. In the anaerobic digestion of piggery and poultry, solid waste streams under the mesophilic condition (37 °C) and pH of 7.0 ± 0.2, less than 10% of the total phosphorous, calcium, and magnesium were in a soluble form in the digestate [45]. A bioavailability test using citric acid extraction indicated the total availability of the phosphorus, calcium, and magnesium. It was also apparent that most of the nitrogen and potassium remained soluble in the digestate [45]. The study further indicated that complete solubilization of phosphorus, calcium, and magnesium occurred below the threshold pH of 5.5, emphasizing that these elements are either bound to form inorganic compounds or were adsorbed on solid surfaces in the digestate, reducing the feasibility of a post-digestion nutrient recovery through struvite crystallization. In their conclusion, it was suggested that nutrient solubilization and recovery during anaerobic digestion could be accomplished by adding complexing chemicals and operating at a lower pH or modifying the operating conditions [45].

A study aimed at exploring the agronomic potential of sewage sludge digestate from various treatments indicated variations in the composition of nitrogen, despite having a similar source (feed) [63]. According to the study, the primary liquid digestate (P), the secondary liquid digestate (S), the centrifuged solid digestate (C) derived from a mix of the primary and secondary digestates, and the dried pulverulent digestate (D) obtained by thermal treatment at 200 °C had 7.4, 7.5, 6.3, and 5.0% nitrogen contents, respectively [63]. Similarly, the phosphorus content was 4.2, 5.7, 6.7, and 6.3% respectively while the potassium content was 0.5, 0.7, 0.4, and 0.2% respectively [63].

Co-digestion of thickened waste activated sludge (TWAS) with macroalgal residue (MAR) (*Gelidium sesquipedale*) from a company located in Morocco, which extracts agar-agar from red macroalgae in a 1 : 1 ratio under mesophilic conditions (37 °C) with a hydraulic retention time of 20 days and an organic loading rate of 1 gVS l^{-1}.d, led to a slight decrease in the N content of TWAS from 6.1 ± 0.1 to $5.6 \pm 0.2\%$ [64].

Composting is a natural process of decomposing organic matter by microorganisms under controlled conditions. Composting processes could be aerobic or anaerobic with aerobic composting producing a very stable product, favored for agricultural land use [65]. Composting is associated with the loss of nitrogen in the form of NH_3 volatilization as the pH and temperature increases, particularly during the thermophilic stage [66]. It is, however, possible to minimize such losses by dissolving and immobilizing ammonia through the addition of wood chips and composted green waste of a higher C : N ratio [67].

The impact of composting on the TKN content of the final product varied among studies. Some studies reported a significant reduction in the TKN after composting compared to the initial nitrogen content of the feedstock mixture [41, 68, 69]. Other studies [20, 24] reported no change in TKN, although most studies reported an increase in total nitrogen [70–74]. The increase in the total nitrogen of the compost relative to the nitrogen content of the initial mixture is attributed to the net loss of dry mass, resulting from CO_2 and water loss through evaporation during the oxidation of organic matter [70].

The observed variation on the impact of composting in TKN among composting studies is attributable to various factors, including the raw materials used for composting [28], C : N ratio of the feedstock [24], and the moisture content during composting [24]. For instance, the composting of a mixture of broccoli waste, laying hen manure, and sawdust reduced the initial total nitrogen content of the mixture (1.23%) by 25% at maturity (0.92%) [28]. In contrast, the replacement of broccoli waste by tomato waste while maintaining the rest (tomato + laying hen manure + sawdust) resulted in an increase in the total nitrogen content of the mature compost (1.60%) compared to the initial nitrogen content of the mixture (1.29%). Similarly, two different total nitrogen responses to composting were reported with no or an insignificant change (a combination of low moisture (45%) and a low C : N ratio (16.2) or high moisture (60%) with a high C : N ratio (19.6)) and a significant increase in total nitrogen (a combination of higher moisture content (60 and 75%) and lower C : N ratio (16.2)), despite having similar initial ingredients [28].

There are limited studies on the phosphorus and potassium response to composting, such as that by [69] and [20] who reported an increase in the total phosphorus from $1.0 \pm 0.1\%$, to $1.2 \pm 0.1\%$ [69] and 0.3 to 0.5% [20]. The potassium content of the compost increased from 1.6 to 2.1% (mean value) [20].

Vermicomposting of human feces reduced the total nitrogen content of human feces from $4.0 \pm 0.4\%$ to 3.1 ± 0.2, leading to a 21% loss of nitrogen compared with the initial N content of the feces [39].

15.7 Conclusion

The nitrogen, phosphorus, and potassium content of household bio-waste is relatively higher than in urban biowaste. It is also apparent that the nitrogen content of bio-waste from restaurants is higher than that of households and in urban areas. Poultry manure is reported to have the highest nitrogen, phosphorus, and potassium content compared to cattle and swine manure. Bio-waste from a slaughterhouse had a high nitrogen content of 10%; 69–80% of the total nitrogen, 40–45% of the total phosphorus, and 60% of the total potassium in municipal waste water originate from urine, despite urine accounting for merely 1% of the domestic waste water. Waste water treatment methods and technologies affect the waste nutrient composition and content. Aerobic and anaerobic digestion of sludge increases the nutrient composition of the digestate compared to raw sludge. Thermal hydrolysis pre-treatment of sludge before anaerobic digestion increases the nitrogen and phosphorus total while reducing the potassium content of sludge compared to lime stabilized class B sludge of a similar origin. Composting of bio-waste was reported to have varying impacts on the nitrogen content of the final product (compost). Some studies indicated a decrease in the total nitrogen content of the compost compared with the initial feedstock mixture, while others indicated an increase in total nitrogen and the rest did not observe any difference. This variation is attributed to various factors, including, but not limited to, the composition of the feedstock, moisture content during composting, and the C : N ratio of the feedstock.

References

1 United Nations (2017). *World Population Prospects the 2006 Revision*, vol. 1, 932 pp. New York: United Nations.

2 Balatsky, A.V., Balatsky, G.I., and Borysov, S.S. (2015). Resource demand growth and sustainability due to increased world consumption. *Sustainability* 7 (3): 3430–3440.

3 United Nations. (1994). Report of the International Conference on Population and Development, Cairo, Egypt, 5–13 September 1994. A/CONF.171/13/Rev.l, New York, p. 15, 1995.

4 Tesfamariam, E.H., Badza, T., Demana, T. et al. (2018). Characterising Municipal Wastewater Sludge for Sustainable Beneficial Agricultural use. WRC Report No. TT 756/18.

5 Alvarez-Campos, O. and Evanylo, G.K. (2019). Biosolids improve urban soil properties and vegetable production in urban agriculture. *Urban Agriculture and Regional Food Systems* 4 (1): 1–11.

6 Aggelides, S.M. and Londra, P.A. (2000). Effects of compost produced from town wastes and sewage sludge on the physical properties of a loamy and a clay soil. *Bioresource Technology* 71: 253–259.

7 Bary, A., Hummel, R.L., and Cogger, C. (2016). Urban highway roadside soils and shrub plantings enhanced by surface-applied and incorporated organic amendments. *Arboriculture and Urban Forestry* 42: 418–427.

8 Garcia-Orenes, F., Guerrero, C., Mataix-Solera, J. et al. (2005). Factors controlling the aggregate stability and bulk density in two different degraded soils amended with biosolids. *Soil and Tillage Research* 82: 65–76. doi: 10.1016/j.still.2004.06.004.

9 Mekki, A., Aloui, F., and Sayadi, S. (2019). Influence of biowaste compost amendment on soil organic carbon storage under arid climate. *Journal of the Air and Waste Management Association* 69 (7): 867–877.

10 Wells, A., Chan, K., and Cornish, P. (2000). Comparison of conventional and alternative vegetable farming systems on the properties of a yellow earth in New South Wales. *Agriculture, Ecosystem and Environment* 80: 47–60. doi: 10.1016/S0167-8809(00)00133-X.

11 Badza, T., Tesfamariam, E.H., and Cogger, C.G. (2020). Agricultural use suitability assessment and characterization of municipal liquid sludge: based on South African survey. *Science of the Total Environment* 721: 135678.

12 Ma, J. and You, F. (2019). Superstructure optimization of thermal conversion based poultry litter valorization process. *Journal of Cleaner Production* 228: 1111–1121.

13 Zając, G., Szyszlak-Barglowicz, J., Golebiowski, W. et al. (2018). Chemical characteristics of biomass ashes. *Energies* 11: 2885. doi: 10.3390/en11112885.

14 Chew, K.W., Chia, S.R., Yen, H.W. et al. (2019). Transformation of biomass waste into sustainable organic fertilizers. *Sustainability* 11: 2266. doi: 10.3390/su11082266.

15 Ahmad Sanadi, N.F., Lee, C.T., Sarmidi, M.R. et al. (2019). Characterisation of liquid Fertiliser from different types of bio-waste compost and its correlation with the compost nutrients. *Chemical Engineering Transactions* 72: 253–258. doi: 10.3303/CET1972043.

16 Schröder, J.J., Cordell, D., Smit, A.L. et al. (2010). Sustainable use of phosphorus. EU Tender ENV.B.1/ETU/2009/0025. Plant Research International, Wageningen, UR, Report 357, p. 71.

17 Cordell, D., Drangert, J.-O., and White, S. (2008). The story of phosphorus: global food security and food for thought. *Global Environ Chang* 19: 292–305.

18 Abdel-Shafy, H. and Mansour, M.S.M. (2018). Solid waste issue: sources, composition, disposal, recycling, and valorization. *Egyptian Journal of Petroleum.* 27: 1275–1290.

19 Hanc, A., Novak, P., Dvorak, M. et al. (2011). Composition and parameters of household bio-waste in four seasons. *Waste Management* 31 (7): 1450–1460.

20 Andersen, J.K., Boldrin, A., Christensen, T.H. et al. (2011). Mass balances and life cycles of home composting of organic waste. *Waste Management* 31 (9–10): 1934–1942.

21 Eisted, R. and Christensen, T.H. (2011). Characterization of household waste in Greenland. *Waste Management* 31 (7): 1461–1466.

22 Borowski, S., Boniecki, P., Kubacki, P. et al. (2018). Food waste-codigestion with slaughterhouse waste and sewage sludge: digestate conditioning and supernatant quality. *Waste Management* 74: 158–116.

23 Idowu, I., Li, L., Flora, J.R.V. et al. (2017). Hydrothermal carbonization of food waste for nutrient recovery and reuse. *Waste Management* 69: 480–491.

24 Kumar, M., Ou, Y., and Lin, J. (2011). Co-composting of green waste and food waste at low C:N ratio. *Waste Management* 30 (4): 602–609.

25 Schwalb, M., Rosevear, C., Chin, R. et al. (2011). Food waste treatment in a community center. *Waste Management* 31 (7): 1570–1575.

26 Chen, M., Wang, C., Wang, B. et al. (2019). Enzymatic mechanism of organic nitrogen conversion and ammonia formation during vegetable waste composting using two amendments. *Waste Management* 95: 306–315.

27 Margaritis, M., Psarras, K., Panaretou, V. et al. (2018). Improvement of home composting process of food waste using different minerals. *Waste Management* 73: 87–100.

28 Gavilanes-Terán, I., Jara-Samaniego, J., Idrovo-Novillo, J. et al. (2016). Windrow composting as horticultural waste management strategy – a case study in Ecuador. *Waste Management* 48: 12–134.

29 Boldrin, A. and Christensen, T.H. (2010). Seasonal generation and composition of garden waste in Aarhus (Denmark). *Waste management* 30 (4): 551–555.

30 Cieślik, B. and Konieczka, P. (2017). A review of phosphorus recovery methods at various steps of wastewater treatment and sewage sludge management. The concept of "no solid waste generation" and analytical methods. *Journal of Cleaner Production* 142: 1728–1740.

31 Ye, Y., Ngo, H.H., Guo, W. et al. (2017). Insight into chemical phosphate recovery from municipal wastewater. *Science of the Total Environment* 576: 159–171.

32 Abbasy, H.N., Xie, J., Hussain, S.I. et al. (2019). Nutrient removal in hybrid constructed wetlands: spatial-seasonal variation and the effect of vegetation. *Water Science and Technology* 79 (10): 1985–1994.

33 Pennsylvania Department of Environmental Protection (2016). *Wastewater Operator Certification Training Module 1: Introduction to Wastewater Treatment*. Pennsylvania Department of Environmental Protection (Pa. DEP), The Pennsylvania State Association of Township Supervisors (PSATS), Gannett Fleming Inc., Dering Consulting Group, and Penn State Harrisburg Environmental Training Center.

34 Teles, C.C., Mohedano, R.A., Tonon, G. et al. (2017). Ecology of duckweed ponds used for nutrient recovery from wastewater. *Water Science and Technology* 75 (12): 2926–2934.

35 Fumasoli, A., Etter, B., Sterkele, B. et al. (2016). Operating a pilot-scale nitrification/distillation plant for complete nutrient recovery from urine. *Water Science and Technology* 73 (1): 215–222.

36 Larsen, T.A. and Gujer, W. (1996). Separate management of anthropogenic nutrient solutions (human urine). *Water Science and Technology* 34 (3–4): 87–94.

37 Hanæus, J., Hellström, D., and Johansson, E. (1997). A study of a urine separation system in an ecological village in northern Sweden. *Water Science and Technology* 35 (9): 153–160.

38 Kujawa-Roeleveld, K. and Zeeman, G. (2006). Anaerobic treatment in decentralised and source-separation-based sanitation concepts. *Reviews in Environmental Science and Biotechnology* 5 (1): 115–139.

39 Yadav, K.D., Tare, V., and Ahammed, M. (2011). Vermicomposting of source-separated human faeces by Eisenia Fetida: effect of stocking density on feed consumption rate, growth characteristics and vermicompost production. *Waste Management* 31 (6): 1162–1168.

40 Filho, J.L.dP., Tonetti, A.L., Guimarães, M.T. et al. Nutrient recovery from airplane wastewater: composition, treatment and ecotoxicological assay. *Water Science and Technology* 75 (8): 1952–1960.

41 Ravindran, B. and Sekaran, G. (2010). Bacterial composting of animal fleshing generated from tannery industries. *Waste Management* 30 (12): 2622–2630.

42 Quiroga, G., Castrillón, L., Fernández-Nava, Y. et al. (2010). Physico-chemical analysis and calorific values of poultry manure. *Waste Management* 30 (5): 880–884.

43 Henuk, Y.L. and Dingle, J.G. (2003). Poultry manure: source of fertilizer, fuel and feed. *World's Poultry Science Journal* 59 (3): 350–360.

44 Wang, M., Cao, W., Sun, C. et al. (2019). To distinguish the primary characteristics of agro-waste biomass by the principal component analysis: an investigation in East China. *Waste Management* 90: 100–120.

45 Mehta, C.M. and Batstone, D.J. (2013). Nutrient solubilization and its availability following anaerobic digestion. *Water Science and Technology.* 6 (4): 756–763.

46 Materechera, S.A. and Mkhabela, T.S. (2002). The effectiveness of lime, chicken manure and leaf litter ash in ameliorating acidity in a soil previously under black wattle (*Acacia mearnsii*) plantation. *Bioresource Technology* 85 (1): 9–16.

47 Cavalli, D., Cabassi, G., Borrelli, L. et al. (2016). Nitrogen fertilizer replacement value of undigested liquid cattle manure and Digestates. *European Journal of Agronomy* 73: 34–41.

48 Zhang, F., Gu, W., Xu, P. et al. (2011). Effects of alkyl Polyglycoside (APG) on composting of agricultural waste. *Waste Management* 31: 1333–1338.

49 Benke, M.B., Hao, X., Caffyn, P. et al. (2010). Using manure from cattle fed dried distillers' grains with solubles (DDGs) as fertilizer: effects on nutrient accumulation in soil and uptake by barley. *Agriculture, Ecosystems and Environment* 139: 720–727.

50 Chen, S., Liao, W., Liu, C. et al. (2004). Value-added chemicals in animal manures. Northwest Bioproducts Research Institute Technical Report. US Department of Energy Contract DE-AC06-76RLO 1830, p. 135.

51 Tilche, A., Bortone, G., Malaspina, F. et al. (2001). Biological nutrient removal in a full-scale SBR treating piggery wastewater: results and modelling. *Water Science and Technology* 43 (3): 363–371.

52 Meng, L., Xie, L., Riya, S. et al. (2019). Impact of turning waste on performance and energy balance in thermophilic solid-state anaerobic digestion of agricultural waste. *Waste Management* 87: 183–191.

53 Galliou, F., Markakis, N., Fountoulakis, M.S. et al. (2018). Production of organic fertilizer from olive mill wastewater by combining solar greenhouse drying and composting. *Waste Management* 75: 305–311.

54 Taddeo, R. and Lepistö, R. (2015). Struvite precipitation in raw and co-digested swine slurries for nutrient recovery in batch reactors. *Water Science and Technology* 71 (6): 892–897.

55 Jiang, T., Schuchardt, F., Li, G. et al. (2011). Effect of C/N ratio, aeration rate and moisture content on ammonia and greenhouse gas emission during the composting. *Journal of Environmental Sciences* 23 (10): 1754–1760.

56 Lal, R. (1995). *Sustainable Management of Soil Resources in the Humid Tropics*. United Nations University Press. Online. http://archive.unu.edu/unupress/unupbooks/uu27se/uu27se00.htm# (contents accessed 9 July 2020).

57 Moukazis, I., Pellera, F., and Gidarakos, E. (2018). Slaughterhouse by-product treatment using anaerobic digestion. *Waste Management* 71: 652–662.

58 Kirby, M.E., Theodorou, M.K., Brizuela, C.M. et al. (2018). The anaerobic digestion of pig carcass with or without sugar beet pulp, as a novel on-farm disposal method. *Waste Management* 75: 251–260.

59 López-Cano, I., Roig, A., Cayuela, M.L. et al. (2016). Biochar improves N cycling during composting of olive mill wastes and sheep manure. *Waste Management* 49: 553–559.

60 Barber, W.B.F. (2015). Influence of wastewater treatment on sludge production and processing. *Water Practice and Technology* 10 (1): 178–186.

61 Černe, M., Palčič, I., Paskovič, I. et al. (2019). The effect of stabilization on the utilization of municipal sewage sludge as a soil amendment. *Waste Management* 94: 27–38.

62 Wang, X., Andrade, N., Shekarchi, J. et al. (2018). Full scale study of class A biosolids produced by thermal hydrolysis pretreatment and anaerobic digestion. *Waste Management* 78: 43–50.

63 Cristina, G., Camelin, E., Pugliese, M. et al. (2019). Evaluation of anaerobic digestates from sewage sludge as a potential solution for improvement of soil fertility. *Waste Management* 99: 122–134.

64 Elalami, D., Monlau, F., Carrere, H. et al. (2020). Evaluation of agronomic properties of Digestate from macroalgal residues anaerobic digestion: impact of pretreatment and co-digestion with waste activated sludge. *Waste Management* 108: 127–136.

65 Ahmed, M., Ahmad, S., Hassa, F. et al. (2019). Innovative processes and technologies for nutrient recovery from wastes: a comprehensive review. *Sustainability* 11: 4938.

66 Zhang, L. and Sun, X. (2016). Influence of bulking agents on physical, chemical, and biological properties during the two-stage composting of green waste. *Waste Management* 48: 115–126.

67 Jolanun, B. and Towprayoon, S. (2010). Novel bulking agent from clay residue for food waste composting. *Bioresource Technology* 101: 4484–4490.

68 Malinowski, M., Wolny-Koladka, K., and Vaverková, M.D. (2019). Effect of biochar addition on the OFMSW composting process under real conditions. *Waste Management* 84: 364–372.

69 Hait, S. and Tare, V. (2011). Optimizing Vermistabilization of waste activated sludge using vermicompost as bulking material. *Waste management* 31 (3): 502–511.

70 Kalamdhad, A., Singh, Y.K., Ali, M. et al. (2009). Rotary drum composting of vegetable waste and tree leaves. *Bioresource Technology* 100: 6442–6450.

71 Yañez, R., Bueno, P., Rivera, A. et al. (2010). Selective organic compounds degradation under controlling composting conditions. *Waste Management* 30 (5): 755–763.

72 Vuorinen, A.H. and Saharinen, M.H. (1997). Evolution of microbiological and chemical parameters during manure and straw co-composting in a drum composting system. Agriculture. *Ecosystem and Environment* 66: 19–29.

73 Sánchez-Monedero, M.A., Roig, A., Paredes, C. et al. (2001). Nitrogen transformation during waste composting by the Rutgers system and its effects on pH, EC and maturity of the composting mixtures. *Bioresource Technology* 78: 301–330.

74 Inbar, Y., Hadar, Y., and Chen, Y. (1993). Recycling of cattle manure: the composting process and characterization of maturity. *Journal of Environmental Quality* 22: 857–863.

16

Energy and Biomethane Production

Dagmar Juchelková[1], Helena Raclavská[3], Maw Maw Tun[2], Klaus Koppe[1], Abdulaziz Atabani[4], Barbora Švédová[3], and Marek Kucbel[3]

[1]*Department of Electronics, Faculty of Electrical Engineering and Computer Science, VŠB-TU Ostrava, Ostrava Poruba, Czech Republic*
[2]*Ministry of Education, Naypyidaw, The Republic of the Union of Myanmar*
[3]*Centre ENET VŠB-TU Ostrava, Ostrava Poruba, Czech Republic*
[4]*Alternative Fuels Research Laboratory (AFRL), Energy Division, Department of Mechanical Engineering, Erciyes University, 38039 Kayseri, Turkey*

16.1 Biomethane Origin

Biomethane, also known as renewable natural gas (RNG) or sustainable natural gas (SNG), is methane of a purely biological origin. Biomethane is produced by qualitative processing of raw biogas in several steps, to separate it from other components. Biomethane production consists of a complex sequence of processes. A number of influencing factors play a role in the journey from biomass production to the application of biomethane, while at the same time, different actors are involved.

Following the so-called "upgrade biogas," biomethane can replace fossil fuels in various applications, e.g. it can be fed into the existing gas network and used as a fuel in compressed natural gas (CNG) vehicles and used in heat and power generation.

It is clear from the above mentioned discussion that obtaining high-quality biomethane is one of the main routes to obtain a high-quality biogas. The high quality biogas is not only derived from the technology involved but also from the quality of the feedstocks used. Figure 16.1 shows the logistical chain of biomethane generation and use.

Biogas, alongside hydroelectric, wind, and solar power plants, is one of the major sources of the alternatives fuels, based on the processing of organic materials. In line with the IEA's processed submissions [1–4], the growth in biogas and/or biomethane production can be expected worldwide and the availability of sustainable feedstocks for these purposes is set to grow by 40% over the period to 2040 (Figure 16.2a–c).

16.2 Methane/Biomethane Sources

Biogas is one of the most attractive renewable resources due to its ability to convert waste into energy. Biogas is produced by natural fermentation/digestion of biomass. This occurs in an anaerobic environment (i.e. without the presence of oxygen) in special biogas plants. A wide range of feedstocks can be used as the main biomass [5]. Biogas mainly consists of methane (CH_4) (50–75%),

Biodegradable Waste Management in the Circular Economy: Challenges and Opportunities, First Edition.
Edited by Malgorzata Kacprzak, Eleonore Attard, Kari-Anne Lyng, Helena Raclavska, BalRam Singh, Eyob Tesfamariam, and Franck Vandenbulcke.

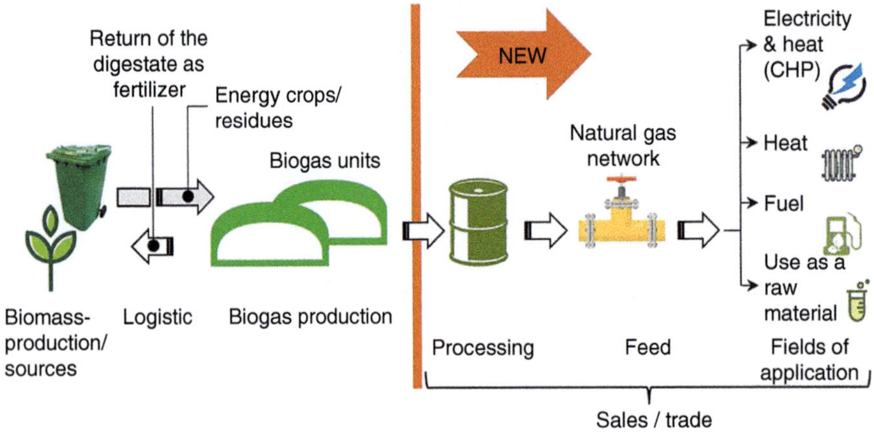

Figure 16.1 Logistical chain of biomethane generation and use.

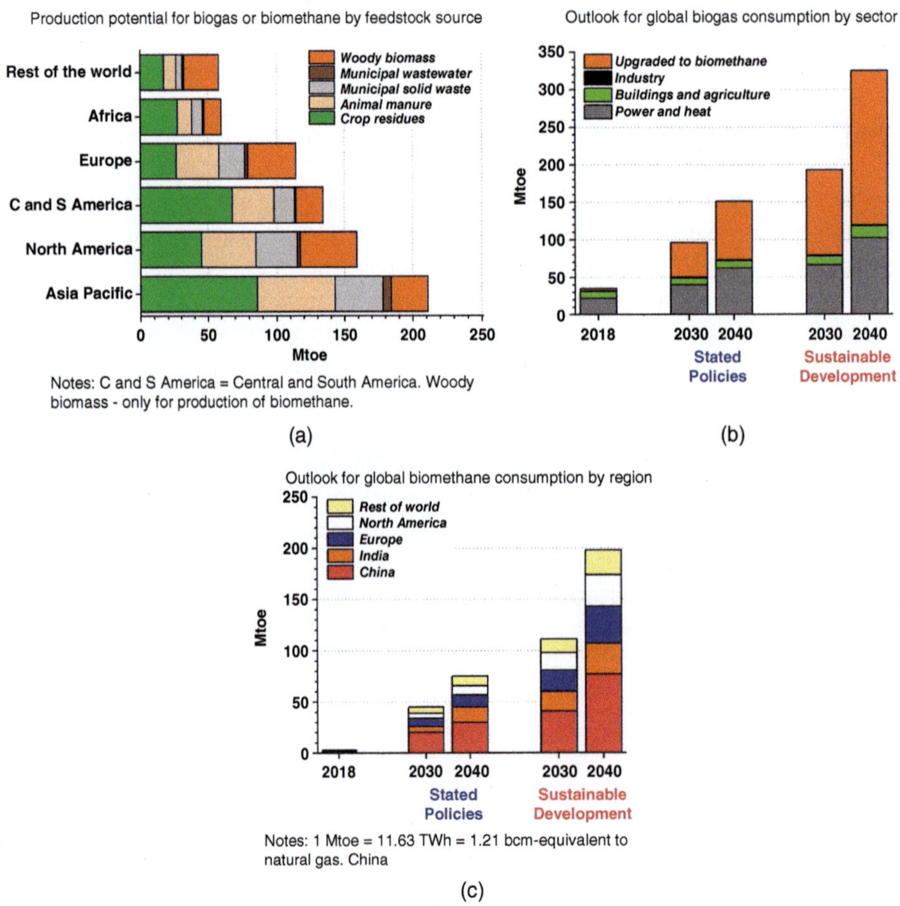

Figure 16.2 Current production and expected trend in biomethane production (Mtoe – million tonnes of oil equivalent); (a) Biogas or biomethane by feedstock source; (b) outlook for global biogas consumption by sector; (c) outlook for global biomethane consumption by region [1].

carbon dioxide (CO_2) (25–50%), hydrogen sulphides (H_2S), hydrogen (H_2), ammonia (NH_3) (1–2%) and traces of other gases such as oxygen (O_2) and nitrogen (N_2). The percentage of these trace gases is related to operating conditions and feedstocks.

16.2.1 The Main Feedstocks for Biogas Production Can Be Divided into Four Groups

- Residues from the agricultural crops: for instance, residues from soya and other oilseeds, rice, sugar beet or wheat harvests, etc. In particular, feedstocks should be produced as by-products so as to maintain sustainable land management (preserve soil fertility, preserve soil carbon, and prevent erosion). There must be no competition with the food chain.
- Manure: organic livestock products.
- Organic fraction of the municipal waste: paper, cardboard, green maintenance (leaves and grass), unusable pulped wood, and waste from the food industry.

  Sewage sludge: organic matter obtained from municipal waste water treatment plants.

Using waste and residues/by-products as feedstocks is more advantageous in terms of value chain sustainability.

Most biomethane production comes from upgrading biogas, so the feedstocks are the same as those described above. However, the biomethane gasification pathway may use wood biomass (excluding municipal solid waste- (MSW) and agricultural residues) as feedstocks, consisting of forest residues and wood processing.

In biogas stations, co-generation (most commonly gas engines) or (less frequently gas turbines) is the most often used. In most processes, heat is also generated as a by-product.

Two temperature ranges are most commonly used in biogas production – mesophilic bacteria (37–38 °C) or thermophilic bacteria (55–57 °C). Waste heat from the process, unless otherwise applied, is used to reduce the humidity of digests and heat biogas reactors. With the electricity generated, loading is simpler and is mostly diverted to the grid. The downside of such systems is that heat has to be used on site and only electricity can be connected to the grid.

16.2.2 Accumulation and Management

In an optimal biomethane production system, it is essential to manage the process efficiently, e.g. using fuzzy-logic and to be able to accumulate the gas produced. With small amounts of treated feedstock, biogas upgrades cannot be realized (Figure 16.3).

16.2.3 By-products of the Biogas Process

Digestate is a stable organic matter rich in various nutrients (phosphorus, nitrogen, potassium). It is a by-product of the anaerobic digestion process. It can be directly used as a fertilizer feedstocks used to produce biogas. In this case, it can replace different types of organic fertilizers available

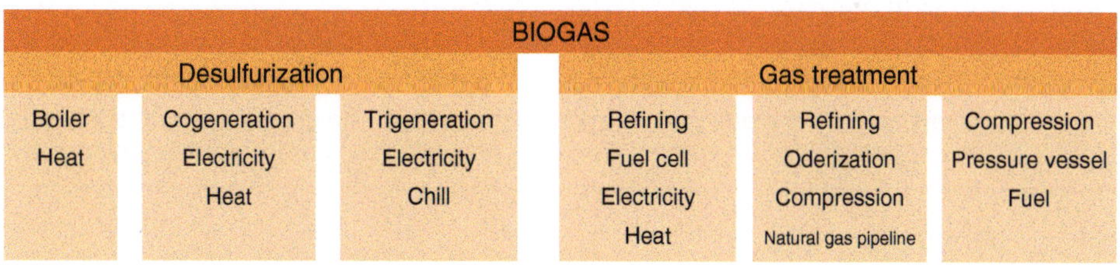

Figure 16.3 Overview of basic uses of biogas.

in the market that are made from manure, compost, or other animal and plant products, but this places high demands on the qualitative composition of feedstocks for biogas production. By passing through higher temperatures, the material neutralizes most pathogenic substances, is relatively homogeneous, and is rapidly absorbed into the soil.

If its quality cannot be assured, the digestate is further processed by industrial processes.

16.3 Biogas Treatment Technologies

In the context of the upgrading process of biogas in general, it is particularly necessary to remove the following ingredients/components.

16.3.1 Water Removal

The biogas should be dried to remove the water (moisture) in order to prevent it from accumulating and condensing in the pipes or reservoirs and producing an acidic corrosive solution when acidic gases are present. In this way, the dew point can also be reduced to prevent water condensation and subsequent freeze-out when biogas is stored under increased pressure. Water is usually removed from the biogas along with other impurities.

16.3.2 Carbon Dioxide Removal

The elimination of carbon dioxide is necessary to upgrade the biogas to natural gas and to be used as a transportation fuel. This will increase the energy content of biogas resulting from the methane content; its content in purified biogas should be greater than 95% vol.

16.3.3 Sulfan Removal

The most commonly used methods of removing sulfan are those already used in the production of biogas, i.e. the dosing of air and oxygen respectively and the dosing of ferrous chloride into fermented organic matter. Biological desulfuration (oxidation), for which microorganisms belonging to the *Thiobacillus* community are used, allows up to 95% desulfuration to be achieved, reducing the sulfan content to below 50 ppm

16.3.4 Removal of Higher Hydrocarbons and Halogenated Hydrocarbon Derivatives

Higher hydrocarbons and halogenated (mainly F and Cl) hydrocarbons present primarily in landfill biogas are responsible for causing corrosion problems. Their removal is carried out under pressure on specially activated charcoal. Substances with a small molecule such as methane, carbon dioxide, nitrogen, and oxygen go through the adsorbent, while large ones get adsorbed.

16.3.5 Removal of Organo-Silicate Compounds

The presence of siloxanes in biogas is undesirable because, when they are burned, they cause the formation of silica, which deposits on the surface of various structural elements (cylinder heads, pistons, valves) and severely damages them as a result of the contour. They are removed using a special absorption agent, which is a mixture of hydrocarbons with the capacity to absorb these substances.

16.3.6 Oxygen and Nitrogen Removal

Both gases are mainly present in landfill biogas. To remove them, the pressure change adsorption (PSA) method is used, i.e. adsorption on molecular networks, the main drawback of this method

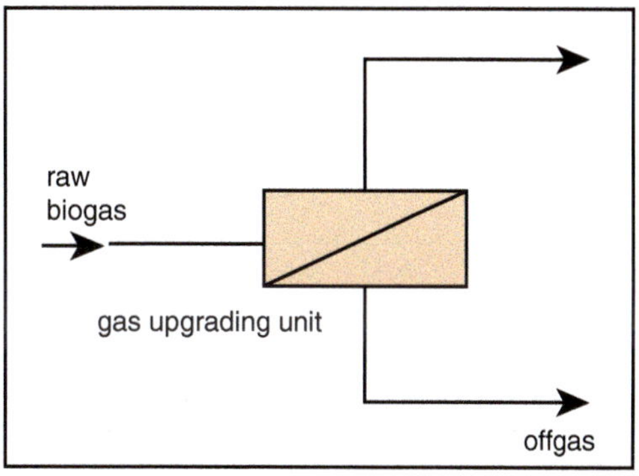

Figure 16.4 Basic principle of biogas refining [6].

is that it is costly (Figure 16.4). It is therefore appropriate to monitor the oxygen concentration in the biogas produced as a precautionary measure to capture any higher increase in time.

16.3.7 Biogas Refinement on Biomethane

Biogas processing methods are mostly based on physico-chemical carbon dioxide removal:

(a) Pressure water scrubber (Figure 16.5)
(b) Diaphragm separation (Figure 16.6)
(c) Pressure change adsorption (PSA)

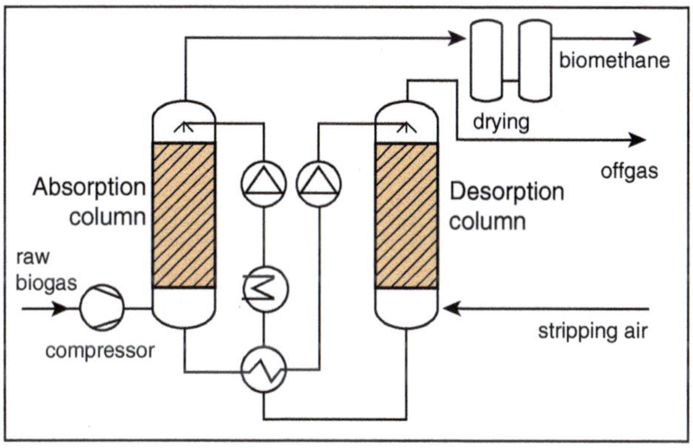

Figure 16.5 Refining biogas to biomethane pressure washing machine [6].

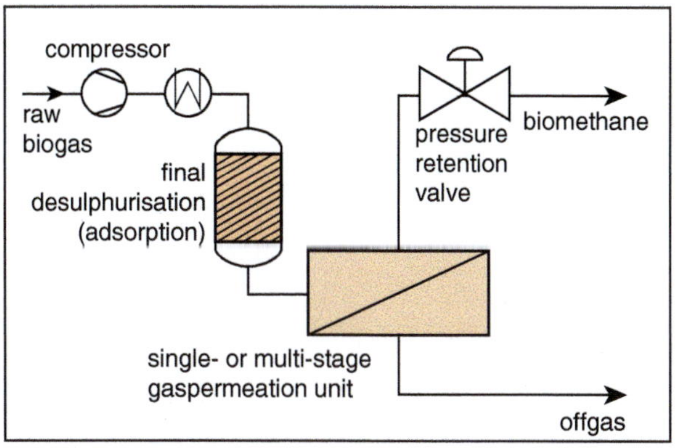

Figure 16.6 Refining biogas to biomethane membrane [6].

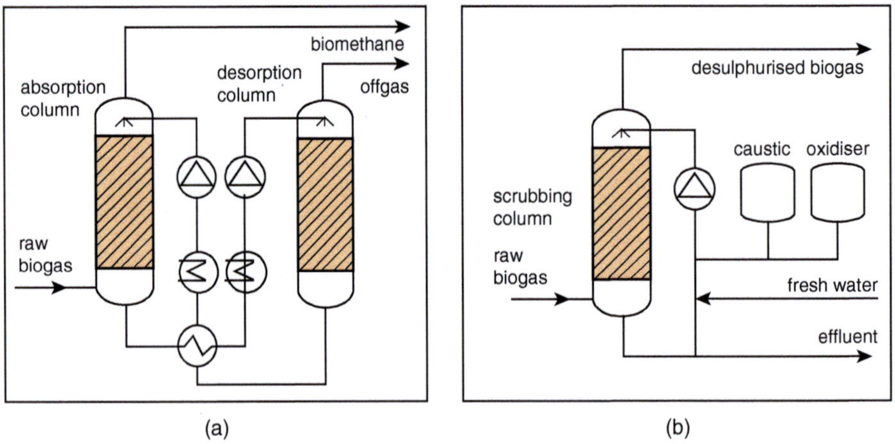

Figure 16.7 Refining biogas to biomethane: (a) ammine wash, (b) chemical oxidation [6].

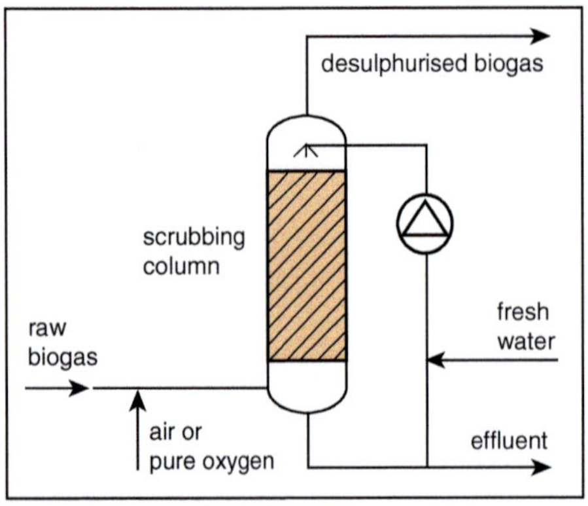

Figure 16.8 Refining biogas to biomethane biologically [6].

(d) Chemical scrubbers – glycol washing machines, amine washing machines (Figure 16.7a, b)
(e) Cryogenic separation
(f) Other (Figure 16.8)

Different technologies are used in different countries, depending on the country's preference for end-uses of biomethane.

16.3.8 Processing CO_2 into Biomethane

The possibility of turning CO_2 into a usable product – methane or various organic substances – is also an emerging technology. Converting CO_2 to methane needs reduction equivalents in the form of hydrogen or directly electrons:

Sabatier reaction $4\,H_2 + CO_2 \rightarrow CH_4 + 2\,H_2O$ $\Delta G° = -130.7\,\text{kJ mol}^{-1}$

Reaction feasibility – needs catalysts and appropriate physical conditions

16.3.9 Conversion of CO_2 with Hydrogen May Occur

- Chemically: high pressures and temperatures necessary, precious metal catalysts, need for purification

Table 16.1 Composition of different gases [6] modified.

Parameter	Biogas	Deponie gas	Natural gas	Biomethane
Methane (vol.%)	60–70	35–65	89	Min 95
H_2 (vol.%)	0	0–3	0	0
CO_2 (vol.%)	30–40	15–50	0.67	<5
N_2 (vol.%)	Up to 1	5–40	0.28	<2
O_2 (vol.%)	Up to 0.5	0–5	0	0.5
H_2S (ppv)	0–4000	0–100	2.9	<5
Ammonia (ppv)	Up to 100	Up to 5	0	0
HCV (kWh m^{-3}N)	6.5	4.4	11.0	9.4–11.8

- Biochemically: using hydrogenotrophic methanogens under anaerobic fermentation conditions, temperatures 40–64 °C
- CO_2 sources for "synthetic biomethane" can come from different sources, e.g.
- capture from flue gas (expensive, nitrogen-polluted from air)
- gases from fermentation plants, e.g. ethanol (contaminated with nitrogen from air)
- CO_2 from biogas – most advantageous source

However, these are very expensive solutions.

16.3.10 Composition of Different Gases

Table 16.1 shows the typical gas composition for biogas and landfill gas and compares these values with typical natural gas. The quality of this type of natural gas appears to be relatively typical of the many different qualities of natural gas available in Europe.

16.4 Circular Economy

The "improvement/upgrade" of biogas to biomethane offers higher added value on many levels: increased use of biomass, logistics, biogas production, biogas processing, feed, and sales and trade.

At each of these stages of value creation, there is a potential for optimization of a technical and economic nature. Industry-specific know-how is needed to realize this potential.

Biomethane is produced by qualitative processing of raw biogas in several steps and by separation from other components.

The presence of carbon dioxide and the low pressure of the biogas produced interfere with the use of the natural gas transport system, so the biogas is refined (upgraded) to natural gas-quality biomethane and can also be used as a fuel in natural gas (CNG) vehicles – agricultural machinery, cars, and buses (trains) (see Figure 16.9).

The digestate obtained contains a high proportion of nutrients and is used as a fertilizer in agriculture and horticulture.

Currently, the biomethane produced is more expensive than natural gas. For biomethane, cost increases by additional connections to pipelines also play an important role.

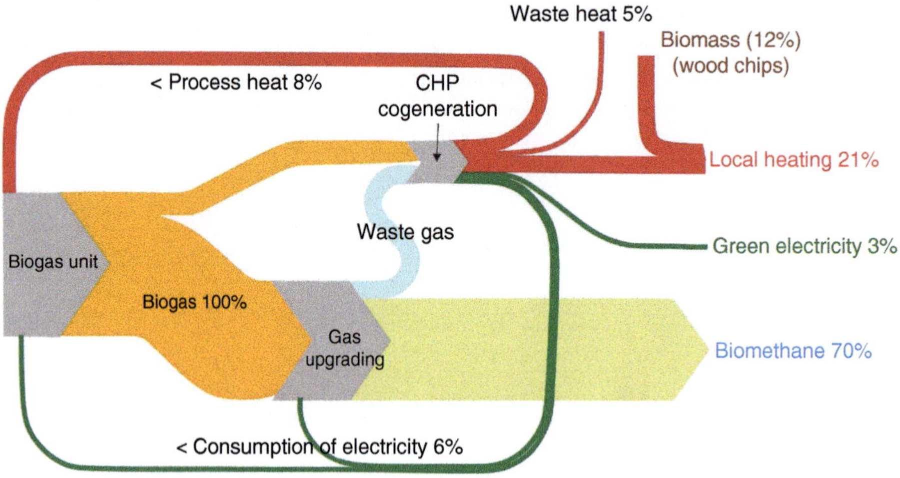

Figure 16.9 Example – Diagram of energy flows to the different parts of biogas processing and utilization.

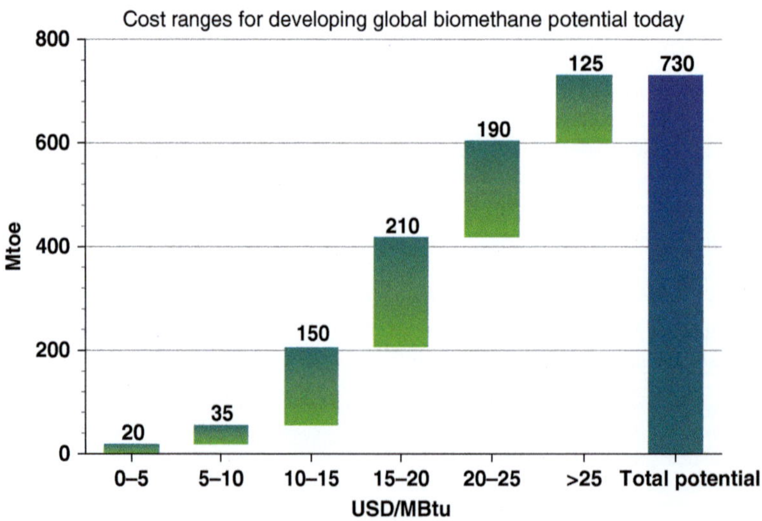

Figure 16.10 Example of biomethane price structure [1].

The "upgrade" of landfill gases to biomethane appears to be the cheapest option yet (Figure 16.10).

Improvements in the biomethane production technologies is highly expected in the future, which will also entail lower prices.

16.4.1 SWOT Analysis of Biomethane Use

Strengths

- A dense network of pipelines
- Quality comparable to natural gas ⇨ possibility of removal to pipeline network
- Improved energy use efficiency over Biogas Station

- A controlled, storable, and permanent energy source
- Forms a closed cycle of nutrients
- A renewable resource, neutral in terms of CO_2 emissions
- The best fuel in terms of GHG emissions/1 carbon in the CH_4 molecule

Weaknesses

- Sometimes inadequate legislation
- Expensive cleaning and treatment of biogas
- Large investment costs

Opportunities

- A new renewable energy source
- The possibility of partial substitution of petroleum products for "domestic" fuels

Threats

- Insufficient knowledge and experience of biomethane production and use technologies
- Difference in EU Member States legislation, inconsistency, sometimes lack of regulations and standards for the use and modification of biomethane, transport
- Investor uncertainty in the "new" field of biogas use.

16.5 Summary

16.5.1 Return – Economy

The decision to build a biomethane plant requires heavy investment. The profitability analysis can be very extensive due to the long duration of operation of the facility (typically at least 20 years).

The annual economic assessment (Figure 16.11) should include at least the following points.

The potential for realizing reductions in CO_2 emissions from the use of biogas or biomethane depends on how these gases are produced and where they are used in the value chain. Politically, it

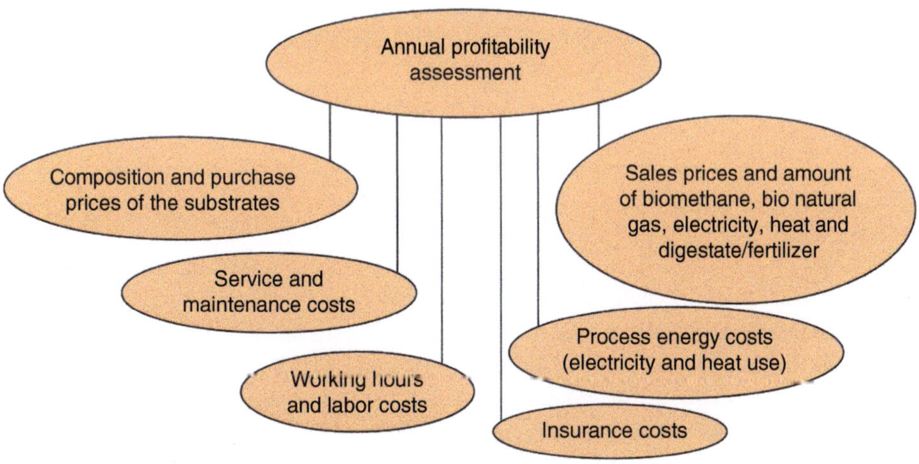

Figure 16.11 Scheme of the annual economic assessment.

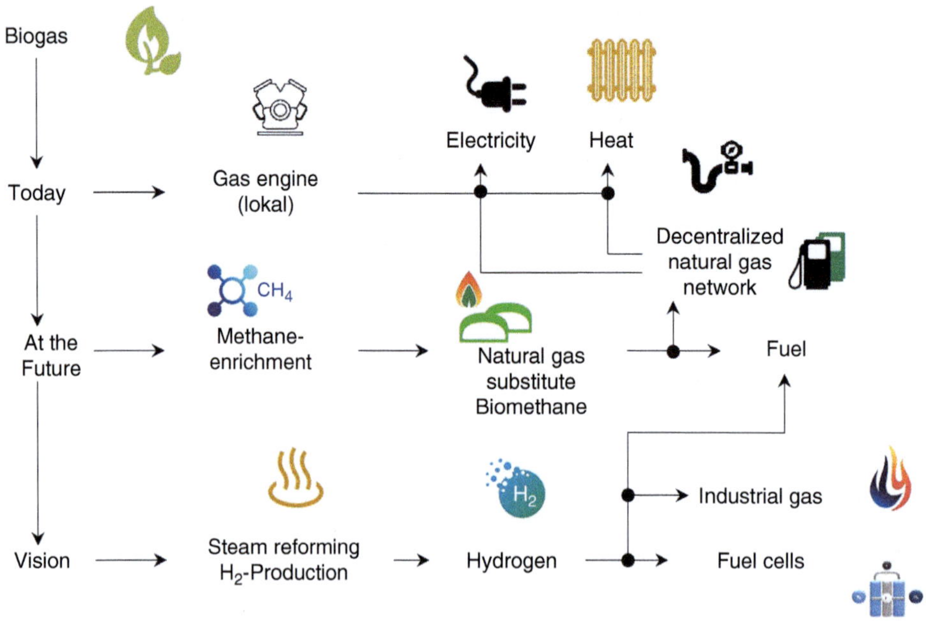

Figure 16.12 Biogas/biomethane use scheme now and in the future.

is crucial that biogas production (and all other forms of bioenergy) actually leads to a net reduction in CO_2 emissions over the life cycle.

One solution is to use 10% of the biomethane in gas distribution pipelines, which could achieve an adequate reduction in CO_2 emissions. In reality, however, CO_2 emissions have to be considered as part of the entire production chain [7] (Figure 16.12).

The situation is similar in the use of low-carbon hydrogen, but here it can be expected that using 10% biomethane in distribution pipelines can reduce CO_2 emissions by 3–4% (with the same energy content/net calorific value). However, the assessment of the whole hydrogen production chain leads to the need for further development of the technologies used to ensure that the technologies used are truly low-carbon [8].

A number of studies have shown that restoring the use of organic residues, carbon and digestate can be one of the drivers of rural development. Sustainable landscape development can contribute to the development of more efficient/sustainable agriculture, which has an important economic function in rural areas.

References

1 International Energy Agency (2020). Outlook for biogas and prospects for organic growth: World energy outlook special report on biomethane (Internet). France: International Energy Agency; March 2020 (cited 1 September 2020), p. 93. https://webstore.iea.org/download/direct/2970?fileName=Outlook_for_biogas_and_biomethane.pdf.

2 International Energy Agency (2019). World energy outlook 2019 (Internet). (cited 3 September 2020). https://www.iea.org/reports/world-energy-outlook-2019.

3 International Energy Agency (2018). World energy outlook 2018 (Internet). https://www.iea.org/reports/world-energy-outlook-2018.

4 International Energy Agency. Renewables (2019). Analysis and forecasts to 2024 (Internet). Paris: International Energy Agency (cited 16 September 2020). doi: 10.1787/b3911209-en.

5 Scholwin F. (2020). Biomethane (Internet). Institut für Biogas (cited 3 September 2020). https://www.biogasundenergie.de/en/page/biomethane.

6 Überblick über biogasaufbereitungstechnologien zur produktion von biomethan, Promotion of bio-methane and its market development through local and regional partnerships (Internet). Institute of Chemical, Environmental and Bioscience Engineering; 2012 (cited 4 September 2020). http://www.lev.at/download/BiomethaneRegions/D_3_1_1_BiogasUpgradingTechnologyReview_DE.pdf.

7 Giuntoli J, Agostini A, Edwards R, Marelli L, European Commission, Joint Research Centre et al. (2015). Solid and gaseous bioenergy pathways: input values and GHG emissions: calculated according to the methodology set in COM(2010) 11 and SWD(2014) 259 (Internet). Luxembourg: Publications Office (cited 16 September 2020). https://op.europa.eu/en/publication-detail/-/publication/ead42a9c-3105-4046-a401-9f1ca27a3bf5/language-en.

8 McDonagh, S., Deane, P., Rajendran, K., and Murphy, J.D. (2019). Are electrofuels a sustainable transport fuel? Analysis of the effect of controls on carbon, curtailment, and cost of hydrogen. *Applied Energy* 247: 716–730. https://linkinghub.elsevier.com/retrieve/pii/S0306261919307068.

17

The Governance and Social Aspects

Malgorzata Kacprzak[1,2] *and Kupich Iwona*[2]

[1] *Warsaw University of Technology, Faculty of Civil Engineering, Mechanics and Petrochemistry, Institute of Civil Engineering, Płock, Poland*
[2] *Czestochowa University of Technology, Department of Environmental Engineering and Biotechnology, Częstochowa, Poland*

17.1 Criteria for Selecting the Best Solutions for a Safe Biodegradable Waste Management in Accordance with the Requirements of Closed-Circular Management

Waste management is a complex problem because it involves legislative (laws, rules, guidelines on management, responsibilities, and competences), environmental (resources use and their optimal management), economic (reuse, disposal, and tariff), and urban planning aspects [1]. Moreover, it may pose serious toxic problem connected with permanent contamination of air, soil, and groundwater and may affect the climate. The World Bank has estimated that the greenhouse gas (GHG) emissions from waste processes achieved 5% of global GHG emissions or 1.6 billion tonnes of CO_2, primarily from open dumping and disposal in landfills without landfill gas capture systems [2]. On the other hand, it could be valuable resource. According to the World Energy Council, it has estimated that the global waste to energy market will reach \$40 billion by 2023 [3]. Moreover, the capital expenditure for waste management in developing Asian countries has been estimated at between \$23.7 and \$90.3 tonnes^{-1} for composting and energy from waste respectively. The overall investment required, taking an average of the global waste forecast to be generated between 2016 and 2050, is between \$2.1 trillion and \$7.8 trillion.

Hence the proper management of waste in accordance to three fields – technology, economy and social aspects – should be implemented.

17.1.1 Definition of Biodegradability and Biodegradable Waste Collection

According to most simple definitions, the term biodegradability means capable of being decomposed by biological agents, especially bacteria. The official definitions are, however, different. "Bio-waste" is defined in the Waste Framework Directive (WFD) as "biodegradable garden and park waste, food and kitchen waste from households, restaurants, caterers, and retail premises, and comparable waste from food processing plants." It does not include forestry or agricultural residues, manure, sewage sludge, or other biodegradable waste (natural textiles, paper, or processed wood). In turn, "biodegradable waste" is a broader concept defined in the Landfill Directive (LD) as any waste that is capable of undergoing anaerobic or aerobic decomposition, such as food and garden waste, and paper and paperboard.

Biodegradable Waste Management in the Circular Economy: Challenges and Opportunities, First Edition.
Edited by Malgorzata Kacprzak, Eleonore Attard, Kari-Anne Lyng, Helena Raclavska, BalRam Singh, Eyob Tesfamariam, and Franck Vandenbulcke.

Types of biodegradable waste:

- biodegradable fractions (BMSW) of municipal solid waste (MSW);
- commercial food waste, not collected as part of the MSW, including: (i) waste from markets, (ii) catering waste;
- forestry residues, including: (i) bark, (ii) wood residues;
- waste from agriculture, including: (i) animal husbandry excrements (solid and liquid manure), (ii) straw residues, (iii) sugar beet and potato haulm, (iv) residues of growing of beans, peas, flax and vegetables, (v) spent mushroom compost;
- wastes from the food and beverage industries, including: (i) breweries and malt houses, (ii) wineries; (iii) fruit and vegetable production industry; (iv) potato industry including starch; (v) residues of beet sugar production: (vi) slaughterhouse residues, (vii) meat production, (viii) whey;
- sewage sludge (derived from biological treatment of municipal wastewater).

Some artificially produced organic compounds are similar to naturally occurring ones that are slowly degraded by microorganisms, while others had such complex molecular structures that they are never exposed to microorganisms and are even not recognized by them. We have to remember, however, that the ability to biodegrade does not guarantee low toxicity of intermediates or final products. Some chemicals known to be rapidly biodegradable could also be identified as carcinogens, mutagens, or even reproductive toxicants. Usually the waste substrates are built from many compounds and their biodegradability may be very different. That is why some of the bio-wastes are so important when choosing the best criteria for a safe biodegradable waste management in accordance with the requirements of a closed-circular economy. It is estimated that, depending on local conditions, eating and drinking habits, climate, and the degree of industrialization, between 60 and 70% of MSW consists of BMSW (food and green waste and paper and cardboard waste). All countries and regions surveyed employ a traditional "bagged waste" collection and a separate collection foe everything else (Figure 17.1). Bio-waste is the largest single component of municipal waste in the EU. According to data from 18 European countries, a total of 47.5 million tonnes of bio-waste is separately collected. On average, this equates to 117 kg of bio-waste per capita per year. Taking into consideration the total amount of bio-waste separately collected per year and country, Germany collects more than 14 million tonnes of bio-waste per year, followed by the UK with 8.9 million, Italy with 6.5 million, and France with 4.6 million tonnes [4]. In Poland, according to GUS (Statistic Poland) [5] in 2019, there were 2190 separate collection points for municipal waste. In 2005, separately collected waste accounted for only 3% of collected municipal waste (295 thousand tonnes). In 2019, almost 4 million tonnes were collected selectively (31% of the total municipal waste generated). The structure of selectively collected municipal waste has changed over the years. The dominant waste fractions in 2005, such as paper and cardboard, glass, and plastics (in total 80% of separately collected waste) currently account for 33%, while biodegradable waste has the highest share (30% in 2019). It is interesting that in 2005 biodegradable waste was not collected selectively in Poland. Selective collected waste were processed as follows [5]:

- recycling – 3192.1 thousand tonnes (25.0%),
- biological treatment processes (composting or fermentation) – 1–153.2 thousand tonnes (9.0%),
- thermal transformation with energy recovery – 2741.8 thousand tonnes (21.5%).

Moreover, the fact that waste is biodegradable does not mean that it can be transformed into a useful product in any technological process.

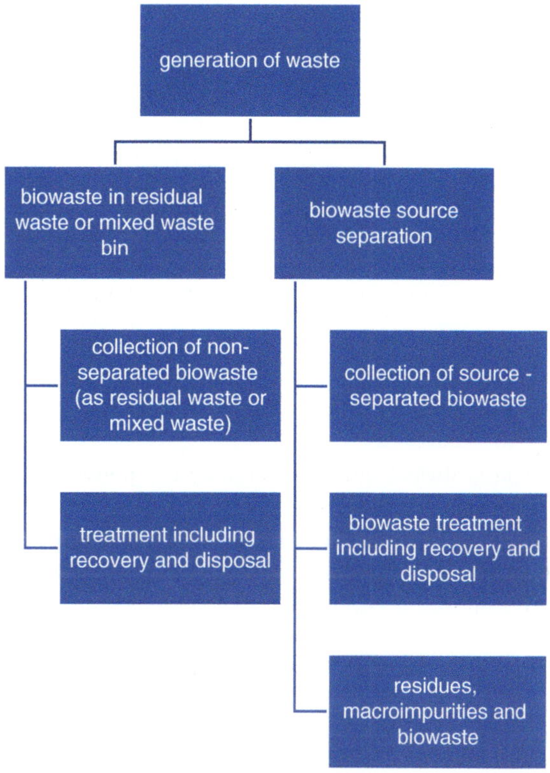

Figure 17.1 The possible pathways of biodegradable waste collection and treatment.

17.1.2 Biodegradable Waste Management System

Over the last decades the linear economy has played a main role in the economic development and many industrial branches have been based on natural resources and improper waste disposal. Whereas most countries have already introduced a waste management system, even the most developed technologies have to meet specific methodology and approach in order to advance the quality of life, meet environmental and legal standards, adopt best practices, and substantially increase public participation and environmental education and the need for efficiency and quality management. The following steps to develop, establish, and monitor a strategy at local and/or central level are now proposed, which must comply with United Nations Developed Programme Goals (UNDPGs) the circular economy (CE), and the industrial symbiosis concept as well as the Water Framework Directive (WFD) targets, concept, and philosophy [6]:

(a) Country Selection/Area Political Decision;
(b) Evaluation of the selected area including evaluation SWOT/PESTEL (Political, Economic, Social, Technological, Environmental, Legal) analysis:
   - demographic data, level of population, size of population, waste production, habits (eating), daily routines, ethics, incomes,existing legislations in place, multicriteria analysis (Eco – Env – Soc);
   - economic activities, main players (polluters – stakeholders), existing waste management plans in place, as well as existing practices and existing infrastructures;
(c) Citizen engagement – public awareness, public participation, stakeholder recognition, SMEs, and industries;

(d) Compositional analysis (per economic activity where possible), including microanalysis;
(e) Environmental performance EIA (Environmental Impact Assessment), RAA (Renin–Angiotensin–Aldosterone system), ERA (European Research Area), LCA (Life-Cycle Assessment), WFD (Water Framework Directive) and set up benchmarking indicators in order, for example, to measure the circularity of the circular economy (including processing, procedures, materials);
(f) Set up priorities having in mind the 10 R concept (per waste category where possible) targets on the circular economy WFD, food waste, plastics, CO_2 reduction organics, etc. and the end of waste criteria;
(g) Targets and goals with a time frame (realistic not ambitions), pilot areas, live labs;
(h) (Optional) political decisions to receive environmental license and become an independent authority to collect, treat, and transport waste according to the implanted legislations;
(i) Activities (i) to engage more citizens in the "game" and (ii) decide which target groups will be implemented;
(j) Educational and informal program educational material motivation, with smart applications such as waste mapping;
(k) Public participation, with an announcement advertising the strategy "pay as you save";
(l) Measuring the strategy;
(m) Monitoring – update – continual improvement.

An important development is the 10 R concept (Refuse, Reduce, Reuse, Repair, Refurbish, Re-manufacture, Re-purpose, Recycle materials, Recovery energy, Remain). According to biodegradable waste management this is correlated mainly with: (i) refuse and reject – people could choose between buying and not buying; (ii) rethink – start to analyze how to dispose of it; (iii) reduce – reducing the amount of waste as the best way to help the environment, i.e. to start home composting to reduce biodegradable waste; (iv) recycle –collect biodegradable waste to produce a new product; and last (v) recover – after collection, the recyclables are remanufactured and/or energy is recovered [6].

Biodegradable wastes can undergo different forms of treatment, such as [7]:

- landfilling;
- incineration and other thermal treatments;
- mechanical biological treatment (MBT) and mechanical biological stabilization (MBS);
- composting;
- anaerobic digestion;
- fermentation.

Fortunately, landfilling of biodegradable waste in many countries has been forbidden due to the high volume of methane production. Methane as a powerful greenhouse gas is 30 times more powerful than carbon dioxide. Efficiency of biodegradable waste incineration as an energy source is strongly dependent upon waste composition (which may exhibit seasonal variations). For example, garden wastes may be best treated through composting, both because of their seasonal nature and due to the fact that much of the material (e.g. grass clippings) may have quite a low calorific value. MBT is used as a valuable tool for pre-treating wastes prior to landfilling. Nowadays, due to the increase in separate collection of wastes, MBT has become less used in biodegradable management.

17.1.3 Goals

According to the circular economy, the management of biodegradable waste should be treated as a large “bioreactor”, which is why two ways are proposed:

- energy recovery – anaerobic digestion with intense biogas production;
- matter recovery – a composting or other anaerobic stabilization process that aims to obtain fertilizers or other materials that should be able to be safely used in the environment.

As mentioned by Saveyn and Eder [7], composting and digestion represent the vast majority of the processes used. The authors defined compost and digestate as follows:

- **Compost** is the solid particulate material that has been sanitized and stabilized by a biological treatment process of which the last step is aerobic composting. Composting is a process of controlled decomposition of biodegradable materials under managed conditions, which are predominantly aerobic and which allow the development of temperatures suitable for thermophilic bacteria as a result of biologically produced heat.
- **Digestate** is the semi-solid or liquid product that has been sanitized and stabilized by a biological treatment process of which the last step is an anaerobic digestion step. It can be presented as whole digestate or separated into a liquor phase and a semi-solid phase. Anaerobic digestion is a process of controlled decomposition of biodegradable materials under managed conditions, predominantly anaerobic and at temperatures sui for mesophilic or thermophilic bacteria.table

eports given by the European Compost Network [8, 9] estimates the potential of Europe to produce compost from bio-waste could achieve 32 million Mg per year (Table 17.1). This was calculated on the base value of 48 million Mg of bio-waste treated in 2018 and as a result estimated that 12 million Mg of compost and 4 million Mg of anaerobic digestate could be produced. Scaling up, if this is taken as a median value of 113–138 million Mg of bio-waste the potential production of compost in Europe then even 32 million Mg per year could be achieved.

According to EPA in 2018, the recycling, composting, combustion with energy recovery, and landfilling of MSW saved over 193 million metric tonnes of carbon dioxide equivalent ($MMTCO_2E$) (EPA 2020). EPA estimates that the total MSW composted was 25 million tonnes. This included approximately 22.3 million tonnes of yard trimmings (more than a fivefold increase since 1990) and 2.6 million tonnes of food waste (4.1% of generation of wasted food) (Figure 17.2). Moreover, in 2018, 17.7 million tonnnes of food (28.1% of the generation of wasted food) was managed through

Table 17.1 The current and potential compost production in Europe [8].

Indicator	Current (million Mg yr^{-1})	Potential (million Mg yr^{-1})
Bio-waste per year	48	122
Compost production per year	12	32
CO_2 equivalents	1.3[a)]	3.5

a) Calculations based on Gilbert et al. [9] – C sequestration rate of 50 kg of soil organic carbon per tonnes of compost (dry matter, or 30 kg on a fresh mass basis).

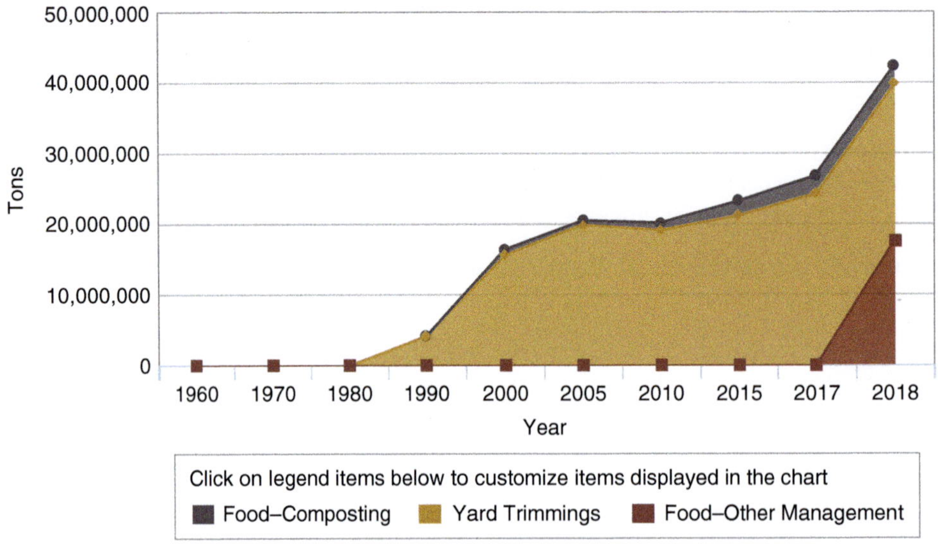

Figure 17.2 Composting and other food waste management in the USA in the years 1960–2018 according to EPA 2020 [10].

animal feed, co-digestion/anaerobic digestion, bio-based materials/biochemical processing, donation, land application, and sewer/wastewater treatment.

17.2 Local Market Demand and the Paradigm "Waste as a Resource"

When countries are drawing up their national strategies, the basic question concerning markets and other outlets needs to be addressed. While it is easy to design and build the infrastructure to provide for separate collection and treatment of separate collected waste/substrate, such as paper waste, garden waste and food waste, there is no guarantee that reliable and stable markets will be available for the products produced. Both national and regional planners should be fully aware of the importance of establishing and maintaining adequate markets and outlets when drawing up plans for the diversion of biodegradable waste away from landfill.

17.2.1 Separate Collection of Biodegradable Waste

At the household level different biowaste subfractions are generated:

- Food waste is defined as any food, and inedible parts of food, removed from the food supply chain. Food waste to be recovered or disposed of include composting, crops plowed in/not harvested, anaerobic digestion, bio-energy production, co-generation, incineration, disposal to sewer, landfill, or discarding to sea. Food waste can be classified according to different properties, e.g. edible and inedible parts (e.g. bones, banana peels), mainly animal-derived food waste (eggs, dairy products, meat, fish, etc.), mainly plant-derived food waste (vegetables, bakery products, rice, pasta, etc.), as well as food waste from meal preparation (e.g. peels from vegetables) and food waste from leftovers (e.g. what people leave on their plates in a restaurant) [11].
- Garden waste includes biodegradable garden waste that is generated in gardens, parks, and other locations, such as cemeteries, road sides, dykes, etc. Garden waste can also be divided into:

(i) green waste, a part that is easily digestible under anaerobic conditions, and (ii) woody waste, which is difficult to digest:

✓ green waste includes grass cuttings, leaves, flowers, herbs, etc.;
✓ woody waste contains the ligno-cellulosic (wood-rich) part of garden waste, which includes twigs, branches, etc., and is rigid, bulky, and difficult to digest by anaerobic microorganisms without a previous special pre-treatment.

In theory, collecting compostable material requires only a little additional work simply to place organic items, such as food waste and scraps, in a collection container separate from other waste. First at all, for safe composting of biodegradable waste in a household area is the choice of container for composting. Several types of containers include: homemade containers, compost bins, and compost tumblers. Organic materials must be placed thoughtfully and carefully in the compost pile. A well laid out compost pile will promote drainage and aeration while deterring clumping and rotting. Many consumers buy a designated indoor compost bin and collect their food scraps and waste in repurposed plastic food containers, trash cans, or compostable bags. To get the compost to an acceptable quality, the pile should include about 50% green waste or manure. These items introduce nitrogen into the pile and serve as a catalyst for the compost process. Table scraps and food waste are just two forms of green manure. Other green waste items include: grass clippings, clover, buckwheat, wheatgrass, coffee grounds, tea leaves, or tea bags. Brown waste should make up the other 50% of the compost pile. Brown waste is a resource of carbon for the compost pile. These items include: shredded newspaper, shredded paper, shredded brown paper bags, dead branches, twigs, leaves, straw, and untreated sawdust. People frequently forget to alternate layers of moist and dry green and brown waste. Moist items include food scraps, food waste, coffee grounds, and tea bags. Dry items include straw, twigs, leaves, and untreated sawdust.

Silvennoinen et al. [12] took into consideration possible sociodemographic, behavioral, and attitudinal factors that might explain food waste in households and found a few statistically significant factors:

- size of household ($p < 0.001$);
- gender of person mainly responsible for grocery shopping (female, male, or both) ($p < 0.05$);
- opinion of potential to reduce food waste ($p < 0.001$);
- appreciation of low food prices (buy one get one free (BOGOF)) and discounted food products bought often) ($p < 0.100$);
- opinion on the ability to reduce food waste by buying smaller packages ($p < 0.001$).

Analyzing the correlation between the waste amounts and the gender responsible for food purchases, the authors found that food waste was considerably higher in those households where a woman did the shopping than in households where a man did the shopping, or where a couple shopped together/took turns shopping [12].

The value of avoidable food waste has been estimated in a number of European countries and ranges between EUR 3.2 and EUR 6.1 per kilogram of waste (Table 17.2).

17.2.2 City/Regional Composting Plan and Management Facilities

Cities/regions in Europe, the United States, and in many other countries (such as Australia and China) have committed to reducing or eliminating their residents' waste. In an effort to achieve this goal, these cities have established curb-side composting programs. Management facilities estimate and inquire about the cost of the service, the possibility to provide indoor and curb-side collection

Table 17.2 Estimated amounts and value of avoidable household food waste (EEA 2020 [8]).

Country and reference year	Avoidable food waste generated (kg/person per year)	Cost of food waste (EUR kg^{-1})	Cost of avoidable food waste per year and capita (EUR/person per year)	References
Netherlands, 2013	47	3.2	150	van Dooren et al. [13]
Netherlands, 2019	34	3.5	120	van Dooren [14]
Germany, 2012	53	4.4	235	Braun [15]
Norway, 2015	42	6.1	256	Stensgård and Hanssen [16]
Finland, 2011	23	4.0	92	Silvennoinen et al. [12]
Belgium, 2016	37	4.3	159	Criel and Fleurbaey [17]

bins, and frequency of the picks-up of compostable material (collections twice a week, once a week, every two weeks, etc.). Local municipalities should provide consumers/households with the equipment and educational resources needed to start composting. The equipment provided may include an indoor collection bin, an outdoor bin, and/or compostable bags. It may also provide a set of instructions, a list of approved biodegradable materials, and a list of unaccepted organic materials.

- The indoor collection bin is typically intended for a kitchen. Many people store this bin under the kitchen sink or on their kitchen counter.
- The outdoor collection bin is emptied by city workers. Some people who live in an apartment building may be required to share this outdoor collection bin with other residents.

17.2.3 Compost/Digestate Feedstocks

The biodegradable waste management should take into consideration the impacts associated with all stages of the system as well as the level of resources consumed (e.g. electricity, fuel, machinery, etc.) within the system and the avoided production of the goods (material and energy) substituted by the outputs generated in the bio-waste management system (bio-based products and energy), such as mineral fertilizers [18]. A comparable overview of the breakdown of compost/digestate according to the feedstocks used is shown in Table 17.3 [19].

The compost/digestate is usually used as an organic fertilizer and/or soil improver. As shown in Figure 17.3, there are several market sectors: agriculture and horticulture, landscape and green areas, and growing media. An average of 50% of compost produced in Europe is used in agriculture and horticulture – in Switzerland this value amounts to almost 90%, but in Flanders (Belgium) it is less than 20%. The increase in use of compost across Europe can also be connected with the decrease in use of peat for growing media production. Peatland Strategies in Europe recognize the cross-cutting effect of sustainable peatland management to reach national commitments to EU regulations and international agreements, such as the Paris Agreement, the Water Framework Directive, and the Convention on Biodiversity. In many countries (Norway, Finland, UK) targets were set for peat-free policies in professional sectors by 2030.

Interestingly, in the case of emerging markets, local solutions could be easier for wide implementation than national ones. The compost sales rapidly grown during spring 2020 were due to a

Table 17.3 Estimated quantity of compost/digestate in the EU28 by feedstock used, in tonnes.

Type	Production quantity (t)	Feedstocks type	Feedstocks quantity (t)
Compost			
Green waste compost	8 000 000 (7 000 000–9 000 000)	Source separated garden and park waste	16 500 000
Vegetable, fruit, and garden waste compost (VFG compost)	100 000 (100 000–150 000)	Only separately collected biodegradable materials according to the specific input list of ECN 20144 (ECN-QAS Part CII, Annex CII[a)]) and the input materials listed in Table 14 of JRC 2014[b)])	300 000
Bio-waste compost	6 000 000 (5 000 000–7 000 000)	Only separately collected biodegradable materials according to the specific input list of ECN 20144 (ECN-QAS Part CII, Annex CII[a)]) and the input materials listed in Table 14 of JRC 2014[b)])	12 500 000
Sewage sludge compost	ca. 800 000[b)]	Sewage sludge; sometimes mixed with green waste.	ca. 1 700 000
Digestate			
Digestate from source separated biowaste from MSW	At least 7 000 000	Commonly biodegradable kitchen waste from individual housing, restaurants, caterers, offices, canteens, hospitals, and schools. In some municipalities, organic material from landscaping may be used such as grass clippings.	8 300 000
Agricultural digestate	ca. 120 000 000	The most common inputs are one or a mixture of the following: livestock manure, energy crops[c)], cover crops or secondary crops for energy use (triticale, miscanthus), straw left over from harvest. Other in-farm organic residues are not expected to be used in significant amounts.	144 000 000
Sewage sludge digestate	At least 1 700 000	Sewage sludge. In a few cases, source separated bio-waste from nearby MSW may be added.	2 000 000
Digestate from agroindustry by-products	At least 1 700 000	A large proportion of sewage water from households and commerce in urban areas is digested. Industrial sludge may be treated alongside urban sludge or separately.	2 000 000

a) Assuming a weight reduction of 50% of the input quantity during composting.
b) Excluded: organic fraction from mixed municipal household waste, separated through mechanical, physicochemical, biological, and/or manual treatment; sewage sludge; sludges from the paper industry and non-biodegradable materials.
c) Plants grown with the specific purpose of being used for energy generation (e.g. to produce biogas or biofuel), mostly maize.

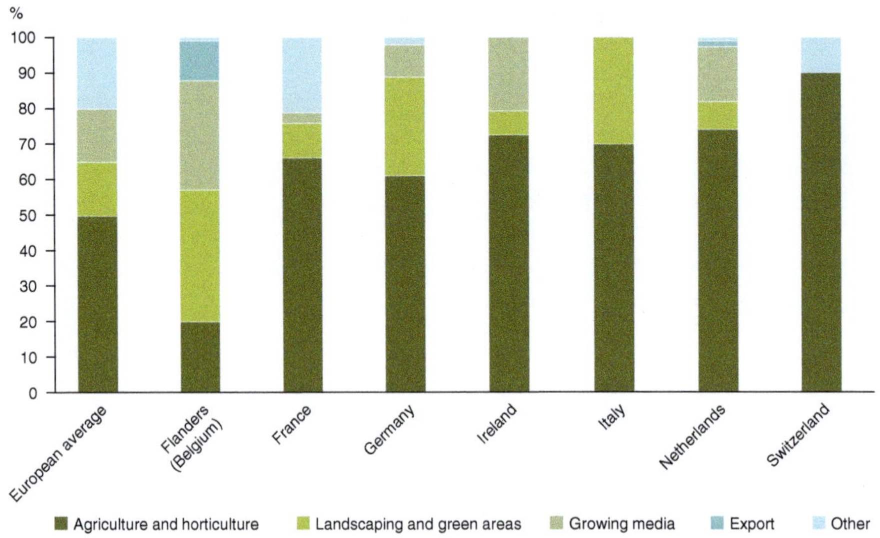

Sources: Data for France, Ireland, Italy, Netherlands fromETC/WMGE (2019a), for the European average from ECN (2019), for Germany from BGK (2018), for Flanders from VLACO (2019), and for Switzerland from Fuchs (2016).

Figure 17.3 Market for compost from bio-waste in selected countries/regions [8].

dramatic increase in the number of people closed in during lockdown in the pandemic who were planting gardens and landscaping.

17.3 Legal Pluralism

17.3.1 European Law

According to the European Commission, the management of bio-waste/biodegradable waste depends on a number of local factors, inter alia collection systems, waste composition and quality, climatic conditions, and the potential of use of various waste-derived products such as electricity, heat, methane-rich gas, or compost. Therefore, the EC suggested that national strategies for the management of this waste should be determined in a transparent manner and be based on a structured and comprehensive approach such as Life Cycle Thinking (LCT). In order to fulfill the waste hierarchy and to assist decision-makers in making the best use of biodegradable waste, the Commission prepared a set of guidelines on how to apply Life Cycle Assessment (LCA) and LCT to planning the management of bio-waste.

A number of EU laws address the problem of treatment of bio-waste. On their website, the Commission gave waste management requirements, such as environmental and human health protection during waste treatment and priority for waste recycling, laid down in the revised WFD, which also contains specific bio-waste related elements (new recycling targets for household waste, which can include bio-waste) and a mechanism allowing quality criteria to be set for compost (end-of-waste criteria). In turn, landfilling of bio-waste is addressed in the Landfill Directive, which requires the diversion of biodegradable municipal waste from landfills. The IPPC Directive (soon to be replaced by the Industrial Emissions Directive) lays down the main principles for permitting and control of bio-waste treatment installations of a capacity exceeding 50 tonnes day^{-1}. The incineration of bio-waste is regulated in the Waste Incineration Directive, while the health rules for composting and biogas plants that treat animal by-products (ABPs) are laid down in the Animal By-products Regulation (ec.europa.eu/environment/waste/compost/index.htm).

According to European Law the main elements of bio-waste management include the following goals:

- recycling and preparing for re-use of municipal waste (including bio-waste) to be increased to 70% by 2030;
- phasing out landfilling by 2025 for recyclable waste (including plastics, paper, metals, glass, and bio-waste) in non-hazardous waste landfills – corresponding to a maximum land-filling rate of 25%;
- measures aimed at reducing food waste generation by 30% by 2025;
- introduction of a separate collection of bio-waste.

The European Waste Framework Directive (2008/98/EC), among other amendments, introduces a new procedure for defining end-of-waste (EoW) criteria, which are criteria that a given waste stream has to fulfill in order to cease to be waste. Waste streams that are candidates for this procedure must have undergone a recovery operation and comply with a set of specific criteria. These criteria are yet to be defined for each specific waste stream, but the general conditions that a waste material has to follow are defined by Article 6 of the WFD in the following terms: "certain specified waste shall cease to be waste (within the meaning of point (1) of Article 3) when it has undergone a recovery, including recycling, operation and complies with specific criteria to be developed in accordance with the following conditions:

(a) the substance or object is commonly used for a specific purpose;
(b) a market or demand exists for such a substance or object;
(c) the substance or object fulfills the technical requirements for the specific purpose referred to in (a) and meets the existing legislation and standards applicable to products;
(d) the use of the substance or object will not lead to overall adverse environmental or human health impacts."

In 2018, the revised WFD [8] introduced several substantial changes relevant for bio-waste:

- an obligation for all EU Member States to collect bio-waste separately or ensure recycling at source from the end of 2023 onwards;
- new targets for the preparation for reuse and recycling of municipal waste which, in combination with the Landfill Directive's landfill reduction targets for municipal waste, are unlikely to be met without proper management of bio-waste;
- an aspirational target to reduce food waste in line with Sustainable Development Goal 12.3 to halve food waste by 2030, and a mandate for the European Commission to propose a binding food waste reduction target by the end of 2023;
- a requirement for EU Member States to measure and report food waste generation annually, starting in 2020, and to adopt specific food waste prevention programmes.

Fertilizing Products Regulation and the Single-Use Plastics Directive contain provisions concerning biodegradable waste management. The EU Fertilizers Regulation, published in July 2019, determines the rules of placing secondary raw materials on the EU fertilizer market in order to protect primary raw materials. It aims to enable recycled organic fertilizers and soil improvers (composts and digestate products) access to the EU internal market so that they can compete on an equal level with mineral fertilizers.

The main goals regulations are:

- boosting organic matter recycling from bio-waste;
- integration of organic fertilizing products into the scope of the NFR;

- introducing harmonized EU rules for products diverting from organic waste materials;
- creating access to CE (Conformité Européenne) marking and free trade for organic fertilizing products across the EU.

Food waste is also included in the European Commission's monitoring framework of indicators for the circular economy.

17.3.2 United States Law

In the United States, the Environmental Protection Agency (EPA) regulates household, industrial, manufacturing, and commercial solid and hazardous wastes under the 1976 Resource Conservation and Recovery Act (RCRA). Effective solid waste management is a cooperative effort involving federal, state, regional, and local entities. Federal guidelines have provided state, tribal, and some local governments with regulatory responsibility for ensuring proper management of wastes generated from each source in their region, but the programs might vary considerably in their guidelines and implementation. As of 2011, the EPA has authorized 48 states, except Alaska and Iowa (Hawaii was added in 2001), to implement the RCRA, meaning that the states' regulations must meet at least the requirements set at the federal level and may be more stringent.

California's SB 1383 regulations require jurisdictions to buy compost, mulch, and biogas from the state's organics recycling facilities. The law will be implemented in January 2022. In turn, Washington state passed a Compost Procurement and Use Law in June 2020. The intent is to ensure that compost created from municipal recycling programs moves onto state, county, and city projects. It also encourages cities with organics collection to enter into compost buy-back agreements with their processors, closing the loop on their organics recycling programs. In the Washington State House Bill 2713 it is written that state agencies and local governments are required to consider whether compost products can be used in government-funded projects.

In the Clean Water Act, land application, surface disposal, and combustion of biosolids sewage sludge as well as biosolids composting are specified. EPA published federal standards for the use or disposal of sewage sludge in the form of the Code of Federal Regulations (CFR) in part 503.

17.3.3 China Law

China' Solid Waste Environmental Pollution Prevention and Control Law ("Solid Waste Law") has been amended four times (this is the Fifth Amendment) since 1995. The Solid Waste Law covers the prevention and control of pollution from industrial waste, household waste, construction waste, agricultural waste, and hazardous waste. The new revision strengthens the solid waste handling principles of reduction, recycling, and harmlessness. The government and its relevant departments are responsible for supervision and management, furthers efforts to achieve "zero" solid waste imports, and integrates solid waste management into current environmental programs.

According to Wei et al. [20], the Chinese government has established national policies and regulations to promote source separation of MSW, including the Classification Symbols for MSW (GB/T19095–2008), and the Classification and Evaluation Standard of MSW (CJJ/T102–2004). Moreover, the Technical Guide for Treatment of Domestic Waste (MOHURD [2010] No. 61), released in 2010, noted that source-separated MSW collection should be implemented in more cities, and that biodegradable organic material should be recycled as a resource [21]. The consultation paper System Plan to Compel Separation of Domestic Waste (NDRC [2016] No. 1467), published in 2016, set quantitative targets for separate collection of MSW [21]. The authors noted that waste separation systems should cover at least 90% of the city area and the waste recycling rate, including organic waste recycling, should reach 35%.

17.4 Access Barriers

If biodegradable waste processing led to obtaining valuable products, such as compost, digestate, fertilizers or soil improvers and requirements to design quality standards and quality assurance processes, this could help to build a good bio-waste management system. The compost produced by households or small communities would be used at the local level. This is of course a sample of how to use the circular economy locally as a de-centralized system. Home composting requires people to have some knowledge of good composting practices in order to avoid negative environmental impacts and to obtain good-quality compost. However, frequently the management of the such products creates more and more problems. The protection of soil (and therefore of agricultural crops) against pollution forced the introduction of very strict requirements regarding the content of harmful substances in all materials, including compost, applied to the soil (especially for fertilization or reclamation). This is especially true for the content of heavy metals and some organic pollutants. The introduced material must of course also be safe in terms of sanitary and epidemiology aspects.

Requirements for these materials in individual countries are different. They also depend on the type of institutions introducing them, as well as on the intended use of the composting product.

According to the regulations, EU fertilizing products bearing the "CE marking" will have to fulfill certain requirements to benefit from free circulation in the EU's internal market. The EU Fertilizing Regulation defines in Annex II a list of component material categories (CMC), of which compost is CMC 3 and non-energy crop digestate is CMC 5. Specific labeling requirements are also set out for product function categories (PFC), which are subdivided into fertilizers, soil improvers, growing media, liming materials, and biostimulants (Table 17.4).

Table 17.4 Maximum level of contaminations according to EU Regulations (2019).

Criteria	PFC 1(A)(I)/(II) Organic fertilizer	PFC 3(A) Organic soil improver
Cd (mg kg^{-1} dm)	1, 5	2
Cr VI/Cr (mg kg^{-1} dm)	2/–	2/–
Hg (mg kg^{-1} dm)	1	1
Ni (mg kg^{-1} dm)	50	50
Pb (mg kg^{-1} dm)	120	120
Cu (mg kg^{-1} dm)	300	300
Zn (mg kg^{-1} dm)	800	800
As (mg kg^{-1} dm)	40	40
$C_2H_5N_3O_2$ (g kg^{-1} dm)	absent	–
Salmonella spp.	absent	absent
E.coli/*Enterococcaceae* CFU (Colony Forming Units) g^{-1}	≤1000	≤1000

In the final report for a project on "Digestate and compost as organic fertilizers – Risk assessment and risk management options" (reference FC/2015/0010 – SR3 under Framework Contract ENV.A.3/FRA/2015/0010), prepared by Wood with partners Peter Fisk Associates and Ramboll for the European Commission, DG Environment [19], based on 54 relevant publications, 17 substance groups were identified, covering more than 94 single substances and four sub-groups, among others:

- heavy metals – represented by cadmium, nickel, lead, copper, zinc, and mercury.
- bisphenol-A;
- phthalates – represented by di(2-ethylhexyl)phthalate (DEHP);
- pharmaceuticals – represented by 17α-Ethinylestradiol (EE2);
- hexabromocyclododecane (HBCDD);
- dioxins, furans, and dioxin-like PCBs – represented by 2,3,7,8-tetrachlorodibenzodioxin (2,3,7,8-TCDD) and 2,3,4,7,8-pentachlorodibenzofuran (2,3,4,7,8-PCDF);
- other PCB – represented by 2,4,4′-Trichlorobiphenyl (PCB-28);
- nonylphenol isomers and ethoxylates – represented by nonylphenol;
- polycyclic aromatic hydrocarbons (PAH) – represented by benzo[a]pyrene, chrysene, benz[a]anthracene, benzofluoranthene, and indeno[1,2,3-cd]pyrene;
- perfluoralkyl substances (PFAs) – represented by perfluoroctanesulfonic acid (PFOS) and undecafluorohexanoic acid (PFHxA);
- physical impurities – represented by microplastics.

For the component materials "compost" and "digestate" only specific input materials are allowed and specific process requirements for composting and anaerobic digestion are described. As suitable input materials source separated bio-waste, ABP category 2 and 3 materials and residues from food processing industries can be used (Table 17.5).

Table 17.5 The general process requirements for compost and digestate [22].

	Fertilizer regulation
Input material	– bio-waste source separated, animal by-products (ABPs) cat 2&3, excluding sewage sludge and mixed municipal waste – liquid or no-liquid microbial or no microbial extract made out of compost – residues from food processing production except ABPs
Process criteria for digestate	Termophilic at 55 °C 24 h^{-1} for hydraulic retention time of 20 h
	Termophilic at 55 °C including pasteurization step 70 °C by one hour
	Termophilic at 55 °C followed by composting
	Mesophilic at 37–40 °C including pasteurization step 70 °C by one hour
	Mesophilic at 37–40 °C followed by composting
Process criteria to compost	70 °C ≥ 3 days
	65 °C ≥ 5 days
	60 °C ≥ 7 days
	55 °C ≥ 14 days

In the case of all animal by-product input material, like catering waste, the EU Fertilizer Regulation accepts only the ABP transformation standard of 70 °C for at least one hour at a maximum particle size of 12 mm, which is not suitable for composting [23].

Additionally, in the Fertilizing Regulation some specific criteria for compost and digestate are required:

- organic pollutants,
- impurities (glass, metals, and plastics), and
- stability.

17.5 Setting Institutional Systems Promoting the Circular Economy

The European Fertilizing Regulation foresees an "optional harmonization" in order to avoid creating additional barriers to the marketing compost or digestate in national markets. This means that compost and digestate manufacturers can decide whether to label their products with the CE mark: they can either freely trade their CE-marked products on the internal market in accordance with common European rules or simply market their non-CE marked products according to national rules.

Such quality management systems can also help to protect against the use of poor-quality compost or digestate. An EU-wide requirement to put in place quality management systems for the entire bio-waste management system could therefore help to optimize the use of bio-waste as a resource based on the principles of a circular economy and a better level of environmental protection. The EEA, in a report published in 2020, reviewed the separate collection of municipal bio-waste, national compost quality standards, and quality management systems in EU countries (Table 17.6). It has been found that there are differences in the degree of advancement of these systems. Eleven countries have already reported that they have implemented a quality management system, the last step following the implementation of separate bio-waste collection at national level and the introduction of compost quality standards into national legislation [8].

In the USA some local governments run their own organics collection programs, but most are run in conjunction with private hauliers and composting facilities [21].

The revision of the China Solid Waste Environmental Pollution Prevention and Control Law ("Solid Waste Law") introduces several product-related programs to further the development of Chinese circular economy policies and regulations.

17.6 Conclusions

Biodegradable waste management includes a variety of measures that can be employed at national, regional, and company levels. Municipalities can increase participation in composting programs by educating residents and businesses about the benefits of composting food scraps. This can be done through an increase in the educational activity and through social marketing such as advertisements, public announcements, and social media to help promote the concept of composting.

Table 17.6 Quality management systems for the production of compost from bio-waste for selected European countries and regions, 2016–2018 [8].

	Separate bio-waste collection	National standards for compost quality	Quality management system
Albania	–	–	–
Austria	+++	+++	+++
Belgium	+++	+++	+++
Bosnia and Herzegovina	–	–	–
Bulgaria	–	–	–
Croatia	+	+	–
Cyprus	–	–	–
Czechia	++	++	++
Denmark	++	+	–
Estonia	+	+++	+++
Finland	+++	+++	+++
France	+	+++	+++
Germany	+++	+++	+++
Greece	–	–	–
Hungary	+	+	–
Iceland	+	–	–
Ireland	+++	+++	+++
Italy	+++	+++	+++
Kosovo[a)]	–	–	–
Latvia	–	–	–
Liechtenstein	–	–	–
Lithuania	+	+	–
Luxembourg	+	+++	–
Malta	+++	–	–
Montenegro	–	–	–
Netherlands	+++	+++	+++
North Macedonia	–	–	–
Norway	+++	+	–
Poland	++	++	–
Portugal	+	++	–
Romania	–	–	–
Serbia	–	–	–
Slovakia	–	–	–

Table 17.6 (Continued)

	Separate bio-waste collection	National standards for compost quality	Quality management system
Slovenia	+++	+++	–
Spain	+	+++	–
Sweden	+++	+++	+++
Switzerland	+	++	–
Turkey	+	+	+
United Kingdom	+++	++	+++

+++, a mature system embedded in national legislation; ++, part way between starting out and reaching maturity; +, just starting out or on a voluntary basis or developing local/regional or national standards; – not started.
a) Under UN Security Council Resolution 1244/99.
Sources: ECN (2018) [24]. Information for Spain from Puyuelo et al. (2019) [25]. Information for Croatia, Czechia, Estonia, Lithuania, Malta, Poland, Portugal, Slovenia, and Turkey provided during Eionet review.

Acknowledgements

The research leading to these results was possible due to a project financed by the Polish National Agency for Academic Exchange the EnviSafeBioC project – Contract No. PPI/APM/2018/1/00029/U/001.

References

1 Bamonti, S., Bonoli, A., and Tondelli, S. (2011). Sustainable waste management criteria for local urban plans. *Procedia Engineering* 21: 221–228.

2 Kaza, S., Lisa, Y., Bhada-Tata, P., and Van Der Woerden, F. (2018). *What a Waste 2.0: A Global Snapshot of Solid Waste Management to 2050*. Overview booklet. Washington, DC: World Bank.

3 Waste to Energy (2016). World Energy Council.

4 ECN (2019). Status Report 2019 – European Biowaste Management.

5 Statistics Poland (2020). Environment 2020, Warsaw.

6 Zoorpas, A.A. (2020). Strategy development in the framework of waste management. *Science of the Total Environment* 716: 137088.

7 Saveyn, H. and Eder P. (2014). End-of-waste criteria for biodegradable waste subjected to biological treatment (compost & digestate): Technical proposals, Final Report, December 2013, IPTS, Sevilla, Spain, European Commission.

8 EEA Report (2020). Bio-waste in Europe – turning challenges into opportunities. 04.2020. ISSN:1977–8449.

9 Gilbert, J., Ricci-Jurgensen, M., and Ramola, A. (2020). *Quantifying the Benefits to Soil of Applying Quality Compost*. Rotterdam: ISWA.

10 EPA United States Environmental Protection Agency National Overview: Facts and Figures on Materials, Wastes and Recycling Available from: https://www.epa.gov/facts-and-figures-about-materials-waste-and-recycling/national-overview-facts-and-figures-materials (accessed 20 December 2020).

11 FUSIONS (2014). Definitional Framework for Food Waste, Full Report. http://www.eu-fusions.org/phocadownload/Publications/FUSIONS%20Definitional%20Framework%20for%20Food%20Waste%202014.pdf.

12 Silvennoinen, K., Katajajuuri, J.-M., Hartikainen, H. et al. (2014). Food waste volume and composition in Finnish households. *British Food Journal* 116 (6): 1058–1068. https://doi.org/10.1108/BFJ-12-2012-0311.

13 van Dooren, C. (2019). Synthesis report on food waste in Dutch households in 2019, Netherlands Nutrition Centre Foundation, The Hague, Netherlands (access 10.12.2020).

14 van Dooren, C. et al. (2019). Measuring food waste in Dutch households: a synthesis of three studies. *Waste Management* 94: 153–164.

15 Braun, S. (2012). Food waste – report on the situation and recent activities in Germany, Working Group Meeting on Food Losses and Food Waste, Brussels. https://ec.europa.eu/food/sites/food/files/safety/docs/adv-grp_wg_20121005_pres_3_susanne_braun.pdf (accessed 7 August 2020).

16 Stensgård, A. and Hanssen, O. (2016). Food waste in Norway 2010–2015 – final report from the ForMat Project, Commissioned Report No. OR.17.16, Kråkerøy, Norway.

17 Criel, P. and Fleurbaey, F. (2019). Voedselverlies en consumenten-gedrag bij Vlaamse huishoudens, Departement Omgeving. http://www.voedselverlies.be/sites/default/files/atoms/files/Voedselverlies%20\ignorespacesen%20consumentengedrag%20bij%20Vlaamse%20\ignorespaceshuishoudens%20Dossier_0.pdf (accessed 7 July 2020).

18 Decisive Project Report (2018). A Decentralized Management Scheme for Innovative Valorization of Urban Biowaste. https://cordis.europa.eu/project/id/689229/reporting.

19 DG Environment (2019). Digestate and compost as organic fertilisers – risk assessment and risk management options (reference FC/2015/0010 – SR3 under Framework Contract ENV.A.3/FRA/2015/0010), prepared by Wood with partners Peter Fisk Associates and Ramboll for the European Commission.

20 Wei, Y., Li, J., Shi, D. et al. (2017). Environmental challenges impeding the composting of biodegradable municipal solid waste: a critical review. *Resources, Conservation and Recycling* 122: 51–65.

21 Bradford, A., Sundby, J., Truelove, A., and Andre, A. (2019). *Composting in America, a Path to Eliminate Waste, Revitalize Soil and Tackle Global Warming*. U.S. PIRG Education Fund.

22 Economic analysis of options for managing biodegradable municipal waste. Final report to the European Commission. https://ec.europa.eu/environment/pdf/waste/compost/econanalysis_appendices.pdf (accessed 20 December) 2020).

23 Regulation of the European Parliament and of the Council laying down rules on the making available on the market of EU fertilising products and amending Regulations (EC) No. 1069/2009 and (EC) No. 1107/2009 and repealing Regulation (EC) No. 2003/2003. https://eur-lex.europa.eu/legal-content/PL/TXT/?uri=CELEX%3A32019R1009 (accessed 20 December 2020).

24 ECN (2018). Bio-waste management in Europe: capturing of bio-waste in cities and municipalities. Presentation given at: The Road to an Urban Bioeconomy: Barriers and Solutions to Closing the Loop of Bio-Resources, Brussels.

25 Puyuelo, B. et al. (2019). Quality assessment of composts officially registered as organic fertilisers in Spain. *Spanish Journal of Agricultural Research* 17 (1): e1101. https://doi.org/10.5424/sjar/2019171-13853.

18

Biofuels – More than Electricity, Heat, and Biomethane

Biodegradable Waste Management in the Circular Economy: Challenges and Opportunities, First Edition.
Edited by Malgorzata Kacprzak, Eleonore Attard, Kari-Anne Lyng, Helena Raclavska, BalRam Singh, Eyob Tesfamariam, and Franck Vandenbulcke.

18.1

The Role of Biofuels in the Circular Economy

Kari-Anne Lyng

NORSUS Norwegian Institute for Sustainability Research, Kråkerøy, Norway

18.1.1 Biofuels and the Circular Economy

Biofuels are produced from cultivated energy crops, and from bio-waste and side streams from other product systems. Biofuels produced from waste resources are categorized as advanced biofuels or second-generation biofuels, which refers to fuels produced from non-food biomass.

According to the European Parliament [1], the circular economy "is a model of production and consumption, which involves sharing, leasing, reusing, repairing, refurbishing, and recycling existing materials and products as long as possible." When bio-waste is used to produce fuels, the valuable components in the bio-waste are not returned to their natural cycles, as described in Chapter 3.2. Fuels are normally consumed during the use phase of their life cycle, and can thus not be recycled in loops, in contrast to some materials. According to the waste hierarchy, re-use and material recycling should be preferred over energy recycling. As a consequence, the production of fuel from bio-waste should only be used for waste resources that cannot be used for higher value purposes.

Efficient and sustainable use of biological resources is, nevertheless, important when it comes to phasing out non-renewable resources in the energy and transport sector. The share of renewables in the global total final energy production is increasing, and is estimated to be around 18% [2]. According to REN21 [2], the progress relating to increasing the share of renewables remains concentrated in the power sector, while far less growth has occurred in the heating, cooling, and transport sectors. In 2016, modern renewable energy constituted around 10% of the global heating and cooling demand. In the transport sector, the share of renewable resources is estimated to be 3.3% globally [2] and 7.6% in the European Union [3]. This indicates a large potential for increase and represents an important motivation for utilizing available bio-waste resources to increase the share of renewables and reduce environmental impacts in the transport sector. In the European Union, the revised Renewable Energy Directive (RED II) includes the target for a minimum share of advanced biofuels of 3.5% in 2030 [4]. According to a survey performed by the International Renewable Energy Agency (IRENA), the stakeholders consider the target realistic, but appropriately ambitious [5].

Biofuels can be divided into three categories: liquid fuels, gaseous fuels, and solid fuels [6]. Liquid fuels are bioethanol and biodiesel, gaseous fuels are biogas and biohydrogen, and solid fuels are firewood, charcoal, and wood pellets. This chapter focuses on liquid and gaseous biofuels for transport, produced from bio-waste resources. Electric vehicles can also be considered as an important part of future mobility, but are not discussed in detail in this chapter. In the following sections, each of the liquid and gaseous biofuels are described with emphasis on how they can be produced from waste resources.

18.1.2 Biodiesel from Waste Resources

Biodiesel can be used in combination with conventional diesel, and thus requires little change in the fuel distribution infrastructure or the transport fleet. It can therefore be rapidly deployed [5]. Biodiesel can be produced from edible oils, non-edible sources, waste cooking oil, and animal fats [7]. The advantage of producing biodiesel using waste cooking oil as a feedstock is that it does not directly conflict with food accessibility. There are two types of waste cooking or frying oil, yellow and brown grease, and the components in cooking oil are dependent on the prepared foods [7]. The oils can contain animal fat, fish fat, or vegetable oil, depending on the foods prepared.

Several studies have documented environmental advantages of using bio-waste resources as feedstock for biodiesel production. An assessment of the environmental impacts of hydrotreated vegetable oils (HVO) in Sweden concluded that feedstocks grown for biofuel production have larger emissions of greenhouse gases (GHGs) than residual feedstocks in a life cycle perspective [8]. Another study found clear environmental benefits associated with biodiesel produced from used cooking oil compared to imported soybeans, when including impacts from land use change [9].

18.1.3 Bioethanol from Waste Resources

Bioethanol is normally recovered from biomass feedstocks such as sugarcane, sugar beet, and starch crops (mainly corn and wheat). Second generation bioethanol is typically produced from lignocellulosic materials, such as residues from agriculture, forestry, and industry and dedicated lignocellulosic crops [10]. Cellulosic ethanol is primarily used as a blend with gasoline [5]. In addition to biofuels, a biorefinery can produce biomaterials and biochemicals [11, 12]. The production of several high-value products may reduce the impact of each product, as the impacts from the production process can be divided between the co-products [12].

As first-generation biofuels from agricultural crops are dependent on feedstock costs and the energy markets, they compete with other uses of biomass. In sourcing the cheapest feedstock, it is thus not unusual to import from developing countries with the risk of causing social, economic, and environmental problems [13]. It has also been pointed out that the production of bioethanol from food crops should be regulated to ensure food security [14]. Second-generation biofuels, on the other hand, often use low-value raw materials which means that these types of problems can be avoided.

The environmental impacts from ethanol produced from bio-waste will depend on the waste type, the production process, and the amounts of co-products. One study assessing bioethanol production from cattle manure found that it had a low impact in most of the environmental categories assessed and found that the impact could be further reduced by improving the energy carriers used in the production [15]. A life cycle assessment study comparing a waste to an ethanol biorefinery system with conventional waste management found that the biorefinery had a good performance and that the net emissions were quite low for all the impact categories examined [16].

Biorefinery concepts based on bio-waste is, however, less developed than applying homogeneous feedstock from industry and agriculture [17]. This may be explained by the non-homogenous and variable nature of many types of waste streams. Information about the characteristics and properties of feedstocks is required to be able to choose the most adequate processing and desired products [11, 18]. Several technologies are thus under development and at pilot-scale, and will in the future require upscaling to be commercialized [18].

18.1.4 Biomethane (Upgraded Biogas) from Waste Resources

The main difference between natural gas and upgraded biogas, which is also called biomethane, is that biogas is produced from a renewable resource while natural gas is produced from a non-renewable resource. Biogas can be harvested from landfills (landfill gas) or produced through an anaerobic digestion process using organic resources as feedstocks. Examples of common biogas substrates are food waste, sewage sludge, and livestock manure.

The raw biogas that exits the digestion tank consists of methane (approximately 60%) and carbon dioxide and can also contain small amounts of hydroxide. To obtain the same quality as natural gas, the methane content of the biogas must be increased to over 97%, which is done through an upgrading process. Biomethane can be produced as compressed gas or liquified. Compressed biogas is commonly applied in city buses and waste collection vehicles. Liquified biogas increase the energy density, which opens up possibilities of utilizing it in heavy duty vehicles and long-distance transport, and in ships and ferries.

Treatment of food waste through anaerobic digestion has been found to be more beneficial in terms of environmental impacts compared with energy recovery or landfill [19, 20] The results are, however, highly dependent on what the biogas and the digestate or energy from incineration is used for and whether they replace fossil alternatives [19–21]. The upgrading of biogas as a fuel for transport requires a certain scale due to high investment costs and can thus be assumed to be less preferable for farm scale biogas plants [22, 23].

When applying a life cycle perspective, studies indicate that biomethane produced from waste or manure resources is one of the most environmentally beneficial fuels currently available on the market [24].

Despite the fact that production of biogas is considered as a matured technology, its full potential has currently not been exploited [25]. This may be due to the cross-sectoral nature of the biogas value chain, which can involve the waste, agriculture, transport, and energy sectors. A large range of sector integration can result in reductions of greenhouse gas emissions, but can also represent considerable costs [26]. An article describing the barriers of the use of biogas as a fuel for transport found conflicting political priorities and shifting strategic objectives, which has resulted in mixed signals concerning the role and viability of biogas for transportation [27].

18.1.5 Biohydrogen

Biohydrogen is hydrogen produced from biological resources, which can potentially represent a reduction in greenhouse gas emissions compared with diesel and fossil hydrogen pathways [28]. Biohydrogen can be produced from a large range of feedstocks. An analysis of the environmental impacts of biohydrogen used as a fuel for transport in Germany produced from woody biomass from forestry or short rotation coppice, herbaceous biomass, energy crops, bio-waste, and organic by-products found that the lowest emissions of greenhouse gases could be achieved by gasification of forest residues or short rotation coppice [29].

Hydrogen and fuel cell vehicles have a long history, but are still not seen as a realistic alternative in a short-term perspective due to high costs. The costs relate to the requirement for a new and expensive infrastructure [30]. To obtain commercialization of fuel cell vehicles, the most important is thus to reduce investment costs.

It has also been pointed out that the most suitable role of hydrogen and fuel cells in mobility are in large capacity vehicles and that the future role of hydrogen and fuel cell applications in the transport sector is largely dependent on political targets and policy frameworks [30].

18.1.6 Environmental Impacts from Biofuels

The transition from a linear and fossil economy to a bio- and circular economy is not without risks and potential negative impacts. Most biofuels are first-generation biofuels, produced from crops such as sugarcane, sugar beet, wheat, corn, rapeseed, and sunflower [31]. These types of biofuels create a demand for cultivated plants, which in consequence can increase the use of arable land. Production of energy crops on land that could have been used for food production can lead to adverse social impacts and is a threat toward producing enough feed for the global population.

The link to the agri- and horticultural sectors also causes concerns regarding direct and indirect land use change (DLUC and ILUC). Direct land use change (DLUC) can be defined as the man-made conversion of a piece of land from a previous land use to a biofuel crop [6]. Indirect land use change (ILUC) can occur when land use change caused by increased biofuel production drives conversion of marginal land elsewhere to meet the demand for producing other types of crops. Land use change is often discussed in relation to greenhouse gas emissions, but arable land does also represent a scarce resource on earth and land use change can also represent a severe threat to biodiversity.

According to Hjuler and Hansen [6] LUC and ILUC will in most cases contribute significantly to the life cycle impacts of first-generation biofuels when taken into account, and in some cases, LUC impacts can be higher than the impacts of the rest of the life cycle impacts combined. As a consequence, focusing only in GHG emissions without considering LUC can be misleading [9]. Biofuels do, nevertheless, generally perform better than fossil fuels when climate change impacts are assessed from a life cycle perspective. They do, however, not necessarily perform better when it comes to other impact categories. For example, many reviews find that biofuels perform poorly when it comes to the potential impact on eutrophication and stratospheric ozone depletion [6].

The use of second-generation biofuels can avoid some of the challenges related to first generation biofuels. Lyng and Brekke [24] evaluated the life cycle environmental impacts of biomethane from food waste and manure with other available fuels on the market. The spider diagram in Figure 18.1.1 shows the estimated potential impacts of various fuels available on the market compared with conventional diesel for five different environmental impact categories. The non-renewable alternatives are represented by red and orange lines, biofuels from waste by blue lines, and first-generation biofuels by green lines. The lines representing electric transport are shown by dashed lines.

The results show that following the methodology commonly used in environmental product declarations (EPDs), the fuels based on bio-waste resources (blue lines) overall perform better than the first-generation biofuels in most of the impact categories assessed. Biodiesel from palm oil and rapeseed oil has a considerably larger impact than the other fuels in the environmental product categories of photochemical oxidant formation, eutrophication, and acidification. Lyng and Brekke [24] point out that there are reasons to believe that data quality is sufficient for emissions

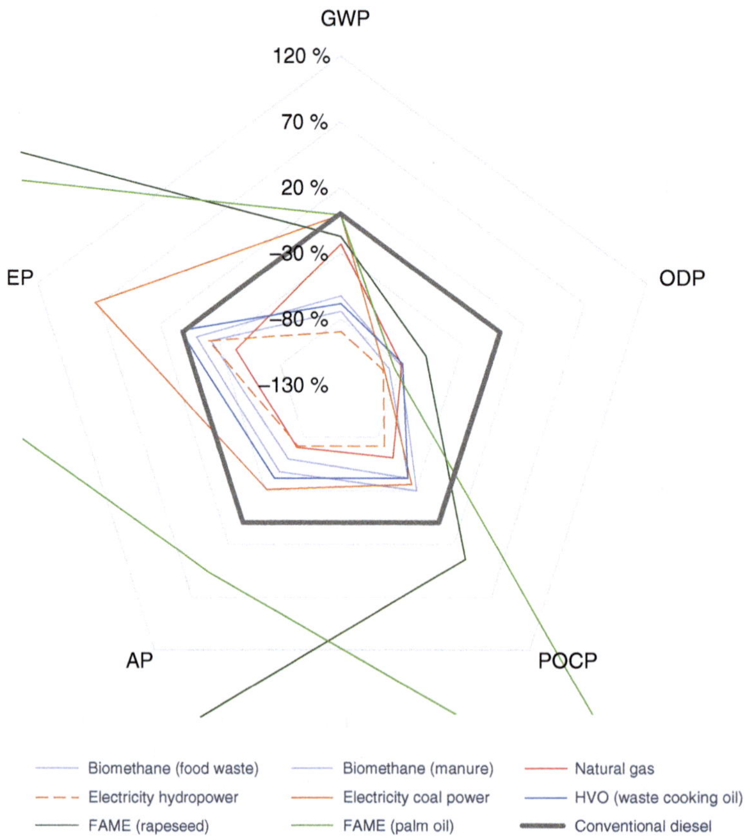

Figure 18.1.1 Estimated environmental impacts for different fuels in a life cycle perspective, compared with conventional diesel. GWP = global warming potential, ODP = ozone depletion potential, POCP = photochemical oxidant formation potential, AP = acidification potential, EP = eutrophication potential (derived from reference [24]).

contributing to climate change, but are poorer for other impact categories. Similarly, the life cycle data from traditional fuels can be assumed to be more robust than for the newer technologies, as there are more data collection points and more operational data available.

18.1.7 Conclusion and the Way Forward

There are political ambitions to increase the use of biofuels in the transport sector to reduce the environmental impact [5]. According to IRENA, three major legislative changes have taken place since 2009 with regards to advanced biofuels: the enactment of the Renewable Energy Directive (RED) I (2009), the Indirect Land-Use Change (ILUC) Directive (2015), and approval of the political agreement on RED II in 2018. Despite this, worldwide investments in advanced biofuel production have been on a declining trend since 2011 [5].

Three major challenges that have to be solved in the future have been described for hydrogen to be regarded as a realistic option: reduction of investment costs of cars, development of infrastructure, and stable policy framework conditions that give predictable support to emerging technologies and can reduce risks related to long-term investments [30]. The same barriers can be assumed to be true also for the other biofuel types. Some barriers are also related to consumer preferences. There is, for example, some skepticism among potential users of biomethane toward gas vehicles and filling stations based on issues of safety, reliability, and availability [27]. Biofuels and transportation industries have a preference for drop-in fuels that require little changes in vehicle or fueling infrastructure, leading to reduced life cycle efficiencies for the overall alternative-fuel production and utilization system [32]. This indicates that sharing experience between users could contribute to reduce the potential resistance to change and to ensure knowledge-based decisions.

Organic wastes should be treated to avoid greenhouse gas emissions and unwanted runoffs to water and soil. Although waste minimization and material recycling as a general rule should be preferred over energy recovery, it is likely that there are waste flows and side streams that are difficult to avoid. No treatment or alternative treatment methods for the bio-waste may cause impacts that can be avoided if using the resource to produce fuels. For example, when taking into consideration that food waste will be sent to energy recovery if not source separated and sent to anaerobic digestion to produce biomethane, the benefits could be considerable when comparing it with alternative fuels on the market [24]. This should be taken into consideration when it is compared with biofuels produced from non-waste biological resources and is also highly relevant when considering goals for increasing the amount of source separated food waste.

Despite the fact that fuels do not contribute to recycling the valuable components in bio-waste, as described in Chapter 3.2, fuel production can be an important contributor to recycling of nutrients. For example, production of biomethane from food waste will also generate digestate, which can be used as a biofertilizer to substitute mineral fertilizer. The income of industrial anaerobic digestion plants is often related to the input (for example the treatment of food waste and sewage sludge) and the sales of gas [23, 26]. The digestate has in many cases a negative value and thus represents a cost for the plant [21, 33]. This can be a barrier for producing biomethane, but it does also imply that biogas as a fuel for transport can be a driver for recycling nutrients in the digestate by providing income.

A life cycle assessment can be a useful tool for indicating which activities are more or less sustainable when developing policies for the transition toward a circular bioeconomy [34]. Despite this, the concept of zero emissions has been central in policy development of mobility. A zero emission vehicle refers to vehicles that have no emissions during combustion, such as electric and hydrogen vehicles. In other words, does the definition only include the use phase and not the production and distribution of the fuel, or the production and end of life of the vehicle. In a life cycle perspective, however, zero emission vehicles do not exist as all human activities will have a small impact on the environment [24].

The different types of biofuels (biodiesel, bioethanol, biomethane, and biohydrogen) can be produced from different types of wastes and have differences in application. This should be explored further to make sure that fossil fuels are substituted by biofuels instead of biofuels competing with and phasing out each other. The shortage of biofuel feedstocks [35] and the risk of creating environmental and social problems related to land use change and occupation of arable land that could have been used for food production are large incentives for increasing the use of available bio-waste resources for fuel production.

References

1 European Parliament (2018). Circular economy: definition, importance and benefits. https://www.europarl.europa.eu/news/en/headlines/economy/20151201STO05603/circular-economy-definition-importance-and-benefits.

2 REN21 (2019). Renewables 2019. Global Status Report.

3 Eurostat (2019). Energy, transport and environment statistics. 2019 edition.

4 European Comission. Renewable Energy – Recast to 2030 (RED II) 2019. Available from: https://ec.europa.eu/jrc/en/jec/renewable-energy-recast-2030-red-ii.

5 IRENA (2019). Advanced biofules. What holds them back?.

6 Hjuler, S.V. and Hansen, S.B. (2018). LCA of biofuels and biomaterials. In: *Life Cycle Assessment: Theory and Practice* (ed. M.Z. Hauschild, R.K. Rosenbaum and S.I. Olsen), 755–782. Cham: Springer International Publishing.

7 Ambat, I., Srivastava, V., and Sillanpää, M. (2018). Recent advancement in biodiesel production methodologies using various feedstock: a review. *Renewable and Sustainable Energy Reviews* 90: 356–369.

8 Soam, S. and Hillman, K. (2019). Factors influencing the environmental sustainability and growth of hydrotreated vegetable oil (HVO) in Sweden. *Bioresource Technology Reports* 7: 100244.

9 Escobar, N., Ribal, J., Clemente, G., and Sanjuán, N. (2014). Consequential LCA of two alternative systems for biodiesel consumption in Spain, considering uncertainty. *Journal of Cleaner Production* 79: 61–73.

10 Cherubini, F. (2010). The biorefinery concept: using biomass instead of oil for producing energy and chemicals. *Energy Conversion and Management* 51 (7): 1412–1421.

11 Carmona-Cabello, M., Garcia, I.L., Leiva-Candia, D., and Dorado, M.P. (2018). Valorization of food waste based on its composition through the concept of biorefinery. *Current Opinion in Green and Sustainable Chemistry* 14: 67–79.

12 Modahl, I.S., Brekke, A., and Valente, C. (2015). Environmental assessment of chemical products from a Norwegian biorefinery. *Journal of Cleaner Production* 94: 247–259.

13 Valente, C., Brekke, A., and Modahl, I.S. (2018). Testing environmental and social indicators for biorefineries: bioethanol and biochemical production. *The International Journal of Life Cycle Assessment* 23 (3): 581–596.

14 Roy, P. and Dutta, A. (2019). Chapter 19 – Life Cycle Assessment (LCA) of bioethanol produced from different food crops: economic and environmental impacts. In: *Bioethanol Production from Food Crops* (ed. R.C. Ray and S. Ramachandran), 385–399. Academic Press.

15 de Azevedo, A., Fornasier, F., da Silva Szarblewski, M. et al. (2017). Life cycle assessment of bioethanol production from cattle manure. *Journal of Cleaner Production* 162: 1021–1030.

16 Papadaskalopoulou, C., Sotiropoulos, A., Novacovic, J. et al. (2019). Comparative life cycle assessment of a waste to ethanol biorefinery system versus conventional waste management methods. *Resources, Conservation and Recycling* 149: 130–139.

17 Vea, E.B., Romeo, D., and Thomsen, M. (2018). Biowaste valorisation in a future circular bioeconomy. *Procedia CIRP* 69: 591–596.

18 Matharu, A.S., de Melo, E.M., and Houghton, J.A. (2016). Opportunity for high value-added chemicals from food supply chain wastes. *Bioresource Technology* 215: 123–130.

19 Lyng, K.-A., Modahl, I.S., Møller, H. et al. (2015). The BioValueChain model: a Norwegian model for calculating environmental impacts of biogas value chains. *The International Journal of Life Cycle Assessment* 20 (4): 490–502.

20 Bernstad, A. and la Cour Jansen, J. (2012). Review of comparative LCAs of food waste management systems–current status and potential improvements. *Waste Management* 32 (12): 2439–2455.

21 Lyng, K.-A. (2018). Reduction of environmental impacts through optimisation of biogas value chains. Drivers, barriers and policy development. PhD thesis 2018:6, Norwegian University of Life Sciences.

22 Lyng, K.-A., Bjerkestrand, M., Stensgård, A.E. et al. (2018). Optimising anaerobic digestion of manure resources at a regional level. *Sustainability* 10 (1): 286.

23 Lyng, K.-A., Skovsgaard, L., Jacobsen, H.K., and Hanssen, O.J. (2020). The implications of economic instruments on biogas value chains: a case study comparison between Norway and Denmark. *Environment, Development and Sustainability* 22: 7125–7152. https://link.springer.com/article/10.1007/s10668-019-00463-9.

24 Lyng, K.-A. and Brekke, A. (2019). Environmental life cycle assessment of biogas as a fuel for transport compared with alternative fuels. *Energies* 12 (3): 532.

25 Mustafi, N.N. and Agarwal, A.K. (2020). Biogas for transport sector: current status, barriers, and path forward for large-scale adaptation. In: *Alternative Fuels and their Utilization Strategies in Internal Combustion Engines* (eds. A.P. Singh, Y.C. Sharma, N.N. Mustafi and A.K. Agarwal), 229–271. Singapore: Springer Singapore.

26 Lyng, K.-A., Stensgård, A.E., Hanssen, O.J., and Modahl, I.S. (2018). Relation between greenhouse gas emissions and economic profit for different configurations of biogas value chains: a case study on different levels of sector integration. *Journal of Cleaner Production* 182: 737–745.

27 Fenton, P. and Kanda, W. (2017). Barriers to the diffusion of renewable energy: studies of biogas for transport in two European cities. *Journal of Environmental Planning and Management* 60 (4): 725–742.

28 Djomo, S.N. and Blumberga, D. (2011). Comparative life cycle assessment of three biohydrogen pathways. *Bioresource Technology* 102 (3): 2684–2694.

29 Wulf, C. and Kaltschmitt, M. (2013). Life cycle assessment of biohydrogen production as a transportation fuel in Germany. *Bioresource Technology* 150: 466–475.

30 Ajanovic, A. and Haas, R. (2020). Prospects and impediments for hydrogen and fuel cell vehicles in the transport sector. *International Journal of Hydrogen Energy* 4 (16): 10049–10058. https://www.sciencedirect.com/science/article/abs/pii/S0360319920310843.

31 Menten, F., Chèze, B., Patouillard, L., and Bouvart, F. (2013). A review of LCA greenhouse gas emissions results for advanced biofuels: the use of meta-regression analysis. *Renewable and Sustainable Energy Reviews* 26: 108–134.

32 Bergthorson, J.M. and Thomson, M.J. (2015). A review of the combustion and emissions properties of advanced transportation biofuels and their impact on existing and future engines. *Renewable and Sustainable Energy Reviews* 42: 1393–1417.

33 Dahlin, J., Herbes, C., and Nelles, M. (2015). Biogas digestate marketing: qualitative insights into the supply side. *Resources, Conservation and Recycling* 104: 152–161.

34 Brekke, A., Lyng, K.-A., Olofsson, J., and Szulecka, J. (2019). Life cycle assessment: a governance tool for transition towards a circular bioeconomy? *Waste to Value*, 1st edn.

35 Oh, Y.-K., Hwang, K.-R., Kim, C. et al. (2018). Recent developments and key barriers to advanced biofuels: a short review. *Bioresource Technology* 257: 320–333.

18.2

Metal Recovery

Agata Rosińska and Anna Grobelak

Faculty of Infrastructure and Environment, Czestochowa University of Technology, Częstochowa, Poland

18.2.1 Introduction

Industrialization and urbanization have led to an increased release of metals, mainly heavy metals, into the environment (soils, lakes, rivers, seas, oceans, underground waters). The occurrence of metals in water, soil, and the atmosphere depends on both natural factors and the economic activity of humans [1]. Metals enter into the environment as a result of geochemical processes caused by volcanic eruptions and as a result of rock weathering. Forms of metals in the environment depend on many factors, such as the properties of a given metal, pH, oxidoreductive potential, and the presence of other metals and ligands. The presence of metals in soils, waters, bottom sediments, and living organisms is not sufficient evidence of contamination. Contamination means the presence of metals in a high concentration relative to their background level. Heavy metals are of particular interest because of their toxicity. Toxicity of heavy metals and their bioaccumulation in the food chain are one of the environmental and health problems [2, 3]. The main anthropogenic sources of metals include combustion processes in power plants, combined heat and power plants, local boiler rooms and industrial plants, road transport, non-combustion production processes, metal ore mining, and waste utilization. Cd, Cr, Cu, Hg, Ni, Pb, and Zn are considered as particularly dangerous elements in the environment [4, 5].

18.2.2 Metal Sources in Biodegradable Waste

Many industrial processes produce aqueous solutions containing various metal ions. Metals in waste water may originate from the metallurgical, mining, chemical, electroplating, tanning, dyeing, chemical fertilizers, plant protection, textile and electrochemical industries, and oil refineries [6]. The above solutions are created both in the metallurgical industry and in other industries as a side effect of the technological process used. In the metallurgical industry, these solutions are intentionally obtained as one of the process stages, while in other industries they are most often an element of the waste stream requiring disposal. Appropriate processing of this waste enables the recovery of valuable metallic raw materials and prevents the introduction of toxic metal ions into the environment [7]. Permissible levels of individual toxic metal ions in waste water are regulated by law [8, 9]. Cadmium is present in waste water from the electroplating industry, the production of dyes, or the production of nickel cadmium batteries. The source of copper comes from the metallurgy, dyeing, textile, crop protection, and fertilizer industries. Chromium occurs in waste water from the electroplating, tanning, textile, wood impregnation, and dye industries. Mercury is found mainly in waste water from the production of phosphoric acid, caustic soda, and pulp mill, as well as from the production of plant protection products and the production of metallic mercury. Mercury in the form of organic compounds is used in fungicides. Lead is present in the waste water from the production of batteries, dyes, and brass, and the nickel source is the galvanizing industry, paper production, refineries, steel mills, and fertilizer factories [10].

In order to implement the policy of sustainable development, one should strive to obtain basic physico-chemical parameters of treated waste water, which are better than the parameters of water in their natural receiver. Fulfilling this criterion is not easy, so multistage purification processes are most often used, combining the advantages of conventional methods: mechanical, chemical, and biological [7, 11, 12]. In industrial plants where waste water contains heavy metals, there should be separate technological systems for their pre-treatment. This is not only for safety reasons, but also for economic reasons. The need to protect the environment and economic considerations caused the removal of heavy metals from industrial waste water and the recovery of precious metals became a very important practical issue. It was found that the purification from metals of smaller amounts of waste water from individual departments is more economically profitable than of mixed waste water in a collective waste water treatment plant [10, 13].

Metal recovery from biodegradable waste is carried out using physical, chemical, or biological methods (Table 18.2.1). Physical methods mainly use membrane filtration processes, ion exchange, etc., in chemical precipitation, electrochemical processes, and in biological methods, biosorption, and bioremediation. These methods have shown different efficiencies in the recovery of various metals. For example, membrane filtration, including microfiltration (MF), ultrafiltration (UF), nanofiltration (NF), and reverse osmosis (RO), use pressures and different pore-sized membranes to physically retain metals while allowing water to pass through. In addition, membranes can be modified by chemical sorbents to selectively remove or retain metal ions [14]. Ritchie et al. [15] have shown that polycysteine functionalized MF membranes are highly effective in the sorption of Hg (II) and Cd (II), which occur in high concentrations. However, membrane technologies are expensive due to high energy consumption and membrane fouling. Ion exchange uses solid resins to exchange metals from waste water with other cations (e.g. H^+). Resins are not selective,

Table 18.2.1 Common methods of metal recovery.

Category Methods	Methods
Physical	Membrane filtration (ultrafiltration (UF), nanofiltration (NF), reverse osmosis (RO), and electrodialysis (ED)); ion exchange and ion flotation; adsorption
Chemical	Precipitation; cementation; electroextraction (EE); electrocoagulation (EC); photocatalysis; membrane electrolysis (ME)
Biological	Biosorption; bioremediation

Source: Wang and Ren [14].

but they can sequentially bind metal ions sequentially based on binding preferences. The ion exchange process is characterized by a high metal removal efficiency and fast kinetics, but is not suitable for waste water with high metal concentration due to resin saturation [16]. The metal adsorption process on activated carbon is expensive. The use of activated carbon for this purpose has been found to pose problems, such as costly recovery of purified water, difficulties in re-use of regenerated activated carbon, and restrictions on the use of activated carbon for non-polar particles. Chemical methods are the most traditional and are widely used for removing metals from aqueous systems, but chemical precipitation and electrochemical methods are not very effective when removing from waste water metals that occur in low concentrations. Chemical methods are easy to use and are characterized by a low capital cost. However, one problem with chemical precipitation is the large amount of toxic sludge generated, which requires additional processing for disposal.

Literature data suggest applying hybrid systems. Zou et al. [17] proposed an innovative and efficient technology for the separation and recovery of precious metals from waste water containing trace amounts of Pd and Ag through a combination of agglomeration and adsorption. The authors conducted experimental research on the impact of working conditions, such as sewage pH value, adsorption time, additional amount of flocculant and adsorbent, and the efficiency of metal recovery. Optimal technological conditions for the recovery process were obtained at pH 8–9, the volume of flocculant to waste water ratio of approximately 1: (2000–4000), the adsorbent to wastewater mass ratio of 1: (30–40), and processing time two to four hours. In determined optimal technological conditions, field tests were carried out, which showed that the total concentration of Pd and Ag (below 11 $mg\,l^{-1}$) in waste water can be reduced to less than 1 $mg\,l^{-1}$. Another solution for recovering metals from biodegradable waste was proposed by Tavakoli and Yoshida [18]. The authors used sub-critical water (sub-CW) technology to recover heavy metal ions, Cd (II), Zn (II), Cu (II), Fe (II), Mn (II), and Ni (II), in the waste of the Japanese scallop *Patinopecten yessoensis*. When used, it provides environmental and financial benefits for the recovery of harmful heavy metal ions. The research proposed a new method using sub-CW treatment to recover the metal ions from scallop waste and the simultaneous production of harmless and valuable materials. The reactions were carried out in the temperature range of 473–653 K and for reaction times of 1–60 minutes. Four phases were obtained after the sub-CW reaction: oil phase, metal-soap phase, water phase, and solid residue (unreacted waste). Part of the oil was hydrolyzed in the sub-CW process and converted into free fatty acids and glycerin. Free fatty acids reacted with metal ions to form a metal-soap phase. Both the metal-soap phase and the oil phase captured almost all metal ions from the primary waste. Metallic soap and oil in this process are chelating agents used to catch almost all metal ions by forming complexes. Moreover, oil and metal-soap phases are important energy sources for the production of heat and electricity. When the phases are burned, energy is released, which can be used to regenerate the ions. The two phases could also be processed to produce biodiesel as a renewable fuel after removal and recovery of metals using a chelate ion exchanger. This process was effective at low and medium reaction temperatures (473–573 K). As the temperature increased, these two phases decomposed and the metal ions concentrated in unreacted waste. Literature data indicate that subcritical water technology can play a significant role, not only in saving energy but also in transforming waste into valuable resources and/or energy and production of invaluable materials in an environmentally friendly manner [19, 20].

Much attention is paid worldwide to methods of biological removal of inorganic impurities. Biotechnological approaches cover all cutting-edge areas of biohydrometallurgy, which include bioleaching, bioprecipitation, bioflotation, bioflocculation, biooxidation, biosorption, bioreduction, bioaccumulation, and the application of biosensors in analytics [21–23]. Processes of great use are biosorption and bioaccumulation, which rely on binding metals from aqueous solutions by dead or living biological material respectively. Biosorption is a process of fast, reversible binding of metal ions from aqueous solutions by biomass. In addition, the advantage of the biosorption process, in contrast to conventional methods, is the lack of economic restrictions resulting from low concentrations of heavy metals in solution [10]. The mechanism of the process is passive and is not controlled by cell metabolism [24]. Binding of metal ions can occur in accordance with the mechanism of physical or chemical adsorption (ion exchange, chemisorption, complexing, chelation, and microcrystallization). It has been shown that metals can be effectively removed by "green" biosorbents, such as fungal biomass [25–27], marine algae [28], bacteria [3, 29], agricultural waste and residues, yeast [27, 30], as well as biosorbents, which include chitosan produced from the shells of crustaceans. Some biosorbents can bind many metals, but parts of them are only compatible with particular heavy metals. Biomass used in the biosorption process can come from industrial waste made available free of charge, and can also be specially bred and propagated. Many agricultural wastes, such as fertilizers, bark, and compost, contain large amounts of lignocellulosic substances that have been used to remove heavy metals from waste water [10]. The use of yeast seems to be the preferred method of removing low concentrations of heavy metals from waste water, using inexpensive carriers and simple fermentation techniques. Moreover, fermentation processes are common in the main industrial plants and waste from them is a source of inexpensive biosorbent. Kordialik-Bogacka [2] found that the simultaneous greater hydrophobicity and lower negative charge of the yeast cell wall cause less efficiency in removing copper, lead, and cadmium from waste water. This shows that knowledge of these parameters can contribute to the selection of the best biosorbent for individual heavy metals.

The process of removing metals from waste water with the participation of microorganisms is very fast. Studies have shown that in 10–15 minutes sorption of 85–90% of the total metal concentration occurs, of which a significant amount is bound in a few seconds. Biosorption efficiency depends on the cell wall structure of microorganisms. It defines the nature of the interaction between organisms and heavy metals. Cell walls of different species of microorganisms, or even their strains, differ in their overall composition, which affects different adsorption capacities. The charge of cell walls depends on the presence of phosphates and carboxyl groups. The cell wall hydrophobicity, which depends on the presence of polysaccharides, proteins, and lipids, also affects the ability of biosorption [21, 31]. Metals can be removed from wastewater as a result of:

- interactions between metal and reactive groups of polymers and macromolecules that make up the microbial cell covers,
- chemical reactions with secreted metabolites, leading to precipitation of insoluble deposits (most often sulfides, carbonates, phosphates, and oxalates),
- transport across cell membranes and intracellular metal accumulation,
- production of insoluble or volatile compounds as a result of metal redox processes or formation of organometallic bonds [32].

18.2.2.1 The Use of Algae in Heavy Metal Sorption

An alternative to the biosorption process using dead organic matter can be natural biomass of microorganisms, especially algae. The concept of using microorganisms, especially algae, as heavy metal biosorbents deserves special attention due to their biological origin and significantly lower costs of their acquisition compared to conventional methods [21]. Algae have been found to be an inexpensive renewable resource found in large quantities in seas and oceans. Algae quickly increase biomass and are characterized by a relatively constant composition and high biosorptive capacity with respect to heavy metals. The use of these organisms in industrial waste water treatment technologies is the answer to the need to search for new, environmentally beneficial, and economical solutions. Due to the constant tendency to tighten standards regulating the permissible values of heavy metal concentrations in waste water, conventional methods of treatment prove to be insufficient and too expensive. This applies especially to waste water in which the metal concentration does not exceed $100\,mg\,l^{-1}$.

Different physicochemical mechanisms involved in biosorption include microprecipitation, chelation, complexation, ion exchange, and physical adsorption (Figure 18.2.1) [21]. The mechanism and efficiency of the heavy metal biosorption process with the use of algae is influenced by the structure of algae and the associated taxonomic affiliation to a given cluster or even species, depending on growth conditions [23, 28, 33]. Photosynthetic dyes of algae have different colors, depending on their systematic affiliation (clusters): red (redweed – *Rhodophyta*), blue (cyanobacteria – *Cyanophyceae*), green (green algae – *Chlorophyta*), and brown (brown algae – *Phaeophyta*).

Due to the content of polysaccharides, the most effective biosorbents are brown algae, especially the *Sargassum* group [21]. Acid polysaccharides, occurring in the cell walls of brown algae, are

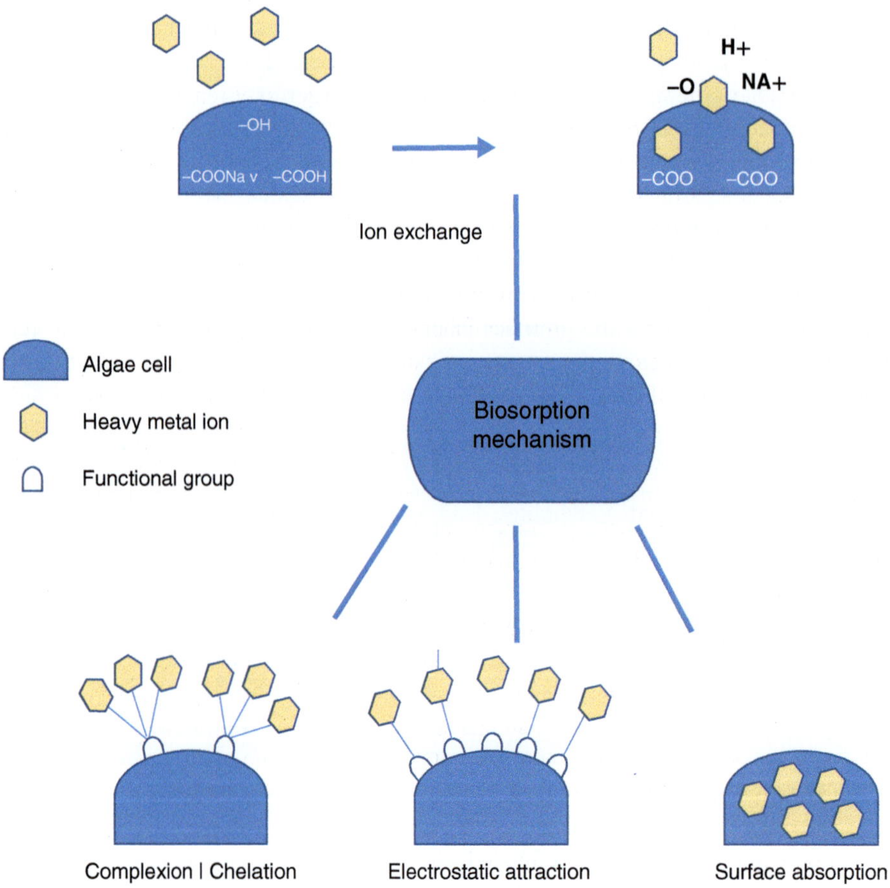

Figure 18.2.1 Physicochemical mechanisms in biosorption. Source: Cheng et al. [21].

indicated as a factor determining the biosorption capacity toward metals. Due to the high content of polysaccharides in the cell walls, brown algae are considered the most efficient biosorbents [3]. Aravindhan et al. [34] demonstrated that brown seaweed *Sargassum wightii* can remove as much as 83% of chromium from aqueous solutions in the pH range of 3.5–3.8. As a result of the biosorption process, the concentration of chromium ions in the solution was reduced to a level that met the standards for tanning wastewater. The confirmation of the high efficiency of brown algae in metal removal is the research by Hashima and Chu [35]. These authors in the process of Cd (II) biosorption obtained the best results for the species *Sargassum baccularia* (q_{max} = 83.00 m g^{-1}) classified as brown algae and then *Chaetomorpha linum* (q_{max} = 53.8 mg g^{-1}) belonging to the green cluster, while the worst was in relation to the *Gracilariasalicornia* species (q_{max} = 17.9 mg g^{-1}) belonging to the redweed cluster. Alginic acids isolated from brown macroalgae are linear carboxylated copolymers, consisting mainly of β-D-mannuronic acids (M-block) and α-L-guluronic acids (G-block), joined in different proportions with a 1,4-glycosidic bond. Alginates, i.e. salts of alginic acid, are seen as relatively inexpensive biopolymers from which sorbent granules can be obtained by appropriate treatment, which can be used in process and apparatus solutions identical to those of ion exchangers. Due to the possibility of regeneration as a result of desorption, the sorbent can be used repeatedly to remove heavy metals. According to Zan et al. [36], copper and zinc can be removed from waste water using dried sea algae *C. linum*, copper, cadmium, lead, and zinc by *Caulerpa lentillifera*, chromium by *Ulva lactuca,* and lead by *Cladophora fascicularis.* Nastaj and Roik [37] conducted a literature review in the field of theoretical and experimental research on the process of biosorption of Cu (II) ions from aqueous solutions using various biosorbents. The authors evaluated the efficiency of removing copper ions from contaminated aqueous solutions. For this purpose, the maximum biosorptive capacities of bacteria, yeast, and fungi were compared with the capacity of biopolymers such as alginates and chitosan. It was demonstrated that these biopolymers are competitive materials in the sorption of copper ions compared to microorganisms as well as conventional adsorbents (activated carbon, polymer adsorbents, and zeolites) due to non-toxicity, easy availability, low price, and the possibility of easy regeneration. The sorption capacity of microalgae cells is influenced by culture parameters, i.e. light intensity and initial glucose concentration. It has been demonstrated that *Spirulina* sp. microalgae biomass grown using auto- or heterotrophic metabolism is characterized by a greater cation binding capacity than biomass grown using mixed (mixotrophic) metabolism. Increasing the light intensity promotes the synthesis of cells with higher biosorption capacity, and under conditions of increased concentration of organic substrate (glucose) cells with lower biosorption capacity are formed. In the literature, there can be found numerous comparisons of the biosorption capacities of various species of micro- and macroalgae (Table 18.2.2).

Research in the field of waste water treatment technology proves the competitiveness of reactors with biomass immobilized on mobile carriers compared to conventional solutions using activated sludge or a biological bed. Ozer et al. [46] showed that green microalgae *Pediastrum boryanum* in a system immobilized with a gelatin-alginate composite has a greater ability to remove chromium (VI) (29.6 mg g^{-1}) compared to free biomass (17.3 mg g^{-1}). Wang and Lu [47] have developed a novel method that uses the aqueous sodium alginate solution for direct gelation with metal ions, which is developed for effective removal and recovery of heavy metals from industrial waste water. Experimental studies were carried out for a model solution containing selected heavy metals: Pb^{2+}, Cu^{2+}, and Cd^{2+}. The results showed that gels can be formed rapidly between the metals and alginate in less than 10 minutes and the gelation rates fit well with the pseudo second-order kinetic model. According to the latest research conducted by Shen and Chirwa [48], application of a fungi–algae pelletized reactor works very well in the bioremediation of waste water contaminated with precious

Table 18.2.2 Adsorption capacity of algal biosorbents for the removal of metal ions.

Algal biosorbent	Metals	Adsorption capacity ($mg\,g^{-1}$)	References
Caulerpa fastigiata	Pb (II)	16.11	[38]
Chara aculeolata	Pb, Cd, Zn	105.3, 23.0, 15.2	[39]
Cystoseira indicia	Cu, Co (II)	103.093, 59.524	[40]
Enteromorpha sp.	Cr (VI)	5.3475	[41]
Eucheuma denticulatum	Pb, Cu, Fe, Zn	81.97, 66.23, 51.02, 43.48	[42]
Laminaria digitata	Cr (VI)	125.99	[43]
Nitella opaca	Pb, Cd, Zn	104.2, 20.5, 13.4	[39]
Scenedesmus quadricauda	Cd	135.1	[44]
Ulva lactuca	Cu^{2+}, Cd^{2+}, Cr^{3+}, Pb^{2+}	64.5, 62.5, 60.9, 68.9	[45]

Source: Cheng et al. [21].

metals as well as their recovery and regeneration in situ of biosorbents. The authors assessed the potential of live and lyophilized fungus–algae deposits as biosorbents for gold recovery and their regenerative capacities. It was demonstrated that the porous characteristics of fungi–algae pellets played an important role in gold adsorption. Lyophilized co-pellets were more effective than live co-pellets due to the greater number of cell wall polysaccharides involved in interaction with Ag. As a result of the process, 97.77% of gold was selectively adsorbed by freeze-dried co-pellets from multimetal waste water in a column reactor.

In assessing the effectiveness of algae in metal removal, studies on the kinetics of the process of biosorption of metal ions from aqueous solutions are necessary. They are essential to determine the time at which the state of chemical equilibrium is established, as well as to determine the optimal conditions for the process. Therefore, it is important to assess the basic parameters of the process, i.e. the impact of pH, initial concentrations of metal ions, and biosorbent, the time of biomass contact with the solution, and the temperature. Deng et al. [49] evaluated the efficiency of chromium (VI) ion biosorption using dried and ground biomass of green algae *Cladophora albida* in a cyclic system. The authors examined the effects of: pH (0.5–12), initial concentration of Cr (VI) ions (20–154 $mg\,l^{-1}$, biomass concentration (0.2–10 $g\,l^{-1}$), temperature (25 °C, 35 °C, 45 °C), and the presence of nitrate, chloride, and sulfate anions (1–10 $mmol\,l^{-1}$) on the kinetics of the biosorption process. The highest Cr (VI) ion removal efficiency was obtained at pH = 0.5, biomass concentration = 2 $g\,l^{-1}$, and temperature = 25 °C. A correlation was found between an increase in the efficiency of ion removal from the solution and increasing the initial concentration of chromium (VI) at a constant process temperature. It was demonstrated that the efficiency of Cr(VI) ion removal from the solution increases with increasing temperature from 25 to 45 °C, with the same initial concentration, constant pH = 2 and biomass concentration = 2 g l. The effect of nitrate, chloride, and sulfate anion presence in the solution was minimal. The binding rate of Cr (VI) ions was the highest in the first 60 minutes of the process, afterwards gradual reduction was observed. A confirmation of the influence of pH on the heavy metal biosorption process is the research by Esposito et al. [50]. According to the authors, the pH determines the amount of functional groups available on the surface of the biosorbent, as well as the processes occurring in the solution (hydrolysis, redox reactions, precipitation, complexation by organic/inorganic ligands). The optimal for the removal of Cr (VI) ions is pH = 2, when carboxylic and amine functional groups are protonated. Under these conditions, the carboxyl groups do not repel anions from the biomass surface and amines are the chromate binding site. When the optimal for the removal of Cr (III) ions is pH = 4.5, the concentration of hydrogen ions decreases, being competitive ions with an affinity for carboxyl groups.

The use of algae biomass as a biosorbent, however, requires further experimental research aimed at obtaining full process characteristics and selecting the best conditions for conducting the biosorption process and biomass regeneration. Biological material binding metal ions can also be a waste product from many industries: pharmaceutical (biomass after the production of antibiotics) and biotechnology of food (fermentation processes).

A different example of the use of biosorption for the removal and recovery of metal ions from waste water is the use of agricultural by-products, such as rice husks, yam peels, cassava peels, orange peels, banana peels, plantain peels, corncobs, sawdust, potato skins. The biosorption techniques consists of a group of applications that involve the detoxication of hazardous chemicals, namely lead, cadmium, zinc, among others, instead of transferring them from one medium to another using these agricultural by-products [51, 52].

18.2.2.2 Bioleaching

Biological methods are used to recover metals not only from liquid waste such as sewage but also from solid waste. Solid waste generation is a global problem that requires environmentally friendly treatment. Some solid waste contains valuable metals. The biotechnological approach is optimal in terms of the ecosphere in an economically feasible manner [53]. In recent years, biohydrometallurgical processes have been used to leach metals from off-balance ores, waste, and waste water. Conducted research is focused on the use of microorganisms in metal removal/recovery from industrial waste water, replacing traditional methods of ore processing with biological ones, which is safer from the point of view of environmental protection and microbiological leaching of metals from industrial waste and sewage sludge. This biotechnological approach can not only reduce the toxicity of various waste streams but can also discover the potential of waste as secondary materials for the recovery of precious metals [22, 54–57]. Literature data confirm the possibility of their recovery using microorganisms and other technologies supporting this process. For the recovery of metals from solid waste by biological methods, the knowledge about interactions between microorganisms and the different types of waste is important. The bioleaching process was proposed for the recovery of metals from solid waste. The major microorganisms used in the recovery of heavy metals are acidophiles, which thrive at acidic pH values ranging from 2.0–4.0. Some of the well-known acidophilic bacteria such as *Acidithiobacillus ferrooxidans*, *Acidithiobacillus thiooxidans*, *Leptospirillum ferrooxidans*, and *Sulfolobus* spp. have been well studied for bioleaching activity, as also have the fungal species such as *Penicillium* spp. and *Aspergillus niger*. Good results were obtained using mesophilic and moderately thermophilic acidophiles, which can play a useful role. Ilyas et al. [58] used *Sulfobacillus thermosulfidooxidans*, a moderate thermophile, to bioleach electronic scrap, and reported recovery of more than 80% of Ni, Cu, Zn, and Al. Pilot studies have shown that with proper adaptation and pre-treatment, in the bioleaching process more than 90% of the metal can be recovered. Involved microbial consortia including acidophilic bacteria and fungi (*A. niger*, *Penicillium simpleicissimum*) are referred to as computer-munching microbes [59]. These consortia dissolve heavy metals such as Cu, Ni, and Zn very well in a one- or two-stage bioleaching process [60]. Xiang et al. [61] reported that the application of mixed bacteria isolated from an acid mine drainage site resulted in the leaching of 95% of the Cu in a two-step process using ferrous (II) iron as the major substrate for bacteria. In most cases, the metallic gold/silver recoveries achievable from tailings by conventional cyanidation methods are low. Applying a pre-treatment by biooxidation/bioleaching of gold- and silver-bearing sulfide tailings before cyanidation has significantly increased the efficiency of recovering these metals from waste [62]. Thanks to the biooxidation process, locked-up precious metals can be released from the sulfide matrix. It has been shown that ruthenium found in industrial waste water is recovered in the biosorption process using purple sulfur-free bacteria (*Rhodopseudomonas palustris* strains) [63]. Some microorganisms can reduce noble metal ions to their metallic forms. For example, sulfate-reducing bacteria can reduce Pt (II) to Pt (0) [64, 65]. *Magnetotactic bacterium*, *Stenotrophomonas* sp. can remove Au (III) from contaminated waste by reducing Au (III) to Au (0), and then nanocrystals of Au (0) particles settle on the cell surface [66]. Yet another application of microbial biotechnology is the bioprocessing of industrial and other wastes to first recover precious metals from the waste and then to add value to a "clean" or at least a less toxic waste product [58, 67, 68].

To date, biotechnological leaching is practiced in few industrialized countries, but due to the huge potential of this method, it is believed that this technology will be implemented in other countries. For sustainable environmental protection, both developed and developing countries should introduce new technologies for recovering metals from waste, including developing the biotechnology leaching process.

18.2.2.3 Bioelectrochemical System (BES)

The recent development of bioelectrochemical technology has influenced a new approach to effective metal recovery because it offers a flexible platform for processes focused on oxidation and reduction reactions [3, 14, 69–71]. The bioelectrochemical system (BES) is a platform technology that employs microorganisms to covert the chemical energy stored in biodegradable materials to electric current and chemicals and can be a new option for removing metals from wastewater due to effectiveness and energy efficiency. In this context, BES, often used in the literature to represent both microbial fuel cells (MFCs) and microbial electrolysis cells (MECs), have emerged as a method of choice because they not only combine the treatment of organic waste waters with that of metal laden waste waters but also offer the opportunity to recover metals. Most BES reactors consist of an anode, a cathode, and optionally a separator. Biodegradable waste, such as wastewater, is oxidized in the anode chamber by microbial bacteria to generate an electron (current) flow to the cathode. The cathode chamber may contain electrons used for direct electricity production (microbial fuel cells, MFCs) or are used to reduce water or oxidized chemicals, such as metal ions, CO_2, or organic chemicals (microbial electrolysis cells, MECs, or microbial electrosynthesis, MES) [70–75]. The application of BES for waste water treatment has shown that this method has many advantages, mainly the savings of aeration energy and sludge disposal. Metal recovery by means of BES is associated with the reduction of metal on the cathode, but there are four mechanisms that need to be distinguished (Figure 18.2.2):

- direct metal recovery using abiotic cathodes;
- metal recovery using abiotic cathodes supplemented by external power sources;
- metal conversion using biocathodes;
- metal conversion using biocathodes supplemented by external power sources.

The first mechanism involves direct reduction of metals on an abiotic cathode for those with redox potential higher than the anode potential (Figure 18.2.2a). These metals include Au (III), V (V), Cr (VI), Ag (I), Cu (II), Fe (III), Hg (II), etc. The reduction is thermodynamically favorable and the metals can be used directly as electron acceptors without external energy consumption [72, 73, 76–83]. The second mechanism requires an external power supply, to force the electrons

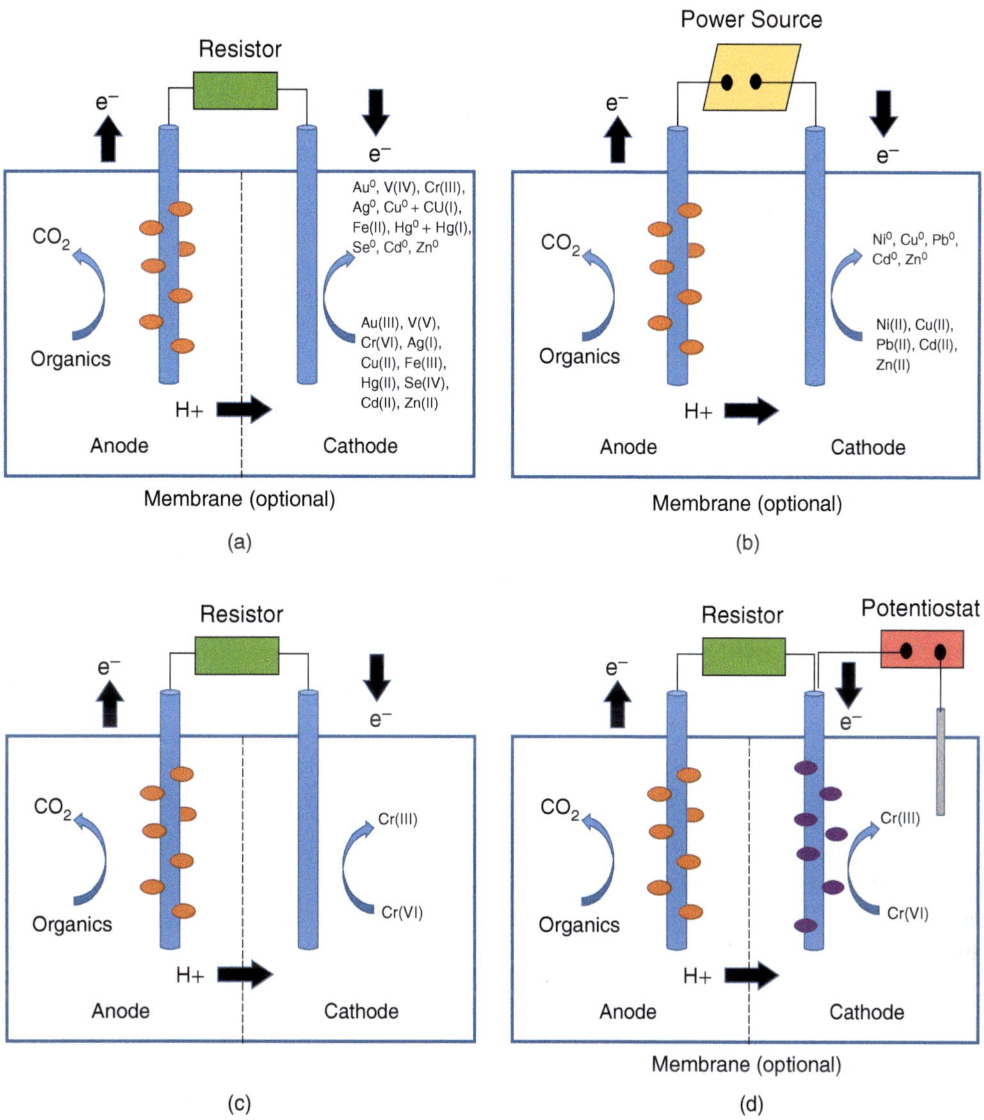

Figure 18.2.2 Four reported mechanisms of bioelectrochemically assisted metal recovery in bioelectrochemical systems. (a) Direct metal recovery using abiotic cathodes; (b) metal recovery using abiotic cathodes supplemented by external power sources; (c) metal conversion using biocathodes; and (d) metal conversion using biocathodes supplemented by external power sources. The membrane is optional. Source: Wang and Ren [71].

to travel from the anode to the abiotic cathode, so metals with lower redox potentials than the anode potentials can be reduced. This process is particularly effective for metals such as Ni (II), Pb (II), Cd (II), and Zn (II) (Figure 18.2.2b) [84, 85]. The third mechanism consists of microbiological reduction of metal oxides on the cathode (Figure 18.2.2c). The fourth mechanism consists of microbially assisted metal reduction with a poised potential (Figure 18.2.2d) and is a combination of processes 2 and 3. In the case of mechanisms 3 and 4, so far test studies have been conducted only for Cr (VI), where a significant reduction was obtained on the biocathode (Huang et al. [86]), but it is believed that similar results can be obtained for other transition metals. Research has confirmed that the BES system is effective in removal and recovery mainly of Cr (VI), Cu (II), Hg (II), Ag (I), Se (IV), and Cd (II) from aqueous solutions [87]. In the BES, microbial communities need to use either a solid electron acceptor (bioanode) or an electron donor (biocathode) as part of their metabolism. However, it is likely that microorganisms that receive electrons from a solid electron donor (cathode) may not gain energy [87, 88]. Bacteria are facing a major challenge of moving or receiving electrons, respectively, and a solid terminal electron acceptor (TEA) or a solid electron donor. It is known that they use various mechanisms of extracellular electron transport (EET) to achieve this.

The BES method is new and most of the research on its use for metal recovery has been carried out so far on a laboratory scale, so further experiments in real conditions for real waste water streams containing various metals are necessary. This will allow the selection of optimal parameters of the electrical potential, pH, retention time, and others for the selective and effective recovery of metals from waste water. Research should also focus on developing new reactor configurations, materials, and system integration processes to improve the efficiency and recovery rate, especially for waste water with a low metal concentration. The use of biocathodes in BES has shown good results in the removal and recovery of metals, but solutions with a high concentration of metals inhibit microbial activity and reduce the effectiveness of the system. Therefore, it is necessary to test more effective and tolerant biofilms. These activities are very important for the development of a specific scenario in the field of metal removal and recovery from waste water of varying quantitative and qualitative compositions.

18.2.3 Metal Sorption on Nanoparticles

A field of science and industry that has been developing rapidly in recent years is nanotechnology. Nanomaterials attract considerable attention as adsorbents in waste-water treatment from heavy metals and rare earths metals, among others [89, 90]. Physical, chemical, and biological properties of particles smaller than 100 nm (nanoparticles, NPs) are different from the properties of the same particles occurring at the macro scale [91]. In the process of metal sorption on nanoparticles, their large specific surface area, lesser flocculent production, and availability of large numbers of active groups for binding of heavy metal ions are important [92]. Another important feature is the ability to re-use and re-cycle nanostructured adsorbents, which makes them cost effective. However, their inconstancy, inconvenience in removal from treated water, and their regeneration limited their commercial use [93]. Therefore, the research focused on seeking solutions to problems through functionalization of NPs because proper surface functionalization improves adsorption capacity and provides easier separation. Scientists are trying to solve this problem by coating NPs with inorganic materials [94], polymers [95, 96], carbon [97], biomolecules [98], etc. Research into the sorption potential of nanoparticles and the efficiency of removing heavy metals from waste water also apply to the use of various nanoadsorbent categories. Research is conducted on the use of various nanostructured materials such as metal, metal oxide, magnetic, chitosan, polymer based, carbonaceous, and zeolite, used as adsorbents for elimination of heavy metal ions. Another area of research concerns the impact of various parameters, i.e. pH, dose of adsorbent, contact time, temperature, and initial ion concentration on heavy metal removal. In the latest literature, there are examples of the metallic NPs, such as Fe^0, i.e. nano zero valent iron (NZVI), used to remove heavy metal ions due to their high stability, greater surface area, non-toxicity, reducing nature, and higher adsorption capacity. It was demonstrated that iron-based NPs synthesized from $FeCl_3$ using

Syzygium jambos (SJA) are highly effective in the elimination of Cr (VI). After using biogenic NP SJAFe, the maximum efficiency of Cr ion removal was achieved at the level of 983.2 mg g^{-1} [99]. Similar results for synthetic NZVI were demonstrated by Kanet et al. [100]. The authors achieved a high 99% As (V) removal rate. NZVI efficiency can be improved by its entrapment or functionalization with some stabilizing agent, e.g. entrapping a non-toxic and biodegradable stabilizer such as chitosan carboxymethyl β-cyclodextrin complex. The synthesized material was used for the complete removal of Cr (VI) and Cu (II), which involves physisorption followed by reduction of Cr (VI) to Cr (III) and also Cu (II) to Cu (0) whilst oxidizing Fe^0 to Fe (III). Then, synthesized bimetallic Fe/Ni NPs embedded in kaolinite (K-Fe/Ni) were used for the simultaneous elimination of Cu (II) and NO_3^- ions, with removal efficiency equal to 99.7 and 40.4% respectively [101]. Another group of nanoadsorbents are magnetic and non-magnetic metal nanoxides. Among non-magnetic NPs, Zn, Ce, Al, Fe, Cu, Mn, etc. oxides were used to remove heavy metal ions. For example [92, 102, 103], synthesized CuO nanoparticles using magnetron sputtering were used to remove harmful Cr (VI) and Pb (II) ions from waste water [104–106]. Another cost-effective NP, i.e. volcanic rock with an α-Fe_2O_3 coating, was used for effective adsorption of Cd (II) ions [104], with a maximum adsorption efficiency of 146.41 mg g^{-1} at a temperature of 30 °C. Hallaji et al. [105] synthesized a nanofiber adsorbent based on ZnO coated with polyvinyl alcohol (PVA) for the adsorption of Cu (II), Ni (II), and U (VI) ions, with a maximum adsorption efficiency for U (VI) and then Cu (II) and Ni (II).

Magnetic nanomaterials (MNPs) are another group of advanced nanomaterials that combine the benefits of magnetic separation and nanotechnology for better removal of heavy metal ions. MNPs have a comparatively greater surface area, biocompatibility, chemical inertness, less toxicity, and easy dispersion ability. Iconaru et al. [106] synthesized magnetite NPs (nFe_3O_4) and compared their removal efficiency with commercial magnetite toward As (II) and Cu (II). Giraldo et al. [107] used Fe_3O_4 MNPs to remove Pb (II), Cu (II), Zn (II), and Mn (II) ions. It was found that the Fe_3O_4 NP adsorption capacity was the highest for Pb (II) and that for Mn (II) was minimal. Among the various nanostructures, nanotubes, nanowires, and nanorods have a greater specific area and exhibit a high capacity to eliminate heavy metal ions from water. Karami [108] used magnetite nanorods to effectively remove Fe (II), Pb (II), Zn (II), Ni (II), Cd (II), and Cu (II) ions from waste water. The maximum adsorption capacities amounted to 127, 113, 107, 95, 88, and 76 mg g^{-1} for Fe (II), Pb (II), Zn (II), Ni (II), Cd (II), and Cu (II) respectively.

Biomolecule functionalized MNPs are another group of modified nanoparticles. Functionalization of MNPs with biomolecules enables their safe and ecological application in the process of metal ion remediation. Many researchers have investigated the use of this group of MNP nanoparticles for adsorption of heavy metal ions. An example of a biomolecule functionalized MNP for Pb (II) ion removal can be glycine modified magnetic Fe_3O_4 NPs (GF-MNPs) enclosed in an alginate polymer in the form of beads [92, 103], which is adsorbent, containing iron oxide MNPs and Ca-alginate embedded on the surface of a biomass strain, i.e. *Phanerochaete chrysosporium* pellets [109] or carboxy methyl cellulose-immobilized Fe_3O_4 MNPs (CMC-Fe_3O_4) [110]. For the above biomolecule functionalized MNP, an adsorption capacity for Pb (II) at the level of 152.0–167.36 mg g^{-1} was obtained. In addition, pH is an important factor in the process of heavy metal adsorption on nanoparticles, which has a large impact on the removal of cations from the solution. It was observed that maximum metal ion adsorption is obtained at moderate pH values. The adsorbent adsorption capacity depends upon the metal to binding sites ratio. It was also demonstrated that temperature has a significant effect on the adsorption process. With the exothermic adsorption process, the adsorption rate decreases with increasing temperature, while endothermic adsorption leads to an increase in adsorption capacity at high temperatures.

18.2.4 Summary

The review concludes that the described methods have great potential for recovery of heavy metals from various types of biodegradable waste. The need for further research in order to improve the effectiveness of these methods was emphasized. It is believed that microbiological methods for cleaning the environment of heavy metals are methods of the future because of the environmental benefits, the possibility of purification in place of pollution (in situ), high metal binding efficiency, metal desorption allowing the recovery of valuable metals, multiple use of metal binding biomass, and the ability to multiply biomass on inexpensive media. The application of bioprocessing is very beneficial for the environment and the main benefits are:

- Waste treatment using these bioprocesses can reduce the total amount of waste. As a consequence, land space for waste storage and disposal could be saved for other human activities. Microorganisms can be used to destroy or degrade toxic compounds (e.g. cyanide) [111] or by initiating bioprocesses that are more environmentally friendly than no or less toxic chemicals.
- The possibility of recovering valuable metals concomitantly to detoxifying the processed wastes [112].
- The use of certain solid wastes as secondary materials for construction purposes, because their toxicity has been reduced by microbiological processing. It can be waste that is used in a more sustainable way, without creating serious environmental risks [22].

The comprehensive description and comparison of the use of different methods to remove heavy metal ions from biodegradable waste will not only provide an insight into various aspects previously researched by scientists but will also help them design their own research problems, taking into account the usefulness and implications of practical work in science and industry. The proposal of new and innovative methods of recovering metals from waste is intended to encourage the extension of research on recovery of metals from biodegradable waste.

References

1 Olaniran, A.O., Balgobind, A., and Pillay, B. (2013). Bioavailability of heavy metals in soil: impact on microbial biodegradation of organic compounds and possible improvement strategies. *International Journal of Molecular Sciences* 14 (5): 10197–10228.

2 Kordialik-Bogacka, E. (2011). Surface properties of yeast cells during heavy metal biosorption. *Central European Journal of Chemistry* 9 (2): 348–351.

3 Vijayaraghavan, K. and Yun, Y.S. (2008). Bacterial biosorbents and biosorption. *Biotechnology Advances* 26: 266–291.

4 Kaplan, O., Yildirim, N.C., Yildirim, N., and Cimen, M. (2011). Toxic elements in animal products and environmental health. *Asian Journal of Animal and Veterinary Advances* 6: 228–232.

5 Sardar, K., Ali, S., Hameed, S. et al. (2013). Heavy metals contamination and what are the impacts on living organisms. *Greener Journal of Environmental Management and Public Safety* 2 (4): 172–179.

6 Wan Ngah, W.S. and Hanafiah, M.A.K.M. (2008). Removal of heavy metal ions from wastewater by chemically modified plant wastes as adsorbents: a review. *Bioresource Technology* 99: 3935–3948.

7 Wei, Y., Li, J., Shi, D. et al. (2017). Environmental challenges impeding the composting of biodegradable municipal solid waste: a critical review. *Resources, Conservation and Recycling* 122: 51–65.

8 Kupper, T., Bürge, D., Bachmann, H.J. et al. (2014). Heavy metals in source-separated compost and digestates. *Waste Management* 34: 867–874.
9 Zheljazkov, V. and Warman, P.R. (2004). Source-separated municipal soil waste compost application to Swiss chard and basil. *Journal of Environmental Quality* 33: 542–552.
10 Mudhoo, A., Garg, V.K., and Wang, S. (2012). Removal of heavy metals by biosorption. *Environmental Chemistry Letters* 10: 109–117.
11 Malik, O.A., Hsu, A., Johnson, L.A., and Sherbinin, A. (2015). A global indicator of wastewater treatment to inform the Sustainable Development Goals (SDGs). *Environmental Science & Policy* 48: 172–185.
12 Zhou, X., Zhang, H., Long, Y. et al. (2020). Sustainable and clean oilfield development: optimal operation of wastewater treatment and recycling system. *Journal of Cleaner Production* 252: 119819.
13 Fu, F. and Qi, W. (2011). Removal of heavy metal ions from wastewaters: a review. *Journal of Environmental Management* 92: 407–418.
14 Wang, H. and Ren, Z.J. (2014). Bioelectrochemical metal recovery from wastewater: a review. *Water Research* 66: 219–232.
15 Ritchie, S.M.C., Kissick, K.E., Bachas, L.G. et al. (2001). Polycysteine and other polyamino acid functionalized microfiltration membranes for heavy metal capture. *Environmental Science & Technology* 35 (15): 3252–3258.
16 Chiarle, S., Ratto, M., and Rovatti, M. (2000). Mercury removal from water by ion exchange resins adsorption. *Water Research* 34 (11): 2971–2978.
17 Zou, H.-S., Z-g, C., and Lin, G. (2007). A novel recovery technology of trace precious metals from waste water by combining agglomeration and adsorption. *Transactions of Nonferrous Metals Society of China* 17 (4): 858–863.
18 Tavakoli, O. and Yoshida, Y. (2008). Application of sub-critical water technology for recovery of heavy metal ions from the wastes of Japanese scallop *Patinopecten yessoensis*. *Science of the Total Environment* 398 (1–3): 175–184.
19 Laria, D., Marti, J., and Guardia, E. (2004). Protons in supercritical water: a multistate empirical valence bond study. *Journal of the American Chemical Society* 126: 2125–2134.
20 Tomita, K. and Oshima, Y. (2004). Stability of manganese oxide in catalytic supercritical water oxidation of phenol. *Industrial and Engineering Chemistry Research* 43 (24): 7740–7743.
21 Cheng, S.Y., Show, P.-L., Lau, B.F. et al. (2019). New prospects for modified algae in heavy metal adsorption. *Trends in Biotechnology* 37 (11): 1255–1268.
22 He, J. and Kappler, A. (2017). Recovery of precious metals from waste streams. *Microbial Biotechnology* 10 (5): 1194–1198.
23 Sivasubramanian, V. (2016). *Environmental Sustainability Using Green Technologies*. Boca Raton, FL: CRC Press.
24 Davis, T.A., Volesky, B., and Mucci, A. (2003). A review of the biochemistry of heavy metal-biosorption by brown algae. *Water Research* 37: 4311–4330.
25 Aftab, K., Akhtar, K., Kausar, A. et al. (2017). Fungal strains isolation, identification and application for the recovery of Zn(II) ions. *Journal of Photochemistry and Photobiology B: Biology* 175: 282–290.
26 Ianis, M., Tsekova, K., and Vasileva, S. (2006). Copper biosorption by Penicillium cyclopium: equilibrium and modelling study. *Biotechnology & Biotechnological Equipment* 20: 195–120.
27 Wang, J. and Chen, C. (2006). Biosorption of heavy metals by *Saccharomyces cerevisiae*: a review. *Biotechnology Advances* 24: 427–451.
28 Romera, E., González, F., Ballester, A. et al. (2007). Comparative study of biosorption of heavy metals using different types of algae. *Bioresource Technology* 98 (17): 3344–3353.

29 Esposito, A., Pagnanelli, F., Lodi, A. et al. (2001). Biosorption of heavy metals by Sphaerotilus natans: an equilibrium study at different pH and biomass concentrations. *Hydrometallurgy* 60: 129–141.

30 Vasudevan, P., Padmavath, Y.V., and Dhingra, S.C. (2003). Kinetics of biosorption of cadmiumon Baker's yeast. *Bioresource Technology* 89: 281–287.

31 Houari, B., Louhibi, S., Tizaoui, K. et al. (2019). New synthetic material removing heavy metals from aqueous solutions and wastewater. *Arabian Journal of Chemistry* 1: 5040–5048.

32 Barakat, M.A. (2011). New trends in removing heavy metals from industrial wastewater. *Arabian Journal of Chemistry* 4: 361–377.

33 Pandey, L.K. (2020). In situ assessment of metal toxicity in riverine periphytic algae as a tool for biomonitoring of fluvial ecosystems. *Environmental Technology & Innovation* 18: 10067.

34 Aravindhan, R., Madhan, B., Rao, J.R. et al. (2004). Bioaccumulation of chromium from tannery wastewater: an approach for chrome recovery and reuse. *Environmental Science & Technology* 38: 300–306.

35 Hashima, M.A. and Chu, K.H. (2004). Biosorption of cadmium by brown, green, and red seaweeds. *Chemical Engineering Journal* 97: 249–255.

36 Zan, F., Huo, S., Xi, B., and Zhao, X. (2012). Biosorption of Cd^{2+} and Cu^{2+} on immobilized *Saccharomyces cerevisiae. Frontiers of Environmental Science and Engineering* 6 (1): 51–58.

37 Nastaj, J. and Roik, M. (2011). Biosorption of Cu (II) ions on selected biosorbent beds: a literature review. *Chemical Industry* 90 (3): 412–416.

38 Sarada, B.M., Prasad, K., Kumar, K.K., and Murthy, C.V.R. (2014). Cadmium removal by macro algae *Caulerpa fastigiata*: characterization, kinetic, isotherm and thermodynamic studies. *Journal of Environmental Chemical Engineering* 2: 1533–1542.

39 Sooksawat, N., Meetam, N., and Inthorn, D. (2016). Equilibrium and kinetic studies on biosorption potential of charophyte biomass to remove heavy metals from synthetic metal solution and municipal wastewater. *Bioremediation Journal* 20: 240–251.

40 Akbari, M., Hallajisani, A., Keshtkar, A.R. et al. (2015). Equilibrium and kinetic study and modeling of Cu(II) and Co(II) synergistic biosorption from Cu (II)–Co (II) single and binary mixtures on brown algae *C. indica. Journal of Environmental Chemical Engineering* 3: 140–149.

41 Rangabhashiyam, S., Suganya, E., Lity, A.V., and Selvaraju, N. (2016). Equilibrium and kinetics studies of hexavalent chromium biosorption on a novel green macroalgae *Enteromorpha* sp. *Research on Chemical Intermediates* 42: 1275–1294.

42 Rahman, M.S. and Sathasivam, K.V. (2016). Heavy metal biosorption potential of a Malaysian-rhodophyte (*Eucheuma denticulatum*) from aqueous solutions. *International journal of Environmental Science and Technology* 13: 1973–1988.

43 Dittert, I.M., Brandão, H.L., Pina, F. et al. (2014). Integrated reduction/oxidation reactions and sorption processes for Cr(VI) removal from aqueous solutions using *Laminaria digitata* macro-algae. *Chemical Engineering Journal* 237: 443–454.

44 Mirghaffari, N., Ehsan Moeini, E., and Farhadian, O. (2015). Biosorption of Cd and Pb ions from aqueous solutions by biomass of the green microalga *Scenedesmus quadricauda. Journal of Applied Phycology* 27: 311–320.

45 Ibrahim, W.M., Hassan, A.F., and Azab, Y.A. (2016). Biosorption of toxic heavy metals from aqueous solution by *Ulva lactuca* activated carbon. *Egyptian Journal of Basic and Applied Sciences* 3: 241–249.

46 Ozer, T.B., Erkaya, I.A., Udoh, A.U. et al. (2012). Biosorption of Cr(VI) by free and immobilized *Pediastrum boryanum* biomass: equilibrium, kinetic, and thermodynamic studies. *Environmental Science and Pollution Research International* 19: 2983–2993.

47 Wang, F. and Lu, X. (2016). Selective removals of heavy metals (Pb^{2+}, Cu^{2+}, and Cd^{2+}) from wastewater by gelation with alginate for effective metal recovery. *Journal of Hazardous Materials* 308: 75–83.

48 Shen, N. and Chirwa, E.M.N. (2020). Live and lyophilized fungi-algae pellets as novel biosorbents for gold recovery: critical parameters, isotherm, kinetics and regeneration studies. *Bioresource Technology* 306: 123041.

49 Deng, L., Zhang, Y., Qin, J. et al. (2009). Biosorption of Cr(VI) from aqueous solutions by nonliving green algae *Cladophora albida*. *Minerals Engineering* 22: 372–377.

50 Esposito, A., Pagnanelli, F., and Veglio, F. (2002). pH related equilibria models for biosorption in single metal system. *Chemical Engineering Science.* 57 (3): 307–313.

51 Igwe, J.C. and Abia, A.A. (2006). A bioseparation process for removing heavy metals from waste water using biosorbents. *African Journal of Biotechnology* 5 (12): 1167–1179.

52 Okoro, I.A. and Okoro, S.O. (2011). Agricultural by products as green chemistry absorbents for the removal and recovery of metal ions from waste-water environments. *Continental Journal of Water, Air and Soil Pollution.* 2 (1): 15–22.

53 Mishra, D. and Ha Rhee, Y.H. (2014). Microbial leaching of metals from solid industrial wastes. *Journal of Microbiology* 52 (1): 1–7.

54 Awasthi, A.K., Zeng, X., and Li, J. (2016). Integrated bioleaching of copper metal from waste printed circuitboard—a comprehensive review of approaches and chal-lenges. *Environmental Science and Pollution Research* 23: 21141–21156.

55 Mishra, D., Kim, D.J., Ahn, J.G., and Rhee, Y.H. (2005). Bioleaching: a microbial process of metal recovery; a review. *Metals and Materials International* 11: 249–256.

56 Palomo-Briones, R., Ovando-Franco, M., Razo-Flores, E. et al. (2016). An overview of reclaimed wastewater reuse in gold heap leaching. *Mineral Processing and Extractive Metallurgy Review* 37: 274–285.

57 Shin, D., Park, J., Jeong, J., and Kim, B.S. (2015). A biologi-cal cyanide production and accumulation system and the recovery of platinum-group metals from spent automotivecatalysts by biogenic cyanide. *Hydrometallurgy* 158: 10–18.

58 Ilyas, S., Ruan, C., Bhatti, H. et al. (2010). Column bioleaching of metals from electronicscrap. *Hydrometallurgy* 101: 135–140.

59 Brandl, H., Bosshard, R., and Wegmann, M. (2001). Computer-munching microbes: metal leaching from electronic scrap by bacteria and fungi. *Hydrometallurgy* 59: 319–326.

60 Khatri, B.R., Sodha, A.B., Shah, M.B. et al. (2018). Chemical and microbial leaching of base metals from obsolete cell-phone printed circuit boards. *Sustainable Environment Research* 28 (6): 333–339.

61 Xiang, Y., Wu, P., Zhu, N. et al. (2010). Bioleaching of copper from waste printed circuit boards by bacterial consortium enriched from acid mine drainage. *Journal of Hazardous Materials* 184: 812–818.

62 Marsden, J. and House, I. (2006). The Chemistry of Gold Extraction Littleton, CO:Mining, Metallurgyand Exploration, Inc, 2nd edition.

63 Colica, G., Caparrotta, S., and De Philippis, R. (2012). Selective biosorption and recovery of *Ruthenium fromindustrial* effluents with *Rhodopseudomonas palustrisstrains. Applied Microbiology and Biotechnology* 95: 381–387.

64 Ito, R., Kuroda, K., Hashimoto, H., and Ueda, M. (2016). Recovery of platinum (0) through the reduction of platinumions by hydrogenase-displaying yeast. *AMB Express* 6: 88.

65 Riddin, T., Govender, Y., Gericke, M., and Whiteley, C. (2009). Two different hydrogenase enzymes from sul-phate-reducing bacteria are responsible for the bioreductive mechanism of platinum into nanoparticles. *Enzyme and Microbial Technology* 45: 267–273.

66 Song, H.P., Li, X.G., Sun, J.S. et al. (2008). Application of a magnetotactic bacterium, *Stenotro-phomonas* sp. to the removal of Au (III) from contaminated wastewater with a magnetic separator. *Chemosphere* 72: 616–621.

67 Cui, J. and Zhang, L. (2008). Metallurgical recovery of metals from electronic waste: a review. *Journal of Hazardous Materials* 158: 228–256.

68 Maes, S., Claus, M., Verbeken, K. et al. (2016). Platinum recovery from industrial process streams by halophilic bacteria: influence of salt species and platinum speciation. *Water Research* 105: 436–443.

69 Li, W., Yu, H., and He, Z. (2014). Towards sustainable wastewater treatment by using microbial fuel cells-centered technologies. *Energy & Environmental Science* 7: 911–924.

70 Logan, B.E. and Rabaey, K. (2012). Conversion of wastes into bioelectricity and chemicals by using microbial electrochemical technologies. *Science* 337 (6095): 686–690.

71 Wang, H. and Ren, Z.J. (2013). A comprehensive review of microbial electrochemical systems as a platform technology. *Biotechnology Advances* 3 (8): 1796–1807.

72 Choi, C. and Cui, Y. (2012). Recovery of silver from wastewater coupled with power generation using a microbial fuel cell. *Bioresource Technology* 107: 522–525.

73 Heijne, A.T., Liu, F., van der Weijden, R. et al. (2010). Copper recovery combined with electricity production in a microbial fuel cell. *Environmental Science & Technology* 44 (11): 4376–4381.

74 Logan, B.E., Call, D., Cheng, S. et al. (2008). Microbial electrolysis cells for high yield hydrogen gas production from organic matter. *Environmental Science & Technology* 42 (23): 8630–8640.

75 Lohner, S.T., Becker, D., Mangold, K.-M., and Tiehm, A. (2011). Sequential reductive and oxidative biodegradation of chloroethenes stimulated in a coupled bioelectro-process. *Environmental Science & Technology* 45 (15): 6491–6497.

76 Choi, C. and Hu, N. (2013). The modeling of gold recovery from tetrachloroaurate wastewater using a microbial fuel cell. *Bioresource Technology* 133: 589–598.

77 Lefebvre, O., Neculita, C.M., Yue, X., and Ng, H.Y. (2012). Bioelectrochemical treatment of acid mine drainage dominated with iron. *Journal of Hazardous Materials* 241-242: 411–417.

78 Li, Z., Zhang, X., and Lei, L. (2008). Electricity production during the treatment of real electroplating wastewater containing Cr^{6+} using microbial fuel cell. *Process Biochemistry* 43 (12): 1352–1358.

79 Tao, H., Li, W., Liang, M. et al. (2011a). A membrane-free baffled microbial fuel cell for cathodic reduction of Cu(II) with electricity generation. *Bioresource Technology* 102: 4774–4778.

80 Tao, H., Liang, M., Li, W. et al. (2011b). Removal of copper from aqueous solution by electrodeposition in cathode chamber of microbial fuel cell. *Journal of Hazardous Materials* 189 (1–2): 186–192.

81 Tao, H., Zhang, L., Gao, Z., and Wu, W. (2011c). Copper reduction in a pilot-scale membrane-free bioelectrochemical reactor. *Bioresource Technology* 102: 10334–10339.

82 Tao, H., Gao, Z., Ding, H. et al. (2012). Recovery of silver from silver(I)-containing solutions in bioelectrochemical reactors. *Bioresource Technology* 111: 92–97.

83 Wang, G., Huang, L., and Zhang, Y. (2008). Cathodic reduction of hexavalent chromium [Cr(VI)] coupled with electricity generation in microbial fuel cells. *Biotechnology Letters* 30 (11): 1959–1966.

84 Modin, O., Wang, X., Wu, X. et al. (2012). Bioelectrochemical recovery of Cu, Pb, Cd, and Zn from dilute solutions. *Journal of Hazardous Materials* 235-236: 291–297.

85 Qin, B., Luo, H., Liu, G. et al. (2012). Nickel ion removal from wastewater using the microbial electrolysis cell. *Bioresource Technology* 121: 458–461.

86 Huang, L., Chai, X., Chen, G., and Logan, B.E. (2011a). Effect of set potential on hexavalent chromium reduction and electricity generation from biocathode microbial fuel cells. *Environmental Science & Technology* 45 (11): 5025–5031.

87 Nancharaiaha, Y.V., Mohan, S.V., and Lens, P.N.L. (2015). Metals removal and recovery in bioelectrochemical systems: a review. *Bioresource Technology* 195: 102–114.

88 Rosenbaum, M., Aulenta, F., Villano, M., and Angenent, L.T. (2011). Cathodes as electron donors for microbial metabolism: which extracellular electron transfer mechanisms are involved? *Bioresource Technology* 102: 324–333.

89 Kegl, T., Košak, A., Lobnik, A., and Novak, Z. (2020). Adsorption of rare earth metals from wastewater by nanomaterials: a review. *Journal of Hazardous Materials* 386: 121632.

90 Wadhawan, S., Ayushi Jain, A., and Nayyar Jand Mehta, S.K. (2020). Role of nanomaterials as adsorbents inheavy metal ion removal from waste water: a review. *Journal of Water Process Engineering* 33: 101038.

91 Masarovičová, E. and Kráľová, K. (2013). Metal nanoparticles and plants. *Ecological Chemistry and Engineering S* 2 (1): 9–22.

92 Verma, M., Tyagi, I., Chandra, R., and Gupta, V.K. (2017a). Adsorptive removal of Pb (II) ions from aqueous solution using CuO nanoparticles synthesized by sputtering method. *Journal of Molecular Liquids* 225: 936–944.

93 Chen, M.-H. and Duan, Y.-P. (2016). Efficient removal/recovery of Pb onto environmentally friendly fabricated copper ferrite nanoparticles. *Journal of the Taiwan Institute of Chemical Engineers* 000: 1–9.

94 Islam, M.S., Choi, W.S., Nam, B. et al. (2017). Needle-like iron oxide@$CaCO_3$ adsorbents for ultrafast removal of anionic and cationic heavy metal ions. *Chemical Engineering Journal* 307: 208–219.

95 Hasanzadeh, R., Moghadam, P.N., Bahri-Laleh, N., and Sillanpaa, M. (2017). Effective removal of toxic metal ions from aqueous solutions: 2-bifunctional magnetic nanocomposite base on novel reactive PGMAMAn copolymer @Fe_3O_4 nanoparticles. *Journal of Colloid and Interface Science* 490: 727–746.

96 Martínez-Cabanas, M., López-García, M., Barriada, J.L. et al. (2016). Green synthesis of iron oxide nanoparticles: development of magnetic hybrid materials for efficient As (V) removal. *Chemical Engineering Journal* 301: 83–91.

97 Danesha, N., Hosseinib, M., Ghorbanib, M., and Marjania, A. (2016). Fabrication, characterization and physical properties of a novel magnetite graphene oxide/lauric acid nanoparticles modified by ethylenediaminetetraacetic acid and its applications as an adsorbent for the removal of Pb (II) ions. *Synthetic Metals* 220: 508–523.

98 Park, S., Flores, A.G., Chung, Y.S., and Kim, H. (2015). Removal of cadmium and lead from aqueous solution by hydroxyapatite/chitosan hybrid fibrous sorbent: kinetics and equilibrium studies. *Journal of Chemistry* 2015: 1–12.

99 Xiao, Z., Zhang, H., Xu, Y. et al. (2017). Ultra-efficient removal of chromium from aqueous medium by biogenic iron based nanoparticles. *Separation and Purification Technology* 174: 466–473.

100 Kanet, S.R., Greneche, J.M., and Choi, H. (2006). Arsenic (V) removal from groundwater using nano scale zero-valent iron as a colloidal reactive barrier. *Environmental Science and Technology* 40: 2045–2050.

101 Cai, X., Gao, Y., Sun, Q. et al. (2014). Removal of co-contaminants Cu (II) and nitrate from aqueous solution using kaolin-Fe/Ni nanoparticles. *Chemical Engineering Journal* 244: 19–26.

102 Gupta, V.K., Chandra, R., Tyagi, Y., and Verma, M. (2016). Removal of hexavalent chromium ions using CuO nanoparticles for water purification applications. *Journal of Colloid and Interface Science* 478: 54–62.

103 Verma, R., Asthan, A., Singh AK Prasad, S., and Susan, M.A.B.H. (2017b). Novel glycine-functionalized magnetic nanoparticles entrapped calcium alginate beads for effective removal of lead. *Microchemical Journal* 130: 168–178.

104 Zhu, X., Song, T., Lv, Z., and Ji, G. (2016). High-efficiency and low-cost α-Fe_2O_3 nanoparticles coated volcanic rock for Cd (II) removal from wastewater. *Process Safety and Environmental Protection* 104: 373–381.

105 Hallaji, H., Keshtkar, A.R., and Moosavian, M.A. (2015). A novel electrospun PVA/ZnO nanofiber adsorbent for U(VI), Cu(II) and Ni(II) removal from aqueous solution. *Journal of the Taiwan Institute of Chemical Engineers* 46: 109–118.

106 Iconaru, S.L., Guégan, R., Popa, C.L. et al. (2016). Magnetite (Fe_3O_4) nanoparticles as adsorbents for As and Cu removal. *Applied Clay Science* 134: 128–135.

107 Giraldo, L., Erto, A., and Carlos Jand Piraján, M. (2013). Magnetite nanoparticles for removal of heavy metals from aqueous solutions: synthesis and characterization. *Adsorption* 19 (2): 465–474.

108 Karami, H. (2013). Heavy metal removal from water by magnetite nanorods. *Chemical Engineering Journal* 219: 209–216.

109 Xu, P., Zeng, G.M., Huang, D.L. et al. (2012). Adsorption of Pb (II) by iron oxide nanoparticles immobilized *Phanerochaete chrysosporium*: equilibrium, kinetic, thermodynamic and mechanisms analysis. *Chemical Engineering Journal* 203: 423–431.

110 Fan, H., Ma, X., Zhou, S. et al. (2019). Highly efficient removal of heavy metal ions by carboxymethyl celluloseimmobilized O nanoparticles prepared via high-gravity technology. *Carbohydrate Polymers* 213: 39–49.

111 Akcil, A. and Mudder, T. (2003). Microbial destruction of cyanide wastes in gold mining: process review. *Biotechnology Letter* 25: 445–450.

112 Natarajan, K., Modak, J., and Raichur, A. (2001). Bioreactor engineering for treating refractory gold-bearing concentrates: an Indian experience. *Process Metallurgy* 11: 183–190.

18.3

Biosorbents and Biochar Production

Anna Grobelak and Agata Rosińska

Faculty of Infrastructure and Environment, Czestochowa University of Technology, Poland

18.3.1 Introduction

Advanced thermal methods of waste biomass processing are now becoming more and more popular. Moreover, the decrease in the usage of non-renewable energy resources is strongly recommended and needed in the longer perspective. The main factor stimulating such a trend seems to be the legislation, also EU, which changed the significant rules in the field of organic waste storage, including sewage sludge, inscribing strongly in the principle of 3 R-reuse, recycle, revenue [1]. Environmental problems related to greenhouse gas emissions as well as carbon depletion of soils necessitate action in this direction [2]. The sustainable strategies for the management of sewage sludge and bio-wastes, their treatment and disposal, must take into account the current state of knowledge, legislation process, characteristics and ecotoxicology. Decision-making tools such as end-of-waste criteria (EWC) and Life Cycle Assessment (LCA) allow a comprehensive approach to waste management systems, and enable the environmental, technical, and economic aspects to be taken into account. Defining the best technologies for organic waste treatment and examining their ecotoxicity in the long term is currently one of the main research topics [3]. The majority of EU members implementing legislative solutions apply them at the national level, but there is no integrated management system. The activities establishing criteria for selecting the best solutions in accordance with the requirements of the circular economy in the context of activities "from waste to resources" are necessary [4]. Organic wastes are being widespread and arise in many industries [5]. However, we should remember that not all innovative technologies will find their place on the market. Apart from the satisfactory effects of waste management it is also necessary to take into account the economic factor, which often decides about the inability to use modern technology. Biochar production and utilization is undoubtedly one of the most interesting research areas at present, concerning not only environmental sciences (39%), engineering environmental issues (17%), and soil science (13%), but also more of other categories of Web of Science (Figure 18.3.1). The main reason for such great research attention comes from biochar's potential impact on global warming mitigation and on soil remediation properties. The European Biochar Foundation (EBC) recommended a definition of biochar: "Biochar is a charcoal-like substance that is pyrolysed from sustainable obtained biomass under controlled conditions and which is used for any purpose which does not involve its rapid mineralization to CO_2" [6].

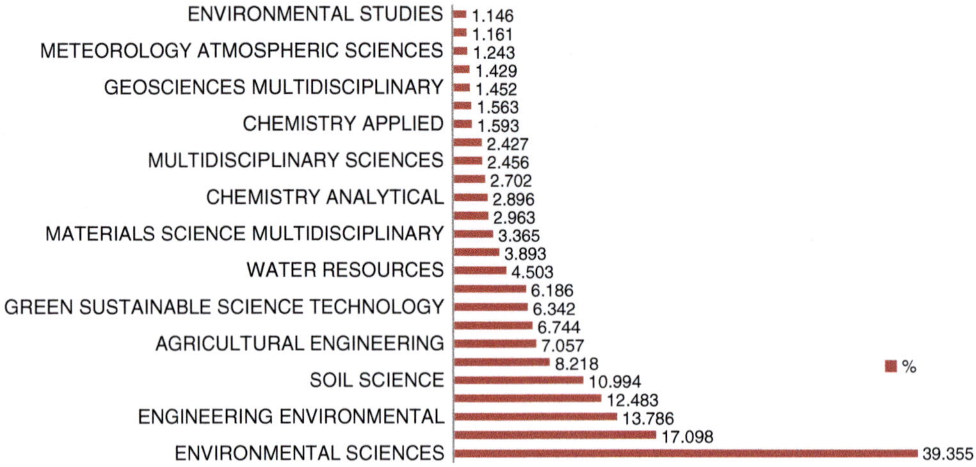

Figure 18.3.1 Percentage of research papers according to Web of Science categories. Source: Modified from Web of Sciences.

Biochar is a porous and fine-grained material with parameters similar to charcoal, which is obtained as a result of biomass thermal conversion without oxygen in a process called pyrolysis [7]. According to the guidelines of the European Biochar Certificate (EBC), biochar is a heterogeneous material, rich in aromatic carbons and mineral compounds, formed as a result of biomass pyrolysis, in which biomass is obtained in a sustainable manner, under controlled process conditions (350–1000 °C) and using pure technologies. According to EBC guidelines, biochar is not a material that is formed as a result of torrefaction or hydrothermal carbonization [8]. The input material for biochar production can be specifically grown biomass as well as waste biomass, which is currently more recommended. Biochar can be used in agriculture and soil remediation as a soil improver or natural fertilizer, but it can also be used as a sorbent or biosorbent [9]. One of the most promising new strategies for restoring carbon to depleted soils and sequestering significant amounts of CO_2 is the use of biochar or other bio-waste and sewage sludge as soil additives [10]. The assessment of the biochar to applications in areas where agricultural crops and energy crops are produced and in degraded soils is especially recommended. Biochar waste, as a sorbent, can be used in the restoration of degraded soil. The important current scientific goal is to access the possibility and efficiency of sequestration of organic carbon on different types of soils using different bio-wastes. It is necessary to have an appropriate dialogue and literature search to answer the question about the most effective and stable indicators for assessing and modeling the degree of soil organic carbon (SOC) sequestration in order to give an unambiguous answer about the effects of additives on SOC [11, 12].

18.3.2 Process of Biochar Production

The chemical composition of the final products is the result of the diverse variety of feedstocks used in the production of biochar, as well as the various thermal processes applied. For these reasons some regulations have been made to standardize the quality requirements connected with the pyrolysis process. Thus some global biochar organizations developed their own procedures and certifications. In Europe the most well-known is the European Biochar Certificate (EBC), defined by the European Biochar Foundation, and in the United Kingdom it is the Biochar Quality Mandate, developed by the British Biochar Foundation. The most known worldwide is the International Biochar Initiative (IBI) in the USA [7].

Biochar production usually includes the following steps:

- Collection, transport, and processing of biomass feedstocks,
- Characterization and testing of biochar,
- Production and utilization of energy co-products: gas, oil, or heat,
- Biochar transport and handling for soil applications,
- Monitoring of biochar applications for carbon accounting,
- Life Cycle Assessment and full system monitoring for sustainability assessment (for IBI certification).

There are many different technologies used to make biochar, but all of them include heating biomass with a limited oxygen supply to extract volatile gasses. This process is named thermal decomposition and is usually from pyrolysis or gasification. These procedures can yield clean energy in the form of gas or oil along with the biochar. Moreover, this energy may be supplied for another use or it may basically be consumed and released as heat. This is one of the few technologies that is reasonably inexpensive, generally applicable, and swiftly scalable [13]. Biochar technology necessarily consists of entire integrated systems that can contain numerous components that may or may not be part of any specific system (Figure 18.3.2).

Equipment for making biochar can be as simple as a primitive campfire or as complex as a modern biorefinery [14]. The basic process for biochar production is pyrolysis, which can be prepared using very simple fire sources, as provided by ancient Amazon Indians who called it "Terra Peta," or

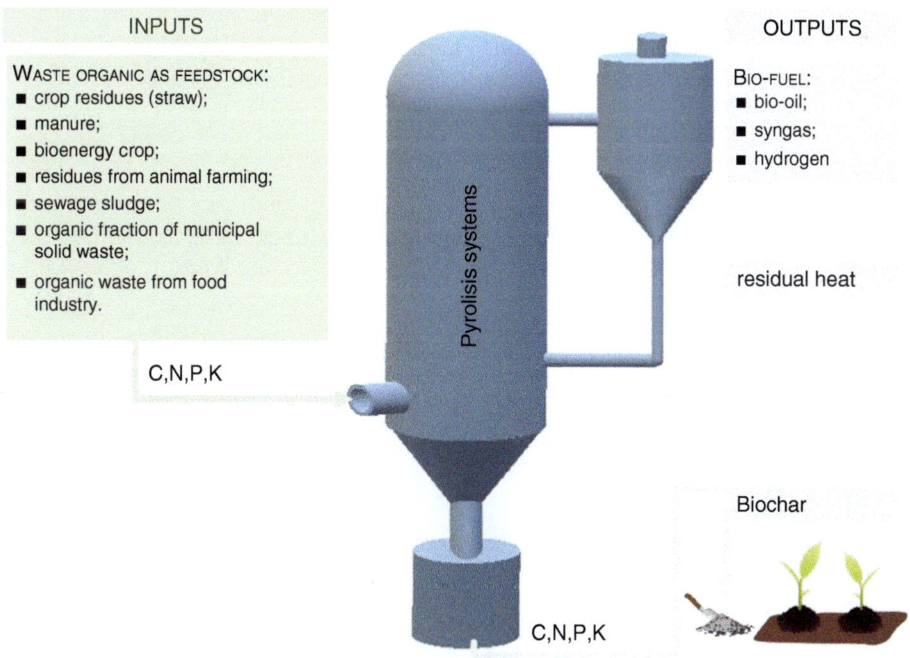

Figure 18.3.2 Overall biochar production technology scheme.

requires more or significant (portable machines for agricultural usage) or sophisticated equipment and facilities like a biorefinery [7]. The pyrolysis process is the breaking down (lysis) of a biomass by heat (pyro) with release of the gas. This is the crucial point in the biomass combustion or gasification process. The biochar production is usually categorized as either a pyrolysis (slow or fast) or a gasification system. Depending on the conditions of the pyrolysis process, i.e. the type of biomass used, temperature, heating rate, degree of biomass fragmentation, used catalysts, etc., biochar with different properties is obtained [15]. Due to the very variable characteristics of waste biomass, it is very difficult to compare various types of biochar and the process of their preparation. The most frequent temperatures of pyrolysis for biochar production are in the range of 260–700 °C [7].

18.3.3 Organic Residual Components for Biochar Feedstock and CO_2 Emission

Biochar should generally be produced from biomass, more specifically from waste biomass. The point is that the production of biomass for biochar should not compete with the production of biomass for other purposes, e.g. for fibers or food. Until now, the most common substrate for biochar production was broadly understood to be biomass derived from wood, which is the easiest way to produce biomass. However, due to the huge demand for wood material, this source should be abandoned. Biochar can generally be obtained from crop residues (both field residues and processing residues such as nut shells, fruit pits, bagasse, etc.), as well as from yard, food, and forestry wastes, and animal manures and sewage sludge [16]. Using these materials to make biochar not only removes it from a pollution cycle but enables it to be obtained as a by-product of producing energy from biomass. Production of biochar from agricultural, municipal, and forestry biomass can avoid releasing huge amounts of CO_2 and other greenhouse gases (GHG) to the atmosphere as well as preventing leaching potentially euthrophogenic elements to groundwater during its decomposition process. It was found that emissions from residual agricultural biomass are significant, and due to increasing food production they are also increasing (Figure 18.3.3).

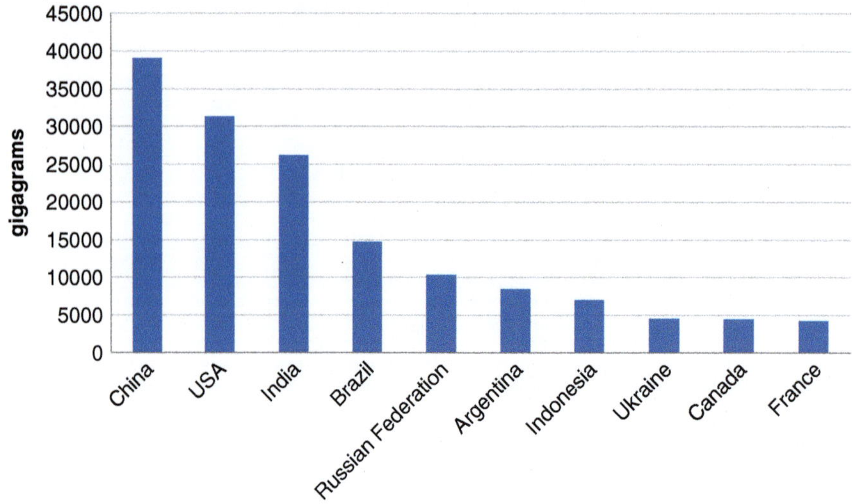

Figure 18.3.3 Top 10 emitters (CO_2 equivalent from crop residues) in 2017, all Crops + (Total) Average 1990–2017. Source: Based on the data from http://www.fao.org/faostat/en/#data/GA/visualize.

Biochar production is also important due to GHG emissions reduction from biomass aerobic decomposition. The problems of CO_2 emissions are also increasing due to increasing crop production and thus crop residue amounts. However, the natural carbon cycles are not adequate to handle the billions of metric tonnes emitted to the atmosphere. The significant amount of CO_2 (CO_2 equivalent kg^{-1}) can be captured and stored in biochar that would have been emitted into the atmosphere if the crop wastes had decomposed naturally [17]. The amount of CO_2 (CO_2 equivalent kg^{-1}) that would have been released to the atmosphere if the crop residues had decomposed naturally or openly burned can be assessed [17]. Firstly, the biochar yield (Equation 18.3.1) is calculated as the ratio of the weight of biochar (M2) to the weight of pyrolysis input (M1) [18]:

$$\text{Biochar yield (Bw\%)} = (\text{M2}/\text{M1}) \times 100 \tag{18.3.1}$$

Next the total potential carbon (TPC) is calculated by multiplying of the biochar yield (Bw%) and the total carbon content achieved from test examination [19] (Equation 18.3.2):

$$\text{TPC} = \text{biochar yield} \times \text{carbon content of biochar} \tag{18.3.2}$$

Finally, the CO_2 reduction potential (CO_2 equivalent kg^{-1}) can be estimated by multiplying the total potential carbon by the carbon stability coefficient (80%) and CO_2 fraction (molecular mass of CO_2, 44 g $mole^{-1}$) and the mass of carbon (12 g $mole^{-1}$) [18] (Equation 18.3.3):

$$\text{Carbon dioxide reduction potential} = \text{TCP} \times 80/100 \times 44/12 \tag{18.3.3}$$

The CO_2 emissions depend also on the residue type. For crop residues the highest emissions come from wheat, rice, and maize (Figure 18.3.4).

Moreover, the feedstocks must not consist of high levels of contamination as heavy metals, usually found in different industrial sources of biomass. The biochar production operations will be considered including the sustainability rules, especially local conditions and potential biomass availability as well as biochar utility and other relative pros and cons. Moreover, feedstock obtainability can fluctuate between years and within years. For areas where feedstock biomass is commonly used as a heating fuel, the expansion with biochar production facilities can be an

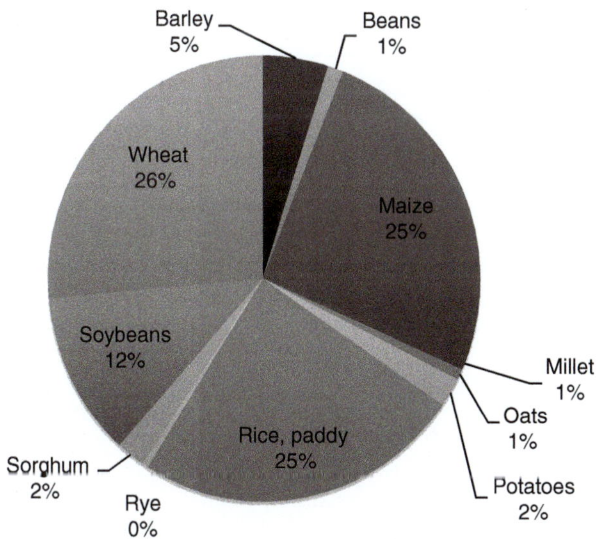

Figure 18.3.4 Worldwide emissions from the decomposition of crop residues on cultivated soils by crop type (CO_2 equivalent) in 2017). Source: Based on the data from: http://www.fao.org/faostat/en/#data/GA/visualize.

essential extra advantage if prepared sustainably [20]. The obtained biochar can then be used on fields to actually raise soil productivity and in consequence food production [6].

The chemical composition of biochar (carbon, nitrogen, potassium, trace elements, etc.) depends on the feedstock applied and the length and temperature of pyrolysis. Biochar has a neutral/alkaline pH and in soil is very stable and resistant to microbiological decomposition [7]. Moreover, the beneficial effects of biochar come from its porous structure, which significantly improves the sorption capacity of soils. The chemical composition is also an important issue, especially in the production of safe biochar from residue biomass. Pyrolysis feedstocks greatly affect nutrient contents of biochar [20]. As an example, biochar made from feedstocks that have higher contents of potassium (such as animal manures of 30 g kg^{-1}) often have a greater potassium content than biochar made fully from straw (6.3 g kg^{-1}). Biochar made of straw frequently has a higher carbon content (73%) compared to that made of sewage sludge (42%) or poultry manure (44%). The highest content of N was obtained for biochar made of poultry manure (3.6%) compared to straw (1.25%) or sewage sludge (0.3%). When considering phosphorus content the values were as follows: sewage sludge (5.25%) and poultry manure (290 mg kg^{-1}) [15]. Some other examples of biochar produced from organic residues and biosolids are provided in Table 18.3.1.

In a few countries the standards and recommendations related to the substrates and processes applied in the biochar production and requirements for biochar applied into the soil are implemented (biochar standards defined by the IBI, USA; European Biochar Certificate defined by the European Biochar Foundation (EBC); and Biochar Quality Mandate developed by the British Biochar Foundation in the United Kingdom (BQM)) [7]. Special attention in this regard is given to the permissible content of contaminants as heavy metals (Pb, Cd, Zn, Hg, Ni, Cr) and metalloids

Table 18.3.1 Selected properties of biochar, depending on the feedstocks applied and the temperature of the process.

Material applied	C	N	P	K	Pb	Cd	Hg	Temperature	pH	References
Unit		%		g kg^{-1}		mg kg^{-1}		°C	-	
Rye straw	73	1.25	0.29	30	1.95	0.58	2.5	500	10	[15]
Rice straw	49	1.32	–	–	–	–	–	500	9.54	[21]
Rice husk	55	0.5	0.12	3.4	–	–	–	450	8	[17]
Coconut coir	64.4	1.01	–	–	–	–	–	550	9.44	[22]
Corn cob	78	0.4	0.1	2	–	–	–	650	11	[17]
Ground nut shell	71.8	0.79	–	–	–	–	–	550	8.83	[22]
Coffee	77	0.6	0.15	4.3	–	–	–	490	8.5	[17]
Green waste	68	0.17	0.2	1	–	–	–	450	6.2	[23]
Poultry manure	44	3.6	0.29	30	1.95	0.58	2.5	500	10	[15]
Compost	83.6	1	–	–	9.3	nd	nd	600	9	[24]
Organic fraction of municipal solid waste	43	1.3	0.53	0.6	166	11	–	500	9.35	[25]
Municipal sewage sludge	42	0.4	5.25	1.56	46.2	2.31	0.0045	500	6.3	[15]

(As, Se) or organic contaminants (WWA, PCB, dioxins, and furans) in biochar, based on the present quality criteria.

Moreover, processing parameters significantly affect biochar properties. Gasification systems produce smaller quantities of biochar in a directly heated reaction vessel with introduced air. The more oxygen a production unit can exclude, the more biochar it can produce. Biochar production is optimized in the absence of oxygen [26].

The biochar may have diverse properties and thus may have many different applications: soil additive, water retention product, soil conditioner, nutrients carrier, soil remediation assistant, precursor for activated carbon, filtration/decontamination material, and animal feed additive [27].

18.3.3.1 Biochar as a Sorbet

The potential of biochar is thus far to be utilized in additional areas than farming. It is used as a textile colorant, as a raw material in construction materials, especially against electromagnetic radiation, in water treatment processes, which is a significant application in the aspect of the growing need for water treatment facilities, and particularly in the emerging treatment of contaminants [7].

The major biochar application that finally provides biochar to soils is at present the cascading use in animal husbandry. In particular, in Germany, Switzerland, and Austria, the use of biochar in animal husbandry is implemented very quickly by small-to-medium scale planters, although scientific research is deficient [28]. It was found that husbandry biochar application reduces the odor, ammonia, and methane emissions when applied as bedding material, or as a manure and slurry conditioner. In animal feeding it was found that feed efficiency increased and animal condition was enhanced. Biochar can be applied as a feed additive: an ingredient in probiotic animal feed ("carbon feed") and silage additive [29]. Biochar is well known as a traditional feed additive that was often used to cure digestive complications of animals. Since 2010, biochar is used as an additive to standard feed mixtures. The biochar application as a feed additive has been authorized by the EU Feed Regulation 2002/32/EC of 7 May 2002 on undesirable substances in animal feed and EU Regulation 396/2005 of 23 February 2005 on maximum residue levels of pesticides in or on food and feed of plant and animal origin. Biochar can only be certified under EBC feed when all conditions for EBC premium quality are met and the production was consequently certified. Moreover, only biochar made from natural and untreated wood is allowable [6].

Due to very high sorbent properties, biochar can efficiently immobilize wide groups of contaminants like heavy metals, pharmaceuticals, and emerging contaminants, from different media like gas, liquid, or solid. It was found that the higher pyrolysis temperature the higher sorption properties can be achieved, mainly as a result of larger specific surfaces and microporosity [7]. Activated carbons can be produced from biochar but require additional steps of physical (oxidizing agent such as water vapor, carbon monoxide (IV), or a mixture of both) or chemical activation (mainly potassium and sodium hydroxide, sodium and potassium carbonates, zinc chloride, and phosphoric acid (V)) [7]. In the biochar production process, by adjusting the pyrolysis temperature and other conditions certain properties can be directed to a variety of diverse applications. In comparison to raw biochar, the activated carbon made of biochar has a wider usage in many branches of the industry.

Biochar is a material that is gaining more and more attention as a cost-effective sorbent for the purification of aqueous solutions from environmentally toxic substances. The properties of biochar differ significantly depending on the type of input material and the pyrolysis temperature used.

However, there is a lack of information on sorption mechanisms for various pollutants, including key factors and adsorption mechanisms. Various types of anthropogenic effects significantly increased the amount of waste water containing toxic substances, mainly in the aquatic environment. It is estimated that bioabsorbents will find application in this area [30]. Biochar has many properties, such as the capacity to bind pollutants, retain nutrients and water, and therefore has many potential applications. In the case of metal ions, specific sorption mechanisms are variable depending on the properties of the biochar. Biochar, which is rich in negative charges on the surface, has been applied to decrease plant uptake of heavy metals and metalloids from contaminated soils by an adsorption process. Moreover, the application of biochar also decreases the uptake of other substances such as PAHs, pesticides, herbicides, nanoparticles, emerging contaminants, and pharmaceuticals by plants [31]. Biochar-mediated adsorption of organic contaminants is the result of electrostatic interactions with polar or non-polar groups [7]. Surface functional groups, mainly those containing oxygen and found on the surface of biochar, can interact strongly with heavy metals such as electrostatic attraction, ion exchange, and complexation at the surface [30]. Five mechanisms regulating metal sorption by biochar from aqueous solutions are described, i.e. complexation, cation exchange, precipitation, electrostatic interactions, and chemical reduction (Table 18.3.2). However, the role of each mechanism differs depending on the target metal ions to be sorbed [32]. Research on biochar sorption mainly concerns heavy metal removals from aquatic solutions (Table 18.3.2).

Table 18.3.2 Selected biochar contaminant adsorption and mechanisms.

Sorbed contamination	Functional groups or sorption mechanism	Biochar feedstock	References
Pb, Cu, Zn i Cd	CO_3^{2-}, PO_4^{3-}	Rice hulls and manure	[32]
Cr	Electrostatic attraction, reduction of Cr (VI) to Cr (III) and complexation	White beet	[33]
Heavy metals	Oxygen functional groups	Palm oil and rice husk	[30]
Pb	Functional groups, exchange with K^+- and Na^+ due to external electrostatic complexation, complexation of the surface with free carboxylic functional groups and free functional hydroxyl of the group, complexation of the inner sphere with free hydroxyl groups	Sewage sludge	[33]
Cd	Similar to above	Corn straw	[34]
Pb, Cd	Magnetization, cation exchange	Waste agro-materials	[35]
As, F	Chemisorption process	Perennial grass	[36]
U (VI)	Surface carboxylic moieties on oxidized biochars	Foodwaste agro-materials	[37]
Atrazine and nicosulfuron	Hydrophobic partition, π–π electron donor–acceptor interactions, H-bonding, and pore-filling mechanism	Peanut shell	[38]
Cu (II)	Dominated by functional groups	Municipal sewage sludge	[39]

Biochar was also applied as an immobilizing carrier for plant growth-promoting bacteria (PGPB) [40] for soil remediation purposes. It was found that biochar is the most suitable carrier for bacteria among vermiculite and perlite. Biochar was also applied as a carrier for bacteria-producing siderophores [41] and for bacteria with ACC-deaminase activity [42]. Moreover, when bacteria were immobilized in the biochar, the high rate production of important biological compounds after months was still noted in heavy metal contaminated soil during the assisted bioremediation process.

18.3.4 Conclusions

The worldwide biochar demand in 2018 was estimated at 395.3 kilo tonnes (https://www.grandviewresearch.com/industry-analysis/biochar-market). Increasing consumption of biochar in organic food production and the properties to enhance the soil productiveness and crop production are estimated to be the crucial growth drivers. The economy of a closed cycle assumes continuous improvement of the cycle of matter and energy cycles, focuses on saving natural resources, and uses raw materials and recycling. The challenge for the coming years is to transform organic waste (bio-waste) into useful resources through an interdisciplinary approach to the life cycle. The scientific assessment of the impact of various methods of sewage sludge/bio-waste management on the environment using decision tools, e.g. LCA, is defining the best available technologies (BAT), effective recovery of raw materials adapted to local needs, ecotoxicological evaluation (the fate of selected pollutants in the short and long term), comparison of scenarios for the management of bio-waste in the technological, economic, and sociological aspect, and encourage policymakers to update legislation are all strongly needed. The biodegradable fraction of products, wastes, and residues from agriculture forestry and related food processing wastes, as well as the biodegradable fraction of industrial and municipal wastes, could be used to produce a safe and valuable biochar. Biochar contains crucial plant nutrients such as potassium, carbon, and magnesium as well as properties (alkaline pH and high CEC) that improve the fertility of poor and acidic soils and raise crop yields. In the aspect of sorption purposes, biochar seems to be a very promising material. The biochar's sorption properties depend on many factors. In addition to the structure, type of raw material, as well as the degree of fragmentation and the properties of the removed ion, external conditions such as the pH of the solution and the temperature of the pyrolysis process are also very important. Currently, research confirms that biochar can be used as a demanding sorbent, especially in the waste water treatment sector. Predominantly responsible for binding water pollutants are adsorption mechanisms, which reveal that different types of interaction exist, including electrostatic attraction, ion exchange, physical adsorption, and chemical bonding. The high carbon content further indicates that biochar has the potential to decrease CO_2 emissions in the atmosphere. Therefore, recycling of crop wastes to biochar is intensely suggested to mitigate climate change through carbon sequestration.

For the feedstock choice and biochar production the main drivers should be taken into account: feedstock production, collection, and transport, use tradeoff, feedstock storage, and pre-processing. These factors will most often affect the choice of locally available biomass for biochar production and its price, and thus its availability.

The main aim of European researchers in the biochar field is to recognize the function of biochar as a soil additive for fertility and crop yield improvement and for conserving environmental resources, while other biochar applications like sorbent or in animal husbandry feed additive are very lacking. Moreover, it should be pointed out that due to the lack of detailed law regulations for biochar, its marketable use is restricted; thus the uniform legal and quality regulations must be provided in national law regulations.

Acknowledgements

This work was supported by an internal research grant of Czestochowa University of Technology. The research leading to these results has received funding from the EnviSafeBioC project – ontract No. PPI/APM/2018/1/00029/U/001. The project is financed by the Polish National Agency for Academic Exchange (NAWA).

References

1 Smol, M., Adam, C., and Preisner, M. (2020). Circular economy model framework in the European water and wastewater sector. *Journal of Material Cycles and Waste Management* 22: 682–697.

2 Placek, A., Grobelak, A., Hiller, J. et al. (2017). The role of organic and inorganic amendments in carbon sequestration and immobilization of heavy metals in degraded soils. *Journal of Sustainable Development of Energy, Water and Environment Systems* 5 (4): 509–517.

3 Jaskulak, M., Rorat, A., Grobelak, A., and Kacprzak, M. (2018). Antioxidative enzymes and expression of rbcL gene as tools to monitor heavy metal-related stress in plants. *Journal of Environmental Management* 218: 71–78.

4 Grobelak, A., Grosser, A., Kacprzak, M., and Kamizela, T. (2019). Sewage sludge processing and management in small and medium-sized municipal wastewater treatment plant-new technical solution. *Journal of Environmental Management* 234: 90–96.

5 Suleiman, H., Rorat, A., Grobelak, A. et al. (2017). Determination of the performance of vermicomposting process applied to sewage sludge by monitoring of the compost quality and immune responses in three earthworm species: *Eisenia fetida*, *Eisenia andrei* and *Dendrobaena veneta*. *Bioresource Technology* 241: 103–112.

6 EBC (2012). European Biochar Certificate – Guidelines for a Sustainable Production of Biochar. European Biochar Foundation (EBC), Arbaz, Switzerland. http://www.europeanbiochar.org/en/download. Version 8.3E of 1 September 2019. doi: https://doi.org/10.13140/RG.2.1.4658.7043.

7 Saletnik, B., Zaguła, G., Bajcar, M. et al. (2019). Biochar as a multifunctional component of the environment – a review. *Applied Sciences* 9 (6): 1139.

8 Poluszyńska, J., Ślęzak, E., and Wieczorek, P.P. (2019). Biowęgiel jako środek polepszający właściwości gleby. *Przemysł Chemiczny* 98 [in Polish].

9 Sładeczek, F. and Głodek-Bucyk, E. (2017). Badania wykorzystania niskotemperaturowej pirolizy do przetwarzania biomasy odpadowej na biowęgiel w instalacji testowej. *Prace Instytutu Ceramiki i Materiałów Budowlanych* 10.

10 Kacprzak, M., Grobelak, A., Grosser, A., and Napora, A. (2014). The potential of biosolid application for the phytostabilisation of metals. *Desalination and Water Treatment* 52 (19–21): 3955–3964.

11 Placek, A., Singh, B.R., Grobelak, A. et al. (2018). Phytosequestration of carbon in *Miscanthus* × *giganteus* and *Pinus sylvestris* L. in degraded zinc smelter and post-mining soil. *Pedosphere* 28 (3): 555–560.

12 Were, K., Bui, D.T., Dick, Ø.B., and Singh, B.R. (2015). A comparative assessment of support vector regression, artificial neural networks, and random forests for predicting and mapping soil organic carbon stocks across an Afromontane landscape. *Ecological Indicators* 52: 394–403.

13 Růžičková, J., Kucbel, M., Raclavský, K., and Raclavská, H. (2016). Chemical composition of biochar from combustion in local heating.

14 Godlewska, P., Schmidt, H.P., Ok, Y.S., and Oleszczuk, P. (2017). Biochar for composting improvement and contaminants reduction. A review. *Bioresource Technology* 246: 193–202.

15 Sładeczek, F. and Głodek-Bucyk, E. (2017). Badania wykorzystania niskotemperaturowej pirolizy do przetwarzania biomasy odpadowej na biowęgiel w instalacji testowej. Prace Instytutu Ceramiki i Materiałów Budowlanych 10 (in Polish).

16 Malińska, K., Golańska, M., Caceres, R. et al. (2017). Biochar amendment for integrated composting and vermicomposting of sewage sludge–the effect of biochar on the activity of *Eisenia fetida* and the obtained vermicompost. *Bioresource Technology* 225: 206–214.

17 Billa, S.F., Angwafo, T.E., and Ngome, A.F. (2019). Agro-environmental characterization of biochar issued from crop wastes in the humid forest zone of Cameroon. *International Journal of Recycling of Organic Waste in Agriculture* 8 (1): 1–13.

18 Allyson S (2011) Biochar production for carbon sequestration. Dissertation. Worcester Polytechnique Institute.

19 Sandip, G. and Harsha, M.W. (2013). Generation of biochar from crop residues. *International Journal of Emerging Technology and Advanced Engineering* 3 (3): 566–570.

20 Wystalska, K., Malińska, K., Włodarczyk, R., and Chajczyk, O. (2018). Effects of pyrolysis parameters on the yield and properties of biochar from pelletized sunflower husk. In *E3S Web of Conferences* (Vol. 44, p. 00197). EDP Sciences.

21 Tsai, W.T., Lee, M.K., and Chang, Y.M. (2006). Fast pyrolysis of rice straw, sugarcane bagasse and coconut shell in an induction-heating reactor. *Journal of Analytical and Applied Pyrolysis* 76: 230–237.

22 Suman, S. and Gautam, S. (2018). Biochar derived from agricultural waste biomass act as a clean and alternative energy source of fossil fuel inputs. *Energy Systems and Environment* 207–219.

23 Chan, K.Y., Van Zwieten, L., Meszaros, I. et al. (2007). Agronomic values of green waste biochar as a soil amendment. *Australian Journal of Soil Research* 45: 629–634.

24 Ayiania, M., Terrell, E., Dunsmoor, A. et al. (2019). Characterization of solid and vapor products from thermochemical conversion of municipal solid waste woody fractions. *Waste Management* 84: 277–285.

25 Benavente, I., Gasco, G., Plaza, C. et al. (2018). Choice of pyrolysis parameters for urban wastes affects soil enzymes and plant germination in a Mediterranean soil. *Science of the Total Environment* 634: 1308–1314. https://doi.org/10.1016/j.scitotenv.2018.04.120.

26 Panwar, N.L., Pawar, A., and Salvi, B.L. (2019). Comprehensive review on production and utilization of biochar. *SN Applied Sciences* 1: 168. https://doi.org/10.1007/s42452-019-0172-6.

27 Jeffery, S., Verheijen, F.G., van der Velde, M., and Bastos, A.C. (2011). A quantitative review of the effects of biochar application to soils on crop productivity using meta-analysis. *Agriculture, Ecosystems & Environment* 144 (1): 175–187.

28 Conte, P., Schmidt, H.P., and Cimò, G. (2016). Research and application of biochar in Europe. *Agricultural and Environmental Applications of Biochar: Advances and Barriers* 63: 409–422.

29 Hagemann, N., Joseph, S., Schmidt, H.P. et al. (2017). Organic coating on biochar explains its nutrient retention and stimulation of soil fertility. *Nature Communications* 8 (1): 1–11.

30 Ślęzak, E. and Poluszyńska, J. (2018). Możliwości sorpcyjne biowęgla w usuwaniu zanieczyszczeń ze środowiska wodnego. *Prace Instytutu Ceramiki i Materiałów Budowlanych* 11: 62–72.

31 Kavitha, B., Reddy, P.V.L., Kim, B. et al. (2018). Benefits and limitations of biochar amendment in agricultural soils: a review. *Journal of Environmental Management* 227: 146–154.

32 Li, H., Dong, X., da Silva, E.B. et al. (2017). Mechanisms of metal sorption by biochars: biochar characteristics and modifications. *Chemosphere* 178: 466–478.

33 Dong, X., Ma, L.Q., and Li, Y. (2011). Characteristics and mechanisms of hexavalent chromium removal by biochar from sugar beet tailing. *Journal of Hazardous Materials* 190 (1–3): 909–915.

34 Sun, J., Lian, F., Liu, Z. et al. (2014). Biochars derived from various crop straws: characterization and Cd (II) removal potential. *Ecotoxicology and Environmental Safety* 106: 226–231.

35 Trakal, L., Veselská, V., Šafařík, I. et al. (2016). Lead and cadmium sorption mechanisms on magnetically modified biochars. *Bioresource Technology* 203: 318–324.

36 Saikia, R., Goswami, R., Bordoloi, N. et al. (2017). Removal of arsenic and fluoride from aqueous solution by biomass based activated biochar: optimization through response surface methodology. *Journal of Environmental Chemical Engineering* 5 (6): 5528–5539.

37 Paschalidou, P., Pashalidis, I., Manariotis, I.D., and Karapanagioti, H.K. (2020). Hyper sorption capacity of raw and oxidized biochars from various feedstocks for U (VI). *Journal of Environmental Chemical Engineering*: 103932.

38 Wang, P., Liu, X., Yu, B. et al. (2020). Characterization of peanut-shell biochar and the mechanisms underlying its sorption for atrazine and nicosulfuron in aqueous solution. *Science of the Total Environment* 702: 134767.

39 Fan, J., Li, Y., Yu, H. et al. (2020). Using sewage sludge with high ash content for biochar production and Cu (II) sorption. *Science of the Total Environment* 713: 136663.

40 Grobelak, A., Kokot, P., Hutchison, D. et al. (2018). Plant growth-promoting rhizobacteria as an alternative to mineral fertilizers in assisted bioremediation-sustainable land and waste management. *Journal of Environmental Management* 227: 1–9.

41 Grobelak, A. and Hiller, J. (2017). Bacterial siderophores promote plant growth: screening of catechol and hydroxamate siderophores. *International Journal of Phytoremediation* 19 (9): 825–833.

42 Grobelak, A., Kokot, P., Świątek, J. et al. (2018). Bacterial ACC deaminase activity in promoting plant growth on areas contaminated with heavy metals. *Journal of Ecological Engineering* 19 (5): 150–157.

18.4

Other Perspectives (e.g. Chitin Recovery, Carbon-Coated Magnet-Sensitive Nanoparticles, Proteins, Carbohydrates, and Humic Acid)

Agata Rosińska

Faculty of Infrastructure and Environment, Czestochowa University of Technology, Częstochowa, Poland

18.4.1 Chitin Recovery

18.4.1.1 Introduction

Chitin (from the Greek chiton – a covering) is the second most common biopolymer in nature – polycarbohydrate. However, the poor solubility of chitin practically prevents its processing. Despite this, the unique biological properties of chitin and its huge and renewing raw material base have been for years encouraging scientists from around the world to look for ways of utilizing it, and especially to develop methods for its chemical modification, aimed at facilitating processing while maintaining its unique properties. Therefore, chitin has become a valuable raw material and Jaganathan [1] reports that on the world market a kilo of chitin costs USD 220 [2].

Chitin is mainly produced by arthropods: crustaceans, insects, and arachnids. It also occurs in fungal cell walls and can be found in insects, sponges [3, 4], corals [5], and some unicellular eukaryotes [6]. In all living organisms it is combined into complexes with proteins and other polymers produced by these organisms. Chitin is mainly a waste product of the shellfish industry. The main source of obtaining chitin are shrimps, with annual global production amounting to over 2 600 000 Mt. According to Ioelovich, the production of chitin and its derivatives is estimated at 100 billion tonnes year^{-1} [2, 7]. The natural production of chitin amounts to between 1010 and 1011 tonnes year^{-1} [2]. In the United States, 750 000 000 pounds of crab, shrimp, and lobsters are harvested annually [6]. Crabs represent 47% of the catch, lobsters 16%, and shrimps 35%. After removing the meat, the shells are often sent to a landfill. The content of chitin in shells ranges from 10 to 40% [6, 8–11]. According to Ameh et al. [12], these carapaces have about 20–30% chitin, varying according to species and season. In the case of chitin production from crab shells, from over 1 300 000 Mt of raw material, 2.15% of chitin is obtained.

18.4.1.2 Chemical Structure of Chitin

The chemical structure of chitin is very similar to cellulose, which consists of several hundred to over a thousand β-(1-4) linked D-glucose units [13]. In the chitin structure hydroxyl at position C-2 of cellulose has been replaced by an acetamide group. Change of the oxygen atom to the nitrogen atom in chitin results in greater mechanical strength of chitin compared to cellulose. This results directly from stronger intermolecular interactions between chains. Chitin is therefore a linear polysaccharide, consisting of 2-acetylamino-2-deoxy-D-glucose units connected by β-glycosidic bonds at the 1,4 position. In nature, chitin occurs in complexes, bound by hydrogen bonds with proteins, other polycarbohydrates, and with multimolecular organic pigments (melanins). These complexes are often strengthened with calcium carbonate molecules that form additional chelate bonds. Chitin without foreign additives is a polymer with an ordered crystal structure whose macromolecules are bonded with hydrogen bonds. Native chitin is a linear polymer found in three polymorphs, types α, β, and γ (Figure 18.4.1), depending on the source of origin [14, 15]. Each of these types differs primarily in the arrangement of the chains in the crystalline area. In the case of α-chitin, we deal with the anti-parallel arrangement of polymer chains. In β-chitin, the polymer chains are arranged in parallel, so that the unit cell consists of only one chain. The last type, γ-chitin, has a unit cell composed of three chains. The central chain is directed anti-parallel to the other two. It is the rarest polymorph of chitin in nature.

Each of these forms has a different molecular weight. The largest is for α-chitin and the smallest for γ-chitin. Type α-chitin is the most common form in the natural world and occurs in many forms of living organisms, such as fungi, yeast cell walls, krill, crab shells, and also insect skeletons [16]. Although β-chitin is not a commonly found form, it occurs in places where, next to high strength, body components of a given species must be characterized by a considerable flexibility.

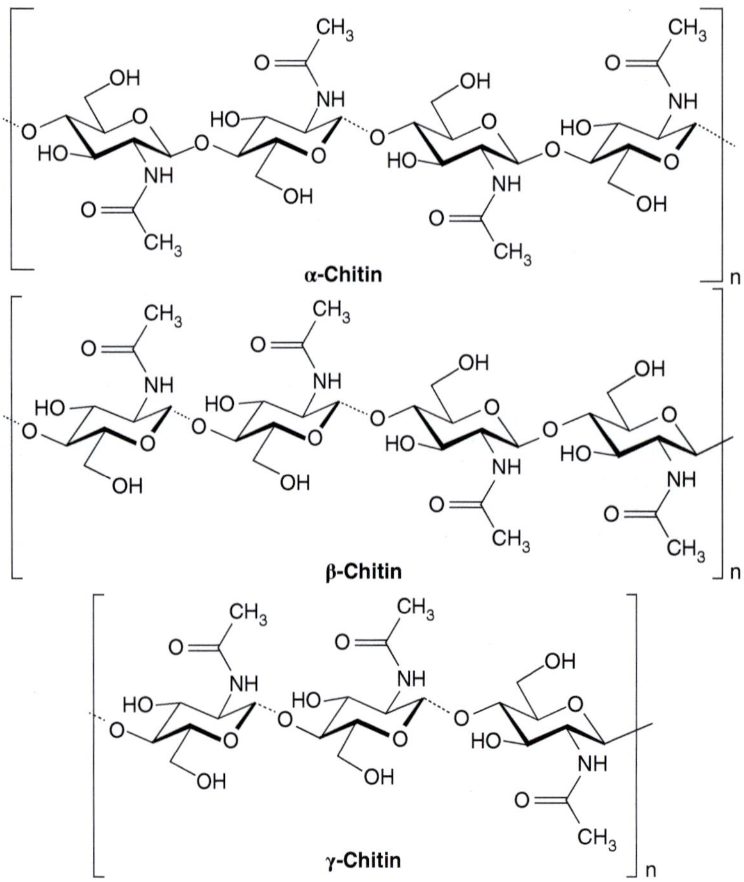

Figure 18.4.1 Three polymorphs, types α, β, and γ, of chitin.

This type of chitin can be extracted from squids and the skeletons of marine tube worms. The γ-type chitin crystal form occurs sporadically and it is assumed that this form is a derivative of the α form. In both of the above-mentioned α- and β-type structures, chitin polymer chains are arranged in layers, with numerous hydrogen bonds that integrate them, forming highly stable structures. The hydrogen bonds formed between the C—O···NH groups are characterized by a length of approximately 0.47 nm, corresponding to the "a" size of the unit cell of the crystal structure [16, 17]. In the α-chitin structures, in comparison to the β-chitin structure, the occurrence of numerous intermolecular hydrogen bonds between the layers forming the crystal structure are additionally observed. The length of these bonds is the same as the size "b" of the unit cell. The different structures of these two types affect the diversity of solubility in typical organic solvents. Polar solvents, such as water and alcohols, are able to diffuse into the β-type structure, causing its loosening, where such a phenomenon is not observed in the α-chitin [16].

18.4.1.3 Chitin Properties

Chitin has a hydrophobic character. Due to strong intermolecular interactions it is practically insoluble in ordinary organic and inorganic solvents. It dissolves in concentrated acid solutions, i.e. sulfuric (VI) acid, nitric (V) acid, or concentrated formic acid. During the dissolution of chitin, an acid hydrolysis process occurs, which causes strong degradation of chitin. Most often, chitin is dissolved in toxic, corrosive, and environmentally harmful organic solvents, i.e. sets of:

- lithium chloride (LiCl) – N-methylpyrrolidone,
- lithium chloride (LiCl) – N, N-dimethylacetamide1,2-chloroethanol – H_2SO_4 [16].

Poor solubility of chitin affects the limitations if its applications and hence it is used only in strictly defined areas. Recently, research has been conducted on the use of ionic liquids as alternative chitin solvents. Ionic liquids are a new generation of solvents, which are referred to as green. Due to a very limited amount of chitin solvents, its molar mass is most often determined by measuring the viscosity of very diluted solutions of chitin in dimethylacetamide (DMAc) containing 5–7% of lithium chloride (LiCl). This solvent creates stable solutions and does not cause the degradation of chitin [37]. Chitin has a relatively high molecular weight. In the case of viscometric average weight, it ranges from 100 000 to 600 000 (Da). The most valuable properties of chitin include: bioactivity, biodegradability, biocompatibility, non-toxicity, film-forming and fiber-forming, and good chelating properties of heavy metal ions. It has bioactivity, antibacterial [2] and antimicrobial [18] activities, anticancer activity, and anticoagulant and molecular adsorption properties [37], [2, 19]. It is physically versatile, can be processed in the form of powder, fibers, membranes, sponges, hydrogels, and scaffolds [2].

18.4.1.4 Recovery and Use of Chitin

The properties of chitin resulted in it finding wide application in various fields (Table 18.4.1). It is used to remove ions from waste water and drinking water, i.e. Hg^{2+}, Cu^{2+}, Pb^{2+}, Ni^{2+}, As^{3+}, Fe^{2+}, Cr^{3+}, as well as radioactive and toxic organic substances [40]. Chitin can be used as an additive in cosmetics [41], antibacterial agents [42], pharmaceutical support agents [11, 32, 37, 39], for the production of paper, finishing textiles, photo products, cement, heavy metal chelating agents, membranes, hollow fibers, waste removal [43], and in biomedical applications such as tissue engineering scaffolds, wound dressings, separation membranes, antibacterial coatings, stent coatings, and sensors [44], since they are harmless for the human body [2]. Moreover, like cellulose, chitin

Table 18.4.1 Applications of chitin and its derivatives as a biomaterial.

Area	Applications	References
Agriculture	Soil modifier; fungicide [20]; defensive mechanism in plants; seed coating; time release of fertilizers [21]; agro-bioscience [22]	[23–26]
Biocatalysis	Scaffolds as a support for metals in order to produce catalysts	[27, 28]
Bioengineering	Reduced susceptibility to echinocandin with elevated chitin levels in *C. albicans* cells	[28, 29]
Biomedical	Enzyme immobilization and purification chelator [11, 30]; emulsifier; flocculen [20]; blood cholesterol control [31]; lectin affinity chromatography; biosensor; immobilization of antibody in the presence of alginate [30]; haemostatic agents [28]	[22, 31, 32]
Biotechnological products	Packaging; manufacture of film and sponge sheet materials; hydrogels [33]	
Cosmetic	Lotions; hair additives; body creams [20]	[24, 26, 34, 35]
Energy production	*Clostridium paraputrificum* M-21 to produce hydrogen gas as a potential source of alternative energy	[36]
Food industries	Stabilizing agent [11]; dietary supplements; antioxidant [20]; emulsifying agent [27]; food preservation such as for edible films [37]; weight loss agent; food and feed additives	[22, 34, 35]
Medical	Fibers; membranes; artificial organs [20] and skin; surgical sutures [28]; bone and cartilage regeneration [37]; wound healing and dressings; cancer diagnosis [30, 38]; aid in cataract surgery; periodontal disease treatment; collagen synthesis [33]; contact lenses [28]; tumor therapy; stem cell technology [38]	[19, 33, 37]
Pharmaceutical	Manganese supplement complex; drug release [20]; gene [38]	[11, 32, 37, 39]
Pollutants removal	Copper removal capability to obtain more stable diesel oil	[36]
Water treatment	Dye removal [21]; absorbent for heavy and radioactive metals	[25]

Source: Pighinelli et al. [2].

is also the starting substrate for the preparation of various chitin derivatives by chemical reactions using the acetylamino group and hydroxyl groups. Hydrolysis in an alkaline environment is a reaction used to obtain the most important chitin derivative – chitosan [4]. Acid hydrolysis is used much less frequently than alkaline hydrolysis, which uses hydrochloric, nitric, and acetic acids. Hydrolysis in an acid environment leads to breaking of glycosidic bonds in the polymer chain. In this way oligochitin is obtained, which has a high biological activity, including anticancer, bactericidal, or fungicidal activities. The esterification of chitin is carried out using acid anhydrides such as carboxylic acid anhydrides with the usage of catalysts. This reaction produces the second important derivative of chitin, di-O-dibutyrylchitin. The esterification reaction is also used to regenerate chitin, i.e. for its recovery. The last group of reactions involving chitin are etherification reactions, which also produce soluble derivatives. Chitin ethers, however, are not completely known and studied [14].

In recent years, there has been a great interest in the application of chitin in biofuels, primarily due to the fact that chitin is also an important source of nitrogen and carbon. Blank et al. [45] proposed using a renewable nitrogen source, which is a waste product of the shellfish industry, to produce valuable biomass from algae. In an aqueous environment, chitin is effectively degraded during microbiological processes using chitinase, chitosanase, and chitin deacetylase enzymes [46, 47]. Currently, to produce biofuels from algae, nitrogen fertilizers obtained in the Haber–Bosch process are used [45]. However, the Haber–Bosch synthesis of conventional nitrogen fertilizers is highly unsustainable [48]. This is because the production of 1 kg of nitrogen fertilizer costs 57 MJ of energy, consuming up to 5% of the global energy supply [49] and accounts for up to 26% of the carbon footprint for algae biofuels [50]. It has a large carbon footprint [50] and life cycle analyses show that it accounts for up to 21% of the greenhouse gas footprint in the production of biomass from algae. Therefore, the authors proposed the use of chitin as a source of nitrogen. The energy footprint of chitin is minimal and it does not require much processing to make it accessible to algae. Chitin is made from renewable resources and the carbon footprint is minimal. Finally, chitin is obtained in a sustainable way from industries producing crustacean, insect armor, fungus, or yeast. The usage of chitin to produce biomass from algae can also significantly reduce production costs [51]. Nutrients make up a significant portion of the operating costs of algae production. The production of 1 tonne of algae biomass using conventional nitrogen requires 96 kg of ammonia and 1.9 tonnes of CO_2, which represents 7.2 and 15.7% of the production costs respectively, i.e. in total 22.9% [52]. For comparison, producing the same amount of biomass would require 1.9 tons of chitin from cost-free production out of the shellfish industry waste products. Due to the fact that carbon in chitin is oxidized by heterotrophic bacteria or is absorbed directly by algae, this may reduce the need to buy commercial CO_2. According to Blank et al. [45], the use of chitin can reduce the cost of producing biomass from algae by up to 22.9%. The application of chitin as the source of nitrogen is also confirmed by research by Zhaia et al. [53]. Chitin is a kind of renewable N-containing biomass resource achieved from the shell of crustaceans. In the recently proposed concept of shell biorefinery [54, 55], chitin and its derivatives have been identified as key feedstock to make N-containing chemicals [56, 57], polymers [58] and carbon-based materials [59]. Zhaia et al. [53] described a new type of Co—Ni bimetallic catalyst (Co_8Ni_2/NC) accompanied with nitrogen-doped carbon materials as support. The N-doped carbon support is derived from sustainable chitin. Such a catalyst shows excellent catalytic activity, high selectivity, and is a promising way to improve biofuel.

A different proposal for the use of chitin for the production of biofuels is presented by Aranda-Martinez et al. [60]. The authors demonstrated that chitin can be a substrate for the production of bioethanol with the participation of fungi. The results of this research are the starting point for the development of biofuels based on chitin and chitosan as the main components of crustacean waste. Fungal parasites of invertebrates such as the nematophagous *Pochonia chlamydosporia* (*Pc*) or the entomopathogens *Beauveria bassiana* (*Bb*) and *Metarhizium anisopliae* (*Ma*) are biocontrol agents of plant parasitic nematodes (e.g. *Meloidogyne* spp.) or insect pests such as the red palm weevil (*Rhynchophorus ferrugineus*). These fungi degrade chitin-rich barriers for the host. The authors demonstrated the possibilities of *Pc*, *Bb,* and *Ma* to reduce sugars using chitosan as the only nutrient. The best chitosan degrader was *Pc*, not only in aerobic but also in anaerobic conditions.

Another proposal for the use of chitin is presented by Wang et al. [61]. It is a simple and sustainable way to directly recover precious metal ions from oil/water emulsions by chitin nanofibrous membrane. The process of recovering precious metals from electronic waste is complex and consists of many stages: leaching noble metals from electronic wastes, extracting noble metal ions from the leaching solution, and reducing noble metal ions. In the first stage, leaching of precious metal ions from electronic materials is carried out using surfactants and organic solvents [62]. The resultant heterogeneous organic/water mixture and surfactant-stabilized emulsion, however, cannot be used to directly extract noble metal ions, as the surface of normal extracting materials are easily contaminated by surfactants or organic solvents [63]. Therefore, additional operations are introduced [64] that undoubtedly complicate the process. In order to simplify the process and increase efficiency, intelligent strategies for recovering precious metals are applied. An integrative material combining oil/water emulsions by chitin nanofibrous membrane is such a solution. The chitin nanofibrous membrane (CNFM) has a nanoporous structure with a superhydrophilic and underwater superoleophobic surface and could effectively separate oil/water emulsion. To endow oil/water separation materials with the ability of extracting/reducing noble metal ions, complicated fabrication procedures involving adding functional polymers containing specific ligands for noble metals have to be used.

The chitin content percentage in shrimp wastes (dry basis) varies from 14 to 30%. The percentage in weight of protein and mineral salts can be up to 40 and 35%, respectively [65, 66]. In addition, these residues contain large amounts of pigments, mainly astaxanthin [67] and the recovery and applications of such added value products is of increasing interest [68]. Astaxanthin is a xanthophyll that has been described as an excellent antioxidant and anticarcinogen as well as a retina precursor that can be used in a wide range of applications such as in aquaculture, food dyes, and cosmetics [67, 69]. Due to these valuable ingredients and chitin properties, a number of chitin recovery studies are being conducted. However, during chitin recovery, its quality can be diminished by depolymerization and astaxanthin can undergo degradation [66]. Therefore, efforts have been made to reduce chemical treatments during chitin recovery using more environmentally friendly processes such as bacterial fermentation, e.g. lactic acid fermentation (LAF) [68, 69]. Literature data confirm that LAF combined with chemical treatments is an alternative to chemical extraction for chitin recovery, reducing the amount of alkali and acid required [68]. LAF proved successful for CH recovery, as well as protein hydrolysates and pigments from crustacean wastes [70]. The pH and acid production have been clearly established as crucial factors in LAF for inhibition of pathogens, spoilage of microorganisms, and demineralization (DM) of these wastes. According to Pacheco et al. [65], the maximum protein removal was determined at the optimum lactic acid bacteria (LAB) growth, which proves the key role that LAB proteases play in the extraction of chitin using homofermentative *Lactobacillus plantarum* as a starter. According to Aranday-García et al. [70], the deproteinization is a major limitation of chitin recovery from crustacean waste. The authors used the combination of a two-stage solid state culture of *Lactobacillus brevis* and *Rhizopus oligosporus* to remove high protein. The protein removal was attributed to acid and neutral proteolytic activities, where the highest activity during culture was determined at pH 5. The extracted chitin presented an ash content of 10.24% and residual protein of 1.09%. The studies have confirmed the feasibility to utilize *R. oligosporus*, a GRAS microorganism, for in situ production of acid proteases that work efficiently on the chitin–protein network.

Usage of the lactic fermentation process of shrimp waste on solid substrates for chitin recovery was proposed by Cira et al. [68]. For this, the authors inoculated shrimp waste with lactobacilli as well as different types and levels of carbohydrate and subjected the mixture to fermentation. It was demonstrated that the efficiency of the fermentation process is influenced by a carbohydrate source. Better lactic acid production was obtained using sucrose compared to lactose and whey powder. Lactic acid fermentation was increased from 2 to 30 kg. The use of a column reactor with an appropriate design was an efficient way of shrimp waste ensilation, producing high protein and mineral removal and facilitating chitin extraction.

Crab waste can be used for the simultaneous production of chitin and L (+) – lactic acid [71]. Flores-Albino et al. [71] submerged fermentation of *Lactobacillus* sp. B2 using sugar cane molasses as the carbon source. Response surface methodology was applied to design the culture media considering demineralization. As a result of the fermentation process, 56% deproteinization with 34% yield of chitin and 19.5 g l^{-1} of lactic acid (77% yield) was obtained. Chitin purified after the fermentation showed 95% acetylation degree and 0.81 and 1 ± 0.125% residual ash and protein content, respectively.

18.4.2 Summary

Spanning into the first decade of the twenty-first century, the roughly 50 years can be considered the Golden Era of research in chitin, chitosan, and their derivatives. The valuable properties of chitin and a significant amount of waste from the shellfish industry was the impetus for research carried out by scientists around the world, which led to the mastery over this intractable issue. These studies covered the following areas:

1. The recovery of chitin primarily from shellfish sources.
2. Structural studies that led to the advent of many procedures and techniques to elucidate the
3. properties of chitin.
4. The chemical derivatization of chitin and chitosan.
5. The investigations into the myriad of potential applications for chitin material.

Some conclusions will have to be verified in future technological and economic research. To date, the generation of chitin-based products has not made the desired impact that was anticipated. There is still a huge potential for chitin in its applications, so further commercial use of chitin will have to be more thoughtful. For each product, the reproducibility of material and manufacturing at acceptable costs will have to be worked out. It is known that the proposed solutions are compatible with the circular economy because they are reducing the amount of waste from the shellfish industry and provide an alternative procedure for the production of many valuable products.

18.4.3 Carbon-Coated Magnet-Sensitive Nanoparticles

For several years there has been a rapid development of nanotechnology, meaning techniques for producing materials and structures with nanometric sizes, i.e. up to ~100 nm. Currently, nanotechnology is one of the most dynamically developing fields of science and technology, combining selected areas of solid-state physics, chemistry, material science, and even molecular biology. Reducing the size of a selected material to the nanometric scale creates the possibility of numerous potential applications. Nanostructures (which allow unparalleled mechanical properties to be obtained), artificial fibers (with very precise molecular structures), carbon nanotubes (CNTs), nanoparticles (NPs), thin layers and superstructures of various materials are currently being produced in scientific laboratories around the world, where their properties are also studied [72].

Potential applications of modern nanomaterials or nanocomposites include various areas of life, so the key issue is a beneficial combination of chemical, physical, and strength properties of the phases that make up these materials. In many scientific centers around the world, NPs of various materials are being intensively researched, in the hope that they will revolutionize many areas of life [73]. Nanomaterials have also begun to revolutionize the world around us; they have been used in various scientific fields such as biotechnology, engineering, biomedical, environmental, catalysis, drug delivery, and materials science [73, 74]). Nanotechnology has become an important technique that has made it possible to work at the molecular and cellular levels to produce many recent advances in the field of research and development. High hopes are also associated with the use of NPs in modern medicine, e.g. in cancer therapy. Over the past decade, much research has focused on magnetic nanoparticles (MNPs) [75–78]. MNPs are very sensitive to oxidation and agglomeration due to having a large specific surface area and are also high in its chemical reactivates as well as its magnetic dipole interaction [79]. Under ambient conditions, a rapid oxidation process occurs on the surface of NPs with the formation of a thin oxide layer, which radically changes the properties of the particles. Another problem is natural agglomeration of nanoparticles into larger clusters. In order to preserve specific magnetic properties and protect nanoparticles from oxidation and agglomeration, the application of the encapsulation procedure has been proposed. Encapsulation of nanoparticles has been successfully employed by using carbon, silica, precious metals, metal oxides, organic polymers, and surfactants. NMs contain carbon and are found in morphologies such as hollow tubes, ellipsoids, or spheres. Fullerenes (C_{60}), CNTs, carbon nanofibers, carbon black, graphene (Gr), and carbon onions are included under the carbon-based NMs category [72]. Carbon encapsulated MNPs (CEMNPs) have received considerable attention in recent years because of their potential as small magnetic clusters for high density magnetic data storage, in ferrofluid applications, and magnetic resonance imaging [80, 81]. Carbon is one of the best solutions for encapsulation of metal NPs, because carbon is more thermally stable, cheap, lightweight, and biocompatible. Due to these advantages, various methods of preparation of carbon capsulated NP have been developed, such as arc discharge techniques, chemical vapor depositions, combustion, pulse laser decomposition, and pyrolysis of metal complexes. Interest in carbon–metal systems is due to the possibility of obtaining nanocrystalline metal particles encapsulated by crystalline or amorphous carbon. Park et al. [82] observed that graphite coatings have a better chemical stability in acids with regard to a material without this coating. Due to the potential benefits of MNPs, researchers have shown the possibility of their application in industry, medicine, and environmental applications. In biomedics, there are currently two strategies for producing multifunctional magnetic nanostructures based on nanoparticles. The first is molecular functionalization involving the attachment of antibodies, proteins, and dyes to MNPs. The other method integrates the MNPs with other functional nanocomponents, such as quantum dots (QDs) or metallic NPs. MNPs and nanoparticle-based materials offer attractive opportunities for the development of new therapeutic strategies in targeted drug delivery, magnetic separation-based blood purification, or as magnetic field-driven artificial muscles. The synthesis of MNPs with a metal core protected by a carbon shell has only recently been achieved and gives access to magnetic carriers with improved magnetic properties (saturation magnetizations up to three times higher than commonly used metal oxide particles), extending both targeting and hyperthermia applications [83–85]. Due to their unique properties, NMPs can also be used in other areas. Catalysts supported on MNPs are known. The surface of MNPs can be functionalized to accommodate a wide variety of organic and organometallic catalysts. An example can be the research by Dálaigh et al. [86]. The authors developed the first magnetic NPs-supported 4-N,N-dimethylaminopyridine (DMAP) analog, for use as a robust heterogeneous nucleophilic catalyst with unprecedented activity and recyclability. Also, Hu et al. [87] designed novel magnetite NPs supported by chiral Ru complexes that catalyze heterogeneous asymmetric hydrogenation of aromatic ketones to remove inorganic impurities. Pure MNPs can be used for this purpose, but it is generally necessary to modify the surfaces of the synthesized MNPs in order to attach some functional groups. An example is the use of magnetic Fe_2O_3

nanoparticles as an adsorbent for the removal and recovery of Cr (VI) from wastewaters. Studies have shown that the adsorption capacity of NMPs is very high. MNPs have been synthesized with a number of different compositions and phases, including pure metals Fe, Co, and Ni; metal oxides, such as Fe_3O_4 and γ-Fe_2O_3; ferrites, such as MFe_2O_4 (M = Cu, Ni, Mn, Mg, etc.); and metal alloys, such as FePt, CoPt [87, 88]. A number of studies have been conducted in recent years on the use of magnetic carbon nanotubes (MCNTs) in medicine, diagnostics, cell biology, analytical chemistry, and environmental technology [89]. FeCo nanoparticles that have a core diameter of 10 nm and bear a graphitic carbon shell decorated with poly(ethylene glycol) provide an MPI signal intensity that is sixfold and fifteenfold higher than the signals from the superparamagnetic iron oxide tracers VivoTrax and Feraheme respectively, at the same molar concentration as iron. As sensitive multifunctional and multimodal imaging tracers, carbon-coated FeCo nanoparticles may confer advantages in cancer imaging and hyperthermia therapy [90]. The magnetic Fe_3O_4 NPs coated with carbon (Fe_3O_4@C NPs) were expected to have good biocompatibility and chemical stability, as well as ease of further surface functionalization and modification to introduce additional moleculars or ligands due to the thermodynamic stability of carbon in various multiatomic structures under room conditions. The functionalized or modified core–shell magnetic Fe_3O_4@C NPs have been used as adsorbents for the high-efficient adsorption and removal of heavy metal ions or antibiotics from aqueous media [91, 92]. The novel process for fabricating Fe_3O_4@1C NPs involves the formation of ultrasmall Fe_3O_4 NPs core and in situ growth of one single carbon layer. Moreover, the one single carbon layer is very homogeneously coated on the surface of the Fe_3O_4 core. A possible formation process of the Fe_3O_4@1C NPs is schematically shown in Figure 18.4.2 [61].

The introduction of MCNTs paved the way for the emergence of new approaches in nanobiotechnology and biomedicine as a result of their multifarious properties embedded within either the

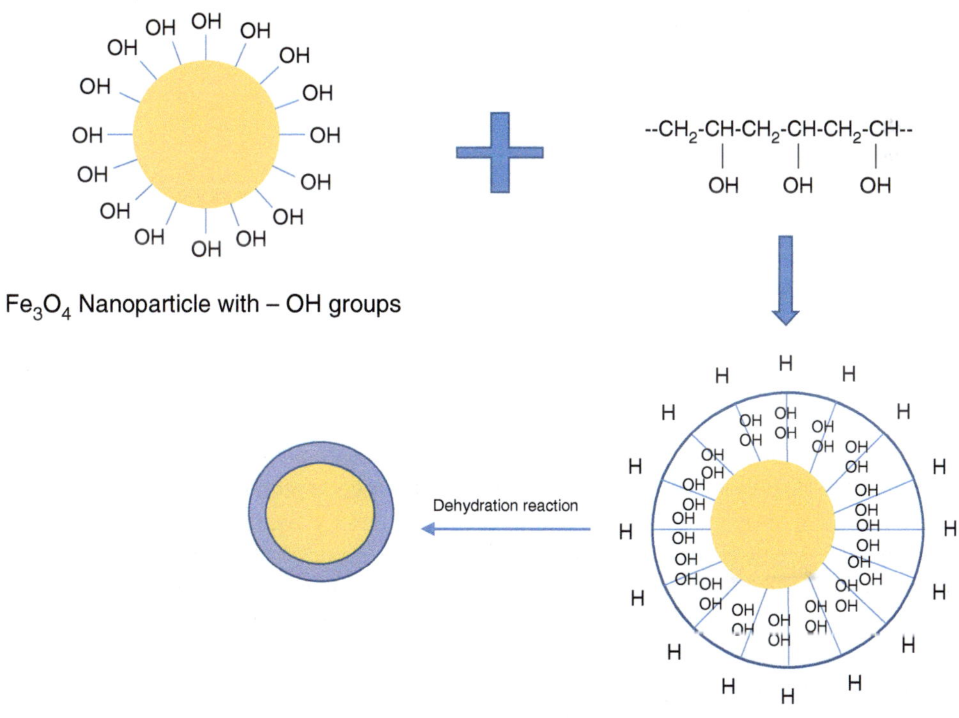

Figure 18.4.2 Formation process of the Fe_3O_4@1C NPs. Source: Wang et al. [61].

CNTs or magnetic parts. In the biomedical field, the most promising approach is the development of non-cytotoxic vectors with targeting ability exploited by the presence of MNPs that can be guided through selected tissues by an external magnetic field. In the industrial field, they are unique nanostructures that adapt and show the desired properties of any other material. They have special electronic and mechanical properties that lead to incredible forms of strength and conductivity. Due to these features, the range of MNP applications is almost unlimited, ranging from reinforcements in composites, sensors, and probes to energy storage, electrochemical devices, and nanometer sized electronics. It is believed that CNT can revolutionize the world. Therefore, there is no doubt that the future of magnetic CNT is very promising.

18.4.4 Proteins, Carbohydrates, and Humic Acid

It is known that sewage sludge usually consists of many valuable organic substances such as, for example, nucleic acids, humic acids, proteins, and polysaccharides. These components can be recovered, which gives a different perspective on waste management [93]. According to Garcia et al. [94], hydrothermal treatment is currently one of the most attractive alternatives to sludge management. Much research has been conducted into the use of hydrothermal treatment processes as a pre-treatment or post-treatment technology to stabilize and/or minimize sludge. However, there is still a lack of knowledge about products produced during these processes. This information is a key step in assessing the recovery of valuable products from sludge, mainly proteins, humic acids, and carbohydrates, which can significantly improve the economic sustainability of hydrothermal treatment.

Recovering protein from sludge is a simple approach to convert "wasted" sludge into useful "product." For this purpose, sludge pre-treatment methods are used, the release/solubilization of protein from sludge, including physical, thermal, chemical, biological treatments, and also combined processes. Protein recovered from sewage sludge can be used as fertilizers, adhesives, or food for animals [95] and can be a new renewable source for both energy and resources recovery [96]. The protein thus obtained constitutes alternatives for poultry feedstuff, wood adhesive, and fire extinguishing foam. The separation of protein from the sewage sludge reduces the nitrogen content in the sewage and therefore in the environment. The process of protein recovery from sewage sludge consists of several steps, starting from screening, then pre-treatments for protein solubilization, subsequently exposure to filtration, and precipitating protein from protein solution. Ultimate drying of protein precipitation and hence protein product (crude protein) can be separated and recovered. According to the literature data, protein release/solubilization from sludge is the step determining protein recovery. The best protein solubilization effects were obtained after an alkaline pre-treatment at pH = 12. Pre-treatment can change the protein properties (e.g. molecular weight, conformation, type, etc.). Proper selection of chemicals for protein separation is also important for effective protein recovery, and a purification process is required to separate the protein from non-protein components of the mixture. The principle for protein separation is to disrupt the hydration shell of proteins and precipitate proteins from the liquid phase to the solid phase [97]. The main reagents used for protein precipitation are: hydrochloric acid, sulfuric acid, sodium lignosulphonate, trichloroacetic acid, and ammonium sulphate. García et al. [94] demonstrated that the addition of ammonium sulfate was the best separation method, with a protein recovery of approximately 87%. According to Pervaiz and Sain [98] and Hwang et al. [99], during protein precipitation at pH 3, its recovery can be as high as 80–92%. After the solubilization, further protein separation and purification from the complex solution takes place. The best protein purification effects were obtained using pH graded gel, an ion exchange column, size exclusion chromatography, and high-performance liquid chromatography.

Another alternative of protein recovery from waste was proposed by Lamp et al. [100]. According to the authors, biogenic residues offer a high potential for protein recovery. The authors investigated protein recovery from bioethanol decoction by treatment with liquid hot water at pH 3–11 and 110–210 °C for 10–90 minutes. The highest protein recovery efficiency was achieved at the level of 75%.

Shrimp waste is also a valuable source of protein. Deng et al. [101] used an enzyme process to recover most of the shrimp shell waste components, including protein, chitin, astaxanthin, and minerals. The process used two recombinant aspartic proteases for protein hydrolysis, recombinant chitinase for chitin hydrolysis, and ethyl acetate for astaxanthin extraction. The proposed enzyme process promotes sustainable development by introducing circular economy goals in shrimp processing, thus offering economic and environmental benefits. Research on protein recovery from shrimp waste was also conducted by Forghani et al. [102]. For this purpose, the authors subjected a boiling biomass from waste to flocculation and sedimentation using food polysaccharides (carrageenan, alginate, and carboxymethylcellulose (CMC)) as flocculants. Protein recovery ranging from 60 to 78.5% on a dry weight basis was obtained.

According to the literature data, further studies are needed to achieve complete purification of the protein. The new perspective of sludge management is open to the recovery of valuable compounds, but the lack of operational/environmental standards as well as an appropriate incentive policy for protein recovery from waste may make it difficult to transfer research from the laboratory scale to the commercial market. Future research directions in this area should focus on the recovery of a specific protein by modifying the protein structure with a specific pre-treatment method.

In the case of humic substances (HSs) occurring in municipal waste, Gua et al. [103] proposed their recovery from municipal leachate. For this purpose, for the recovery of resources from leachate, the authors used an innovative process two-stage anaerobic digestion comprising an expanded granular sludge bed reactor and an anaerobic membrane. The recovery of HS is particularly important in terms of their fertilizing values because the use of HS increases the content of organic matter in soil, improves the structure of soil aggregates, improves fertilizer efficiency, and stimulates plant growth. Therefore, HSs are widely used as a new type of fertilizer in agriculture [104]. Currently, in order to recover HA from the leachate, mainly chemical precipitation methods and membrane filtration are used. However, the chemical precipitation process is carried out at low pH (pH = 2), and thus the recovered HS is not suitable for use as a fertilizer [105]. Membrane filtration is a method for efficient HS recovery. HS particles found in leachates are mainly in the size classes of less than 1 kDa and between 1 and 3.5 kDa [106]. The molecular weight cut-off (MWCO) of the nanofiltration (NF) membrane is 200–1000 Da. Although NF membranes are detrimental to the fractionation and desalination of leachate concentrate due to the deterioration of membrane flux caused by the retention of all divalent salts [107], the use of the loose NF membrane (MWCO 860 Da) allows the recovery of HS at the level of 95.7%. The leachates always contain toxic heavy metals, which may limit the re-use of the recovered HS as fertilizer. However, according to Xu et al. [108], the use of a two-step process with a sealed UF membrane allows the concentration of heavy metals in the concentrate to be reduced below standard limits for water-soluble fertilizers. The efficiency of HS recovery from leachate depends on many factors, e.g. pressure, temperature and pH [109]. Under high operating pressures, the so-called dilution effect and concentration polarization promote the separation behavior of the solute [110]. In a two-stage tight ultrafiltration membrane system [108], an HS recovery efficiency of over 90% can be obtained. Temperature influences the permeate transport behavior and the film structure of the membrane [111]. With increasing temperatures, the pore size of the membrane increases and the solute diffusion coefficient increases [109]. In general, the temperature and pH of the leachate are favorable for the recovery of the HS by membrane filtration. However, more energy is required to manage flux by increasing the operating pressure. Therefore, the optimal applied pressure must be selected when designing the UF process conditions for HS recovery from municipal waste leachate.

References

1 Jaganathan, K. (2016). Extraction and characterization of chitin from marine bycatch crustaceans employing fermentation method. *World Journal of Pharmacy and Pharmaceutical Sciences* 5 (1): 1290–1301.

2 Pighinelli, L., Broquá, J., Zani, B.G. et al. (2019). Methods of chitin production. *American Journal of Biomedical Science & Research* 3 (4): 307–314.

3 Kaya, M., Seyyar, O., Baran, T. et al. (2014). Physicochemical characterization of the fully acetylated chitin structure isolated from two spider species. *International Journal of Biological Macromolecules* 65: 553–558.

4 Merzendorfer, H. and Zimoch, L. (2003). Chitin metabolism in insects: structure, function and regulation of chitin synthases and chitinases. *The Journal of Experimental Biology* 206 (24): 4393–4412.

5 Juárez-de la Rosa, B.A., Quintana, P., Ardisson, P.L. et al. (2012). Effects of thermal treatments on the structure of two black coral species chitinous exoskeleton. *Journal of Material Science* 47 (2): 990–998.

6 Blank, C.E. and Hinman, N.W. (2016). Cyanobacterial and algal growth on chitin as a source of nitrogen; ecological, evolutionary, and biotechnological implications. *Algal Research* 15: 152–163.

7 Jafari, A.M., Gharibi, S., Farjadmand, F., and Sadighara, P. (2012). Extraction of shrimp waste pigments by enzymatic and alkaline treatment: evaluation by inhibition of lipid peroxidation. *Journal of Material Cycles and Waste Management* 14: 411–413.

8 Boβelmann, F., Romano, P., Fabritius, H. et al. (2007). The composition of the exoskeleton of two crustacea: the American lobster *Homarus americanus* and the edible crab *Cancer pagurus*. *Thermochimica Acta* 463: 65–68.

9 Hsieh, C.-H. and Wu, W.-T. (2009). Cultivation of microalgae for oil production with a cultivation strategy of urea limitation. *Bioresource Technology* 100: 3921–3926.

10 Xu, Y., Gallert, C., and Winter, J. (2008). Chitin purification from shrimp wastes by microbial deproteination and decalcification. *Applied Microbiology and Biotechnology* 79: 687–697.

11 Arbia, W. (2013). Extraction from crustacean shells using biological methods a review. *Food Technology and Biotechnolog* 51 (1): 12–25.

12 Ameh, A.O., Isa, M.T., Adeleye, T.J., and Adama, K. (2013). Kinetics of demineralization of shrimp exoskeleton in chitin and chitosan synthesis. *Journal of Chemical Engineering and Materials Science* 4 (3): 32–37.

13 Islam, S., Bhuiyan, M.A.R., and Islam, M.N. (2017). Chitin and chitosan: structure, properties and applications in biomedical engineering. *Journal of Polymers and the Environment* 25: 854–866.

14 Elieh-Ali-Komiand, D. and Hamblin, M.R. (2016). Chitin and chitosan: production and application of versatile biomedical nanomaterials. *International Journal of Advanced Research (Indore)* 4 (3): 411–427.

15 Namboodiri, M.M.T. and Pakshirajan, K. (2020). Chapter 10 – Valorization of waste biomass for chitin and chitosan production. In: *Waste Biorefinery Integrating Biorefineries for Waste Valorisation*, 241–266. Elsevier.

16 Rinaudo, M. (2006). Chitin and chitosan: properties and applications. *Progress in Polymer Science* 31: 603.

17 Pillai, C.K.S., Paul, W., and Sharma, C.P. (2009). Chitin and chitosan polymers: chemistry, solubility and fiber formation. *Progress in Polymer Science* 34: 641.

18 Kim, M., Han, Y., Jo, Y. et al. (2016). Extraction of chitin and chitosan from housefly, *Musca domestica*, pupa shells. *Entomological Research* 46 (5): 324–328.

19 Younes, I. (2012). Chitin and chitosan preparation from shrimp shells using optimized enzymatic deproteinization. *Process Biochemistry* 47 (12): 2032–2039.

20 Cao, W., Tan, C., Zhan, X. et al. (2014). Ultraviolet irradiation and gradient temperature assisted autolysis for protein recovery from shrimp head waste. *Food Chemistry* 164: 136–141.

21 Khor, E. and Lim, L.Y. (2003). Implantable applications of chitin and chitosan. *Biomaterial* 24 (13): 2339–2349.

22 Ofem, M. (2017). Properties of chitin reinforces composites: a review. *Nigerian Journal of Technology* 36 (1): 57–71.

23 Azuma, K. (2014). Preparation and biomedical applications of chitin and chitosan nanofibers. *Journal of Biomedical Nanotechnology* 10 (10): 2891–2920.

24 Carvalho, L.B., Nader, H.B., and Bezerra, R.S. (2012). Recovery of protein, chitin, carotenoids and Glycosaminoglycans from Pacific white shrimp (*Litopenaeus vannamei*) processing waste. *Process Biochemistry* 47 (4): 570–5779.

25 Jalal, A.F., Risheed, C.M., and Ibrahim, B.I. (2012). Optimization of chitin extraction from chicken feet. *Journal of Analytical & Bioanalytical Techniques* 3 (05): 1–5.

26 Younes, I., Hajji, S., Rinaudo, M. et al. (2015). Optimization of proteins and minerals removal from shrimp shells to produce highly acetylated chitin. *International Journal of Biological Macromolecules* 84: 246–253.

27 Puvvada, Y.S., Vankayalapati, S., and Sukhavasi, S. (2012). Extraction of chitin from chitosan from exoskeleton of shrimp for application in the pharmaceutical industry. *International Current Pharmaceutical Journal* 1 (9): 258–263.

28 Mao, X., Zhang, J., Kan, F. et al. (2013). Antioxidant production and chitin recovery from shrimp head fermentation with *Streptococcus thermophilus*. *Food Science and Biotechnology* 22 (4): 1023–1032.

29 Pachapur, V.L., Guemiza, K., Rouissi, T. et al. (2016). Novel biological and chemical methods of chitin extraction from crustacean waste using saline water. *Journal of Chemical Technology and Biotechnology* 91 (8): 2331–2339.

30 Hongkulsup, C., Khutoryanskiy, V.V., and Niranjan, K. (2016). Enzyme assisted extraction of chitin from shrimp shells (*Litopenaeus vannamei*). *Journal of Chemical Technology and Biotechnology* 91 (5): 1250–1256.

31 Mao, X. (2017). Comprehensive utilization of shrimp waste based on biotechnological methods: a review. *Journal of Cleaner Production* 143: 814–823.

32 Liu, P., Liu, S., Guo, N. et al. (2014). Cofermentation of *Bacillus licheniformis* and *Gluconobacter oxydans* for chitin extraction from shrimp waste. *Biochemical Engineering Journal* 91: 10–15.

33 Yang, G. (2016). Chitin and chitosan: production and application of versatile biomedical nanomaterials. *International Journal of Advanced Research (Indore)* 4 (3): 411–427.

34 Muzzarelli, R.A.A. (2011). Biomedical exploitation of chitin and chitosan via mechanochemical disassembly. Electrospinning, dissolution in imidazolium ionic liquids, and supercritical drying. *Marine Drugs* 9 (9): 1510–1533.

35 Kandra, P., Challa, M.M., and Jyothi, H.K.P. (2012). Efficient use of shrimp waste: present and future trends. *Applied Microbiology and Biotechnology* 93 (1): 17–29.

36 Kaur, S. and Dhillon, G.S. (2014). The versatile biopolymer chitosan: potential sources, evaluation of extraction methods and applications. *Critical Reviews in Microbiology* 40 (2): 155–175.

37 Kumari, S. and Rath, P. (2014). Extraction and characterization of chitin and chitosan from (*Labeo rohit*) fish scales. *Procedia Materials Science* 6: 482–489.

38 Hamed, I., Ozogul, F., and Regenstein, J.M. (2016). Industrial applications of crustacean by-products (chitin, chitosan, and chitooligosaccharides): a review. *Trends in Food Science and Technology* 48: 40–50.

39 Ameh, A.O. (2014). Kinetics of demineralization of shrimp shell using lactic acid. *Leonardo Electronic Journal of Practices and Technologies* 13 (24): 13–22.

40 Mincea, M., Negrulescu, A., and Ostaf, V. (2012). Preparation, modification, and applications of chitin nanowhiskers: a review. *Reviews on Advanced Materials Science* 30: 225–242.

41 Gautier, S., Xhauflaire-Uhoda, E., Gonry, P., and Piérard, G.E. (2008). Chitin-glucan, a natural cell scaffold for skin moisturization and rejuvenation. *International Journal of Cosmetic Science* 30: 459–469.

42 Potara, M., Jakab, E., Damert, A. et al. (2011). Synergistic antibacterial activity of chitosan-silver nanocomposities *Staphylococcus aureus*. *Nanotechnology* 22 (13): 135101.

43 Zargar, V., Asghari, M., and Dashti, A. (2015). A review on chitin and chitosan polymers: structure, chemistry, solubility, derivatives, and applications. *ChemBioEng* 2 (3): 204–226.

44 Jayakumar, R., Nwe, N., Tokura, S., and Tamura, H. (2007). Sulfated chitin and chitosan as novel biomaterials. *International Journal of Biological Macromolecules* 40 (3): 175–181.

45 Blank, C.E., Parks, R.W., and Hinman, N.W. (2016a). Chitin: a potential new alternative nitrogen source for the tertiary, algal-based treatment of pulp and paper mill wastewater. *Journal of Applied Phycology* 28: 753–2766.

46 Azuma, K., Izumi, R., Osaki, T. et al. (2015). Chitin, chitosan, and its derivatives for wound healing: old and new materials. *Journal of Functional Biomaterials* 6 (1): 104–142.

47 Wan, A.C.A. and Tai, B.C.U. (2013). CHITIN – a promising biomaterial for tissue engineering and stem cell technologies. *Biotechnology Advances* 1: 1776–1785.

48 Peccia, J., Haznedaroglu, B., Gutierrez, J., and Zimmerman, J.B. (2013). Nitrogen supply is an important driver of sustainable microalgae biofuel production. *Trends in Biotechnology* 31: 134–138.

49 Erisman, J.W., Sutton, M.A., Galloway, J. et al. (2008). How a century of ammonia synthesis changed the world. *Nature Geoscience* 1: 636–639.

50 Huo, Y.X., Wernick, D.G., and Liao, J.C. (2012). Toward nitrogen neutral biofuel production. *Current Opinion in Biotechnology* 23: 406–413.

51 Pate, R., Klise, G., and Wu, B. (2011). Resource demand implications for US algae biofuels production scale-up. *Applied Energy* 88: 3377–3388.

52 Davis, R., Aden, A., and Pienkos, P.T. (2011). Techno-economic analysis of autotrophic microalgae for fuel production. *Applied Energy* 88: 3524–3531.

53 Zhaia, Y., Chua, M., Shang, N. et al. (2020). Bimetal Co_8Ni_2 catalyst supported on chitin-derived N-containing carbon for upgrade of biofuels. *Applied Surface Science* 506: 144681.

54 Chen, X., Yang, H., and Yan, N. (2016). Shell biorefinery: dream or reality? *Chemistry – A European Journal* 22: 13402–13421.

55 Yan, N. and Chen, X. (2015). Don't waste seafood waste. *Nature* 524: 155–157.

56 Chen, X., Chew, S.L., Kerton, F.M., and Yan, N. (2014). Direct conversion of chitin into a N-containing furan derivative. *Green Chemistry* 16: 2204–2212.

57 Gao, X., Chen, X., Zhang, J. et al. (2016). Transformation of chitin and waste shrimp shells into acetic acid and pyrrole. *ACS Sustainable Chemistry & Engineering* 4: 3912–3920.

58 Dai, J., Gözaydın, G., Hu, C., and Yan, N. (2019). Catalytic conversion of chitosan to glucosaminic acid by tandem hydrolysis and oxidation. *ACS Sustainable Chemistry & Engineering* 7: 12399–12407.

59 Zhang, Z., Jung, J.C., and Yan, N. (2016). Designed synthesis of MOx (M = Zn, Fe, Sn, Ni, Mn Co, Ce, Mg, Ag, Pt, and Au) nanoparticles supported on hierarchical CuO hollow structures. *Nanoscale* 8: 19684–19695.

60 Aranda-Martinez, A., Ortiz, M.A.N., García, I.S.A. et al. (2017). Ethanol production from chitosan by the nematophagous fungus *Pochonia chlamydosporia* and the entomopathogenic fungi *Metarhizium anisopliae* and *Beauveria bassiana*. *Microbiological Research* 204: 30–39.

61 Wang, Z., Li, P., Fang, Y. et al. (2019). One-step recovery of noble metal ions from oil/water emulsions by chitin nanofibrous membrane for further recycling utilization. *Carbohydrate Polymers* 223: 115064.

62 Sun, Z., Cao, H., Xiao, Y. et al. (2016). Toward sustainability for recovery of critical metals from electronic waste: the hydrochemistry processes. *ACS Sustainable Chemistry & Engineering* 5 (1): 21–40.

63 Shah, A., Shahzad, S., Munir, A. et al. (2016). Micelles as soil and water decontamination agents. *Chemical Reviews* 116 (10): 6042–6074.

64 Andrade, L.N., Amorim, C.C., Santos, S.V. et al. (2015). Efficient demulsification of wastewater by steel furnace dust with amphiphilic and surface charge properties. *Chemical Engineering Journal* 271: 281–286.

65 Pacheco, N., Garnica-González, M., Ramírez-Hernández, J.Y. et al. (2009). Effect of temperature on chitin and astaxanthin recoveries from shrimp waste using lactic acid bacteria. *Bioresource Technology* 100 (11): 2849–2854.

66 Synowiecki, J. and Al-Khateeb, N.A. (2003). Production, properties, and some new applications of chitin and its derivatives. *Critical Reviews in Food Science and Nutrition* 43: 145–171.

67 Gimeno, M., Ramírez-Hernández, J.Y., Mártinez-Ibarra, C. et al. (2007). One-solvent extraction of astaxanthin from lactic acid fermented shrimp wastes. *Journal of Agricultural and Food Chemistry* 55: 10345–10350.

68 Cira, L.A., Huerta, S., Hall, G.M., and Shirai, K. (2002). Pilot scale lactic acid fermentation of shrimp wastes for chitin recovery. *Process Biochemistry* 37: 1359–1366.

69 Bhaskar, N., Sures, P.V., Sakhare, P.Z., and Sachindra, N.M. (2007). Shrimp biowaste with *Pediococcus acidolactici* CFR2182: optimization of fermentation conditions by response surface methodology and effect of optimized conditions on DP and carotenoid recovery. *Enzyme and Microbial Technology* 40: 1427–1434.

70 Aranday-García, R., Guerreroa, A.R., Ifukub, S., and Shiraia, K. (2017). Successive inoculation of *Lactobacillus brevis* and *Rhizopus oligosporus* on shrimp wastes for recovery of chitin and added-value products. *Process Biochemistry* 58: 17–24.

71 Flores-Albino, B., Arias, L., Gómez, J. et al. (2012). Chitin and *L*(+)-lactic acid production from crab (*Callinectes bellicosus*) wastes by fermentation of *Lactobacillus* sp. B2 using sugar cane molasses as carbon source. *Bioprocess and Biosystems Engineering* 35: 1193–1200.

72 Jeevanandam, J., Barhoum, A., Chan, Y.S. et al. (2018). Review on nanoparticles and nanostructured materials: history, sources, toxicity and regulations. *Beilstein Journal of Nanotechnology* 9: 1050–1074.

73 Khan, I., Saeed, K., and Khan, I. (2019). Properties, applications and toxicities. *Arabian Journal of Chemistry* 12 (7): 908–931.

74 Tartaj, P., Morales, M.P., Veintemillas-Verdaguer, S. et al. (2003). The preparation of magnetic nanoparticles for applications in biomedicine. *Journal of Physics D: Applied* 36: R182–R197.

75 Faivre, D. and Bennet, M. (2016). Materials science: magnetic nanoparticles line up. *Nature* 535: 235–236.

76 Faraji, M. (2016). Recent analytical applications of magnetic nanoparticles. *Nanochemistry Research* 1 (2): 264–290.

77 Mohammed, L., Gomaa, H.G., Ragab, D., and Zhu, J. (2017). Magnetic nanoparticles for environmental and biomedical applications: a review. *Particuology* 30: 1–14.

78 Wierucka, M. and Biziuk, M. (2014). Application of magnetic nanoparticles for magnetic solid-phase extraction in preparing biological, environmental and food samples. *TrAC Trends in Analytical Chemistry* 59: 50–58.

79 Lu, A.-H., Salabas, E.L., and Schüth, F. (2007). Magnetic nanoparticles: synthesis, protection, functionalization, and application. *Angewandte Chemie International Edition* 46 (8): 1222–1244.

80 Faraji, M., Yamini, Y., and Rezaee, M. (2010). Magnetic nanoparticles: synthesis, stabilization, functionalization, characterization, and applications. *Journal of the Iranian Chemical Society* 7 (1): 1–37.

81 Kang, J., Kim, Y., Kim, H.-M. et al. (2017). In-situ one-step synthesis of carbon-encapsulated naked magnetic metal nanoparticles conducted without additional reductants and agents. *Scientific Reports* 6: 1–9.

82 Park, J.B., Jeong, S.H., Jeong, M.S. et al. (2008). Synthesis of carbon-encapsulated magnetic nanoparticles by pulsed laser irradiation of solution. *Carbon* 46: 1369–1377.

83 Bircher, L., Theusinger, O.M., Locher, S. et al. (2014). Characterization of carbon-coated magnetic nanoparticles using clinical blood coagulation assays: effect of PEG-functionalization and comparison to silica nanoparticles. *Journal of Materials Chemistry B* 2: 3753–3758.

84 Grass, R.N., Athanassiou, E.K., and Stark, W.J. (2007). Covalently functionalized cobalt nanoparticles as a platform for magnetic separations in organic synthesis. *Angew. Chem. Int. Ed. Engl.* 46 (26): 4909–4912.

85 Herrmann, I.K., Grass, R.N., Mazunin, D., and Stark, W.J. (2009). Synthesis and covalent surface functionalization of nonoxidic iron core–shell nanomagnets. *Chem. Mater.* 21 (14): 3275–3281.

86 Dálaigh, C.O., Corr, S.A., Guňko, Y., and Connon, S.J. (2007). A magnetic-nanoparticle-supported 4-N,N-dialkylaminopyridine catalyst: excellent reactivity combined with facile catalyst recovery and recyclability. *Angewandte Chemie International Edition* 46: 4329–4332.

87 Hu, A., Yee, G.T., and Lin, W. (2005). Magnetically recoverable chiral catalysts immobilized on magnetite nanoparticles for asymmetric hydrogenation of aromatic ketones. *Journal of the American Chemical Society* 127: 12486–12487.

88 Jeong, U., Teng, X., Wang, Y. et al. (2007). Superparamagnetic colloids: controlled synthesis and niche applications. *Advanced Materials* 19 (1): 33–60.

89 Samadishadlou, M., Farshbaf, M., Annabi, N. et al. (2018). Magnetic carbon nanotubes: preparation, physical properties, and applications in biomedicine. *Artificial Cells, Nanomedicine, and Biotechnology – An International Journal* 46 (7): 1–17.

90 Song, G., Kenney, M., Chen, Y.-C. et al. (2020). Carbon-coated Fe_3O_4 nanoparticles as sensitive magnetic-particle-imaging tracers with photothermal and magnetothermal properties. *Nature Biomedical Engineering* 4: 325–334.

91 Bao, X., Qiang, Z., Chang, J.H. et al. (2014). Synthesis of carbon-coated magnetic nanocomposite (Fe_3O_4@C) and its application for sulfonamide antibiotics removal from water. *Journal of Environmental Sciences* 26 (5): 962–969.

92 Kakavandi, B., Esrafili, A., Mohseni-Bandpi, A. et al. (2014). Magnetic Fe_3O_4@C nanoparticles as adsorbents for removal of amoxicillin from aqueous solution. *Water Science & Technology* 69 (1): 147–155.

93 Xiao, K. and Zhoud, Y. (2020). Protein recovery from sludge: a review. *Journal of Cleaner Production* 249: 119373.

94 García, M., Urrea, J.L., Collado, S. et al. (2017). Protein recovery from solubilized sludge by hydrothermal treatments. *Waste Management* 67: 278–287.

95 Adebayo, O.T., Fagbenro, O.A., and Jegede, T. (2004). Evaluation of *Cassia fistula* meal as a replacement for soybean meal in practical diets of *Oreochromis niloticus* fingerlings. *Aquaculture Nutrition* 10 (2): 99–104.

96 Suáreziglesias, O., Urrea, J.L., Oulego, P. et al. (2017). Valuable compounds from sewage sludge by thermal hydrolysis and wet oxidation. A review. *Science of the Total Environment* 584–58: 921–934.

97 Xu, Q., Wang, Q., Zhang, W. et al. (2018). Highly effective enhancement of waste activated sludge dewaterability by altering proteins properties using methanol solution coupled with inorganic coagulants. *Water Research* 138: 181–191.

98 Pervaiz, M. and Sain, M. (2011). Protein extraction from secondary sludge of paper mill wastewater and its utilization as a wood adhesive. *Bioresource* 6 (2): 961–970.

99 Hwang, J., Zhang, L., Seo, S. et al. (2008). Protein recovery from excess sludge for its use as animal feed. *Bioresource Technology* 99 (18): 8949–8954.

100 Lamp, A., Kaltschmitt, M., and Lüdtke, O. (2020). Protein recovery from bioethanol stillage by liquid hot water treatment. *The Journal of Supercritical Fluids* 15: 104624.

101 Deng, J.-J., Mao, H.-H., Fang, W. et al. (2020). Enzymatic conversion and recovery of protein, chitin, and astaxanthin from shrimp shell waste. *Journal of Cleaner Production* 271: 122655.

102 Forghani, B., Bordes, R., Ström, A., and Undeland, I. (2020). Recovery of a protein-rich biomass from shrimp (*Pandalus borealis*) boiling water: a colloidal study. *Food Chemistry* 302: 125299.

103 Gua, N., Liua, J., Yea, J. et al. (2019). Bioenergy, ammonia and humic substances recovery from municipal solid waste leachate: a review and process integration. *Bioresource Technology* 293: 122159.

104 Nardi, S., Pizzeghello, D., Muscolo, A., and Vianello, A. (2002). Physiological effects of humic substances on higher plants. *Soil Biology and Biochemistry* 34 (11): 1527–1536.

105 Iskander, S.M., Novak, J.T., and He, Z. (2019). Reduction of reagent requirements and sludge generation in Fenton's oxidation of landfill leachate by synergistically incorporating forward osmosis and humic acid recovery. *Water Research* 151: 310–317.

106 Xu, Y.-D., Yue, D.-B., Zhu, Y., and Nie, Y.-F. (2006). Fractionation of dissolved organic matter in mature landfill leachate and its recycling by ultrafiltration and evaporation combined processes. Fractionation of dissolved organic matter in mature landfill leachate and its recycling by ultrafiltration and evaporation combined processes. *Chemosphere* 64 (6): 903–911.

107 Wang, Y.-N. and Tang, C.Y. (2011). Protein fouling of nanofiltration, reverse osmosis, and ultrafiltration membranes – the role of hydrodynamic conditions, solution chemistry, and membrane properties. *Journal of Membrane Science* 376 (1–2): 275–282.

108 Xu, Y., Chen, C., Li, X. et al. (2017). Recovery of humic substances from leachate nanofiltration concentrate by a two-stage process of tight ultrafiltration membrane. *Journal of Cleaner Production* 161: 84–94.

109 Ye, W., Liu, H., Jiang, M. et al. (2019). Sustainable management of landfill leachate concentrate through recovering humic substance as liquid fertilizer by loose nanofiltration. *Water Research* 157: 555–563.

110 Lin, J., Tang, C.Y., Huang, C. et al. (2016). A comprehensive physico-chemical characterization of superhydrophilic loose nanofiltration membranes. *Journal of Membrane Science* 501: 1–14.

111 Rahimpour, A., Rajaeian, B., Hosienzadeh, A. et al. (2011). Treatment of oily wastewater produced by washing of gasoline reserving tanks using self-made and commercial nanofiltration membranes. *Desalination* 265 (1–3): 190–198.

18.5

Biofuel Production from Agricultural Waste

Esra Meşe Erdoğan[1], Pınar Karagöz[2], Enes Yılmaz[1], and Melek Özkan[1]

[1] *Environmental Engineering Department, Gebze Technical University, Gebze, Kocaeli, Turkey*
[2] *Department of Biochemical Engineering, University College London, London, UK*

18.5.1 Introduction

The concern for food security, water scarcity, and land consumption is increasing each day and these aspects are far from being independent from energy production and use. Nonetheless, the need for energy is increasing tremendously, which triggers environmental concerns. Production of liquid biofuels, such as bioethanol, has been advocated as a sustainable option to tackle the problems associated with rising crude oil prices, global warming, and diminishing petroleum reserves [1]. Today, the primary energy consumption in the world is mainly dominated by fossil energy sources (85.3%, of which 33.1% for oil, 28.1% for coal and 24.1% for gas), as well as hydraulic (6.9%), nuclear (4.4%), and other renewable energies (3.2%) [2]. These percentages may show changes with respect to the country. For example, in 2018, the Indian energy mix was different from the global energy stage, as 49% of its primary energy consumption was coming from coal [3]. In Europe, the heterogeneity of the energy mix is very important, which has a strong effect on energy policies [2].

The negative impact of the greenhouse gases caused by the use of fossil energy sources to the environment has led governments and scientists to look for alternative energy sources. In a globalized World, the demand for finite resources continues to increase, and the current production and consumption strategies seems to be not sufficient enough to achieve sustainability with regard to the protection of the environment and the climate [4]. In addition, gasoline prices increased remarkably during the last decades. Following the Iraq war, a significant increase in oil prices shortened the time required for global production of biofuels and led to their rapid entry to the energy market [5].

Alternative energy sources stand as an option for countries' strategies to increase the range and diversity of their energy resources and not to be largely dependent on certain types of energy sources. Although there are numerous types of alternative energy sources including solar, thermonuclear, hydraulic, geothermal, tidal, and wave energy, they cannot be used as direct fuel in the transport sector [6]. The storage and transportation of biofuels, on the other hand, is much easier and simpler than other renewable energy sources. Moreover, it is reported that the use of biofuels can be based directly on today's oil and gas infrastructure. This situation significantly reduces the cost of infrastructure works in this area and makes the use of biofuels very attractive. If the oil crises continue to increase, biofuels can replace gasoline for tomorrow's transportation sector. Nevertheless, the cost of biofuel production is still remarkably higher as compared to fossil fuels.

Utilization of underused biowaste for energy production is one of the most exciting areas of research, which has a great potential for supporting the circular economy. The low cost of raw materials and elimination of waste disposal will contribute not only to the cost of bioenergy production but also the sustainability. Bio-based waste includes a considerable amount of material at the global level, in the order of magnitude of gigatonnes per year [7]. Recent researches and developments have enabled the conversion of bio-waste into either bioenergy, biofuels, or valuable secondary raw materials [8]. The selection of the appropriate feedstock for biofuel production is an important aspect for more economical production. Selection of feedstocks highly depends on the local conditions. For example, North American and European countries have based their ethanol industry mostly on starchy materials, due to their agro-economic conditions. The world leader USA produces ethanol with starch hydrolyzed from corn, and in the second order in Brazil, where sugar is extracted and ethanol is produced from sugarcane derivatives [9]. However, the use of edible crops as feedstock for energy conversion is in direct competition with food production [10, 11]. For current technologies employed at the commercial level, the main share in the cost structure corresponds to the feedstock [9, 12]. From a sustainability point of view, bioethanol or other biofuels should be produced from secondary biomass mainly from lignocellulosic waste, which brings the advantages of both utilizing the waste through resource recovery and avoiding using biomass useful as food for humans [13, 14].

The most important biofuels in the World today are regarded to be bioethanol, methane, and biodiesel. Other biofuels with less significant impact and diffusion include biogas, biomethanol, biobutanol, firewood, vegetable oil, bioethers, dried manure, and agricultural waste [15]. Liquid biofuels cover bioethanol, biodiesel, pyrolysis bio-oil, and drop-in transportation fuels. World trends in renewable energy production and usage are well documented in the Renewables Global Status Report [16]. In Asia, China, Thailand, Indonesia, Singapore, and India are top countries in biofuel production. The United States was reported to be the world's largest producer and user of biofuels, with the achievement of a record average blend rate of 10.8%. The European Union is the major biodiesel producer. USA, Brazil, Argentina, Indonesia, and Thailand along with the EU together produce 85% of all biodiesel worldwide [17].

18.5.2 Major Biofuels and Potential Use of Agricultural Waste for Production

18.5.2.1 Bioethanol

Ethanol (C_2H_5OH), a promising energy source, has unique advantages compared to oil (C_7H_{17}). One liter of ethanol gives 66% of the energy supplied by the same amount of gasoline, since ethanol (106–110) has a higher octane number than gasoline (91–96), which improves the performance of gasoline when blended with ethanol and allows the use of smaller, high-performance engines [18]. Bioethanol is the highly preferred biofuel, especially in the transportation sector, and provides more efficient combustion in the vehicle engine due to its 35% oxygen content. This results in a decrease in NO_x, hydrocarbons, and particle emission values [19, 20]. Furthermore, ethanol can be produced by renewable sources. Due to its advantages over other biofuels there have been a huge number of researches on the production of ethanol from lignocellulosic waste materials. Commercial production of bioethanol from lignocellulosic materials has just started. However, bioethanol is still produced from fermentation of edible parts of crops in the leading countries. The annual production of 22 billion gallons predominantly comes from food crops [15]. The USA and Brazil are leading countries in bioethanol production. Due to the high production and global usage of bioethanol in the transportation sector, processes for pre-treatment of lignocellulosic waste and fermentation of released sugars have remarkably been developed during the last few decades. Bioethanol is one of the major fuels that can be produced from lignocellulose. The use of agricultural waste for bioethanol production and the steps used to produce lignocellulosic ethanol are presented in detail in related titles throughout this chapter. As can be seen under the title of pre-treatment of lignocellulosic waste, many crop wastes, including wheat and rice straws, sugarcane baggase, corn, and cotton stalks, have been used as source of sugar for bioethanol fermentation.

18.5.2.2 Biodiesel

Biodiesel is a mixture of long-chain fatty acid alkyl esters, which are produced via the transesterification reaction of triglycerides and alcohols [21]. Biodiesel is an alternative fuel for diesel engines with some properties being that they are non-toxic, environmentally friendly, renewable, sulfur free, oxygenated, sustainable, biodegradable, and economic [22]. It can be produced from edible and non-edible resources.

Currently, more than 95% of biodiesel is produced by transesterification reaction of oily food crops and approximately 84% of it is met from rapeseed oil. Oil crop plantation for biodiesel production caused large amounts of deforestation in some countries like Malesia, Brazil, etc. [23], and also caused fluctuations in food prices and global food insecurity [24]. In order to overcome these problems, some researchers tried to find alternative materials for biodiesel production and used non-edible crops and agricultural residues [25–27]. Second and third generation sources for biodiesel production are waste cooking oils, animal oils, non-edible crops (jatropha, karanja, macaw, lignocellulosic waste biomass, etc.) [25]. Researches are being performed for production of biodiesel from agricultural waste. Biodiesel was produced from waste cotton-seed oil by the microwave assisted KOH catalyzed transesterification reaction in the presence of methanol and biodiesel; the yield observed in this process was $96.55 \pm 0.23\%$ [28]. Similarly high biodiesel production yield (97%) was achieved from the transesterification reaction of cotton seed oil with methanol in the presence of KOH catalyst in a batch mode [29].

18.5.2.3 Biomethane

Biomethane is a natural gas produced from anaerobic digestion (AD) of organic substances such as manure, sewage, agricultural, and municipal wastes [30]. AD is a biochemical process that converts organic biomass to biogas through four steps (hydrolysis, acidogenesis, acetogenesis, and methanogenesis) of anaerobic conversion [31]. Biogas composition is generally 50–70% methane, 30–50% carbon dioxide, 1–5% water vapor, 0–5% nitrogen, and other trace concentrations of ammonia, hydrogen, carbon monoxide, and hydrogen sulfide [32]. Although agricultural crops are currently used for biomethane production, they are also used as food or feed for animals. Lignocellulosic biomasses are not only cheap but also do not have any negative effects on the food chain [33]. Agricultural wastes including corn [34], cotton [35], rice [36], and wheat [37] were intensely investigated in biomethane production. The biomethane production potential of agricultural wastes has been determined by researchers and average methane yields of corn, wheat, sugarcane, and rice wastes were reported to be 338, 290, 278, and 232 l CH_4 kg^{-1}-VS respectively [33, 36, 38–42].

In Europe, Germany is in the leader position in anaerobic digestion deployment with a large scale of energy crops production and the UK and Italy are between the top biogas producer countries and produce biogas mainly from landfill gas and agro wastes [43].

18.5.2.4 Biohydrogen

Hydrogen is the most abundant element in the universe. As a fuel, hydrogen gas is an environmentally friendly, clean alternative energy source. Biohydrogen with a high energy content (142 kJ g^{-1}) can be used directly in internal combustion engines and can also be used to generate electricity in fuel cells after purification. As a result of the use of biohydrogen as a fuel, the formation of only water vapor as a by-product has enabled it to be seen as an alternative, ideal, and environmentally friendly fuel to fossil fuels. Currently 88% of global hydrogen production is supplied by fossil fuels. Hydrogen can be produced by various processes. Hydrogen production by the electrolysis of water constitutes only 4% of the total production. Also, this method is not economical because it consumes high energy [44]. Biohydrogen production from agricultural wastes is an alternative cost-effective technology [45]. Biohydrogen production is divided into three main categories: thermochemical routes, electrochemical routes, and biological routes. Biomass gasification is a thermochemical biohydrogen production method that is performed at high temperatures without air. Syngas generated from gasification of biomass consists of hydrogen and CO [46].

Biological routes to produce biohydrogen are direct and indirect biophotolysis, microbial electrolysis, dark fermentation (anaerobic process), and photofermentation (photosynthesis) [47]. The efficient and beneficial hydrogen productions by fermentation were influenced by various factors such as substrate composition, nutrient availability, reactor mode-operations, and bacterial consortium and its yield [48]. Substrate selection is one of the important factors that are necessary for the metabolism of microbes involved in biohydrogen production. Zhang et al. used corn stover as a carbon source and the hydrogen yield was 51.9 L H_2 kg^{-1} of solids [49].

Hydrogen fermentation is greatly enhanced by effective pre-treatment methods of lignocellulosic biomass. There is a considerable amount of research in the literature on hydrogen production via fermentation using different bacteria. For example, *Bacillus cereus* yielded 156.6 l H_2 kg^{-1}-VS on fermentation of alkaline hydrolyzed wheat straw [50]. In another study, it was observed that 58.78 l H_2 kg^{-1}-VS production yield was obtained from fermentation of CaOH and enzymatically hydrolyzed wheat straw [51]. Rice straw can also be used for H_2 fermentation and 4.22 times more glucose was obtained from organosolv pre-treated rice straw biomass than an untreated one. Organosolv pre-treated rice straw was used as a substrate and 19.73 l H_2 kg^{-1} dry biomass was obtained from H_2 fermentation with *Enterobacter aerogenes* strain [52]. Vasconcelos de Sa et al. produced H_2 from hydrothermally and acid hydrolysis sugarcane bagasse that was inoculated with anaerobic sludge and the yield was 231 l H_2 kg^{-1} carbohydrate [53]. Li et al. observed higher than 90 l H_2 kg^{-1} substrate production yield by *Klebsiella* sp. WL1316 with cotton stalk hydrolysate [54].

18.5.2.5 Biobutanol

Butanol (C_4H_9OH) is a four-carbon alcohol that can be produced chemically or biologically and is used in various industries. Biobutanol is the name given to butyl alcohol produced by acetone–butanol–ethanol (ABE) fermentation in a biological way [55]. In recent years, the physicochemical properties of butanol and its usability as a fuel have been investigated intensively. These properties are a high energy content, low corrosion, and is similar to gasoline, and thus it can be used directly without the need of any modifications in gasoline engines [56, 57].

The history of butanol production by the microbial route is very old. During the First World War, the ABE fermentation process was developed to produce acetone, and butanol was produced as a by-product. After that, with the developments in the petrochemical industry, the ABE fermentation process became unable to compete for butanol production because of the high substrate cost and low production rates [58]. Due to the depletion of fossil fuels and damage to the environment, interest in the ABE fermentation process has increased again [57]. Clostridia strains were identified to produce butanol and the most popular commercial clostridia strains are *Clostridium acetobutylicum* and *Clostridium beijerinckii* [59, 60]. These strains can ferment both pentoses and hexoses [61]. However, biobutanol production by clostridia strains also have some drawbacks: being strictly anaerobic, low butanol titter, by-products formation, butanol inhibition effect on strain, a low biomass growth rate, etc. [56].

Hence, researchers now focused on developing one by product forming, not by strictly anaerobic strains resistant to high concentrations of butanol [62]. Biobutanol production studies were also conducted from non-clostridial strains using synthetic biology. *Escherichia coli* and *Saccharomyces cerevisiae* are the most studied host strains for biobutanol production [63, 64].

Finding a low-cost sustainable substrate is another important issue and various researchers also focused on the lignocellulosic-based substrate investigations. Agricultural waste is a high cellulose–hemicellulose containing biomass and a potential substrate for biobutanol production [65]. However, lignin inside it can hinder the enzymatic hydrolysis of lignocellulosic biomass, so delignification of biomass is important [66]. Cai et al. used NaOH hydrolyzed corn stover hydrolysate as a substrate for biobutanol fermentation and achieved 18.6 $g\,l^{-1}$ of butanol with an ABE yield of 83.2 g ABE kg^{-1} of corn stover bagasse [67]. In another study, it was estimated that 169 l of ABE can be produced per ton of hydrothermally pretreated sugarcane straw (or 47.5 l butanol per ton raw sugarcane straw) [68]. Zong-Wen Pang et al. also used alkali pre-treated sugarcane bagasse and the production yields for butanol and ABE were found to be 15.4 g and 22.9 g per 100 g of raw sugarcane bagasse respectively [69]. Another agricultural waste material, barley straw, was used as a substrate for butanol production and 10.8 $g\,l^{-1}$ ABE and 7.9 $g\,l^{-1}$ butanol was produced with a 135 g ABE kg^{-1} pre-treated barley straw production yield [70].

18.5.3 Agricultural Waste Potential for Biofuel Production

In nature, sugars such as glucose, fructose, and sucrose are the main energy sources of microorganisms. When microorganisms convert these sugars to cellular energy, by-products including ethanol, butanol, hydrogen, and carbon dioxide are produced. Biofuels are commonly produced by fermentation of sugar from high sugar and starch containing plants (First Generation) or non-edible parts of lignocellulosic biomass (Second Generation). The development of new technologies to extract sugars from non-food crops or non-edible parts of agricultural products (wastes) solved the conflict of using land for food or fuel production.

The use of renewable biomass resources for the production of second-generation biofuels has gained great importance in the world. The renewable plant biomass especially includes cheap and abundant lignocellulose-rich materials that are not used as food. It is the most abundant and promising resource, which can be used to produce biofuel without interfering with the production of food and feed crops without the need for additional land, although waste generated from processing of agro products, managing livestock, and distribution of fruits and vegetables are also regarded as agricultural waste [71]. However, the crop biomass has high production worldwide and mainly waste produced from crop production is considered as agricultural waste in this chapter. It is reported that global production of plant biomass is 200×10^9 tonnes $year^{-1}$ and 8×10^9–20×10^9 tonnes can be used for biofuel production [1, 72]. Considering six major crops grown globally (barley, maize, rice, soybean, sugarcane, and wheat), Bentsen et al.

reported that global potential production of these crops annually is 3.7 Pg of dry matter (3700 million tons) [73]. In energy terms, the total amount of residue is theoretically equivalent to 65 EJ year^{-1}, corresponding to 15% of global primary energy consumption or 66% of energy consumed for transportation [74]. However, most of the agro-industrial biomass is used as feeding stock for animals or is burned in the fields.

Lignocellulosic biomass is commonly found in four main sources: (i) agricultural residues, (ii) forest residues, (iii) energy plants, and (iv) cellulosic waste [75]. Agro-industial biomass is a large category of biomass consisting of non-edible or unused parts of edible and non-edible plants. This intense activity produces a large amount of lignocellulosic waste. Various types of residues such as straw, bagasse, foliage, bunches, and shells are separated. Part of these discarded wastes is removed in the first stage of the process and left in the field, used as animal feed, or incinerated. Although the structure and composition of these residues is diverse, they are formed basically by cellulose, hemicellulose, and lignin [12, 76].

The most common and the largest quantities available as agricultural crop residues for biomass generation in the world are wheat and rice straw. Wheat is the world's most produced grain and is produced in 115 countries in a wide variety of environmental conditions. Consequently, wheat straw is the lignocellulosic waste that is produced in the highest quantities in the world. It is reported that complete removal of wheat straw in the field may reduce organic matter in the soil and cause soil erosion. The ratio of wheat straw to be left in the field depends on the weather, crop rotation, current soil fertility, slope of the area, and tillage practices [77]. According to Kim and Dale, approximately 1.7 Mg of wheat residue per hectare should be left in the area to prevent soil erosion [78]. The global average yield of wheat is approximately 2.4 Mg ha^{-1}. As a result, approximately 430 Tg (Teragram) of wheat straw is available annually for the production of about 120 Gl of bioethanol. This ethanol volume can replace approximately 93 ($\pm$ 3) Gl of gasoline when the differences between volumetric energy contents and octane numbers are accounted for [79]. Due to diversity in the industry base, resources, living conditions, and other factors, wheat straw is used differently in different areas, such as burning, animal nutrient, and fertilizers. For example, in a city with a relatively large amount of coal, the use of straw as fuel will be low [80].

Rice straw, including stems, leaf blades, leaf sheaths, and the remains of the panicle after threshing, has global production rates reported to be between 499 and 731 million tonnes year^{-1} [1, 81]. The highest global production of rice and wheat straw is in Asian countries, while the majority of corn stalks and sugarcane bagasse are produced in America [82]. Sugarcane pulp is an inexpensive residue from sugarcane processing, with estimated production often reported differently, changing between 300 and 1841 million tonnes year^{-1} [81, 83]. High sugar and starch containing plants such as barley, wheat, corn, sugar beet, sugarcane, potatoes, and sweet sorghum have been widely used for biofuel production. Therefore, the non-edible parts of these crops can be considered for biofuel production. Corn stalk is one of the most promising agricultural wastes with a high potential for bioethanol production. Corn stover is the leftover residue after harvesting corn kernel and comprises stalks, leaves, cobs, and husks. Its annual production is approximately 1 kg kg^{-1} of corn grain or 4 tonnes acre^{-1} [78].

The estimated annual production amount of biofuels from agricultural wastes are given in Table 18.5.1. Technical bioethanol, biobutanol, biohydrogen, and biomethane production potential from main agricultural wastes were calculated through equations given by Bidart et al. [95]. As can be seen from Table 18.5.1, the highest production potential from agricultural residues of all the crops belongs to bioethanol. Compared to other liquid biofuels, bioethanol is produced at a higher quantity in the world and is used globally, especially for the transportation sector. Although the

Table 18.5.1 Estimated annual biofuel production potential of agricultural wastes.

Waste material	Annual production (million tonnes year^{-1}) [81]	Residue to product ratio [84]	Estimated biofuel production potential				References
			Bioethanol (Gl year^{-1})	Biobutanol (Gl year^{-1})	Biohydrogen ($\times10^{-3}$Gl year^{-1})	Biomethane (Gl year^{-1})	
Maize (corn)	1134.75	1	310.6	141.6	58.9	275.2	[49, 85–87]
Rice	499.16	1.76	246	161.6	14.73	117.9	[52, 87–89]
Wheat	765.41	1.75	388.4	205.05	59.56	88.2	[51, 87, 88, 90]
Cotton	25.72	3.50–4.0	26.3	13.01	8.94	6.4	[35, 69, 91, 92]
Sugarcane	1841.53	0.1–0.3	54.6	17.49	40.36	17.5	[68, 87, 93, 94]

production processes from lignocellulosic residues is well established for bioethanol, industrial production and usage of lignocellulosic ethanol is very limited. In this chapter, the steps of lignocelluosic bioethanol production process are discussed in detail since these pre-treatment processes that are applied for releasing sugars from waste are common for the other biofuels produced via fermentation.

18.5.4 Processes for Converting Lignocellulosic Waste to Bioethanol

Lignocellulose, one of the most abundant agricultural residues, contains cellulose, hemicellulose, lignin, and minor components such as protein and other extractives. The composition varies depending on the type of plant. Moreover, within a single plant it varies with respect to age and growing conditions. Cellulose, the main component of the lignocellulose, is a beta (1–4)-linked chain of glucose molecules. Cellulose content and accessibility play an important role in biofuel production efficiency. The cellulose content of major agricultural wastes is shown in Table 18.5.2. Hemicellulose generally found on the outer surface of the cellulose fibers is also diffused into inter-fibrillar space, which acts as a physical barrier limiting cellulose accessibility [107]. Lignin, a non-water soluble amorphous heteropolymer provides structural support and resistance to the plant, is the main factor limiting the cellulose accessibility. Delignification of biomass enhances enzyme effectiveness and biofuel production yield. Lignocellulosic agricultural

Table 18.5.2 Cellulose content of various agricultural wastes.

Type of agricultural waste	Composition (%, dry basis)			References
	Cellulose	Hemicellulose	Lignin	
Barley straw	34–36	19–33	20–24	[70, 96, 97]
Corn stover	33–38	20–23	19–20	[98–100]
Rice straw	35–38	18–25	19–21	[101, 102]
Sugarcane baggase	37–42	21–26	20–28	[69, 96]
Wheat straw	30–42	18–29	15–23	[103–106]

wastes have cellulose as the main component, but their chemical composition varies considerably (Table 18.5.2).

Conversion of biomass to bioethanol or other biofuels obtained by fermentation requires many steps, as depicted in Figure 18.5.1. In the first step, the recalcitrant lignin–hemicellulose–cellulose backbone structure is broken down by physical or chemical (sometimes both) processes. After the pre-treatment, the accessibility of cellulose and hemicellulose is increased and becomes more suitable for hydrolysis. In the second step, hydrolysis, cellulose and hemicellulose macromolecules

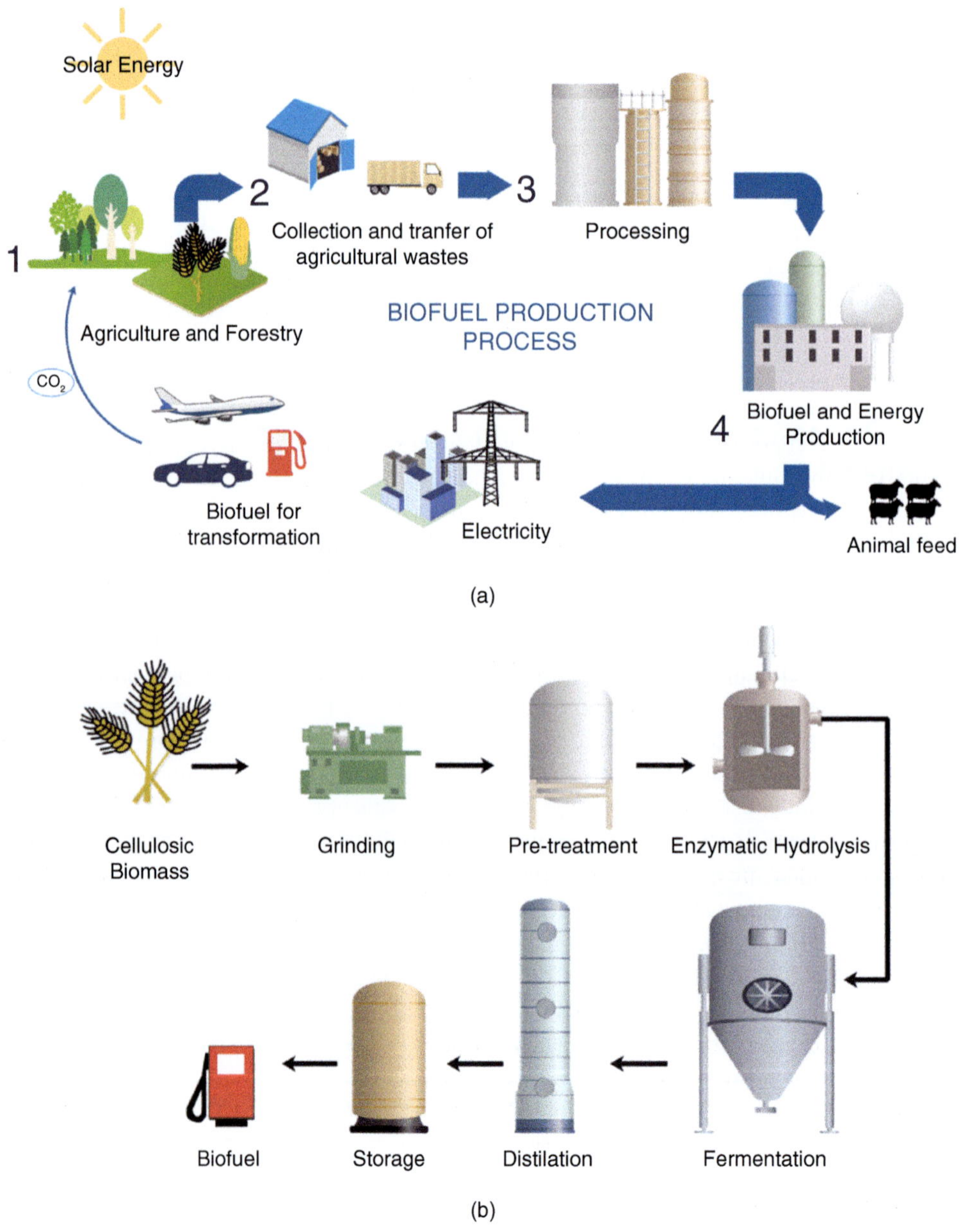

Figure 18.5.1 Biofuel production and usage (a) and steps of production (b).

are broken down into their monomeric forms by enzymatic activities. Depending on the biofuel to be produced, fermentation and separation of biofuel steps may differ. For example, for bioethanol production, produced sugars are converted to bioethanol by yeast fermentation. Depending on the biomass, sugar profile, and used microorganism, different amounts of bioethanol can be produced from per tonne of biomass. The final step of bioethanol production is product recycling and condensation with distillation [108]. These steps are schematized in Figure 18.5.1.

18.5.4.1 Pre-treatment of Lignocellulosic Waste

Steps of bioethanol production is depicted in Figure 18.5.1b. Every step of the biofuel production process has many challenges and affects the final production cost and of course the product price. Among the processes, pre-treatment has the highest cost and affects the total bioethanol production cost, excluding the investment cost [109–111]. Different pre-treatment methods and techniques have been developed to remove lignin and make cellulose and hemicellulose macromolecules more accessible for enzymes. These methods can involve physical, chemical, physicochemical, and biological processes. In general, physical methods refer to milling, extrusion, microwave irradiation, while chemical methods usually apply to acid and alkaline substances and sometimes to organic solvents or ionic liquids. Ammonia fiber explosion, liquid hot water, steam explosion, and supercritical CO_2 explosion are physicochemical methods that have been used on the pre-treatment of lignocellulosic biomass. Biological pre-treatment employs microorganisms, typically white-rot fungi, that can degrade lignocellulosic materials due to their highly efficient extracellular enzymatic system. The advantages and the limitations of pre-treatment methods are summarized in Table 18.5.3. The table not only shows pre-treatment methods used for delignification and hydrolysis but also lists various agricultural waste materials used for different biofuel productions, especially for bioethanol. The main aim in the pre-treatment step is to reach easily fermentable sugars in the lignocellulose structure. These sugars can then be utilized for production of bioethanol, biobutanol, or biogases by using suitable microorganisms for fermentation. The most preferred agricultural waste for bioethanol production researches are wheat straw, rice straw, corn stover, and baggase. Efficiency of wheat and rice straw for biohydrogen and bioethane production were also studied (Table 18.5.3).

18.5.4.2 Enzymatic Hydrolysis of Pre-treated Biomass

Enzymes responsible for the degradation of polymers such as endogluconases, exogluconases, and β-glucosidase are added into pre-treated biomass to release sugars from cellulose and hemicellulose macromolecules. Higher enzyme loadings are necessary when the recalcitrance of the biomass is greater. Development of cellulosic ethanol plants and biofuel industry increases the demand of these enzymes. Due to the high cost of the enzymes, enzymatic hydrolysis is regarded as one of the bottlenecks in the biofuel process. However, the success of the enzymatic hydrolysis step has a direct and significant effect on the biofuel yield. For example, Bernier-Oviedo et al. tested different amounts of enzyme mixture (endoglucanase and cellobiohydrolase) on pre-treated biomass and showed that increasing the enzyme amount from 7.5 FPU (Filter Paper Unit) to 15 FPU per gram of dried biomass pre-treated by steam explosion, increases the sugar production from 211.66 to 402.97 mg glucose equivalent g^{-1} dry solid (mgGE g^{-1} DS) [130]. It is obvious that pre-treatment and hydrolysis processes needs optimization for almost every type of biomass in order to find the most economical way of production.

Table 18.5.3 Most common pre-treatment methods applied to agricultural wastes for biofuel and bioethanol production.

Pre-treatment method	General procedure	Main effects	Advantages	Limitations	Examples		References
					Pre-treated biomass	General purpose of the study	
Ammonia fiber explosion (AFEX)	1–2 kg ammonia kg^{-1} dry biomass applied on biomass at high temperatures (>90 °C)	Increases accessible surface area	Low inhibitor production	Method is not efficient at high lignin contents	Corn stover, prairie cord grass, switch grass,	Pre-treatment assessment	[112]
		Removes lignin and hemicellulose			Agave bagasse	Pre-treatment assessment	[113]
Alkali	Dilute NaOH, or $Ca(OH)_2$ applied on the biomass at high temperatures (90–120 °C)	Increase accessible surface area	High lignin removal	Long residence time	Wheat straw	Bioethanol production	[104]
		Removes lignin and hemicellulose		Salt formation	Jerusalem artichoke stem	Pre-treatment assessment	[114]
					Rice straw	Biohydrogen production	[115]
Biological (fungal) treatment	Brown-, white-, or soft-root fungi grow on the solid material	Degrades lignin and hemicellulose	Low energy consumption	Very slow hydrolysis	Cotton stalks	Bioethanol production	[116]
	Cellulase and hemicellulose produced by fungi degrade biomass				Corn stover	Bioethanol production	[117]
Concentrated-acid hydrolysis	Hydrolysis of biomass by concentrated (10–30%) sulfuric or phosphoric acid	Hydrolysis hemicellulose and cellulose	High glucose yield	Equipment corrosion	Bagasse	Bioethanol production	[118]
			Less or no enzyme requirement	Formation of inhibitory compound	Woody core	Sugar production	[119]
				Acid recovery is required			

Dilute-acid hydrolysis	Disruption of biomass with dilute (0.5–5%) acid concentrations at high (>90 °C) temperatures	Hydrolysis hemicellulose Renders cellulose more accessible for further enzymatic hydrolysis	Low inhibitor production Less corrosion problems than concentrated-acid method	High temperature is required	Rapeseed straw	Succinic acid and bioethanol production	[120]
					Rice straw	Biomethane production	[121]
					Hard wood, soft wood	Sugar production	[122]
Liquid hot water (LHW)	Pressurized (>5 Mpa) hot water applied on biomass at very high (>160 °C) temperatures	Hemicellulose removal Partial solubilization of lignin	Low or no formation of inhibitors	High amount of energy is required	Poplar	Pre-treatment assessment	[123]
					Hard wood, soft wood	Pre-treatment assessment	[124]
Steam explosion	Saturated steam applied on biomass at very high (>160 °C) temperatures	Lignin transformation Hemicellulose solubilization	Cost effective	Partial hemicellulose degradation Incomplete disruption of cellulose-hemicellulose-l ignin matrix	Wheat straw	Sugar production	[125]
					Poplar, wheat straw	Pre-treatment assessment	[126]
					Rice husk, elephant grass	Sugar production	[127]
Wet air oxidation	Pre-treatment of biomass with water and air or oxygen at high temperatures (>120 °C)	Removes lignin	Low inhibitor formation	High cost of catalysts and oxygen	Wood chips	Pre-treatment assessment	[128]
					Sugarcane bagasse, rice hulls, cassava stalks, peanut shells	Bioethanol production	[129]

18.5.4.3 Fermentation

Fermentation is the step where sugars released during hydrolysis of biomass are converted to biofuel by microbial activity. As mentioned before, different microorganisms are being used for production of different biofuels by fermentation. For biobutanol fermentation, *Clostridium* spp. are used whereas the yeast *S. cerevisiae* is the most productive one for bioethanol production. Biobutanol is not produced on an industrial scale as hat of bioethanol. There are some drawbacks for commercialization of biobutanol as a transportation fuel. Low yield through fermentation, sensitivity of the fermenting microbes to end products, and cost-intensive downstream processing are the major challenges [131]. In addition, the single species used for biobutanol (i.e. *Clostridium* spp.) produces considerable amounts of other compounds (acetone, bioethanol, and organic acids) [132]. Due to less drawbacks and high productivity of yeast fermentation, ethanol is the top fermentation biofuel used in the transportation sector. The maximum theoretical conversion efficiency of glucose to ethanol by *S. cerevisiae* is 51% (w/w), which comes from the basic stoichiometric calculation of conversion of 1 mol of glucose to 2 mol of ethanol and 2 mol of CO_2. However, in practice some glucose is used by the yeast cells for the production of cell mass and other metabolic products, and hence the actual conversion efficiency is slightly lower than 51% w/w. Even though wild-type *S. cerevisiae* can ferment glucose, it cannot ferment xylose and other 5-carbon sugars extracted from hemicellulose. *Zymomonas mobilis* is another popular ethanologenic microbe with high ethanol yield and tolerance [133]. Its higher specific rate of sugar uptake and lower biomass production make this bacterium an ideal platform for commercial-scale production of biofuel [134]. Similar to *S. cerevisiae*, wild-type *Z. mobilis* cannot ferment 5-carbon sugars. In nature, there are some microorganisms such as *Pichia stipitis* and *Candida shehatae* that can ferment 5-carbon sugars efficiently. However, these organisms need microaerophilic conditions, and they are more sensitive to high ethanol concentrations and inhibitors [135]. For this reason, a number of studies focused on development of engineered organisms capable of fermenting both 5- and 6-carbon sugars [136–138]. In co-fermentation, a microorganism capable of fermenting 5-carbon sugars can be used with other organisms fermenting 6-carbon sugars [104, 139, 140].

For bioethanol production from lignocellulosic biomass, based on the production structure, there are three main process configurations: separate hydrolysis and fermentation (SHF), simultaneous saccharification and fermentation (SSF), and consolidated bioprocess (CBP). In general, the temperature required for enzyme activity is higher than the temperature required for microbial cells. In SHF, hydrolysis of pre-treated biomass and fermentation takes place at their optimum temperature and pH [141], although glucose and cellobiose accumulation in this step inhibits the activity of the cellulolytic enzymes [111, 142]. In SSF, both saccharification and fermentation take place under the same conditions in a single vessel and released sugars can be rapidly consumed by microorganisms. Even though the suboptimal conditions for saccharification and fermentation process efficiency of this process was proved by different studies [111, 143, 144], the combination of saccharification and fermentation decreases the number of vessels needed and thereby capital investment costs [145]. CBP refers to combinations of four bioprocesses required for conversion of lignocellulose to fuel in one reactor mediated by a microbial consortium [135]. These bioprocesses are (i) production of saccharolytic enzymes; (ii) hydrolysis of polysaccharides in pre-treated biomass, and fermentation of (iii) 6-carbon and (iv) 5-carbon sugars. This process is an alternative to the traditional pure-culture strategies and utilizes recombinant or natural mixed cultures. Due to no sterilization requirements, at-site enzyme production, simultaneous saccharification, and co-fermentation, CBP offers significantly lower cost than other configurations [146]. However, in a CBP microbial growth, enzymatic hydrolysis and fermentation steps are carried out synchronically and it is very difficult to find culture conditions that are optimal for all these processes and, furthermore, microorganisms used in a CBP must display high levels of sugar/alcohol tolerance. There are number of studies working on improving the capacities of these microorganisms by genetic approaches [147].

The thermophilic, anaerobic bacterium, *Clostridium thermocellum*, is a promising candidate for CBP because it can rapidly solubilize and ferment cellulosic biomass without pre-treatment or added enzymes. Furthermore, this organism naturally produces ethanol, hydrogen, acetate, and isobutanol. Thermally stable biomass-degrading enzymes and high-value products of *C. thermocellum* make this organism of great interest to CBP development [148]. To improve its productivity, *C. thermocellum* has been engineered [149, 150]. In a previous study, by engineering its pathway and protein sequence of key enzymes, an engineered *C. thermocellum* strain able to produce 357 mg l of n-butanol from cellulose within 120 h was successfully developed [151]. Similarly, *S. cerevisiae* was genetically engineered to assemble cellulosomes to develop cellulolytic ethanol producing *S. cerevisiae* strains for CBP [152, 153]. Furthermore, the white rot fungus *Phelebia radiata* was adopted for CBP to convert wood waste and barley straw into bioethanol without any pre-treatment and up to 32.2 g l of ethanol was accumulated [154].

18.5.5 Economical Feasibility of Lignocellulosic Biofuel Production

The food for fuel debate encountered in the production of first-generation biofuels from starchy foods and sugar is regarded as an important problem in terms of food safety. In order to reduce the risk of competition with food, non-edible parts of plants like agricultural residues or wastes produced in large quantities has started to be used as feedstock for bioethanol production. However, biofuel production from lignocellulosic residues is still not economically feasible as compared to the low oil prices. Over the last decade, research in the 2G market has been searching for a significant breakthrough that will lead second-generation biofuels being cost competitive with first-generation biofuels. Unfortunately, the development of the lignocellulosic ethanol market has been slower than expected due to a perception of high technological risk and intensive capital costs [155, 156]. It is believed that lignocellulosic-based biofuels could replace about 30% of the petroleum if materials such as forest residues (e.g. wood barks, sawdust), agricultural residues (e.g. corn stover), herbaceous grass (e.g. switchgrass), as well as municipal solid waste are used as feedstock [108, 157].

It is estimated that lignocellulosic biomass can be used to produce about 442 billion liters of bioethanol annually and production can be increased to 491 billion liters if total crop residues and wasted crops are also taken into account. This figure is about 16 times higher than actual global production [78, 158].

In 2008, global biofuel production reached 83 billion liters, corresponding to 1.5% of global transport fuel consumption [159]. Biofuels are expected to provide roughly 9% of total transport fuel demand by 2030. A comprehensive analysis estimates that biofuels can supply 26% of total transport fuel and second-generation biofuels can account for 90% of all biofuels by the year 2050 [160]. However, an international energy agency report of 2019 underlines that global biofuel production is not increasing quickly enough to meet sustainable development scenario demand, which requires 10% growth through to 2030 [16].

Although production varies from country to country, ethanol production by chemical synthesis in the world accounts for as little as 7% of total ethanol production in the world. Ethanol production from ethylene is more expensive due to the high ethylene price. Despite this, the rise in food prices can sometimes lead to an increase in chemical production in some countries [161]. In the cost analysis performed by Denčić et al. [9], the cost of production of 1000 l bioethanol from corn and wheat was calculated as 333 and 360 dollars respectively. These values are very low compared to production from agricultural wastes.

Bioethanol production cost from agricultural wastes depends on key parameters such as type of waste, field output value of waste, size of ethanol production facility, production efficiency, and investment costs. The investment cost and production cost of a plant with an annual production capacity of 200 million liters of ethanol from wheat straw was analysed by Gnansounou and Dauriat [162]. This cost analysis was performed according to the bioethanol production process given by Wooley et al. [163]. This process is an ideal process for cost reduction and includes physical, chemical, and enzymatic hydrolysis steps as a pre-treatment. The facility is supposed to include both celullase enzyme production from waste hydrolysate and energy production from lignin. The waste water released from such an ideal process is used for the production of methane in the anaerobic grinder. Methane is also used to obtain electrical energy. The two products produced in this ideal bioethanol production facility are ethanol and electricity. According to the cost analysis of [162], an expenditure of $1580 is required to produce 1000 l of bioethanol from wheat straw. This value is about 4.5-fold of the cost of bioethanol production from wheat or corn crops. However, agricultural products such as corn and wheat are also needed as food, and their use for bioethanol production is expected to increase product prices. The use of food products in bioethanol production is considered to be a process that should be limited as it is thought to cause food–energy competition. This problem will be overcome when bioethanol production from agricultural wastes is realized.

It has been stated that the collection and transportation of agricultural waste constitutes 50% of the cost of cellulosic ethanol production [164]. For this reason, it will be more economical to install bioethanol production facilities in the vicinity of regions where agricultural waste is produced.

The cost of biofuel production from agricultural waste should be reduced to a reasonable level in order to expand its usage and compete with fossil fuels. There is huge literature on usage of different pre-treatments or fermentation technologies aiming to reach more efficient biofuel production processes. Although fermentation processes are different for bioethanol and biobutanol production, waste hydrolysis and pre-treatment processes include similar steps, which have a remarkable effect on process efficiency and the cost of production. The researches mainly concentrated on pre-treatment of waste for bioethanol production since bioethanol is already used as a blend with fossil fuel in motor vehicles. Various bioethanol production efficiencies can be reported with respect to the applied process. For example, Wang et al. [165] analyzed the effect of different pre-treatment technologies including liquid hot water, steam explosion, acid catalyzed steam explosion and dilute acid hydrolysis on efficiency of ethanol production from wheat straw. When wet oxidation is preferred, conversion of dry matter to bioethanol was found to be around 20%, whereas acid catalyzed steam explosion increases this yield to 25% [165]. A similar study was performed by Zhao et al. for corn stover [14]. According to their analysis, bioethanol production efficiency from corn stover ranges between 19 and 31% depending on the pre-treatment process, including dilute acid, alkali, solvent based ionic liquids or organic solvents, steam explosion, liquid hot water, ammonia based fungal hydrolysis, and combined pre-treatment processes. When fungal pre-hydrolysis is applied, bioethanol production efficiency was 19%. On the other hand, when the ammonia fiber expansion (AFEX) process is preferred for hydrolysis of the waste, production efficiency increases to 31%, which will decrease the cost of production at the end.

We know that the more environmentally friendly process is selected, the costlier will be the production. In a microbial pre-treatment, chemical products that are harmful to the environment are produced at a minimum level. However, to break down lignin microbiologically using fungi is a long-lasting and slow process. Nonetheless, environmental issues should also be noted during waste management for biofuel production. In addition, when considering the factors such as limited oil reserves and geopolitical problems worldwide, biofuel production from agricultural wastes cannot be regarded as an expensive process to be put aside.

References

1 Saini, J.K., Saini, R., and Tewari, L. (2015). Lignocellulosic agriculture wastes as biomass feedstocks for second-generation bioethanol production: concepts and recent developments. *3 Biotech* 5 (4): 337–353.

2 Hache, E. (2018). Do renewable energies improve energy security in the long run? *Journal of International Economics* 156: 127–135.

3 World Energy Primary Production (2018). Energy Production, Enerdata.

4 Corrado, S. and Sala, S. (2018). Bio-economy contribution to circular economy. In: *Designing Sustainable Technologies, Products and Policies: From Science to Innovation* (eds. E. Benetto, K. Gericke and M. Guiton), 49–59. Cham: Springer International Publishing.

5 Adigüzel, A.O. (2013). Biyoetanolün Genel Özellikleri ve Üretimi İçin Gerekli Hammadde Kaynakları. *BEU Journal of Science* 2 (2): 204–220.

6 Valdez-Vazquez, I., Sparling, R., Risbey, D. et al. (2005). Hydrogen generation via anaerobic fermentation of paper mill wastes. *Bioresource Technology* 96 (17): 1907–1913.

7 Pleissner, D., Qi, Q., Gao, C. et al. (2016). Valorization of organic residues for the production of added value chemicals: a contribution to the bio-based economy. *Biochemical Engineering Journal* 116: 3–16.

8 Venkata Mohan, S., Nikhil, G.N., Chiranjeevi, P. et al. (2016). Waste biorefinery models towards sustainable circular bioeconomy: critical review and future perspectives. *Bioresource Technology* 215: 2–12.

9 Denčić, S., Pejin, D., Zekić, V. et al. (2012). A comparison of the costs of bioethanol production from triticale, wheat and maize. *Romanian Agricultural Research* 29: 261–269.

10 Harvey, M. and Pilgrim, S. (2011). The new competition for land: food, energy, and climate change. *Food Policy* 36: 40–51.

11 Moioli, E., Salvati, F., Chiesa, M. et al. (2018). Analysis of the current world biofuel production under a water–food–energy nexus perspective. *Advances in Water Resources* 121: 22–31.

12 Cardona, C.A. and Sánchez, Ó.J. (2007). Fuel ethanol production: process design trends and integration opportunities. *Bioresource Technology* 98 (12): 2415–2457.

13 Agostinho, F., Bertaglia, B.A.B., Almeida, C., and Giannetti, B.F. (2015). Influence of cellulase enzyme production on the energetic-environmental performance of lignocellulosic ethanol. *Ecological Modelling* 315: 46–56.

14 Zhao, Y., Damgaard, A., and Christensen, T.H. (2018). Bioethanol from corn Stover – a review and technical assessment of alternative biotechnologies. *Progress in Energy and Combustion Science* 67: 275–291.

15 Guo, M., Song, W., and Buhain, J. (2015). Bioenergy and biofuels: history, status, and perspective. *Renewable and Sustainable Energy Reviews* 42: 712–725.

16 REN21 (2019). Renewables 2014 Global Status Report (Paris: REN21Secretariat), pp. 1–335.

17 Bušić, A., Kundas, S., Morzak, G. et al. (2018). Recent trends in biodiesel and biogas production. *Food Technology and Biotechnology* 56 (2): 152–173.

18 Nigam, P.S. and Singh, A. (2011). Production of liquid biofuels from renewable resources. *Progress in Energy and Combustion Science* 37 (1): 52–68.

19 Hsieh, W.D., Chen, R.H., Wu, T.L., and Lin, T.H. (2002). Engine performance and pollutant emission of an SI engine using ethanol-gasoline blended fuels. *Atmospheric Environment* 36 (3): 403–410.

20 Bayrakci Ozdingis, A.G. and Kocar, G. (2018). Current and future aspects of bioethanol production and utilization in Turkey. *Renewable and Sustainable Energy Reviews* 81: 2196–2203.

21 Abdelhady, H.H., Elazab, H.A., Ewais, E.M. et al. (2020). Efficient catalytic production of biodiesel using nano-sized sugar beet agro-industrial waste. *Fuel* 261: 116481.

22 Singh, D., Sharma, D., Soni, S.L. et al. (2020). A review on feedstocks, production processes, and yield for different generations of biodiesel. *Fuel* 262: 116553.

23 Gui, M.M., Lee, K.T., and Bhatia, S. (2008). Feasibility of edible oil vs. non-edible oil vs. waste edible oil as biodiesel feedstock. *Energy* 33 (11): 1646–1653.

24 Tegtmeier, E.M. and Duffy, M.D. (2004). External costs of agricultural production in the United States. *International Journal of Agricultural Sustainability* 2 (1): 1–20.

25 Rezania, S., Oryani, B., Park, J. et al. (2019). Review on transesterification of non-edible sources for biodiesel production with a focus on economic aspects, fuel properties and by-product applications. *Energy Conversion and Management* 201: 112155.

26 Shafiei Alavijeh, R., Karimi, K., and van den Berg, C. (2020). An integrated and optimized process for cleaner production of ethanol and biodiesel from corn stover by *Mucor indicus*. *Journal of Cleaner Production* 249: 119321.

27 Zhang, K., Xu, R., Abomohra, A.E.F. et al. (2019). A sustainable approach for efficient conversion of lignin into biodiesel accompanied by biological pretreatment of corn straw. *Energy Conversion and Management* 199: 111928.

28 Sharma, A., Kodgire, P., and Kachhwaha, S.S. (2019). Biodiesel production from waste cotton-seed cooking oil using microwave-assisted transesterification: optimization and kinetic modeling. *Renewable and Sustainable Energy Reviews* 116: 109394.

29 Onukwuli, D.O., Emembolu, L.N., Ude, C.N. et al. (2017). Optimization of biodiesel production from refined cotton seed oil and its characterization. *Egyptian Journal of Petroleum* 26 (1): 103–110.

30 Anderson, L.G. (2015). Effects of using renewable fuels on vehicle emissions. *Renewable and Sustainable Energy Reviews* 47: 162–172.

31 Hosseini Koupaie, E., Johnson, T., and Eskicioglu, C. (2017). Advanced anaerobic digestion of municipal sludge using a novel and energy-efficient radio frequency pretreatment system. *Water Research* 118: 70–81.

32 Qyyum, M.A., Haider, J., Qadeer, K. et al. (2020). Biogas to liquefied biomethane: assessment of 3P's–production, processing, and prospects. *Renewable and Sustainable Energy Reviews* 119: 109561.

33 Frigon, J.C., Mehta, P., and Guiot, S.R. (2012). Impact of mechanical, chemical and enzymatic pre-treatments on the methane yield from the anaerobic digestion of switchgrass. *Biomass and Bioenergy* 36: 1–11.

34 Paul, S., Dutta, A., and Defersha, F. (2018). Biocarbon, biomethane and biofertilizer from corn residue: a hybrid thermo-chemical and biochemical approach. *Energy* 165 (Part B): 370–384.

35 Civelek Yoruklu, H., Koroglu, E.O., Ozdemir, O.K. et al. (2020). Bioenergy production from cotton straws using different pretreatment methods. *International Journal of Hydrogen Energy* 46 (50): 34720–34729.

36 Kainthola, J., Shariq, M., Kalamdhad, A.S., and Goud, V.V. (2019). Electrohydrolysis pretreatment methods to enhance the methane production from anaerobic digestion of rice straw using graphite electrode. *Renewable Energy* 142: 1–10.

37 Romagnoli, F., Dorella, M., Gruduls, A. et al. (2019). Anaerobic co-digestion of Baltic seaweeds with wheat straw and straw pellets: synergetic effects on biomethane yield and kinetic biodegradability constant. *Energy Procedia* 158: 854–860.

38 Chandra, R., Takeuchi, H., and Hasegawa, T. (2012). Methane production from lignocellulosic agricultural crop wastes: a review in context to second generation of biofuel production. *Renewable and Sustainable Energy Reviews* 16 (3): 1462–1476.

39 Huang, W., Yuan, H., and Li, X. (2020). Multi-perspective analyses of rice straw modification by Pleurotus ostreatus and effects on biomethane production. *Bioresource Technology* 296: 122365.

40 Li, J., Wachemo, A.C., Yu, G., and Li, .X. (2020). Enhanced anaerobic digestion performance of corn stalk pretreated with freezing-thawing and ammonia: An experimental and theoretical study. *Journal of Cleaner Production* 247: 119112.

41 Ecem Öner, B., Akyol, Ç., Bozan, M. et al. (2018). Bioaugmentation with Clostridium thermocellum to enhance the anaerobic biodegradation of lignocellulosic agricultural residues. *Bioresource Technology* 249: 620–625.

42 Arimi, M.M., Knodel, J., Kiprop, A. et al. (2015). Strategies for improvement of biohydrogen production from organic-rich wastewater: a review. *Biomass and Bioenergy* 75: 101–118.

43 Zhu, T., Curtis, J., and Clancy, M. (2019). Promoting agricultural biogas and biomethane production: lessons from cross-country studies. *Renewable and Sustainable Energy Reviews* 114: 109332.

44 Lin, C.Y., Nguyen, T.M.L., Chu, C.Y. et al. (2018). Fermentative biohydrogen production and its byproducts: a mini review of current technology developments. *Renewable and Sustainable Energy Reviews* 82 (Part 3): 4215–4220.

45 Yukesh Kannah, R., Kavitha, S., Sivashanmugham, P. et al. (2019). Biohydrogen production from rice straw: effect of combinative pretreatment, modelling assessment and energy balance consideration. *International Journal of Hydrogen Energy* 44 (4): 2203–2215.

46 Reaño, R.L. (2020). Assessment of environmental impact and energy performance of rice husk utilization in various biohydrogen production pathways. *Bioresource Technology* 299: 122590.

47 Mirza, S.S., Qazi, J.I., Liang, Y., and Chen, S. (2019). Growth characteristics and photofermentative biohydrogen production potential of purple non sulfur bacteria from sugar cane bagasse. *Fuel* 255: 115805.

48 Mota, V.T., Ferraz Júnior, A.D.N., Trably, E., and Zaiat, M. (2018). Biohydrogen production at pH below 3.0: is it possible? *Water Research* 128: 350–361.

49 Zhang, S.C., Lai, Q.H., Lu, Y. et al. (2016). Enhanced biohydrogen production from corn Stover by the combination of *Clostridium cellulolyticum* and hydrogen fermentation bacteria. *Journal of Bioscience and Bioengineering* 122 (4): 482–487.

50 Saleem, A., Umar, H., Shah, T.A., and Tabassum, R. (2020). Fermentation of simple and complex substrates to biohydrogen using pure *Bacillus cereus* RTUA and RTUB strains. *Environmental Technology and Innovation* 18: 100704.

51 Reilly, M., Dinsdale, R., and Guwy, A. (2014). Mesophilic biohydrogen production from calcium hydroxide treated wheat straw. *International Journal of Hydrogen Energy* 39 (30): 16891–16901.

52 Asadi, N., Karimi Alavijeh, M., and Zilouei, H. (2018). Biological hydrogen production by Enterobacter aerogenes: structural analysis of treated rice straw and effect of substrate concentration. *International Journal of Hydrogen Energy* 43 (18): 8718–8728.

53 de Sá, L.R.V., Faber, M., d.O., da Silva, A.S.A. et al. (2020). Biohydrogen production using xylose or xylooligosaccharides derived from sugarcane bagasse obtained by hydrothermal and acid pretreatments. *Renewable Energy* 146: 2408–2415.

54 Li, Y., Zhu, Q., Ding, P. et al. (2020). Effects of Fe0 and Ni0 nanoparticles on hydrogen production from cotton stalk hydrolysate using *Klebsiella* sp. WL1316: evaluation of size and concentration of the nanoparticles. *International Journal of Hydrogen Energy* 45 (11): 6243–6253.

55 Pugazhendhi, A., Mathimani, T., Varjani, S. et al. (2019). Biobutanol as a promising liquid fuel for the future – recent updates and perspectives. *Fuel* 253: 637–646.

56 Li, Y., Tang, W., Chen, Y. et al. (2019). Potential of acetone–butanol–ethanol (ABE) as a biofuel. *Fuel* 242: 673–686.

57 Dürre, P. (2007). Biobutanol: an attractive biofuel. *Biotechnology Journal* 2 (12): 1525–1534.

58 Ndaba, B., Chiyanzu, I., and Marx, S. (2015). N-Butanol derived from biochemical and chemical routes: a review. *Biotechnology Reports* 8: 1–9.

59 Ibrahim, M.F., Ramli, N., Kamal Bahrin, E., and Abd-Aziz, S. (2017). Cellulosic biobutanol by clostridia: challenges and improvements. *Renewable and Sustainable Energy Reviews* 79: 1241–1254.

60 Jang, Y.-S., Malaviya, A., Cho, C. et al. (2012). Butanol production from renewable biomass by *Clostridia*. *Bioresource Technology* 123: 653–663.

61 Kudahettige-Nilsson, R.L., Helmerius, J., Nilsson, R.T. et al. (2015). Biobutanol production by *Clostridium acetobutylicum* using xylose recovered from birch Kraft black liquor. *Bioresource Technology* 176: 71–79.

62 Al-Shorgani, N.K.N., Kalil, M.S., Yusoff, W.M.W., and Hamid, A.A. (2015). Biobutanol production by a new aerotolerant strain of *Clostridium acetobutylicum* YM1 under aerobic conditions. *Fuel* 158: 855–863.

63 Atsumi, S., Cann, A.F., Connor, M.R. et al. (2008). Metabolic engineering of *Escherichia coli* for 1-butanol production. *Metabolic Engineering* 10 (6): 305–311.

64 Steen, E.J., Chan, R., Prasad, N. et al. (2008). Metabolic engineering of *Saccharomyces cerevisiae* for the production of n-butanol. *Microbial Cell Factories* 7 (1): 36.

65 Huzir, N.M., Aziz, M.M.A., Ismail, S.B. et al. (2018). Agro-industrial waste to biobutanol production: eco-friendly biofuels for next generation. *Renewable and Sustainable Energy Reviews* 94: 476–485.

66 Kumar, S. and Banerjee, R. (2018). Enzymatic delignification and saccharification of *Bambusa bambos* for biobutanol production. *Industrial Crops and Products* 125: 386–394.

67 Cai, D., Chen, C., Zhang, C. et al. (2017). Fed-batch fermentation with intermittent gas stripping using immobilized *Clostridium acetobutylicum* for biobutanol production from corn stover bagasse hydrolysate. *Biochemical Engineering Journal* 125: 18–22.

68 Pratto, B., Chandgude, V., de Sousa, R. et al. (2020). Biobutanol production from sugarcane straw: defining optimal biomass loading for improved ABE fermentation. *Industrial Crops and Products* 148: 112265.

69 Pang, Z.W., Lu, W., Zhang, H. et al. (2016). Butanol production employing fed-batch fermentation by *Clostridium acetobutylicum* GX01 using alkali-pretreated sugarcane bagasse hydrolysed by enzymes from *Thermoascus aurantiacus* QS 7-2-4. *Bioresource Technology* 212: 82–91.

70 Yang, M., Zhang, J., Kuittinen, S. et al. (2015). Enhanced sugar production from pretreated barley straw by additive xylanase and surfactants in enzymatic hydrolysis for acetone–butanol–ethanol fermentation. *Bioresource Technology* 189: 131–137.

71 Pattanaik, L., Pattnaik, F., Saxena, D.K., and Naik, S.N. (2019). Biofuels from agricultural wastes. In: *Second and Third Generation of Feedstocks* (ed. A. Basile and F. Dalena), 103–142. India: Elsevier.

72 Kuhad, R.C. and Singh, A. (1993). Lignocellulose biotechnology: current and future prospects. *Critical Reviews in Biotechnology* 13 (2): 151–172.

73 Bentsen, N.S., Felby, C., and Thorsen, B.J. (2014). Agricultural residue production and potentials for energy and materials services. *Progress in Energy and Combustion Science* 40 (1): 59–73.

74 Hiloidhari, M., Bhuyan, N., Gogoi, N. et al. (2020). Agroindustry wastes: biofuels and biomaterials feedstocks for sustainable rural development. In: *Refining Biomass Residues for Sustainable Energy and Bioproducts* (ed. R. Praveen Kumar, E. Gnansounou, J.K. Raman and G. Baskar), 357–388. USA: Academic Press.

75 Salehi Jouzani, G. and Taherzadeh, M. (2015). Advances in consolidated bioprocessing systems for bioethanol and butanol production from biomass: a comprehensive review. *Biofuel Research Journal* 5: 152–195.

76 Magalhães, N., Cavalcante, A.V., Andrade, L.S. et al. (2019). Citric acid production by *Aspergillus niger* an 400 from agroindustry waste. *Engenharia Sanitária e Ambiental* 24 (1): 101–107.

77 Kerstetter, D.J. and Lyons, J.K. (2001). Wheat straw for ethanol production in Washington: a resource, technical, and economic assessment. Washington State University: Cooperative Extension Energy Program.

78 Kim, S. and Dale, B.E. (2004). Global potential bioethanol production from wasted crops and crop residues. *Biomass and Bioenergy* 26 (4): 361–375.

79 Otero, J.M., Panagiotou, G., and Olsson, L. (2007). Fueling industrial biotechnology growth with bioethanol. *Advances in Biochemical Engineering/Biotechnology* 108: 1–40.

80 Li, Q., Hu, S., Chen, D., and Zhu, B. (2014). System analysis of grain straw for centralised industrial usages in China. *Biomass and Bioenergy* 47: 277–288.

81 Food and Agriculture Organization of the United Nations (FAOSTAT) (2019). Statistical Data. Rome, Italy.

82 Zabed, H., Sahu, J.N., Suely, A. et al. (2017). Bioethanol production from renewable sources: current perspectives and technological progress. *Renewable and Sustainable Energy Reviews* 71 (December 2016): 475–501.

83 Sánchez, C. (2009). Lignocellulosic residues: biodegradation and bioconversion by fungi. *Biotechnology Advances* 27 (2): 185–194.

84 Fleming, T.H., Kress, W.J., Fleming, T.H., and Kress, W.J. (2015). The Resource Base. *Ornaments Life* (January 1997): 63–106.

85 Zhao, Y., Damgaard, A., Liu, S. et al. (2020). Bioethanol from corn stover – integrated environmental impacts of alternative biotechnologies. *Resources, Conservation and Recycling* 155: 104652.

86 Karimi Alavijeh, M. and Karimi, K. (2019). Biobutanol production from corn stover in the US. *Industrial Crops and Products* 129: 641–653.

87 Tian, Q.Q., Liang, L., and Zhu, M.J. (2015). Enhanced biohydrogen production from sugarcane bagasse by *Clostridium thermocellum* supplemented with CaCO3. *Bioresource Technology* 197: 422–428.

88 Mostafaeipour, A., Sedaghat, A., Hedayatpour, M., and Jahangiri, M. (2020). Location planning for production of bioethanol fuel from agricultural residues in the south of Caspian Sea. *Environmental Development* 33: 100500.

89 Chi, X., Li, J., Leu, S.Y. et al. (2019). Features of a staged acidogenic/solventogenic fermentation process to improve butanol production from rice straw. *Energy and Fuels* 33 (2): 1123–1132.

90 Qi, G., Xiong, L., Li, H. et al. (2019). Hydrotropic pretreatment on wheat straw for efficient biobutanol production. *Biomass and Bioenergy* 122: 76–83.

91 Dimos, K., Paschos, T., Louloudi, A. et al. (2019). Effect of various pretreatment methods on bioethanol production from cotton stalks. *Fermentation* 5 (1): 5.

92 Wang, J.-H., Liu, Y.-P., Wang, J.-J., and Zhang, M. (2017). Effect of acid pretreatment on hydrogen production from cotton stalk by anaerobic fermentation. *Energy Sources, Part A: Recovery, Utilization, and Environmental Effects* 39 (4): 383–388.

93 Sharma, V., Nargotra, P., and Bajaj, B.K. (2019). Ultrasound and surfactant assisted ionic liquid pretreatment of sugarcane bagasse for enhancing saccharification using enzymes from an ionic liquid tolerant *Aspergillus assiutensis* VS34. *Bioresource Technology* 285: 121319.

94 Ghaleb, A.A.S., Kutty, S.R.M., Ho, Y.-C. et al. (2020). Response surface methodology to optimize methane production from mesophilic anaerobic co-digestion of oily-biological sludge and sugarcane bagasse. *Sustainability* 12 (5): 2116.

95 Bidart, C., Fröhling, M., and Schultmann, F. (2014). Livestock manure and crop residue for energy generation: macro-assessment at a national scale. *Renewable and Sustainable Energy Reviews* 38: 537–550.

96 Murciano Martínez, P., Bakker, R., Harmsen, P. et al. (2015). Importance of acid or alkali concentration on the removal of xylan and lignin for enzymatic cellulose hydrolysis. *Industrial Crops and Products* 64: 88–96.

97 Lara-Serrano, M., Morales-delaRosa, S., Campos-Martín, M.J., and Fierro, L.G.J. (2019). Fractionation of lignocellulosic biomass by selective precipitation from ionic liquid dissolution. *Applied Sciences* 9 (9): 1862.

98 Zhong, L., Zhang, X., Tang, C. et al. (2018). Hydrazine hydrate and organosolv synergetic pretreatment of corn stover to enhance enzymatic saccharification and co-production of high-quality antioxidant lignin. *Bioresource Technology* 268: 677–683.

99 Jin, C., Yang, M., Shuang, E. et al. (2020). Corn Stover valorization by one-step formic acid fractionation and formylation for 5-hydroxymethylfurfural and high guaiacyl lignin production. *Bioresource Technology* 299: 122586.

100 Li, W.-C., Zhang, S.-J., Xu, T. et al. (2020). Fractionation of corn stover by two-step pretreatment for production of ethanol, furfural, and lignin. *Energy* 195: 117076.

101 Castro, R.C.d.A. and Roberto, I.C. (2015). Effect of nutrient supplementation on ethanol production in different strategies of saccharification and fermentation from acid pretreated rice straw. *Biomass and Bioenergy* 78: 156–163.

102 Tsegaye, B., Balomajumder, C., and Roy, P. (2020). Organosolv pretreatments of rice straw followed by microbial hydrolysis for efficient biofuel production. *Renewable Energy* 148: 923–934.

103 Seguin, A., Marinkovic, S., and Estrine, B. (2012). New pretreatment of wheat straw and bran in hexadecanol for the combined production of emulsifying base, glucose and lignin material. *Carbohydrate Polymers* 88 (2): 657–662.

104 Karagöz, P. and Özkan, M. (2014). Ethanol production from wheat straw by *Saccharomyces cerevisiae* and *Scheffersomyces stipitis* co-culture in batch and continuous system. *Bioresource Technology* 158: 286–293.

105 Xie, X., Feng, X., Chi, S. et al. (2018). A sustainable and effective potassium hydroxide pretreatment of wheat straw for the production of fermentable sugars. *Bioresource Technology Reports* 3: 169–176.

106 Tsegaye, B., Balomajumder, C., and Roy, P. (2019). Alkali pretreatment of wheat straw followed by microbial hydrolysis for bioethanol production. *Environmental Technology* 40 (9): 1203–1211.

107 Meng, X. and Ragauskas, A.J. (2014). Recent advances in understanding the role of cellulose accessibility in enzymatic hydrolysis of lignocellulosic substrates. *Current Opinion in Biotechnology* 27: 150–158.

108 Hu, G., Heitmann, J.A., and Rojas, O.J. (2008). Feedstock pretreatment strategies for producing ethanol from wood, bark, and forest residues. *BioResources* 3 (1): 270–294.

109 da Costa Sousa, L., Chundawat, S.P.S., Balan, V., and Dale, B.E. (2009). "Cradle-to-grave" assessment of existing lignocellulose pretreatment technologies. *Current Opinion in Biotechnology* 20 (3): 339–347.

110 Shi, J., Sharmashivappa, R., Chinn, M., and Howell, N. (2009). Effect of microbial pretreatment on enzymatic hydrolysis and fermentation of cotton stalks for ethanol production. *Biomass and Bioenergy* 33 (1): 88–96.

111 Tomas-Pejo, E., Dominguez, J., and Ballesteros, M. (2008). Realistic approach for full-scale bioethanol production from lignocellulose: a review. *Journal of Scientific and Industrial Research (India)* 67: 874–884.

112 Sundaram, V., Muthukumarappan, K., and Kamireddy, S.R. (2015). Effect of ammonia fiber expansion (AFEX™) pretreatment on compression behavior of corn Stover, prairie cord grass and switchgrass. *Industrial Crops and Products* 74: 45–54.

113 Perez-Pimienta, J.A., Flores-Gómez, C.A., Ruiz, H.A. et al. (2016). Evaluation of agave bagasse recalcitrance using AFEX™, autohydrolysis, and ionic liquid pretreatments. *Bioresource Technology* 211: 216–223.

114 Li, M., Wang, J., Yang, Y., and Xie, G. (2016). Alkali-based pretreatments distinctively extract lignin and pectin for enhancing biomass saccharification by altering cellulose features in sugar-rich Jerusalem artichoke stem. *Bioresource Technology* 208: 31–41.

115 Dong, L., Cao, G., Zhao, L. et al. (2018). Alkali/urea pretreatment of rice straw at low temperature for enhanced biological hydrogen production. *Bioresource Technology* 267: 71–76.

116 Shi, J., Sharma-Shivappa, R.R., and Chinn, M.S. (2009). Microbial pretreatment of cotton stalks by submerged cultivation of *Phanerochaete chrysosporium*. *Bioresource Technology* 100 (19): 4388–4395.

117 Wan, C. and Li, Y. (2010). Microbial pretreatment of corn stover with *Ceriporiopsis subvermispora for* enzymatic hydrolysis and ethanol production. *Bioresource Technology* 101 (16): 6398–6403.

118 Cuzens, J.C. and Miller, J.R. (1997). Acid hydrolysis of bagasse for ethanol production. *Renewable Energy* 10 (2): 285–290.

119 Yoon, S.-Y., Han, S.-H., and Shin, S.-J. (2014). The effect of hemicelluloses and lignin on acid hydrolysis of cellulose. *Energy* 77: 19–24.

120 Kuglarz, M., Alvarado-Morales, M., Dąbkowska, K., and Angelidaki, I. (2018). Integrated production of cellulosic bioethanol and succinic acid from rapeseed straw after dilute-acid pretreatment. *Bioresource Technology* 265: 191–199.

121 Peng, J., Abomohra, A.E.-F., Elsayed, M. et al. (2019). Compositional changes of rice straw fibers after pretreatment with diluted acetic acid: towards enhanced biomethane production. *Journal of Cleaner Production* 230: 775–782.

122 Smits, J., Bevers, L., van Haastert, M. et al. (2019). Fast screening of optimal acid-pretreatment conditions in the conversion of wood to lignocellulosic sugars. *Bioresource Technology Reports* 5: 220–229.

123 Li, M., Cao, S., Meng, X. et al. (2017). The effect of liquid hot water pretreatment on the chemical–structural alteration and the reduced recalcitrance in poplar. *Biotechnology for Biofuels* 10 (1): 237.

124 Nitsos, C.K., Choli-Papadopoulou, T., Matis, K.A., and Triantafyllidis, K.S. (2016). Optimization of hydrothermal pretreatment of hardwood and softwood lignocellulosic residues for selective hemicellulose recovery and improved cellulose enzymatic hydrolysis. *ACS Sustainable Chemistry & Engineering* 4 (9): 4529–4544.

125 Alvira, P., Negro, M.J., Ballesteros, I. et al. (2016). Steam explosion for wheat straw pretreatment for sugars production. *Bioethanol* 2 (1): 66–75.

126 Auxenfans, T., Crônier, D., Chabbert, B., and Paës, G. (2017). Understanding the structural and chemical changes of plant biomass following steam explosion pretreatment. *Biotechnology for Biofuels* 10 (1): 36.

127 Montipó, S., Ballesteros, I., Martins, A.F. et al. (2020). Optimisation of Uncatalysed steam explosion of lignocellulosic biomasses to obtain both C6- and C5-sugars. *Waste and Biomass Valorization* 11 (1): 231–244.

128 Palonen, H., Thomsen, A.B., Tenkanen, M. et al. (2004). Evaluation of wet oxidation pretreatment for enzymatic hydrolysis of softwood. *Applied Biochemistry and Biotechnology* 117 (1): 1–17.

129 Martín, C. and Thomsen, A.B. (2007). Wet oxidation pretreatment of lignocellulosic residues of sugarcane, rice, cassava and peanuts for ethanol production. *Journal of Chemical Technology and Biotechnology* 82 (2): 174–181.

130 Bernier-Oviedo, D.J., Rincón-Moreno, J.A., Solanilla-Duqué, J.F. et al. (2018). Comparison of two pretreatments methods to produce second-generation bioethanol resulting from sugarcane bagasse. *Industrial Crops and Products* 122: 414–421.

131 Gottumukkala, L.D., Haigh, K., and Görgens, J. (2017). Trends and advances in conversion of lignocellulosic biomass to biobutanol: microbes, bioprocesses and industrial viability. *Renewable and Sustainable Energy Reviews* 76: 963–973.

132 Martín-Juárez, J., Markou, G., Muylaert, K. et al. (2017). Breakthroughs in bioalcohol production from microalgae: solving the hurdles. In: *Microalgae-Based Biofuels and Bioproducts* (eds. C. Gonzalez-Fernandez and R. Muñoz), 183–207. Woodhead Publishing.

133 Todhanakasem, T., Salangsing, O.-L., Koomphongse, P. et al. (2019). *Zymomonas mobilis* biofilm reactor for ethanol production using rice straw hydrolysate under continuous and repeated batch processes. *Frontiers in Microbiology* 10: 1777.

134 He, M.X., Wu, B., Qin, H. et al. (2014). *Zymomonas mobilis*: a novel platform for future biorefineries. *Biotechnology for Biofuels* 7 (1): 101.

135 Refaat, A.A. (2012). Biofuels from waste materials. *Comprehensive Renewable Energy* 5: 217–261.

136 Xiong, M., Chen, G., and Barford, J. (2014). Genetic engineering of yeasts to improve ethanol production from xylose. *Journal of the Taiwan Institute of Chemical Engineers* 45 (1): 32–39.

137 de Vilela, L., F., de Araujo, V.P.G., de Paredes, R., S. et al. (2015). Enhanced xylose fermentation and ethanol production by engineered *Saccharomyces cerevisiae* strain. *AMB Express* 5 (1): 16.

138 Cunha, J.T., Soares, P.O., Romaní, A. et al. (2019). Xylose fermentation efficiency of industrial *Saccharomyces cerevisiae* yeast with separate or combined xylose reductase/xylitol dehydrogenase and xylose isomerase pathways. *Biotechnology for Biofuels* 12 (1): 20.

139 Nguyen, D.T.T., Praveen, P., and Loh, K.-C. (2019). Co-culture of *Zymomonas mobilis* and *Scheffersomyces stipitis* immobilized in polymeric membranes for fermentation of glucose and xylose to ethanol. *Biochemical Engineering Journal* 145: 145–152.

140 Wang, L., York, S.W., Ingram, L.O., and Shanmugam, K.T. (2019). Simultaneous fermentation of biomass-derived sugars to ethanol by a co-culture of an engineered *Escherichia coli* and *Saccharomyces cerevisiae*. *Bioresource Technology* 273: 269–276.

141 Loaces, I., Schein, S., and Noya, F. (2017). Ethanol production by *Escherichia coli* from *Arundo donax* biomass under SSF, SHF or CBP process configurations and in situ production of a multifunctional glucanase and xylanase. *Bioresource Technology* 224: 307–313.

142 Xiao, Z., Zhang, X., Gregg, D.J., and Saddler, J.N. (2004). Effects of sugar inhibition on cellulases and β-glucosidase during enzymatic hydrolysis of softwood substrates. *Applied Biochemistry and Biotechnology* 115 (1): 1115–1126.

143 Ask, M., Olofsson, K., Di Felice, T. et al. (2012). Challenges in enzymatic hydrolysis and fermentation of pretreated *Arundo donax* revealed by a comparison between SHF and SSF. *Process Biochemistry* 47 (10): 1452–1459.

144 Dahnum, D., Tasum, S.O., Triwahyuni, E. et al. (2015). Comparison of SHF and SSF processes using enzyme and dry yeast for optimization of bioethanol production from empty fruit bunch. *Energy Procedia* 68: 107–116.

145 Olofsson, K., Bertilsson, M., and Lidén, G. (2008). A short review on SSF – an interesting process option for ethanol production from lignocellulosic feedstocks. *Biotechnology for Biofuels* 1 (1): 7.

146 Lynd, L.R., van Zyl, W.H., McBride, J.E., and Laser, M. (2005). Consolidated bioprocessing of cellulosic biomass: an update. *Current Opinion in Biotechnology* 16 (5): 577–583.

147 Ali, S.S., Nugent, B., Mullins, E., and Doohan, F.M. (2016). Fungal-mediated consolidated bioprocessing: the potential of Fusarium oxysporum for the lignocellulosic ethanol industry. *AMB Express* 6 (1): 13.

148 Mbaneme-Smith, V. and Chinn, M. (2015). Consolidated bioprocessing for biofuel production: recent advances. *Energy and Emission Control Technologies* 3: 23–44.

149 Lin, P.P., Mi, L., Morioka, A.H. et al. (2015). Consolidated bioprocessing of cellulose to isobutanol using *Clostridium thermocellum*. *Metabolic Engineering* 31: 44–52.

150 Hon, S., Holwerda, E.K., Worthen, R.S. et al. (2018). Expressing the Thermoanaerobacterium saccharolyticum pforA in engineered Clostridium thermocellum improves ethanol production. *Biotechnology for Biofuels* 11 (1): 242.

151 Tian, L., Conway, P.M., Cervenka, N.D. et al. (2019). Metabolic engineering of Clostridium thermocellum for n-butanol production from cellulose. *Biotechnology for Biofuels* 12 (1): 186.

152 Tsai, S.L., Oh, J., Singh, S. et al. (2009). Functional assembly of minicellulosomes on the *Saccharomyces cerevisiae* cell surface for cellulose hydrolysis and ethanol production. *Applied and Environmental Microbiology* 75 (19): 6087–6093.

153 Fan, L.-H., Zhang, Z.-J., Yu, X.-Y. et al. (2012). Self-surface assembly of cellulosomes with two miniscaffoldins on <em> *Saccharomyces cerevisiae*</em> for cellulosic ethanol production. *Proceedings of the National Academy of Sciences* 109 (33): 13260–13265.

154 Mattila, H., Kačar, D., Mali, T., and Lundell, T. (2018). Lignocellulose bioconversion to ethanol by a fungal single-step consolidated method tested with waste substrates and co-culture experiments. *AIMS Energy* 6 (5): 866–879.

155 Stephen, J.D., Mabee, W.E., and Saddler, J.N. (2012). Will second-generation ethanol be able to compete with first-generation ethanol? Opportunities for cost reduction. *Biofuels, Bioproduct and Biorefining* 6 (2): 159–176.

156 Valdivia, M., Galan, J.L., Laffarga, J., and Ramos, J.-L. (2016). Biofuels 2020: biorefineries based on lignocellulosic materials. *Microbial Biotechnology* 9 (5): 585–594.

157 Wyman, C.E., Dale, B.E., Elander, R.T. et al. (2005). Coordinated development of leading biomass pretreatment technologies. *Bioresource Technology* 96 (18 SPEC. ISS): 1959–1966.

158 Sarkar, N., Ghosh, S.K., Bannerjee, S., and Aikat, K. (2012). Bioethanol production from agricultural wastes: an overview. *Renewable Energy* 37 (1): 19–27.

159 Campodónico, H. (2011). An Energy Panorama of Latin America. In: *The Oxford Handbook of Latin American Economics* (eds. J.A. Ocampo and J. Ros). Oxford University Press.

160 International Energy Agency (IEA) (2017). Technology Roadmap: Delivering Sustainable Bioenergy – Summary Paper. IEA, Paris.

161 Roozbehani, B., Mirdrikvand, M., Moqadam, S.I., and Roshan, A.C. (2013). Synthetic ethanol production in the Middle East: a way to make environmentally friendly fuels. *Chemistry and Technology of Fuels and Oils* 49 (2): 115–124.

162 Gnansounou, E. and Dauriat, A. (2010). Techno-economic analysis of lignocellulosic ethanol: a review. *Bioresource Technology* 101 (13): 4980–4991.

163 Wooley, R., Ruth, M., Sheehan, J. et al. (1999). *Lignocellulosic Biomass to Ethanol Process Design and Economics Utilizing Co-Current Dilute Acid Prehydrolysis and Enzymatic Hydrolysis Current and Futuristic Scenarios*. Golden, CO: US Department of Energy Laboratory.

164 Hess, J.R., Wright, C.T., and Kenney, K.L. (2007). Cellulosic biomass feedstocks and logistics for ethanol production. *Biofuels, Bioproduct and Biorefining* 1 (3): 181–190.

165 Wang, L., Littlewood, J., and Murphy, R.J. (2013). Environmental sustainability of bioethanol production from wheat straw in the UK. *Renewable and Sustainable Energy Reviews* 28: 715–725.

Index

Biodegradable Waste Management in the Circular Economy: Challenges and Opportunities, First Edition.
Edited by Malgorzata Kacprzak, Eleonore Attard, Kari-Anne Lyng, Helena Raclavska, BalRam Singh, Eyob Tesfamariam, and Franck Vandenbulcke.

d

e

f

g

h

n